Selectivity in Chemical Reactions

W0246202

NATO ASI Series

Advanced Science Institutes Series

A Series presenting the results of activities sponsored by the NATO Science Committee, which aims at the dissemination of advanced scientific and technological knowledge, with a view to strengthening links between scientific communities.

The Series is published by an international board of publishers in conjunction with the NATO Scientific Affairs Division

A	Life Sciences	Plenum Publishing Corporation
B	Physics	London and New York
C	Mathematical and Physical Sciences	Kluwer Academic Publishers Dordrecht, Boston and London
D	Behavioural and Social Sciences	
E	Applied Sciences	
F	Computer and Systems Sciences	Springer-Verlag
G	Ecological Sciences	Berlin, Heidelberg, New York, London,
H	Cell Biology	Paris and Tokyo

Series C: Mathematical and Physical Sciences - Vol. 245

Selectivity in Chemical Reactions

edited by

J. C. Whitehead

Department of Chemistry,
University of Manchester,
Manchester, U.K.

Kluwer Academic Publishers

Dordrecht / Boston / London

Published in cooperation with NATO Scientific Affairs Division

Proceedings of the NATO Advanced Research Workshop on
Selectivity in Chemical Reactions
Bowness-on-Windermere, U.K.
7–11 September 1987

Library of Congress Cataloging in Publication Data

NATO Advanced Research Workshop on Selectivity in
   Chemical Reactions (1987 : Windermere, England)
   Selectivity in chemical reactions.

   (NATO ASI series. Series C, Mathematical and
physical sciences ; vol. 245)
   "Published in cooperation with NATO Scientific
Affairs Division."
   Includes index.
   1. Chemical reaction, Rate of--Congresses. 2. Chemi-
cal reaction, Conditions and law of--Congresses.
I. Whitehead, J. C., 1947-   . II. North Atlantic
Treaty Organization. Scientific Affairs Division.
III. Title. IV. Series: NATO ASI series. Series C,
Mathematics and physical sciences ; no. 245.
QD502.N35 1987   541.3'94   88-23102

ISBN-13: 978-94-010-7870-2   e-ISBN-13: 978-94-009-3047-6
DOI: 10.1007/ 978-94-009-3047-6

Published by Kluwer Academic Publishers,
P.O. Box 17, 3300 AA Dordrecht, The Netherlands.

Kluwer Academic Publishers incorporates the publishing programmes of
D. Reidel, Martinus Nijhoff, Dr W. Junk, and MTP Press.

Sold and distributed in the U.S.A. and Canada
by Kluwer Academic Publishers,
101 Philip Drive, Norwell, MA 02061, U.S.A.

In all other countries, sold and distributed
by Kluwer Academic Publishers Group,
P.O. Box 322, 3300 AH Dordrecht, The Netherlands.

All Rights Reserved
© 1988 by Kluwer Academic Publishers.
Softcover reprint of the hardcover 1st edition 1988
No part of the material protected by this copyright notice may be reproduced or utilized
in any form or by any means, electronic or mechanical, including photocopying, recording
or by any information storage and retrieval system, without written permission from the
copyright owner.

This book contains the proceedings of a NATO Advanced Research Workshop held within the programme of activities of the NATO Special Programme on Selective Activation of Molecules running from 1983 to 1988 as part of the activities of the NATO Science Committee.

Other books previously published as a result of the activities of the Special Programme are

BOSNICH, B. (Ed.) - *Asymmetric Catalysis* (E103), 1986

PELIZZETTI, E. and SERPONE, N. (Eds.) - *Homogeneous and Heterogeneous Photocatalysis* (C174) 1986

SCHNEIDER, M. P. (Ed.) - *Enzymes as Catalysts in Organic Synthesis* (C178) 1986

SETTON, R. (Ed.) - *Chemical Reactions in Organic and Inorganic Constrained Systems* (C165) 1986

VIEHE, H. G., JANOUSEK, Z. and MERÉNYI, R. (Eds.) - *Substituent Effects in Radical Chemistry* (C189) 1986

BALZANI, V. (Ed.) - *Supramolecular Photochemistry* (C214) 1987

FONTANILLE, M. and GUYOT, A. (Eds.) - *Recent Advances in Mechanistic and Synthetic Aspects of Polymerization* (C215) 1987

LAINE, R. M. (Ed.) - *Transformation of Organometallics into Common and Exotic Materials: Design and Activation* (E141) 1988

BASSET, J.-M., et al. (Eds.) - *Surface Organometallic Chemistry: Molecular Approaches to Surface Catalysis* (C231) 1988

CONTENTS

viii

PREFACE

The aim of this Workshop on "Selectivity in Chemical Reactions" was to examine the specific preferences exhibited by simple chemical reactions with regards to reagents having particular energy states, symmetries, alignment and orientation and the resulting formation of certain products with their corresponding energies, states, alignment and polarisation. Such problems come close to the ultimate goal of reaction dynamics of being able to determine experimentally and theoretically state-to-state cross sections and stereochemical effects under well defined and characterised conditions. There are many examples of highly selective and specific processes to be found in atmospheric and combustion chemistry and the production of population inversions amongst vibrational and electronic states lies at the heart of the development of chemical laser systems. Only when we can understand the fundamental processes that underlie the selectivity in the formation of products in a chemical reaction and the specific requirements of initial states of the reagents, can we expect to be able to develop the explanatory and predictive tools necessary to apply the subject to the development of new laser systems, efficient combustion schemes and specific methods of chemical synthesis, to the control of atmospheric pollution and to all problems in which it is necessary to direct the outcome of a chemical reaction in a specific way.

The brief given to the Workshop was to critically review the field, to discuss the present limitations and difficulties and to identify new directions. This was achieved by having a small group of 34 participants and a mixture of nine review lectures, three panel discussions and a poster session. The topics that were covered were grouped under the headings of *"Specificity of Reagent States in Chemical Reactions"*, *"Spin-Orbit Effects in Chemical Reactions"*, *"Orientation and Alignment in Chemical Reactions"*, *"Reactions of van der Waals Complexes and Clusters"*, , *"Accuracy and Availability of Potential Energy Surfaces"*, *"Multiple Pathways in Chemical Reactions"* and *"Applied Aspects of Chemical Selectivity"*. As far as is possible, the various contributions to this book have been grouped in the same way. Inevitably, this book cannot record the discussion that took place at the Workshop or the collaborations and future projects that have originated from it. There were many themes which ran through the Workshop that connected various contributions in a way that is restricted by the above categorisations and could not have been predicted in advance. One example is the HOCO system that was discussed in the context of the photo-initiated reaction $H + CO_2$ within the van der Waals complex $HBr \cdot CO_2$, the kinetics of the atmospheric and combustion reaction $OH + CO$ and the relaxation of OH by CO in flames and the associated potential energy surfaces.

The success of the Workshop and its ultimate record in this book is due to many people. Firstly, Dr. C. Sinclair of the NATO Scientific Affairs Division for his help in the organisation of the Workshop and Prof. J.P. Simons who was the member of the NATO "Selective Activation of Molecules" Panel who suggested this Workshop. I was greatly assisted in the scientific organisation and choice of the participants by the organising committee, Jonathan Connor, Paul Dagdigian and Steven Stolte. All the manuscripts were retyped with speed and efficiency by Debra Acton and proof-read by Tim Watkinson. To all these people and all the authors and participants, I express my grateful thanks.

J.C. Whitehead,
Manchester,
March 1988.

SELECTIVITY IN ELEMENTARY CHEMICAL REACTIONS

Richard B. Bernstein
Chemistry Department
University of California
Los Angeles
CA 90024
USA

ABSTRACT.   This introductory paper presents a brief version of those areas of molecular reaction dynamics relevant to the subject of the Workshop.   These include the following: influence of reagent's electronic, vibrational and rotational states upon reactivity; spin-orbit effects in elementary reactions, and dynamical aspects of stereochemistry, with special emphasis on orientation and alignment effects in chemical reactions.

Selectivity with respect to reagents leads to specificity with respect to products, exemplified by non-statistical branching ratios, anisotropy of angular distribution and "surprising" product state distributions.   The importance of vector correlations (from crossed beam experiments with polarized laser-induced fluorescence detection) is touched upon, as well as the subject of polarized laser photofragmentation dynamics (half-collisions) leading to (non-statistical) specificity in product polarizations.

An important subject of this paper is that of reagent preparation and stereoselectivity.   Experimental techniques for production of oriented (and of aligned) molecules and their theoretical description are outlined.   Reactive asymmetry experiments using oriented molecule beams are briefly reviewed, and new approaches described.   Results are presented of recent experiments which take advantage of the inherent mutual orientation of the molecules in a van der Waals complex, i.e., "precursor geometry-limited" bimolecular reactions.   Finally, a new technique is described for real time, picosecond, clocking of the collision complex in such bimolecular reactions.   Results are reported for the time-dependent birth of OH from $H + CO_2$.

1.   INTRODUCTION

State-selectivity is indeed a central theme in molecular reaction dynamics [1].   This was already clear by 1974, the date of publication of "Molecular Reaction Dynamics" [2].   In 1977 it became "official" with the publication of the American Chemical Society Symposium Volume "State-to-State Chemistry" [3].   The writer presented the introductory paper at the symposium, entitled "State-to-State Cross Sections and Rate Constants for Reactions of Neutral Molecules" [4].   The symposium offered a truly impressive array of speakers and topics, covering many of the subjects of the present Workshop.   Reading over the volume a decade later is not a waste of time.

J. C. Whitehead (ed.), Selectivity in Chemical Reactions, 1–21.
© *1988 by Kluwer Academic Publishers.*

It was already fully understood that state-selectivity with respect to reagents leads to state-specificity of products, *via* the principle of microscopic reversibility (which relates state-to-state cross sections and rates for the forward and reverse reactions at the same total energy [5]). Most of the attention was directed to the product side, e.g. angular and velocity distributions of reactive scattering of crossed molecular beams [6], product state distributions *via* ir, vis and uv chemiluminescence [7], and laser-induced fluorescence [8], and both ground- and excited-electronic state product branching ratios [9], but the importance of reagent energy and its effect on product energy partitioning was recognized [10]. However, the main excitement was in the "surprising" energy partitioning in elementary exoergic reactions typified by pronounced population inversions, of importance in connection with chemical lasers (both ir and visible) [11]. The interweaving of the fields, chemical lasers and laser chemistry, was already accomplished [12], *via* the bell-wether reaction of $F + H_2$ [13], and soon laser-induced selective chemistry took precedence over chemical lasers [14].

Already familiar a decade ago was the information-theoretic approach [15] to the characterization of disequilibrium population distributions and the applicability of surprisal analysis to non-statistical quantum state distributions and branching ratios [16].

With regard to the reactants' side, there was a considerable body of results on the translational energy dependence of the reaction cross section [17], vibrational energy dependence [18], rotational energy dependence [19], and electronic state selectivity [20], including spin-orbit effects [21].

There were also significant experiments [22] on state-to-state rotational excitation using the electrostatic quadrupole lens for state selection and analysis (a technique confined to polar diatomic molecules), and, subsequently inelastic state-to-state cross sections measured by the more general time-of-flight (TOF) method [23].

From the chemical viewpoint, an interesting development was the introduction in 1965 of the electrostatic hexapole lens [24] for producing beams of oriented symmetric top molecules, suitable for use as reagents in crossed beam reactions. The first such "reactive asymmetry" experiments succeeded in demonstrating the orientation dependence of the reactivity of CH_3I with alkali atoms [25]:

$$CH_3I + M$$
$$ICH_3 + M \longrightarrow MI + CH_3 \qquad (1)$$

Large differences in reactivity were observed for "heads" vs. "tails" orientations. However, the results were only qualitative and no useful theoretical framework for describing the reactive asymmetry was yet available.

A significant breakthrough on the alignment effect (on reactivity) came from the introduction of polarized laser excitation techniques in 1978 [26]. Linearly polarized laser light can selectively excite a homonuclear (as well as a heteronuclear) diatomic molecule so as to align its rotational angular momentum $\mathbf{j}$, i.e. achieve $\pm m_j$ quantum number selectivity. Of course since the populations of the $+$ and $-$ m_j states are equal, there is no true orientation, only alignment of the angular momentum vector $\mathbf{j}$ in space. Nevertheless there is found to be a significant influence of this ($\mathbf{j}$) alignment (with respect to the relative velocity vector $\mathbf{v}$) upon reactivity; e.g. for the reaction

$$Sr + HF \ (v=1, j=1; m_j) \rightarrow SrF + H \qquad (2)$$

the reaction cross section is greater for $j \mathbin{\|} v$ ("broadside" collisions) than for $j \perp v$ ("edge-on" collisions).

Thus, by the late 1970's the "basics" of reagent state selectivity and product state specificity in bimolecular reactions were well established, both experimentally and theoretically [27]. In addition to the above-mentioned extensive work on reactive (as well as inelastic) collisions, the chemical dynamics of "half-collisions", i.e. laser-induced photofragmentation [28], had already reached a relatively mature state by the end of the 1970's. The effect of laser polarization on the angular distribution of photofragments could be expressed by the simple equation

$$I(\theta) = I_0[1 + \beta P_2(\cos\theta)] \qquad (3)$$

where θ is the angle between the velocity vector of the detected fragment and the laser polarization axis. For a parallel transition dipole moment and a prompt dissociation of the excited state of the molecule undergoing photodissociation $\beta=2$, i.e. $I(\theta) \propto \cos^2\theta$. For a perpendicular transition moment with prompt dissociation $\beta=-1$, i.e. $I(\theta) \propto \sin^2\theta$. For a long-lived excited state, i.e. if $\tau \gg t_{rot}$ (the classical rotational period of the molecule), the angular distribution of photofragments is isotropic ($\beta=0$). From measured β values estimates of excited state lifetimes could be made, with greatest sensitivity in the range from $ca.$ 0.1 - 10 ps, a region then inaccessible to direct measurement.

2. WHAT'S NEW?

All that has been discussed above was known by the end of the 1970's. What developments of consequence have taken place since then? One way to answer this question is to look at the programs of papers presented at a few important recent conferences. In November 1986 a Workshop on Dynamical Aspects of Stereochemistry [29] was held at the Hebrew University in Jerusalem, and in July 1987 the XIth International Symposium on Molecular Beams [30] was held in Edinburgh. The Faraday Discussion on Dynamics of Elementary Gas Phase Reactions [31] is scheduled to follow the present Workshop, in Birmingham. (The Faraday Discussion is especially favoured by the presence of, and presentations by, the three 1986 Nobel Laureates in Chemistry, Professors D.H. Herschbach, Y.T. Lee and J.C. Polanyi, whose Award served to "legitimize" the field of chemical dynamics of elementary reactions).

Examination of these programmes reveals an ever-greater emphasis on selectivity and specificity. On the product side, there is much more emphasis on the more detailed attributes such as product rotational alignment [32], Doppler-resolved laser-induced fluorescence of nascent products of inelastic collisions, reaction or photodissociation [33], as well as lambda doublet and fine structure analysis of such products as OH and CN radicals [34]. There is a preponderance of studies of bimolecular reactions involving excited reagents (sometimes as a function of collision energy), e.g. the elementary reactions

$$D + H_2(v=1) \rightarrow HD + H \quad [35] \qquad (4)$$

$$O(^1D) + H_2(v) \rightarrow OH(A^2\Sigma^+) + H \quad [36] \qquad (5)$$

$$N(^2D,{}^2P) + CS_2,\ OCS \rightarrow NS(B^2\Pi) + CS,\ CO \quad [37] \qquad (6)$$

4

$$Ca(^3P, ^1D) + SF_6 \rightarrow CaF(A,B) + SF_5 \quad [38] \tag{7}$$

as well as the reaction

$$Cs(7p) + H_2 \rightarrow CsH + H \quad [39a]. \tag{8}$$

For (8), the fine-structure dependence of the reaction cross section has been measured [39b,c]; the lower $(7P_{1/2})$ spin-orbit state of Cs is an order of magnitude more reactive with H_2 than the upper $(7P_{3/2})$ state! Spin-orbit effects in reactions of $Ba(^3D)$ atoms have also been observed [40].

Very detailed reactive scattering studies of electronically excited alkali atoms with HCl, O_2, NO_2, etc. have been reported [41], and the reactions of polarized laser-excited rare gases with halogenated molecules to yield aligned excimer products studied [42].

Reactions of highly vibrationally [43] and rotationally excited [44] molecules have also been studied. In Ref. 44b the influence of the alignment of the rotationally and vibrationally excited reagent molecule upon reactivity was measured, for the reaction

$$HF(v=1,j) + K \rightarrow KF + H. \tag{9}$$

Reaction cross sections decrease with j at high E_{tr}. The alignment effect, defined as $(\sigma_{||} - \sigma_\perp)/\frac{1}{2}(\sigma_{||} + \sigma_\perp)$, is found to be as high as 17% at the lowest E_{tr} (0.46 eV). These results [44b] are consistent with analogous earlier, less extensive results (Ref. 19e, j-dependence of the K + HCl $(v=1,j)$ reaction, and Ref. 26, alignment dependence and j-dependence of the Sr + HF $(v=1,j)$ reaction). A review on the subject of polarization of reaction products is available [42c].

An important new development of the 1980's is the use of orbitally aligned atoms to study inelastic [45] and reactive [46] atom-molecule collisions. Here an atomic beam is excited by polarized laser radiation, which produces a laboratory alignment of the appropriate atomic orbital. The beam of excited, "aligned" atoms is crossed by a molecular beam and the angle between the polarization axis and the relative velocity vector v_r varied. For example, the cross section for the inelastic energy-transfer collision

$$Na(3p) + H_2(v=0) \rightarrow Na(3s) + H_2(v=1,2,3) \tag{10}$$

varies with the initial alignment of the Na p-orbital with respect to v_r [45a,b]. This is especially interesting since one might have expected "orbital following" to severely damp this alignment effect [45c,d]. Experiments on the *orientation* dependence of the scattering of m_j-selected alkali beams with rare gases have been reported [45c]. Orbital alignment effects have also been observed in near-resonant electronic energy transfer of $Ca(^1P_1)$ atoms with rare gases [45f].

For the elementary chemical reaction

$$Ca(^1P_1) + HCl \rightarrow CaCl(A,B) + H \tag{11}$$

it has been found [46a] that an initially perpendicular alignment of the Ca p-orbital favours the formation of the $A^2\Pi$ product state, while an initially parallel alignment leads preferentially to the $B^2\Sigma^+$ state, this in spite of any orbital following [46a].

More extensive crossed beam experiments for the reactions of excited, aligned Na atoms,

$$e.g. \quad Na(3p,4d,5s) + HCl \rightarrow NaCl + H \qquad (12)$$

have revealed even more detail, showing that not only the reaction cross sections, but also the product angular and recoil velocity distributions are influenced by the (initial) orbital alignment.

The considerations underlying the problem of "orbital following" are similar to those which apply to the interpretation of reactive asymmetry experiments with oriented molecules [47] in which the dependence of the reactive scattering cross section upon the initial mutual orientation of the reagents is measured. Since the earliest days, and right up to the present [25,48], the question of "reorientation effects" has been recognized as an important issue. The experimenter can, at best, specify the initial orientation or alignment of a reagent; what happens thereafter during the course of a collision is not under control. Fortunately one is accumulating considerable experience on this problem (largely from classical trajectory simulations [48]), and the relative importance of the variables (both extrinsic and intrinsic) such as relative velocity, initial orientation and rotational state, intermolecular attractive forces and impact parameter have been ascertained.

As will be discussed below, there have been a number of new developments in the field of oriented molecule scattering involving the venerable hexapole lens focusing technique [49]. However such experiments (as well as those involving laser alignment of reagents) suffer potentially from the reorientation effects. One recent new approach [50], using jet-cooled van der Waals complexes, offers the possibility of bypassing this problem, and thus has considerable potential as a significant new tool for the study of orientational selectivity. There is an inherent mutual orientation of the component molecules in a van der Waals complex [51], e.g. FH···OCO is known [52] to have a linear equilibrium geometry (although the FHO bending force constant is very weak so there is a wide, zero-point angular amplitude).

In the prototype experiment [50c] the 1:1 van der Waals adduct of HBr·CO_2 was used as the precursor whose geometry is restricted to near-collinear; thus the name precursor geometry-limited (PGL) method. The HBr moiety in the van der Waals molecule is photolyzed, thus ejecting a translationally hot H atom toward the O atom of the CO_2 and initiating the endoergic "near-bimolecular" reaction of H + CO_2. Schematically,

$$BrH \cdots OCO \xrightarrow[h\nu]{1} Br\ H \rightarrow OCO \xrightarrow{2} Br + OH(v',j') + CO \qquad (13)$$

where step 1 (unimolecular photodissociation) is the hot H atom ejection and step 2 the "bimolecular" reaction

$$H + OCO \longrightarrow [HOCO] \longrightarrow OH + CO \qquad (14)$$

(to be discussed below). The rotational state distribution in the OH product was found [50c-e] to be significantly different for the PGL reaction than for the gas phase reaction, and this difference was attributed to the restricted range of impact parameters and angles of attack of the H atom. Recently, the PGL method has been applied to the branching reaction [50f]

$$D + \begin{cases} OCS \\ SCO \end{cases} \longrightarrow \begin{cases} OD + CS \\ SD + CO \end{cases} \tag{15}$$

based on the precursor BrD···OCS. There are many exciting possibilities for further stereo-selective dynamical experiments using this powerful PGL technique.

The remainder of this presentation is devoted to discussion of very recent activities, mainly of the writer, in the title area, i.e. selectivity in elementary chemical reactions (more specifically dynamical stereochemistry). A more general overview of this subject (i.e., that of the Jerusalem Workshop, Ref. 29) appears elsewhere [53].

3. NEW EXPERIMENTS ON ORIENTED MOLECULE BEAMS

A new molecular beam machine has been constructed at UCLA with a 3m electrostatic hexapole field for the production of intense, pulsed beams of essentially pure |JKM> states of symmetric top molecules [54]. Figure 1 shows a couple of so-called "focusing curves" for CH_3Cl molecules

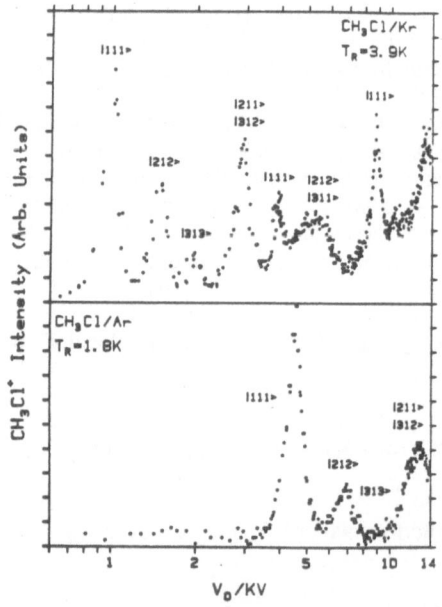

FIGURE 1. Typical hexapole focusing curves for pulsed, rotationally cold seeded beams of CH_3Cl, showing the resolved |JKM> rotational states |111>, |212>, |313>, |211>, (+|312>), the "first overtone" of |111>, that of |212> (+|311>), the "second overtone" of |111>, etc. Rotational temperature T_R from J-state population analysis. [Reprinted from Ref. 54a, with permission].

Figure 2 shows calculated [55] orientational probability density functions P(cosθ), for several of the readily obtained |JKM> states. (Here θ is the angle between the dipole moment μ and the electric field ε).

Similarly resolved rotational states (in oriented molecule beams) have been reported independently [56]. Elegant reactive asymmetry measurements have been carried out with oriented NO and N_2O molecules, as reviewed in Ref. 47. In the chemiluminescent (CL) reaction of oriented N_2O with Ba[57a], i.e.,

$$N_2O(J\ell M = 111) + Ba \rightarrow BaO(A^1\Sigma^+) + N_2. \qquad (16)$$

The polarization of the CL emission of the $BaO(X^1\Sigma^+ \leftarrow A^1\Sigma^+)$ was measured as function of mutual orientation, $<\cos\theta> = +0.5$, i.e. "favourable"; 0; and -0.5 ("unfavourable"), and the CL spectrum itself found to shift as a function of reagent

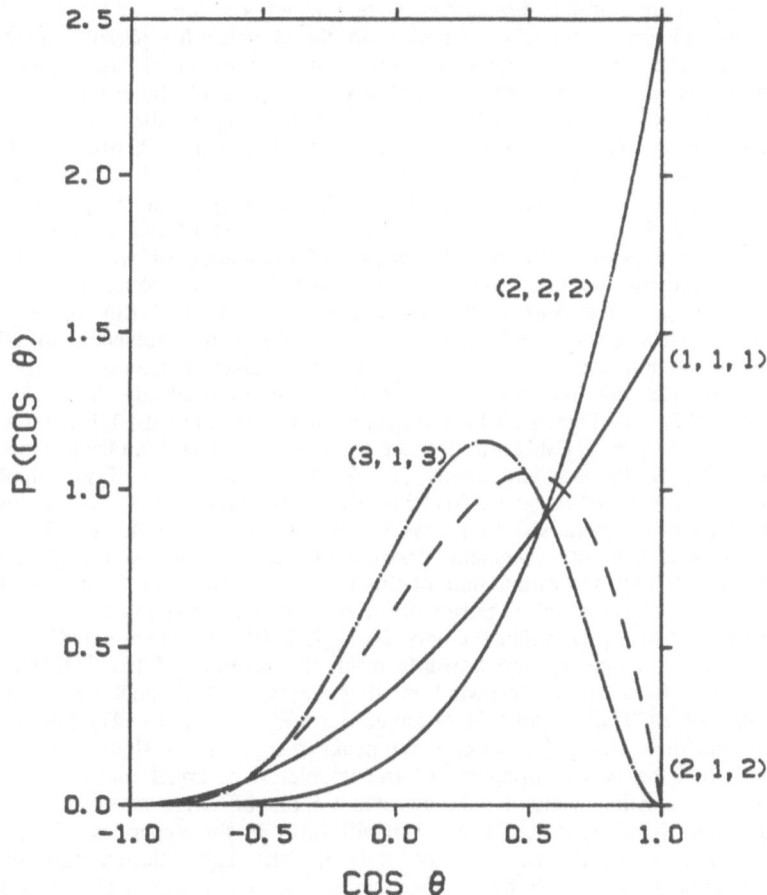

FIGURE 2. Calculated orientational probability density functions for the selected JKM states of a symmetric top molecule [adapted from Ref. 55]. Average values, $<\cos\theta>$, are as follows: |222>, 0.667; |111>, 0.500; |212>, 0.333; |313>, 0.250.

orientation [57b]. For the "favourable" orientation, i.e. for the NNO pointing toward the Ba, there was found to be strong, in-plane product alignment. For the opposite initial orientation, both the CL cross section and its polarization approach zero [57].

Recently, orientation-dependent CL from the CF_3 product of the reaction of metastable Ar atoms with hexapole-oriented CF_3H has been reported [57c]. The orientation dependence of reaction (1) of oriented CH_3I with Rb [25a] has been re-measured [58] and the results treated with more refined analysis [59].

With the advent of extensive oriented CH_3I reactive scattering experiments [58] it has become important to ascertain the actual degree of orientation of the reagent beam of rotationally state-selected molecules, as prepared by the electrostatic hexapole technique. The appropriate theory for JKM-selected symmetric top molecules (in the absence of hyperfine effects) suggested [55] that polarized laser photofragmentation could yield information on the degree of orientation, at least for the case of prompt dissociation. For CH_3I, recent direct, real-time, sub-picosecond laser measurements of the lifetime of the dissociative state yielded a value $\tau < 0.5$ ps [60], confirming the promptness previously inferred from the β value for photofragmentation, i.e. $\beta \simeq 2$ [61]. Thus, for CH_3I the direction of emission of the fragment I atoms should be an indication of the orientation in space of the parent CH_3I molecules at the instant of photoexcitation. A strong asymmetry of the angular distribution of fragments is expected. The probability density function $P(\cos\theta)$ for the emitted I atoms should be that of the parent molecules modulated by the $\cos^2\theta$ (or $\sin^2\theta$) factor due to the absorption of the parallel (or perpendicular) polarized laser radiation.

New UCLA experiments [62] represent a test of this concept, and demonstrate by a purely physical method the degree of orientation of an "oriented molecule beam", hitherto inferred from crossed molecular beam reactive asymmetry experiments. (The only related observations are those of Ref. 63 in which a partially state-selected and oriented beam of CH_3I was subjected to UV photoionization and the anisotropy of the photoelectron angular distribution measured.)

In Ref. 62 the pulsed, state-selected oriented molecule beam is generated as before [54], but here an added capacitor field assembly established the direction (vertical) of the ϵ-field with respect to which the CH_3I molecules precess. According to the standard theory [24,29] the negative end of the dipole (here the I atom) points to the negative plate of the orientation field (here the upper plate), so the I atom fragment should be ejected preferentially into the upper hemisphere. Figure 3 is a highly schematic drawing of the experimental arrangement, which employs a laser ionization time-of-flight mass spectrometer (TOFMS) detector.

The polarized laser pulse which induces the one-photon photofragmentation also ionizes the $I(^2P_{1/2})$ product atoms *via* a 2+1 REMPI process [64] at 281.74 nm. Ions from I atoms ejected upwards reach the aperture of the TOFMS at an earlier time than those from downward moving atoms, so for parallel laser polarization a well-resolved TOF doublet is observed for the I^+ (m/z = 127) ions (cf. Ref. 65). (For perpendicular polarization these peaks merge into a single one). Under ideal conditions the two components of the doublet have equal integrated intensities. This has been verified using a bulk gaseous sample of CH_3I at a pressure of *ca.* 1 x 10^{-7} Torr in the ionizer of the TOFMS (also in the apparatus of Ref. 64). This is also found to be the case for randomized, |JK> state-selected molecules with loss of M orientation (i.e. with the voltage across the orientation field set at zero). TOF spectra were recorded (*via* a fast waveform recorder) using a CH_3I selected state with and without voltage on the orientation capacitor. This allowed a comparison of the "up" vs "down" intensities of the I atoms from the photolysis of the oriented |JKM> states vs. the randomized |JK> molecules.

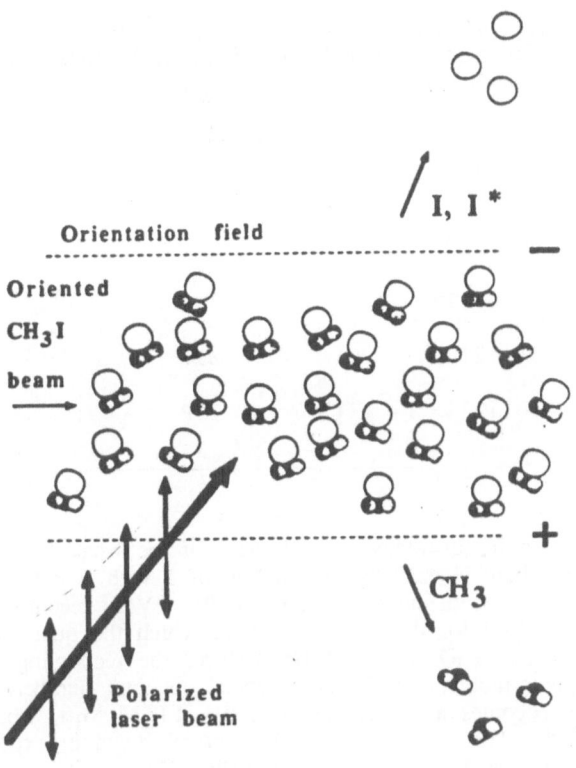

FIGURE 3. Highly schematic (!) portrayal of the experimental arrangement [62] for the polarized laser photofragmentation of oriented CH_3I molecules. The iodine atoms formed from the 1-photon photodissociation are detected by a 2+1 resonance-enhanced MPI process at an appropriate I atom resonance wavelength. [Reprinted from Ref. 53, with permission].

Experiments were carried out for four selected $|JKM\rangle$ parent states: $|111\rangle$, $|212\rangle$ and $|211\rangle$ (the latter containing a small admixture of $|312\rangle$), all from rotationally cold Ar- of Kr-seeded beams, as well as $|222\rangle$ (from a neat, "warm" beam). A strong up/down asymmetry was observed, indicating substantial orientation. Table 1 summarizes the experimental results.

TABLE 1. Experimental results (Ref. 62) on the asymmetry of the polarized laser photofragmentation of oriented CH_3I molecules in selected ⏐JKM> "parent" rotational states. (The measured up/down ratios have an uncertainty of *ca.* 10%).

⏐JKM>	Up/Down Ratio
⏐222>	3.9
⏐111>	2.4
⏐212>	2.7
⏐211> (+ ⏐312>)	1.9

Analysis of these results indicates [62] that, for the weak fields used (in the orientation capacitor), there is extensive recoupling of **J** with **I** and a consequent lowering of the degree of molecular orientation [47b]. Very recently, calculations have been carried out [66] for the realistic case in which the quadrupole of the I atom (nuclear spin = 5/2, eqQ ≃ 1940 MHz) induces the recoupling **J** + **I** = **F**, where the total angular momentum F is the "good" quantum number, so a given ⏐JKM> parent state becomes an ensemble of states ⏐FJKM$_F$M$_I$>, whose average orientation is reduced. This effect has already been observed in experiments with aligned molecules, and has been treated theoretically [67].

Not included in the analysis of Refs. 55, 62 or 66 is the possible effect of the change in J (or F) of the parent ⏐JKM> state upon photoexcitation; however, under the assumption of prompt dissociation, the framework of the molecule cannot reorient appreciably in the few hundred femtoseconds before the excited state falls apart, so the experimental results should be compared to calculations without change in quantum numbers.

The experiments of Ref. 62 have demonstrated by a purely physical (i.e., non-chemical) technique the high degree of molecular orientation that can be provided by the hexapole method, even for the case of CH_3I which, of all the methyl halides, has the most severe hyperfine (quadrupole coupling) disorientation effect. Experiments are now in progress to extend these studies on CH_3I and investigate related molecules. As a byproduct of these measurements of the orientation of selected states, one obtains the sign of the dipole moment of the (symmetric top) molecule under study.

The influence of the orienting field strength ϵ upon the hyperfine coupling is also of great importance. The transition from the weak field (fully coupled **J**, **I**) to the strong field (decoupled **J**, **I**) occurs at modest values of ϵ, and thus should be observable. The deleterious effect of the hyperfine coupling upon molecular orientation in weak fields is clear. How does this relate to the conditions used to perform the "chemical" experiments (e.g., Refs. 58)? Re-examination of this question is in progress.

Returning to the issue raised earlier, namely, the "uncontrollable" effect of reorientation (in the course of the encounter) of the initially oriented reagents,

11

precursor geometry-limited (PGL) reactions [50] offer unusual opportunities in chemical dynamics. In addition to the attributes discussed earlier, there is another advantage to the use of "potential reagents" combined within a van der Waals adduct. By the use of an initiating "pump" laser pulse, one can effectively establish the zero of time and then, with a probe laser pulse delayed with respect to the pump pulse, detect the formation and decay of the collision complex and measure the time-evolution of the products of a bimolecular reaction [68]. The first experiment of this kind dealt with the original PGL reaction (of Refs. 50c,d)

$$H + OCO \rightarrow HOCO \rightarrow OH(v',j') + CO, \tag{14}$$

but here the precursor van der Waals molecule was IH⋯OCO (prepared by jet expansion of a He-seeded mixture of HI and CO_2). The sequence is as follows: a picosecond laser pulse initiates the hot atom reaction by dissociating the HI; the H atom projectile strikes the OCO with sufficient translational energy to overcome the barrier to form the collision complex, HOCO, which decays to yield the OH product, detected by a second picosecond laser pulse via laser induced fluorescence (LIF). Figure 4 shows the scheme [69].

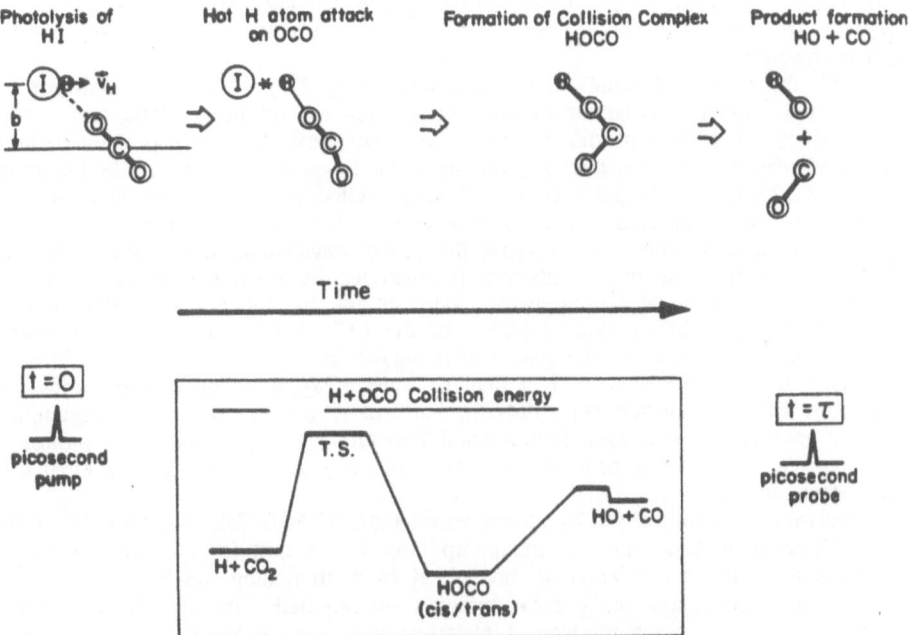

FIGURE 4. Course of the overall PGL reaction of Ref. 68. Upper: time development, from ps laser photolysis of HI to hot atom attack forming the collision complex which dissociates to CO + OH (detected by the ps probe laser, delayed by time τ). Lower: schematic of the minimum energy path from an *ab initio* calculated potential energy surface [69]. Here T.S. stands for transition state; this barrier is dependent upon the "angle of attack" of the H on the OCO. [Adapted from Ref. 68].

From the wavelength of the dissociation (pump) laser, it can be calculated that the velocity of the H atom was ca. 200 Å/ps and the initial relative translational energy E_{tr} of the H with respect to the OCO was ca. 200 kJ mol^{-1}. At this energy, the characteristic "rise" time τ_r for the formation of the OH is ca. 5 ps [68]. Classical trajectory simulations at different E_{tr} on the *ab initio* potential surface have been carried out [69], but initial results appear to under-estimate the lifetime of the HOCO complex. Further experiments are in progress by the authors of Ref. 68 to extend the photolysis wavelength range and thus the initial translational energy range of the "clocked" reaction, i.e. the birth of the OH from the H + CO_2 reaction.

4. CONCLUDING REMARKS

This paper cannot be construed to be a review [70], but rather a kaleidoscopic view of the subject of selectivity in elementary chemical reactions. Without editorializing, it is clear that after a decade the field has become rather mature [71]; the exciting "breakthroughs" in experiments and theory are coming less frequently now. (Nevertheless, there are many more on the horizon!). What's next?

The writer was recently asked (by the National Science Foundation) to identify one or more "promising areas" in the future of physical chemistry. Although predictions are always dangerous, here is the author's response: *femtosecond, real-time chemical dynamics*.

With the advent of femtosecond laser technology [72], it is now possible to study, in *real time*, the dynamics of elementary chemical reactions. The first example of an application of this technique to a chemical dynamical problem comes from sub-picosecond experiments [73] on the photodissociation of ICN. In the most recent work [73b], a 40-fs pulse ($\lambda \equiv 307$ nm) excited the ICN molecule to a repulsive state and a delayed, second femtosecond pulse probed the CN fragment (via LIF) as a function of time. By varying the probe wavelength to the red of the free CN bandhead, they were able to observe fluorescence from the CN in the I–CN molecule in the process of falling apart. They found that the transition state lives for about four times the vibrational period of the I-CN bond. In this time interval there is negligible rotation of the parent ICN molecule.

There has already been a great deal of preliminary work on the dynamics and spectroscopy of the transition state [74-76], but Ref. 73b now signals the beginning of a new era, that of *real time* femtosecond transition-state spectroscopy/dynamics. It is important to note that progress here does not require heroic new developments in laser technology.

For transform-limited pulses at the wavelength of Ref. 73b, the FWHM of the wavelength-intensity distribution of the pump laser is 3.5 nm; for the probe laser 4.4 nm. Thus there is *no advantage* to be gained from shortening the laser pulse, if well-resolved spectroscopic probe measurements are required. In fact, the uncertainty principle makes it uninteresting from a chemical viewpoint to work with pulses much shorter than those presently available (e.g. 7 fs corresponds to 20 nm at 308 nm; thus E = 4.02 ± 0.13 eV, i.e. an energy FWHM of 0.26 eV, or 6 kcal mol^{-1}, large by chemical standards). Thus it is already necessary to compromise between temporal and energy resolution. From a chemical dynamical point of view, since for inter- and intramolecular motions atomic velocities are usually < 1 km s^{-1}, a time interval of 1 fs corresponds to an atomic displacement of < 0.01 Å, certainly "slow motion"!

Clearly the dynamics of a great many photon-induced unimolecular processes (dissociations and isomerizations) can now be studied, in real time, via the femtosecond pump-probe technique. But at the same time, a new area is opening up in which the real-time dynamics of bimolecular reactions can be similarly observed [68], using van der Waals complexes as precursors, i.e. the PGL technique [50]. One molecular component in a binary van der Waals complex is photodissociated by a ps or fs laser pulse, initiating a "hot atom" reaction. This pulse establishes the zero of time for the bimolecular reaction, following which the time-evolution of the reaction is observed by a delayed probe laser pulse. With femtosecond techniques one can now study the real-time dynamics not only of so-called complex-mode bimolecular reactions [1], with relatively long-lived collision complexes (*ca.* 1-10 ps), but also of "direct-mode" [1] bimolecular reactions in which the transition state lifetime is less than 100 fs. By the use of improved techniques and instrumentation, e.g. REMPI with OMA and MCP detection, imaging [77] etc., this new area of physical chemistry is bound to yield exciting new results and deeper insights into reaction dynamics [1].

ACKNOWLEDGMENTS

The author's research programme at UCLA is supported by NSF Grant CHE86-15286, hereby gratefully acknowledged. Sincere appreciation and thanks are expressed for the many contributions to this research made by the members of his UCLA research group: S.E. Choi, T.J. Curtiss, S.R. Gandhi, J.F. Garvey, M. Giorgi-Arnazzi, P. Grunberg, Y.-F. Jiang and Q.-X. Xu, as well as to his Caltech collaborators N.F. Scherer and L.R. Khundkar, and especially to Prof. A.H. Zewail for many stimulating discussions. The author also thanks Dr. S. Stolte (Nijmegen) for valuable remarks on many aspects of oriented molecule beam research.

REFERENCES

[1] R.D. Levine and R.B. Bernstein, Molecular Reaction Dynamics and Chemical Reactivity, Oxford University Press, NY (1987).

[2] R.D. Levine and R.B. Bernstein, Molecular Reaction Dynamics, Clarendon Press, Oxford (1974).

[3] P.R. Brooks and E.F. Hayes, eds. State-to-State Chemistry, ACS Symposium Series 56, American Chemical Society, Washington, DC (1977).

[4] R.B. Bernstein, in Ref. 3, p. 3.

[5] (a) J. Ross, J.C. Light and K.E. Shuler, in Kinetic Processes in Gases and Plasmas, ed. A.R. Hochstim, Academic, N.Y. (1965), p. 281;
(b) K.G. Anlauf, D.H. Maylotte, J.C. Polanyi and R.B. Bernstein, J. Chem. Phys., 51 (1969) 5716;
(c) R.D. Levine and J. Manz, J. Chem. Phys., 63 (1975) 4280.

14

[6] (a) G.H. Kwei, J.A. Norris and D.R. Herschbach, J. Chem. Phys.,
34 (1961) 1842; 52 (1970) 1317.
(b) D.R. Herschbach, Far. Disc. Chem. Soc., 33 (1962) 149.
(c) T.P. Schafer, P.E. Siska, J.M. Parson, F.P. Tully, Y.C. Wong
and Y.T. Lee, J. Chem. Phys., 53 (1970) 3385.
(d) J.D. McDonald, P.R. LeBreton, Y.T. Lee and D.R. Herschbach,
J. Chem. Phys., 56 (1972) 769;
(e) D.R. Herschbach, Far. Disc. Chem. Soc., 55 (1973) 233.
(f) J.M. Parson, K. Shobatake, Y.T. Lee and S.A. Rice, Far. Disc.
Chem. Soc., 55 (1973) 344.

[7] (a) J.K. Cashion and J.C. Polanyi, J. Chem. Phys., 29 (1958) 455;
(b) C.D. Jonah, R.N. Zare and Ch. Ottinger, J. Chem. Phys., 56
(1972) 263;
(c) J.C. Polanyi, Acc. Chem. Res., 5 (1972) 161;
(d) A.M.G. Ding, L.J. Kirsch, D.S. Perry, J.C. Polanyi, and
J.L. Schreiber, Far. Disc. Chem. Soc., 55 (1973) 252.
(e) J.C. Polanyi, Far. Disc. Chem. Soc., 55 (1973) 389;
(f) K. Tamagake and D.W. Setser, in Ref. 3, p. 124;
(g) U.C. Sridharan, D.M. McFadden and P. Davidovits, in Ref. 3,
p. 136.

[8] (a) H.W. Cruse, P.J. Dagdigian and R.N. Zare, Far. Disc. Chem.
Soc., 55 (1973) 277;
(b) R.N. Zare, in Ref. 3, p. 50;
(c) R.N. Zare and P.J. Dagdigian, Science, 185 (1974) 739.

[9] (a) J.D. McDonald, P.R. LeBreton, Y.T. Lee and D.R. Herschbach,
J. Chem. Phys., 56 (1972) 769;
(b) A. Persky, J. Chem. Phys., 59 (1973) 5578;
(c) C.A. Mims, S.M. Lin and R.R. Herm, J. Chem. Phys., 58 (1973)
1983;
(d) J. Grosser and H. Haberland, Chem. Phys., 2 (1973) 342;
(e) D.M. Manos and J.M. Parson, J. Chem. Phys., 69 (1978) 231.

[10] A.M.G. Ding, L.J. Kirsch, D.S. Perry, J.C. Polanyi and
J.L. Schreiber, Far. Disc. Chem. Soc., 55 (1973) 252.

[11] (a) J.H. Parker and G.C. Pimentel, J. Chem. Phys., 51 (1969) 91;
(b) M.J. Berry, J. Chem. Phys., 59 (1973) 6229;
(c) K.L. Kompa, Chemical Lasers, Springer, Berlin (1973).

[12] M.J. Berry, Ann. Rev. Phys. Chem., 26 (1975) 259.

[13] (a) J.H. Parker and G.C. Pimentel, Ref. 11a;
(b) T.P. Schafer et al., Ref. 6c;
(c) M.J. Berry, Ref. 11b;
(d) J.C. Polanyi and J.L. Schreiber, Far. Disc. Chem. Soc.,
62 (1977) 267.

[14] A.H. Zewail, editor, <u>Advances in Laser Chemistry</u>, Springer-Verlag, Berlin (1978).

[15] (a) R.D. Levine and R.B. Bernstein, in Ref. 3, p. 100;
      (b) R.D. Levine and R.B. Bernstein, <u>Acc. Chem. Res.</u>, 7 (1974) 393;
      (c) R.B. Bernstein and R.D. Levine, <u>Adv. At. Mol. Phys.</u>, 11 (1975) 215;
      (d) R.D. Levine, <u>Annu. Rev. Phys. Chem.</u>, 29 (1978) 59.

[16] (a) R.D. Levine and R. Kosloff, <u>Chem. Phys. Lett.</u>, 28 (1974) 300;
      (b) U. Dinur and R.D. Levine, <u>Chem. Phys. Lett.</u>, 31 (1975) 410;
      (c) R.B. Bernstein, <u>Intl. J. Quant. Chem. Symp. No. 9</u>, (1975) 385.

[17] (a) M.E. Gersh and R.B. Bernstein, <u>J. Chem. Phys.</u>, 56 (1972) 6131;
      (b) J.G. Pruett, F.R. Grabiner and P.R. Brooks, <u>J. Chem. Phys.</u>, 63 (1975) 1173;
      (c) J.J. Valentini, M.J. Coggiola and Y.T. Lee, <u>Far. Disc. Chem. Soc.</u>, 62 (1977) 232;
      (d) M.W. Geis, H. Dispert, T.L. Budzynski and P.R. Brooks, in Ref. 3, p. 103;
      (e) T.M. Mayer, B.E. Wilcomb and R.B. Bernstein, <u>J. Chem. Phys.</u>, 67 (1977) 3507.

[18] (a) T.J. Odiorne, P.R. Brooks and J.V. Kasper, <u>J. Chem. Phys.</u>, 55 (1971) 1980;
      (b) J.G. Pruett and R.N. Zare, <u>J. Chem. Phys.</u>, 64 (1976) 1774;
      (c) D. Arnoldi and J. Wolfrum, <u>Ber. Bunsenges. Phys. Chem.</u>, 80 (1976) 892;
      (d) D.J. Douglas, J.C. Polanyi and J.J. Sloan, <u>J. Chem. Phys.</u>, 13 (1976) 15;
      (e) Z. Karny and R.N. Zare, <u>J. Chem. Phys.</u>, 68 (1978) 3360.

[19] (a) A.M.G. Ding et al., Ref. 10;
      (b) S. Stolte, A.E. Proctor, W.M. Pope and R.B. Bernstein, <u>J. Chem. Phys.</u>, 66 (1977) 3468;
      (c) L. Zandee and R.B. Bernstein, <u>J. Chem. Phys.</u>, 68 (1978) 3760;
      (d) B.A. Blackwell, J.C. Polanyi and J.J. Sloan, <u>J. Chem. Phys.</u>, 30 (1978) 299;
      (e) H.H. Dispert, M.W. Geis and P.R. Brooks, <u>J. Chem. Phys.</u>, 70 (1979) 5317.

[20] (a) F. Engelke, J.C. Whitehead and R.N. Zare, <u>Far. Disc. Chem. Soc.</u>, 62 (1977) 222;
      (b) F.J. Van Itallie, L.J. Doemeny and R.M. Martin, <u>J. Chem. Phys.</u>, 56 (1972) 3689;
      (c) J.W. McGowan, ed. 'The Excited State in Chemical Physics', <u>Adv. Chem. Phys.</u>, 28 (1975).

[21] (a) J.T. Muckerman and M.D. Newton, J. Chem. Phys., **56** (1972)
          3191;
      (b) H.F. Krause, S.C. Johnson, S. Datz and F.K. Schmidt-Bleek,
          Chem. Phys. Lett., **31** (1975) 577;
      (c) S. Hayashi, T.M. Mayer and R.B. Bernstein, Chem. Phys. Lett,
          53 (1978) 419.

[22] (a) J.P. Toennies, Far. Disc. Chem. Soc., **33** (1962) 96;
      (b) J.P. Toennies, Z. Physik, **182** (1965) 257;

[23] (a) A.R. Blythe, A.E. Grosser and R.B. Bernstein, J. Chem. Phys.,
          41 (1964) 1917;
      (b) H.E. van den Bergh, M. Faubel and J.P. Toennies, Far. Disc.
          Chem. Soc., **55** (1973) 203;
      (c) W.R. Gentry and C.F. Giese, J. Chem. Phys., **67** (1977) 5389;
      (d) V. Buck, F. Huisken and J. Schleusener, J. Chem. Phys., **68**
          (1978) 5654.

[24] K.H. Kramer and R.B. Bernstein, J. Chem. Phys., **42** (1965) 767.

[25] (a) P.R. Brooks and E.M. Jones, J. Chem. Phys., **45** (1966) 3449;
      (b) R.J. Beuhler, R.B. Bernstein and K.H. Kramer, J. Amer. Chem.
          Soc., **88** (1966) 5331.

[26] Z. Karny, R.C. Estler and R.N. Zare, J. Chem. Phys., **69** (1979)
      5199; see also Z. Karny and R.N. Zare, J. Chem. Phys., **68** (1978)
      3360.

[27] See e.g., R.B. Bernstein, ed., Atom-Molecule Collision Theory
      A guide for the Experimentalist, Plenum, N.Y. (1979).

[28] (a) R.N. Zare and D.R. Herschbach, Proc. I.E.E.E., **51** (1963) 173;
      (b) S.H. Lin and R. Bersohn, Adv. Chem. Phys., **16** (1969) 67;
      (c) R.N. Zare, Mol. Photochem., **4** (1972) 1;
      (d) S.J. Riley and K.R. Wilson, Far. Disc. Chem. Soc., **53** (1972)
          132;
      (e) S.C. Yang and R. Bersohn, J. Chem. Phys., **61** (1974) 4400;
      (f) M.J. Dzvonik, S.C. Yang and R. Bersohn, J. Chem. Phys.,
          61 (1974) 4408.

[29] A special issue of J. Phys. Chem., **91** (1987) No. 21 is
      devoted to the subject of this Workshop, namely, Dynamical
      Stereochemistry.

[30] Abstracts, XIth International Symposium on Molecular Beams,
      Univ. of Edinburgh, Royal Society of Chemistry (July, 1987).

[31] Faraday Discussion No. 84 on Dynamics of Elementary Gas Phase
      Reactions, Univ. of Birmingham, Royal Society of Chemistry
      (Sept. 1987).

[32] (a) P. Andresen and E.W. Rothe, J. Chem. Phys., **78** (1983) 989;
      (b) R. Vasudev, R.N. Zare and R.N. Dixon, J. Chem. Phys., **80** (1984) 4863:
      (c) M.G. Prisant, C.T. Rettner and R.N. Zare, J. Chem. Phys.,
          81 (1984) 2699;
      (d) G.E. Hall, N. Sivakumar and P.L. Houston, J. Chem. Phys.,
          84 (1986) 2120;
      (e) M.S. deVries, G.W. Tyndall, C.L. Cobb and R.M. Martin,
          J. Chem. Phys., **84** (1986) 3753;
      (f) R.M. Martin, C.L. Cobb, G.W. Tyndall and M.S. deVries, in
          Ref. 30.

[33] (a) K. Bergmann, U. Hefter and J. Witt, J. Chem. Phys., **72**
          (1980) 4777;
      (b) J.A. Serri, A. Morales, W. Moskowitz, D.E. Pritchard,
          C.H. Becker and J.L. Kinsey, J. Chem. Phys., **72** (1980) 6304;
      (c) J.A. Serri, C.H. Becker, M.B. Elbel, J.L. Kinsey,
          W.P. Moskowitz and D.E. Pritchard, J. Chem. Phys., **74** (1981)
          5116;
      (d) R. Schmiedl, H. Dugan, W. Meier and K.H. Welge, Z. Phys. A,
          304 (1982) 137;
      (e) I. Nadler, D. Mahgerefteh, H. Reisler and C. Wittig,
          J. Chem. Phys., **82** (1985) 3385.

[34] (a) E.J. Murphy, J.H. Brophy, G.S. Arnold, W.L. Dimpfl, and
          J.L. Kinsey, J. Chem. Phys., **74** (1981) 324;
      (b) P. Andresen, G.S. Ondrey, B. Titze and E.W. Rothe, J. Chem.
          Phys., **80** (1984) 2548;
      (c) I. Nadler, J. Pfab, H. Reisler and C. Wittig, J. Chem. Phys.,
          81 (1984) 653;
      (d) J.L. Kinsey, J. Chem. Phys., **81** (1984) 6410;
      (e) P. Andresen, V. Beushausen, D. Häusler, H.J.W. Lülf and
          E.W. Rothe, J. Chem. Phys., **83** (1985) 1429.

[35] V. Herrero, J.P. Toennies and M. Vodegel, in Ref. 30.

[36] A. Lebéhot, J. Marx, F. Aguillon, S. Drawin and R. Campargue,
      in Ref. 30.

[37] K. Tabayashi and K. Shobatake, in Ref. 30.

[38] E. Verdasco, V. Sáez Rábanos, F.J. Aoiz and A. González Ureña,
      in Ref. 30.

[39] (a) G. Rahmat, J. Vergès, R. Vetter, F.X. Gadea, M. Pelissier
          and F. Spiegelmann, in Ref. 30;
      (b) G. Rahmat, F. Spiegelmann, J. Vergès and R. Vetter,
          Chem. Phys. Lett., **135** (1987) 459;
      (c) F.X. Gadea and J. Durup, Chem. Phys. Lett., **138** (1987) 43.

[40] M.L. Campbell and P.J. Dagdigian, in Ref. 31.

18

[41] (a) M.F. Vernon, H. Schmidt, P.S. Weiss, M.H. Covinsky and
Y.T. Lee, <u>J. Chem. Phys.</u>, **84** (1986) 5580;
(b) J.M. Mestdagh, B.A. Balko, M.H. Covinsky, P.S. Weiss,
M.F. Vernon, H. Schmidt and Y.T. Lee, in Ref. 31.

[42] (a) K. Johnson, J.P. Simons, P.A. Smith, C. Washington and
A. Kvaran, <u>Mol. Phys.</u>, **57** (1986) 255;
(b) R.J. Donovan, P. Greenhill, A. Hopkirk, W.S. Hartree,
K. Johnson, C. Jouvet, A. Kvaran and J.P. Simons,
in Ref. 31 (1987);
(c) J.P. Simons in Ref. 29, 5378.

[43] M. Becker, U. Gaubatz and K. Bergmann, in Ref. 30.

[44] (a) H.-J. Loesch, <u>Chem. Phys.</u>, **104** (1986) 213; **112** (1987) 85;
(b) M. Hoffmeister, R. Schleysing and H.-J. Loesch, in Ref. 30;
also in Ref. 29, 5441.

[45] (a) P. Botschwina, W. Meyer, I.V. Hertel and W. Reitland,
<u>J. Chem. Phys.</u>, **75** (1981) 5438;
(b) W. Reitland, H.U. Tittes and I.V. Hertel, <u>Phys. Rev. Lett.</u>,
46 (1982) 1389;
(c) M. Hale, I.V. Hertel and S.R. Leone, <u>Phys. Rev. Lett.</u>,
53 (1984) 2296;
(d) I.V. Hertel, H. Schmidt, A. Bähring and E. Meyer,
<u>Rep. Prog. Phys.</u>, **48** (1985) 375;
(e) R. Düren and E. Hasselbrink, in Ref. 29, 5445;
(f) W. Bussert, D. Neuschäfer and S.R. Leone, <u>J. Chem. Phys.</u>,
87 (1987) 3833.

[46] (a) C.T. Rettner and R.N. Zare, <u>J. Chem. Phys.</u>, **77** (1982) 2416;
(b) M.F. Vernon *et al.*, Ref. 41a.

[47] (a) D.H. Parker, S. Stolte and H. Jalink in Ref. 29, 5427;
(b) S. Stolte, <u>Ber. Bunsenges. Phys. Chem.</u>, **86** (1982) 413.

[48] (a) M. Karplus and M. Godfrey, <u>J. Amer. Chem. Soc.</u>, **88** (1966)
5332;
(b) N.C. Blais, R.B. Bernstein and R.D. Levine, <u>J. Phys. Chem.</u>,
89 (1985) 10.
(c) N.C. Blais and R.B. Bernstein, <u>J. Chem. Phys.</u>, **85** (1986)
7030;
(d) N. Agmon, <u>Int. J. Chem. Kin.</u>, **18** (1986) 1047;
(e) M. Janssen and S. Stolte, in Ref. 29, 5480;
(f) G.T. Evans, <u>J. Chem. Phys.</u>, **87** (1987) 3865.

[49] (a) K.H. Kramer and R.B. Bernstein, Ref. 24;
(b) R.B. Bernstein, <u>Chemical Dynamics via Molecular Beam and
Laser Techniques</u>, Oxford Univ. Press, N.Y. (1982), Chap. 3;
(c) S. Stolte, chapter in <u>Atomic and Molecular Beam Methods</u>,

[50] (a) C. Jouvet and B. Soep, Chem. Phys. Lett., **96** (1983)
          426;
      (b) C. Jouvet and B. Soep, J. Chem. Phys., **80** (1984) 2229;
      (c) S. Buelow, G. Radhakrishnan, J. Catanzarite and C. Wittig,
          J. Chem. Phys., **83** (1985) 444;
      (d) G. Radhakrishnan, S. Buelow and C. Wittig, J. Chem. Phys.,
          84 (1986) 727;
      (e) S. Buelow, M. Noble, G. Radhakrishnan, H. Reisler, C. Wittig
          and G. Hancock, J. Phys. Chem., **90** (1986) 1015;
      (f) D. Häusler, J. Rice and C. Wittig, in Ref. 29, 5413;
      (g) S. Buelow, G. Radhakrishnan and C. Wittig, in Ref. 29, 5409.

[51] A.C. Legon and D.J. Miller, Acc. Chem. Res., **20** (1987) 39.

[52] F.A. Baiocchi, T.A. Dixon, C.H. Joyner and W. Klemperer,
      J. Chem. Phys., **74** (1981) 6544.

[53] R.B. Bernstein, D.R. Herschbach and R.D. Levine, in Ref. 29,
      5365.

[54] (a) S.R. Gandhi, T.J. Curtiss, Q.-X. Xu, S.E. Choi and
          R.B. Bernstein, Chem. Phys. Lett., **132** (1986) 6;
      (b) S.R. Gandhi, Q.-X. Xu, T.J. Curtiss and R.B. Bernstein,
          in Ref. 29, 5437.

[55] S.E. Choi and R.B. Bernstein, J. Chem. Phys., **85** (1986) 150.

[56] (a) H. Jalink, M. Janssen, F. Harren, D. Van den Ende,
          K.H. Meiwes-Broer, D.H. Parker and S. Stolte, in
          Proc. Conf. on Recent Advances in Molecular Reaction
          Dynamics, eds. R. Vetter and J. Vigué, C.N.R.S. Paris
          (1986), p. 41;
      (b) D.H. Parker, S. Stolte and H. Jalink, Ref. 47.

[57] (a) H. Jalink, D.H. Parker and S. Stolte, J. Chem. Phys., **85**
          (1986) 5372;
      (b) H. Jalink, D.H. Parker and S. Stolte, Chem. Phys. Lett.,
          in press (1987);
      (c) H. Ohoyama, T. Kasai, K. Ohashi and K. Kutwata, Chem.
          Phys. Lett., **136** (1987) 236.

[58] (a) D.H. Parker, K.K. Chakravorty and R.B. Bernstein,
          J. Phys. Chem., **85** (1981) 466;
      (b) D.H. Parker, K.K. Chakravorty and R.B. Bernstein,
          Chem. Phys. Lett., **86** (1982) 113.

[59] (a) K.K. Chakravorty, D.H. Parker and R.B. Bernstein,
          Chem. Phys., **68** (1982) 1;
      (b) S. Stolte, K.K. Chakravorty, R.B. Bernstein and D.H. Parker,
          Chem. Phys., **71** (1982) 353;
      (c) R.B. Bernstein, J. Chem. Phys., **82** (1985) 3656;

(d) S.E. Choi and R.B. Bernstein, J. Chem. Phys., 83 (1985) 4463;
(e) R.B. Bernstein, in Proc. Conf. on Recent Advances in
    Molecular Reaction Dynamics, eds. R. Vetter and J. Vigué,
    C.N.R.S., Paris (1984), p. 51;
(f) N.C. Blais and R.B. Bernstein, Ref. 48c.

[60] J.L. Knee, L.R. Khundkar and A.H. Zewail, J. Chem. Phys., 83
      (1985) 1996.

[61] (a) M.J. Dzvonik, S.C. Yang and R. Bersohn, in Ref. 28f;
      (b) M.D. Barry and P.A. Gorry, Mol. Phys., 52 (1984) 461;
      (c) G.N. Van Veen, T. Baller, A.E. deVries and N.J. Van Veen,
          Chem. Phys., 87 (1984) 405;
      (d) P.A. Gorry, F.G. Godwin, P.M. Hughes and C. Paterson, in
          Ref. 30.

[62] S.R. Gandhi, T.J. Curtiss and R.B. Bernstein, Phys. Rev. Lett.,
      submitted (1987).

[63] S. Kaesdorf, G. Schönhense and U. Heinzmann, Phys. Rev. Lett.,
      54 (1985) 885.

[64] Y.-F. Jiang, M. Giorgi-Arnazzi and R.B. Bernstein, Chem. Phys.,
      106 (1986) 171.

[65] (a) G.E. Hall, N. Sivakumar, R. Ogorzalek, G. Chawla,
          H.-P. Haerri, P.L. Houston, I. Burak and J.W. Hepburn,
          Far. Disc. Chem. Soc., 82 (1986) 13;
      (b) P.L. Houston, in Ref. 29, 5388.

[66] S.E. Choi, Ph.D. thesis, UCLA (1987).

[67] R. Altkorn, R.N. Zare and C.H. Greene, Mol. Phys., 55 (1985) 1.

[68] N.F. Scherer, L.R. Khundkar, R.B. Bernstein and A.H. Zewail,
      J. Chem. Phys., 87 (1987) 1451.

[69] (a) L.B. Harding, to be published (1987);
      (b) G.C. Schatz and M.S. Fitzcharles, paper in this Volume.
      (c) G.C. Schatz, M.S. Fitzcharles and L.B. Harding, in Ref. 31.

[70] A conscious omission is the fascinating story of the $H + H_2$
      reaction, which could be the subject of a full workshop
      of its own! See, e.g.,
      (a) D.G. Truhlar and R.E. Wyatt, Ann. Rev. Phys. Chem., 21
          (1976) 1;
      (b) D.P. Gerrity and J.J. Valentini, J. Chem. Phys., 81 (1984)
          1298; 82 (1985) 1323;
      (c) E.E. Marinero, C.T. Rettner and R.N. Zare, J. Chem. Phys.,
          80 (1984) 4142;

(d) R. Götting, H.R. Mayne and J.P. Toennies, J. Chem. Phys., **80** (1984) 2230; **85** (1986) 6396;

(e) S.A. Buntin, C.F. Giese and W.R. Gentry, J. Chem. Phys., **87** (1987) 1443.

[71] (a) J.C. Polanyi, Science, **236** (1986) 680;
(b) Y.T. Lee, Science, **235** (1986) 793.

[72] C.V. Shank, Science, **233** (1986) 1376.

[73] (a) N.F. Scherer, J.L. Knee, D.D. Smith and A.H. Zewail, J. Phys. Chem., **89** (1985) 5141;
(b) M. Dantus, M.J. Rosker and A.H. Zewail, J. Chem. Phys., **87** (1987) 2395.

[74] (a) J.C. Polanyi, Far. Disc. Chem. Soc., **67** (1974) 129;
(b) H.-J. Foth, J.C. Polanyi and H.H. Telle, J. Phys. Chem., **86** (1982) 5027.

[75] (a) P. Hering, P.R. Brooks, R.F. Curl, R.S. Judson and R.S. Lowe, Phys. Rev. Lett., **44** (1980) 687;
(b) P.R. Brooks, R.F. Curl and T.C. Maguire, Ber. Bunsenges. Phys. Chem., **86** (1982) 401;
(c) P.R. Brooks, R.F. Curl, S. Kaesdorf and T.C. Maguire, in Ref. 31.

[76] D. Imre, J.L. Kinsey, A. Sinha and J. Krenos, J. Phys. Chem., **88** (1984) 3956.

[77] D.W. Chandler and P.L. Houston, J. Chem. Phys., **87** (1987) 1445.

LASER STUDIES ON THE SELECTIVITY OF ELEMENTARY CHEMICAL REACTIONS: PRODUCTS, ENERGY, ORIENTATION

Jürgen Wolfrum
Physikalisch-Chemisches Institut der Universität Heidelberg
Im Neuenheimer Feld 253
D-6900 Heidelberg
FRG

1. INTRODUCTION

The multivalence structure of chemical elements opens numerous possibilities for chemical product formation, energy utilization and distribution, stereo-selectivity and product orientation. Experiments on the effect of selective product formation and vibrational, translational and orientational excitation of reactants in bimolecular reactions can give important insight into their microscopic dynamics. The information obtained in such experiments can be compared with the results of theoretical calculations of the reaction dynamics based on *ab initio* potential energy surfaces and is also of fundamental interest for improving the kinetic data used in detailed chemical kinetic modelling.

2. REACTIONS AND PATHWAYS IN THE SELECTIVE REDUCTION OF NO WITH NH_3

Inhibition and sensitization of chemical processes by nitric oxide are well-known phenomena. Recently the role of the nitrogen oxides in the formation of acid rain, photochemical smog and the possible depletion of the stratospheric ozone layer has stimulated the interest in chemical reactions which can selectively remove nitrogen oxides. An interesting elementary chemical reaction in this respect is the reaction of nitric oxide with NH_2 radicals. First direct studies of the rate and products of this reaction showed a very fast complex addition rearrangement sequence which forms nitrogen molecules and highly vibrationally excited water molecules in the single step [1]

$$NH_2 + NO \rightarrow N_2 + H_2O^*. \tag{1}$$

Later investigations with various experimental methods showed different amounts of OH radicals formed in this reaction [2-8]. Recent studies [5-8] restrict this channel to about 10% at room temperature. Modelling the NH_3-NO-O_2 reaction system at 1000-1500 K [9,10] requires higher branching ratios in this temperature range to explain the selective reduction of NO by NH_3 [11]. Theoretical calculations of the potential energy barriers for the various channels are in agreement with the two

J. C. Whitehead (ed.), Selectivity in Chemical Reactions, 23–45.
© 1988 by Kluwer Academic Publishers.

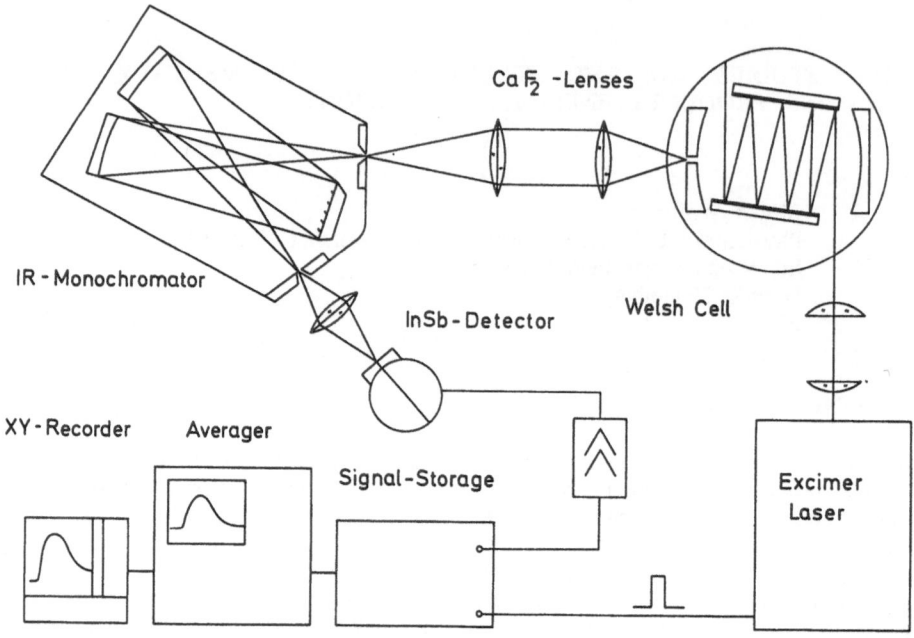

FIGURE 1. Experimental set-up for detection of time and spectral resolved infrared fluorescence signals from water molecules formed in Reaction (1).

possibilities [12]. However, since this branching ratio is determined by very small energy differences no quantitative answer can be obtained in this way. With the experimental arrangement depicted in Fig. 1 direct measurements of the distribution of reaction energy in the water molecule formed in reaction (1) can be carried out. An enlarged photolysis volume for the production of NH_2 radicals by photolysis of NH_3 is created by multireflection of an ArF-Exciplex laser beam combined with a Welsh-mirror light gathering system for the effective collection of infrared fluorescence from excited H_2O molecules. The spectrally resolved infrared emission shown in Fig. 2 can approximately be simulated with a vibrational temperature for the H_2O molecules formed of 10^4 K. Significantly lower vibrational temperatures are found in the nitrogen molecules by CARS spectroscopy [13]. Thus, the reaction (1) is very selective in channelling the available reaction energy preferably into the stretching vibrations of one reaction product. Therefore, as shown in Fig. 3 many simultaneous and competing pathways have to be considered in a simulation of the selective reduction of NO by NH_2 radicals in the presence of various amounts of O_2 (see Fig. 4).

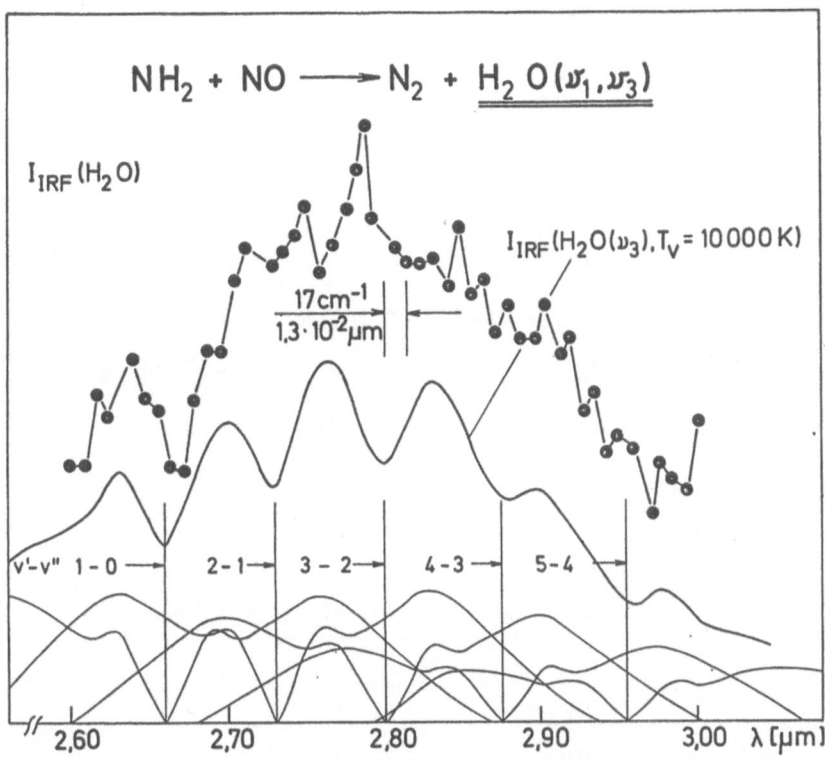

FIGURE 2. Measured (dots) and simulated (solid line) spectra of H_2O^* (v_1, v_3) 25 μs after laser photolysis of a NH_3/NO gas mixture. Shown on the bottom in the figure are the individual H_2O-vibrational transitions that constitute the simulated spectrum.

3. REACTIVE AND INELASTIC CHANNELS IN THE REACTIONS OF ATOMS WITH VIBRATIONALLY EXCITED MOLECULES

The simplest systems in which the effect of a selective vibrational excitation can be studied are those of reactions of free atoms with vibrationally excited diatomic molecules. The various channels for removal of the vibrationally excited molecules BC (v) may be written as

$$
\begin{aligned}
A + BC\ (v) &\to A + BC\ (v') \\
&\to AB\ (v') + C \\
&\to AC\ (v') + B \qquad \Delta H^0 \gtrless 0 \\
&\to (ABC)^*
\end{aligned}
$$

$$NH_3 + H \longrightarrow NH_2 + H_2$$

$$NH_3 + O \longrightarrow NH_2 + OH$$

$$NH_3 + OH \longrightarrow NH_2 + H_2O$$

$$NH_2 + OH \longrightarrow NH + H_2O$$

$$NH_2 + O_2 \longrightarrow HNO + OH$$

$$NH_2 + HNO \longrightarrow NH_3 + NO$$

$$NH_2 + NO \longrightarrow N_2 + H_2O^*(\nu_i)$$

$$H_2O^*(\nu_i) + M \longrightarrow H_2O + M$$

$$\longrightarrow OH + H + M$$

$$H_2O^*(\nu_i) + O_2 \longrightarrow HO_2 + OH$$

$$H_2O^*(\nu_i) + NO \longrightarrow HNO + OH$$

$$NH_2 + NO \longrightarrow N_2 + H + OH$$

$$HNO + H \longrightarrow NO + H_2$$

$$HNO + M \longrightarrow NO + H + M$$

$$HNO + OH \longrightarrow NO + H_2O$$

$$H_2, CO, C_nH_m \text{ Oxidation}$$

FIGURE 3. Elementary steps in the selective reduction of NO by NH_3-radicals.

As model systems for the competition between energy transfer processes and chemical reactions under non-equilibrium conditions one can use simple thermoneutral halogen and hydrogen atom exchange reactions. Vibrationally excited HCl (v) molecules can be consumed by H or D atoms in electronically adiabatic processes either by thermoneutral hydrogen atom exchange, the slightly exothermic hydrogen atom abstraction reaction, or in non-reactive collisions. Figure 5 shows an experimental arrangement for the observation of these elementary processes [14].

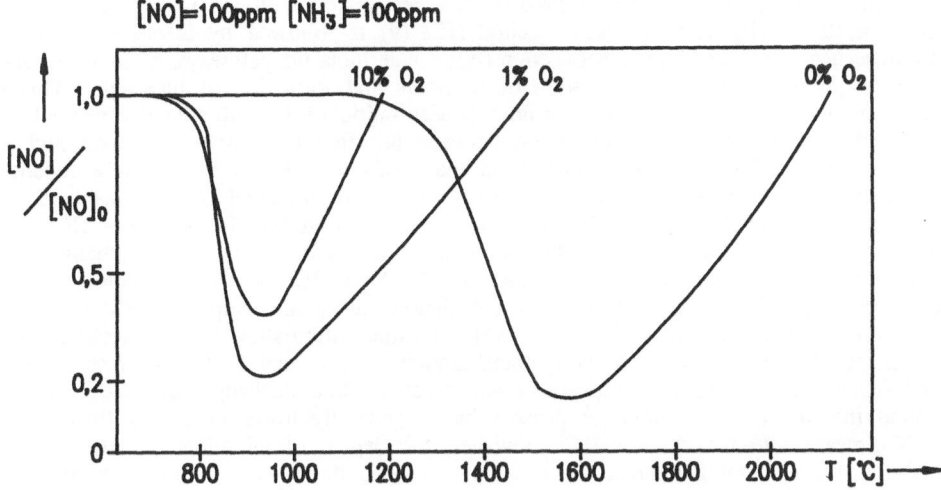

FIGURE 4. Selective reduction of NO with NH$_3$ in the presence of O$_2$.

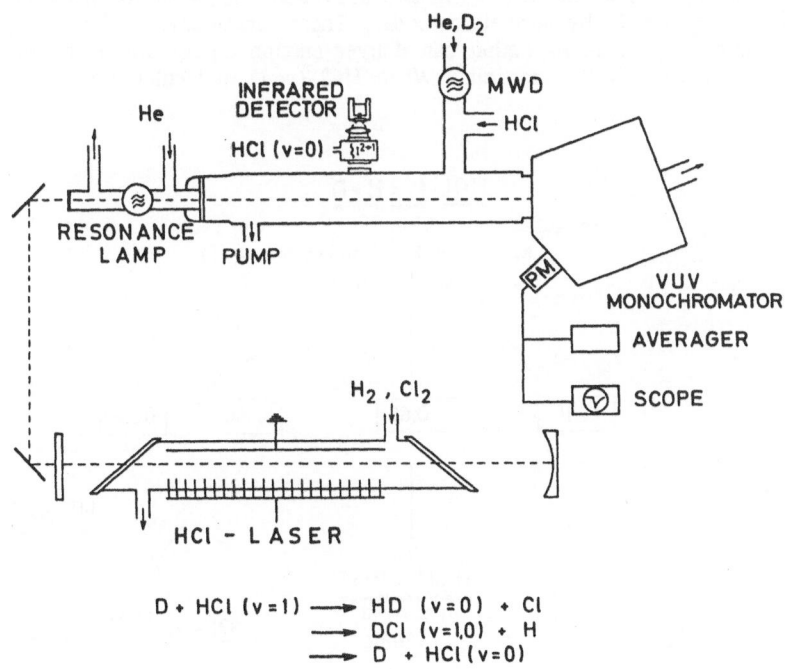

FIGURE 5. Discharge-flow system for simultaneous time-resolved detection of the concentration of reacting atoms and vibrationally excited HCl (v) molecules [14].

HCl molecules in the vibrational ground state are mixed with atoms in a discharge flow reactor. The decay of laser excited HCl (v) is followed by infrared fluorescence. To distinguish between reactive and inelastic pathways, it is necessary to measure the absolute consumption of reactants and formation of products. This is achieved here by measuring the absolute concentration of the vibrationally excited molecules using the rapid equilibration between the HCl (v) vibrational levels and a measurement of the relative population in the levels v = 1 and v = 2 as a function of time. The concentrations of the reacting atoms are followed by time-resolved atomic resonance absorption. Figure 6 summarizes the results for the D + HCl (v=1) system. The non-reactive relaxation and not the hydrogen atom exchange or abstraction reaction is mainly responsible for the high HCl (v=1) deactivation rate in contrast to predictions from theoretical calculations using semiempirical [15] as well as *ab initio* potential energy surfaces [16]. Further information on this system can be obtained from experiments using translationally hot H and D atoms to study the individual roles of inelastic excitations and reactive atom exchange processes. It is found that the reactive exchange process has a generally lower efficiency than the T-V process. In the case of HCl, however, the degree of vibrational excitation in the reactive channel is higher. It appears that once the system is following the potential energy surface for reactive exchange and has entered the transition state for reaction, the deposition of energy into higher vibrational states is much more facile. However, the system does not enter the reactive surface as readily, perhaps because of some geometrical constraints to overcome the reactive barrier [17].

Ground state oxygen atoms react relatively slowly with thermal HCl at room temperature. The rate and Arrhenius activation energy of the reaction has been measured directly by several methods. These measurements show that a single vibrational quantum excitation can deliver enough energy for overcoming the potential energy barrier of the reaction. When HCl (v=1) molecules are generated in the flow

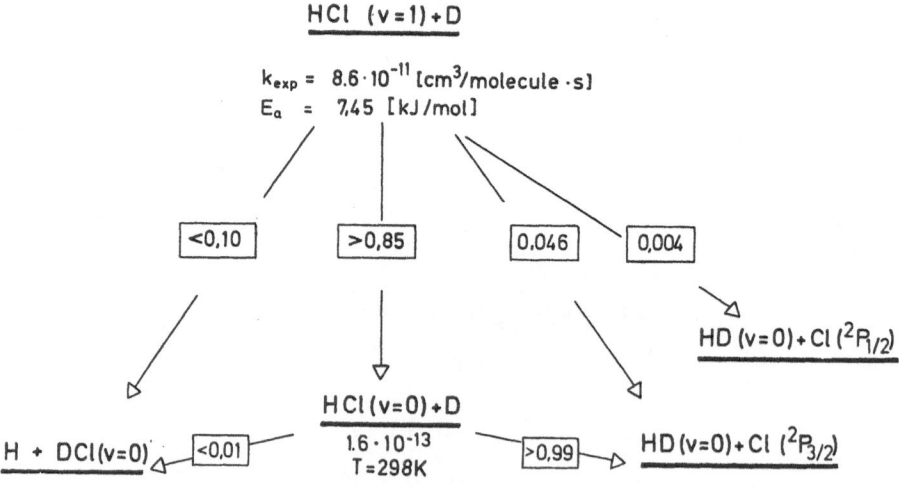

FIGURE 6. Experimental results for different channels in the D + HCl (v=0,1) reaction [14].

system by absorption of the laser pulse, the decay of HCl (v=1) is significantly accelerated in the presence of oxygen atoms. However, the data given in Fig. 7 indicate that the reactive channel to form OH + Cl gives only a small contribution to the rapid removal of HCl (v=1) by $O(^3P)$. The rate enhancement is much less than the factor exp $(E_v(v=1)/RT)$. Since the Arrhenius pre-exponential factor is not changed significantly by vibrational excitation, the contribution of HCl (v=1) molecules to the thermal reaction is small for most temperatures of interest.At 200 K thermal excited HCl (v=1) molecules contribute less than 10^{-3} % and at 2000 K about 10% to the total consumption of HCl by $O(^3P)$ atoms. This is also true for HCl (v=2) molecules. As shown by quasiclassical trajectory calculations, the remaining thermal activation energies for HCl (v=1,2) are very similar [18]. A theoretical model to explain the effective energy transfer in collisions involving P-state atoms as a result of electronically nonadiabatic curve crossing was given by Nikitin and Umanski [19]. As shown in Fig. 8, several potential energy surfaces exist for the interaction of $O(^3P)$ atoms with HCl (v). At a certain distance a nonadiabatic coupling between the different vibronic states is possible. The approach of the reactants $O(^3P)$ and HCl on a triplet surface followed by a nonadiabatic transition to the singlet HOCl surface as an intermediate complex has been discussed as the possible origin of the potential energy barrier in this reaction. However, the fact that this crossing point appears to be necessarily lower than the saddle point of the lowest triplet surface is of course an artifact of the single coordinate correlation

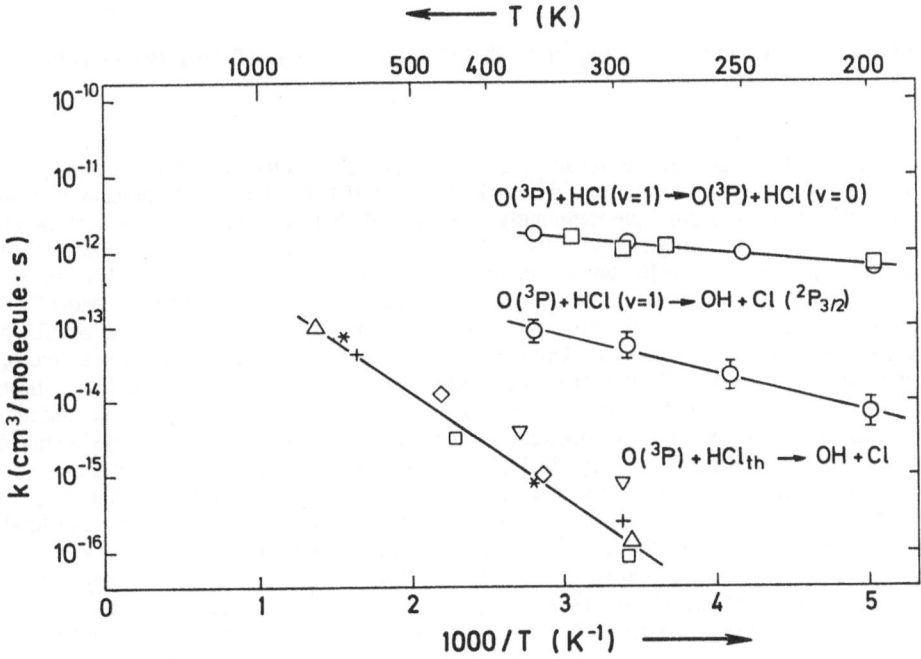

FIGURE 7. Experimental data for the temperature dependence of the rates for vibrational relaxation and reaction of HCl (v=1,0) with $O(^3P)$ atoms [14].

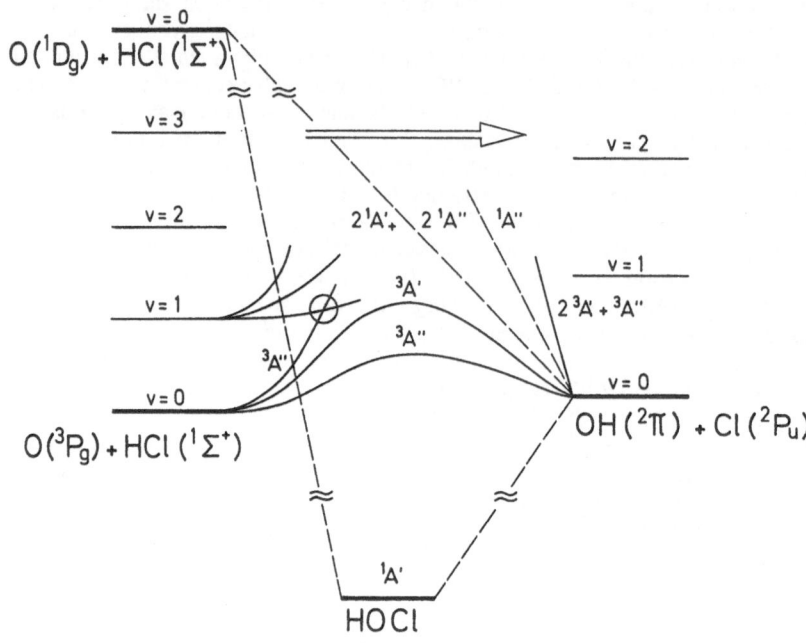

FIGURE 8. Chemical reaction and vibrational deactivation of HCl (v) in collisions with $O(^3P)$ atoms.

diagram. The experimental results on the reverse Cl + OH (v ⟨ 9) reaction [20] and the observed formation of OH (v=1) from $O(^3P)$ + HCl (v=2) indicate that the chemical reaction occurs predominantly vibronic adiabatically on a triplet surface and does not proceed through a long-lived HOCl complex. However, such interpretation of the experimental results on the competition of reactive and inelastic channels in the reactions of vibrationally excited HCl molecules are still qualitative. More quantitative *ab initio* calculations including electronic nonadiabaticprocesses should be carried out for these systems. On the other hand the interaction between a hydrogen atom and hydrogen molecule provides a theoretically simple system which has been studied now for more than half a century. Owing to the lack of a dipole moment and an electronic absorption spectrum in the vacuum ultraviolet for the H_2 molecule, state-selective studies were very difficult to perform experimentally before laser methods became available. As shown in Fig. 9 single quantum vibrational excitation of the H_2 molecule exceeds the Arrhenius activation energy (E_a) and the threshold energy (E_0) as well as the classical barrier height (E_c) of the reaction D + H_2. A CARS detection system shown in Fig. 10 provides a method for monitoring directly reactants and products in the D + H_2 (v=1) reaction. The reaction is followed in a discharge flow system, where the atoms and H_2 (v=1) molecules were generated by microwave discharges [22]. As shown in Fig. 11, HD (v=1) and HD(v=0) molecules are formed in adiabatic and non-adiabatic reaction pathways. Information on the competition of reactive and inelastic channels can be obtained by monitoring the decrease of H_2 (v=1) in the presence of D atoms (see Fig. 12) corrected for the energy transfer process HD (v=1) + H_2 (v=0) → HD (v=0) + H_2 (v=1) [23]. The

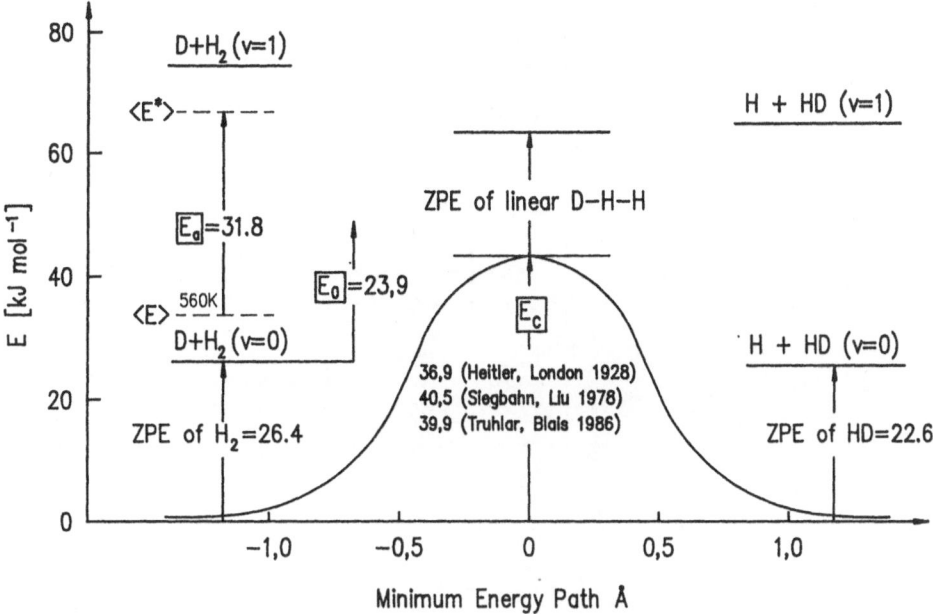

FIGURE 9.   Characteristic energies for the D + H$_2$ (v=0,1) reaction.

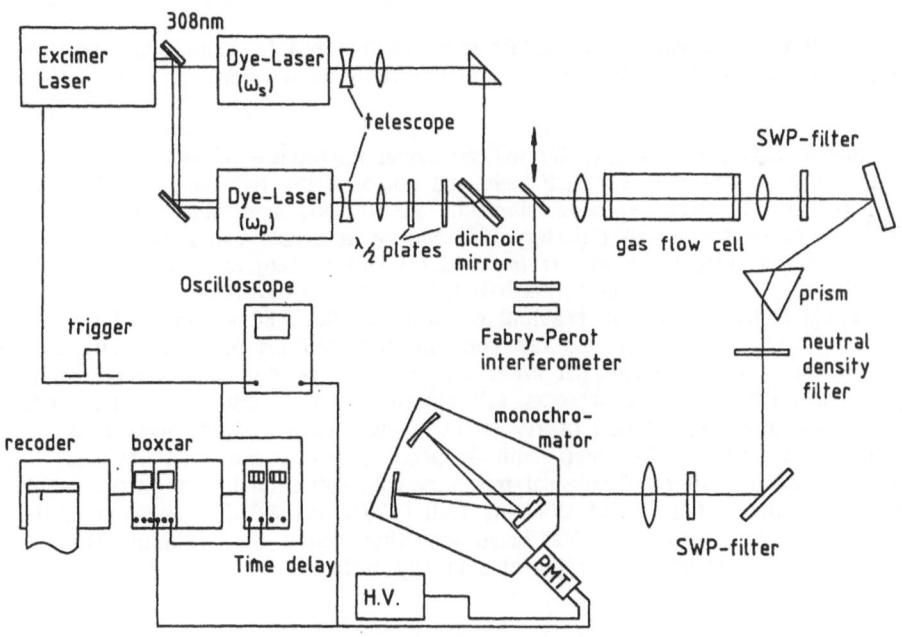

FIGURE 10a.   Schematic of the discharge-flow system used for monitoring reactants and products in the D + H$_2$ (v=1) → HD (v=0,1) + H reaction.

32

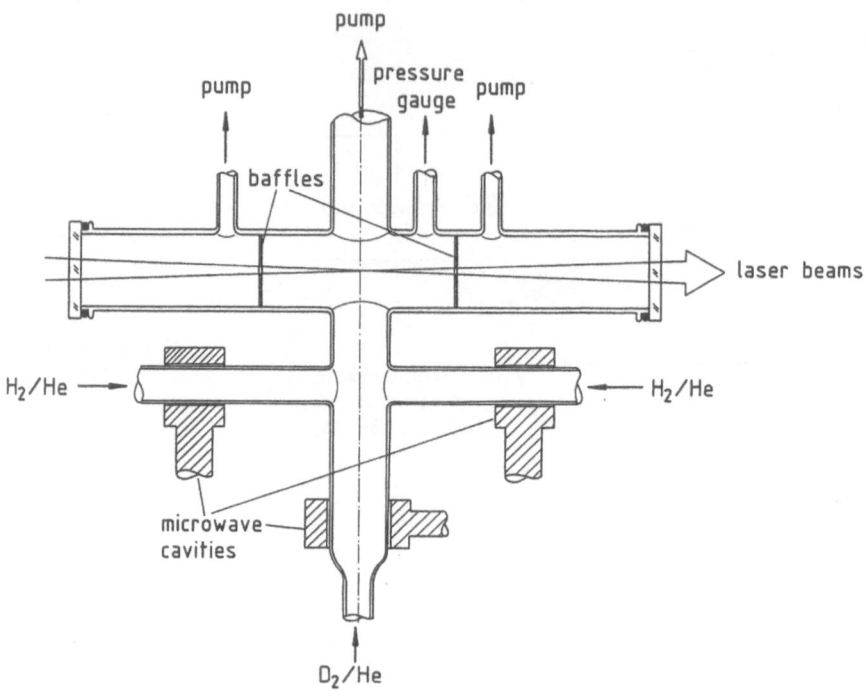

FIGURE 10b. Schematic of the CARS-spectrometer used for monitoring reactants and products in the D + H$_2$ (v=1) → HD (v=0,1) + H reaction.

experimental results obtained so far indicate equal importance of inelastic and reactive channels as well as of adiabatic and non-adiabatic reactive channels. As shown in Fig. 13, these experimental results are in good agreement with the predictions of quasi-classical trajectory [24] and semiclassical variational transition state calculations [25] using the *ab initio* LSTH surface [26] or a new surface based on double-many-body-expansions (DMBE) [25]. The new DMBE surface has a lower classical barrier of 9.65 kcal/mol compared to the 9.80 kcal/mol of the LSTH surface. However, the calculated rate constants for reactions of vibrationally excited hydrogen molecules are somewhat lower than previous results, in agreement with the experimental data. In some respects, this agreement with quasi-classical trajectories is surprising since the reaction of H$_2$ (v=1) involves only a small number of state-to-state processes. However, similarly good agreement has been obtained between predictions of quasi-classical trajectory calculations and experimental results using translationally hot H and D atoms with CARS and REMPI detection of the reaction products [27] or molecular beam scattering studies [28] sampling also regimes of higher energy at the potential energy surface.

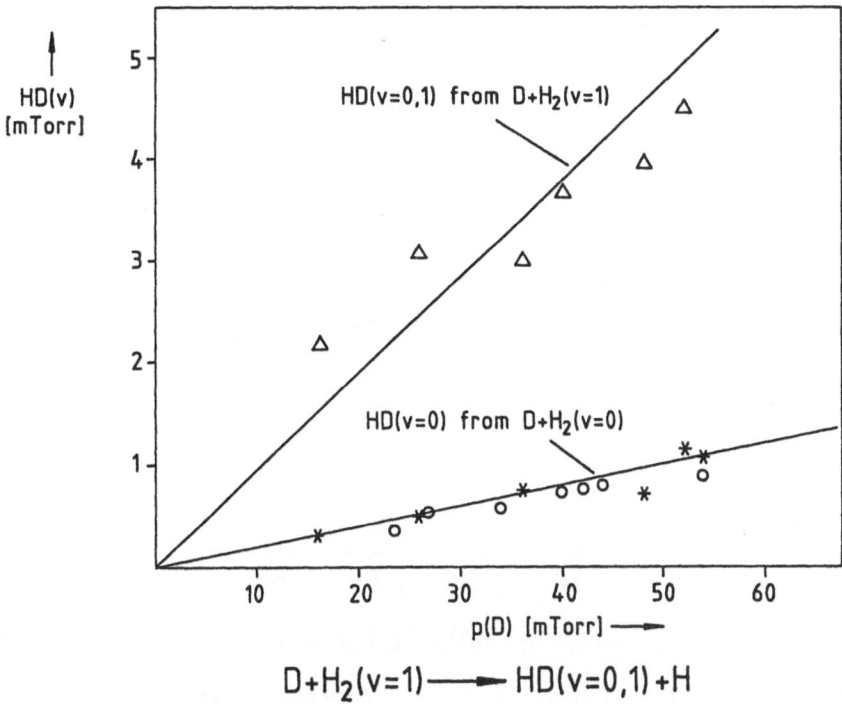

$$D + H_2(v=1) \longrightarrow HD(v=0,1) + H$$

FIGURE 11.   CARS-signals of HD (v=0,1) formed in the D + H_2 (v=1) reaction.

4.   SELECTIVE ENERGY REQUIREMENTS IN THE REACTION CO + OH → CO_2 + H

The reaction

$$OH + CO \rightarrow CO_2 + H \tag{2}$$

is the dominant step for conversion of CO into CO_2 in combustion systems [29] and in the chemistry of the upper atmosphere [30].   Various measurements of the thermal rate constants over a wide temperature range (200K < T < 2500K) show an upward curvature of the Arrhenius plot [31].   Transition-state theory explains the curvature in the ln k_2 versus 1/T plot equally well by using quite different activated complex configurations and barrier heights in the exit channel [30-34]. Measurements with state-selected reactants should give more microscopic details of the reaction.

Experiments using vibrationally excited hydroxyl radicals do indicate that OH (v=1) is quenched by CO at least as rapidly as it reacts with it.   A rate enhancement of less than a factor of two was measured at room temperature [35]. Vibrational excitation of the CO molecule should provide another clue to the microscopic mechanism of this reaction.   The energy of the vibrational quanta in CO

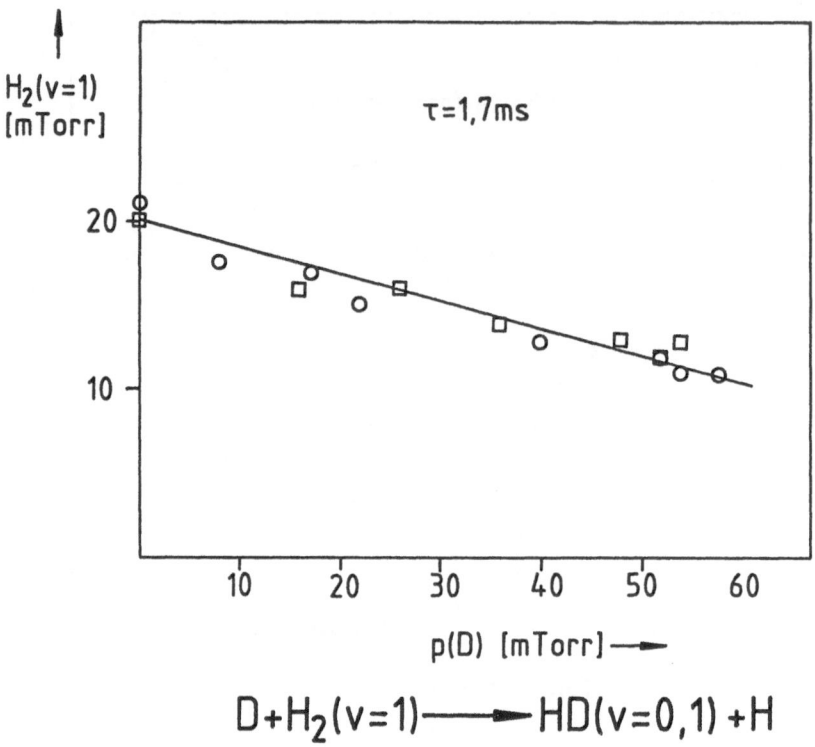

$$D + H_2(v=1) \longrightarrow HD(v=0,1) + H$$

FIGURE 12. Relative decrease of the H_2 (v=1) concentration in the D + H_2 (v=1) reaction as a function of the D-atom partial pressure.

is comparable to the apparent activation energy observed at higher temperatures. Compared with OH radicals and other diatomic hydrides, the CO molecule has a much smaller dipole moment in the vibrational ground state. The change due to the dynamic dipole moment after vibrational excitation should therefore be more dramatic compared to those cases. Figure 14 shows an experimental set up for investigation of the CO (v) + OH reaction [36]. The experiments were performed using a discharge flow system coupled to a quadrupole mass spectrometer by a nozzle beam sampling system. Vibrational excitation of CO was achieved by energy transfer from vibrationally excited N_2 molecules. The vibrational and rotational temperatures of the reacting CO molecules could be determined directly by a stabilized cw-CO-laser as the light source for infrared resonance absorption measurements. From the measured absolute population in the v=1 - 4 CO vibrational levels the vibrational temperature of CO is monitored at different excitation levels of the N_2 molecules. At the same time the rotational and translational temperature of CO was kept at room temperature. With constant $T_T = T_R = 298$ K, k_2 decreases with increasing vibrational temperature of CO from $k_2 = 9 \times 10^{10}$ cm^3/mol sat $T_v^{CO} = 298$ K to 7.8×10^{10} cm^3/mol s at $T_v^{CO} = 1800$ K.

Such a decrease may be explained by the formation of an intermediate reaction

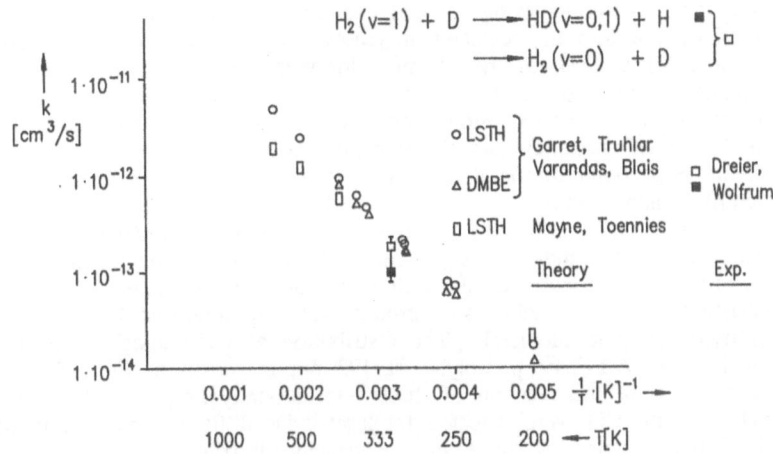

FIGURE 13. Comparison of rate constants for D + H_2 (v=1) in inelastic and reactive channels (□) and vibrational adiabatic and non-adiabatic channels (■) from experiments [22] and theoretical calculations [24,25].

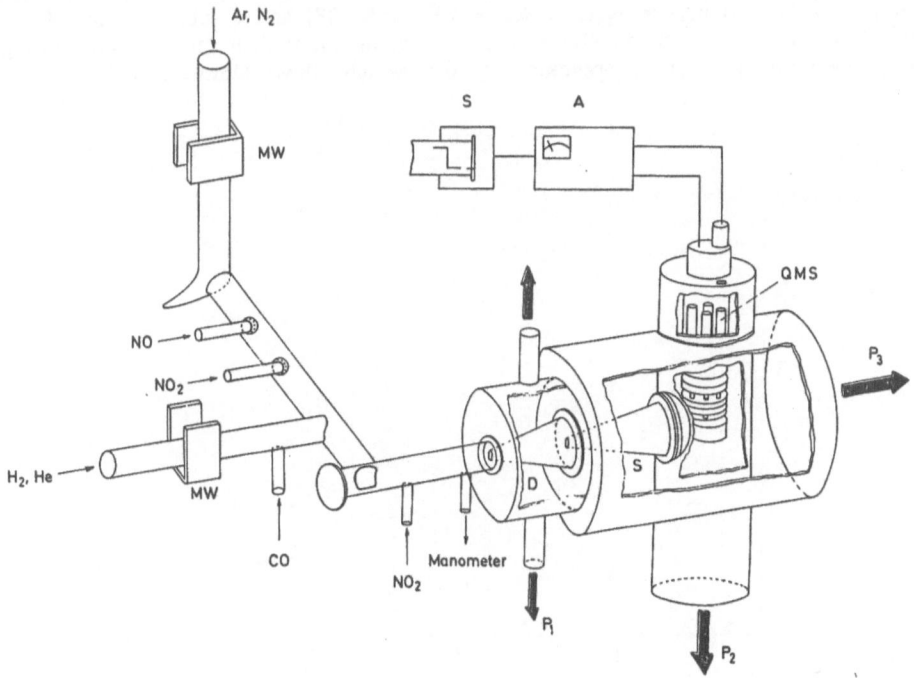

FIGURE 14. Schematic of the experimental arrangement for investigation of the reaction CO (v) + OH → CO_2 + H.

complex. With increasing internal excitation of the complex the rate of redissociation into the reactants increases, which results in a decreasing reaction rate if the cross section for the complex formation is not significantly enhanced by the vibrational excitation of the reactants. If the additional energy carried by the vibrationally excited CO molecules is equipartitioned among the degrees of freedom in the HOCO complex and the transition state, the measured rate for CO (v=1) should correspond to the thermal rate at a temperature of 830 K. The experiments described above show that k_2 (T = 830 K) $\geqslant$ 4 $\times$ k_2 (v=1).

In contrast, however, selective excitation of the relative translational energy of the reactants can induce a significant enhancement of the reaction rate in both directions. Hydroxyl radicals with high translational energy in well-defined rotational energy states in the vibrational ground state are produced by the photodecomposition of hydrogen peroxide [37]. The distribution of OH radicals over the rotational levels obtained on ArF-laser photolysis at 193 nm is shown in Fig. 15. Hydrogen atoms produced from the reaction of these translationally hot, yet vibrationally cold OH radicals with CO were detected by laser-induced fluorescence using tunable VUV light in the region of the Lyman-α wavelength (Fig. 16). The VUV light was generated by frequency tripling [12,13] the emission of an excimer laser-pumped dye laser tunable around 364 nm. The dye laser pulse of 20 mJ/cm^2 was focused by a quartz lens with f = 10 cm into a cell containing 400 Torr of a krypton-argon mixture. The cell was separated from the reactor by a MgF$_2$ window. A 90° arrangement of the two laser beams was chosen to reduce scattering from as well as damage of the MgF$_2$ window at the tripling cell by the photolysis laser. At a total cell pressure of 90 mTorr, v_{OH} = 4.7 $\times$ 10^5 cm/s [38] and a cross section of 60 Å^2, the probability of an OH radical undergoing more than one collision during the probe time of 60 ns is approximately 0.1, which allows almost collision free

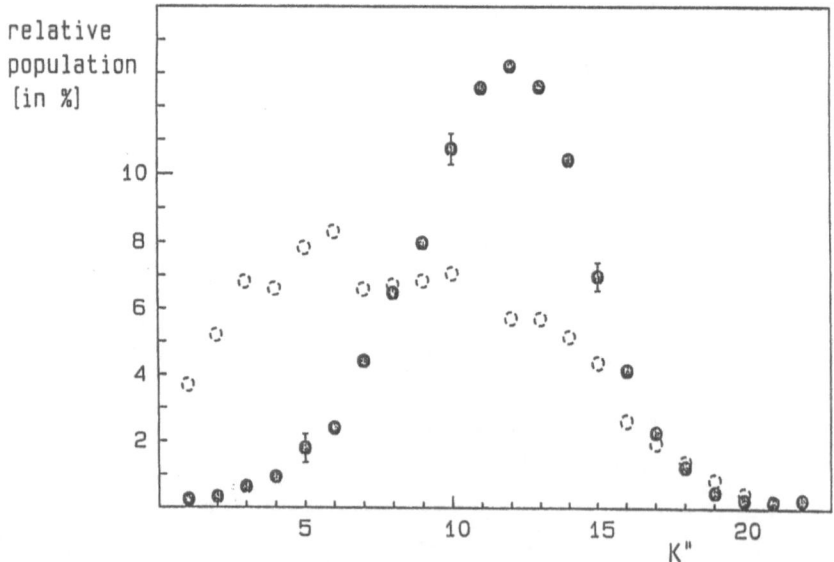

FIGURE 15. Rotational energy distribution of OH ($^2\Pi$, v"=0) radicals formed in the laser photolysis of H$_2$O$_2$ at 193 nm.

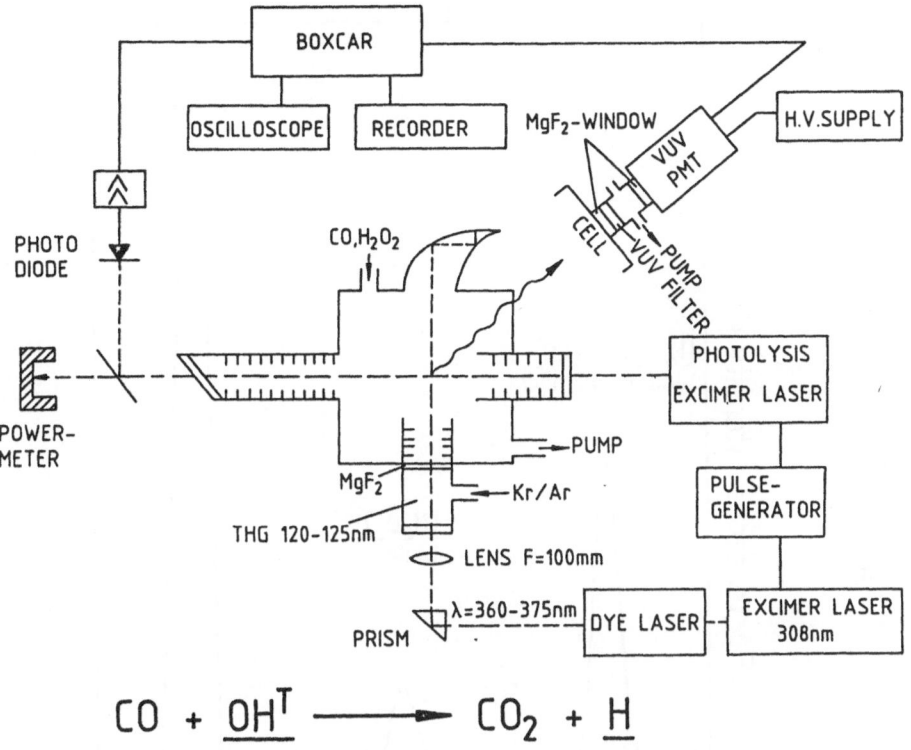

$$CO + \underline{OH}^T \longrightarrow CO_2 + \underline{H}$$

FIGURE 16. Experimental arrangement for the time-resolved detection of H atoms formed in the reaction of translationally hot OH radicals with CO by LIF in the VUV spectral region.

spectra to be recorded. The reaction cross section S is determined according to

$$S(v_{OH}) \approx [H]_t/[OH]_{t=0}[CO]v_{OH}t$$

where v_{OH} is the velocity of the OH radicals after photolysis and t the probe time [39]. The initial density of the OH reactant $[OH]_{t=0}$ is evaluated from the H_2O_2 partial pressure in the cell and its absorption cross section at 193 nm $S_{H_2O_2}$ = 1.2 × 10^{-18} cm^2 [40].

The amount of H atoms produced in the reaction at time t $[H]_t$ is obtained by calibrating the H atom detection sensitivity *via* photolysis of HBr at 193 nm using S_{HBr} = 1.8 × 10^{-18} cm^2 [41]. Only the ratio [H]/[OH] is needed, therefore the photolysis laser intensity cancels.

Figure 17 shows H atom fluorescence excitation spectra from a typical experiment. The LIF signal of the product H atoms is displayed and compared to the calibration signal of H atoms from photolysed HBr. The H atom signal from

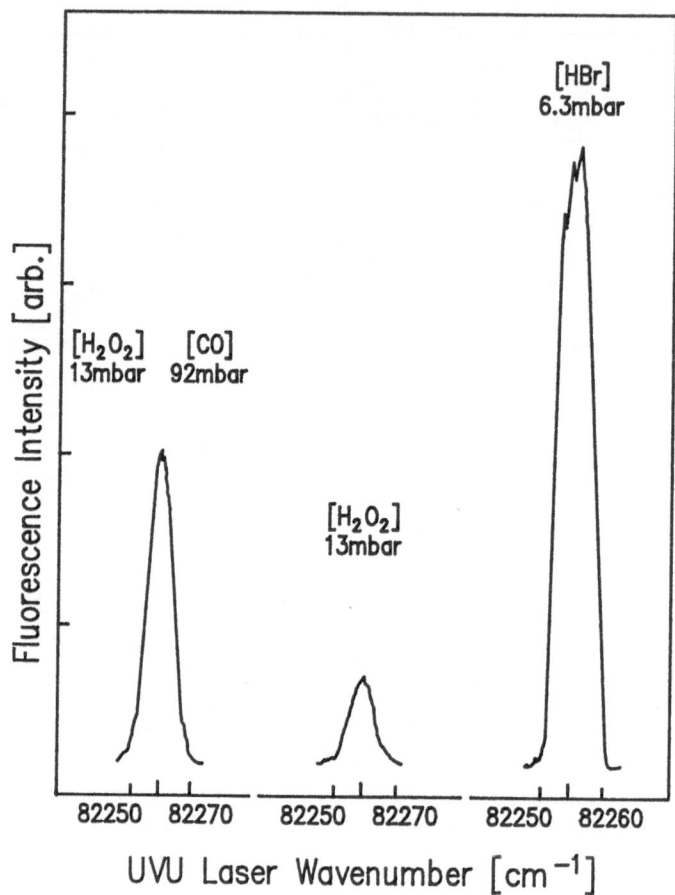

FIGURE 17. H atoms formed in the reaction CO + OH → CO$_2$ + H at E$_{C.M}$ = 1.32 eV, and in the photolysis of H$_2$O$_2$ and HBr at 193 nm.

the photolysis of H$_2$O$_2$ at 193 nm [42] has to be subtracted from the reaction signal. Evaluation of the nine best measurements yields a total reaction cross section of 19 ± 10 Å^2 at the centre of mass collision energy of E$_{CM}$ = 1.32 eV. First calculations on the dynamics of the reaction (-2) using fast hydrogen atoms are now available [43]. Such calculations should allow microscopic interpretations of the experimental results described here.

5. VECTOR SELECTIVE PROPERTIES OF THE REACTION H + O$_2$ → OH + O

Despite the large number of elementary reactions taking place in the oxidation of hydrocarbons, the important parameters of the combustion process are controlled by relatively few elementary reactions. Sensitivity analysis shows that the important parameters such as flame velocity are controlled to a large extent by the reaction of hydrogen atoms with oxygen molecules [29]. This endothermic reaction leads to the formation of two reactive radicals and is therefore the most important chain-branching step. The dynamics of such an elementary reaction with a high energy barrier can be studied in microscopic detail by combining translationally hot atom formation by laser photolysis with time-, state- and orientation-resolved product detection with laser-induced fluorescence spectroscopy.

The apparatus is shown in Fig. 18. Two antiparallel laser beams are directed coaxially through a flow reactor equipped with a baffle system to reduce the scattered light from the laser photolysis pulse and from the dye laser analysis pulse. The dye laser operates with rhodamine 640 and a frequency doubling KDP crystal to generate a pulse in the 306-311 nm region to probe OH radicals by laser-induced fluorescence. Fluorescence light is then detected as a function of the dye laser wavelength through emission optics and a filter transmitting between 240 and 390 nm and by a photomultiplier. For experiments using polarized photolysis and analysis laser beams both lasers were linearly polarized (*ca.* 95% polarization) by using 10 Brewster quartz plates (rack-polarizer). Both light beams are then directed through $\lambda/2$ plates so that the electric vectors of both lasers can be adjusted independently to any desired angle. The polarization experiments are based on measuring the distribution of orientations of the OH angular momentum vector J by using the polarized dissociation and analysis laser. OH fluorescence intensity is observed with the electric vectors of both lasers E_D and E_E parallel and perpendicular to each other. As illustrated in Fig. 19 the cases (a) $E_E||E_D||Z$ and (b) $E_E \perp E_D||Z$ were used with the lasers propagating along the X and the phototube along the Y axes. The polarization ratio I_{rel} (ratio of the fluorescence intensity in case (a) to that in case (b)) can be measured as a function of J for Q branch transitions, because here μ_{OH} lies along J_{OH} and larger polarization effects are expected. For P or R transitions, the μ_{OH} is perpendicular to J_{OH} and rotates in the OH plane of rotation [44].

Dissociation of HBr at 193 nm to H + Br $(^2P_{3/2})$ is induced by a perpendicular transition, so that the H atom flight direction is aligned with a $\sin^2$-distribution along E_D, i.e. $v_{H} \perp E_D$ preferentially. Figure 20 shows the variation

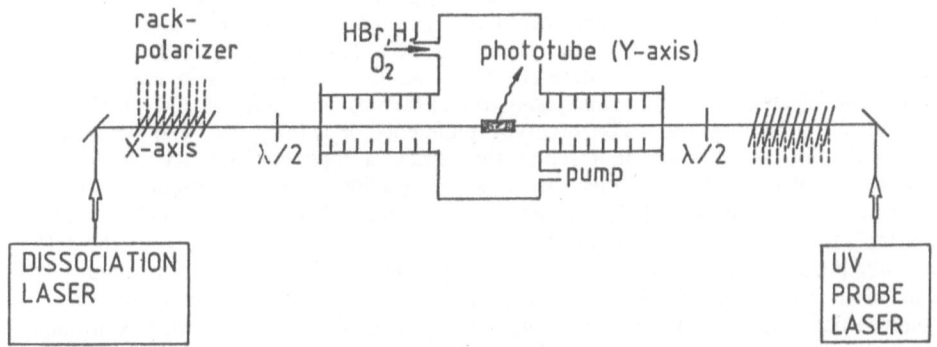

FIGURE 18. Experimental arrangement for the investigation of the H + $O_2 \rightarrow$ OH + O reaction product spatial rotational alignment as a function of the H atom flight direction.

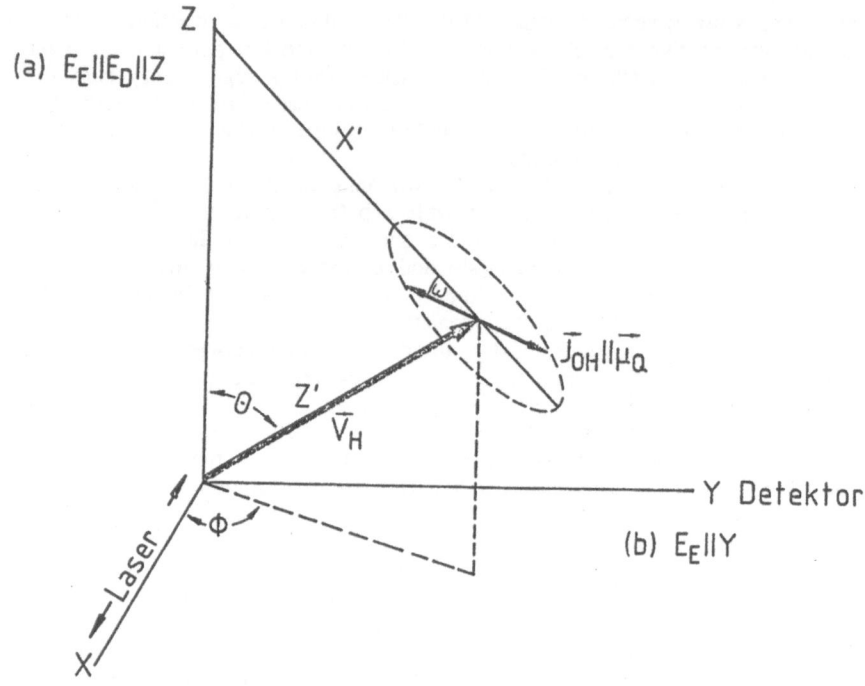

FIGURE 19. Schematic of the polarization analysis of the H + O_2 → OH + O reaction.

of the OH-$Q_1$16 (v''=0) fluorescence intensity with polarization of the dissociation laser E_D relative to analysis laser E_E. Experimental polarization effects at 251 kJ/mol collision energy for different rotational product states of the OH-radical are depicted in Fig. 21. Within the error limits similar ratios of the fluorescence signal intensity for the polarization of the photolysis laser parallel and perpendicular to the analysis laser are obtained for different rotational lines as well as in experiments where the electric vectors of the photolysis (193 nm) and analysis (307 nm) laser are rotated independently. The observed preference $J_{OH}||E_D\perp v_H$ can be explained by restrictions in the possible reaction geometries at high collision energies. Trajectory calculations show that the H + O_2 reaction occurs essentially in a plane at high collision energies [45]. From that we expect $J_{OH}\perp v_H$ for randomly oriented O_2 molecules. The transition moment μ_E of Q-lines is perpendicular to the OH rotation plane (||J_{OH}) for high OH rotational states. Thus we get maximum OH excitation probability |$E_E.\mu_E$|^2 for $\mu_Q||E_E||J_{OH}\perp v_H$, resulting in higher fluorescence intensity for $E_E||E_D$ than for $E_E\perp E_D$ [46]. This is also confirmed by analysing the λ-doublet excitation of the OH-radicals. The physical difference between the two λ-doublet components Π^+ (A') and Π^- (A") arises from interaction of the electronic spin-orbit momentum with the rotation of the molecule. For fast rotation of the OH radical, the unpaired electron in the p orbital of the oxygen is

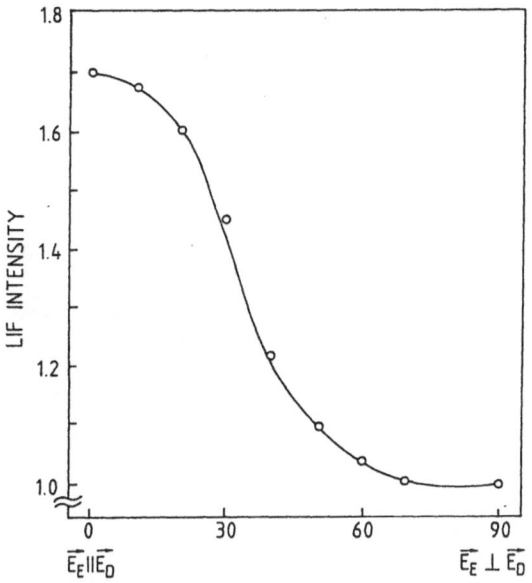

FIGURE 20. Variation of the OH (Q_1, 16 ($v''=0$)) fluorescence intensity from the H + O_2 → OH + O reaction as a function of the relative polarization of the dissociation laser E_D relative to the analysis laser E_E.

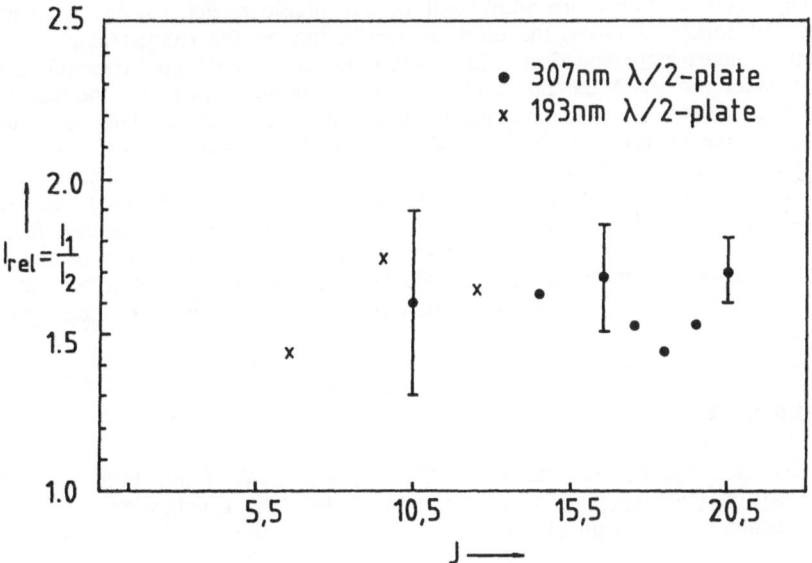

FIGURE 21. Experimental polarization effects in the H + O_2 → OH + O reaction at 251 kJ/mol collision energy.

42

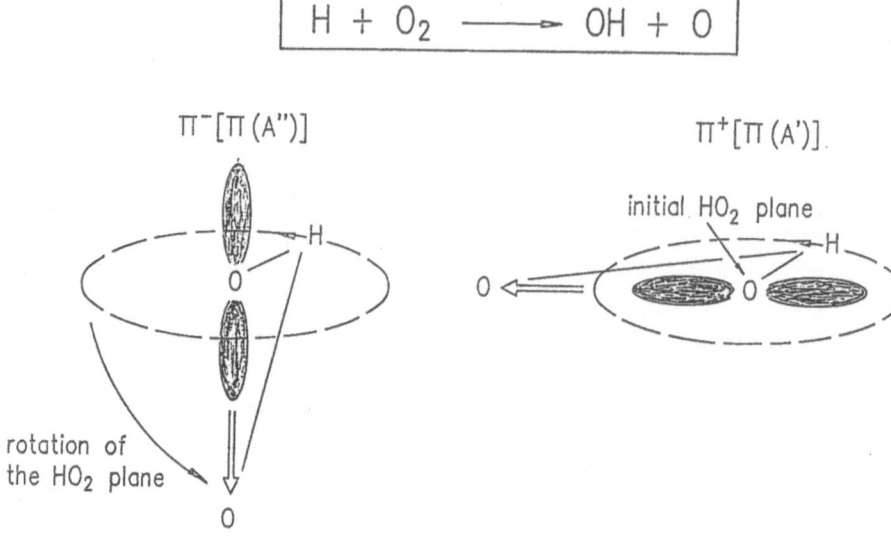

$$H + O_2 \longrightarrow OH + O$$

$\Pi^-[\Pi\,(A")]$ $\Pi^+[\Pi\,(A')]$

initial HO_2 plane

H

O

rotation of
the HO_2 plane

O

nonplanar reaction pathway   planar reaction pathway

FIGURE 22.   Vector properties of the $H + O_2 \rightarrow OH + O$ reaction (see text).

no longer able to follow the movement of the atomic nuclei.   If the p orbital lies
in the OH rotational plane, the electron distribution on the oxygen atom changes,
becoming increasingly spherical.   In contrast, for a Π^- (A") configuration, the
oxygen atom moves in the nodal plane of the p orbital and thus continues to "see"
a dumbbell-shaped electron environment, even for fast rotation.   This leads to a
splitting of the energies of the Π^+ (A') and Π^- (A") configurations, which
selectively increases with increasing rotational energy.   Experimentally, at 251 kJ
mol^{-1} collision energy, three OH radicals were found in the Π^+ (A') state for each
OH radical in the Π^- (A") state.   This shows that the unpaired electron formed
after bond cleavage of O_2 stays in an orbital in the rotational plane of the OH
radical.   During the reaction, most of the HO_2 complexes do not rotate out of the
initial plane, because of the short reaction time at high collision energies (Fig. 22).

REFERENCES

[1]   M. Gehring, K. Hoyermann, H. Schacke and J. Wolfrum, 14th
      Symp. (International) on Combustion, p. 99, The Combustion
      Institute, Pittsburgh (1973).

[2]   J.A. Silver and C. Kolb, J. Phys. Chem., 86 (1982) 3240.

[3]   P. Andresen, K. Kleinermanns, A. Jacobs and J. Wolfrum,
      19th Symp. (International) on Combustion, p. 11 (1982).

[4]   L.J. Stief, W.B. Brobst, D.F. Nava, R. Borkowski and
      J.V. Michael, J. Chem. Soc., Faraday. Trans. 2, **78** (1982) 1391.

[5]   J.L. Hall, D. Zeitz, J. Kasper, G.P. Glass, R.F. Curl,
      F.K. Tittel, J. Phys. Chem., **90** (1986) 2501.

[6]   D.A. Dolson, J. Phys. Chem., **90** (1986) 6714.

[7]   J.A. Silver and C. Kolb, J. Phys. Chem., **91** (1987) 3713.

[8]   R. Weller and J. Wolfrum, Periodic Report of the EC Energy
      Conservation Programme EN, 3F-0092 (1986).

[9]   J.A. Miller, M.C. Branch and R.J. Kee, Combustion Flame, **43**
      (1981) 81.

[10] M.A. Kimmball-Linne and R.K. Hanson, Combustion Flame, **64** (1986)
      337.

[11] R.K. Lyon, U.S. Patent No. 3900.554.

[12] C.F. Melius and J.S. Binkley, 20th Symp. (International) on
      Combustion, p. 575, The Combustion Institute, Pittsburgh (1984).

[13] Th. Dreier and J. Wolfrum, 20th Symp. (International) on
      Combustion, p. 695, The Combustion Institute, Pittsburgh (1984).

[14] D. Arnoldi and J. Wolfrum, Ber. Bunsenges. Phys. Chem., **80**
      (1976) 892; J. Wolfrum, 'Reactions of Vibrationally Excited
      Molecules' in Reactions of Small Transient Species, p. 105-156
      (A. Fontijn and M.A.A. Clyne Eds.), Academic Press, London
      (1983).

[15] R.E. Weston Jr., J. Phys. Chem., **83** (1979) 61; J.C. Miller
      and R.J. Gordon, J. Chem. Phys., **78** (1983) 3713 and references
      therein.

[16] P. Botschwina and W. Meyer, Chem. Phys., **20** (1977) 43;
      A.F. Voter and W.A. Godard III, J. Chem. Phys., **75** (1981) 3638.

[17] C.A. Wight, F. Magnotta and S.R. Leone, J. Chem. Phys., **81**
      (1984) 3951.

[18] R.D.H. Brown and I.W.M. Smith, Int. J. Chem. Kinet., **10** (1978)
      1; A. Persky and M. Broida, J. Chem. Phys., **81** (1984) 4352.

[19] E.E. Nikitin and S.Y. Umanski, Faraday Discuss. Chem. Soc.,
      53 (1972) 1.

[20]   B.A. Blackwell, J.C. Polanyi and J.J. Sloan, Chem. Phys., 24 (1977) 25.

[21]   J.E. Butler, J.W. Hudgens, M.C. Lin and G.K. Smith, Chem. Phys. Lett., 58 (1978) 216.

[22]   Th. Dreier and J. Wolfrum, Int. J. Chem. Kinet., 18 (1986) 919.

[23]   J. Arnold, D. Chandler, Th. Dreier and J. Wolfrum, J. Chem. Phys. (to be published).

[24]   H.R. Mayne and J.P. Toennies, J. Chem. Phys., 75 (1981) 179. N.C. Blais and D.G. Truhlar, Chem. Phys. Lett., 102 (1983) 120.

[25]   B.C. Garrett, D.G. Truhlar, A.J.C. Varandas and N.C. Blais, Int. J. Chem. Kinet., 18 (1986) 1065.

[26]   B. Liu, J. Chem. Phys., 58 (1973) 1925; 80 (1984) 581; P. Siegbahn and B. Liu, J. Chem. Phys., 68 (1978) 2457; D.G. Truhlar and C.J. Horrowitz, J. Chem. Phys., 68 (1978) 2466; 71 (1979) 1514.

[27]   E.E. Marinero, C.T. Rettner and R.N. Zare, J. Chem. Phys., 80 (1984) 4142; J.J. Valentini and D.P. Gerrity, Int. J. Chem. Kinet., 18 (1986) 937.

[28]   R. Götting, J.P. Toennies and M. Vodegel, Int. J. Chem. Kinet., 18 (1986) 949; R. Götting, H.R. Mayne, J. Chem. Phys., 85 (1986) 6396; R. Götting, V. Herrero, J.P. Toennies, M. Vodegel, Chem. Phys. Lett., 137 (1987) 524.

[29]   J. Wolfrum, 20th Symp. (International) on Combustion, p. 559, The Combustion Institute, Pittsburgh (1984).

[30]   I.W.M. Smith and R. Zellner, J. Chem. Soc., Faraday Trans. 2, 69 (1973) 1617.

[31]   R. Zellner, J. Phys. Chem., 83 (1979) 18.

[32]   A.R. Ravishankara and R.L. Thomson, Chem. Phys. Lett., 99 (1983) 377.

[33]   M. Morzukewich and S.W. Benson, J. Phys. Chem., 88 (1984) 6429; M. Morzukewich, J.J. Lamb and S.W. Benson, J. Phys. Chem., 88 (1984) 6441.

[34]   S. Buelow, M. Noble, G. Radhakrishnan, H. Reisler, C. Wittig and G. Hancock, J. Phys. Chem., 90 (1986) 1015.

[35] J.E. Spencer, H. Endo and G.P. Glass, 16th Symp. (International) on Combustion, p. 829, The Combustion Institute, Pittsburgh (1977).

[36] Th. Dreier and J. Wolfrum, 18th Int. Symp. on Combustion, p. 801, The Combustion Institute, Pittsburgh (1981).

[37] A. Jacobs, M. Wahl, R. Weller and J. Wolfrum, Appl. Phys. B, 42 (1987) 173.

[38] G. Ondrey, N. van Veen and R. Bersohn, J. Chem. Phys., 78 (1983) 3732.

[39] K. Kleinermanns and J. Wolfrum, J. Chem. Phys., 80 (1984) 1446.

[40] C.L. Lin, H.K. Rohatgi and W.B. DeMore, Geophys. Res. Lett., 5 (1978) 113.

[41] B.J. Huebert, R.M. Martin, J. Phys. Chem., 72 (1968) 3046.

[42] U. Gerlach-Meyer, E. Linnebach, K. Kleinermanns and J. Wolfrum, Chem. Phys. Lett., 133 (1987) 113.

[43] G.C. Schatz, M.S. Fitzcharles and L.B. Harding, Faraday Discuss. Chem. Soc., 84 (1987).

[44] M.P. Sinka, C.D. Caldwell and R.N. Zare, J. Chem. Phys., 61 (1974) 491; D.A. Case, G.M. McClelland and D.R. Herschbach, Mol. Phys., 35 (1978) 541.

[45] K. Kleinermanns and R. Schinke, J. Chem. Phys., 80 (1984) 1440.

[46] K. Kleinermanns and E. Linnebach, Appl. Phys. B, 36 (1985) 203.

ROTATIONAL AND STERIC EFFECTS IN THREE CENTRE REACTIONS

W. Grote, M. Hoffmeister, R. Schleysing,
H. Zerhau-Dreihöfer, and H.J. Loesch
Fakultät für Physik
Universität Bielefeld
D-4800 Bielefeld 1
Fed. Rep. of W. Germany

ABSTRACT. The current information on integral cross sections and rates for three centre reactions measured as a function of the molecular rotational quantum number is presented and the employed preparation methods briefly discussed. A simple dynamical model is described relating the most common rotational effects (initial decline, minimum and subsequent rise of cross sections and rates) to properties of the potential energy surface (anisotropy of the reagent valley and position of the barrier). Model predictions compare well with QCT-results on $O + HCl(v=0,1;j)$ and $H + H_2(v=0,j)$. Reaction cross sections for $K + HF(v=1,j=2)$ and $Xe^* + IBr$ measured at various alignments of the molecular angular momentum with respect to the relative velocity of the reagents are presented. The different approach geometries resulting from these alignments influence the cross sections markedly. These "steric effects" are interpreted in terms of the preferred reaction geometry using a simple formula relating the steric opacity function and the preparation of dependent cross sections. The reliability of the analysis procedure is confirmed by QCT-results for $Li + HF$ and $O + HCl$. From the dependence of the steric effects on reagent translational energy measured for $K + HF$ crude information on the anisotropy of the potential energy at the reaction shell is extracted.

1. INTRODUCTION

The outcome of a three centre reaction taking place on a single potential energy surface (PES) is governed by four quantities which specify the reagent motion completely: the relative translational energy (E_{tr}), the molecular vibrational $(E_{vib}(v))$ and rotational energy $(E_{rot}(j))$ as well as the quantum number m of the magnetic sublevels describing the component of the angular momentum vector $\vec{j}$ with respect to a quantization axis. These four quantities are the only parameters available for an external control of such a chemical process.

Triggered by the discovery of substantial vibrational enhancement of reactivity considerable experimental and theoretical effort was devoted to the exploration of the role of vibrational energy in bimolecular reactions. On the other side the role of rotational energy and the vector properties of $\vec{j}$ were only marginally investigated. One obvious reason for this is that only minute effects due to the usually much smaller energy of rotation were expected and first experimental results supported this

J. C. Whitehead (ed.), Selectivity in Chemical Reactions, 47–78.
© 1988 by Kluwer Academic Publishers.

conclusion [1]. However, recent results demonstrate that there do exist substantial effects which range from an inhibition of reaction to a moderate enhancement. Furthermore, j and m-dependent reaction cross sections provide detailed information on steric properties of the intermolecular interaction and give new insight into the dynamics of elementary chemical reactions.

The integral reaction cross section of the completely specified process

$$A + BC(v,j,m) \rightarrow AB(v',j',m') + C$$

can be formally denoted as a function of all initial and final (primed) quantum numbers and the translational energy E_{tr}

$$\sigma(v',j',m'/v,j,m;E_{tr});$$

where m, m' refer to a suitable quantization axis. All observable quantities except for angular distributions of products can be deduced from this fundamental state-to-state cross section.

2. EXPERIMENTAL METHODS

In Table 1 we have compiled a complete list of reactions for which j or m specific cross sections or rates are now available. Reagents and products are characterized by the prepared and observed states only while the non-observed summed up states are omitted. The reactions listed are subdivided into the following three groups according to the method used in the investigation:

a) Detailed balance: The basic experimental informations are rate constants of exoergic reactions leading from a thermal mixture of initial states to a specific final state of products. The principle of detailed balance then provides rates for the reverse endoergic reactions starting at specific v', j' states and ending in the thermal state distribution of the products, the former reagents [2].

b) Chemiluminescence depletion: The most extensive set of data with respect to both the range of v and j was obtained for Na + HF,HCl using the chemiluminescence depletion method. In a precursor reaction the reagents in various v, j-states are generated and their concentration monitored by measuring the infrared luminescence. The atomic reagent was then admitted to the reactor and the correlated decrease of molecular reactants measured via the depletion of the luminescence [3].

c) Laser preparation methods: Various laser schemes like infrared pumping, selective photodissociation, selective stimulated emission pumping [4a] and stimulated Raman pumping [4b] appear to be qualified for the preparation of molecules.

Among these techniques only the first two have been applied to date in experimental studies on the j or m-dependence of reaction cross sections. Infrared pumping (IRP) was applied in crossed beam [5-9] and beam/gas [10-13] experiments with a variety of reactions of HF and HCl with alkali and alkaline earth atoms. The obvious restriction of the method to hydrogen-halides is a consequence of the restricted availability of suitable high power ir lasers. Several years ago only

TABLE 1. Experimental methods and reactions for which j and/or m-dependent cross sections are available (Me=alkaline earth).

Experimental Method	Reactions	authors/reference
Application of detailed balance to chemiluminescence (and LIF) data from reverse reactions	$I + HCl(v,j) \to IH + Cl$ $Cl + ClH(v,j) \to Cl_2 + H$ $H + HF(v,j) \to H_2 + F$ $D + DF(v,j) \to D_2 + F$	Bernstein, Polanyi *et al.* [2a,b] Kaplan, Levine, Manz [2c]
Chemiluminescence depletion	$Na + ClH(v,j) \to NaCl + H$ $Na + FH(v,j) \to NaF + H$	Blackwell, Polanyi, Sloan [3]
Laser preparation: infrared pumping	$Me + FH(1,j) \to MeF(v') + H$ $Me + ClH(1,j) \to MeCl(v') + H$ $Sr + FH(1,j) \to SrF(v') + H$ $K + ClH(1,j) \to KCl + H$ $K + FH(1,j) \to KF + H$ $Na + FH(1,j) \to NaF + H$ $Li + FH(1,j) \to LiF + H$ $K + FH(1,2,m) \to KF + H$	Zare et al. [10,11,12] Man, Estler [13] Dispert, Geis, Brooks [5] Grote, Hoffmeister, Potthast, Schleysing, and Loesch [6,7,8,9]
Selective photo-dissociation	$Xe^* + IBr(m) \to$ $\qquad XeI^* + Br$ $\qquad XeBr^* + I$	de Vries, Srdanov, Hanrahan, Martin [16]

50

chemical lasers whose active media are gases of exactly these molecules were available and could be applied. But recently, tunable ir lasers such as OPO, colour centre laser, Raman shifted visible dye lasers and others were developed or became available for standard use in a molecular beam lab. Thus it is very likely that in the near future the method will be increasingly used also for the preparation of other molecules and it has good chances to become a general and widely applied tool.

For a more detailed characterization of the IRP-method we will briefly discuss apparatus and experimental technique recently used to measure m- and j-dependent cross sections for the reaction K + HF(v=1,j,m) → KF + H [7,9]. Figure 1 shows a schematic drawing of the crossed molecular beams machine together with the optical set up employed to prepare HF via IRP.

The two reagent beams cross perpendicularly; the products emerging from the intersection volume are detected by surface ionization on a Re ribbon which can be rotated in the plane of the beams. Due to the special kinematics of this reaction the flux of products measured at the centroid angle is proportional to the total flux of products [17].

The source of ir radiation is a continuous wave colour centre laser operated in single mode. The light has a fixed polarization plane but the polarization vector $\bar{E}$ can be rotated in the plane of the two molecule beams by means of the $\lambda/2$-plate.

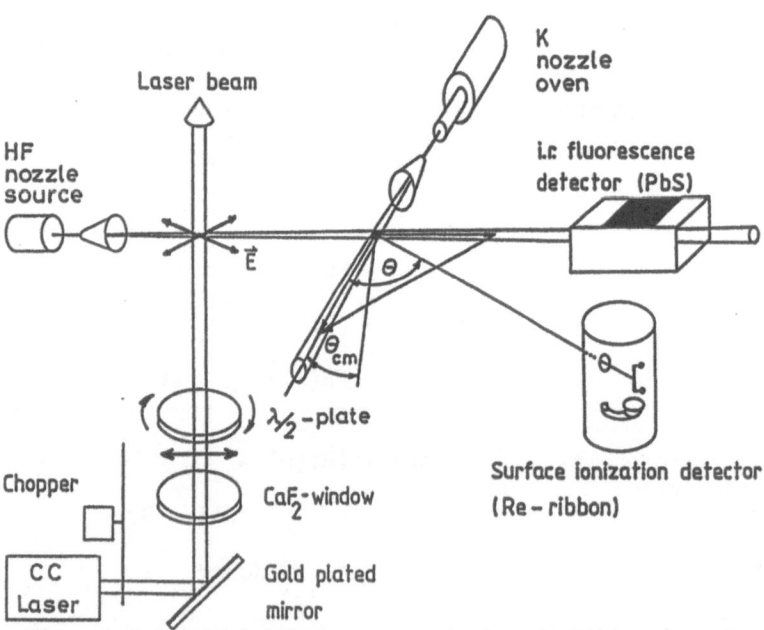

FIGURE 1. Schematic drawing of molecular beam apparatus and optical set-up of Ref. 9.

This feature is used to rotate the quantization axis of the optically populated m-states with respect to the relative velocity of the reagents when m-dependent cross sections are measured (c.f. Section 5). The laser beam intersects the HF-molecule beam perpendicularly 1.5 cm before the scattering volume. With the ir laser off, all molecules are in the vibrational ground state while the rotational states are distributed according to

$$f(j') \quad \text{with} \quad \sum_{j'} f(j') = 1.$$

When the laser is tuned to one of the $v=0 \to v=1$ transition frequencies like one of the $R_1(j-1)$ lines a fraction ϵ of the molecules in the relevant ground state $v=0$, $j-1$ is excited to the specific vibrational-rotational state $v=1,j$. In the absence of saturation effects ϵ is proportional to the spectral power density $\delta(\omega)$ of the radiation, Einstein's B coefficient and the period of time τ the molecule senses the radiation field

$$\epsilon(j-1) \propto \delta(\omega)B_{v=0,j-1 \to v=1,j} \cdot \tau.$$

ϵ amounts at a laser power of 300 mW/cm^2 and a bandwidth of 5 MHz to the order of 1%.

The population of v,j'-states is then

$$v=0: \quad f(j') - \epsilon(j-1).f(j-1).\delta_{j-1,j'}$$

$$v=1: \qquad \epsilon(j-1).f(j-1).\delta_{j,j'}.$$

The number density of molecules in the excited state which is proportional to $\epsilon(j-1).f(j-1)$ is monitored by the infrared radiation detector (a PbS-element) which measures the intensity of the fluorescence $F(v,j)$ from the spontaneous $v=1 \to v=0$ transition (Fig. 1).

The total flux of products with the ir laser off is

$$N_{off} \propto \sum_{j'} f(j').\sigma(v=0,j';E_{tr})$$

and with the laser on

$$N_{on} \propto \sum_{j'} [f(j').\sigma(v=0,j';E_{tr})]$$

$$+ \epsilon(j-1).f(j-1).[\sigma(v=1,j;E_{tr}) - \sigma(v=0,j-1;E_{tr})].$$

The difference of these fluxes which is proportional to the modulated detector signal divided by the infrared fluorescence intensity eventually furnishes the desired state specific reaction cross section

$$S(j) = \frac{N_{on} - N_{off}}{F(v,j)} \propto [\sigma(v=1,j;E_{tr}) - \sigma(v=0,j-1;E_{tr})], \qquad (1)$$

however, not isolated but diminished by the corresponding $v=0$ cross section.

For reactions with strong vibrational enhancement of reactivity like K + HF or at translational energies below the threshold of the v=0 reaction the second term is negligible or zero, respectively, and the method directly provides the state specific cross section.

Furthermore, use of polarized radiation permits the preparation of polarized rotational states even of specific magnetic sublevels [14]. To illustrate this we assume linearly polarized radiation. With respect to the polarization vector $\bar{E}$ the selection rule $\Delta m=0$ applies. If the molecules are irradiated, for example, with the $R_1(1)$ line only the substates m=-1,0,+1 of the v=1,j=2-level are excited and in the absence of saturation effects, the relative m-state population indicated by the

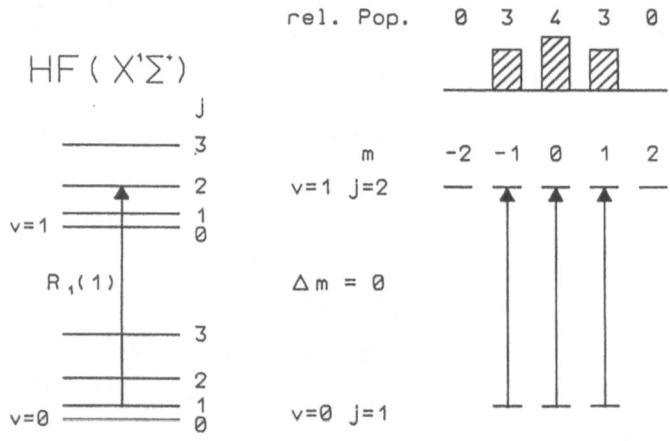

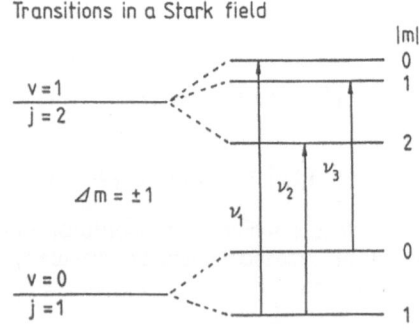

FIGURE 2. Level diagrams for the preparation of specific states of HF by infrared pumping (IRP).

histogram in Fig. 2 is proportional to the square of the transition matrix elements.

If the molecule in question is polar the Stark effect can be exploited to select one specific m-transition. The lower level diagram of Fig. 2 shows a typical splitting of the rotational levels in a homogeneous electric field. With the radiation directed parallel to the Stark field (quantization axis) the selection rule $\Delta m = \pm 1$ applies in this case, that is, depending on the frequency of the laser, $j=2, |m|=2$ or $|m|=1$ or $m=0$ can be prepared.

In general, molecules posess nuclear spin which couples to the rotation. This coupling may lead to a complete or partial loss of the prepared polarization [15]. With a Stark field these angular momenta can be decoupled and thus the loss of polarization avoided.

The other laser method applied to prepare reagent molecules is based on photodissociation. It is useful to generate an alignment of the angular momentum vector, however, it is not specific with respect to v and j-states. The technique exploits the dependence of the photodissociation probability for polarized light on the initial m-state. For example, using linearly polarized radiation, the dissociation probability of a parallel transition is much higher for states near m=0 than for those near $|m|=j$. Consequently, the radiation depletes the states near m=0 and after the laser pulse the state distribution is characterized by a maximum at $|m|=j$ and a minimum at m=0. The first and only application of the method in this field so far is by de Vries et al. who investigated the reaction Xe* + IBr → XeI* + Br/XeBr* + I [16].

3. RESULTS

Several typical examples for the effect of rotational energy on the integral reaction cross sections or rates of three centre reactions are depicted in Fig. 3.

Detailed thermal rates were measured for Na + HCl, HF → NaCl, NaF + H [3] as a function of j at various vibrational states using the chemiluminescence depletion method. Figure 3a shows the results for v=2. The rates first drop, reach a minimum and then increase slowly as j grows.

A quite similar behaviour has been found for Sr + HF(v=1,j) (Fig. 3b) and K + HCl (v=1,j) (Fig. 3c). The cross sections of the former reaction were measured in a beam/gas experiment using IRP (pulsed chemical HF laser) to prepare the HF molecules and LIF for product detection [13]. Cross sections of the latter system were obtained in a crossed beam experiment using the same preparation technique (cw chemical HCl-laser) but surface ionization for product detection [5]. The data feature an initial descent, which is particularly steep for K + HCl, to some j and then rise again. For K + HCl the latter property is not yet well established experimentally.

For the first time the effect of translational energy E_{tr} on the j-dependence of σ was studied for the reaction K + HF(v=1,j) → KF + H(j=5,6,7) [6]. In this crossed beam experiment the HF-beam was prepared via IRP using a pulsed chemical HF-laser. With increasing j the cross sections rise steeper for the two lower energies than for the two higher ones (Fig. 3d). Very recently measurements for the states j=0 to j=3 were performed [7]. Also in this j-range a marked effect of E_{tr} was found (Fig. 3e). At the lowest E_{tr} (0.49 eV) the cross section exhibits the initial descent followed by a minimum and subsequent rise. At the highest energy (1.21 eV) the minimum has disappeared making room for a continuous increase. At medium E_{tr} (0.75 eV) the cross section is essentially constant between

54

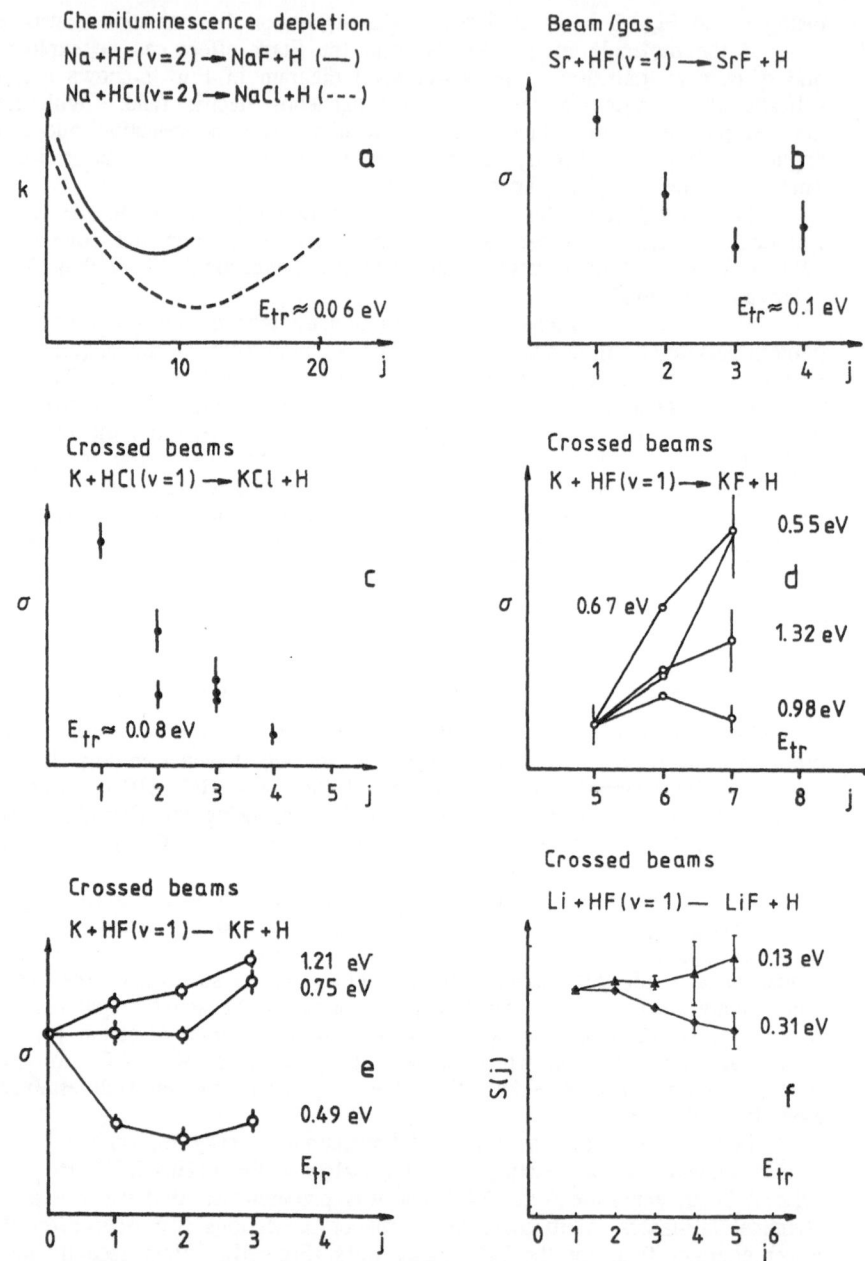

FIGURE 3. Measured j-dependence of cross sections (σ), rates (k) or signals (S) for various reactions. The data are taken from Ref. 2 (a), Ref. 13 (b), Ref. 5 (c), Ref. 6 (d), Ref. 7 (e), Ref. 8 (f).

j=0 and 2 and rises at j=3.

New data also exist on the reaction Li + HF(v=1,j) → LiF + H [8]. This reaction is of particular interest as *ab initio* surfaces exist [18,19,20] for which j-dependent cross sections were calculated [20,21]. The measurements were carried out in a crossed beam apparatus similar to the one shown in Fig. 1 except for the use of a pulsed chemical HF-laser and a pulsed HF-nozzle source. The results are displayed in Fig. 3f. As E_{tr} = 0.14 eV is only slightly above threshold the signal S(j) is essentially proportional to $\sigma(v=1,j)$. However, at E_{tr} = 0.32 eV, contributions to the signal arising from the v=0 reaction (c.f. Section 2) are certainly not negligible. Thus the observed decrease of S(j) might be caused by $\sigma(v=0,j)$ rising steeper with j than $\sigma(v=1,j)$ (for more details see Ref. 8).

The j-dependence of rate constants calculated via detailed balance from the strongly exoergic reverse reaction [2] all feature a similar shape. The rates start with a rise at small j, reach a maximum and descend steeply towards large j. No significant data could be deduced for low j. A considerable disadvantage of the method with respect to an investigation of pure rotational effects is that the total energy is a constant. As a consequence an increase in rotational energy at a fixed vibrational energy is possible only at the expense of the translational energy. Due to this fact and the thermal final state distribution the results of detailed balance calculations are difficult to compare with data from experiments with prepared reagents and differently weighted detection of products and are excluded from further consideration in this paper.

Summary:

(i)   Most cross sections or rates feature an initial decline, pass a minimum and rise again with increasing j.

(ii)   A few cross sections rise monotonously or are more or less constant.

(iii)   Translational energy has a marked effect on the shape of $\sigma(j)$.

In the following section we will try to give qualitative explanations for this variety of j-dependencies in terms of characteristic attributes of the PES.

4. INTERPRETATION OF j-DEPENDENT REACTION CROSS SECTIONS

A strong motivation for the great experimental effort to measure such detailed quantities like j-dependent cross sections is that they certainly also contain detailed information on the microscopic motion of the particles which is useful to improve our insight into the dynamics of elementary chemical reactions. There are two well established ways to extract clues about the microscopic events from cross sections. The one requires the knowledge of a PES and the computational simulation of data. The desired information is then deduced from the degree of agreement achieved and from the analysis of the scattering dynamics. The other way is to conceive a transparent model of the microscopic motion which directly relates the characteristic properties of the interaction with the observed features. In the subsequent Section 4.1 we touch on the first way whilst in Section 4.2 we concentrate on the second.

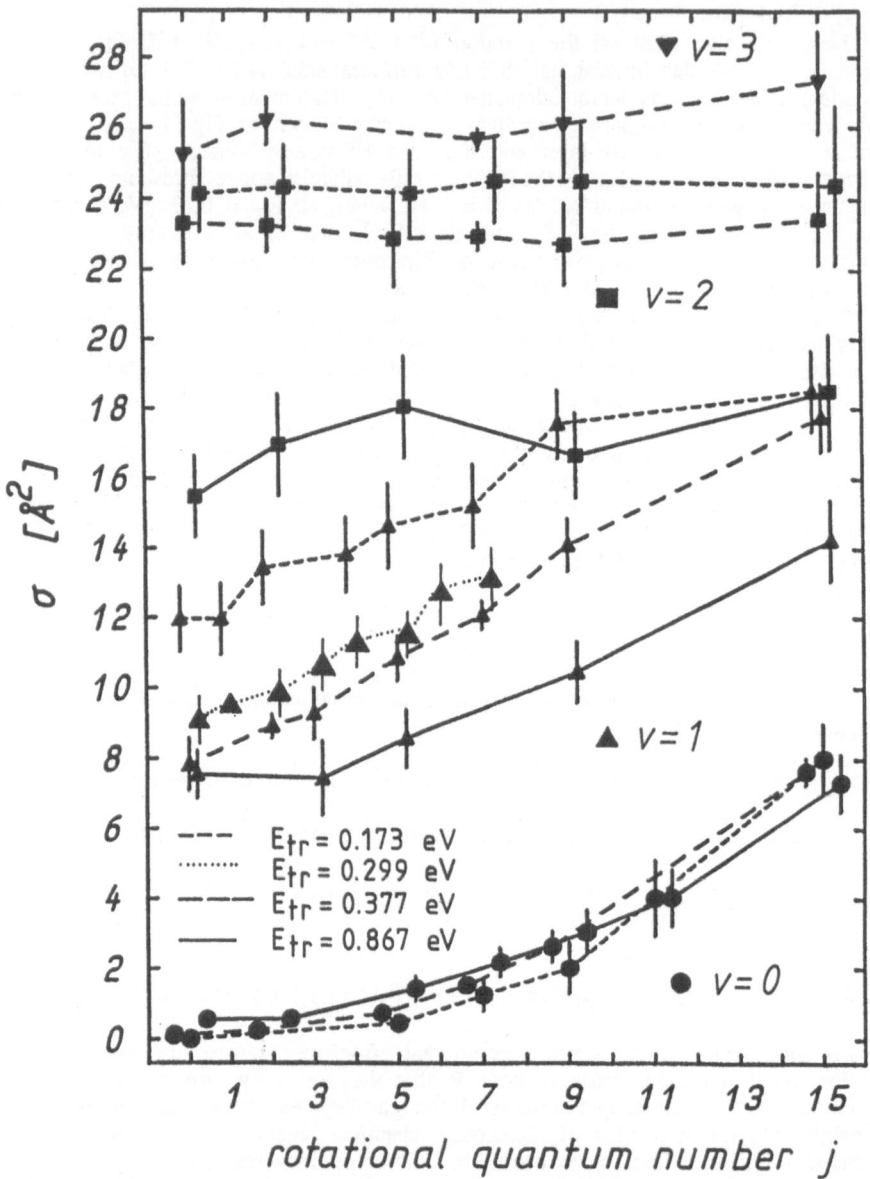

FIGURE 4. QCT-reaction cross sections for Li + HF(v,j) → LiF + H, calculated for the modified CSCM-PES (from Ref. 8).

4.1 A Case Study: Li + HF

The most reliable *ab initio* surface for this system was calculated by Chen and
Schaefer [18] and fitted to an analytic function by Carter and Murrell [19]. Since
then this surface - the CSCM-PES - has been used in a great number of
quasi-classical trajectory (QCT) calculations (e.g. Refs. 20,21). Among these
investigations there was also a detailed study on the influence of rotational energy
on reaction cross sections by Noorbatcha and Sathyamurthy [21]. These authors
found a striking structure in the shape of $\sigma(j)$. However, we had to realize later
that the structure was an artifact. We could clearly show on the basis of a simple
dynamical model [23] - the rotational sliding mass model (RSMM) - that the marked
structure was due to spurious oscillations of the fit polynomials of the CSCM-PES at
large distances between reagents. We removed these artifacts by using additional
damping functions and received a surface - the CSCM-mod PES- with a proper
asymptotic behaviour towards both the reagents and products [20]. The results of
our QCT-calculations on the modified surface exhibit a completely different shape
(Fig. 4): the cross sections for v=0 and v=1 rise monotonously with j while it is
nearly constant for v=2. The QCT results based on the CSCM-mod surface compare
well with our data. For more details see Ref. 8.

4.2 A Simple Model to Rationalize $\sigma(j)$

The model is based on the following reaction mechanism. We assume the existence
of a reaction shell $R_c=R_c(\gamma_c)$ which is defined by the condition that a reaction can
occur only if the distance R between A and the centre of mass of BC falls below
$R_c(\gamma_c)$ during the approach (c.f. Fig. 5). Here, γ_c (the reaction angle) denotes the
orientation angle γ at R_c. We then make the rather general statement that for the
reaction cross section:

$$\sigma(v,j,E_{tr}) = \sigma_0(v).P_c(j,E_{tr}).P_r(v,j,E_{tr}), \tag{2}$$

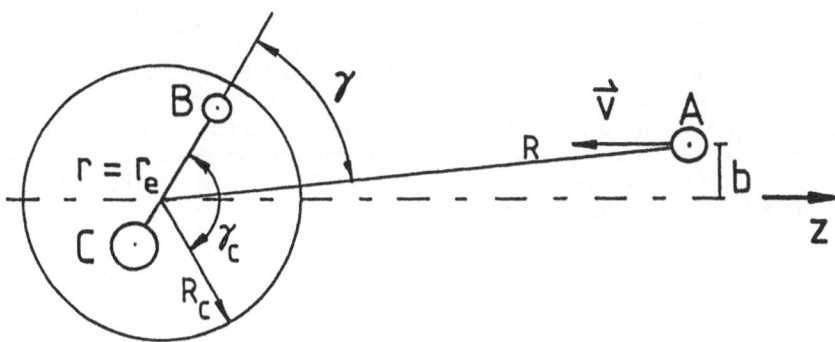

FIGURE 5. Collision geometry and reaction shell $R_c(\gamma_c)$.

where σ_0 is a normalization factor. $P_c(j,E_{tr})$ denotes the probability for the reagents to reach the shell and $P_r(v,j,E_{tr})$ the probability of the complex to react. A encounters BC usually in a sterically confined region and thus P_c is certainly not only dependent on the radial forces but to a great extent also on the orientation dependent angular forces (torques). It is for this reason that P_c is frequently referred to as describing an "orientational effect". On the other side there is evidence as shown below that P_r is dependent on the rotational energy and thus this factor describes an "energy effect".

4.3 The "Orientational Effect"

The probability P_c can be determined on the basis of exact three dimensional rigid rotor QCT calculations or by using the rotational sliding mass model (RSMM). The results of the latter are of course more qualitative but the method has the striking advantage that the trajectories and thus the probabilities are easily interpretable in terms of PES attributes. It should be mentioned that for all systems studied so far in this context the model provided nearly quantitative results. For these reasons the RSMM will be exclusively employed. The model has been published elsewhere [23,24] so only a short introduction suffices here.

The basic idea is to reduce the 6 spatial coordinates of the full problem to the essential two, the R and γ-coordinate, by assuming that

(i)   BC is a rigid rotor,

(ii)   A (and the centre-of-mass of BC) move on a straight line,

(iii)   central collisions with impact parameter b=0 are representative.

The classical Hamiltonian is then given by

$$H = \frac{1}{2\mu_{A,BC}} \; (P_R^2 + P_\delta^2) + V_{eff}(R,\delta), \qquad (3)$$

which describes the two dimensional (2D) motion of a mass $\mu_{A,BC}$ in cartesian R,δ coordinates sliding on a surface $V_{eff}(R,\delta)$, where $\delta = \gamma \cdot r_e \cdot [\mu_{BC}/\mu_{A,BC}]^{1/2}$ is the mass weighted orientation angle γ. The combined R, γ motion of the approaching system and its relation to the forces can be easily visualized by superimposing the path of the sliding mass on the R, γ contour diagram of the potential.

The term $P_\delta^2/2\mu_{A,BC}$ denotes the fraction of the rotational energy due to δ (or γ) - motion; the energy due to the azimuthal motion around the z-axis (c.f. Fig. 5) is constant as $V(R,\delta)$ is independent of the azimuthal angle and can be added to V to give the effective potential energy

$$V_{eff}(R,\delta) = E_{rot} \cdot \cos^2\chi/\sin^2\gamma + V(r \equiv r_e,R,\gamma). \qquad (4)$$

χ denotes the helicity angle between $\bar{j}$ and $\bar{v}$ and thus describes the initial polarization of the molecule.

Eventually $P_c(j,E_{tr})$ is determined by solving the 2D equations of motion for adequately distributed χ and initial conditions for R and δ. Then

$$P_c(j,E_{tr}) = N_c/N_{tot},$$

where N_c and N_{tot} denote the number of paths which crossed R_c and the total number of paths, respectively. For the purpose of illustrating how j-dependent effects evolve it is sufficient in many cases to plot planar trajectories with $\chi = 90^0$. Then the effective and the real potentials are equal and the initial angles γ_i are distributed uniformly between 0^0 and $(\pm)$ 180^0. P_c is then roughly given by the length of γ_i-intervals where all those trajectories start which lead to R_c.

4.3.1. <u>An Application of the RSMM</u>. The RSMM was first applied to reactions for which the j-dependence of cross sections is dominated by the "orientational" effect. For example, the model proved to be very useful to rationalize the striking structure of the theoretical $\sigma(j)$ for Li + HF found by Noorbatcha and Sathyamurthy mentioned earlier [21]. Here, we will illustrate the RSMM in an application to the system $O(^3P) + HCl \rightarrow OH + Cl$ [24]. For this reaction two LEPS PES's exist both equally qualified to explain temperature dependencies of experimentally known rate constants.

Figure 6 shows the computational results of Persky and Broida [25] for LEPS I and II. For LEPS I, $\sigma(j)$ increases slowly with growing j but decreases dramatically for LEPS II. This finding raises immediately the question of the dynamical origin for this contradictory behaviour. A comparison of the usually presented collinear surfaces (Fig. 7, upper panel) gives absolutely no hint of a possible explanation. The contours are nearly equal, the position and height of the saddle point are also similar.

Application of the RSMM to the reaction furnishes reaction probabilities (solid lines in Fig. 6) which recover nearly quantitatively the j-dependence of $\sigma(j)$. This is clear evidence for the predominance of "orientational" effects in both cases and a justification to rationalize the rotational effect by an inspection of the R,γ contour maps of the surface and traces of the sliding mass.

As shown in Fig. 7 (middle and lower panel) these contour maps indeed exhibit substantial deviations. The contours for LEPS I are weakly curved towards larger R while they are rather strongly curved into the opposite direction for LEPS II. This means that there are strong torques exerted onto the molecule (the derivatives of V with respect to γ) which are always directed towards $\gamma = 0^0$ in case of LEPS II, while the rather weak torques are directed the opposite way for LEPS I.

The reaction shell is for both PES's a slim parabolic curve and thus the potential energy at R_c, $V_c(\gamma_c)$, rises steeply with γ_c demonstrating the strong steric requirements of this reaction. Inspection of the few trajectories calculated for j=0 and j>0 and superimposed on the contour maps of Fig. 7 reveals now clearly the tight correlation between the shape of $P_c(j;E_{tr})$ and the anisotropy of the PES.

LEPS I

j=0: As there is no angular motion the trajectories enter the contour diagram on straight lines parallel to the R-axis and due to the weak angular forces they essentially remain straight lines up to R_c or the turning point. Energy conservation demands that paths can cross R_c only when $|\gamma_c| < \gamma_{c,max}$ where $\gamma_{c,max}$ is given by $V_c(\gamma_{c,max}) = E_{tr}$; this γ_c-interval is indicated by the two arrows on the γ-axis

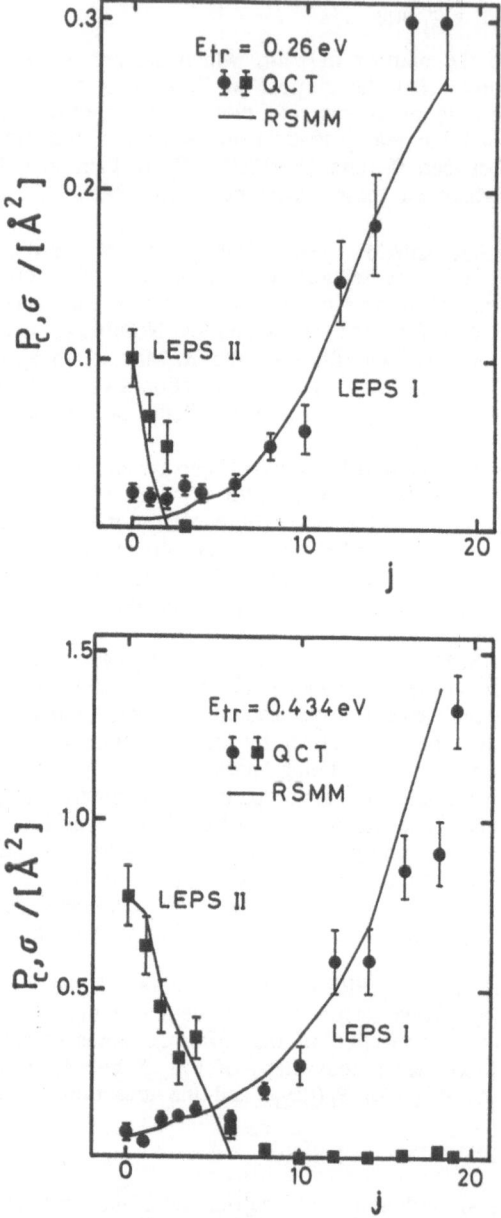

FIGURE 6. Reaction cross sections (σ) [25] and probabilities P_C for the reaction O + HCl(v=0,j) → OH + Cl calculated using LEPS I and II PES's (from Ref. 24).

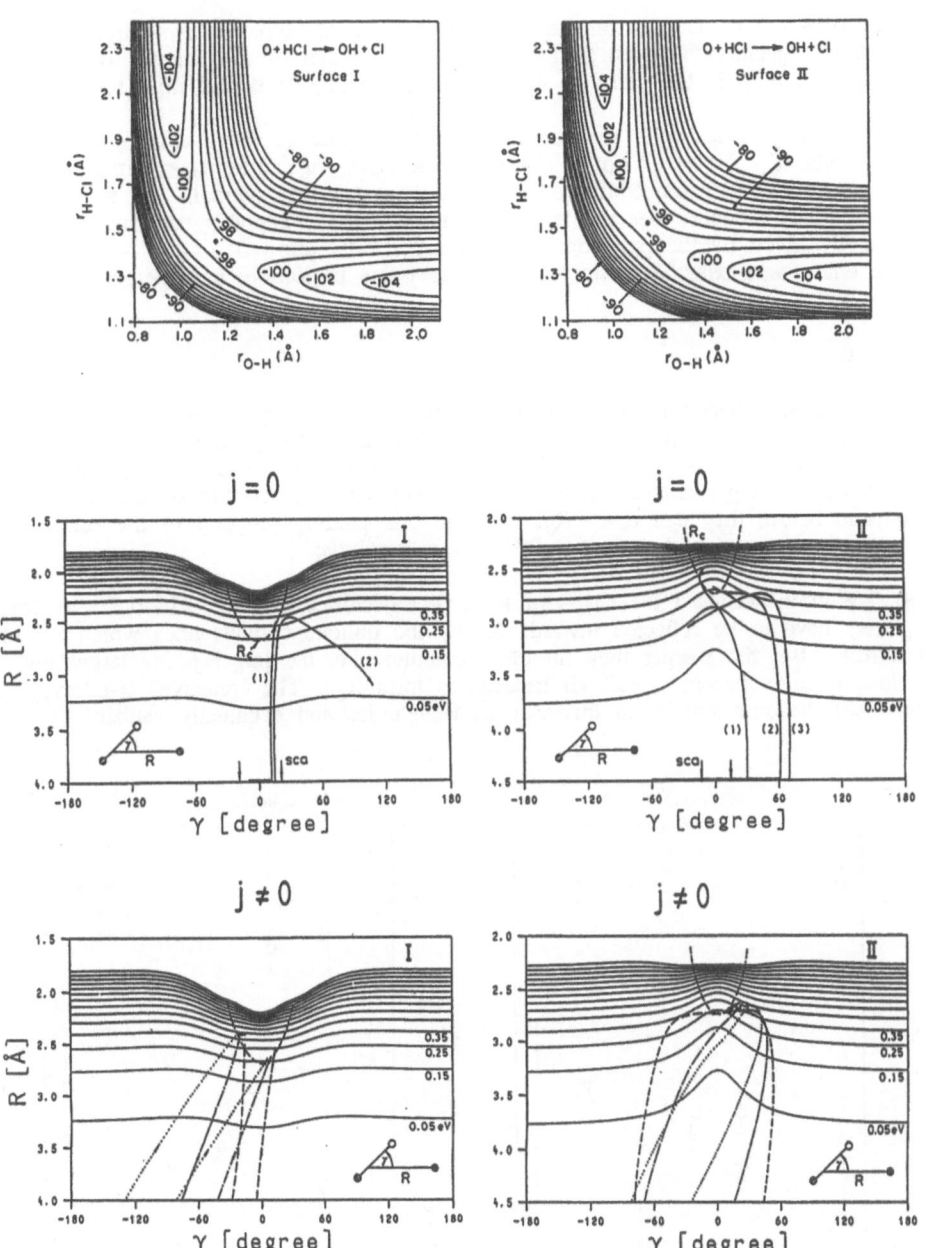

FIGURE 7. Collinear (upper panel) and R,γ-contour maps for the LEPS I (left) and LEPS II (right) PES of the reaction O + HCl → OH + Cl. The trajectories were calculated for E_{tr} = 0.434 eV with j=0 (—), 1(---), 3(–.–), and 5 (...) (upper panel from Ref. 25, lower panels from Ref. 24).

62

(E_{tr} = 0.434 eV). Due to the fact that the torques are directed away from $\gamma = 0^0$ a reactive trajectory (1) must start within the centre part of this interval (heavy line) to meet R_c, if γ_i is too close to the boundaries (path (2)) it will miss R_c.

j>0: The trajectories enter the contour diagram inclined to the R-axis by the angle B $\propto$ arctan $(E_{rot}/E_{tr})^{1/2}$. For a given j all reactive trajectories start within the γ_i intervals enclosed by trajectories drawn with the same line symbol. As the paths are essentially straight lines the lengths of these reactive intervals and thus $P_c(j;E_{tr})$ is roughly given by the projection of the energetically allowed portion of the $R_c(\gamma_c)$ curve onto the γ-axis. With increasing B or j this projection, and therefore also P_c, grows.

LEPS II

j=0: Even the trajectories (1) and (2) starting outside the γ-interval defined by energy conservation (arrows) are reactive as they are strongly curved by the marked angular forces towards $\gamma=0$ and guided to R_c. Trajectory (3) and all the others which start at larger angles are bent too strongly and are rejected by the radial repulsion before they can reach R_c. Due to this guiding property of the surface the "reactive" γ_i-interval and thus $P_c(j=0;E_{tr})$ is rather large.

j>0: With increasing j the shape of reactive trajectories changes. In order to cross R_c they have to be reflected towards R_c by the oblique contour lines which act like a mirror. But the steeper they hit these contours, i.e. the larger j, the larger this deflection angle becomes until all trajectories miss R_c. The "reactive" γ_i-interval and thus $P_c(j)$ become smaller in this way as j increases and eventually vanish.

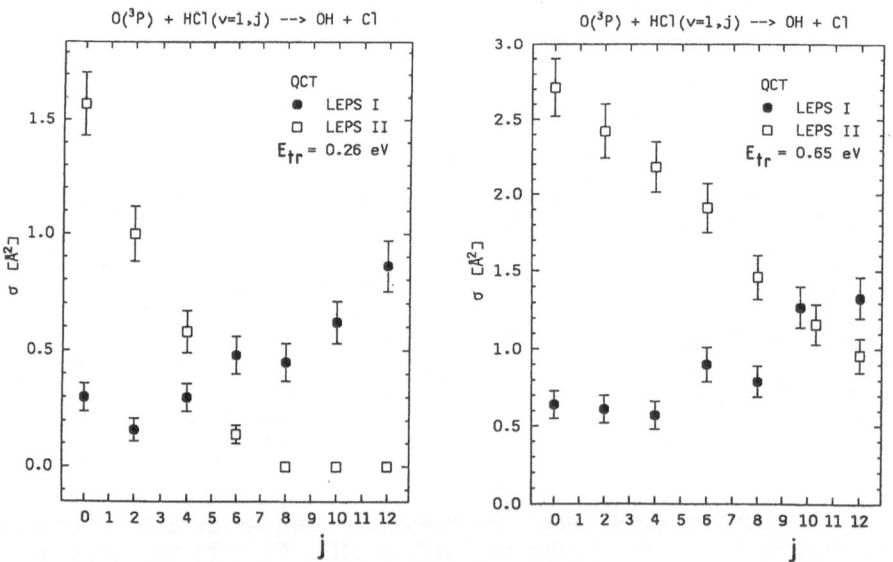

FIGURE 8. QCT-reaction cross sections for O + HCl(v=1,j) → O + HCl calculated from LEPS I and II.

The RSMM incorporates the effect of vibrational energy of BC only in the shape of the reaction shell. As this shape is not essentially different for O + HCl(v=1) one would expect similar j-dependencies for the reaction with vibrationally excited HCl. The results of our recent QCT-study on this reaction are shown in Fig. 8. As expected LEPS I and II furnish again ascending and descending cross sections, respectively, however with a somewhat flatter slope.

The results of the above and of numerous other RSMM-calculations can be summarized as follows: $P_C(j,E_{tr})$

(i)   decreases steeply to zero if the (positive) reactive γ_C-interval defined by energy conservation, $\Delta\gamma_c$, is small ($< \pi/4$) and the anisotropy of the PES large, that is $<\delta V/\delta\gamma>/E_{tr} >> \Delta\gamma_C \cdot (\mu_{BC}/\mu_{A,BC})$ ($<\delta V/\delta\gamma>$ is an average torque acting on BC during the approach),

(ii)   decreases initially and becomes constant for large $\Delta\gamma_C$ ($> \pi/2$) or weak anisotropy, that is $<\delta V/\delta\gamma>/E_{tr} << \Delta\gamma_C \cdot (\mu_{BC}/\mu_{A,BC})$,

(iii)   is constant for an isotropic potential with $R_C(\gamma_C) \equiv R_C$,

(iv)   increases for parabolic (or similarly shaped) $R_C(\gamma_C)$ and weakly or non-anisotropic PES.

4.4   The Energy Effect

A prerequisite for reaching the critical distance at an elevated j is that the surface is only weakly anisotropic to avoid inhibition of the reaction. In this case the angular momentum of BC at $R_C = R_C(\gamma_C)$ is roughly conserved and the rotational energy is then still the initial one

$$E_{rot} = \frac{\hbar^2}{2\mu_{BC} \cdot r_e^2} j(j + 1).$$

First we consider a reaction characterized by a surface with a late barrier and a rather sudden transition from reagents to products. These properties are typical for reactions with strong vibrational enhancement. At the instant A reaches R_c the BC bond stretches rapidly from r_e to r_b and j remains constant until the saddle configuration with $r=r_b$ is reached. There the rotational energy is then

$$E'_{rot} = \frac{\hbar^2}{2\mu_{BC} r_b^2} j(j + 1).$$

The difference

$$\Delta E_{rot} = E_{rot} - E'_{rot} = E_{rot}(1 - r_e^2/r_b^2)$$

appears as kinetic energy along the r-coordinate. With the assumptions that (i) ΔE_{rot} acts like vibrational energy and (ii) the fraction $\Delta E_{rot}/\hbar\omega$ of the vibrational quantum $\hbar\omega = E_{vib}(v=2) - E_{vib}(v=1)$ leads to an increase of the cross section by the same fraction of the cross section enhancement caused by the full quantum, one obtains the expression

$$\sigma_\epsilon(v=1,j;E_{tr}) = \sigma(v=1,0;E_{tr})(1 + [\eta(E_{tr}) - 1].\frac{\Delta E_{rot}}{\hbar\omega}), \quad (5)$$

where $\eta(E_{tr})$ denotes the vibrational enhancement of reaction

$$\eta(E_{tr}) = \sigma(v=2;E_{tr})/\sigma(v=1;E_{tr}).$$

The expression contains two parameters, $\eta(E_{tr})$ and r_b. Fundamentally, $\eta(E_{tr})$ can be determined independently in another experiment and thus r_b can be obtained by fitting Eq. 5 to an ascending j-dependent cross section.

For systems governed by an early barrier with insignificant vibrational enhancement the model predicts, of course, also an insignificant energy effect.

To test Eq. 5 it is easiest to select reactions for which at given translational energies the shape of $\sigma(j)$ is dominated by the "energy" effect. According to extensive QCT calculations the reactions Li + HF → LiF + H and H + H$_2$ → H + H$_2$ comply with this requirement.

Li + HF:

Inspection of the cross sections in Fig. 4 shows at least qualitative agreement with Eq. 5. For v=1 and particularly for v=0 the cross sections rise roughly quadratically in j. For v=2 σ is nearly independent of j in accord with the absence of a significant vibrational enhancement between v=2 and v=3 [8].

H + H$_2$:

A quantitative recovery of QCT results is obtained for H + H$_2$ as shown in Fig. 9. Plotted are the QCT results of Boonenberg and Mayne [26] which are based on the SLTH-PES; at low E_{tr} they obviously manifest a competition between the orientational and the energy effect, but at the two elevated energies the energy effect appears undisturbed. The model functions (dashed lines) are calculated using r_e/r_b=0.796 deduced from the SLTH-PES and η=2.6 (v=0 → v=1); in a QCT study Toennies et al. [27] find an enhancement of η=2 at E_{tr}=0.8 eV.

Summary:

The combined j dependence of P_c and P_r suggests that

(i)   A steeply descending $\sigma(j)$ is due to a strongly anisotropic PES.

(ii)   A monotonously increasing $\sigma(j)$ is caused either by a late barrier (energy effect) and/or by a parabolic reaction shell (orientational effect), both in connection with a weakly anisotropic PES.

(iii)   An initial descent followed by a minimum and a subsequent rise is a consequence of a competition between the orientational and the energy effect in reactions governed by a PES with moderate anisotropy and late barrier.

(iv)   All of the presently available data can be categorized according to this
         scheme which provides a first rough picture about the anisotropy of the
         PES and the position of the barrier.

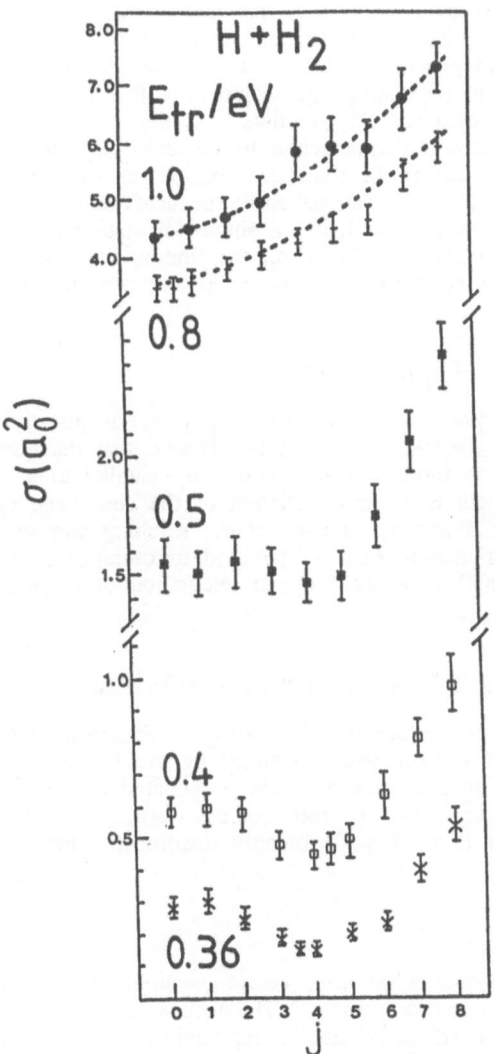

FIGURE 9.   j-dependent reaction cross sections for H + $H_2(v=0,j)$ (from Ref. 26).
The dashed lines are calculated using Eq. 5.

5. STERIC EFFECTS

5.1 Preparation of Reaction Geometry

In a scattering experiment the best one can do to prepare the geometry of the reagents is to fix the angle of attack, γ_a, between the molecular axis and the relative velocity $\bar{v}$ (approach geometry). In the case of a $^1\Sigma$ molecule the orientation of the molecular axis is possible only in strong electric fields where the free rotational motion of the molecule is converted to hindered rotations or oscillations around the field vector. However, this method has not yet been used probably due to the extremely high field strengths involved.

The usual procedure to generate at least a non-uniform distribution $A_p(\gamma_a)$ of the angle of attack consists in populating one single or a few j,m-states (alignment); the index p denotes the prepared state distribution. As the angular momentum vector of a $^1\Sigma$-molecule is always perpendicular to the axis the symmetry relation $A_p(\gamma_a) = A_p(180^0-\gamma_a)$ holds: that is the molecular axis is not orientable in this case.

Neglecting nuclear spin, $A_p(\gamma_a)$ can be easily deduced from the rigid rotor eigenfunctions $Y_{j,m}$ of BC. After populating a single j,m-state where m refers to a quantization axis parallel to $\bar{v}$ the probability A_p for finding the axis at the polar angles γ_a, φ_a pointing into a solid angle $d\omega$ is the square modulus of the rotor eigenfunctions.

$$A_p(\gamma_a,\varphi_a) = |Y_{j,m}(\gamma_a,\varphi_a)|^2.$$

If an incoherent mixture of j,m-states is populated $A_p(\gamma_a,\varphi_a)$ is just the sum of the adequately weighted $|Y_{j,m}(\gamma_a,\varphi_a)|^2$. It should be stressed that the quantization axis with respect to the preparation process is not necessarily parallel to $\bar{v}$. In case $\bar{v}$ and this axis include the angle B a simple rotation of the coordinate system by the angle B has to be carried out and the squares of the resulting coherent superpositions of rotor eigenfunctions have to be used to calculate $A_p(\gamma_a,\varphi_a)$. The desired distribution $A_p(\gamma_a)$ is then obtained by an integration of $A_p(\gamma_a,\varphi_a)$ over the azimuthal angle φ_a:

$$A_p(\gamma_a)d\gamma_a = \int A_p(\gamma_a,\varphi_a)d\varphi_a \cdot \sin(\gamma_a)d\gamma_a. \qquad (6)$$

Figure 10 shows two illustrative examples of approach geometries. Plotted is $A_p(\gamma_a)$ versus γ_a for the mixture of j=2, m-states populated by the IRP process (upper curves) and for j=3, m=j a state characteristic for a preparation by selective photodissociation (lower curves). For the mixture, the distribution peaks sharply at $\gamma_a=90^0$ for $\bar{E} \perp \bar{v}$ while for $\bar{E} \mathbin{\mathrm{II}} \bar{v}$ γ_a is broadly distributed. For j=3, m=j the opposite holds.

5.2 Results

The number of experimental studies on three centre reactions with molecules in prepared m states is rather scarce at present. The results of the first one [10], a beam/gas experiment on Sr + HF could not be reproduced [12] so we have to deal only with two recent crossed beam investigations: one is on the reaction Xe* + IBr → XeI*, XeBr* + Br,I [16] and the other on K + HF → KF + H [9]. Further studies did not come to our attention.

In the K + HF experiment the molecules were rotationally aligned by

IRP as discussed earlier. This technique leads to a predominant population of the $m=0$-state and thus the angular momentum $\vec{J}$ of the molecule is roughly <u>perpendicular</u> to the polarization vector $\vec{E}$ (quantization axis). The classical analogue of the

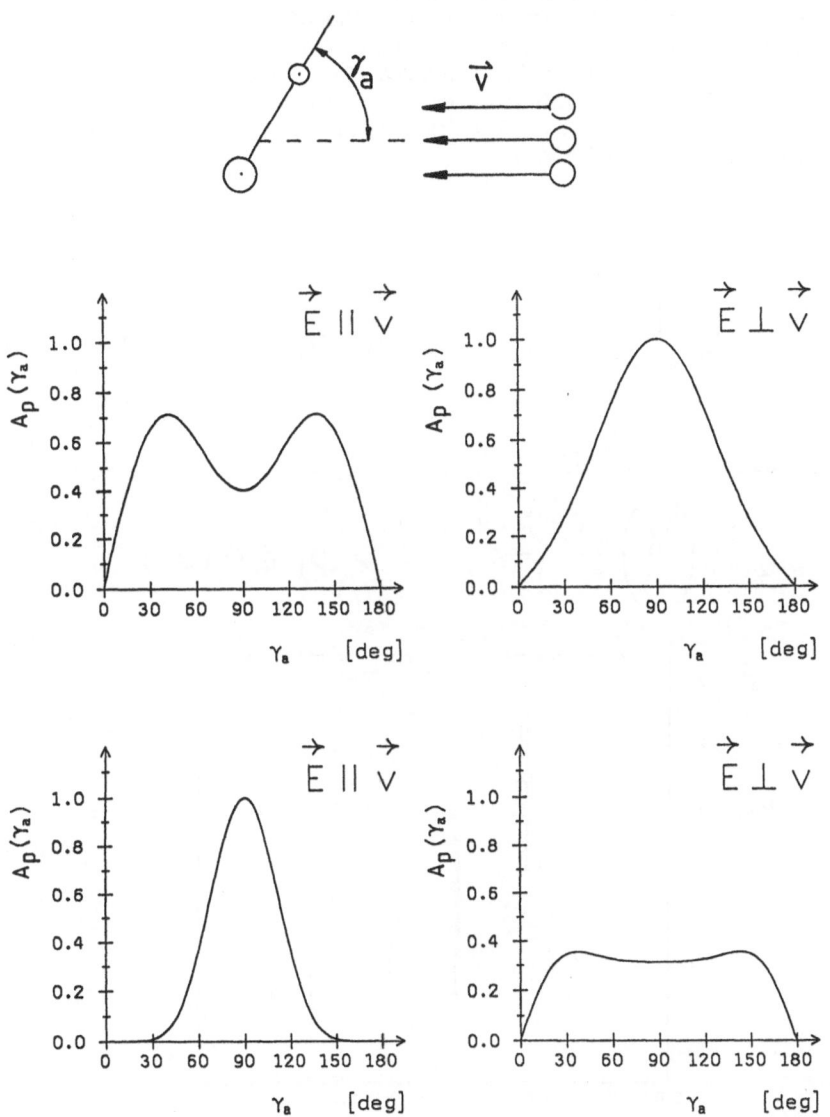

FIGURE 10. Probability distribution $A_p(\gamma_a)$ as a function of the angle of attack γ_a for optically excited $j=2$-states (R(1)-line) (upper curves) and for the $j=3$ $m=3$-state of a $^1\Sigma$-molecule (lower curves). No nuclear spin is considered.

68

prepared state corresponds to HF-molecules rotating in planes which have the polarization vector as a common axis. Therefore, if $\vec{E} \parallel \vec{v}$ the approaching atoms face the small side of the rotation disk and with $\vec{E} \perp \vec{v}$ also the broad side as depicted in Fig. 11. The probability distribution for the angle of attack is displayed in Fig. 10 (upper curve) for both directions of $\vec{E}$.

The experimental result is given in Fig. 11. Plotted is the difference of cross sections for $\vec{E} \parallel \vec{v}$ and $\vec{E} \perp \vec{v}$ divided by the mean cross section versus the translational energy E_{tr}. The data exhibit a marked effect on the direction of the polarization vector. The relative difference of cross sections is largest at $E_{tr} = 0.42$ eV where it amounts to 17%. The effect descends monotonously with rising E_{tr}, and reaches zero near $E_{tr} = 1.1$ eV.

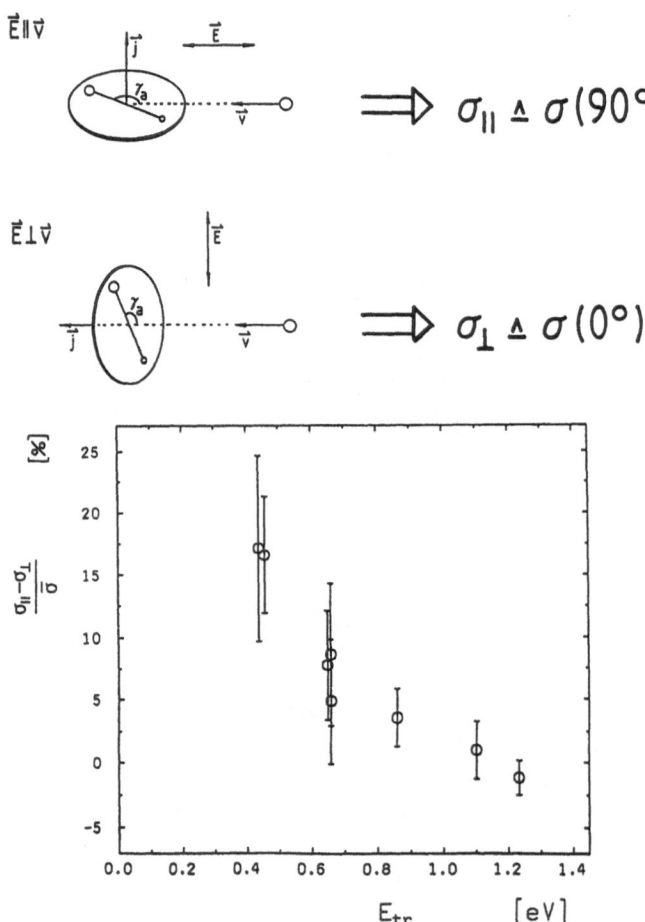

FIGURE 11. Prepared approach geometry and results for the reaction K + HF(v=1,j=2) → KF + H (from Ref. 9).

In the Xe* + ICl experiment the molecules were prepared by photodissociation using linearly polarized laser light. Here the |m| = j states are predominantly populated (c.f. Section 2) and thus the angular momentum of the molecule is roughly parallel to Ē. Due to this fact the classical analogues shown in Fig. 12 are obtained just by interchanging the two rotation disks drawn on Fig. 11 for the IRP-case. For example, with Ē || v̄ (B=0⁰) the atoms see the broad side of the rotation disk and for Ē ⊥ v̄ (B=90⁰) the small side.

The experimental results are given in Fig. 12. Plotted is the total reaction cross section (intensity of product fluorescence) versus the angle B between Ē and v̄. The signal is significantly dependent on B with a maximum at B=90⁰ and a minimum at B=0⁰. That is, the difference of cross sections for j̄ roughly perpendicular and parallel to v̄ ($\sigma(90^\circ) - \sigma(0^\circ)$) is larger than zero as it is also for K + HF.

In the following final section we will try to rationalize these results in terms of the favoured reaction geometry.

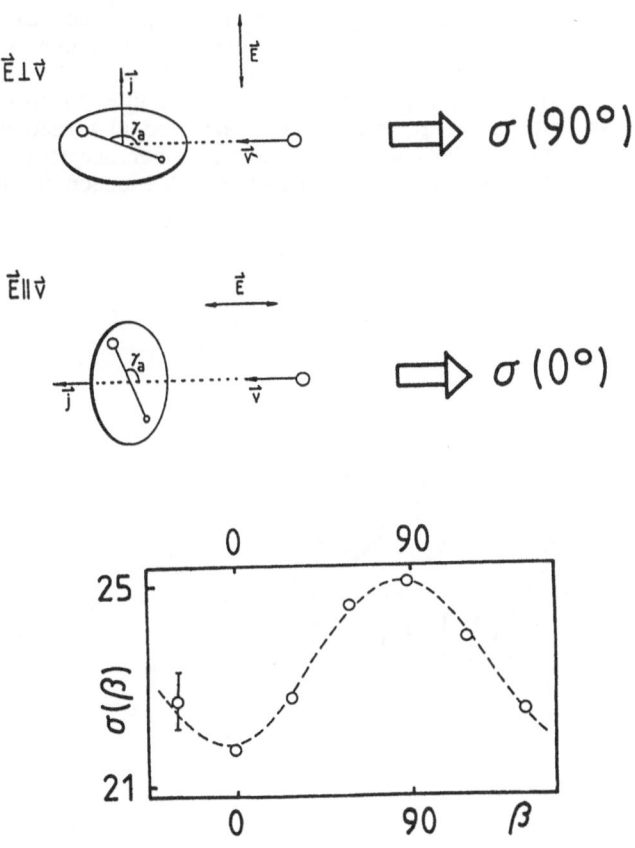

FIGURE 12. Prepared approach geometry and results for the reaction Xe* + IBr → XeI*/XeBr* + Br,I (from Ref. 16).

5.3   Interpretation of m-dependent Reaction Cross Sections

The preparation of j,m-states leads primarily to a non-uniform distribution of the angle of attack γ_a.   However, what is important for a reaction to be initiated is into what distribution of reaction angles γ_c (angle at the reaction shell, c.f. Fig. 5) the prepared distribution is transformed during the approach.   For example, the large range of impact parameters and the action of torques certainly lead to a deterioration of the prepared geometry.   A general treatment of this problem requires of course the consideration of these torques and the resulting angular motion of the molecules.   But for the present systems this is likely not to be necessary.   In the case of K + HF the j-dependent cross sections suggest a rather weak anisotropy of the PES and for Xe* + IBr the reaction shell is likely to be located at large distances where all forces are not yet essential.   We, therefore, assume that the angular motion is not disturbed by torques and the γ_a distribution transforms into the corresponding γ_c distribution in a purely geometric way.

The γ_c-distribution can be found in the following way.   As illustrated in Fig. 13, the position of the molecular axis in a coordinate frame with z as the polar axis (∥ $\vec{v}$) is given by the polar angles γ_a, φ_a. Atom A approaching BC with impact parameter $b \leqslant R_c$ and azimuthal angle φ hits the shell at point t. In a frame with polar axis z connecting the center-of-mass of BC and t the position of the molecular axis is given by γ_c, φ_c. With the help of a little trigonometry γ_a and φ_a are expressed as functions of γ_c, φ_c, b and φ and used to transform A_p (γ_a, φ_a) into the corresponding distribution in the z'-frame.   After integration over φ_c and averaging over φ one eventually obtains the desired distribution for fixed b

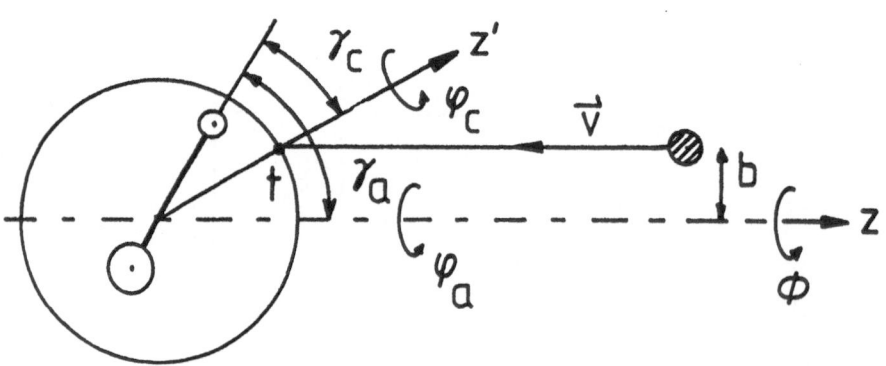

FIGURE 13.   Geometric relation between γ_a and γ_c at given b and φ.

$$B_p(\gamma_c, b) = \sin \gamma_c \cdot \frac{1}{2\pi} \int \int d\varphi d\varphi_c A_p(\gamma_a, \varphi_a).$$

With known $B_p(\gamma_c, b)$ the cross section σ_p measured for preparation A_p and the steric properties of the reaction can now be correlated. For this purpose we use the angle dependent opacity function $P(b, \gamma_c)$ introduced by Levine and Bernstein [28] which describes the probability that a reaction occurs if a collision with impact parameter b leads to a reaction angle γ_c. The cross section is then given by

$$\sigma_p = 2\pi \int_0^\pi \int_0^{b_{max}} d\gamma_c db \; b.P(b, \gamma_c).B_p(\gamma_c, b) \tag{7}$$

where b_{max} is the maximum impact parameter leading to reaction.
   To evaluate this integral we employ for the sake of simplicity two approximations:

(i) $P(b, \gamma_c)$ factorizes into

$$P(b, \gamma_c) = P(b).S(\gamma_c), \tag{8}$$

where $P(b)$ and $S(\gamma_c)$ denote the usual (b-dependent) and the steric opacity functions, respectively. $S(\gamma_c)$ describes the probability that a reaction occurs at a given reaction angle γ_c on contact with the shell and defines in this way the "cone of-acceptance". It is the aim of this section to extract information about $S(\gamma_c)$ from the observed steric effects.

(ii) $P(b)$ and $S(\gamma_c)$ are step functions defined by

$$P(b) = \begin{cases} 1 & b_{min} < b < b_{max} \\ 0 & \text{otherwise} \end{cases} \tag{9}$$

$$S(\gamma_c) = \begin{cases} 1 & \gamma_{c,min} < \gamma_c < \gamma_{c,max} \\ 0 & \text{otherwise.} \end{cases} \tag{10}$$

Insertion of Eqs. 8,9,10 into Eq. 7 furnishes:

$$\sigma_p = \int_{\gamma_{c,min}}^{\gamma_{c,max}} G_p(\gamma_c) \; d\gamma_c \tag{11}$$

with

$$G_p(\gamma_c) = 2\pi \int_{b_{min}}^{b_{max}} db \; b.B_p(\gamma_c, b). \tag{12}$$

According to Eq. 11 the cross section σ_p is simply given by the area below the $G_p(\gamma_c)$-curve between $\gamma_{c,min}$ and $\gamma_{c,max}$. To calculate $G_p(\gamma_c)$ one has to specify the preparation p of the approach geometry characterized by A_p as well as b_{min} and b_{max}.
   In the case of K + HF we use the A_p distributions which lead to the curves

shown in the upper part of Fig. 10 (IRP-preparation of the $v=1,j=2$-states) and choose $b_{min}=0$ by analogy with other M + HX-reactions [29]. As b_{max} is essentially unknown we computed two sets of G_p-functions one with $b_{max}=R_c/2$ and the other with $b_{max}=R_c$. The results are depicted in Fig. 14. The $b_{max}=R_c/2$ curves still show all the characteristics of the approach geometry while the structures are completely washed out for $b_{max}=R_c$. For the further analysis we use the $b_{max}=R_c/2$-curves; however, all conclusions drawn are not dependent on this.

To interpret the Xe* + IBr-data we employ the $A_p(\gamma_a)$ distributions which lead to the curves displayed in the lower part of Fig. 10 considered typical for a preparation by photodissociation ($m=j=3$); b_{min} and b_{max} are approximately determined by the fact that the reaction proceeds via a harpooning mechanism. Curve crossing into the ionic state, and thus reaction, is most likely to occur at

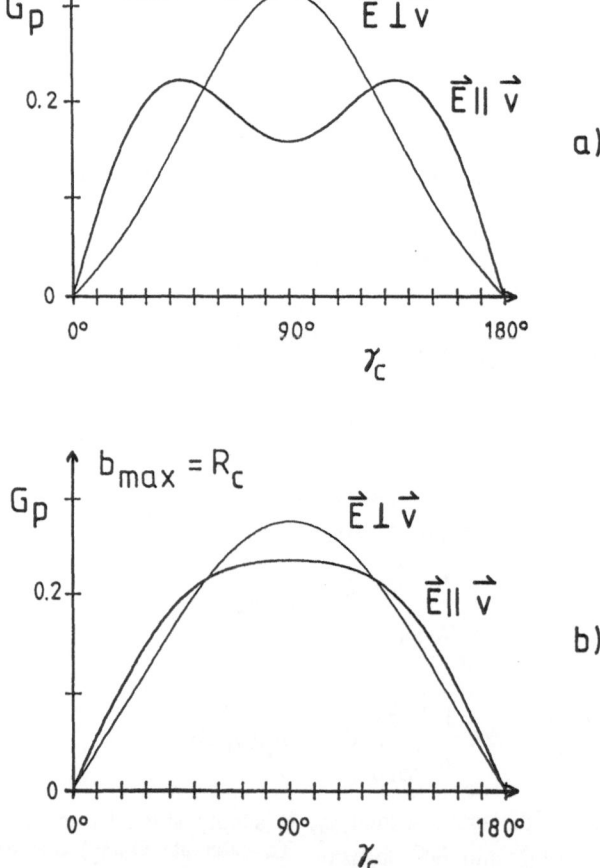

FIGURE 14. Probability distribution of reaction angle γ_c, $G_p(\gamma_c)$ for the optically prepared $j=2$-state of HF.

large reagent separations and at large impact parameters where the radial velocity is small. We take this fact into account by considering only large impact parameters and set b_{min} = 0.75 R_c and $b_{max}=R_c$. The result is depicted in Fig. 16. The curves resemble very much the $b_{max}=R_c$-curves of Fig. 14 but one should bear in mind that due to the different preparation methods $\bar{E}_\perp \bar{v}$ (B=90°), $\bar{E}_{||} \bar{v}$ (B=0°) in Fig. 16 correspond to $\bar{E}_{||} \bar{v}$ and $\bar{E}_\perp \bar{v}$, respectively, in Fig. 14.

The steric properties of both reactions are now determined by using two sets of trial values for $\gamma_{c,min}$ and $\gamma_{c,max}$ which are inserted into Eq. 11. The resulting $\sigma_{||} - \sigma_\perp$ and $\sigma(90°) - \sigma(0°)$ (differences of areas below the $G_{||}(90°)$ and $G_\perp(0°)$ curves) are then compared to the corresponding experimental data. We consider the two cases:

(i) The favoured reaction geometry is bent and reaction occurs only if the atom encounters the shell at an angle near γ_c = 90° that is between $\gamma_{c,min}$ and $\gamma_{c,max}$ as depicted by the cone of acceptance shown in Fig. 15a.

(ii) The favoured reaction geometry is collinear and reaction occurs only if the atom encounters the shell at an angle near γ_c = 0°, that is between $\gamma_{c,min}=0°$ and $\gamma_{c,max}$ as illustrated by the cone of acceptance as shown in Fig. 15b.

For the system K + HF the relevant areas are illustrated in Fig. 15. It immediately follows that cases (i) and (ii) lead to the results $\sigma_{||} - \sigma_\perp$ <0 and $\sigma_{||} - \sigma_\perp$ >0, respectively. As the latter result is in agreement with experiment (Fig. 13) we draw the conclusion that case (ii) is correct and the reaction proceeds via a collinear reaction geometry (transition state). For Xe* + IBr (Fig. 16) case (i) provides $\sigma(90°) - \sigma(0°)$ >0 in agreement with the data (Fig. 12). That is, the reaction proceeds predominantly via a bent reaction geometry in accordance with the interpretation given by de Vries et al. [16].

Besides the sign and magnitude also the E_{tr}-dependence of $(\sigma_{||} - \sigma_\perp)/\sigma$ has been measured for K + HF (c.f. Fig. 11). In the following we show that such data are closely related to the angular dependence of the potential energy at the shell, $V_c(\gamma_c)$, and make an attempt to give a rough estimate of an absolute value of V_c.

From the discussion in Section 4.3 it is clear that a reaction can take place only if γ_c falls into the interval $\Delta\gamma_c$ whose boundaries are defined by energy conservation with $V_c(\gamma_{c,max/min}) = E_{tr}$, (we assume here that always the full translational energy is available to reach the shell). It is known from the conclusion drawn above that the minimum of V_c occurs at a collinear configuration (most likely: K···F···H) with $\gamma_c=0°$. Thus, $V_c(0°)$ is the threshold energy of the v=1-reaction which amounts to ~ 0.3 eV [17]. Provided $V_c(\gamma_c)$ increases monotonously with γ_c similar as for LEPS I and II in Fig. 7 the outer boundary $\gamma_{c,max}$ of $\Delta\gamma_c$ ($\gamma_{c,max}=90°$) grows with E_{tr} starting from $\gamma_{c,max}=0°$ at $E_{tr}=V_c(0°)$. With increasing $\gamma_{c,max}$ the ratio $\sigma_{||} - \sigma_\perp / \sigma$ calculated on the basis of Fig. 15b drops until it reaches zero at $\gamma_{c,max}=90°$ (the full areas below both curves are equal as a spherical shell is considered) which is at least in qualitative agreement with the experimental data. As shown in Fig. 11 the loss of steric effects is observed at E_{tr} ~ 1.1 eV, that means within the framework of this primitive model, that at E_{tr} = 1.1 eV $\gamma_{c,max}=90°$ is reached with $V_c(90°)$ ~ 1.1 eV. Using the threshold energy given above we eventually obtain

$$V_c(90°) - V_c(0°) \sim 0.8 \text{ eV}.$$

The increase of γ_c up to 90^0 at $E_{tr} \sim 1.1$ eV is in agreement with the observed loss of the orientational effect in the j-dependent cross section between $E_{tr} = 0.75$ and 1.21 eV according to the criteria given at the end of Section 4.3.

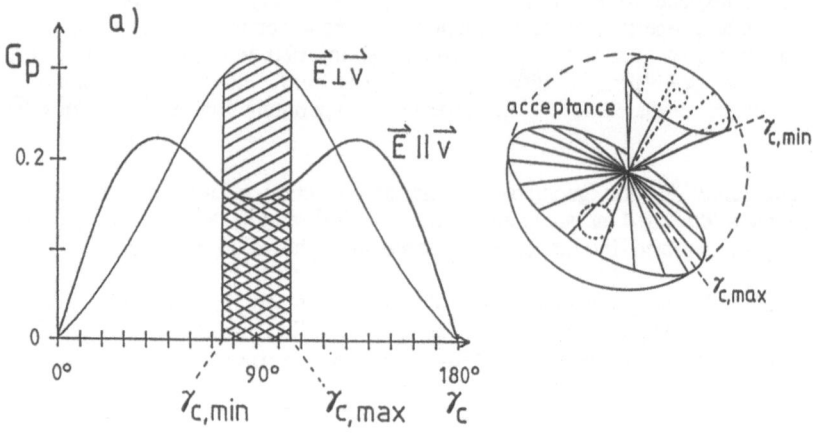

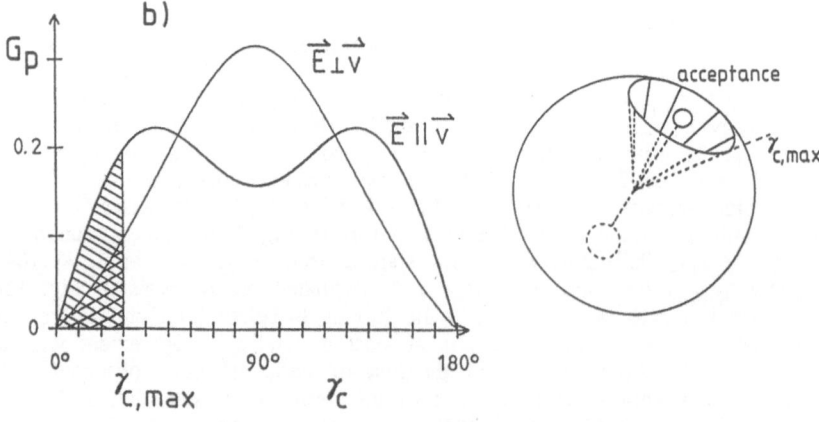

FIGURE 15. The hatched areas between $\gamma_{c,min}$ and $\gamma_{c,max}$ are proportional to the reaction cross section at the indicated position of $\vec{E}$ with respect to $\vec{v}$. (a) lateral attacks lead to products; (b) frontal attacks lead to products.

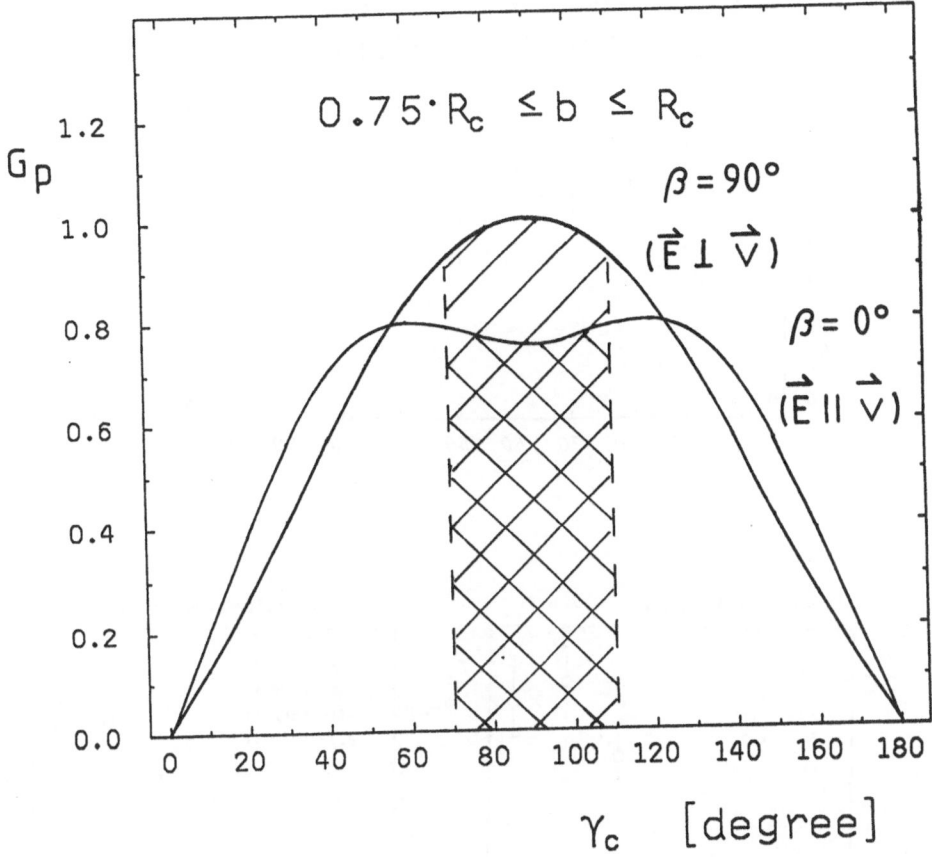

FIGURE 16. Probability distribution of reaction angle γ_c for molecules in the j=m=3-state which is considered typical for a preparation by selective photodissociation. With the assumption that lateral collisions lead favourably to products the experimental results are qualitatively recovered. This interpretation is in accord with the one of Ref. 16.

5.4 Theoretical Results

We have checked this simplistic interpretation of m-dependent cross sections by comparison with QCT-results. As examples for reactions with bent and collinear reaction geometry (transition state) we investigated Li + HF described by the CSCM-mod PES and O + HCl described by LEPS I and II, respectively. Cross sections were calculated as a function of the angle between $\vec{j}$ and $\vec{v}$ (helicity angle).

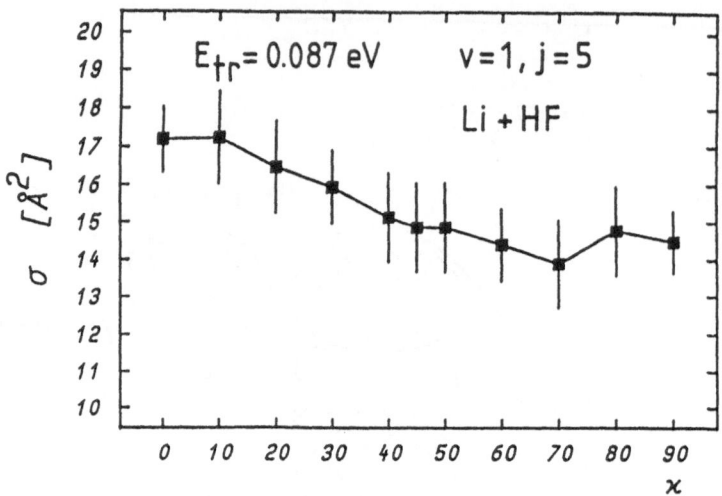

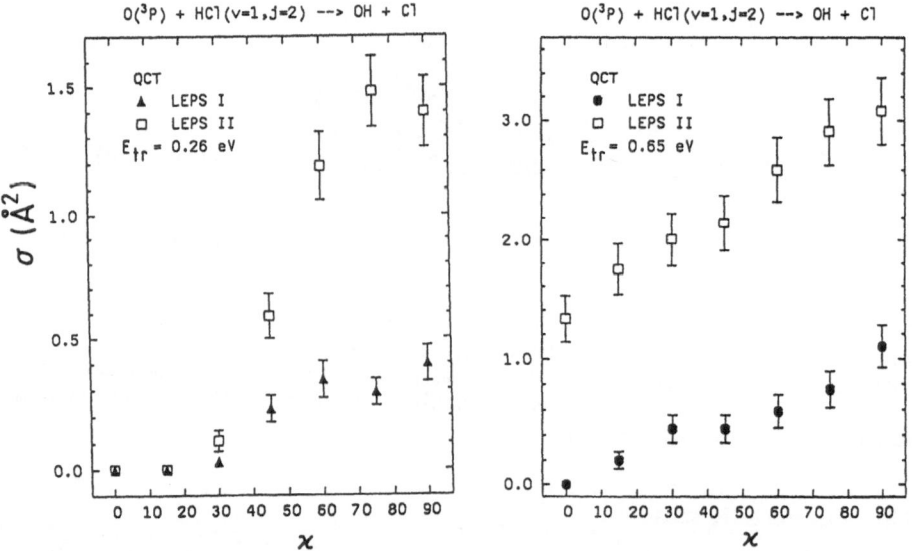

FIGURE 17.   Reaction cross sections as a function of $\chi(\hat{j},\bar{v})$ for Li + HF(v,j) → LiF + H (CSCM-mod-PES [19]) and O + HCl → OH + Cl (LEPS I and II).

Thus $\chi = 0^0$ and 90^0 refer approximately to $\bar{E} \perp \bar{v}$ and $\bar{E} \parallel \bar{v}$, respectively, in an IRP-experiment.

The result for Li + HF is shown in Fig. 17.   The cross section drops with increasing χ.   This is equivalent to $\sigma_{\parallel} - \sigma_{\perp} < 0$ as predicted by the model.

The slow descent is a consequence of the already large apex angle of the cone of acceptance ($2\Delta\gamma_C = 160^0$).

The reaction $0 + HCl$ exhibits exactly the opposite behaviour (Fig. 17). The cross sections for both surfaces increase with rising χ which corresponds to $\sigma_{||} - \sigma_{\perp} > 0$ in accordance with the model. The steep ascent is due to the small apex angle of only $2\Delta\gamma_C = 40^0$ (c.f. Fig. 7).

ACKNOWLEDGEMENT

Support of this work by the Deutsche Forschungsgemeinschaft (SFB 216, P5 and SPP "Dynamik zustandsselektierter chemischer Primärprozesse") is gratefully acknowledged.

REFERENCES

[1]   F.S. Klein and A. Persky, J. Chem. Phys., **61** (1974) 2472.

[2]   a) K.G. Anlauf, D.H. Maylotte, J.C. Polanyi and R.B. Bernstein,
        J. Chem. Phys., **51** (1969) 5716.
      b) J.C. Polanyi and D.C. Tardy, J. Chem. Phys., **51** (1969) 5719.
      c) H. Kaplan, R.D. Levine, and J. Manz, Chem. Phys.,
        12 (1976) 447.

[3]   B.A. Blackwell, J.C. Polanyi and J.J. Sloan, Chem. Phys., **30**
      (1978) 299.

[4]   (a) C.E. Hamilton, J.L. Kinsey and R.W. Field, Ann. Rev. Phys.
        Chem., **37** (1986) 493.
      (b) A.E. De Pristo, H. Rabitz and R.B. Miles, J. Chem. Phys.,
        73 (1980) 4798.

[5]   H.H. Dispert, M.W. Geis and P.R. Brooks, J. Chem. Phys., **70**
      (1979) 5317.

[6]   M. Hoffmeister, L. Potthast and H.J. Loesch, Chem. Phys., **78**
      (1983) 369.

[7]   M. Hoffmeister, Thesis, R. Schleysing, Thesis, Universität
      Bielefeld, FRG (1987) and M. Hoffmeister, R. Schleysing, and
      H.J. Loesch, Chem. Phys., to be published.

[8]   W. Grote, Thesis, Universität Bielefeld, FRG (1987) and
      W. Grote, H. Zerhau-Dreihöfer, and H.J. Loesch, Chem. Phys.,
      to be published.

[9]   M. Hoffmeister, R. Schleysing, and H.J. Loesch, J. Phys. Chem.,
      91 (1987) 5441.

[10] Z. Karny, R.C. Estler and R.N. Zare, J. Chem. Phys., **69** (1978)
      5199.

[11] R. Altkorn, F.E. Bartoszek, J. Dehaven, G. Hancock, D.S. Perry, and R.N. Zare, Chem. Phys. Letters, 98 (1983) 212.

[12] R.N. Zare, K.G. McKendrick and D.J. Rakestraw, private communication.

[13] C-K. Man and R.C. Estler, J. Chem. Phys., 75 (1981) 2779.

[14] R.N. Zare, Ber. Bunsenges. Phys. Chem., 86 (1982) 422.

[15] R. Altkorn, R.N. Zare and C.H. Greene, Mol. Phys., 55 (1985) 1.

[16] M.S. de Vries, V.I. Srdanov, C.P. Hanrahan and R.M. Martin, J. Chem. Phys., 77 (1982) 2688; ibid., 78 (1983) 5582.

[17] F. Heismann and H.J. Loesch, Chem. Phys., 64 (1982) 43.

[18] M.M.L. Chen and H.F. Schaefer III, J. Chem. Phys., 72 (1980) 4376.

[19] S. Carter and J.N. Murrell, Mol. Phys., 41 (1980) 567.

[20] H. Zerhau-Dreihöfer, Diplomarbeit, Universität Bielefeld, FRG, (1986).

[21] I. Noorbatcha and N. Sathyamurthy, Chem. Phys., 77 (1983) 67.

[22] A. Laganá, M.L. Hernandez and J.M. Alvariño, Chem. Phys. Letters, 106 (1984) 41.

[23] H.J. Loesch, Chem. Phys., 104 (1986) 213.

[24] H.J. Loesch, Chem. Phys., 112 (1987) 85.

[25] A. Persky and M. Broida, J. Chem. Phys., 81 (1984) 4352.

[26] C.A. Boonenberg and H.R. Mayne, Chem. Phys. Letters., 108 (1984) 67.

[27] G.-D. Barg, H.R. Mayne and J.P. Toennies, J. Chem. Phys., 74 (1981) 1017.
       H.R. Mayne and J.P. Toennies, J. Chem. Phys., 75 (1981) 1794.

[28] R.D. Levine and R.B. Bernstein, Chem. Phys. Letters, 105 (1984) 467.

[29] C.H. Becker, P. Casavecchia, P.W. Tiedemann, J.J. Valentini and Y.T. Lee, J. Chem. Phys., 73 (1980) 2833.
       J.R. Airey, E.F. Greene, K. Kodera, G.P. Peck and J. Ross, J. Chem. Phys., 46 (1967) 3287.

SELECTIVITY OF TRANSLATIONAL EXCITATION IN REACTIVE COLLISIONS. THE ALKALINE PLUS ALKYL HALIDE REACTION FAMILY

A. Gonzalez Ureña
Departamento de Quimica
Facultad de Quimica
Universidad Complutense
28040 - Madrid
Spain

ABSTRACT. This paper represents recent results on the reaction dynamics of the M + RX → MX + R (M = alkali, X = halogen and R = radical group) family obtained from crossed molecular beam studies. The selectivity of the translational excitation of the reactants as well as of the chemical nature of the M, R and X group is outlined. A comparison of these reactive processes with photofragmentation and electron attachment studies of the same RX molecule is also presented revealing important similarities. This common behaviour seems to indicate the same overall selectivity as a result of the similar main dynamics associated with the same R-X bond rupture.

1. EARLY STUDIES

The reactions of the alkyl halides with alkali atoms are perhaps one of the families best studied so far and have become a venerable example in many books and reviews of the molecular reaction dynamics field [1-5].

Early studies included diffusion flames [6-7] and extensive work on molecular beams measuring total and differential reaction cross-sections from elastic [8] and reactive scattering experiments [9-10]. Among the many measurements that have been carried out one may mention product angular and recoil velocity distributions [9], collision energy dependence of the differential and total reaction cross-section [10], rotational and polarization analysis, via electric deflection of the MX product [11], orientation dependence of the reaction barrier [12], etc.

As a result of this significant body of data, and following the work of Herschbach [4], Bernstein [3] and their co-workers, the electron transfer mechanism is well established for these reactions that belong to the classic rebound or impulsive mechanism [1,4]. The MX product molecules scatter preferentially backward in the centre of mass (c.m.) system and a high fraction of the reaction exoergicity goes into translational recoil. Increasing the reagents' translational energy [9-10] leads to a substantial increase in the products' translational energy. In addition these studies indicated that the reaction cross-sections fall as the electron affinity of RX becomes more negative in accordance with the electron jump mechanism previously cited. A huge number of simple models [5] of a statistical or dynamical nature seem to recover some, but not all, of the reported features, leading to a general need for a

79

J. C. Whitehead (ed.), Selectivity in Chemical Reactions, 79–94.
© 1988 by Kluwer Academic Publishers.

more complete treatment. Trajectory calculations using repulsive surfaces [13] on the other hand, account for most of the gross features, at least in the M + CH$_3$I dynamics.

One of the main goals of the present paper is to review and discuss the most recent data and new findings on the family of reactions of the title. Emphasis will be made on the selectivity of the translational excitation of the reagents, as well as of the chemical nature of the M, X and R groups. A comparison of these reactive processes with those of photofragmentation and electron attachment of the same alkyl halides is also presented.

2. EXPERIMENTAL AND DEFINITIONS

The experimental technique has been widely described elsewhere [10]. Typically a thermal beam (M = K, Na, Rb etc.) crosses at 90^0 the supersonic RX beam (X = halogen, R = organic radical group). The reactively scattered MX is collected over the whole laboratory angular range in order to measure the product's angular distribution and, by integration, the total reaction cross-section. As is well known, for a bimolecular reaction A + BC → AB + C the reaction cross-section is given by [5]

$$\sigma_R(E_T) = \frac{F_{AB}}{n_A \cdot n_{BC} \cdot v_R \cdot \Delta V}$$

where F_{AB} is the total integrated flux over the laboratory system, the n's are the respective densities, v_R the relative velocity of the beams and ΔV is the intersection volume.

On the other hand, if we measure some specific state α of the product, the differential excitation function given by

$$\sigma_R(E_T \cdot \alpha) = \sigma_R(E_T) \; . \; P(E_T/\alpha)$$

could also be obtained. Examples of this function can be the collision energy dependence of the products' angular or energy distributions for which $P(E_T \cdot /\theta)$ and $P(E_T \cdot /E'_T)$ would be the normalized products' distribution (i.e. $\alpha = \theta$ or E'_T, respectively).

3. RESULTS AND DISCUSSION

3.1 Products' Centre of Mass Angular Distributions

The MX product is always scattered in the backward hemisphere in the c.m. corresponding to a rebound mechanism.

In this case the character of the angular and velocity distribution of a direct mode reaction may be well described by simple dynamical models, and one of the most used for the title reaction is that of the electron jump mechanism the basic assumptions of which follow Herschbach's idea of the so called diatomic reflecting approximation [4]. The M atom induces, by an electron jump, a vertical transition from the BC to the BC$^-$ repulsive potential leading to the impulsive separation of the anion, resulting in a high fraction of the reaction exothermicity appearing as product translation. Moreover the requirements of angular and linear momentum conservation which are present in many impulsive models (see Ref. 5 for a more complete description), in a clear analogy with photofragmentation dynamics, allow us to recover the (backward) products' angular distribution.

Figure 1 shows the percentage of backward character of the KI angular distributions obtained in the molecular beam study of the K + RI → KI + R reactions, where R was the methyl, ethyl and propyl radicals. The increase in this backward character as the radical size increases was interpreted as an effect of a larger steric factor as one goes from the methyl to the propyl radical (see below). This reduction in the reaction cross-section, as the radical size increases, would confine the reactive trajectories to the collinear approach, leading to a more repulsive interaction, and so to an increase in the backward distributions of the product.

3.2 Products' Energy Disposal as a Function of Collision Energy

In analysing the reaction dynamics, a well-known correlation comes from the work of Polanyi and co-workers [14] on the dynamical problem of barrier crossing.

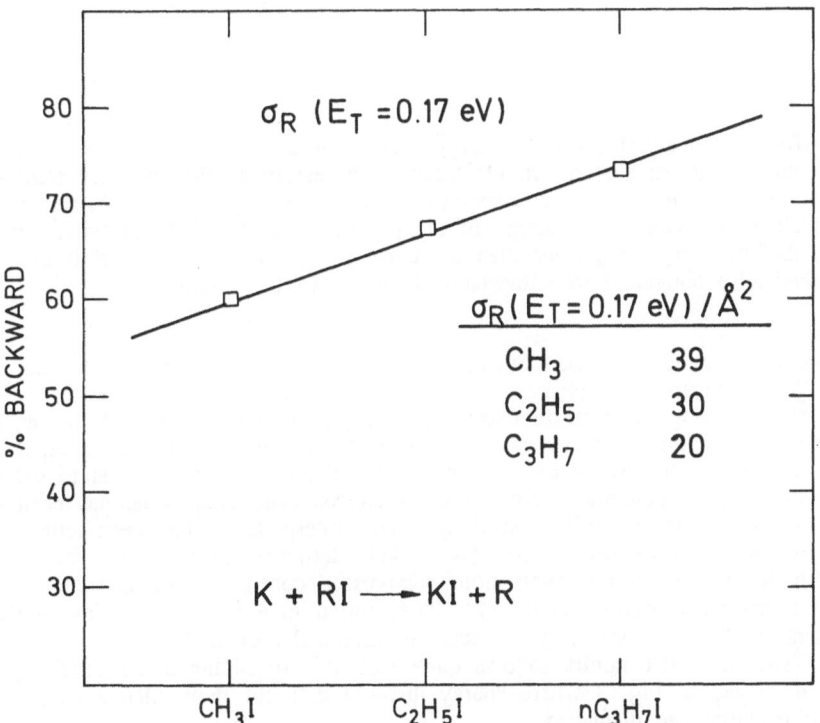

FIGURE 1. Percentage of backward character of the KI formed in the indicated reactions as a function of the radical identity. Total reaction cross-section values are also indicated. Adapted from Ref. 15 and from E. Verdasco, V. Saez Rabanos, F.J. Aoiz and A. Gonzalez Ureña, Mol. Phys., 62 (1987) 1207

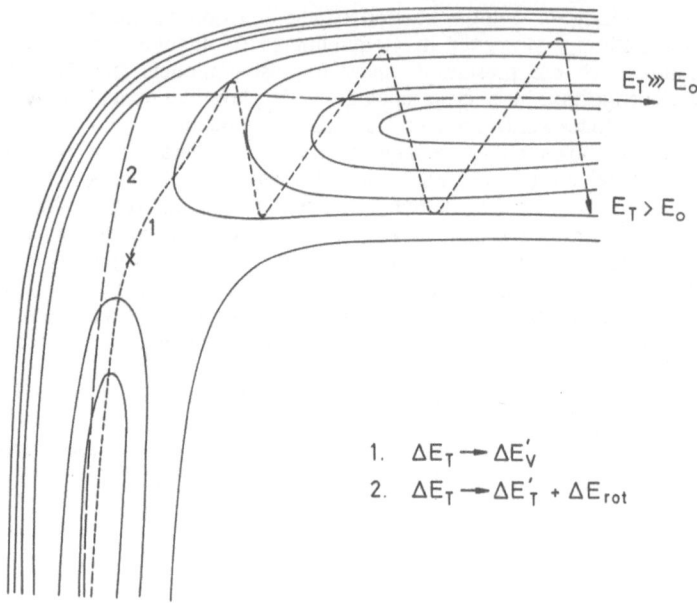

FIGURE 2. Idealised potential energy surface of a repulsive character. Two different trajectories are shown: (1) Example trajectory of the so-called transposition mechanism. In this case the excess of energy with respect to the barrier is channelled into vibrational energy of the product. (2) Example of trajectory where the collision energy is greater than the barrier. A higher fraction of it is (adiabatically) converted into translational energy of the product.

Figure 2 shows the so-called Surface I, representing an early barrier location. In this case a transposition phenomenon has been reported. It occurs when the translational energy is little in excess with respect to the barrier and this energy appears as vibrational energy of the product, as shown by trajectory number 1. On the other hand, the excess of collision energy above the barrier leads to extra translational and rotational energy in the products, originating what has been called (a pronounced) translational adiabaticity. This interpretation has been termed "induced repulsive energy release" [14]. This definition gives a clear indication of a shifting toward reaction through more compressed configurations, as shown in the same figure by trajectory number 2. They might give either recrossing to the reactants valley or, conversely, products translationally excited.

The M + RX family follows quite well this translational adiabaticity pattern as it is displayed in Fig. 3 where energy disposal data for more than a couple of dozen reactions are displayed.

Recently [15] we have measured the alkyl group effect in this energy disposal behaviour, and the result is shown in Fig. 4 where data for the methyl, ethyl and propyl reactions are displayed. Note that whereas F_T, the fraction of the total energy that appears in translational energy of the products, is about 0.6 (0.5) for the methyl (ethyl) reaction, it reduces to 0.3 for the propyl case. This enhancement in

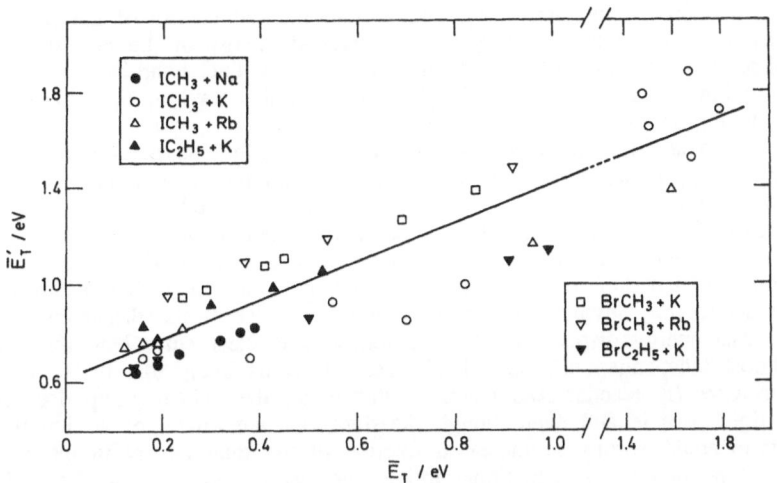

FIGURE 3. Translational energy disposal for several members of the M + RX family as indicated. Note the translational adiabaticity (see text for more details). Adapted from Ref. 5.

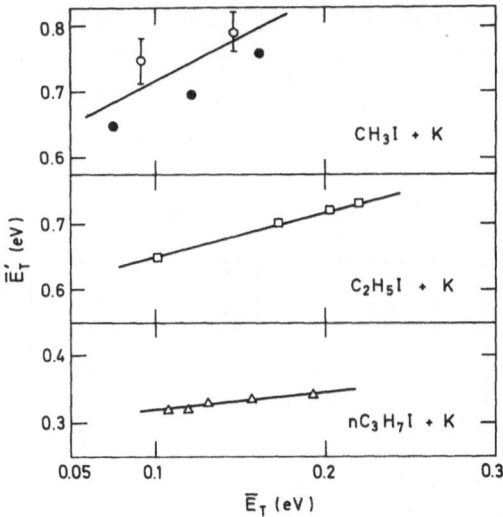

FIGURE 4. Radical effect in the translational energy disposal for the indicated reaction. Adapted from Ref. 15.

the fraction of the total energy going into internal energy of the product, and therefore we might expect going into internal energy of the radical, can only be associated with the presence of the symmetric C-C-C bond which has much lower vibrational quanta than the stiff C-C bond of the ethyl or the C-H bonds of the methyl group.

In addition, it is very interesting to compare this fraction of the total energy appearing in internal energy of the radical with the same fraction obtained from photofragmentation studies of the alkyl iodides, as well as for the Ba + RI reaction family [9b]. Figure 5 shows the comparison among all these reactive scattering and photofragmentation studies. The agreement between the K reactions and the photofragmentation results is quite satisfactory showing a clear analogy and that both processes may involve the same dynamics. However the disagreement with the internal energy disposal for the Ba reactions is clear since here this quantity is almost independent of the radical size. It seems likely that in the above two processes (K reaction and electron attachment) the radical group does not behave as a rigid particle and even though the dynamics are direct, of an impulsive nature, this group could absorb an increased fraction of the total energy in internal motion, as the total number of vibrational modes increases. On the other hand, in the Ba reaction the higher ionization potential for the Ba than for the K atom plus the poor electron acceptor capability of the methyl radical may well produce migratory encounters leading to BaI in highly excited vibrational states. These sort of

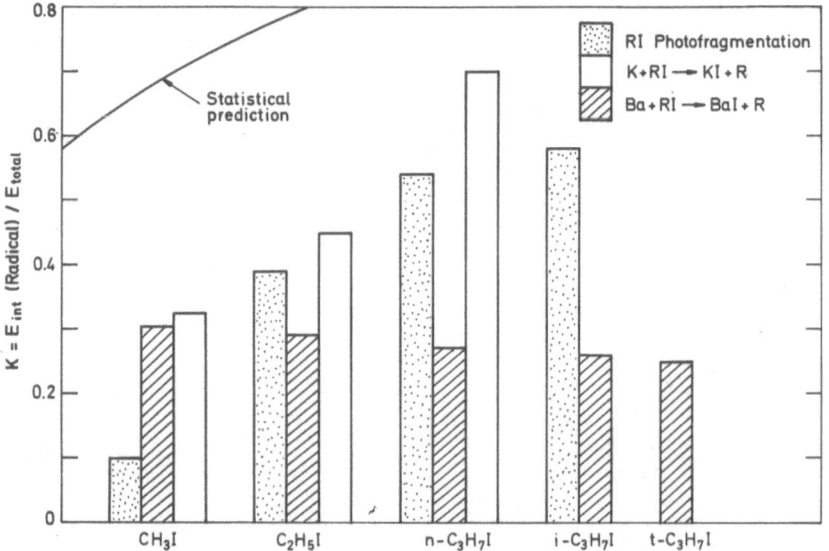

FIGURE 5. Fraction of the total energy available to the products that appears in internal excitation of the radical group plotted against identity of the alkyl iodides. Three processes are displayed, the two K and Ba + RI reactions and the RI photofragmentation. Adapted from Refs. 15 and 9b.

transposition effects originating from "induced attractive energy release" [14] have been reported for several analogous systems [16] at low collision energy. Under these circumstances the radical group will merely act as a rigid particle and the percentage of the total energy appearing in internal modes will be almost independent of its actual size, as was found [9b] to be the case.

3.3   Total Excitation Function

3.3.1 Family trends and comparison. General and common features can be noted. They have been rationalized in terms of the following three effects: halogen, attacking atom and radical (alkyl group) effect. For a better illustration they are summarised in Table 1 where a comparison with electron attachment data is also included. Figures 6, 7 and 8 show some examples of these common features.

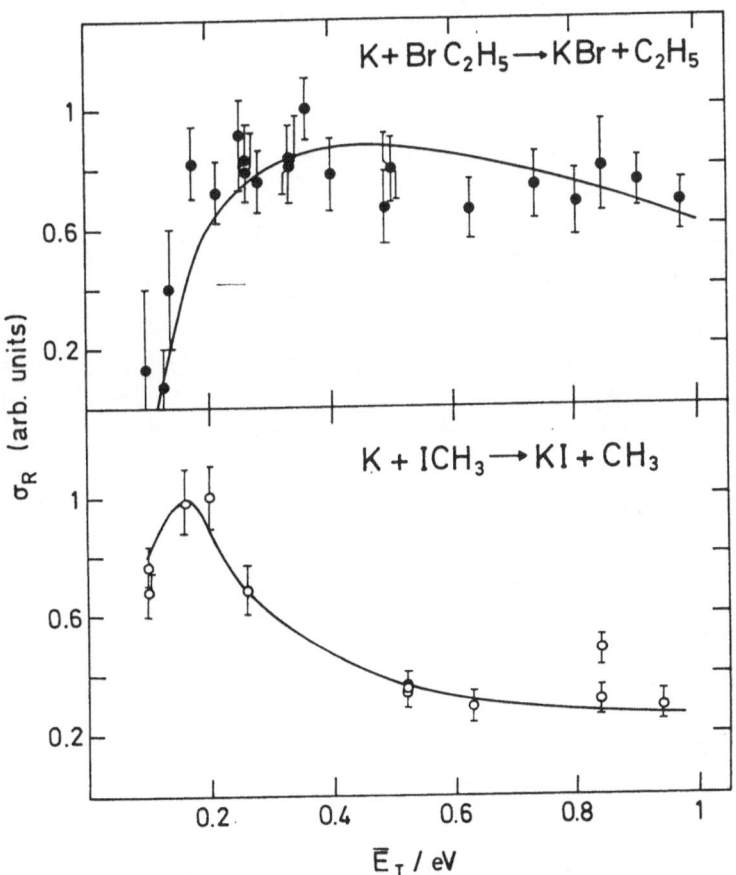

FIGURE 6.   Halogen effect in the shape of the excitation function.   Top: K + C_2H_5Br reaction.   Bottom: K + CH_3I reaction.   Adapted from Ref. 5.

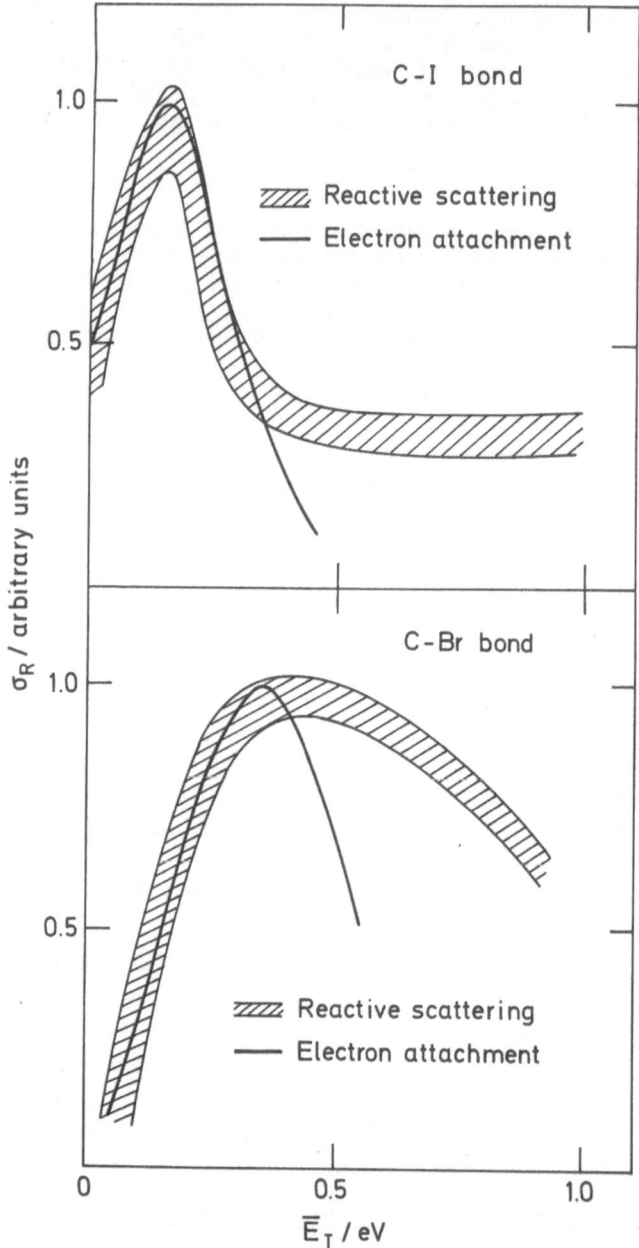

FIGURE 7. Reactive scattering and electron attachment data comparison. Note the close similarity between the two processes, at least at low collision energy. Adapted from Ref. 5.

TABLE 1. Family trends and comparisons in the M + RX → MX + R system. Selectivity and translational excitation effects.

Effect/comparison	Comment
Halogen effect (Fig. 6)	(a) Alkyl iodides: low thresholds, (< 0.1 eV) early maximum in $\sigma_R(E_T)$, sharp post-maximum decline.
	(b) Alkyl bromides: higher thresholds, (> 0.1 eV) weak maximum in $\sigma_R(E_T)$.
Reactive vs electron attachment cross-section (Fig. 7)	Same trends as noted in (a) and (b).
Radical group effect (Figs. 1 and 5)	As the size of the group increases:
	(a) The lower reactivity is found, larger steric factor.
	(b) The backward character of the MX angular distribution increases.
	(c) A higher fraction of the total available energy appears as internal motion of the radical group. The radical group behaves as a soft group, similarly to the behaviour found in the photofragmentation dynamics of the same alkyl iodides.
Attacking atom effect (Fig. 8)	(a) Reaction thresholds with the same target molecule seem to decrease in the Na > Rb > Cs(Xe) sequence according to Magee's prediction (electron transfer model).
	(b) The main change is observed as the electron affinity of the target molecule is changed. Reaction thresholds seem to follow the empirical law $E_0 \sim \exp(b\,\Delta E)$, $\Delta E = I.P - EA$ (ionization potential minus vertical electron affinity).

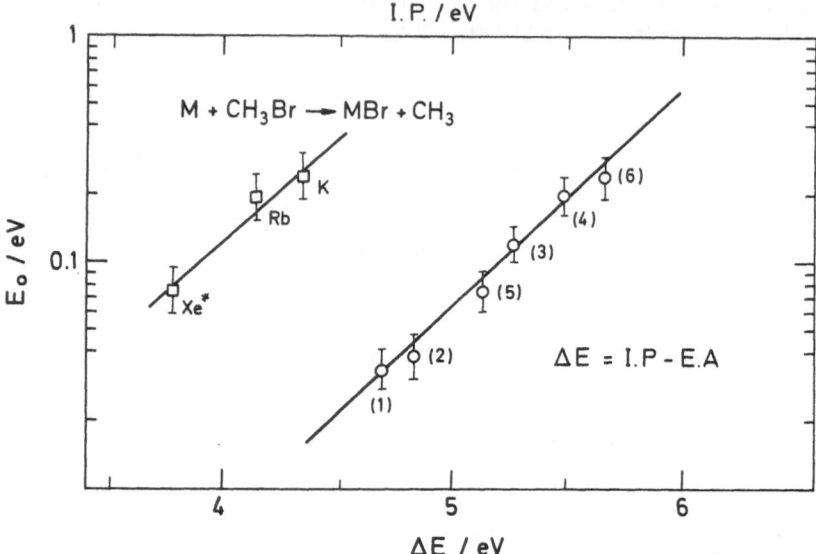

FIGURE 8. Attacking atom dependence of the energy threshold for the indicated reactions. Note how as the attacking atom mass increases (ionization potential decreases) the E_0 values decrease. The numbers correspond to: (1) Rb + CH_3I; (2) K + CH_3I; (3) K + C_2H_5Br; (4) Rb + CH_3Br; (5) Xe* + CH_3Br; and (6) K + CH_3Br. Threshold values from Ref. 5.

3.3.2 <u>Post-threshold behaviour</u> [17-19]. Figure 6 showed several post-threshold cross-section data obtained for some members of the M + RX family. At low collision energy the best agreement is always obtained for the line of centres model [1, 18, 21]. At higher collision energies a maximum in the excitation function may appear. One of the most common interpretations used to explain such a maximum is to consider a recrossing mechanism. From this viewpoint the natural behaviour is to assume an increasing function of $\sigma_R(E_T)$ with E_T, but as the collision energy increases depletion of reactive trajectories occurs because of the onset of the recrossing mechanism due to reflection from the (inner) repulsive potential to the reactant valley [13].

In addition to this recrossing mechanism we can consider a different interpretation for the appearance of the maximum in the reaction cross-section [20]. If we assume an orientational dependence of the reaction barrier of the linear form [12] $E_{th} = E_0 + E_0' (1 - \cos \gamma)$ with respect to the cosine of γ, i.e. the A -- BC angle, one can show that the bending frequency of the "A -- B -- C" transition state ω_B is directly proportional to E_0. In this case at low collision energy the $\omega_B.\tau$ parameter, e.g. the bending frequency times the reaction time, is long enough that the only reaction barrier for any initial angle is exactly the collinear one, E_0. In other words, the reaction time is so long compared with the rotation period of BC that the reaction always takes place through the collinear configuration. At higher collision energies $\omega_B.\tau \ll 1$ and the long-range (alignment forces) no longer operate. The reactive trajectories sample all the reaction barriers depending upon the

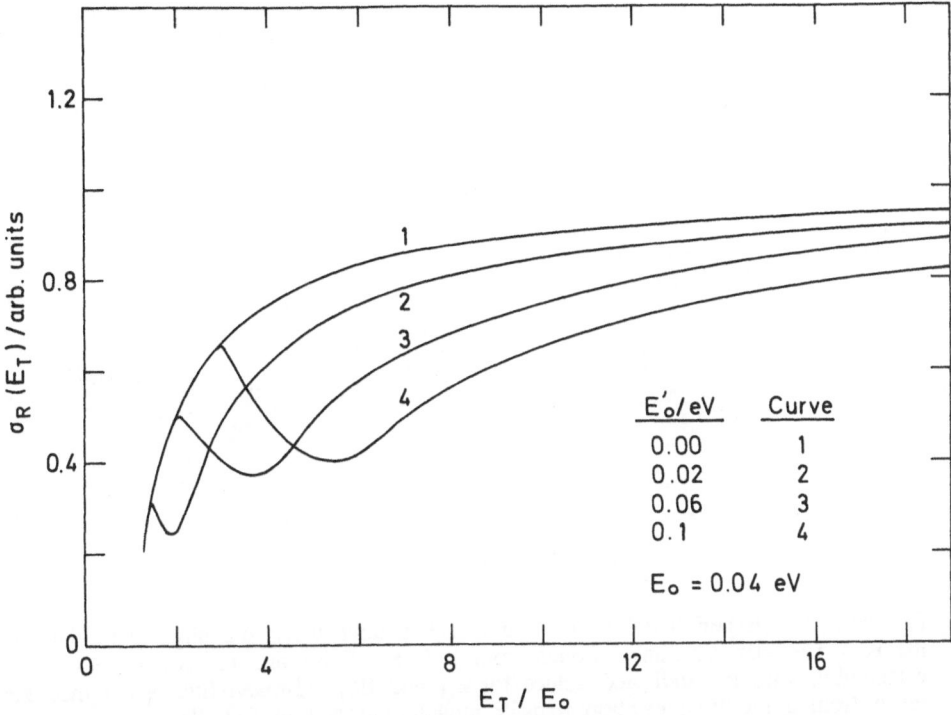

FIGURE 9. Theoretical result using the alignment model described in the text (see also Ref. 20). Different excitation functions are displayed as a function of the E'_0/E_0 value. An orientational angle dependence of the reaction barrier of the linear form

$$E_{th} = E_0 + E'_0 \ (1 - \cos \gamma)$$

was assumed being γ the orientation angle (see Ref. 12).

angle of attack, and so the reaction cross-section increases smoothly. One can show that over an intermediate range of collision energies a maximum appears followed by a shallow minimum whose location is a function of the model parameter, essentially the E_0' value.

Fig. 9 shows a model calculation to illustrate this behaviour in a clear way.

Fig. 10 shows the K + C_2H_5Br excitation function measurements (taken from Ref. 10c). The dotted line is the model calculation obtained with this alignment model for E'_0/E_0 = 3 and E'_0 = 0.012 eV. Even though there is no convolution considering the collision energy spread the agreement between experimental and theoretical calculation is satisfactory.

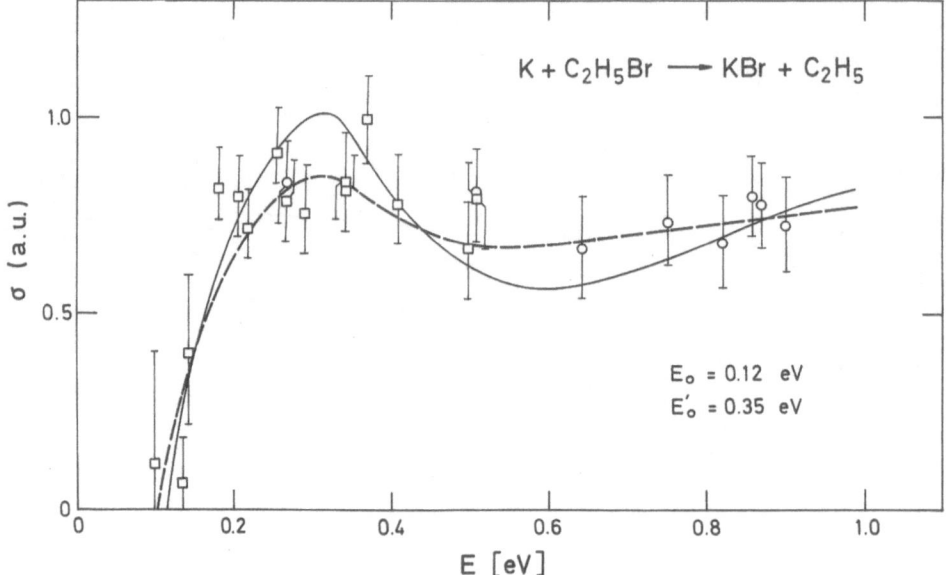

FIGURE 10. Experimental vs theoretical calculations using the alignment model for the K + C_2H_5Br reactions. Points from Ref. 5. Solid line present model calculation, with the indicated values for E_0 and E'_0. Broken line is the theoretical result from a modified electron transfer model described in Ref. 21.

4. CONCLUSIONS

In what follows a summary of well established features about the M + RX → MX + R family are presented.

4.1 Differential Excitation Functions

(1) The backward character and the high translational energy disposal are key features of the rebound, direct mechanism, not only at low collision energy, but also up to 1 eV.

(2) As the collision energy increases (i) the c.m. angular distribution becomes slightly forward peaked, (ii) most of the excess of energy appears as translational energy of the products indicating a high level of translational adiabaticity.

(3) Whereas the attacking atom and halogen do not seem to produce important changes in the differential excitation function, the radical group does. As the radical group increases there is an increase in the backward character of the MX (c.m.) angular distribution as well as a reduction of the translational adiabaticity. The latter has been found to be similar to that of photofragmentation studies of the same alkyl iodides.

4.2 Total Excitation Function

All the cases studied so far showed low reactivity, reaction cross-section, σ_R < 50 $Å^2$, with clear family trends as far as the shape of the excitation function is concerned

(a)   The attacking atom does not seem to play a major part except that energy threshold decreases as the M size increases (as the ionization potential decreases).

(b)   The halogen atom plays a decisive role.   First of all the reactivity increases in the sequence F < Cl < Br < I following the change in E_0 values.   With respect to the evolution of the shape upon halogen substitution, the only available comparison that could be made was for Br and I reactions.   The alkyl iodide reactions showed lower energy thresholds and sharper post-maximum declines than the bromide ones.   These two well established behaviours are also found for the low collision electron attachment cross-section of the respective alkyl halides, confirming that the electron transfer process is the reaction pathway for these reactive collisions.

On the other hand the radical influence upon the total reaction cross-section has been studied showing a definite decrease as the alkyl group size increases, therefore indicating a larger steric factor in that precise sequence.

With respect to the energy threshold and post-threshold evolution, several conclusions can be drawn:

The energy threshold value can be well predicted by the electron transfer mechanism, and it seems to be closely related to the (lower) value of the target molecule electron affinity.

The post-threshold evolution of the reaction cross-section can be well described by the line-of-centres model.   From a transition state point of view that behaviour can be explained by a vibrationally adiabatic character of the transition state motion, i.e. the only active degree of the transition state is the "A.....BC" rotation. Moreover this picture is consistent with the prediction of the alignment model which only considers collinear configurations for the transition state since the long range interactions, at low collision energy, will induce or favour the collinear reaction pathway.   In conclusion a linear rotating complex seems to be the more adequate description for the post-threshold evolution of the reaction dynamics.

As the collision energy increases significant deviations, or pronounced extrema, may appear, indicating the onset of additional dynamical factors governing the collision energy dependence of the reaction dynamics.   New channels, as for example collision-induced dissociation, non-collinear pathways with higher reaction barriers (i.e. insertion mechanism), the onset of repulsive interaction leading to recrossing trajectories, may change significantly the reaction dynamics.   Once again total and differential excitation function studies with oriented and state selected reagents are of great and promising value to elucidate the molecular dynamics governing these reactive collisions.

Finally, it was noted that there were close similarities in the total and differential excitation function between the M + RX reaction and the RX photofragmentation and/or electron attachment studies.   The same overall selectivity was invoked as a result of a similar dynamics associated with the rupture of the same C-X bond.

ACKNOWLEDGEMENT

I wish to thank Professor J.P. Simons for stimulating discussions, as well as the SERC for a visiting fellowship to the Chemistry Department of Nottingham University, during which period of time this work was concluded. The current support of the author's research by the Comision Interministerial de Ciencia y Tecnologia is gratefully acknowledged.

REFERENCES

[1] R.D. Levine and R.B. Bernstein, Molecular Reaction Dynamics, Clarendon, Oxford (1974).

[2] R.B. Bernstein, Chemical Dynamics via Molecular Beam and Laser Techniques, Clarendon, Oxford (1982).

[3] R.B. Bernstein (ed), Atom-Molecule Collision Theory. A Guide for the Experimentalist, Plenum, New York (1979).

[4] D.R. Herschbach, Faraday Disc. Chem. Soc., 55 (1973) 233.

[5] A. Gonzalez Ureña, Adv. Chem. Phys., 66 (1987) 213.

[6] H. Beutler and M. Polanyi, Z. Physik. Chem. B, 1 (1928) 31.

[7] (a) H.V. Hartel and M. Polanyi, Z. Physik. Chem. B, 11 (1930) 97.

     (b) M. Polanyi in Atomic Reactions, Williams & Norgate, London (1932).

[8] J.E. Mosch, S.A. Safrom and J.P. Toennies, Chem. Phys., 8 (1975) 304.

[9] See for example:

     (a) See section 4 of Ref. 5.

     (b) K.K. Chakravorty and R.B. Bernstein, J. Phys. Chem., 88 (1984) 3465.

     (c) M.E. Gersh and R.B. Bernstein, J. Chem. Phys., 55 (1971); 4461; 56 (1972) 6131.

    (d)   H.E. Litvak, A. Gonzalez Ureña and R.B. Bernstein,
J. Chem. Phys., **61** (1974) 738; **61** (1974) 4091.

    (e)   S.A. Pace, H.F. Pang and R.B. Bernstein, J. Chem. Phys., **66** (1977) 3635.

    (f)   H.F. Pang, K.T. Wu and R.B. Bernstein, J. Chem. Phys., **54** (1977) 2088.

    (g)   K.T. Wu, H.F. Pang and R.B. Bernstein, J. Chem. Phys., **61** (1978) 1064.

[10]   See for example:

    (a)   M.E. Gersh and R.B. Bernstein, J. Chem. Phys., **55** (1971); 4461; **56** (1972) 6131.

    (b)   V. Saez Rabanos, E. Verdasco, V.J. Herrero and A. Gonzalez Ureña. J. Chem. Phys., **81** (1984) 5725.

    (c)   V.J. Herrero, F.L. Tabares, V. Saez Rabanos, F.J. Aoiz and A. Gonzalez Ureña, Mol. Phys., **44** (1981) 1239. See also V.J. Herrero, V. Saez Rabanos and A. Gonzalez Ureña, Mol. Phys., **47** (1982) 725.

[11]   (a)   D.S.Y. Hsu and D.R. Herschbach, Faraday Disc. Chem. Soc., **55** (1972) 116.

    (b)   D.S.Y. Hsu, N.D. Weinstein and D.R. Herschbach, Mol. Phys., **29** (1975) 257.

    (c)   See also Ref. 5.

[12]   R.B. Bernstein, J. Chem. Phys., **82** (1985) 3656 and references cited therein.

[13]   (a)   D.L. Bunker, E.A. Goring, Chem. Phys. Lett., **15** (1972) 52.

    (b)   R.A. La Budde, P.J. Kuntz, R.B. Bernstein and R.D. Levine, J. Chem. Phys., **59** (1973) 6286.

[14]   (a)   J.C. Polanyi, Faraday Disc. Chem. Soc., **44** (1967) 293.

    (b)   *ibid.* **55** (1973) 293.

    (c)   J.C. Polanyi and J.L. Schreiber, Physical Chemistry. An Advanced Treatise Vol. VI, A.H. Eyring, D. Henderson and W. Jost (eds). Academic, New York (1974).

[15]   V. Saez Rabanos, F.J. Aoiz, V.J. Herrero, E. Verdasco and A. Gonzalez Ureña, Mol. Phys., **59** (1986) 707. See also reference 10c.

[16]   See for example T. Munakata, Y. Matsuni and T. Kasuya,
       J. Chem. Phys., **79** (1983) 1698.

[17]   P. Pechukas, J.C. Light and C. Rankin, J. Chem. Phys., **44** (1986)
       794.

[18]   A. Gonzalez Ureña, Mol. Phys., **52** (1984) 1145.

[19]   K.T. Wu, J. Phys. Chem., **83** (1979) 1043.

[20]   A. Gonzalez Ureña, in preparation.

[21]   A. Gonzalez Ureña and M. Menzinger, Chem. Phys., **99** (1985) 437.

THE EFFECT OF CHANGING REAGENT TRANSLATION ON REACTION DYNAMICS:
$F + CF_3I \rightarrow IF + CF_3$

M. Trautmann*, K. Wagemann, J. Wanner, X.K. Zeng**
Max-Planck-Institut für Quantenoptik
8046 Garching
Federal Republic of Germany

1. INTRODUCTION

Within the general discussion of the selectivity of energy consumption in chemical reactions the influence of translational energy upon the dynamics of neutral reactions has received significant attention [1]. A first series of investigations of the effect of reagent translational energy on the internal state distribution of reaction products was carried out by Polanyi and co-workers, applying the method of infrared arrested relaxation chemiluminescence. A number of three-atom exchange reactions forming hydrogen halides have been found to obey the general rule that collision energy in excess of a barrier of type I is channelled into translational and rotational energy of the reaction product: $\Delta T \rightarrow \Delta T' + \Delta R'$. This is interpreted in terms of "induced repulsive energy release" [2]. The development of novel techniques, in particular the combination of the molecular beam and the laser-induced fluorescence methods, has permitted study of the detailed dynamics of chemical reactions with selective translational reagent excitation to be extended to a variety of other reaction systems [3-5].

Despite this, the collision energy dependence of the internal product state distributions for complex reactions occurring on a covalent surface has so far barely been obtained owing to the relatively small detailed cross-sections. Yet, the change of complex dynamics with reagent translation has been revealed in product angular velocity distribution measurements. A prominent example is the reaction $F + CH_3I$, studied by Farrar and Lee [6]. At a low collision energy of about 11 kJ mol⁻¹ this system exhibits long-lived complex dynamics, which changes to more direct interaction at a reagent translation of about 59 kJ mol⁻¹.

In this contribution we report the results of a crossed molecular beam study of the dependence of internal IF product state distributions on the collision energy in the reaction $F + CF_3I \rightarrow IF + CF_3$. The laser-induced fluorescence product state analysis of this process at a collision energy of 3.2 kJ mol⁻¹ in conjunction with the analogous hydrogenated system $F + CH_3I$ has been reported earlier [7]. In the present work the analysis of the internal product energy distribution of the $F + CF_3I$

---

* Wilhelmsgymnasium, München, ** Chinese Academy of Sciences, Dalian Institute of Chemical Physics, People's Republic of China.

J. C. Whitehead (ed.), Selectivity in Chemical Reactions, 95–104.
© 1988 by Kluwer Academic Publishers.

reaction was extended to collision energies of up to 14 kJ mol⁻¹ by using a supersonic He-seeded microwave discharge F atom source. The results presented in this paper show that for the F + CF_3I reaction the excess reagent translation is channelled strongly into product vibration and moderately into rotation: $\Delta T \rightarrow \Delta V' + \Delta R'$.

2. EXPERIMENTAL

High-velocity fluorine atoms were produced by supersonic expansion of F atoms seeded in He. The F atom discharge based on the design of Schwalm [8] was incorporated with a differentially pumped stage into our molecular beam apparatus [9]. The fluorine atoms were prepared by a microwave discharge (2450 MHz, 150 W) in a gas mixture of pure F_2 and He (mixing ratio 1:4) at the end of a sufficiently heat-resisting alumina ceramic tube (Coors, Ltd) of 13 mm o.d. and 8 mm i.d. The details of the discharge source are depicted in Fig. 1. At a

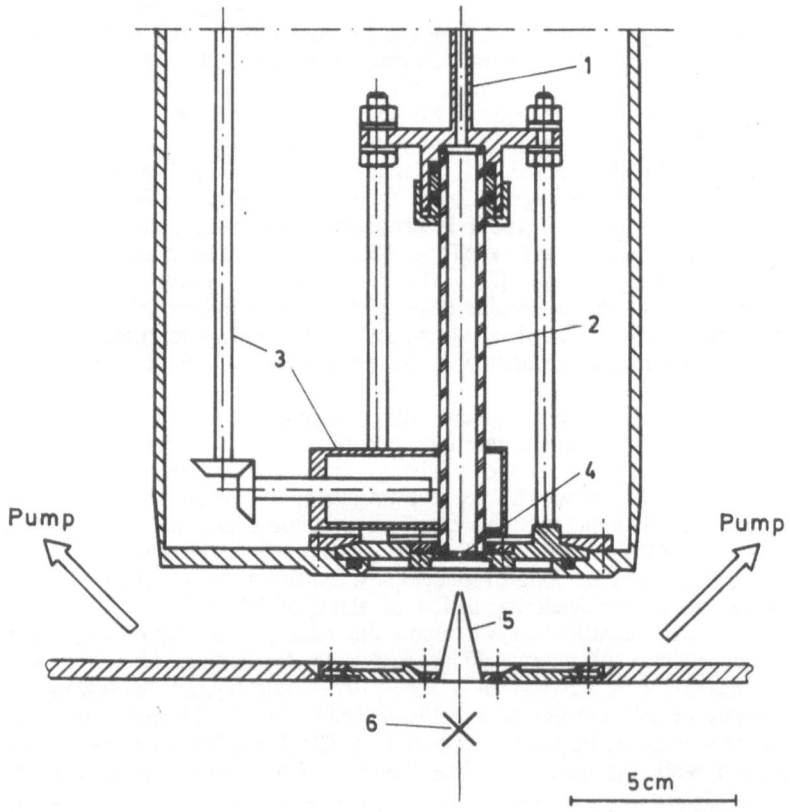

FIGURE 1. Details of the experimental arrangement: 1. gas supply, 2. discharge tube from Al_2O_3, 3. microwave cavity including tuning elements, 4. nickel nozzle, 5. skimmer, 6. scattering center of F atoms with the quasieffusive CF_3I beam.

stagnation pressure of 130 mbar the gas expands through a 0.2 mm nickel nozzle and enters the main chamber through a skimmer (0.93 mm in diameter) placed 10 mm apart from the nozzle. The terminal speed of the F atoms was calculated according to isentropic expansion [10]. In order to determine the average molecular weight of the expanding gas mixture, the degree of dissociation of F_2 was measured by a crossed beam ionizer quadrupole mass spectrometer (Extranuclear) placed 80 mm downstream from the nozzle. Under our experimental gas flow conditions we determined a fraction of dissociation of F_2 of about 0.10 and calculated a terminal velocity of the F atoms of 1200 m/s.

In a preliminary set of experiments the F source was operated with pure fluorine at a stagnation pressure of 30 mbar where a 0.5 mm i.d. nozzle plate made of sapphire was used [9,11]. Under these conditions the resulting F beam velocity was measured to be 850 m/s with 30% FWHM [12].

The comparative experiment at a collision energy of ~3.2 kJ mol^{-1} was again conducted under experimental conditions identical to those in Ref. 13. Here the F beam was prepared from a microwave discharge of F_2 (2450 MHz, 50 W) and effused from a quartz tube narrowed to 3 mm with flows of approximately 7 SCCM. In all cases the quasieffusive CF_3I source described in Ref. 13 was used. Single collision conditions were established for all experiments.

The fluorescence of the IF product molecules originating from the $B^3\Pi(0)^+ \rightarrow X^1\Sigma^+$ band system was observed in the wavelength range 470-675 nm. The relative product vibrational population densities were determined from the measured fluorescence intensities of the individual bands, the relative laser intensities and the transition probabilities, as described in Ref. 12. At slow scanning speed the resolved rotational structure in the IF excitation spectrum was obtained, except in the vicinity of the band heads. It was analysed according to the procedure described below.

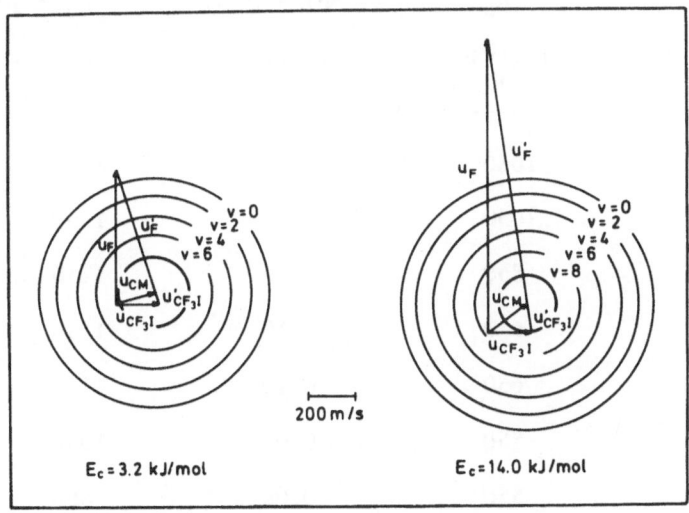

FIGURE 2. Newton diagram for the F + CF_3I reaction at two collision energies applied. In accordance with the favourable kinematics the translational energy in the relatively light attacking F atom is largely retained as translation in the CM frame.

The effective change of reagent energy can be calculated according to a Newton diagram as shown in Fig. 2. Owing to the favourable kinematics the translational energy in the relatively light attacking F atom is largely retained as translation in the CM frame. By contrast, the thermal motion of the CF_3I beam is largely motion of the CM system as a whole. The change of reagent translation from $E_c = 3.2$ kJ mol^{-1} to $E_c = 14$ kJ mol^{-1} corresponds to a change of the total available energy E_{tot} from 55.1 kJ mol^{-1} to 65.9 kJ mol^{-1}.

3. RESULTS

3.1 Rotational Product Excitation

The difference in rotational product state excitation is visualized in Fig. 3 for the (0,4) band of IF for two reagent energies. It can be seen that the rotational excitation of the product molecules is obviously enhanced at higher collision energy. This holds for all vibrational product levels investigated. A Boltzmann rotational distribution and a Gaussian line profile for P and R branch lines can be used to simulate the observed rotational spectra while taking the molecular constants of Ref. 4 for the X and B states of IF. The result of the rotational analysis is given in Table 1.

TABLE 1. Rotational excitation of individual vibrational product states. To a good approximation $<f'_R(v)>$ was obtained from a fit of a "rotational temperature" $T'_R(v)$ to rotationally resolved spectra.

IF(v)	$E_c = 3.2$ kJ mol^{-1}		$E_c = 14.0$ kJ mol^{-1}	
	$T'_R(v)$ (K)	$<f'_R(v)>$	$T'_R(v)$ (K)	$<f'_R(v)>$
0	1050	0.16	1350	0.17
1	1050	0.16	1350	0.17
2	850	0.13	1250	0.16
3	650	0.10	1150	0.15
4	650	0.10	1150	0.15
5	550	0.08	1050	0.13
6	550	0.08	850	0.11
7	350	0.05	550	0.07

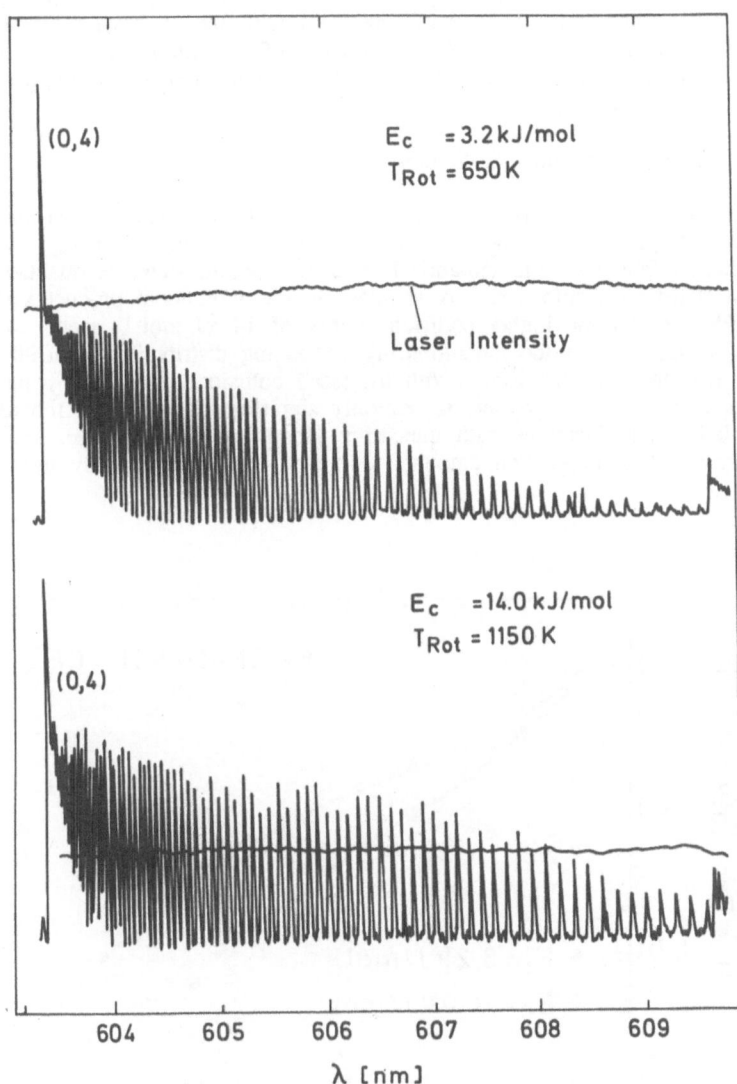

FIGURE 3. Influence of the excess reagent translation on the rotational manifold of the IF(v=4) product state. The effect is visualized by excitation spectra of the (0,4) band.

In summary, 23% of the excess translation is channelled into product rotation (see Table 3 below). This is in agreement with the "Polanyi rule" stated earlier and can be understood in terms of a correlation of J' with the increase in the angular momentum L, which is determined by the relative velocity of the reagents and the impact parameter [2].

3.2   Vibrational Product Excitation

Figure 4 shows the experimentally determined relative number densities N'(v) of IF vibrational product state occupations. At the lower collision energy the vibrational population falls with the vibrational product quantum state, as qualitatively expected for a statistical distribution. A shoulder in the vibrational product state distribution at v=5, e.g. for the higher collision energy of 14 kJ mol⁻¹, seems to mark the onset of a deviation from the monotonically decreasing distribution, indicating the beginning of a change in mechanism. With increased collision energy the population of vibrationally excited products is not only significantly enhanced; the excess vibrational population also increases with increasing vibrational product state. For IF(v=6) the enhancement in population amounts to a factor of approximately three at E_c=14 kJ mol⁻¹.

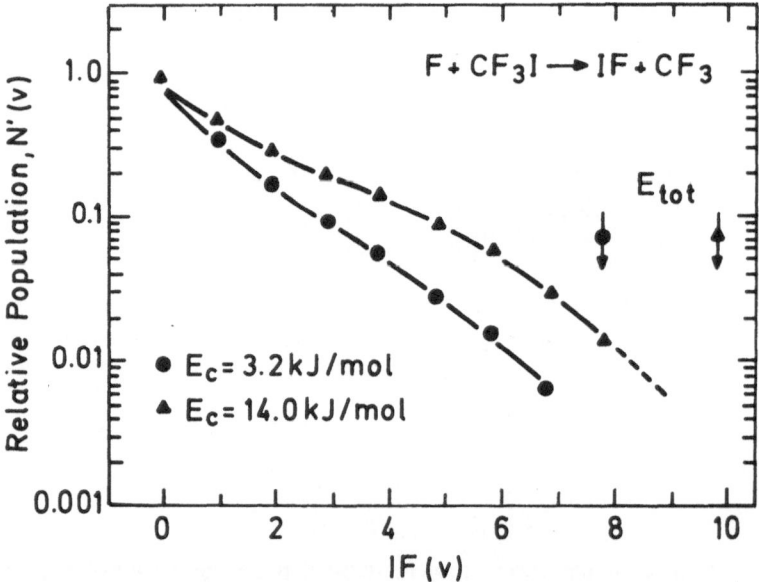

FIGURE 4. Measured vibrational product number densities plotted on a logarithmic scale as a function of the IF vibrational quantum number. The monotonically decreasing distributions reveal essentially long-lived complex dynamics. The shoulder in the upper distribution at a collision energy of 14 kJ mol⁻¹ evolving at v=5 seems to indicate the deviation from a statistical mechanism. The probability of populating higher vibrational product states strongly rises with excess reagent translation.

TABLE 2. Measured relative vibrational occupations $N'(v)$ for the reaction $F + CF_3I$ at two collision energies E_C. The number densities have been converted into detailed rate constants $k'(v)$ by multiplying by the corresponding laboratory velocities.

| | $E_C = 3.2$ kJ mol^{-1} | | $E_C = 14.0$ kJ mol^{-1} | |
IF(v)	$N'(v)$	$k'(v)$	$N'(v)$	$k'(v)$
0	1	1	1	1
1	0.33	0.31	0.44	0.38
2	0.17	0.15	0.29	0.26
3	0.09	0.07	0.21	0.18
4	0.06	0.04	0.17	0.13
5	0.03	0.017	0.09	0.06
6	0.015	0.008	0.06	0.04
7	0.006	0.003	0.03	0.02
8			0.01	0.005

TABLE 3. Summary of the internal product energy partitioning in terms of the reduced variables $<f'_V>$ and $<f'_R>$. $\Delta E'_V/\Delta E_C$ and $\Delta E'_R/\Delta E_C$ denote the fraction of excess reagent translation channelled selectively into product vibration and rotation, respectively.

collision energy (kJ mol^{-1})	E_{tot} (kJ mol^{-1})	$<f'_V>$	$<f'_R>$	$\dfrac{\Delta E'_V}{\Delta E_C}$	$\dfrac{\Delta E'_R}{\Delta E_C}$
3.2	55.1	0.10	0.15		
14.0	65.9	0.14	0.17	0.42	0.23

If one converts the measured number densities into detailed rate constants, assuming forward-backward symmetry, this effect becomes even more pronounced (Table 2). Summarizing the data we find that a total of 42% of the excess reagent translation is channelled into product vibration (Table 3). A more detailed analysis of the measured internal product state distributions will be given elsewhere [15].

4. DISCUSSION

In the following we need to explain why for the polyatomic system F + CF$_3$I, contrary to the "Polanyi rule", excess reagent translation is preferentially transformed into vibration of the diatomic product. It has already been noted that the internal product state distributions exhibit similar features for the F + CF$_3$I and F + CH$_3$I reactions. This led to the conclusion that both reactions at collision energies corresponding to 300 K proceed via an identical, essentially statistical, mechanism [7]. As we are not aware of any other dynamic investigations of the F + CF$_3$I reaction, we shall thus consider in the following discussion two related studies of the F + CH$_3$I system: the angular distribution measurement of Farrar and Lee [6] and the subsequent trajectory calculation of Fletcher and Whitehead [16].

The F + CH$_3$I reaction proceeds via a potential hypersurface with a well. At a collision energy of 10.9 kJ mol^{-1} the IF angular velocity distribution as determined by Farrar and Lee for F + CH$_3$I shows forward-backward symmetry. This corresponds to a complex lifetime exceeding one rotational period. At a higher collision energy of 58.9 kJ mol^{-1} more pronounced forward scattering was observed. This implies a complex lifetime comparable to one rotational period [6]. In terms of the internal product energy distribution the observed changes can be interpreted as follows: At a low collision energy the internal energy transfer is rapid compared with the chemical reaction time. The system then represents unimolecular decay following chemical activation. According to the RRKM theory the internal product state distribution is expected to be statistical. At higher energy the lifetime may be reduced to the extent that energy randomization cannot occur in this time period. Consequently, the reaction will be more direct, especially for systems containing only a few atoms as in our case. Hence one would expect to find a change in dynamics from statistical to nonstatistical as E$_c$ is increased. The lifetime is shortened, preventing complete randomization of the energy. Hence the energy remains to a larger extent in IF product vibration, as seen by the conversion of a fraction of about 40% of the excess collision energy. Consequently, the relative amount of CF$_3$ excitation decreases. The collision-energy-dependent partitioning of the internal energy is visualized in Fig. 5.

It is worthwhile to compare the change in internal product state distributions as a function of E$_c$ with dynamic calculations based on an adequate potential energy surface. Fletcher and Whitehead evaluated a collinear LEPS potential energy surface for the F + CH$_3$I → IF + CH$_3$ reaction [16]. In their approach the well depth was adjusted according to the stability of the FICH$_3$ intermediate. The well extends into the entrance and exit valleys without barriers in either coordinate. However, the authors reduced the problem to a three-particle approximation by considering the methyl group to be a single particle of the appropriate mass. The trajectory calculations showed agreement with the experimental angular and recoil energy distributions of the IF product, the energy dependence of the total reaction cross-section and the values of the thermal rate constant. Owing to the single-particle approximation of the CH$_3$ radical, however, this approach could not

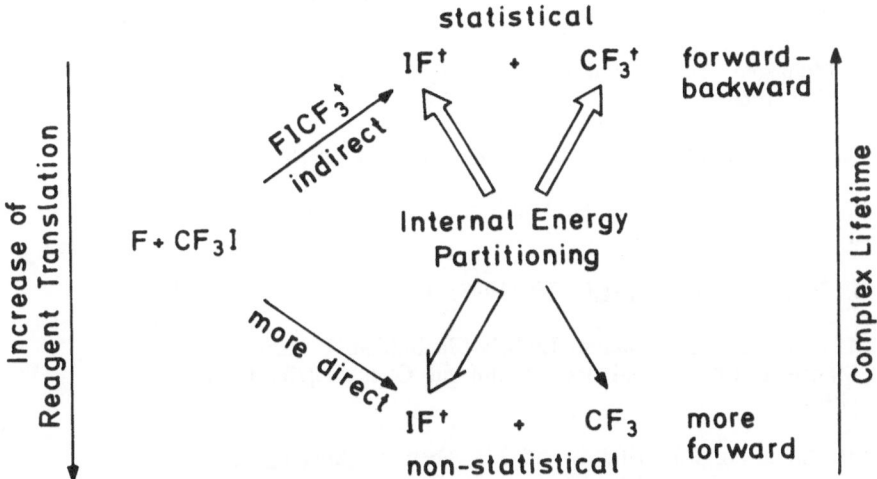

FIGURE 5. Schematic representation of the collision-energy-dependent partitioning of the internal energy between the reaction products IF and CF_3 as observed in this study.

determine the extent of the partitioning of the internal energy between the reaction products IF and CH_3.

The collision-energy-dependent partitioning of the internal energy between the reaction products IF and CF_3 as observed in this study should encourage further investigations. Experimentally it would be desirable to operate the F atom beam source at further increased velocities in order to exploit the change of energy partitioning at higher collision energies. Recent work of the Grice group shows that with a similar design a F beam velocity of 1950 m/s can be obtained [17]. From the theoretical point of view there seems to be a lack of simple theoretical models which account for the observed effect.

The authors are grateful to K.L. Kompa for continual support of this work. One of them, X.K.Z., acknowledges the award of a fellowship by the Max-Planck-Gesellschaft.

REFERENCES

[1]   A.G. Urena, Advances in Chemical Physics, **66** (1987) 213.

[2]   L. Cowley, D.S. Horne and J.C. Polanyi, Chem. Phys. Lett., **12** (1971) 144.
      A.M.G. Ding, L.J. Kirsch, D.S. Perry, J.C. Polanyi and J.L. Schreiber, Faraday Discuss. Chem. Soc., **55** (1973) 252;
      J.C. Polanyi, J.J. Sloan and J. Wanner, Chem. Phys., **13** (1976) 1.

[3]   A. Siegel and A. Schultz, J. Chem. Phys., **72** (1980) 6227.

104

[4] A. Gupta, D.S. Perry and R.N. Zare, J. Chem. Phys., 72 (1980)
     6237.

[5] T. Munakata, Y. Matsumi and T. Kasuya, J. Chem. Phys., 79
     (1983) 1968.

[6] J.M. Farrar and Y.T. Lee, J. Chem. Phys., 63 (1975) 3639.

[7] L. Stein, J. Wanner and H. Walther, J. Chem. Phys., 72 (1980)
     1128.

[8] U. Schwalm, Appl. Phys.B, 30 (1983) 149.

[9] M. Trautmann, Dissertation, Ludwig-Maximilians-Universität,
     München, 1984: Max-Planck-Institut für Quantenoptik, Bericht 90,
     (1984).

[10] J.B. Anderson, R.P. Andres and J.B. Feen, in Advances in
     Atomic and Molecular Physics, D.R. Bates and I. Estermann, Eds.
     Vol. 1, Academic Press, New York, 1965, p. 345.

[11] Gmelin Handbuch der Anorganischen Chemie, "Fluorine" Supplement
     Vol. 2, Springer, Berlin, Heidelberg, New York, 1980, p. 97.

[12] S.T. Collins, M. Trautmann and J. Wanner, J. Chem. Phys., 84
     (1986) 3814.

[13] T. Trickl and J. Wanner, J. Chem. Phys., 78 (1983) 6091.

[14] T. Trickl and J. Wanner, J. Mol. Spectrosc., 104 (1984) 174.

[15] K. Wagemann, J. Wanner and X.K. Zeng, paper in preparation.

[16] I.W. Fletcher and J.C. Whitehead, J. Chem. Soc. Faraday Trans. 2,
     77 (1981) 2329.

[17] N.C. Firth, D.J. Smith and R. Grice, Molec. Physics, 61(1987) 859.
     N.C. Firth, N.W. Keane, D.J. Smith and R. Grice, Faraday Discuss.
     Chem. Soc., 84 (1988), in the press.

INTERNAL ENERGY DISTRIBUTIONS IN AlO GIVEN BY THE REACTIONS OF ALUMINIUM ATOMS WITH SEVERAL OXIDANT MOLECULES

Michel Costes, Christian Naulin, Gérard Dorthe and
Guy Nouchi
U.A. 348: Photophysique Photochimie Moléculaire and
U.A. 283: Centre de Physique Moléculaire Optique
et Hertzienne
Université de Bordeaux I
33405 Talence
France

ABSTRACT. The reactions of ground state Al atoms (^2P_J) with O_2, CO_2, SO_2 and N_2O are studied with a pulsed crossed supersonic molecular beam apparatus. The Al atom beam is generated by laser ablation of the solid at the exit of a pulsed nozzle and seeding in the supersonic expansion of neutral carrier gases. Velocities ranging from 800 to 3400 ms^{-1} are thus obtained. Al atoms are probed, together with the product AlO, by laser-induced fluorescence. The effects of the reagent relative translational energy on the relative reactive cross-sections and AlO internal energy distributions are studied. All the vibrational and rotational distributions found are governed by statistical behaviour at low collision energy and show a lack of excitation at high collision energy.

1. INTRODUCTION

Since the first age of molecular beam scattering there has been a constant interest in the dynamics of metal atom-molecule reactive collisions. The effect of relative translational energy of reagents on reaction dynamics has been extensively studied. However a great majority of experiments has been devoted to alkali and alkali-earth reactions. The insight into the reactivity of other metal elements is comparatively less rich and this can be easily connected with the technical constraints associated with the generation of beams of these atoms.

Advances in the development of pulsed nozzle beams and pulsed U.V. lasers now offer an alternative to the metal-oven effusive source. Metal atom beams can be obtained by laser ablation of the solid at the exit of a pulsed valve. Seeding in different neutral carrier gas mixtures provides a wide, continuous range of velocities, from 800 ms^{-1} with pure argon to 3400 ms^{-1} with pure hydrogen, and narrow velocity spreads (17 - 11% F.W.H.M.). The versatility of the source allows one to study the reactivity of a large number of metal or solid metalloidic atoms including the most refractory ones and the wide velocity range gives an easy access to the study of endoergic reactions characterized by a translational energy threshold [1-4].

Reactions of aluminium atoms with the oxidizers O_2, CO_2, SO_2 and N_2O reported here represent an interesting family, since it contains both exoergic and

105

J. C. Whitehead (ed.), Selectivity in Chemical Reactions, 105–116.
© *1988 by Kluwer Academic Publishers.*

endoergic members. Fontijn and co-workers [5-9], using a high temperature fast flow reactor, determined the rate constants except for Al + N_2O. This latter reaction was reported to give AlO(B-X) and AlO(A-X) chemiluminescences in a flow system [10-11]. Information on the dynamics was also obtained for the exoergic Al + O_2 reaction, in a beam gas configuration, using an aluminium oven effusive source [12-13].

2. EXPERIMENTAL

The scattering chamber is a stainless steel cylinder (400 mm i.d. and 500 mm long) evacuated by an oil diffusion pump of the same diameter. The two pulsed valves are of the Gentry and Giese design [14] in its commercial version (Beam Dynamics). The base pressure which is 10^{-6} mbar with beams off increases to less than 10^{-5} mbar with both valves in operation. The nozzle-crossing point distance is 80 mm for the oxidizer beam and 110 mm for the aluminium beam. The two beams crossing at right angles are collimated to 4^0 and 6.5^0 F.W.H.M. respectively. Fast ionisation gauges (Beam Dynamics) are used to measure the time profiles, velocities and velocity distributions of the beams, as well as the attenuation of one beam by another. This latter factor remains lower than 10% in all experiments thus ensuring reasonably single collision conditions.

The reactant Al and the product AlO are probed by LIF at the crossing point (laser beam diameter $\approx$ 2.5 mm) using a Nd-Yag pumped dye laser (Quantel YG481 + TDLIV) operated with 532 nm pumping of rhodamine dyes. Radiations at shorter wavelengths are generated by stimulated Raman scattering in H_2 and the desired anti-Stokes line is selected through a set of four Pellin-Broca prisms. When the dye laser is operated with a mixture of Rh590 and Rh610, rapid switching from Al to AlO detection is possible. In that case, Al atoms are detected on the $4s^2S_{1/2} \leftrightarrow 3p^2P_{1/2}$ and $4s^2S_{1/2} \leftrightarrow 3p^2P_{3/2}$ transitions at 394.4 and 396.2 nm situated in the second anti-Stokes wavelength range of the dye. AlO is probed on the $(B^2\Sigma^+ \leftrightarrow X^2\Sigma^+)$ $\Delta v = +1$ sequence, between 465 and 472 nm, by scanning over the first anti-Stokes wavelength range of the same dye and the fluorescence is isolated through a coloured glass filter centred at 465 nm (85 nm F.W.H.M., T = 0.60). Some experiments have also been performed with excitation of AlO($X^2\Sigma^+$) on the $\Delta v = -1$ sequence, between 508 and 515 nm, by scanning through the first anti-Stokes wavelength range of the Rh640 dye. AlO excitation spectra are recorded in the intermediate saturation regime keeping the mean laser power constant. The effects of optical pumping are not a factor as long as the variation of absorption coefficients remains small [15]. This is the case for v" = 0-2 states pumped on the $\Delta v = + 1$ sequence or v" = 1-3 states pumped on the $\Delta v = -1$ sequence [16]. Line intensities remain thus linearly proportional to the densities in the ground state manifold.

The aluminium beam is produced by the technique described in previous works of Smalley and co-workers for the generation of supersonic metal clusters [17,18]. However, as our goal is to produce atoms and not clusters, our set up stands out with several particularities: (i) firstly we use an excimer laser (Lambda Physik EMG101E) operated at 249 nm with an unstable resonator, rather than a doubled Nd-Yag laser. A fraction of the output ($\approx$ 7mJ) is selected by spatial filtering and focused to a 0.4 mm diameter circular spot on the rod (96% Al - 4% Mg alloy). The combination of the 5 eV photon energy and the high fluence ensures rupture of all the Al-Al bonds in the ablated material. A higher production rate of atoms versus clusters is thus expected; (ii) secondly, the gas load of our valve is about 20 times less than that used by Smalley and co-workers for their pulsed valve. Cluster growth by three body recombination is thus much lower; (iii) finally, this latter effect is enhanced as we use a short extension channel: the distance between the beam waist and the expansion into the vacuum is only 4 mm. The Al beam obtained is a mixture of the $^2P_{1/2}$ and $^2P_{3/2}$ ($\epsilon = 0.014$ eV) states, and its velocity can be adjusted between 800 (pure Ar) and 3400 m s^{-1} (pure H_2), depending upon the carrier gas mixture used. Throughout this range, its velocity spread (F.W.H.M.) decreases from 17 to 11%, and its pulse duration (F.W.H.M.) decreases from 20 to 3 µs. The time compression of the pulse results in a sevenfold increase of the Al peak density.

Experiments are performed with the whole system synchronized at 10 Hz repetition rate, except the oxidizer beam which is triggered at 5 Hz. AlO excitation spectra are thus recorded as the result of the difference signal in consecutive shots with the oxidizer beam on and off. This procedure was found necessary as the aluminium beam also contains small amounts of AlO, in the low rotational levels of v" = 0.

3. TREATMENT OF THE DATA

The quantitative information contained in the excitation spectra is extracted by comparison with synthetic spectra, calculated with the molecular constants from Coxon and Naxakis [19]. In these calculations, the following distribution function is introduced to model the rotational populations:

$$P(v",K") \propto (2K" + 1) \{\epsilon_{tot} - G(v") - F(K")\}^{1/2} \exp\{-\beta F(K")\} \qquad (1)$$

where $P(v",K")$ is the probability to populate the level v", K" (K" = J" ± 1/2, incorporating the spin splitting), ϵ_{tot} the total energy available to reaction products and $G(v")$ and $F(K")$ the vibrational and rotational energies respectively. In the three atom case the first part of the expression represents the prior expectation given by the information - theoretic approach to the analysis of state-to-state reaction dynamics, and the exponential term the deviance to the prior defining the surprisal parameter β [20]. In the four atom case this is simply a fitting expression because no summation over the density of states of the second diatomic product is carried out.

The rate of production per unit volume for a particular rovibrational state is given by:

$$dN(v",K")/dt = n_{Al} \, n_{ox} \, \sigma_r \, v_r \, P(v",K") \qquad (2)$$

with n_{Al} and n_{ox} the Al and the oxidizer densities, σ_r the total reactive cross-section, v_r the relative translational energy and $P(v",K")$ defined by Eq. 1. The densities $n(v",K")$ measured by LIF at the crossing point are not strictly proportional to the rate; this has been taken into account since early LIF crossed beam works through the use of the density-flux transformation of the product recoiling with laboratory velocity $v_{lab}(v",K")$ [21]:

$$dN(v",K")/dt \propto n(v",K") \, v_{lab}(v",K"). \qquad (3)$$

However, systematic use of this $v_{lab}(v",K")$ correction is not always correct. A

108

steady state has to be reached and this requirement is not necessarily fulfilled with pulsed beams of narrow temporal width. Also the laser beam section has to be greater than the reaction zone, indeed if it is smaller, molecules produced outside the irradiated volume and scattered into it are counted. Clearly in our case the product densities n(v",K") in the irradiated volume are dependant on the outcome of previous outside reactive collisions. We have therefore developed a simple realistic model to evaluate the correction function defined by the spatial and temporal overlap of the molecular beams.

The mathematical description of the model is out of the scope of this paper. Briefly, in this model, each reactant beam density is fitted to gaussian radial and temporal distribution functions, the spread in relative translational energy is neglected and the densities are assumed to be constant within the probed volume, which is smaller than the reaction zone. These assumptions result in a simple analytic expression of the overlap integral. Calculations are carried out for each rovibrational state of the outcoming molecule and for extreme velocity vector orientations, i.e. forwards and backwards. An example of the correction function, Γ, obtained for the Al + O$_2$ reaction at ϵ_{trans} = 0.49 eV is displayed on Fig. 1, together with the corresponding Newton diagram. In this case, the density of molecules recoiling backwards is higher than for those recoiling forwards. This may be related to the asymmetry of the reaction zone which is larger along the O$_2$ beam direction. This effect is enhanced at lower energy as the Al beam pulse duration increases when decreasing its velocity.

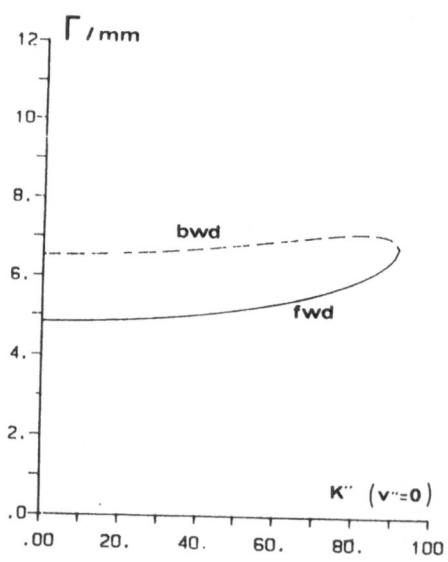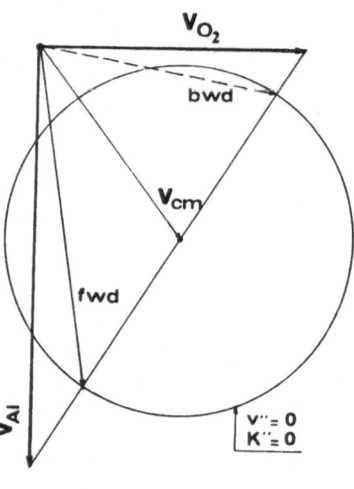

FIGURE 1. Correction function, relating AlO(v"=0, K") densities at the crossing point to the rate, and Newton diagram for Al + O$_2$ reaction at 0.49 eV.

4. RESULTS AND DISCUSSION

$$Al + O_2 \longrightarrow AlO + O$$

$AlO(X^2\Sigma^+)$ excitation spectra of this exoergic reaction ($\Delta\epsilon$ = -0.15 eV) have been recorded at relative translational energies ϵ_{trans} ranging from 0.083 eV to 0.49 eV. At the lowest energy there is close agreement with the synthetic spectrum calculated with statistical vibrational and rotational distributions, and symmetric forward-backward peaking of the angular distribution (Fig. 2). At the highest energy, deviance to the prior expectation occurs (Fig. 3). The whole results are reported on Fig. 4 with the variation of the detailed rate constants k(v") with the collision energy. There is no significant change of the fraction of energy going into AlO rotation $<f_R>$ and vibration $<f_v>$. They remain about 30% and 14% respectively throughout this energy range, and the excess translational energy is channelled mainly into product rotation and translation ($\approx$ 85%).

The $Al + O_2$ reaction was first studied by Dagdigian, Cruse and Zare in a beam-gas arrangement using an unselected effusive Al source at an average translational energy of collision ϵ_{trans} = 0.13 eV [12]. With phase space theory as a basis they concluded that the available energy was not entirely statistically distributed, as vibrational levels above v" = 1 and high rotational levels in v" = 0 appeared less populated than *a priori* expected. A substantial improvement was made in the beam-gas study of Pasternack and Dagdigian [13] with the use of a velocity selected aluminium effusive source. Our experimental results exhibit only

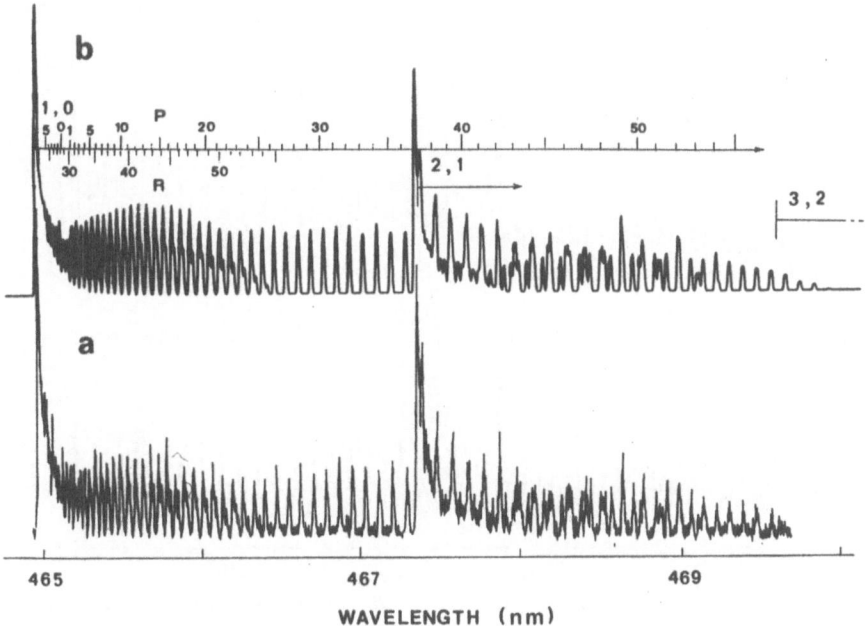

FIGURE 2. Excitation spectrum of $AlO(X^2\Sigma^+)$ from a $Al + O_2$ reaction at ϵ_{trans} = 0.083 eV: experimental (a), and calculated (b).

minor differences from theirs. In particular they found that a high fraction of the excess translational energy (80%) was channelled into product translation and rotation. They compared this behaviour with the general rules established by Polanyi and co-workers for exoergic direct reactions with an early energy barrier [22], and suggested a direct mechanism involving an attractive surface with mixed energy release. However we stress that the preferential channelling of the excess translational energy of reactants into product rotation and translation is not a determining factor for the Al + O_2 reaction because statistical behaviour gives also the same result on this point.

A complex mode mechanism was also suggested by the kinetic studies of Fontijn and co-workers, with a fast flow reactor between 300 and 1700 K [5-7]. Our data are in good agreement with such a mechanism although it does not bring a definitive argument for its support. Even if the potential energy surface has no well, the low amount of energy available to the products can prevent them to give rovibrational distributions markedly different from statistical.

$$Al + CO_2 \rightarrow AlO + CO.$$

This reaction is endoergic and threshold for the detection of the $AlO(X^2\Sigma^+, v'' = 0)$ bandhead occurs at $\epsilon_{trans} = 0.17 \pm 0.02$ eV. At this energy only the $^2P_{3/2}$ state is presumably reactive. Spectra near threshold and at higher energy (Fig. 5) show the growth of the rotational and vibrational excitations. The results on the AlO

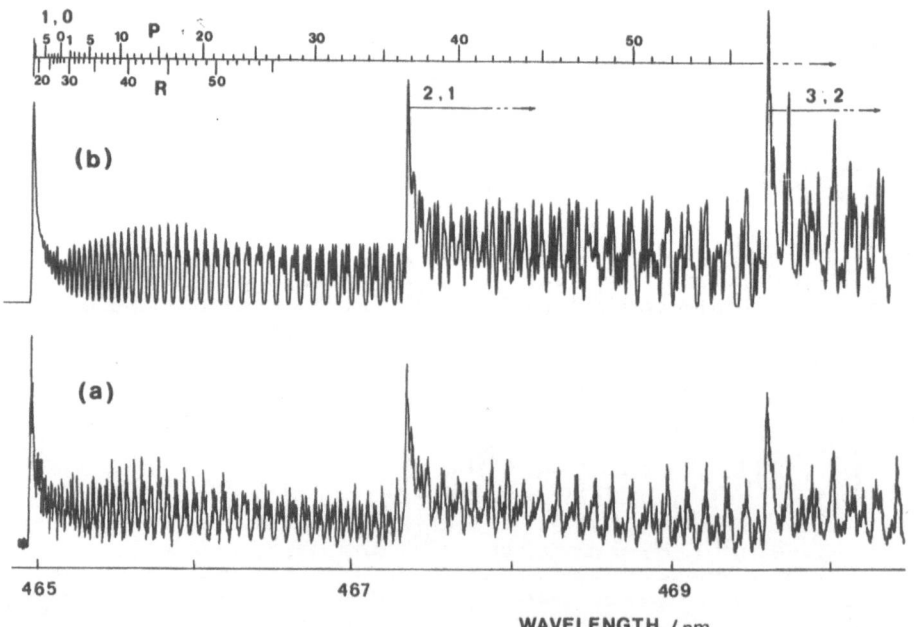

FIGURE 3. Excitation spectrum of $AlO(X^2\Sigma^+)$ from the Al + O_2 reaction at $\epsilon_{trans} = 0.49$ eV: experimental (a), and calculated (b) with a statistical rovibrational distribution.

internal energy partitioning are reported on Fig. 6. Vibrational and rotational distributions remain statistical up to ϵ_{trans} = 0.44 eV and show deficiency to the prior at high energy.

The variation of the relative reactive cross-section is plotted in Fig. 7, with the following modelled excitation functions:

$$\sigma_r = \sigma_0 \ (\epsilon_{th}/\epsilon_{trans})^{2/s} \ (1-\epsilon_{th}/\epsilon_{trans})^{1-2/s} \qquad (4)$$

where σ_0 is the maximum cross section, ϵ_{th} the translational energy at threshold and s takes the value 4, 6, or ∞ for an ionic, Van der Waals, or hard sphere potential, respectively [23]. A good agreement is obtained for s = 4 or 6 in the vicinity of threshold when computing the curves for ϵ_{th} = 0.19 eV which corresponds to the reaction endoergicity. However, the two highest energy points significantly deviate from the corresponding curves: a second threshold seems to occur in the 0.5 eV region. This kind of behaviour was reported for the endoergic Eu + O_2 reaction and was explained by two reaction mechanisms, the first at low energy resulting from insertion of Eu into O_2, and the second from collinear collisions [24]. In our case the second threshold could be due to reactions of CO_2 dimers. It has been shown indeed that the reactive cross-section of CO_2 dimers with Ba atoms was four to eightfold larger than that of the monomers [25]. Reactions of dimers were also found to give much colder rotational distributions of

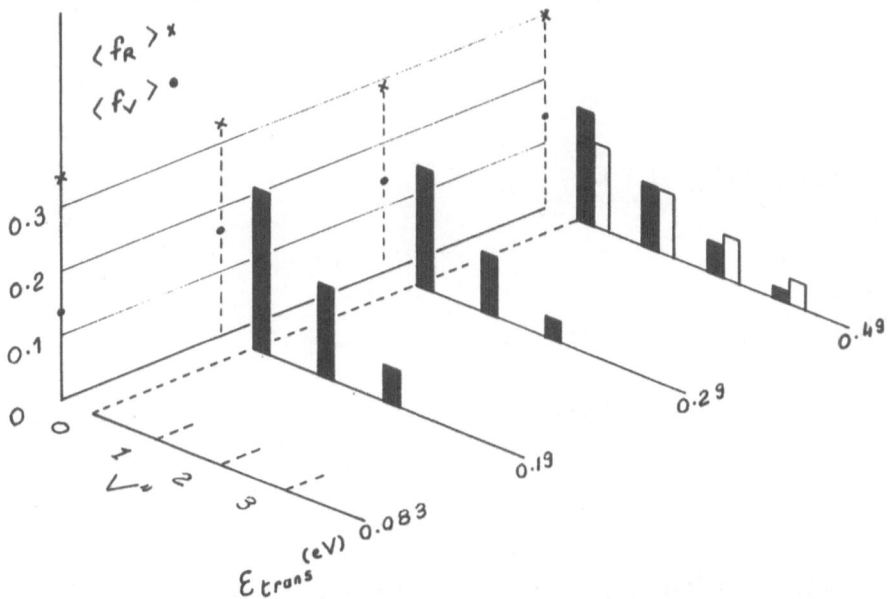

FIGURE 4. Energy partitioning in AlO from the Al + O_2 reaction and histogram of the detailed (normalized) rate constants k(v"). Open bars refer to the prior. O_2 beam: 12.5% O_2 in He at P = 4 bars, velocity v = 1360 ms^{-1}, and 15% velocity spread. The rate constants are not given at ϵ_{trans} = 0.083 eV (k(v"=0)/k(v"=1) = 0.76/0.24) since the experiment was made under different O_2 beam conditions: 12.5% O_2 in Ar at P = 4 bars, v = 660 ms^{-1}, and 17% velocity spread.

the BaO product. However when decreasing the backing pressure of the CO_2-He mixture by a factor of 4 which should result in a dramatic decrease of the dimer density in the molecular beam, we do not detect a difference in the rotational distributions of the spectrum at ϵ_{trans} = 0.53 eV. Finally as a last explanation for the second threshold, we assume that it could correspond to the threshold for the

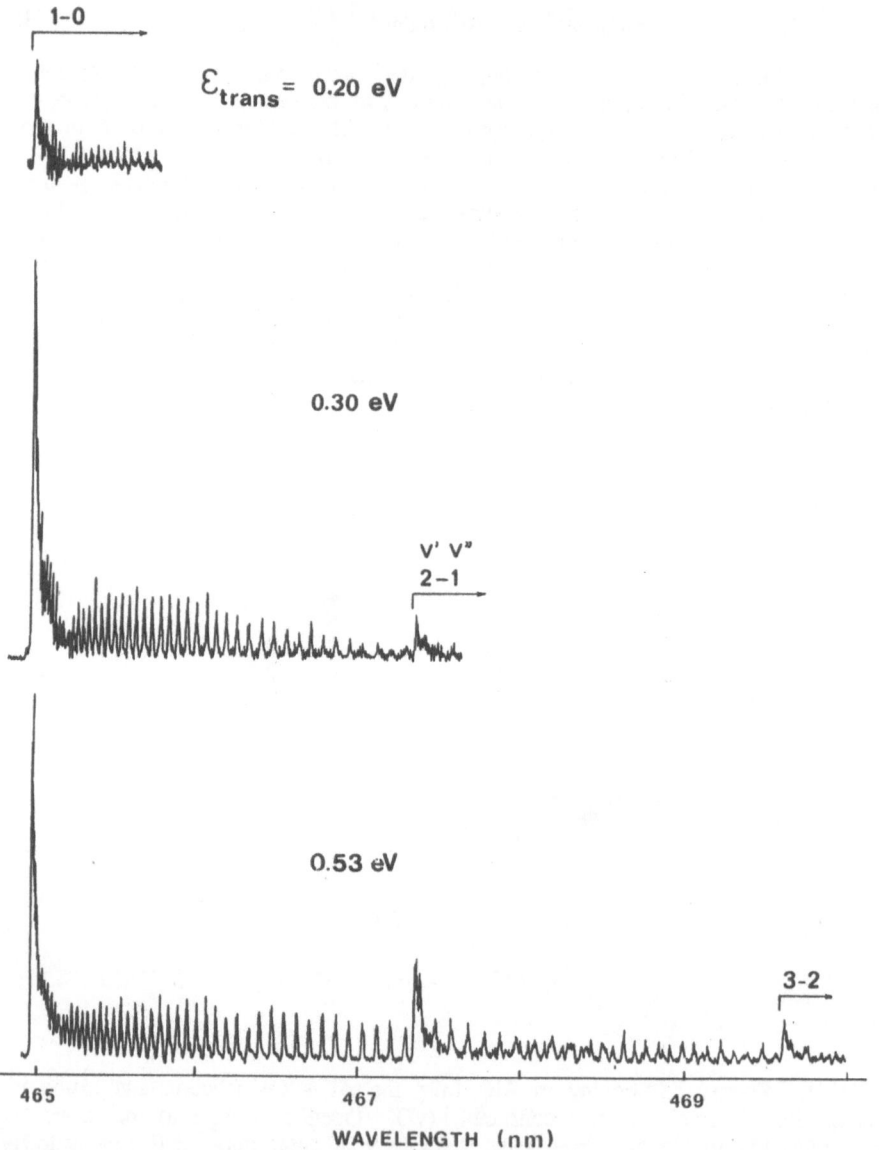

FIGURE 5. Excitation spectra of $AlO(X^2\Sigma^+)$ from the Al + CO_2 reaction at ϵ_{trans} = 0.20, 0.30 and 0.53 eV.

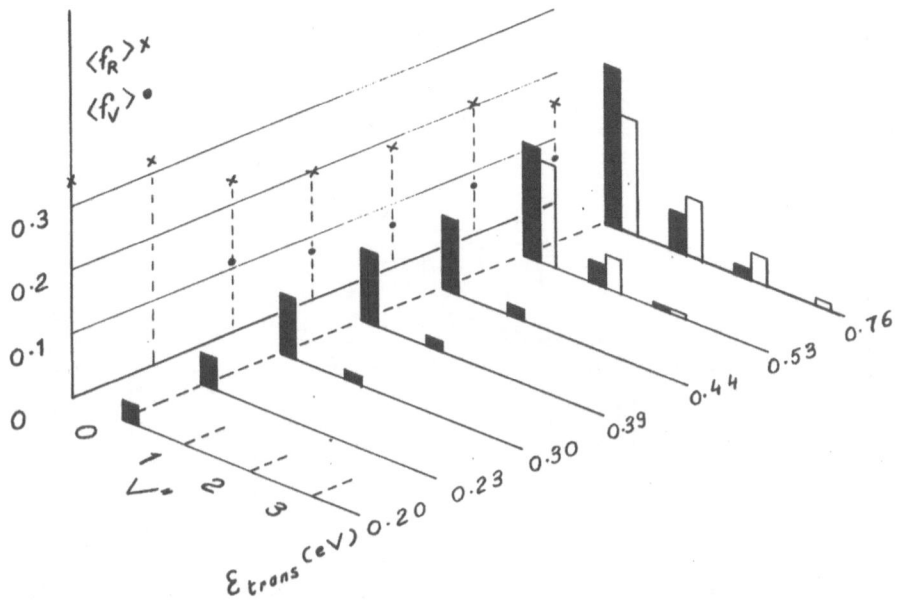

FIGURE 6. Energy partitioning in AlO from the Al + CO_2 reaction and histogram of the detailed (normalized) rate constants k(v"). Open bars refer to the prior. CO_2 beam: 20% CO_2 in He at P = 4 bars, v = 1220 ms^{-1}, and 18% velocity spread.

production of CO(v" = 1) which should occur after ϵ_{trans} = 0.46 eV. In that case the CO(v" = 1) production would be well above the statistical prediction (threefold at 0.76 eV), and the reaction would exhibit a very specific unusual channelling of the energy into the old bond

$$Al + SO_2 \rightarrow AlO + SO.$$

The Al + SO_2 reaction is also endoergic ($\Delta\epsilon_0 \approx 0.32$ eV) and the excitation of the AlO($X^2\Sigma^+$, v" = 0) bandhead starts at ϵ_{trans} = 0.36 ± 0.04 eV. As for Al + O_2 and Al + CO_2 the increase of the translational energy results in gradual rotational and vibrational excitation of the AlO product. Results on the AlO internal energy partitioning are displayed on Fig. 8. The amount of energy going into rotation is fairly constant, $<f_R> \approx 0.24$, up to ϵ_{trans} = 0.66 eV but drops to $<f_R>$ = 0.18 at ϵ_{trans} = 1.19 eV. The fraction going into vibration does not exceed a mere $<f_v>$ = 0.10. The vibrational and rotational distributions remain statistical except at the highest collision energy sampled.

$$Al + N_2O \rightarrow AlO + N_2.$$

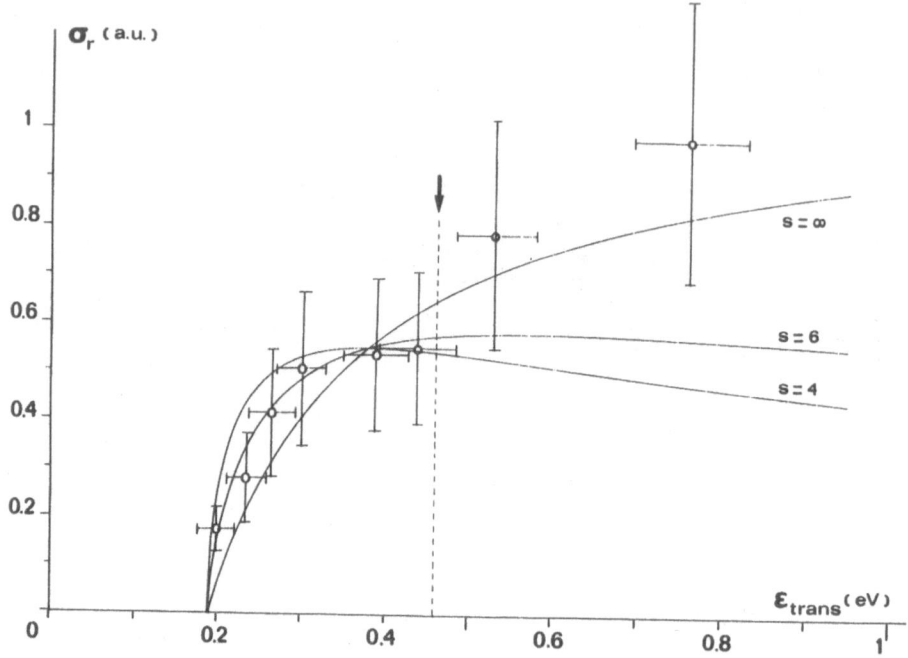

FIGURE 7. Relative reactive cross section, σ_r, of the Al + CO_2 reaction as a function of the relative translational energy, ϵ_{trans} (o: experimental values; solid line: calculated - see text -). The arrow indicates the CO(v" = 1) threshold.

We have not obtained quantitative results for this exoergic reaction ($\Delta\epsilon$ = -3.6 eV). LIF signals of $AlO(X^2\Sigma^+)$ remained low even at a collision energy of 0.52 eV. We also failed to detect by LIF any electronically excited $AlO(A^2\Pi_i)$ product, by excitation on B-A transitions in the fundamental range of rhodamine dyes, and collecting the fluorescence of B-X transitions. Combined factors are certainly responsible for the low LIF signals: the spread of the products on the large manifold of accessible rovibrational states, the high recoil velocities which lower the densities, and perhaps a lower reactive cross-section than for the other reactions.

5. CONCLUSION

The reactions of Al atom with O_2, SO_2 and CO_2 give statistical AlO internal energy distributions at low translational energy. Deviance from the prior expectation at high collision energy occurs at the expense of the AlO internal excitation. Selectivity of the channelling of the incident reagent translational energy into the CO bond vibration might exist for the Al + CO_2 reaction.

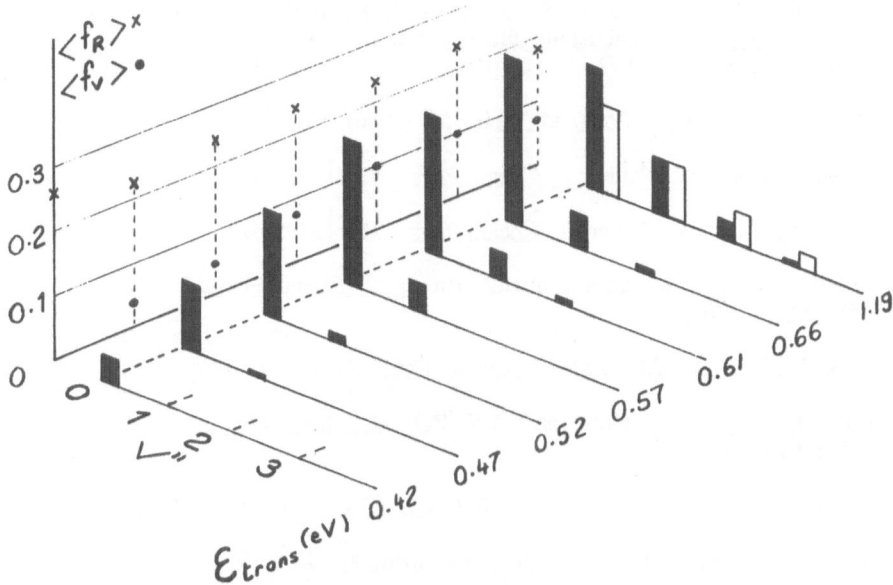

FIGURE 8. Energy partitioning in AlO from the Al + SO$_2$ reaction and histogram of the detailed (normalized) rate constants k(v"). Open bars refer to the prior. SO$_2$ beam: 10% SO$_2$ in He at P = 3 bars, v = 1280 ms^{-1}, and 20% velocity spread.

REFERENCES

[1] G. Dorthe, M. Costes, C. Naulin, J. Joussot-Dubien, C. Vaucamps and G. Nouchi, J. Chem. Phys., 83 (1985) 3171.

[2] M. Costes, G. Dorthe, B. Duguay, P. Halvick, J. Joussot-Dubien, C. Naulin, G. Nouchi, J.C. Rayez, M.T. Rayez and C. Vaucamps, in Recent Advances in Molecular Reaction Dynamics, ed. R. Vetter and J. Vigue (Editions du C.N.R.S., Paris, 1986), p. 97.

[3] M. Costes, C. Naulin, G. Dorthe, J. Marchais and C. Vaucamps, C.R. Acad. Sc. Paris II, 303 (1986) 1279.

[4] M. Costes, C. Naulin, G. Dorthe, C. Vaucamps and G. Nouchi, Faraday Disc. Chem. Soc., 84 in the press.

[5] A. Fontijn, W. Felder and J.J. Houghton, Fifteenth Symposium (International) on Combustion, the Combustion Institute, Pittsburgh, 1975, p. 775.

116

[6]   A. Fontijn, W. Felder and J.J. Houghton, Sixteenth Symposium
      (International) on Combustion, the Combustion Institute,
      Pittsburgh, 1977, p. 871.

[7]   A. Fontijn, Combust. Sci. and Tech., 50 (1986) 151.

[8]   A. Fontijn and W. Felder, J. Chem. Phys., 67 (1977) 1561.

[9]   A. Fontijn and W. Felder, J. Chem. Phys., 71 (1979) 4854.

[10] S. Rosenwaks, R.E. Steele and H.P. Broida, J. Chem. Phys.,
      63 (1975) 1963.

[11] D.M. Lindsay and J.L. Gole, J. Chem. Phys., 66 (1977) 3886.

[12] P.J. Dagdigian, H.W. Cruse and R.N. Zare, J. Chem. Phys.,
      62 (1975) 1824.

[13] L. Pasternack and P.J. Dagdigian, J. Chem. Phys., 67 (1977) 3854.

[14] W.R. Gentry and C.F. Giese, Rev. Sci. Instrum., 49 (1978) 595.

[15] R. Alktorn and R.N. Zare, Ann. Rev. Phys. Chem., 35 (1984) 265.

[16] G.R. Hebert, R.W. Nicholls and C. Linton, J. Quant. Spectrosc.
      Radiat. Transfer, 23 (1980) 229.

[17] D.E. Powers, S.G. Hansen, M.E. Geusic, A.C. Pulu, J.B. Hopkins,
      T.G. Dietz, M.A. Duncan, P.R.R. Langridge-Smith and R.E. Smalley,
      J. Phys. Chem., 86 (1982) 2556.

[18] J.B. Hopkins, P.R.R. Langridge-Smith, M.D. Morse and
      R.E. Smalley, J. Chem. Phys., 78 (1983) 1627.

[19] J.A. Coxon and S. Naxakis, J. Mol. Spect., 111 (1985) 102.

[20] R.B. Bernstein in Chemical Dynamics via Molecular Beam and Laser
      Techniques, Oxford University Press, New York, 1982, and
      references therein.

[21] H.W. Cruse, P.J. Dagdigian and R.N. Zare, Faraday Disc.
      Chem. Soc., 55 (1973) 277.

[22] A.M. Ding, L.J. Kirsch, D.S. Perry, J.C. Polanyi and
      J.L. Schreiber, Faraday Disc. Chem. Soc., 55 (1973) 252.

[23] R.D. Levine and R.B. Bernstein, J. Chem. Phys., 56 (1972) 2281.

[24] R. Dirscherl and K.W. Michel, Chem. Phys. Lett., 43 (1976) 547.

[25] J. Nieman and R. Naaman, Chem. Phys., 90 (1984) 407.

THE EFFECT OF ELECTRONIC EXCITATION IN THE REACTIONS OF OXYGEN ATOMS WITH SIMPLE HYDRIDE MOLECULES

J.J. Sloan, E.J. Kruus and B.I. Niefer
Department of Chemistry
University of Waterloo
Waterloo
Canada N2L3G1
and
Department of Chemistry
Carleton University
Ottawa
Canada K1S5B6

ABSTRACT. On excitation of the lowest electronic state of the oxygen atom, the rates of its reactions with hydride molecules typically increase by about six orders of magnitude and the dynamics of the reactions change from simple hydrogen atom abstractions to more complex processes in which the oxygen inserts into the molecular bond under attack. This results in the transient formation of a strongly-bound singlet intermediate, which may undergo very complex dynamical behaviour before dissociating to products. This behaviour is illustrated by detailed discussions of the oxygen atom reactions with H_2 and the hydrogen halides.

1. INTRODUCTION

The central position occupied by oxygen atom reactions in many high-temperature processes makes it important to understand the behaviour of the excited states, as well as that of the ground state, in these reactions. The excitation energy of the lowest-lying electronically excited state of the atom, $O(^1D)$, is 1.967 eV; and therefore it is not abundant under conditions of common high-temperature processes such as combustion. The cross sections of its reactions, however, typically exceed those of the ground state by many orders of magnitude (*vide infra*), thus making its role in the chemistry of many high-temperature processes much more important than its abundance would suggest.

Many high-energy processes which are less common than combustion, but equally important, involve energies at which $O(^1D)$ atoms may be much more abundant. For example, in certain electronic fabrication procedures such as the deposition or etching of oxide films by plasma processing, the ion energies are several tens or hundreds of eV [1]; and the electron energy distributions extend to similar energies [2]. The (^1D) state is certainly excited under these conditions. A second example involving oxygen atoms with energies substantially in excess of that required to excite the (^1D) state, occurs in collisions between atmospheric oxygen atoms and the surfaces of spacecraft in low earth orbit [3]. These collisions induce

117

J. C. Whitehead (ed.), Selectivity in Chemical Reactions, 117–133.
© *1988 by Kluwer Academic Publishers.*

unusually severe materials degradation by initiating chemical reactions on the exposed surfaces [4]. The importance of the $O(^1D)$ state in many processes such as these is speculative at the present time, largely due to the extreme difficulty in measuring its concentration (its emission lifetime is 150 s). In view of the energies involved, however, it is almost certainly present; and it is therefore important to consider its possible influence. The importance of $O(^1D)$ reactions to the chemistry of the earth's ozone layer, however, has been well established [5]. Its reactions with trace-level species such as hydrocarbons, water and pollutants such as the chlorofluorocarbons, create reactive species such as OH and Cl which are responsible for dramatic reductions in ozone concentration [6].

The kinetics of oxygen atom reactions are strongly dependent on the atom's electronic state; and selective excitation of the $O(^1D)$ state typically causes increases of several orders of magnitude in the rate constants. Rate measurements on $O(^3P)$ [7-9] and $O(^1D)$ [10,11] reactions, reported in several comprehensive reviews, illustrate this. Table 1 shows a comparison of the rate constants for several selected $O(^3P)$ and $O(^1D)$ reactions relevant to the discussion to follow. The differences shown are typical of those found in most chemical reactions involving these species. At the temperature of the rate data (298 K), the $O(^1D)$ reaction is typically six to eight orders of magnitude faster than the $O(^3P)$ reaction. Under stratospheric conditions relevant to the chemistry of the ozone layer (at altitudes from about 15 km to about 40 km, for example) the concentration of $O(^1D)$ is approximately six orders of magnitude smaller than that of $O(^3P)$. In the ozone layer, therefore, the reactions of the electronically-excited species occur with a slightly higher probability than those of the ground state.

The singlet reaction is always more energetic than the triplet (due to the $O(^1D)$ excitation energy); but the exoergicities alone do not account for the kinetic differences shown in Table 1. These stem mainly from differences in the shapes of the potential energy surfaces. Of primary importance to the kinetics are differences in the shapes of the entrance channels of the surfaces: there is usually a substantial energy barrier to the triplet reactions, whereas the singlet reactions proceed with virtually zero activation energy. The surfaces differ at the later stages of the reactions as well. Generally, the triplet surface decreases smoothly from the location of the energy barrier to the product asymptote, with either no minimum, or only a very weakly-bound triplet intermediate; whereas the singlet surface frequently correlates with a very strongly-bound intermediate (usually a stable molecule), and therefore, although the singlet surface begins at a higher energy than the triplet, it crosses the latter and becomes lower in energy for a substantial part of the configuration space in the region of close approach.

The differences in the shapes of the potential energy surfaces, of course, give rise to differences in the dynamics of the reactions; and it is these to which this article is addressed. The vast majority of previous dynamical studies have involved hydride reactions - since their dynamics are particularly simple - and examples of these will be used to illustrate the discussion. Although they begin at very different energies, both the singlet and triplet reactions correlate with the same (ground state) products which are almost always observed in the experiments. In almost all cases, the dynamics of the triplet reactions can be interpreted in terms of simple abstractions [10], in which a single unpaired p electron of the oxygen atom overlaps with an antibonding orbital of the reagent molecule, forming a linear (or slightly bent) triplet intermediate [20], which may dissociate directly on the triplet surface, or cross to the singlet and undergo a more complex interaction. Conversely, existing data on the singlet reactions indicate that these occur preferentially *via* insertion of

TABLE 1. Comparison of Rate Constants for Selected Reactions of $O(^1D)$ and $O(^3P)$ Atoms.

| Reagent | Rate Constant $(cm^3\ molec^{-1}\ s^{-1})$ [a] [Reference] | |
	$O(^1D)$	$O(^3P)$
H_2	1.1(-10) [12]	3.5(-18) [12]
CH_4	1.4(-10) [12]	2.7(-18) [13]
$CHCl_3$	3.0(-10) [15]	8.3(-16) [19]
CCl_4	3.3(-10) [14]	1.7(-16) [17]
CHF_3	1.9(-12) [15]	<3.7(-15) [17]
HCl	1.5(-10) [16]	1.4(-16) [16]
HBr	-	3.7(-14) [16]
NH_3	2.5(-10) [16]	1.1(-16) [18]
H_2S	2.5(-10) [b]	2.2(-14) [14]

[a] Value for T=298 K given.
[b] Estimate, based on the value for $O(^1D)$ + H_2O.

the $O(^1D)$ atom into the molecular bond, followed by a relatively complex interaction which usually involves the transient formation of the strongly-bound singlet intermediate. Evidence to be presented in a subsequent section of this article suggests that, as in the previous example, the reaction may cross to the other surface (in this case the triplet) in the exit channel, providing yet another, dynamically different, pathway to the products.

The reaction with H_2, to be discussed in greater detail below, exemplifies this behaviour clearly. The triplet reaction involves a nearly-linear O-H-H intermediate which, after the energy barrier has been surmounted, dissociates to the products - $OH(^2\Pi)$ + $H(^2S)$ - in a relatively direct process. The singlet reaction, on the other hand, involves the formation of ground-state H_2O in a very complicated interaction which is (dynamically) quite different from the triplet reaction. Clearly, the differences in the physics of the triplet and singlet interactions are profound. That such strong differences should be caused by a relatively small change in the orbital occupation of one of the reagents is one of the most surprising aspects of excited-state chemistry.

The following article will discuss the changes in the chemistry of oxygen atom reactions resulting from the excitation of the (^1D) state. Recent dynamical measurements on these excited-state reactions, carried out in our laboratory, will be used to illustrate the discussion. Before these results are introduced, however, the novel experimental technique used to obtain them will be described. Following this, the dynamics of the reactions of both ground- and excited-state oxygen atoms with several small hydride molecules will be described, beginning with the H_2 reaction.

2. EXPERIMENT

The dynamics of the $O(^1D)$ reactions were inferred from measurements of the internal energy distributions of their products, obtained from the low pressure emission spectra of the products' infrared chemiluminescence. The experiments are carried out in a cylindrical stainless steel reaction chamber. Premixed reagents (ozone and a molecular reagent) are admitted at the top of the chamber *via* concentric tubes; and products are pumped away from the bottom through a gate valve and cryobaffle by a 38 cm diffusion pump. $O(^1D)$ is produced by photodissociation of ozone in the Hartley bands, between about 200 nm and 300 nm. In order to obtain adequate signal, a high intensity laser (Lumonics TE860 excimer, operating at 330 Hz on KrF) is used for the photolysis. The laser light passes just beneath the end of the concentric reagent-inlet tubes, making two passes through a well-defined photolysis zone. A multipass Welsh cell, having its axis perpendicular to the laser direction, is located just downstream from the photolysis region. This collects the chemiluminescence from the reaction products and focuses it into a very sensitive Fourier transform spectrometer.

These measurements, of course, require that the products be observed before their collisional deactivation. This is achieved by carrying out the experiments at low pressure, and removing the products from the observation zone rapidly. For these reactions, the flow requirements are determined mainly by the large rate constants for both the reactions of the $O(^1D)$ atom, and its physical and chemical quenching by ozone - the latter is about 5.0×10^{-10} cm^3 $molec^{-1}$ s^{-1} [21,22] - which dictate that the $O(^1D)$ is removed and the reaction products are created at approximately the first gas kinetic collision after the photolysis pulse. In the present experiments, the residence time of the reagents in the photolysis zone is about 100 microseconds; and in the observation zone it is about 1 millisecond. It will be shown later that these conditions satisfy the requirements for observation of "initial" unperturbed energy distributions.

Since a (pulsed) excimer laser is used to produce the $O(^1D)$ reagent, the products of the reaction are also pulsed. If the repetition rate is sufficiently high, this may be regarded as a quasi-steady-state reagent source, and normal CW Fourier transform spectroscopy may be used. Some of the early experiments were carried out in this way. To maximize the duty cycle, however, it is desirable to make the observations synchronously with the creation of the products, and for this reason, an implementation of very fast time-resolved Fourier transform spectroscopy (TRFTS) has been developed in this laboratory; this was used for the most recent measurements.

Due to the relatively weak infrared emission spectrum of the major reaction product, OH (the Einstein transition probabilities for Δv=-1 emission are approximately 100 s^{-1}), adequate signal cannot be obtained below total reagent pressures of about 10^{-2} Torr. The mean time between collisions at this pressure is about 10 μs. This determines the time-resolution required for the TRFTS experiment, since the rate constants for vibrational deactivation of OH by ozone and many of the molecular reagents are near gas kinetic. The first TRFTS system, which we introduced some time ago [23,24] had approximately this time-resolution.

More recently, we have improved the time-resolution of the system substantially. The present instrument is capable of recording high time- and frequency-resolution spectra of transients having decay times from the nanosecond to the millisecond regime. The minimum time delay between the initiation of the transient and the first spectral observation can be arbitrarily short. (Typically, the first spectrum is recorded just *before* the transient in order to provide a background observation. A maximum of 128 successive time-delayed spectra of a single transient can be recorded; the minimum time delay between each of these is 10 ns. All operational parameters (resolution, sensitivity, etc.) of the commercial Fourier transform spectrometer with which the system is used, are unchanged by time-resolved operation. Variability in the baseline due to amplitude instabilities in the excitation source (usually a pulsed laser) are taken into account, and appropriate corrections are made.

The dynamical results on the $O(^1D)$ reactions with H_2S, NH_3 and CH_4, to be reported in the following Sections, were obtained without time-resolution of the observations. In this case, the photolysis laser was allowed to free-run: it had no phase relationship with the data acquisition of the Fourier transform spectrometer. The data acquisition time was relatively long, therefore, due to the inherently low duty cycle of such an experiment. The data acquisition time for the experiments using the time-resolved instrumentation (fast or slow versions) was more than an order of magnitude shorter.

3. RESULTS

3.1 The $O + H_2$ Reaction

The kinetics of $O(^3P) + H_2 \rightarrow OH + H$ reaction, important to the H_2/O_2 combustion system, have been thoroughly investigated; accurate data exist for temperatures from 300 K to 2500 K [25-27]. Furthermore, theoretical descriptions have been reported for the tunnelling responsible for the curvature of the Arrhenius plot [28,29], the reaction probabilities, state to state rate constants, isotope effects, and many other aspects of the reaction dynamics [30-33]. The reaction is very slightly endoergic ($\Delta H = + 7.5$ kJ/mol) and the barrier to reaction on the triplet surface is about 52 kJ/mol. The transition state has a collinear O-H-H structure [34], and the lowest-energy saddle point is located near the region of maximum curvature in the (collinear) potential energy surface.

The dynamics of the triplet reaction are typical of a simple abstraction in which the major interaction is between the O atom and the nearest H atom; the end-atom interactions are small. Because of its essential collinearity, the surface can be described accurately by a LEPS potential [35]. Quasiclassical trajectory studies using modified LEPS potentials [36] have found, in agreement with many of the other calculations [30-33], that the products are predominantly backward-scattered for low reagent energies. Additionally, the calculations conclude that vibrational excitation of the H_2 reagent increases the cross section by about three orders of magnitude per vibrational level, the precise value of the rate enhancement being strongly dependent on the location of the barrier along the reaction coordinate. Vibrational excitation of the reagent becomes vibrational energy in the product; reagent translational energy, on the other hand, is converted efficiently into product rotational and translational energy. All of these characteristics indicate relatively simple, direct reaction dynamics.

Both the kinetics and the dynamics of the singlet reaction are quite different from those of the triplet. Table 1 shows that the singlet rate constant is about 3×10^7 times larger than that of the triplet at room temperature. The singlet potential energy surfaces are very well-known as a result of extensive computations [37-44]. There is no barrier to reagent approach in the C_{2v} configuration; and the lowest ($^1A'$) surface correlates directly with the $^1A_1$ ground state of H_2O, at an energy of -10 eV with respect to the reagents. There is a slight barrier to collinear O-H-H reagent approach, with the result that the reaction strongly prefers insertion dynamics; and the H_2O intermediate is formed in the majority of the collisions.

Dynamical calculations on this reaction have been carried out by a large number of groups [45-52]; and most of these agree on the broad aspects of the dynamics. Reaction on the lowest potential energy surface almost invariably involves insertion into the H_2 bond and the formation of an H-O-H intermediate complex with extremely high excitation in the bending vibrational mode. This dissociates within a few vibrations to yield OH with strong rotational excitation. These calculations also reproduce the available experimental product energy distribution measurements [46,53-62] reasonably well, especially with respect to the OH rotational [53-56] and angular [57] distributions.

Reactions which form strongly-bound intermediate complexes of this kind, frequently exhibit statistical energy partitioning due to the complete randomization of the reaction exoergicity during the lifetime of the complex. This appears not to occur in this case, however, and this reaction therefore seems to be exceptional. This aspect of the reaction has been discussed in detail elsewhere [52,60,63]. The experimental evidence for non-statistical behaviour is contained in the OH vibrational excitation [59-62], which is greater than that of the statistical distribution, and the OH/OD branching ratio in the $O(^1D)$ + HD reaction, which is larger than the statistical ratio. These non-statistical effects are not large, however, and the dynamical behaviour which they reflect is subtle.

There are a total of five potential energy surfaces accessible to the singlet reagents at moderate energies. Two of these - $^1A'$ and $^1A''$ symmetry - correlate with the products observed in the energy distribution measurements: $OH(^2\Pi)$ + $H(^2S)$. A third ($^1A'$) surface is close to the first two in the entrance channel of the potential, but correlates with the $OH(^2\Sigma^+)$ product. Nonadiabatic behaviour may occur at the resulting hyperconical intersections; and the dynamics of the trajectories which switch potential energy surfaces are quite different from those which remain on the lowest surface for the entire reaction. In a recent quasiclassical surface-hopping trajectory calculation, using a multisurface DIM representation of the potential [52], we found that trajectories which begin on the first excited $^1A'$ surface, and switch to the lower surface in the exit channel, for example, usually avoid the potential minimum completely, and have fairly direct dynamics resembling those of a simple H atom abstraction.

Even those reactions which remain on the lower surface for the entire trajectory exhibit a variety of interesting behaviour. In the same quasiclassical trajectory calculations on this system [52], we identified a substantial difference in the behaviour of those trajectories on the lowest surface which went deep into the H-O-H potential minimum (below an energy of -8.0 eV with respect to the reagents) and those which avoided it or went below this energy only once. The former yielded products having a substantially lower vibrational excitation than the latter.

This is illustrated in Figs. 1 and 2. Figure 1 shows bond-length/potential energy plots for two typical trajectories. The trajectory in Fig. 1(a) went below -8.0

eV once, and created OH(v'=3,J'=8), while the one in Fig. 1(b) went below -8.0 eV three times, and as a result, created OH(v'=0,J'=23). Figures 2(a) and 2(b) show, respectively, the OH vibrational distributions for the entire batch of 5000 trajectories. Figure 2(a) shows that the OH vibrational distributions from those trajectories which did not go below -8.0 eV more than once, are inverted. Figure 2(b) shows that those trajectories which entered the H-O-H well more than once created OH with much lower vibrational excitation. This effect is the result of intramolecular vibrational energy redistribution during the lifetime of the H-O-H intermediate, which is longer for those reactions which sample the potential minimum several times. Consequently, these data can be used to give an (classical) estimate for the probability of IVR in this system. A more complete analysis of this is presented in Ref. 52.

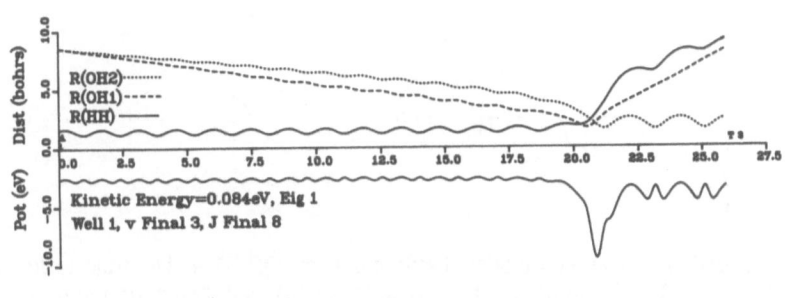

(a)

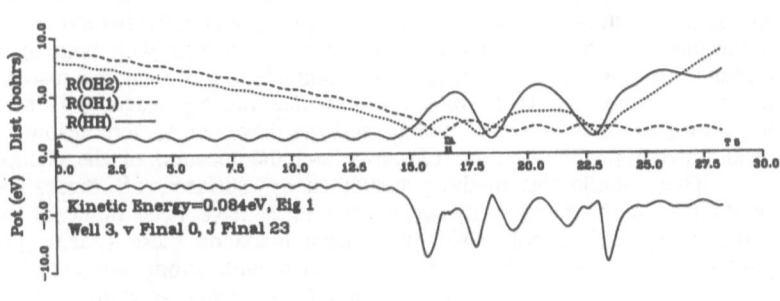

(b)

FIGURE 1. Plots of the potential energy and bond length for typical O(^{1}D) + H$_2$ trajectories which sample the deep H-O-H minimum less than twice (a) and twice or more (b). The former create OH with inverted vibrational distribution, the latter yield vibrationally cold OH.

124

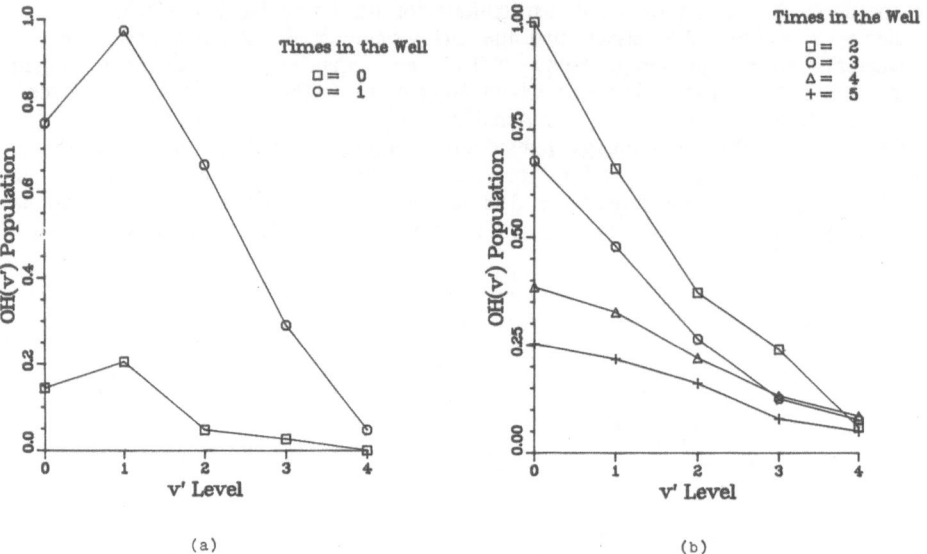

FIGURE 2. The vibrational distributions of O(^{1}D) + H$_2$ trajectories which sample the deep H-O-H minimum less than twice (a) and twice or more (b).

3.2 The O + HCl and HBr Reactions

The behaviour of these reactions is qualitatively similar to that of the H$_2$ reactions. This is not surprising, considering the similarities in the electronic states involved. Although the triplet reactions with HCl [64-68] and HBr [64,69] are considerably faster than the triplet H$_2$ reaction, the singlet reactions with both of the former reagents are still about six orders of magnitude faster than the respective triplet reactions. Since the HCl triplet reaction, like the H$_2$ reaction, is nearly thermoneutral, their potential energy surfaces have even more features in common, including the preference for collinearity and the location of the energy barrier.

These similarities in the potential energy surfaces, of course, lead to similarities in the reaction dynamics; and these have been observed in the energy partitioning measurements which have been made on these systems [70,71]. The OH product of the HBr triplet reaction is created with strong vibrational inversion and substantial rotational excitation, as would be expected of simple, abstraction dynamics, like those described previously for the H$_2$ reaction. The product internal excitation created by the HBr reaction is greater, however, because the mass-combination (a light atom transferred between two heavier masses) causes greater release of energy into the nascent product than the Heavy-Light-Light mass combination [72]. Apart from such obvious mass-combination effects, the dynamics of the O(^{3}P)/HX reactions are the same as those of the H$_2$ reaction.

As a consequence of barrier location, and of the possibility to vibrationally excite the HX reagent using a chemical laser, the O(^{3}P)/HX reactions have been the subject of a great deal of experimental [73-77] and theoretical [78-81] work in which the effect of internal excitation of the HCl reagent was explored. Vibrational excitation enhances the reaction cross section substantially, as would be expected for such a reaction [82]. Recently, the effect of HX rotation has also been examined theoretically [79]. It was also demonstrated in much of this work that an increase in the HX vibrational excitation leads to an increase in the product vibrational excitation, a result which is expected for a potential energy surface of this kind. This "conservation of vibration" as well as the energy partitioning results, imply that the triplet reaction does not cross to the nearby singlet potential energy surface (in the entry valley, for example) because if it did, the energy redistribution attendant on the formation of the HOCl intermediate would destroy this dynamical specificity. This adiabatic behaviour has also been observed for the reverse (Cl + OH → HCl + O) triplet reaction [83]. The singlet reaction behaves differently in this respect, as will be described shortly.

Excitation of the oxygen atom to the (^1D) state increases the reaction rate by six orders of magnitude [84,85]. In addition, the singlet trajectories cross to the triplet surface with a surprisingly high probability. The triplet products $(O(^3P) + HX)$ are created in 10% of the collisions for the HCl case and 20% for the HBr case [86]. For the purely singlet reactions, producing $OH(^2\Pi) + H(^2S)$, energy partitioning measurements [87-89] indicate that the internal excitation of the product molecule is very large. The OH rotational distribution in the lowest vibrational levels (the only ones which have been measured to date) is inverted. This may not carry any dynamical information, however, as it is possibly the result of simple angular momentum constraints. (See Ref. 63 and Refs. 90-93 for a discussion of this in the case of the H_2 reaction).

The OH vibrational distributions from the HCl [89] and HBr [94] reactions are also strongly inverted. The extent of the inversion in these results is surprising in view of the generally-accepted interpretation that these reactions, like the H_2 case, also involve insertion of the $O(^1D)$ atom into the molecular bond. The extent of this inversion is clear, however, from the time-dependent spectra of the products from the $O(^1D)$ + HCl reaction shown in Fig. 3. These were recorded using the TRFTS instrument described in the Experimental Section, under conditions for which the gas kinetic collision time is about 10 μs. The earliest OH spectrum at 20 μs shows that the initial vibrational distribution peaks strongly in OH(v'=3), and has virtually no population in v'=1. (The frequencies of the P-branch lines of each vibrational level are indicated). This excitation is collisionally relaxed in subsequent spectra.

Emission from high vibrational levels of HCl is also evident at the earliest time. These levels are created in the E-V energy transfer process, which results from the singlet-triplet surface crossing. The distribution in this case is also strongly inverted; and has a shape which is virtually identical to that of the OH vibrational distribution [89]. The discussion of this result in Ref. 89 suggests that this is evidence for the participation of an HOCl intermediate, because the available *ab initio* information on this system [95,96] shows that the only region of configuration space where the singlet and triplet surfaces are relatively close corresponds to the HOCl species.

All of the energy distribution data on the $O(^1D)$ + HX reactions, therefore, are consistent with an insertion mechanism, like that of the H_2 reaction. If this interpretation is correct, the discussion of the IVR probabilities given in the previous section may be applied. In this event, the very strong OH vibrational inversion - nearly as extensive as that created by the equally-exoergic triplet reaction with HBr -

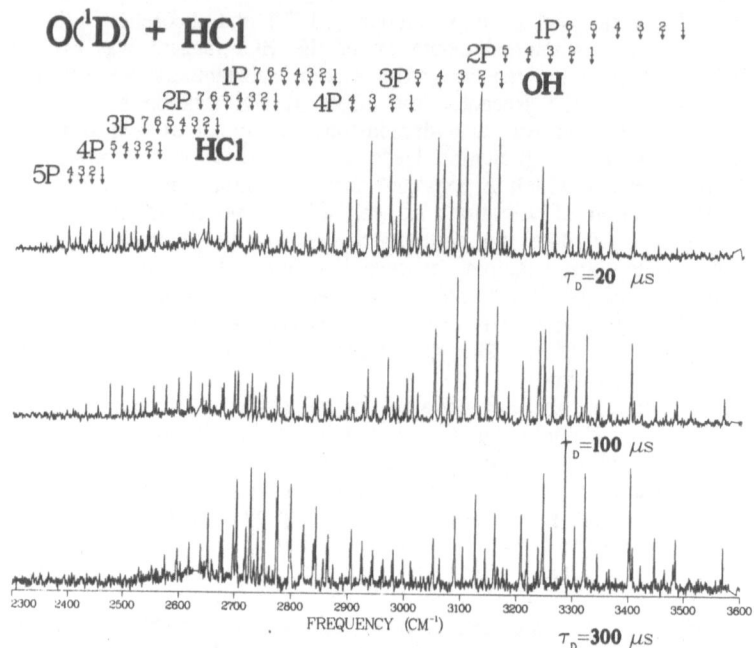

FIGURE 3. The product emission spectra recorded from the reaction of O(^{1}D) with HCl under low-pressure conditions using a fast time-resolved Fourier transform spectrometer. τ_D gives the time between the creation of the O(^{1}D) and the observation of the spectrum.

indicates that virtually no IVR occurs in this case, and the lifetime of the HOCl intermediate is extremely short - not more than one vibrational period of the HOCl. This must be regarded as a speculation at the present time, because the only existing (quasiclassical) trajectory calculation on this system [97] concludes that the HOCl lifetime is relatively long - typically 5×10^{-13} s, which would correspond to about 10 HOCl vibrational inversion periods. This calculation obtained an inverted OH vibrational distribution as well, (although not as strongly inverted as the experimental distribution) despite the long HOCl lifetime. Comparison of these results suggests that the (heavy) Cl atom must introduce an additional factor which inhibits IVR in the HOCl intermediate. This will be the subject of future experimental and theoretical work from this laboratory.

4. CONCLUSION

Excitation of the (^{1}D) state changes both the dynamics and the kinetics of oxygen atom reactions dramatically. For reactions with simple hydrides, the room-temperature rate increases by about six orders of magnitude, and the reaction dynamics change from simple H atom abstraction to insertion into the molecular bond, followed by dissociation of the highly-energetic intermediate in a relatively complex process. The dynamics of the ground state triplet and excited state singlet reactions with H_2 and the hydrogen halides are relatively well-understood, although some aspects of the latter remain to be elucidated.

The reactions with more complex hydride molecules are also the subject of current investigation. The dynamics are correspondingly more complex, although the basic behaviour appears to be consistent with that of the simpler systems discussed here. The dynamics of the singlet reaction with CH_4 have been studied in this laboratory [98] and elsewhere [99]. The major channel in this case is also the insertion-elimination; and the OH created in this process is vibrationally inverted [98,100]. The dynamics appear to be very similar to those of the H_2 reaction. The singlet reaction with H_2S creates vibrationally inverted OH [101], although in this case as well, the added molecular complexity has an influence, since the OH vibrational distribution actually appears to be bimodal [102]. The NH_3 singlet reaction, unlike the others discussed in this article, gives very little OH vibrational excitation [103,104]. The dynamics in this case are not understood.

It is clear that the oxygen atom reactions with simple hydrides can be used as models to understand the effect of electronic excitation on the kinetics and dynamics of the more complex oxygen atom - hydride reactions. The details of the latter, however, appear to be quite complicated. These are the subject of work in many laboratories at present, however, and novel and sophisticated experimental techniques are being used to elucidate the more subtle aspects of their dynamics.

ACKNOWLEDGMENT

The authors are grateful to Dr. P.J. Kuntz for many very helpful discussions. Research from this laboratory, cited in this article, was supported by the National Research Council of Canada, the Natural Sciences and Engineering Research Council of Canada, and NATO.

REFERENCES

[1] Materials Research Society Proceedings, Vol. 29, Plasma Processing and Synthesis of Materials, ed. J. Szekely and D. Apelian, North-Holland, Amsterdam, 1984.

[2] M.J. Kushner, J. Appl. Phys., 53 (1983) 2939; 54 (1983) 4958.

[3] Thirteenth Space Simulation Conference, NASA Conference Publication 2340, NASA Scientific and Technical Information Branch, Springfield, Virginia, 1984.

[4] Second Workshop on Spacecraft Glow, NASA Conference Publication 2391, ed. J.H. Waite Jr. and T.W. Moorehead, NASA Scientific and Technical Information Branch, Springfield, Virginia, 1985.

[5] R.G. Prinn, F.N. Alyea, and D.M. Cunnold, Ann. Rev. Earth Planet. Sci., 6 (1978) 43; J.R. Wiesenfeld, Acc. Chem. Res., 15 (1982) 110.

128

[6] Chlorofluoromethanes and the Stratosphere, NASA Reference
     Publication 1010, ed. R.D. Hudson, NASA Scientific and
     Technical Information Branch, Springfield, Virginia, 1977.

[7] K. Schofield, J. Phys. Chem. Ref. Data, 8 (1979) 25.

[8] R.E. Huie and J.T. Herron, Prog. Reaction Kinetics, 8 (1985) 1.

[9] R.J. Cvetanovic and D.L. Singleton, Rev. Chem. Intermediates,
     5 (1984) 183.

[10] M.C. Lin in Potential Energy Surfaces ed. K.P. Lawley, John
     Wiley and Sons Ltd., New York, 1980.

[11] W.H. Breckenridge in Reactions of Small Transient Species,
     ed. M.A.A. Clyne and A. Fontijn, Academic Press Inc.,
     London, 1983.

[12] D.L. Baulch, R.A. Cox, P.J. Crutzen, R.F. Hampson Jr.,
     J.A. Kerr, J. Troe, and R.T. Watson, J. Phys. Chem. Ref. Data,
     11 (1982) 327.

[13] J.V. Michael, D.G. Keil and R.B. Klemm, Int. J. Chem. Kinet,
     15 (1983) 705.

[14] D.L. Baulch, R.A. Cox, R.F. Hampson Jr., J.A. Kerr, J. Troe,
     and R.T. Watson, J. Phys. Chem. Ref. Data, 13 (1984) 327.

[15] A.P. Force and J.R. Wiesenfeld, J. Phys. Chem., 85 (1981) 782.

[16] W.B. DeMore, J.J. Margitan, M.J. Molina, R.T. Watson,
     D.M. Golden, R.F. Hampson, M.J. Kurylo, C.J. Howard and
     A.R. Ravishankara, JPL Publ. 85-37, Documentation Services,
     Jet Propulsion Laboratory, Pasadena CA, 1985.

[17] D.L. Baulch, J. Duxbury, S.J. Grant and D.C. Montague,
     J. Phys. Chem. Ref. Data, 10 Suppl. 1 (1981).

[18] D.L. Baulch, D.D. Drysdale, D.G. Horne and A.C. Lloyd,
     Evaluated Kinetic Data for High Temperature Reactions, Vol.
     2, Butterworths, London, (1973).

[19] J. Barrassin and J. Combourieu, J. Bull. Soc. Chim. France,
     1 (1974) 10.

[20] R. Grice, Acc. Chem. Res., 14 (1981) 37.

[21] H. Okabe, Photochemistry of Small Molecules, Wiley-Interscience,
     New York, 1978, pg. 153.

[22] S.T. Amimoto, A.P. Force and J.R. Wiesenfeld, Chem. Phys. Lett., **60** (1978) 40.

[23] J.J. Sloan and P.M. Aker in Time-Resolved Vibrational Spectroscopy, Springer Proceedings in Physics, Vol. 4, ed. A. Laubereau and M. Stockburger, Springer-Verlag, Berlin, 1985.

[24] J.J. Sloan, P.M. Aker and B.I. Niefer, Proc. SPIE, **669** (1986) 169.

[25] K. Westberg and N. Cohen, J. Phys. Chem. Ref. Data, **12** (1983) 531.

[26] G.C. Light, J. Chem. Phys., **68** (1978) 2831.

[27] N. Presser and R.J. Gordon, J. Chem. Phys., **82** (1985) 1291.

[28] B.C. Garrett and D.G. Truhlar, Int. J. Quantum Chem., **24** (1986) 1463.

[29] B.C. Garrett and D.G. Truhlar, Int. J. Quantum Chem., **31** (1987) 17.

[30] G.C. Schatz, A.F. Wagner, S.P. Walch and J. Bowman, J. Chem. Phys., **74** (1981) 4984.

[31] J.M. Bowman, A.F. Wagner, S.P. Walch and Thom. H. Dunning, Jr., J. Chem. Phys., **81** (1984) 1739.

[32] G.C. Schatz, J. Chem. Phys., **83** (1985) 5677.

[33] D.C. Clary and J.N.L. Connor, Mol. Phys., **41** (1980) 698.

[34] Thom. H. Dunning, Jr., S.P. Walsh and A.F. Wagner in Potential Energy Surfaces and Dynamics Calculations, ed. D.G. Truhlar, Plenum Press, New York, 1981.

[35] B.R. Johnson and N.W. Winter, J. Chem. Phys., **66** (1977) 4116.

[36] M. Broida and A. Persky, J. Chem. Phys., **80** (1984) 3687.

[37] R.A. Gangi and R.F.W. Bader, J. Chem. Phys., **55** (1971) 5369.

[38] R.E. Howard, A.D. Maclean and W.A. Lester Jr., J. Chem. Phys., **71** (1979) 2412.

[39] K.S. Sorbie and J.N. Murrell, Mol. Phys., **29** (1975) 1387; *ibid*, **31** (1976) 905.

[40] J.N. Murrell and S. Carter, J. Phys. Chem., **88** (1984) 4887.

[41] P.A. Whitlock, J.T. Muckerman and P.M. Kroger in Potential Energy Surfaces and Dynamics Calculations, ed. D.G. Truhlar, Plenum Press, New York, 1981.

[42] G. Durand and X. Chapuisat, Chem. Phys., 96 (1985) 381.

[43] P.J. Kuntz and R. Polak, Chem. Phys., 99 (1985) 405.

[44] R. Polak, K. Paiderova and P.J. Kuntz, J. Chem. Phys., 82 (1985) 2352.

[45] L.J. Dunn and J.N. Murrell, Mol. Phys., 60 (1983) 635.

[46] A.C. Luntz, R. Schinke, W.A. Lester Jr., and Hs.H. Gunthard, J. Chem. Phys., 70 (1979) 5908.

[47] R. Schinke and W.A. Lester Jr., J. Chem. Phys., 72 (1980) 3754.

[48] P.A. Whitlock, J.T. Muckerman and E.R. Fisher, J. Chem. Phys., 76 (1982) 4468.

[49] S.W. Ransome and J.S. Wright, J. Chem. Phys., 77 (1982) 6346.

[50] M.S. Fitzcharles and G.C. Schatz, J. Phys. Chem., 90 (1986) 3634.

[51] P.M. Aker, J.J. Sloan and J.S. Wright, Chem. Phys., 110 (1986) 275.

[52] P.J. Kuntz, B.I. Niefer and J.J. Sloan, J. Chem. Phys., 88 (1988) 3629.

[53] G.K. Smith, J.E. Butler and M.C. Lin, Chem. Phys. Lett., 65 (1979) 115.

[54] A.C. Luntz, J. Chem. Phys., 73 (1980) 1144.

[55] G.K. Smith and J.E. Butler, J. Chem. Phys., 73 (1980) 2243.

[56] G.M. Jurisch and J.R. Wiesenfeld, Chem. Phys. Lett., 119 (1985) 511.

[57] R. Buss, P. Casavecchia, T. Hirooka, S.J. Sibener and Y.T. Lee, Chem. Phys. Lett., 82 (1981) 386.

[58] K. Tsukyama, B. Katz and R. Bersohn, J. Chem. Phys., 83 (1985) 2889.

[59] J.E. Butler, R.G. Macdonald, D.J. Donaldson and J.J. Sloan, Chem. Phys. Lett., 95 (1983) 183.

[60] P.M. Aker and J.J. Sloan, J. Chem. Phys., **85** (1986) 1412.

[61] J.E. Butler, G.M. Jurisch, I.A. Watson and J.R. Wiesenfeld, J. Chem. Phys., **84** (1986) 5365.

[62] Y. Huang, Y. Gu, C. Liu, X. Yang and Y. Tao, Chem. Phys. Lett., **127** (1986) 432.

[63] J.J. Sloan, J. Phys. Chem. (Accepted for publication).

[64] R.D.H. Brown and I.W.M. Smith, Int. J. Chem. Kinet., **7** (1975) 301.

[65] A.R. Ravishankara, G. Smith, R.T. Watson and D.D. Davis, J. Phys. Chem., **81** (1977) 2220.

[66] W. Hack, G. Mex and H. Gg. Wagner, Ber. Bunsenges. Phys. Chem., **81** (1977) 677.

[67] D.L. Singleton and R.J. Cvetanovic, Int. J. Chem. Kinet., **13** (1981) 945.

[68] R.D.H. Brown and I.W.M. Smith, Int. J. Chem. Kinet., **10** (1978) 1.

[69] G.A. Takacs and G.P. Glass, J. Phys. Chem., **77** (1973) 1182.

[70] J.E. Spencer and G.P. Glass, Int. J. Chem. Kinet., **11** (1977) 97.

[71] K.G. McKendrick, D.J. Rakestraw and R.N. Zare, Faraday Discuss. Chem. Soc., **84** in the press.

[72] J.C. Polanyi, Acc. Chem. Res., **5** (1972) 161.

[73] Z. Karney, B. Katz and A. Szoke, Chem. Phys. Lett., **35** (1975) 100.

[74] J.E. Butler, J.W. Hudgens, M.C. Lin and G.K. Smith, Chem. Phys. Lett., **58** (1978) 216.

[75] M. Kneba, R. Stender, U. Wellhausen and J. Wolfrum, J. Mol. Struct., **59** (1980) 207.

[76] R.G Macdonald and C.B. Moore, J. Chem. Phys., **68** (1978) 513.

[77] T.J. Odiorne, P.R. Brooks and J.V.V. Kasper, J. Chem. Phys., **55** (1971) 1980.

[78] A. Persky and M. Broida, J. Chem. Phys., **81** (1984) 4352.

[79] A. Persky and H. Kornweitz, Chem. Phys. Lett., **127** (1986) 609.

132

[80] H. Loesch, Chem. Phys., 112 (1987) 85.

[81] M. Broida, M. Tamir and A. Persky, Chem. Phys. Lett., 110 (1986) 83.

[82] D.J. Douglas, J.C. Polanyi and J.J. Sloan, J. Chem. Phys., 59 (1973) 6679.

[83] B.A. Blackwell, J.C. Polanyi and J.J. Sloan, Chem. Phys., 24 (1977) 25.

[84] J.A. Davidson, C.M. Sadowski, H.I. Schiff, G.E. Streit, C.J. Howard, D.A. Jennings and A.L. Schmeltekopf, J. Chem. Phys., 64 (1976) 57.

[85] J.A. Davidson, H.I. Schiff, G.E. Streit, J.R. McAfee, A.L. Schmeltekopf and C.J. Howard, J. Chem. Phys., 67 (1977) 5021.

[86] P.H. Wine, J.R. Wells and A.R. Ravishankara, J. Chem. Phys., 84 (1986) 1349.

[87] N. Basco and R.G.W. Norrish, Proc. Roy. Soc. A, 260 (1961) 293.

[88] A.C. Luntz, J. Chem. Phys., 73 (1980) 4393.

[89] E.J. Kruus, B.I. Niefer and J.J. Sloan, J. Chem. Phys., (submitted).

[90] K. Rynefors and L. Holmlid, Chem. Phys., 60 (1981) 393.

[91] L. Holmlid and K. Rynefors, Chem. Phys., 60 (1981) 405.

[92] P.-A. Elofson, K. Rynefors and L. Holmlid, Chem. Phys., 100 (1985) 39.

[93] K. Rynefors, P.-A. Elofson and L. Holmlid, Chem. Phys., 100 (1985) 53.

[94] E.J. Kruus, B.I. Niefer and J.J. Sloan, (in preparation).

[95] G. Hirsch, P.J. Bruna, S.D. Peyerimhoff and R.J. Buenker, Chem. Phys. Lett., 52 (1977) 442.

[96] P.J. Bruna, G. Hirsch, S.D. Peyerimhoff and R.J. Buenker, Can. J. Chem., 57 (1979) 1839.

[97] R. Schinke, J. Chem. Phys., 80 (1984) 5510.

[98] P.M. Aker, J.J.A. O'Brien and J.J. Sloan, J. Chem. Phys., **84** (1986) 745.

[99] A.C. Luntz, J. Chem. Phys., **73** (1980) 1143.

[100] P.M. Aker, E.J. Kruus, B.I. Niefer and J.J. Sloan, (in preparation).

[101] P.M. Aker, J.J.A. O'Brien and J.J. Sloan, Chem. Phys., **104** (1986) 421.

[102] S. Klee, K.-H. Gehricke and F.J. Comes, Chem. Phys. Lett., **118** (1985) 530.

[103] P.M. Aker, J.J.A. O'Brien, J.M. Parsons and J.J. Sloan, Can. J. Chem., **64** (1986) 2315.

[104] J.F. Cordova, C.T. Rettner and J.L. Kinsey, J. Chem. Phys., **75** (1981) 2742.

LASER STUDIES OF REACTION DYNAMICS: THE Ca* + HF → CaF(X$^2\Sigma^+$,v) + H AND Ca* + SF$_6$ → CaF(X$^2\Sigma^+$,v) + SF$_5$ REACTIONS

F. Engelke and K.H. Meiwes-Broer
Fakultät für Physik
Universität Bielefeld
D-4800 Bielefeld 1
FRG

ABSTRACT. Laser and molecular beam techniques allow detailed study of many dynamical properties of single reactive collisions. The chemical scope of these methods is now very wide and includes internal state preparation of reactants, change of collision energies, state detection of products, and thus determination of state-to-state reaction rates. The great impact of laser spectroscopy on knowledge in the field of structure, molecular energy transfer and the mechanism of elementary chemical reactions is illustrated by two selected examples, i.e. studies in which laser-induced fluorescence (LIF) has been used to determine the specific impact parameter dependence of the Ca* + HF → CaF(X) + H reaction and the product state distributions for the reaction of metastable Ca with SF$_6$.

1. INTRODUCTION

In order to investigate chemical reactions in a controlled manner, studies of molecular species formed under single collision conditions are carried out in crossed molecular beams [1,2]. The emphasis in such studies has generally been on the measurement of angular and velocity distributions of the products whereas the electronic, vibrational, and rotational excitation of these products has been characterized for only a few systems.

With the discovery of lasers such studies have assumed a new importance. The development of these coherent, intense, highly monochromatic light sources from the far Infrared (IR) to the Vacuum ultraviolet (VUV) has promised to put at the disposal of molecular beam kineticists new and powerful tools that enable the performance of highly refined and sophisticated experiments previously not feasible [3].

It is worthwhile to reexamine molecular beam studies in the light of what laser devices have done and can do to improve these studies of molecular interactions. Basically, the laser has permitted the experimenter to combine molecular spectroscopy with molecular beams [1,3]. Under those felicitous circumstances where both can be utilized effectively, the progeny of this combination is a delight in that it provides some of the most detailed information on dynamical knowledge yet available [3-7].

The observation of bright fluorescence against a dark background makes the laser-induced fluorescence (LIF) technique extremely sensitive; molecular densities in

J. C. Whitehead (ed.), Selectivity in Chemical Reactions, 135–146.
© 1988 by Kluwer Academic Publishers.

the range of 10^4 cm^{-3} have been successfully detected [4]. The best conventional techniques are capable of detecting trace amounts of a substance at a level of about one part per billion (0.001 ppm). LIF is easily capable of detecting small fractions of one part per trillion (0.000001 ppm!), and in some cases has been used to detect a single atom [8].

The use of this highly sensitive laser technique allows experimentalists to study reaction channels leading to individual product states (vibrational, rotational, fine structure and even hyperfine structure). Since we cannot control the choice of the impact parameter b (defined as the distance of closest approach of the incoming reagents if they move in undeflected straight lines [1]), nearly all of our studies so far tend to ignore the detailed dependence of the reaction probability on b. Under certain favourable circumstances we can remove the average over b included in the cross section and gain information on the impact parameter dependence in the state selective outcome of a reactive collision. The key to this is a kinematically constrained reaction having the extreme mass combination, H + HL $\rightarrow$ HH + L (H: heavy atom, L: light atom), which allows us to use a simplified approach to determine a "specific opacity" function $P_v(b)$ for the present reaction Ca* + HF $\rightarrow$ CaF($X^2\Sigma^+$,v) + H, i.e. to predict the reaction probability for a given b and relative velocity v_{rel} to produce the product CaF in a specific vibrational level.

2.   EXPERIMENTAL

2.1. Apparatus and Treatment

Our basic apparatus is shown in Fig. 1, except that the polarization accessories are not immediately required. A primary beam of calcium atoms is produced by resistively heating an oven (nozzle diameter about 0.7 mm) to approximately 1400 K. A DC discharge between a ring electrode in front of the nozzle and the nozzle exit serves to excite approx. 20% of the atoms into metastable states [9]. After collimation, this beam enters the collision zone and impinges on a secondary gas. A pulsed (nitrogen laser pumped) dye laser beam enters through baffle arms to reduce scattered light. A sensitive photomultiplier tube adjacent to the reaction region collects the fluorescence from product molecules so that excitation spectra of the products are generated as the dye laser is scanned. The intensities may then be converted to figures that indicate the relative populations of the various internal energy states of the products. This procedure provides the most detailed description presently available of the nature of a single reactive encounter. Fig. 2 shows an example of the use of laser induced fluorescence taken from Ref. 9. There we carried out a series of beam-gas reactions of metal atoms with various fluorine containing molecules (F_2,HF,NF_3,...) [9,10]. For these single collision processes we used LIF to determine the nascent internal state distribution of the MF product (M: Ca,Sr,Ba). The CaF radical, shown in Fig. 2, has an electronic band system around 530 nm. An analysis of the intensity pattern of the individual bands leads to the conclusion that the unrelaxed CaF product molecule, produced in the highly exothermic reaction Ca + F_2 $\rightarrow$ CaF + F, has a vibrational-rotational distribution in its electronic ground state characterized by a Boltzmann temperature (T$\approx$9000 K!). This result supports the picture that energy is randomized i.e. is statistically redistributed before the decomposition of the CaF_2 collision complex.

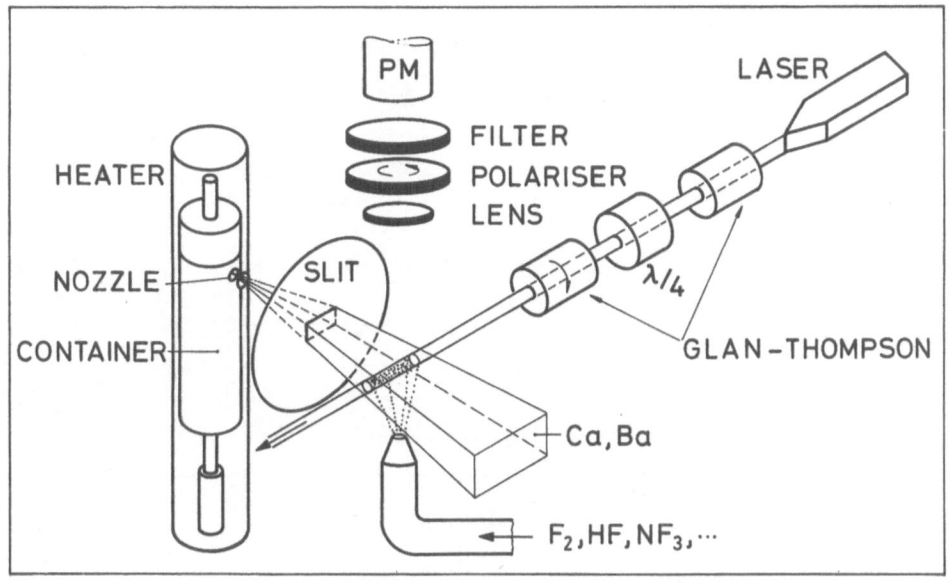

FIGURE 1. Crossed laser and molecular beam apparatus for LIF-studies. In addition, the experimental setup for polarization measurements is given [9].

2.2 Internal State Preparation of Reactants

If we could prepare reactants in known internal states, detection of specific product states would be straightforward and within present capabilities. Such experiments represent an important advance towards the determination of state-to-state reaction rates. The first successful experiments on molecular beam internal state preparation by infrared lasers were carried out by Odiorne, Brooks, and Kasper [11], who excited HCl by means of a primitive pulsed HCl laser. The slightly endothermic reaction K + HCl(v=0) → KCl + H could be accelerated approximately 100-fold in rate by pumping the (1-0) fundamental vibrational transition of the HCl reactant. Using a similar HF laser some of the reagent HF molecules (typically 1%) may be prepared in their first vibrationally excited state. For Ba + HF an intense spectrum due to the reaction with HF(v=0) makes it difficult to distinguish the products resulting from the small fraction of HF(v=1) [12]. However, careful subtraction techniques involving two Ba beams (from the same beam oven!) have been successful in this respect [13] even for the Ba + HF(v=2) reaction [14]. All the foregoing reactions are ground state reactions, that is, electronically ground state reactants evolved into electronically ground state products. However, excitation can

138

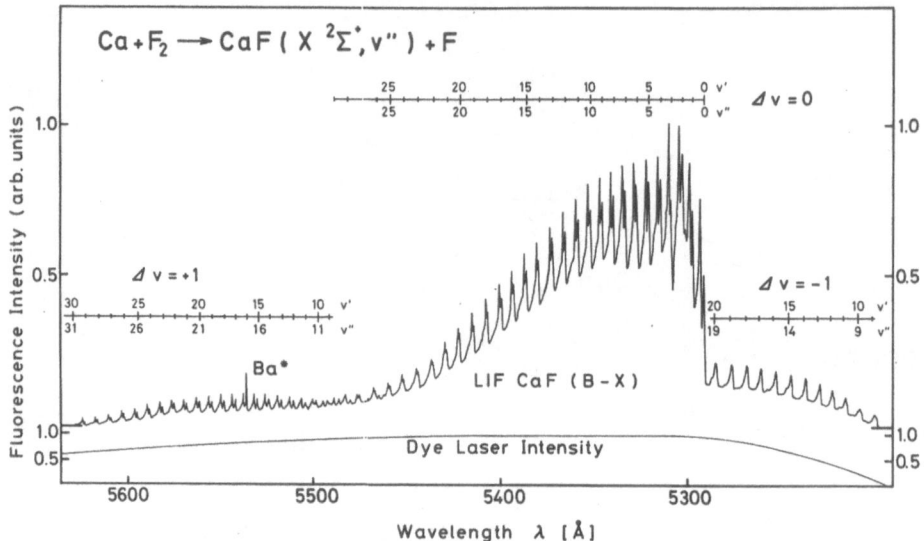

FIGURE 2. LIF-excitation spectrum of the CaF product molecules formed in the reaction $Ca + F_2 \rightarrow CaF(X^2\Sigma^+) + F$. The vibrational bands (v',v'') are noted [9].

be used to prepare electronically excited reactants, here metastable alkaline earth atoms [15]. If they react to form electronically excited products, these products can be readily detected by their characteristic chemiluminescence [15,16]. Here we present some unpublished work [17] of our laboratory, where we use a discharge in front of the metal beam oven to excite Ca atoms to a metastable state $(^3P_{0,1,2})$; then the Ca* beam crosses a beam of gas.

3. RESULTS AND DISCUSSION

3.1 Product State Distribution for $Ca(^3P,^1S)$ with SF_6

Fig. 3 shows the laser-induced fluorescence (LIF) spectrum for the beam-gas reaction of

$$Ca^* + SF_6 \rightarrow CaF + SF_5. \tag{1}$$

With the discharge off, there is no detectable LIF of CaF products from Ca + SF_6. With the discharge on, strong peaks appear which are readily assigned to the CaF product excitation spectrum.

As is characteristic of the visible band systems of the alkaline-earth monohalide, the $\Delta v = 0$ sequence has the highest intensity. In addition, the $\Delta v = -1$ sequence appears also with relatively high intensity which hints at a preferred

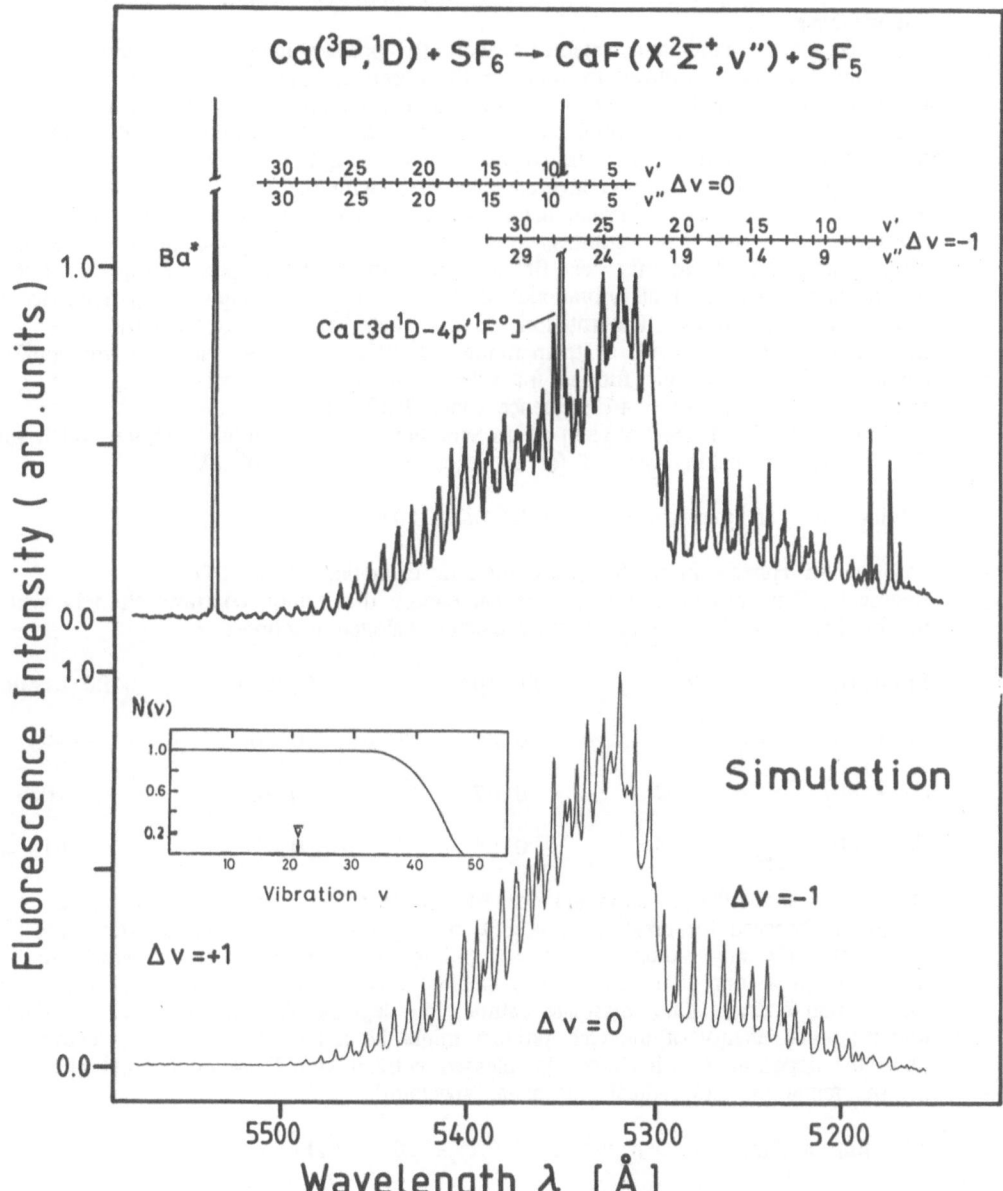

FIGURE 3. Upper part: LIF spectrum of CaF(B-X) for the $Ca(^3P,^1D)$ + SF_6 reaction. The intensities are not corrected for the varying dye laser power. $P(SF_6)$ < 2 x 10^{-4} mbar; discharge current 0.2A.
Lower part: Calculated CaF fluorescence based on the "best fit" for reaction (1).

excitation of CaF product molecules into high vibrational levels. Rotational lines are not resolved.

To make quantitative statements about the product internal distribution a computer program is utilized to simulate the observed excitation spectrum [10]. As input for the calculations we estimate the relative vibrational and rotational populations. Each line is weighted by the population of the initial (v",J") level, by the Franck-Condon factor and the rotational line strength of the pump transition. At each frequency, the program convolutes the lines with the laser bandwidth and power to produce a simulated spectrum; such spectra are compared visually with the observed spectra and new estimates are made for the (v",J") populations. Iteration of this process leads to the "best fit" as shown in the lower part of Fig. 3. For this calculated spectrum all vibrational states v" = 0...35 are equally populated as is shown in the insertion. The rotation, on the other hand, is described by a Boltzmann distribution with a "temperature" of 1200 K. With such low rotational energy no band heads are formed for v" < 5 in the $\Delta v = 0$ sequence and for nearly all v" in the $\Delta v = +1$ sequence (near 5550 Å).

Our simulated spectra yield mean vibrational and rotational energies which give the following branching ratios f for the exoergicity $\Delta E = 3.62$ eV:

f (vibration):   f (rotation) = 0.34 (±0.04):   0.02 (±0.015).

64% of ΔE appears in product translation or excitation of the SF_5 fragment. This pronounced non-statistical energy distribution contrasts strongly with the product state distribution of other calcium-halogen reactions:

Reaction	ΔE	f(vib)	f(rot)	Reference
Ca + F_2	3.95	0.26	0.20	[10]
Ca + NF_3	3.07	0.67	0.14	[10]
Ca + HF	1.54	0.25	0.27	this work

Our results show that reaction (1) does not proceed via a long living complex as has been observed for alkali-atom + SF_6 reactions [20], where the electron jump occurs at a crossing radius $R_c > 5$ Å, leading to a $M^+SF_6^-$ complex which may dissociate on a long time scale. The calcium reaction, however, proceeds directly as the reaction energy is not dissipated among all degrees of freedom. The extremely low rotational energy of the CaF product might be a hint that after the electron jump the departing fluorine atom is released without significant momentum change, i.e. the remaining SF_5-radical acts as a "spectator".

3.2   Impact Parameter Dependence of Ca* + HF → CaF(X,v") + H

Our ability to control the wavelength of lasers allows us to choose the excitation wavelength of the dye laser (Fig. 1), thus it offers the opportunity to probe individual rotational-vibrational states of reaction products. For

$$Ca^* + HF \rightarrow CaF + H, \tag{2}$$

this is illustrated in Fig. 4, a low resolution excitation spectrum of the CaF $B^2\Sigma^+$ - $X^2\Sigma^+$ system, using a nitrogen pumped dye laser with a bandwidth of 1 cm^{-1}.

During the reactive encounter total angular momentum $J + L = J' + L'$ is conserved. Here L and L' are the reagent and product orbital angular momenta and J and J' are the rotational angular momenta, respectively. Since the measured average value of J' for this experiment is in the range of J' $\leqslant$ 110 [17] we may neglect J $\leqslant$ 2 for HF at room temperature. Since one of the products, atomic hydrogen, is so much lighter than either of the two other partners, it will not likely carry away much orbital angular momentum [18-20]. This kinematic constraint necessitates a non-spherical symmetric distribution of CaF transition moments which has been probed by comparing the fluorescence intensity when the dye laser is polarized [21]. As a consequence of the conservation of angular momentum, nearly all of the reagent angular momentum L is channelled into rotational angular momentum J' of the product, i.e.

$$J' \approx L = \mu(v_{rel} \times b). \qquad (3)$$

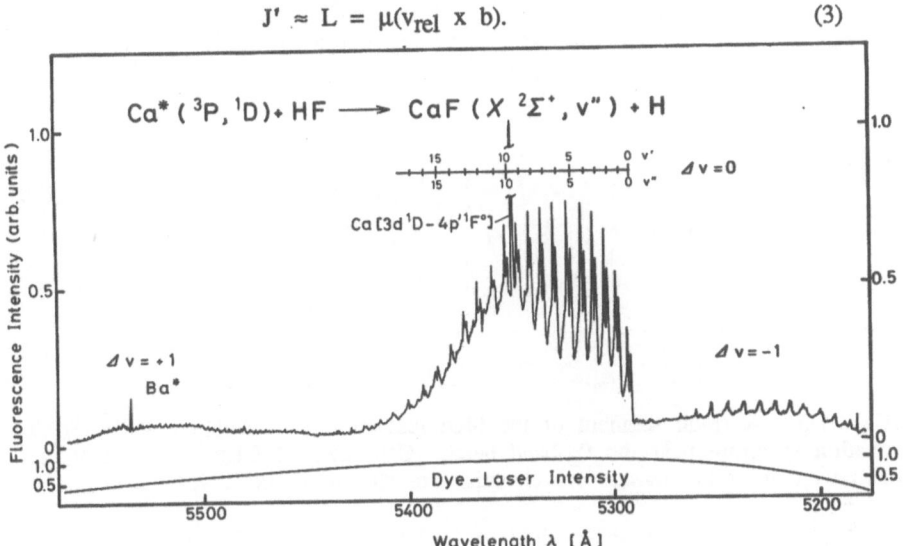

FIGURE 4. LIF-excitation spectrum of the CaF product molecules formed in the reaction of metastable Ca*($^3P_{0,1,2}$,$^1D_2$) and HF(v=0) with an average thermal collision energy of 2.9 kcal/mole. The vibrational bands (v',v") within the $\Delta v = 0$ sequence are noted [17].

For a given v_{rel}, the distribution of reactive impact parameters is mapped into the population of the product rotational levels through the relation (3). Thus, we can derive "specific opacity" functions $P_v(b)$ for the reaction (2) using the fact that in this reaction nearly all the orbital angular momentum L of the reactants Ca* + HF is transferred to product rotational angular momentum J' of CaF($X^2\Sigma^+$) in individual vibrational states v.

An important point could be that, although the departing H atom does not carry away appreciable orbital angular momentum, the CaF vibrational state could depend on the kinetic energy of the departing H atom, as well as the kinetic energy of the reactants.

We have measured the J' distribution for CaF(v), using a single-mode ring dye laser; a small segment of the excitation spectrum is shown in Fig. 5. We then derived $P_v(b)$, see Fig. 6, using Eq. 3, and choosing a full distribution of initial relative velocities $f(v_{rel})$. The choice of $f(v_{rel})$ i.e. a distribution which includes all relative velocities, implies that each collision with initial v_{rel} contributes equally to production of CaF(v), so that $P_v(b)$ is independent of v_{rel}. This assumption may be satisfactory for a substantially exothermic reaction such as Ca* + HF ($\Delta E = 1.7$ eV), where the final vibrational state distribution might not depend strongly on the range of relative kinetic energies which are much smaller than the exothermicity

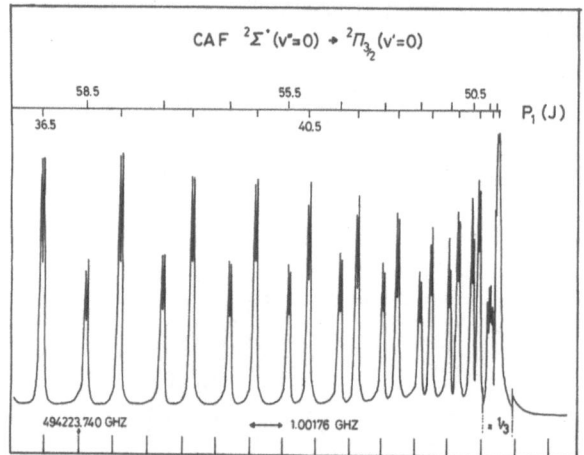

FIGURE 5. A small segment of the high resolution CaF $A^2\Pi_{3/2} \leftarrow X^2\Sigma^+(0,0)$ excitation spectrum near the P_2-band head. Calibrated (1 GHz) Fabry-Perot markers, which were used to measure wavelengths, are shown in the lower trace.

(here about 5%). Note, however, that it would be an unphysical assumption for a nearly thermoneutral reaction. It implies, that a change in the initial translational energy E''_{trans} does not alter the probability of populating CaF(X,v). For increasing E''_{trans} this model predicts that all additional initial translational energy is channelled exclusively into product translational energy, and ignores any coupling between E''_{trans} and E'_{vib}. This lack of coupling has an important implication: Eq. 3 shows that the distribution of impact parameters required to reproduce the J' distribution depends upon $f(v_{rel})$. A wide velocity distribution, like the one we have chosen so far, requires only a small range of impact parameters, and generates narrow $P_v(b)$ opacity functions. In the case of a smaller subclass of v_{rel} which populates CaF(v), this yields a somewhat wider $P_v(b)$. Thus, we are aware that we are making a simplistic assumption by postulating that the specific opacity function is independent of v_{rel}. Our assumption closely follows a computer simulation using a LEPS potential for the reactions Ba + HCl and Ba + HBr [22] as well as experimental work to gather similar information on the impact parameter dependence of the Ba + HI ground state reaction [23].

To obtain the true $P_v(b)$, both energy conservation and the initial velocity dependence of the production of CaF(v) must be considered: One has to measure

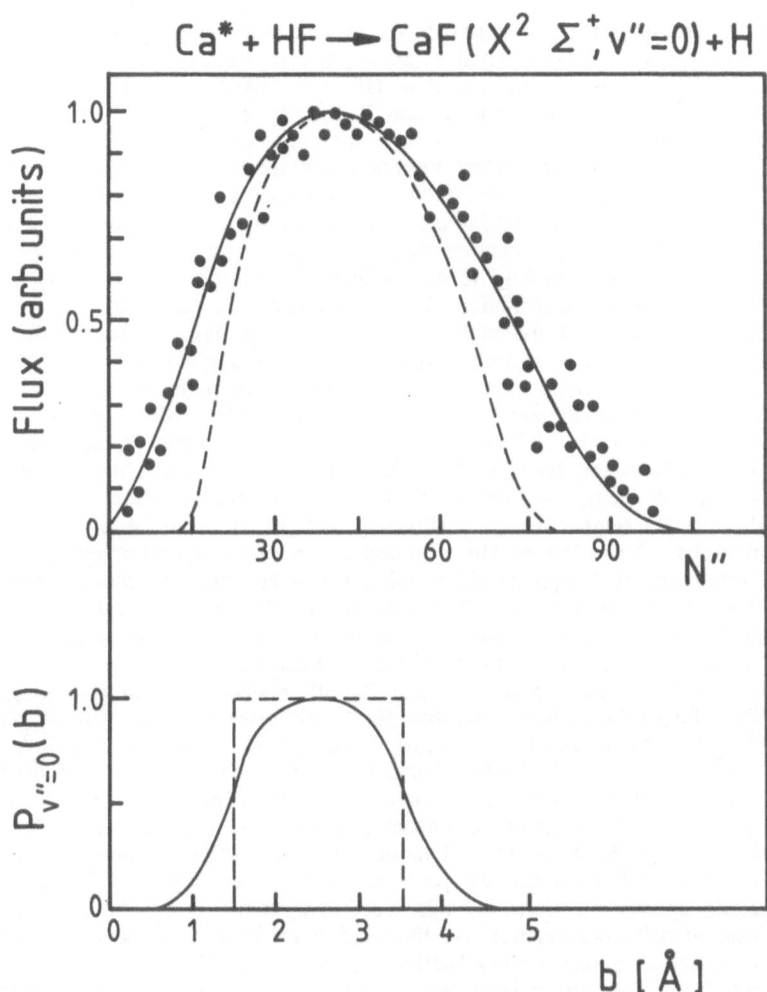

$$Ca^* + HF \longrightarrow CaF (X^2 \Sigma^+, v''=0) + H$$

FIGURE 6

(a) The rotational distribution of the CaF product in v'=0 produced by the Ca* + HF reaction. Calculated rotational distributions using the specific opacity function $P_{v=0}(b)$ shown in (b) are also given.

(b) The specific opacity function $P_{v=0}(b)$ for the Ca* + HF → CaF(X,v=0) + H system, assuming (i) no velocity dependence for the specific reaction cross section and (ii) no activation energy barrier. In addition, a simple step function $P_{v=0}(b)$ is shown, which does not give such a good fit, see (a), dashed line.

the product vibrational state distribution as a function of initial kinetic energy. Thus, the experiments repeated, under low resolution with a well defined initial relative velocity, gave the dependence on collision energy. In order to control the collision energy a crossed molecular beam configuration is employed in which HF gas is seeded with He.

From low resolution spectra similar to the one presented in Fig. 4 it is clear that collision energy has no substantial effect on the distribution of energy among the product modes from the reaction $Ca^* + HF \rightarrow CaF(X^2\Sigma^+,v) + H$. The fraction of energy appearing in vibration and rotation is roughly constant over the range of collision energies with more than half going into translation and the remainder being divided nearly equally between product vibration and rotation. However, while translational energy probably increases with collision energy, this dependence of the final translational energy on the initial translational energy greatly complicates the analysis. The specific opacity functions $P_v(b)$ depend on the coupling between E''_{trans} and E'_{trans}. This coupling is not included in our preliminary analysis. Its influence may give broader distributions $P_v(b)$ than given in Fig. 6(b).

The dramatic effects of the metastable Ca shown in Fig. 3, lead to the question "which reagent mode is more effective in promoting endothermic reactions, rovibration, electronic excitation or translation?" These comparisons are complicated by the fact that the rovibrationally, electronically and translationally excited measurements are done under different experimental configurations [9,10,17-19]. To circumvent this problem the reaction Ba + HF whose cross section depends only modestly on collision energy is used as an internal standard. Ba and Ca can be mixed together in the same oven [19] allowing both reactions to be monitored under identical conditions. Vibration of HF is found to be one order of magnitude more effective in promoting the endothermic reaction Ca + HF than is collision energy, whereas electronic excitation of Ca (40 kcal/mole) accelerates Ca + HF approximately 250-fold in rate compared with an 20-fold increase from threshold to the highest available collision energy of about 12 kcal/mole.

If the reaction probability is the same for all relative collision velocities, then the calculation of specific opacity functions is straightforward. Fig. 6(b) shows the $CaF(X^2\Sigma^+,v'' = 0)$ opacity function $P_{v=0}(b)$, which peaks strongly near 2.5 Å with a full width at half-maximum (FWHM) of about 2.1 Å. The simple step function, also shown in Fig. 6(b), does not give such a remarkably good fit. Thus, we conclude that only a limited range of impact parameters contributes to the formation of CaF(X,v=0). Because of the limited amount of information on higher lying vibrational states of CaF products, we presently cannot give a unique determination of other specific opacity functions for this reaction system. It can be speculated that (i) the intrinsic velocity dependence has little effect on $P_v(b)$; (ii) there is no hint of the existence of an activation energy barrier E_A, which could greatly alter the form of the derived $P_v(b)$ and (iii) experimental work on polarization of the product molecules CaF already shows that for higher vibrational levels v' the departing orbital angular momentum of the hydrogen is getting larger [10,21,24]. The apparatus used to measure this is the same as shown in Fig. 1. Note, that this beam-gas setup brings v_{rel} perpendicular to the probe dye laser beam. The polarization of the dye laser is first fixed by a Glan-Thomson prism, then after circular polarization due to a $\lambda/4$ plate rotated with a second Glan-Thomson (extinction 10^7!).

With these procedures angular momentum polarization in CaF has been observed. The high degree of polarization measured is found to be nearly independent of collision energy; the degree for $v'=0$ is close to the expected value [17,25]. Previous experiments on this system and similar reactions [10] and theoretical [26] and experimental [21,27] work on systems with this mass combination all support the assumption that $L'<<J'$.

We conclude that impact parameter dependence for the reaction Ca* + HF has been demonstrated using LIF, giving direct information on the role of angular momentum of reactant and product molecules. In the future these new techniques may be quite useful in studying state-to-state reacting processes giving us yet another powerful method of exploring reaction dynamics, i.e. providing us with some precise details of atom-molecule collisions.

ACKNOWLEDGEMENTS

It is a pleasure to thank D. Beck and R.N. Zare (Stanford) for their comments and suggestions. We are grateful to J. Heinze for his help in taking spectra shown in Fig. 5, and to H. Ellerbusch for his help in a computing procedure for extracting rotational distributions from our spectra and calculating $P_v(b)$ functions from these distributions. We wish to acknowledge support of our work by the Deutsche Forschungsgemeinschaft (SFB 216) and the Minister für Wissenschaft und Forschung des Landes Nordrhein-Westfalen.

REFERENCES

[1] For a general review see R.B. Bernstein, Chemical Dynamics via Molecular Beam and Laser Techniques, Oxford University Press, New York (1982);
R.D. Levine and R.B. Bernstein, Molecular Reaction Dynamics and Chemical Reactivity, Oxford University Press, New York (1987).

[2] M.R. Levy, 'Dynamics of Reactive Collisions', Progr. React. Kin., 10 (1979) 1.

[3] R.N. Zare and P.J. Dagdigian, Science, 185 (1974) 739.

[4] F. Engelke, Ber. Bunsenges. Phys. Chem., 81 (1977) 135.

[5] J.L. Kinsey, Ann. Rev. Phys. Chem., 28 (1977) 349.

[6] F. Engelke, Comments Atom. Mol. Phys., 11 (1981) 13.

[7] S.J. Silvers, R.A. Gottscho and R.W. Field, J. Chem. Phys., 74 (1981) 6000.

[8] G.S. Hurst, M.H. Nayfeh and J.P. Young, Appl. Phys. Lett., 30 (1977) 229.

[9] D. Beck, H. Ellerbusch, F. Engelke and K.H. Meiwes, Forschungsbericht des Landes Nordrhein-Westfalen, Nr. 3096, Westdeutscher Verlag Opladen (1982).

[10] F. Engelke and K.H. Meiwes-Broer, Z. Phys., A320 (1985) 39.

[11] T.J. Odiorne, R.P. Brooks and J.V.V. Kasper, J. Chem. Phys., 55 (1971) 1980.

[12] J.G. Pruett and R.N. Zare, J. Chem. Phys., 64 (1976) 1774.

[13] A. Torres-Filho and J.G. Pruett, J. Chem. Phys., 72 (1980) 6736.

[14] A. Torres-Filho and J.G. Pruett, J. Chem. Phys., 77 (1982) 740.

[15] F. Engelke, Chem. Phys., 39 (1979) 279.

[16] F. Engelke, Chem. Phys., 44 (1979) 213.

[17] K.H. Meiwes-Broer, Dissertation, Bielefeld (1984).

[18] A. Gupta, D.S. Perry and R.N. Zare, J. Chem. Phys., 72 (1980) 6237.

[19] Z. Karny and R.N. Zare, J. Chem. Phys., 68 (1978) 3360.

[20] S.J. Riley and D.R. Herschbach, J. Phys. Chem., 58 (1973) 27.

[21] F. Engelke and K.H. Meiwes-Broer, Chem. Phys. Letters, 108 (1984) 132.

[22] A. Siegel and A. Schultz, J. Chem. Phys., 72 (1980) 6227.

[23] Ch. Noda, J.S. McKillop, M.A. Johnson, J.R. Waldeck and R.N. Zare, J. Chem. Phys., 85 (1986) 856.

[24] F. Engelke, Naturw., 70 (1983) 594.

[25] M.H.M. Janssen, D.H. Parker and S. Stolte, Chem. Phys., 113 (1987) 357.

[26] N.H. Hijazi and J.C. Polanyi, Chem. Phys., 11 (1975) 1.

[27] D.S.Y. Hsu, G.M. McClelland and D.R. Herschbach, J. Chem. Phys., 61 (1974) 4927.

SPIN-ORBIT EFFECTS IN CHEMICAL REACTIONS

Paul J. Dagdigian
Department of Chemistry
The Johns Hopkins University
Baltimore
Maryland 21218
USA

ABSTRACT. It is well known that forms of reagent energy, for example translational, vibrational, rotational excitation, can have drastically different effects on the rate of a chemical reaction. In this paper, we review investigations of spin-orbit effects in reactions involving atoms with both nonzero electronic orbital and spin angular momenta. There has been increasing interest in reactions of individual atomic spin-orbit levels because of the importance of nonadiabatic transitions in controlling the magnitude of spin-orbit dependences. Experimental methods for the study of reactions of individual reagent spin-orbit levels and for the detection of such states in atomic products will be briefly discussed. Observations for reactions of representative classes of atoms, including halogen atoms, electronically excited and ionized inert gases, and electronically excited alkaline earth and mercury atoms, are presented. The approaches taken for a theoretical modelling of such reactions are also outlined.

1. INTRODUCTION

Experimental and theoretical studies in the past decade and a half have shown that deposition of energy in the different degrees of freedom, e.g., translational, vibrational, rotational, electronic, of the reagents of an elementary gas-phase chemical reaction can influence the rate and outcome of this process drastically differently [1-3]. The measurement of state-specific rate constants (or cross sections) has allowed detailed inferences to be made on the dynamics of simple reactions [4-9]. In this paper, we concentrate on investigations of spin-orbit effects in reactions involving atoms with both nonzero electronic orbital and spin angular momenta, either as reactants or products. By spin-orbit effect, we refer to possible variations in the chemical reactivity of individual states of different total angular momentum J of an atomic multiplet. Reactions of such species necessarily involve multiple potential energy surfaces. The study of reactions of the individual fine-structure levels of an atomic multiplet is increasing, in part because such investigations can reveal the importance of nonadiabatic coupling between the different potential energy surfaces. Of special interest is the possibility of spin-orbit effects for atomic reactants whose fine-structure splittings are small compared with the overall reaction energetics.

J. C. Whitehead (ed.), Selectivity in Chemical Reactions, 147–177.
© 1988 by Kluwer Academic Publishers.

We might naively expect that spin-orbit effects would be manifest in chemical reactions of only those atoms with large spin-orbit splittings because of the expected general increase in reaction rate with total reactant energy. In fact, large spin-orbit dependences of the reaction rate are found for reactions of heavy metals, such as electronically excited $Hg(^3P_J^o)$, for which the spin-orbit spacings are comparable to energy differences between different atomic electronic terms. By contrast, a negligible spin-orbit dependence would be predicted for reactions of atoms with small spin-orbit splittings since the reagent spin-orbit states might be expected to scramble completely as the reactants approach. In spite of these expectations, significant spin-orbit dependences have been found for reactions of light atoms with small spin-orbit splittings, such as $F(^2P_J^o)$ and electronically excited $Ca(^3P_J^o)$. We are thus led to consider the detailed dynamics of these reactions and in particular the role of nonadiabatic mixing.

The simplest theoretical approach for an understanding of spin-orbit dependences in chemical reactions involves consideration of the adiabatic potential energy surfaces correlating to the individual fine-structure states of an atomic multiplet, as has been worked out in detail by Husain [10]. Such adiabatic correlation arguments in the strong spin-orbit coupling limit do predict differing reactivities and product electronic state branching as a function of initial spin-orbit state for many reactions. However, such considerations, in effect, only take account of the energy ordering of states and ignore the detailed dynamics of the reaction. Nevertheless, this approach is often very useful in providing a framework for more sophisticated theoretical modelling.

We have recently written an extensive review article on the spin-orbit dependence of chemical reactivity [11]. The present paper highlights and updates what is known about the most extensively studied classes of atomic reactants and considers in somewhat more detail the theoretical treatment of spin-orbit effects and its relationship to alignment effects, which are also being discussed at this workshop. This article is organized in the following way: The next section summarizes the experimental techniques employed for the study of reactions of individual spin-orbit states of both neutral atoms and ions. Section 3 then reviews the experimental data obtained for reactions of representative classes of atoms for which considerable information is available. These include reactions of halogen atoms, electronically excited and ionized inert gases, and electronically excited alkaline earth and mercury atoms. The spin-orbit energies [12] for these atomic groups are listed in Table 1: it can be seen that the splitting in some cases increases dramatically down the periodic table. Spin-orbit effects for other classes of atoms will not be discussed here as there is considerably less information available; the interested reader may consult our more complete review article on spin-orbit effects [11]. In Section 4 an overview of the approaches taken for a theoretical modelling of reactive collisions of individual atomic spin-orbit states is presented.

2. EXPERIMENTAL METHODS

Experimental techniques which have been employed for the study of reactions of individual spin-orbit states fall into two classes, namely those which monitor the collisional loss of specific reactant states and those which follow the formation of products. The former is simple and widely applicable but suffers from the inability to distinguish between reactive and nonreactive collisional removal processes; this can be a serious limitation when comparing ground and spin-orbit-excited states. The

TABLE 1. Spin-orbit energies (in cm^{-1}) for atomic reactants.[a]

Atom	Energies

Group VIIA [np^5]

Atom	$^2P_{3/2}^o$	$^2P_{1/2}^o$
F	0	404
Cl	0	882
Br	0	3685
I	0	7603

Inert Gases [$np^5(n+1)s$]

Atom	$(n+1)s[3/2]_2^o$ ($^3P_2^o$)	$(n+1)s[3/2]_1^o$ ($^3P_1^o$)	$(n+1)s'[1/2]_0^o$ ($^3P_0^o$)	$(n+1)s'[1/2]_1^o$ ($^1P_1^o$)
Ne	134044	134461	134821	135891
Ar	93144	93751	94554	95400
Kr	79973	80918	85192	85848
Xe	67068	68046	76197	77186

Inert Gas Ions [np^5]

Atom	$^2P_{3/2}^o$	$^2P_{1/2}^o$
Ne+	0	782
Ar+	0	1432
Kr+	0	5371
Xe+	0	10537

Group IIA [$nsnp$ or $ns(n-1)d$]

Atom	$^3P_0^o$	$^3P_1^o$	$^3P_2^o$
Mg	21850	21870	21911
Ca	15158	15210	15316
Sr	14318	14504	14899
	$^3D_1$	$^3D_2$	$^3D_3$
Ba	9034	9216	9597

Group IIB [$nsnp$]

Atom	$^3P_0^o$	$^3P_1^o$	$^3P_2^o$
Zn	32311	32501	32890
Cd	30114	30656	31827
Hg	37645	39412	44043

[a] Energies taken from Ref. 12.

latter technique allows determination of the spin-orbit dependence of individual reaction channels for systems where more than one pathway is available, e.g. formation of ground and electronically excited products. This does require selection of the reagent spin-orbit states by some means, such as optical pumping or selective collisional quenching. Complementary experiments have probed the spin-orbit distributions in product atomic species by one-photon vacuum ultraviolet or two-photon ultraviolet laser fluorescence excitation or by observation of spontaneous infrared emission.

A very widely employed method for the measurement of spin-orbit state-specific rate constants is the time-resolved measurement of the concentrations of individual atomic levels after formation of these species from a suitable precursor, either by flash photolysis [13], or, more recently, by laser photodissociation. The concentrations of the various atomic reactant states are monitored by atomic absorption or fluorescence spectroscopy using atomic emission sources [14], or, for spin-orbit-excited states, by observation of the spontaneous infrared emission [15-18]. Recently, Leone and co-workers have utilized gain/absorption of a colour centre and diode infrared laser to probe the relative populations of ground and spin-orbit excited halogen atoms produced in a chemical reaction [19] and also by photodissociation [20].

This technique of time-resolved monitoring of atomic concentrations has been extensively employed to study reactions of electronically excited atoms [10,21]. This can be applied to the measurement of reaction rate constants of individual spin-orbit states provided that the collisional intramultiplet mixing rate is slow, as would be the case for an atom with reasonably large spin-orbit splitting. Provided this mixing is slow, then the collisional removal rates for the individual spin-orbit reactant states can be measured. However, care must be taken with identifying these rates as chemical reaction rates, in particular for spin-orbit-excited states, since the total collisional removal rate includes processes such as nonreactive collisional de-excitation. The clearest situation is when the ground-state level is removed at a total rate larger than that for a spin-orbit-excited level; in this case, it is clear that chemical reaction is faster for the lower level. Otherwise, the branching ratio between reactive and nonreactive decay of the upper level is required for an unambiguous assessment of the magnitude of the reactive spin-orbit effect [17,22].

Several different methods have been employed for spin-orbit selection of reactant species. The spin-orbit dependence of the rate of nonreactive collisional removal processes may be used in flow experiments to monitor the products from reactions of specific spin-orbit states. For example, for the two metastable states of electronically excited argon, $Ar(^3P_{0,2}^0)$, it is known that krypton much more efficiently quenches the J=2 level [23-25], whilst CO possesses the reverse ordering of quenching efficiency [25,26]. Golde and co-workers have utilized this nonreactive spin-orbit selectivity to prepare nearly pure metastable argon in the J=0 or 2 levels and have investigated the spin-orbit dependence in excimer formation and chemi-ionization processes with halogen-containing compounds [27,28]. The addition of N_2 to a flow of metastable $Hg(^3P_{0,2}^0)$ atoms selectively removes the J=2 level by transfer to the radiating J=1 level [29,30] and has allowed the chemiluminescent reactions of J=0 with halogen-containing compounds to be separately studied [31].

State selection by optical methods is a general and versatile tool which is applicable to the preparation of specific radiating and nonradiating states in both beam and bulb experiments [32-35]. This technique can be used both to prepare excited levels by direct excitation or to deplete a selected nonradiating level by optical pumping. The latter has been employed to study spin-orbit effects in reactions of a number of atoms, including electronically excited metastable Ca(^{3}P$_J^o$) [36-39], Sr(^{3}P$_J^o$) [40], Ba(^{3}D$_J$) [41,42], Ne(^{3}P$_{0,2}^o$) [34,43-46] and Ar(^{3}P$_{0,2}^o$) [24,47,48].

Spin-orbit state selection by optical pumping is exemplified in Fig. 1 by the barium atom. The distribution in the spin-orbit levels of the metastable 6s5d ^{3}D term can be altered by irradiation on 5 of the lines of the 5d6p ^{3}P^o ← 6s5d ^{3}D multiplet near 600 nm. If a cw laser is tuned to one of these atomic lines, then absorption and stimulated emission will occur at a rate proportional to the laser power. The upper level can also decay spontaneously to other lower spin-orbit levels, hence transferring population permanently out of the initially pumped state. (We note that since the ^{3}P$_0^o$ upper level can radiate only to ^{3}D$_1$, the ^{3}P$_0^o$ ← ^{3}D$_1$ line in Fig. 1 is not suitable for state selection). Since a substantial fraction of the upper-state decay (whether spontaneous or stimulated) returns the atoms to the pumped level, many optical pumping cycles are required for essentially complete depletion of the initial level. We note also that the pumped atoms are transferred to the other spin-orbit levels of the ^{3}D multiplet. The power per unit bandwidth of commercially available single-mode cw dye lasers is sufficient for

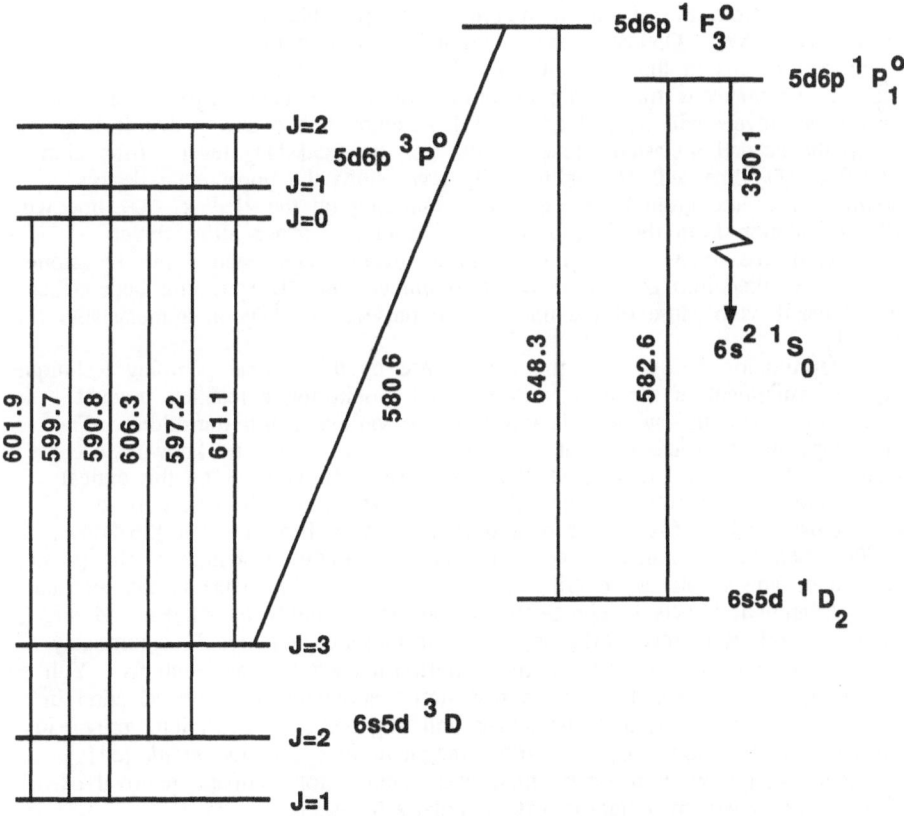

FIGURE 1. Selected low-lying electronic states of the barium atom. The wavelengths of radiative transitions are given in nm.

optical depletion down to the 1% level.

We note that for optical pumping state selection, the pump line width dictates the degree of collimation of the reactant atomic beam. In order to ensure that all atoms in a given spin-orbit level are pumped, the Doppler width must be less than the line width of the pump transition, which will usually be dominated by power broadening. The presence of more than one isotope of an atom will also limit the degree of depletion attainable, both because of isotope shifts and hyperfine splittings for nonzero nuclear spins. In the case of barium, the most common isotope is amu 138, which has nuclear spin I=0 and a natural abundance of 72% [49], while a number of the other isotopes have spin I>0 and hence significant hyperfine splittings [50]. Thus, with a single-mode laser it is not possible to deplete completely a pumped level. Weissmann *et al.* [51] have described a simple standing-wave multimode dye laser with a wide intracavity space for insertion of a molecular beam apparatus. This laser allows simultaneous pumping of all isotopes over a 10 GHz bandwidth with narrow (63 MHz) mode separation. Efficient state selection of $Ne(^3P_{0,2}^0)$, $Ar(^3P_{0,2}^0)$, and $Kr(^3P_{0,2}^0)$ was demonstrated with this arrangement [51]. Pumping with a broadband cw laser outside the laser cavity has also been employed [43].

We note from Fig. 1 that there are several possible pump transitions for a given $Ba(^3D_J)$ level. Optical depletion on these different lines will yield slightly different populations in the other unpumped spin-orbit levels of the multiplet since the relative spontaneous transition probabilities from an excited $^3P_{J'}^0$ level to the various lower levels will depend on J' [41]. Figure 1 also shows that it is possible to pump the second metastable level in Ba, i.e. the 6s5d $^1D_2$ level. Irradiation on the $^1P_1^0 \leftarrow {}^1D_2$ line will deplete the $^1D_2$ level since the upper state decays predominantly to the ground $6s^2$ ^1S state. Pumping on the $^3F_3^0 \leftarrow {}^3D_3$ line will transfer population from the $^3D_3$ to the $^1D_2$ level and hence allow direct comparison of the relative reactivities of these levels. The reaction of $^1D_2$ atoms must also be taken into account in order to unravel the ^3D spin-orbit dependence in chemical reactions because of the unavoidable presence of $^1D_2$ in a metastable Ba beam [41].

Determination of the spin-orbit dependence by this optical pumping technique requires measurement of the change in a signal monitoring a reaction product concentration when the cw laser is tuned to the various pump transitions. The relative populations of the reactant spin-orbit levels must also be known since the populations in the other, unpumped levels are usually modulated by the optical selection process. Several different schemes for extracting spin-orbit dependent reaction cross sections from these data have been described in detail [36,41,47].

One possible complication with this state preparation technique is the possible generation of anisotropic M_J distributions (coherence or alignment) in the reactant atoms. Alignment effects in chemical reactions are certainly of interest and will be discussed by others at this workshop, but their presence here would enormously complicate the extraction of meaningful spin-orbit-dependent cross sections. Yuh and Dagdigian [52] have carried out a density matrix simulation of state selection in $Ca(^3P^0)$. They concluded that coherences and alignment are negligible, principally because of the M_J mixing by the earth's magnetic field. Kroon *et al.* [34] demonstrated that a weak magnetic field was required for complete removal of $Ne(^3P_{0 \text{ or } 2}^0)$ levels by pumping with a polarized laser.

The principal limitation of optical pumping state selection, particularly to light atoms, is the relatively limited wavelength range of cw lasers. However, frequency doubled cw laser radiation is now available [53]. It is interesting to speculate on the use of pulsed laser radiation, which can be generated over a very wide wavelength range by nonlinear optical techniques, for optical depletion. Unfortunately, the state selection attainable would be incomplete because typical laser pulse widths are usually shorter than atomic radiative lifetimes.

Laser techniques have also been extensively employed for the preparation of selected excited radiating atomic states [32], mostly for the study of nonreactive collision processes. Chemical reactions of Ca $4s4p$ $^1P^o$ [54] and Na 3^2P^o, 3^2D, and 4^2S [55] have recently been studied, although spin-orbit effects were not explored in the latter experiments. Two-photon pulsed laser excitation on the J=0 and 2 levels of the $5p^56p$ manifold of xenon has recently been employed for the study of chemiluminescent reactions of these reagent states with halogen donors [56]. We also note that excitation with a resonance lamp has been used to compare the rate constants and branching fractions of the radiating $^3P_1^o$ state of Xe and Kr as compared to those for the corresponding metastable $^3P_2^o$ state [57].

Somewhat different techniques have been used for the study of spin-orbit effects in ion-molecule reactions. Photoionization is the cleanest method for the preparation of reactant ions with well characterized distributions of internal states. A major drawback of a photoionization ion source is its low intensity. This method has been employed for the study of chemical reactions, as well as nonreactive charge transfer processes, involving the $^2P_{3/2,1/2}^o$ states of the argon ion [58-62]. Irradiation of argon atoms with light of wavelengths between the $^2P_{3/2}^o$ and $^2P_{1/2}^o$ ionization thresholds allows preparation of Ar$^+$ solely in the lower $^2P_{3/2}^o$ spin-orbit level. At wavelengths shorter than the second threshold (78.7 nm), production of both levels becomes energetically allowed; the ion spin-orbit state ratio can be determined by photoelectron energy analysis [63].

In the threshold electron-secondary ion coincidence (TESICO) method, the different internal states of an ion are distinguished by detecting only those photoions in coincidence with electrons of low kinetic energy [64]. For a given photoionization wavelength, the internal energy of the ions is then determined by energy conservation. The ions are drawn out of the ionization region, allowed to react with a static target gas, and the product and unreacted reagent ions are extracted and mass analysed. This technique has principally been employed to study the vibrational state dependence of reactivity for molecular reagents. It has also been used to study the spin-orbit dependence of the Ar$^+$ + H$_2$ reaction [58] and several charge-transfer processes [59,60].

Recently, the photoionization state preparation method has been incorporated by Ng and co-workers [61,62] in a crossed-beam apparatus to study ion-molecule collisions. Argon ions generated by photoionization of a supersonic beam are extracted into a separate vacuum chamber and crossed with a second supersonic beam of the neutral reagent. Product ions are then mass analysed and detected. It is found important to maintain a low pressure in the photoionization chamber since the state purity of the reagent ion beam was found to be affected by high background pressures. This also suggests collisional equilibration in the ionization region could affect incident state selectivity in the TESICO method. In their study of the Ar$^+$ + Ar symmetric charge exchange process [61], the product ion spin-orbit state was analysed by utilizing the spin-orbit dependence of the Ar$^+$ + H$_2$ reaction.

The spin-orbit selectivity in nonreactive charge transfer processes has also been employed to investigate spin-orbit effects in ion-molecule reactions involving the heavier inert gas ions Kr$^+$($^2P_J^o$) and Xe$^+$($^2P_J^o$), whose fine-structure splittings are substantial (see Table 1). Near resonant charge transfer between CO$^+$ and Kr has been used to prepare a nearly pure Kr$^{+2}P_{3/2}^o$ beam in order to study the spin-orbit dependence of the Kr^+ + H_2 reaction in an ion cyclotron experiment [65]. The purity of the beam was checked by using the charge-transfer reaction of Kr^+ with CH_4. This process is endoergic for the lower $^2P_{3/2}^0$ state but exoergic for $^2P_{1/2}^0$ so that the amount of ionized methane product is an indication of the state purity of the ion beam. To produce beams of Kr^+ in the excited $^2P_{1/2}^0$ state, the very large difference in charge transfer rate for Kr^+ + N_2O can be utilized [66]: This process occurs rapidly for the $^2P_{3/2}^0$ state, while the excited $^2P_{1/2}^0$ state is depleted much more slowly. These collisional state selection techniques have been used to study spin-orbit effects in ion-molecule reactions in selected ion flow tube [66], drift tube [67], and guided ion beam [68] experiments.

Spectroscopic methods have been applied to the determination of spin-orbit populations of halogen atom products of simple chemical reactions. Such methods are, in principle, applicable to other atoms. Formation of the Br and I spin-orbit-excited $^2P_{1/2}^0$ state has been monitored in both time-resolved bulb [16,17] and steady-state flow [69-75] experiments by observation of the spontaneous $^2P_{1/2}^0$ → $^2P_{3/2}^0$ emission, which is relatively weak because it is electric-dipole forbidden. Because this emission is a measure only of the excited $^2P_{1/2}^0$ state production, this signal must be calibrated in some way. The relative yield of $Br(^2P_{1/2}^0)$ atoms from $I(^2P_{1/2}^0)$ + Br_2 was derived from measurement of relative emission intensities from the excited reagent and product atoms, with correction for detector wavelength sensitivity and radiative transition probabilities [17]. Additionally, correction had to be made for collisional deactivation. The Br or I $^2P_{1/2}^0$ product yield was determined in flow experiments by comparison of the halogen atom emission with that of the accompanying hydrogen halide reaction product [69-73] or from a reference reaction [74,75]. Laser gain/absorption studies with a halogen atom $^2P_{1/2}^0$ → $^2P_{3/2}^0$ chemical laser have also been employed to estimate product halogen atom spin-orbit population ratios [76].

Spin-orbit populations in halogen atom products have also been measured by time-resolved resonance absorption spectroscopy in the vacuum ultraviolet [77,78], in much the same way as the collisional loss of reactant atoms has been monitored. Laser techniques are finding increasing use in the detection of atomic products. Bromine atom product populations in the $^2P_{3/2}^0$ and $^2P_{1/2}^0$ states have been measured in a crossed beam experiment [79] by coherent vacuum ultraviolet radiation generated by nonlinear techniques [80]. There have been developed a number of schemes for the sensitive detection of atoms on the right-hand side of the periodic table by laser 2-photon excitation [81-86]. One such method has been applied to the determination of the $^2P_{1/2}^0$ and $^2P_{3/2}^0$ populations of iodine atom product from the reactions of fluorine atoms with iodine-containing molecules [87,88].

3. RESULTS FOR SPECIFIC ATOMIC REACTANTS

3.1 Halogen Atoms

Spin-orbit effects in the reactions of halogen atoms have been extensively studied, both to investigate the reactivity of ground $^2P_{3/2}^0$ and excited $^2P_{1/2}^0$ reagents as well as the relative formation rates of these states in halogen atom products. Since a halogen atom has a hole in the outer p shell and such a hole can have three orientations, halogen atom reactions necessarily involve three potential energy surfaces. An adiabatic correlation diagram in the weak spin-orbit coupling limit is displayed in Fig. 2 for the reaction with hydrogen atoms. One surface leads directly

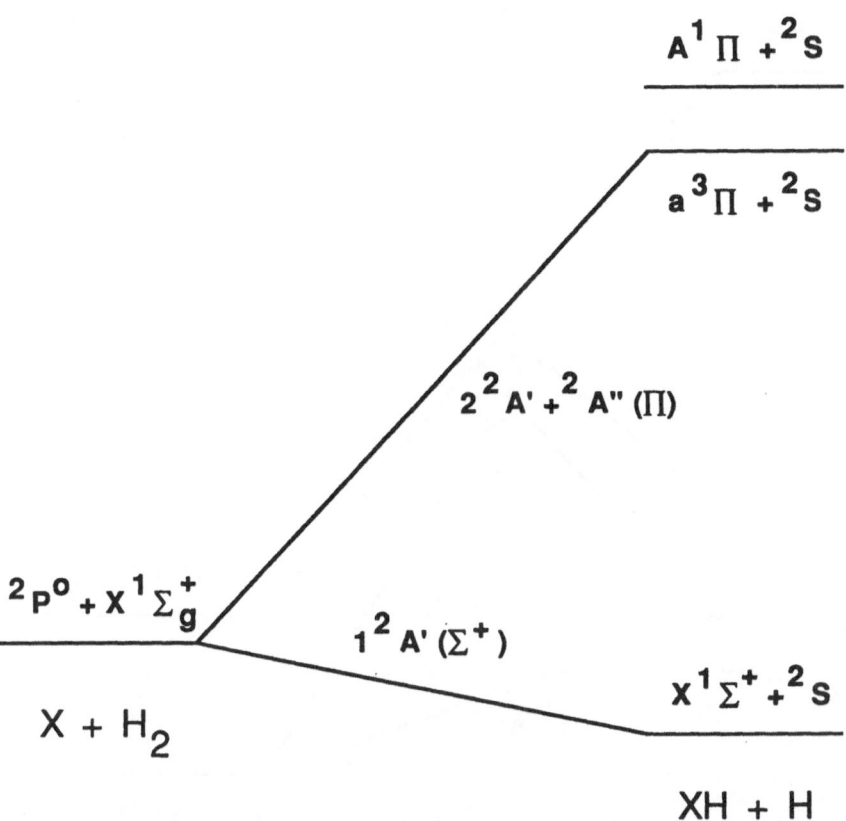

$A^1 \Pi + {}^2S$

$a^3 \Pi + {}^2S$

$2\,{}^2A' + {}^2A''\,(\Pi)$

$^2P^o + X\,{}^1\Sigma_g^+$

$1\,{}^2A'\,(\Sigma^+)$

$X\,{}^1\Sigma^+ + {}^2S$

$X + H_2$

$XH + H$

FIGURE 2. Adiabatic correlation diagram in the weak spin-orbit coupling limit for the reaction of halogen atoms with the hydrogen molecule.

to ground state products, while the other two connect to energetically inaccessible electronically excited products. When the spin-orbit interaction is included, the ground $^2P_{3/2}^o$ state correlates with the reactive $1\,{}^2A'$ and an unreactive $^2A''$ surface, while the excited $^2P_{1/2}^o$ state adiabatically correlates only with an unreactive $(2\,{}^2A')$ surface. This suggests that, if nonadiabatic effects are unimportant, the less energetic $^2P_{3/2}^o$ state will be more reactive. In a high-resolution crossed molecular beam study of the prototype $F + H_2$ reaction, Lee and coworkers [89] found no HF product attributable to the $^2P_{1/2}^o$ reaction. This pathway should have been observable since some portion of the 1.16 kcal/mol reagent spin-orbit energy should appear as product translational energy.

Figure 3 presents an adiabatic correlation diagram in the strong spin-orbit coupling limit for the reaction of a halogen atom with a hydrogen halide. (An analogous diagram can be drawn for the corresponding reaction with halogen

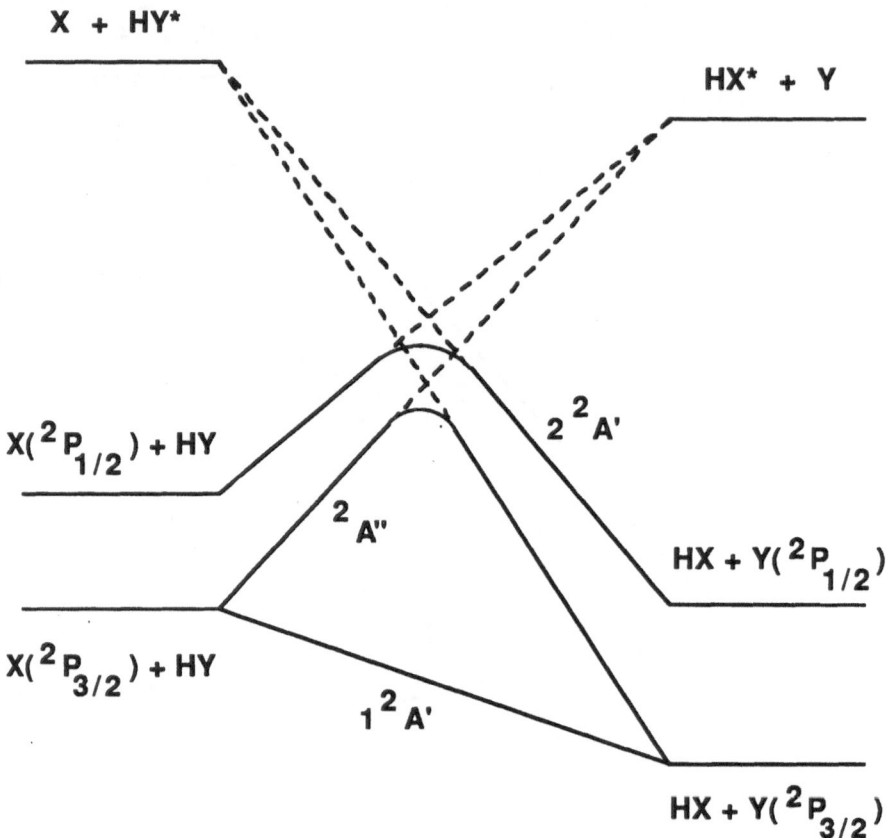

FIGURE 3. Correlation diagram in the strong spin-orbit coupling limit for the reaction of halogen atoms with a hydrogen halide. The $^2A''$ and $2^2A'$ surfaces orbitally correlate with highly excited Π states of HY and HX.

molecules). Here, all three potential energy surfaces correlate between reactant and products in their lowest electronic states. Nevertheless, two of the surfaces orbitally correlate with excited products and reactants and hence may have significant barriers, whose heights will vary with the specific reaction studied. Measurement of the halogen product spin-orbit population can provide significant dynamical information. If the reaction were purely adiabatic and the barriers relatively small, then the product spin-orbit populations should reflect those of the reactant atoms.

Significant differences in the reactivity of halogen $^2P_{3/2}^0$ and $^2P_{1/2}^0$ atoms in reactions with hydrogen halides have, in fact, been found. Bergmann, Leone, and Moore [15] found evidence that the ground state Br $^2P_{3/2}^0$ atoms are at least 4 times more reactive than $^2P_{1/2}^0$ atoms in the Br + HI reaction. Polanyi and co-workers [79] have studied the F + HBr reaction in a crossed beam experiment. The reagent F atom spin-orbit distribution was varied by changing the temperature of its source. From the observed insensitivity of the product Br atom spin-orbit distribution to the estimated F atom $^2P_{1/2}^0$ to $^2P_{3/2}^0$ ratio, they concluded that there is a substantial barrier to the adiabatic process $F(^2P_{1/2}^0)$ + HBr → HF + $Br(^2P_{1/2}^0)$ along the $2^2A'$ surface. They reached similar conclusions for the isotopic F + DBr reaction.

The relative reactivity of ground $^2P_{3/2}^0$ and excited $^2P_{1/2}^0$ halogen atoms has been studied for a number of reactions with halogen molecules by measurement of total collisional removal rates. As discussed previously, the derived bimolecular rate constants include nonreactive processes. Also, in some cases, the $^2P_{3/2}^0$ reaction is endothermic, resulting in a trivial spin-orbit dependence of reactivity. The clearest example of a spin-orbit dependence on reactivity is given by the Br + IBr → Br$_2$ + I reaction. In a recent study of Haugen, Weitz, and Leone [19] using transient F-centre laser absorption spectroscopy, the collisional loss rate for ground-state $Br^2P_{3/2}^0$ atoms was found to be approximately 20 times greater than that for the $^2P_{1/2}^0$ state. Their derived bimolecular rate constants were consistent with those determined in several previous studies [17,90].

The spin-orbit distribution in halogen atom products has also been measured. As can be seen from Fig. 2, the formation of excited $^2P_{1/2}^0$ products will depend on the importance of the adiabatic pathway on the $2^2A'$ surface between spin-orbit-excited reactants and products. Whilst there is considerable variation in the observed product spin-orbit distribution, in general it is found that the lower $^2P_{3/2}^0$ state is preferentially produced, in some cases almost exclusively. In the F + HX reaction, the product atom $^2P_{1/2}^0$ to $^2P_{3/2}^0$ population ratio ranges from 0.10 [91] and approximately 0.07 [72,73,79] for X=Cl and Br, respectively, to <0.01 [69,73,87] for X=I. For the latter reaction, the reported ratio of *ca.* 0.5 by Burak and Eyal [78] appears to be in error since it disagrees with the results of these other experiments [69,73,87]. The observed spin-orbit ratio for the HCl and HBr reactions is close to the room-temperature Boltzmann ratio of F atom spin-orbit states. This similarity has been assumed to imply that the excited $^2P_{3/2}^0$ products were formed on the $2^2A'$ surface from excited $^2P_{1/2}^0$ reactants [76,91]. However, this interpretation disagrees with the observations of Hepburn *et al.* [79] on the F + HBr reaction. They find no dependence of the Br product spin-orbit population ratio on that of the F atom reactant. They conclude that $Br(^2P_{1/2}^0)$ products are formed by nonadiabatic transitions from the $1^2A'$ to the $2^2A'$ surface in the exit channel.

Trickl and Wanner [92] observed bimodal IF vibrational distributions from the F + I$_2$ reaction. The two peaks in the distribution were inferred to be due to branching into the two product iodine spin-orbit states since trajectory calculations [93] on LEPS surfaces could not reproduce the low v component. More recently, the I atom spin-orbit distribution has been directly measured by infrared emission [74,75] and 2-photon laser fluorescence excitation [87]. Whilst there is some disagreement as to the exact $^2P_{1/2}^0$ to $^2P_{3/2}^0$ population ratio, it is clear that it is at most several percent, which is much too small to explain the two peaks in the IF vibrational distribution.

One example of a halogen atom with a halogen molecule reaction is known where adiabatic formation of spin-orbit excited $^2P_{1/2}^0$ halogen atom product on the $2^2A'$ surface appears to be efficient, namely the $I(^2P_{1/2}^0)$ + IBr → IBr + $Br(^2P_{1/2}^0)$ reaction. Houston [16], Wiesenfeld and Wolk [22,77], Spencer and Wittig [94], and Gordon *et al.* [76] demonstrated a high correlation of product $Br(^2P_{1/2}^0)$ with reactant $I(^2P_{1/2}^0)$. Wiesenfeld and Wolk [22,77] have determined branching ratios for the reactive vs. nonreactive channels by vacuum ultraviolet absorption spectroscopy and found that *ca.* 80% of $I(^2P_{1/2}^0)$ reactive collisions yielded

$Br(^2P_{1/2}^{0})$ product. The studies of Spencer and Wittig [94] and Gordon *et al.* [76], both using the laser gain/absorption technique, also found high branching ratios. Gordon *et al.* [76] also presented evidence for a strong correlation between reactant and product halogen atom spin-orbit-excited states for the F + Br_2 reaction. They found that the Br product spin-orbit population ratio was insensitive to the photolytic F atom precursor used.

The I + Cl_2 reaction is endoergic for ground-state I atom reagent. In contrast to the analogous reaction with Br_2, discussed above, the quenching rate constant for $I(^2P_{1/2}^{0})$ atoms is found [95,96] to be extremely small (*ca.* 6×10^{-15} cm^3 molecule^{-1} s^{-1}). The branching ratio for reactive quenching is approximately one half. Lilenfeld *et al.* [95] present evidence that the primary Cl product is the ground $^2P_{3/2}^{0}$ state. These results indicate that for this reaction there is a significant barrier along the $2^2A'$ surface and a small transition probability for nonadiabatic transitions to the $1^2A'$ and $^2A''$ surfaces. An intriguing series of reactions is the gas-phase Walden inversion F + RI $\rightarrow$ FR + I, for which the product I atom spin-orbit distributions have been measured by Bersohn and co-workers [88]. For R=CH_3, the I atom $^2P_{1/2}^{0}$ to $^2P_{3/2}^{0}$ population ratio was found to be 0.42 $\pm$ 0.14, whilst this ratio decreased for larger alkyl groups: 0.32 $\pm$ 0.13 for R=C_2H_5 and 0 for *i*-C_3H_7. Such high $^2P_{1/2}^{0}$ to $^2P_{3/2}^{0}$ product ratios for the lighter alkyl groups are surprising, in view of the results for the homologous hydrogen halide reactions.

3.2 Inert Gas Ions

Because the inert gas ions are isoelectronic with the halogen atoms, it is interesting to compare the collisional behaviour of these two classes of reactants. Spin-orbit effects have been found in the Rg^+ + H_2 $\rightarrow$ RgH^+ + H ion-molecule reaction, as well as in several nonreactive charge transfer processes. Tanaka *et al.* [58] have found that for the Ar^+ + H_2 reaction, the excited $^2P_{1/2}^{0}$ state of the ion has a 50% larger cross section for reaction (independent of the collision energies investigated) than does the ground $^2P_{3/2}^{0}$ state. This confirms the results of an early photoionization study of this reaction by Chupka and Russell [97]. By contrast, an opposite ordering of reactivity is found [65,68] for the Kr^+ + H_2 reaction; here the reaction cross section is about 2.6 times larger for reactant ions in the $^2P_{3/2}^{0}$ state than for the $^2P_{1/2}^{0}$ level. This result for the Kr^+ reaction is consistent with that found for the analogous halogen atom reactions and can be explained by adiabatic correlation arguments in the strong spin-orbit coupling limit. The opposite ordering of reactivity for the Ar^+ reaction has been explained [58], in part, by the mediation of the charge transfer potential energy surface in this ion-molecule reaction: The reactants Ar^+ + H_2 diabatically correlate with Ar^+ + H + H fragments, while the products ArH^+ + H correlate with Ar + H^+ + H [98,99], so that a change in electron occupancy is required for reaction. Similar considerations apply to the Kr^+ reaction; however, because of the difference in the separation of the Rg^+ + H_2 and Rg + H_2^+ asymptotes, the coupling of these charge states will be less efficient for Kr^+ than for Ar^+ [68]. Because of the smaller spin-orbit splitting in Ar^+ than in Kr^+, the asymptotic spin-orbit states would be expected to be mixed significantly by the approach of the H_2 collision partner to the former. Tanaka *et al.* [58] considered such nonadiabatic mixing and developed a surface-hopping classical trajectory model to explain the observed spin-orbit dependence in the Ar^+ + H_2 reaction.

Spin-orbit effects have also been found in charge-transfer collisions involving inert gas ions. In symmetric charge exchange in Ar, Kr and Xe, the $^2P_{1/2}^0$ state of the ion has a slightly smaller cross section than the $^2P_{3/2}^0$ state [60,61,100]. The ratio of these cross sections approaches unity at both high and low collision energies, while the broad minimum in the ratio occurs in the energy region where the kinetic energy is comparable to the spin-orbit splitting. The $^2P_{1/2}^0$ state of Ar+ has been found to have a considerably smaller charge transfer cross section in collisions with N_2 than does the $^2P_{3/2}^0$ state [59,62]. In a recent crossed beam photoionization study [62] the ratio of these cross sections was found to reach a minimum of 0.2 at E_{lab} near 10 eV, with higher values of this ratio at lower and higher collision energies. A similar ordering of reactivity was found in Ar+ + CO charge transfer collisions [59]. A semiclassical calculation [101] of the Ar+ + N_2 charge transfer cross section also finds a smaller reactivity for $^2P_{1/2}^0$ vs. $^2P_{3/2}^0$.

3.3 Electronically Excited Inert Gas Atoms

Because of their high internal electronic energy (see Table 1), many decay channels are available in collisions of electronically excited inert gas atoms with other species [102,103]. These include various nonreactive pathways such as ionization (Penning, associative, and dissociative), excitation transfer, dissociative excitation, as well as a "chemical" deactivation channel, namely excimer formation,

$$Rg* + RX \rightarrow RgX* + R, \tag{1}$$

which forms the basis of operation for the inert gas excimer laser [104]. Spin-orbit effects have been found both in the excimer formation channel and in a number of nonreactive channels. Because of the large number of these decay channels, a comprehensive description of the dynamics of excited inert gas collisions has not been formulated. Most theoretical modelling has focused on the entrance channel.

The dependence on incident spin-orbit level of the excimer formation channel in reactions of halocarbons [27] and molecular fluorine [48] illustrates the role of the inner-shell electrons in influencing the excimer product electronic state distribution. The formation of these excited inert gas halide ionic states is believed to proceed through an electron-jump mechanism involving a Rg^+RX^- intermediate because of the low ionization potential of the excited inert gas reagents [102,105]. This channel competes with direct energy transfer from the excited atom to the RX collision partner. The inert gas halide products can be formed in three molecular electronic states: $B(\Omega=1/2)$ and $C(\Omega=3/2)$, which diabatically correlate to $Rg^+(^2P_{3/2}^0) + X^-$, and the $D(\Omega=1/2)$ state, which dissociates to $Rg^+(^2P_{1/2}^0) + X^-$. Four strong emission bands are expected in the radiative decay of these states to mainly repulsive covalent states: B-X, B-A, C-A, and D-X [106]. The strongest system observed in the reaction of $Ar(^3P_2^0)$ with halide reagents is the B-X continuum, with weaker B-A and C-A emission at longer wavelengths [27,48]. By contrast, significant D-X emission is observed for the corresponding $Ar(^3P_0^0)$ reaction. While both the $^3P_2^0$ and $^3P_0^0$ states have an outer 4s electron, which is transferred to RX by the electron jump, the angular momentum of the ionic core is j=3/2 and 1/2, respectively, for these states. In the absence of core switching effects, $^3P_2^0$ reagents would be expected to form $Rg^+(^2P_{3/2}^0)X^-$ ionic products in the B and C states, as is found [27,48]. Similarly, the $^3P_0^0$ reactions should yield excimer molecules solely in the D state. The experimentally observed accompanying formation of the B state is ascribed to core switching at the crossing of the $Rg+(^2P_{1/2}^0)X^-$ ionic surface with covalent surfaces correlating to the lower-energy $^3P_1^0$ and $^3P_2^0$

reagents. Variation in the production rates of excited $np^4(n+1)s$ $^{2,4}P_J$ states of Cl and Br atoms produced by dissociative excitation has also been observed in $Ar(^3P_{0,2}^0)$-halocarbon collisions [27]. This channel has been postulated as proceeding through predissociation of the RgX* product. The dependence of excited halogen atom production upon incident inert gas spin-orbit state is thus related to differences in the formation rate of the excited RgX* states.

The excimer formation channel has also been compared for Kr and Xe radiating $^3P_1^0$ and metastable $^3P_2^0$ reactants [57,105]. The product emission spectra are fairly similar for these atomic states for the halide donors studied. The principal variations are an extension of the B-X system to shorter wavelengths and evidence of greater C state product vibrational excitation in the C-A system for the $^3P_1^0$ reactant. The additional electronic energy of the $^3P_1^0$ state (see Table 1) can explain these differences. The excimer formation rate constants have also been compared for $Xe^3P_2^0$ and $^3P_1^0$ reactions with several halide donors; some differences in these rate constants were found [107,108]. The excimer formation channel has also recently been investigated for several of the higher excited $Xe(5p^56p)$ states [56]. The branching fraction into this channel was found to be significantly enhanced (by a factor of $ca.$ 40 in the case of HCl) over that observed for the lower $Xe(5p^56s)$ states, even though the total quenching rate constants are only somewhat larger for the former (except for NF_3).

Significant spin-orbit effects have also been observed in a number of the excited inert gas nonreactive collisional decay channels. In excitation transfer collisions between metastable argon and hydrogen atoms,

$$Ar(^3P_{0,2}^0) + H \rightarrow Ar + H^*(n=2), \qquad (2)$$

Sadeghi and Setser [47] found that the higher energy $^3P_0^0$ reactant has a factor of $ca.$ 10 smaller rate constant for this process than does $^3P_2^0$. By considering the Ω-state correlations between the excited ArH states and the separated atomic states, this selectivity was explained [47] as arising from the fact that the $Ar(^3P_2^0)$ + H entrance channel can reach the attractive $D^2\Pi$ and $E^2\Sigma^+$ ArH states, which can lead to H*(n=2) production by curve crossing with the $B^2\Pi$ and $C^2\Sigma^+$ states. By contrast, the $Ar(^3P_0^0)$ + H channel correlates with a repulsive ArH $^4\Pi$ curve.

Significant differences have been found in the vibration- rotation distribution of excited $N_2(C^3\Pi_u)$ molecules formed by excitation transfer from $Ar(^3P_{0,2}^0)$ to nitrogen molecules [109,110]. A somewhat hotter vibrational distribution is found for the $^3P_0^0$ incident level; from variable temperature studies, this is ascribed, at least in part, to energy resonance effects [110]. Unequal $N_2(C)$ spin and Λ doublet populations, which also depended on the Ar* initial level, were deduced from high-resolution studies of the C-B emission [109]. These variations gave rise to an even-odd N alternation in the rotational populations and were explained as arising from dynamical constraints, including the dominance of planar ArNN collisions and the preferential population of Λ doublets whose electronic distribution is symmetric with respect to reflection in the plane of rotation.

Spin-orbit effects have also been investigated in the ionizing excited inert gas collisions. Golde and Ho [111] found that the branching fraction f_{ion} for chemi-ion formation in collisions of $Ar(^3P_J^0)$ with halogen donors was higher for J=0 than for J=2. These differences could be explained by the higher energy available in collisions of the former. It appears that f_{ion} increases sharply as the reagent energy becomes larger than the threshold for Penning ionization. A similar behaviour of the spin-orbit dependence of the ionization cross section was also found in collisions of

$Ne(^3P_{0,2}{}^0)$ with N_2 and the heavier inert gases [44]. A significant spin-orbit dependence has also been observed in the branching fraction f_{AI} for associative ionization in $Ne(^3P_{0,2}{}^0)$ collisions with Ar and has been rationalised by a variation in the formation rate of the $NeAr^+X^2\Sigma_{1/2}{}^+$ and $A^2\Pi_{3/2,1/2}$ states [43]. The J=0 state has a larger f_{AI} because this reagent state apparently produces X state products preferentially. Since the X molecular ionic state is the most attractive [112], formation of this state yields the largest f_{AI}.

The translational energy distribution of electrons produced in Penning and associative ionization collisions has been measured by Hotop and co-workers [45,46,113,114] for state-selected $Ne(^3P_{0,2}{}^0)$ and higher $Ne(2p^5 3p)$ states. The total ionization cross section in collisions of $Ne(^3P_J{}^0)$ with H and D atoms was found to be smaller for J=0 than for J=2, consistent with the previously discussed results [47] for the $Ar(^3P_J{}^0)$ + H excitation transfer channel. As expected for an attractive interaction for J=2 incident level, a very broad electron energy distribution was found for this level, while the distribution was very narrow for the primarily repulsive collisions of J=0. Similar observations were reported for ionizing collisions with alkali atoms [114]. A very interesting anomalous fine-structure branching has been found in $Ar^+(^2P_{3/2,1/2}{}^0)$ products formed in $Ne(^3P_{0,2}{}^0)$-Ar Penning ionizing collisions [45]. The Ar^+ spin-orbit states are easily resolved in the electron energy distribution. The J=3/2 to 1/2 ratio was found to be 3.94 and 1.51 for $^3P_0{}^0$ and $^3P_2{}^0$ reagents, respectively, while a statistical distribution would give a ratio of 2. In an elegant analysis, Morgner [115] showed that, within the exchange model for Penning ionization, the Ar^+ fine-structure branching could be explained by an interference between the dominant Ar $3p\sigma \rightarrow$ Ne $2p\sigma$ amplitude and a minor $3p\pi \rightarrow 2p\pi$ component.

3.4 Metastable Alkaline Earth Atoms

In our laboratory, a series of studies of spin-orbit effects in reactions of alkaline earth atoms has been carried out [36-42]. Optical pumping state selection has been employed in order to determine the relative reactivity of the individual spin-orbit levels of metastable $Ca(^3P^0)$, $Sr(^3P^0)$, and $Ba(^3D)$ atoms. For most of the reactions studied, the products can be formed in the ground electronic state,

$$M^* + RX \rightarrow MX(X^2\Sigma^+) + R, \qquad (3)$$

or in radiating excited states, producing visible chemiluminescence,

$$M^* + RX \rightarrow MX(A^2\Pi, B^2\Sigma^+, C^2\Pi) + R. \qquad (4)$$

In some cases, a chemi-ionization channel, yielding $MX^+ + R^-$, is energetically allowed. Finally, several nonreactive decay pathways, including collisional fine-structure mixing and quenching to the ground ^1S state, are available; the latter channel has not been studied.

A significant spin-orbit dependence was observed for the chemiluminescence channel for just about all the reactions of metastable alkaline earth atoms with diatomic and polyatomic oxidants investigated. Reactions studied include $Ca(^3P^0)$, $Sr(^3P^0)$, $Ba(^3D)$ + Cl_2, Br_2, and several alkyl halides, including CH_3X, CH_2X_2, with X=Br and I, and allyl and benzyl bromide [36,38-41]. Because of the congestion in the spectra of alkaline earth halides, the large number of rovibrational levels populated in these reactions, and the relatively low spectral resolution in these experiments, spin-orbit effects were determined for total production of the A, B, and, in some cases, C states of the excited metal halide products.

The alkaline earth atoms studied all have a second metastable level (^{1}D) whose internal energy is slightly higher than that of the metastable triplet level [12]. For most of the reactions investigated, the ^{1}D reaction yields significant chemiluminescence. (In fact, for some reactions, e.g. M* + CH$_3$Cl, all of the observed chemiluminescence arises from this source [38]). For Ca and Ba, it was possible to correct for the presence of this reaction by optical pumping depletion of the ^{1}D level [38,41].

For the reactions listed above, the highest energy reactant spin-orbit level (^{3}P$_2$^{o} for Ca and Sr, ^{3}D$_3$ for Ba) was found to have the largest chemiluminescence cross sections. In all cases, the lower J reactant atomic levels possessed successively smaller values. Within the relatively large uncertainties, the ratio of the cross section for the highest J level to that for the lowest-energy spin-orbit level was found to be a factor of 5 to 10, independent of reactant metal atom and halide reactant. Several reactions were found to have no detectable spin-orbit dependence of the chemiluminescence cross sections. These include Ba(^{3}D) + N$_2$O, NO$_2$ [41], Ca(^{3}P^o) + SF$_6$ [39], and the CaX(C) channel for Ca(^{3}P^o) + Cl$_2$, Br$_2$ [36]. The chemi-ionization channel for the last two reactions was also investigated and no spin-orbit dependence was found [36].

It is also of greater interest to measure the spin-orbit dependence of the ground state product channel, Eq. 3. For several reactions, the chemiluminescence signal was absent or sufficiently small so that laser fluorescence detection of ground state products could be carried out, and the spin-orbit dependence of this channel determined. An opposite ordering of reactivity from that for the chemiluminescence channel was observed for the Ca(^{3}P^o) + Cl$_2$ [37] and Sr(^{3}P^o) + HBr, CH$_2$Br$_2$ [40] reactions; namely the lowest energy J=0 level had the largest cross section for production of ground state halide product.

Recently, we have been able to determine the spin-orbit dependence for the ground state channel as a function of vibrational quantum number v for the reaction of Ba(^{3}D) with HX and CH$_3$X, where X=Cl and Br [42]. This contrasts with the studies of the above- mentioned Ca(^{3}P^o) and Sr(^{3}P^o) reactions, where the vibrational level(s) detected could not be identified from the excitation spectra. For these Ba(^{3}D) reactions, the spin-orbit level with the lowest energy (^{3}D$_1$) was found to have the largest cross sections. The higher ^{3}D$_J$ levels had successively smaller cross sections. For the HCl and HBr reactions, the spin-orbit effect was found to vary with the product vibrational level detected. For the highest product levels investigated (v=30 and 40, respectively), the ratio of reactivity for the J=1 to J=3 levels was a factor of 3 to 4, whilst essentially no difference in reactivity was observed for the lowest product levels [42].

These Ba reactions fall into a class of kinematically constrained reactions H + H'L $\rightarrow$ HH' + L, where H and L denote heavy and light atoms, respectively [116,117]. One consequence is that initial orbital angular momentum is channelled into rotational angular momentum of the diatomic product. With the assumption of constant product recoil energy, which can be used to interpret the dynamics of a number of Ba(^{1}S) reactions [118], the formation of low and high v product vibrational levels is associated with large and small impact parameters, respectively. Thus, the variation of the spin-orbit effect with product vibrational level for the Ba(^{3}D) reactions provides information on the dependence of the reaction dynamics on incident impact parameter.

Despite the large reaction cross sections expected for these alkaline earth

163

reactions, nonreactive intramultiplet mixing was observed in Ca(^{3}P$_J^o$) collisions [36,38,39]. When the ^{3}P$_J^o$ level was optically pumped, the ^{3}P$_1^o$ → ^{1}S intercombination emission line was found to increase initially with increasing target gas density. These data were analysed with the help of a simple kinetic model to estimate cross sections for collisional transfer from the ^{3}P$_1^o$ to the nonradiating ^{3}P$_{0,2}^o$ levels. For some reactions, these were found to be sizeable [38].

Although adiabatic correlation arguments [10] could, in principle, be used to understand the observed spin-orbit effects, such considerations have been shown [119-123] to be poor predictors for the product electronic state branching in chemiluminescent alkaline earth reactions. Because of the relatively low ionization potentials of the excited alkaline earth atoms, their reactions proceed by charge transfer to an ionic surface [116]. As the reactants approach but before charge

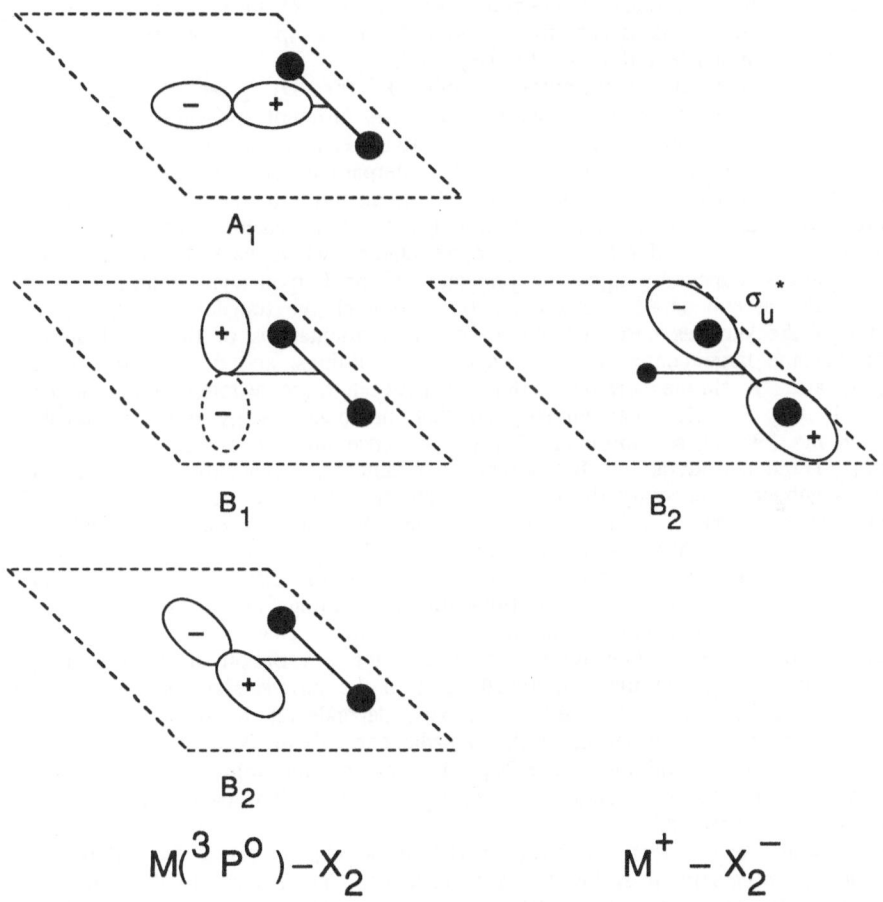

FIGURE 4. Comparison of the p orbital orientation for covalent surfaces arising from the interaction of ^{3}P^o alkaline earth atoms with halogen molecules. Illustrated on the right is the σ_u^* orbital of X$_2^-$ for the lowest M$^+$-X$_2^-$ ionic surface.

transfer has occurred, several potential energy surfaces corresponding to different orientations of the metal valence p or d electron will be accessible. In general, one expects that charge transfer will occur with differing efficiency from these different covalent surfaces. For example, for reactions of $^3P^0$ atoms with the diatomic halogens, the favoured molecular orientation for reaction is a broadside C_{2v} geometry [36,54,123,124]. For this orientation, charge transfer is symmetry allowed only from the $^3B_2$ surface [36,54,123] as can be seen from Fig. 4. We see there is good overlap between the metal atom p orbital and the lowest unoccupied σ_u* halogen molecule orbital. *Ab initio* calculations [124] on the Ca-Cl$_2$ system also indicate that $^3B_2$ is the lowest energy covalent surface correlating with the Ca($^3P^0$) asymptote.

The effect of alignment of the metal atom electron has been probed by Rettner and Zare [54] in somewhat more direct fashion in the reaction of Ca(4s4p $^1P^0$) with Cl$_2$ and other reactants. Aligned Ca($^1P^0$) was prepared by polarized laser excitation. For the Cl$_2$ reaction, chemiluminescence and chemi-ionization cross sections were found to be largest for π approach of the 4p electron, suggesting that these channels are reached through the $^1B_2$ surface.

The correlation of the asymptotic atomic $^3P^0$ (or ^3D) spin-orbit levels with these potential energy surfaces is complicated. The different spin-orbit levels are distinguished by the coupling of the electron orbital **L** and spin **S** angular momenta to form the resultant **J**. The coupling to the interparticle axis is weak, and the p electron asymptotically will have no preferred orientation. In spectroscopic language, this situation is described by a Hund's case (e) coupling scheme [125]. At smaller separations, we can consider a case (c) representation, where each J level is split into components, depending upon the projection Ω of J upon some molecule-frame axis. Finally, at still smaller separations but before charge transfer occurs, the splittings of the surfaces corresponding to different orientations of the metal atom electron become large compared to the spin-orbit splittings, and a case (a) coupling is appropriate. A simple correlation diagram between these different representations is given in Fig. 5. This diagram suggests that the lower energy spin-orbit levels can reach the $^3B_2$ surface more efficiently and hence undergo charge transfer and reaction. These considerations thus provide a qualitative explanation for the observed [37,40,42] enhanced reactivity of the lowest spin-orbit levels to form products in the ground electronic state. In Section 4, we discuss the quantum calculations of Alexander [126], who has developed a quasi-diatomic pseudo-quenching model for a more quantitative understanding of spin-orbit effects in the Ca($^3P^0$) + Cl$_2$ reaction.

Because of the relatively small fine-structure splittings in the alkaline earth atoms (see Table 1), nonadiabatic mixing in the entrance channel will scramble the incident spin-orbit levels. Our recent results showing a variation of the spin-orbit effect as a function of product vibrational level in the ground state product channel for reactions of Ba(^3D) suggests that the mixing depends on the incident impact parameter. Another manifestation of the nonadiabatic mixing is nonreactive collisional intramultiplet mixing [36,38,39]. In view of our original expectations about spin-orbit effects, it is actually surprising that such significant spin-orbit dependences are observed.

We would expect that symmetry restrictions to charge transfer would be less important for polyatomic than for diatomic reactants. Yet essentially the same relative reactivity of incident spin-orbit levels is observed. At least for methyl halide reagents, similar symmetry constraints can be invoked [39] for a collinear M-X-C approach geometry: Of the two covalent surfaces, A_1 and E, which arise in C_{3v} symmetry, charge transfer is symmetry allowed only for the former.

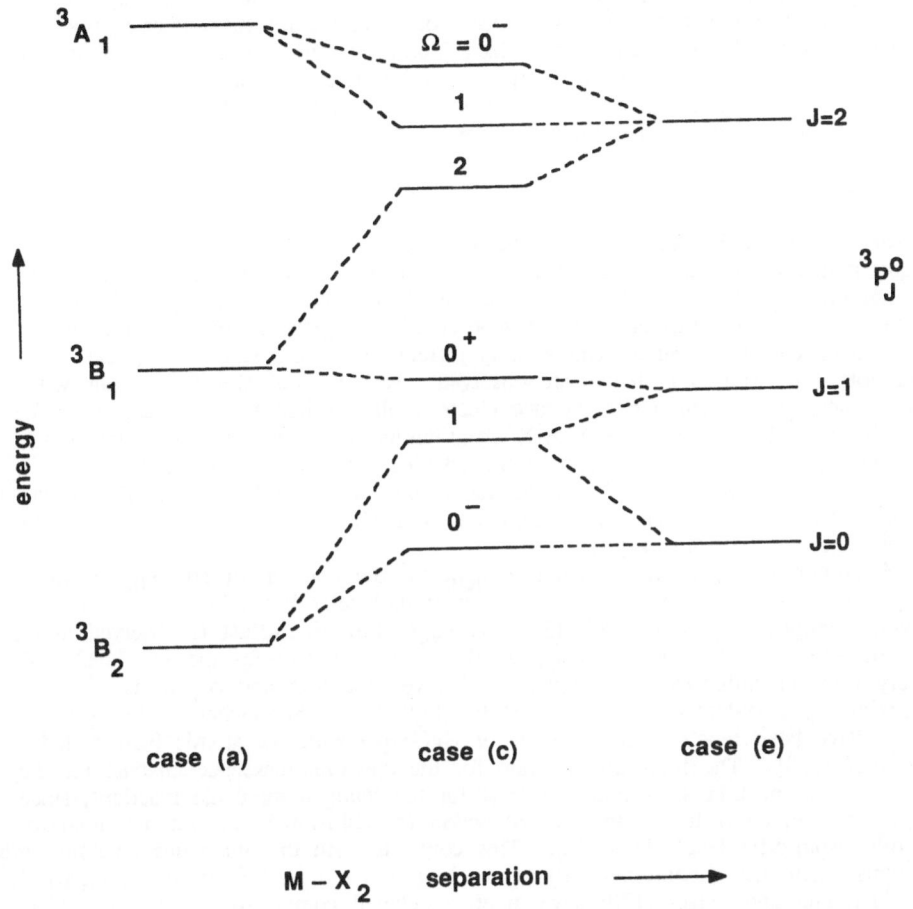

FIGURE 5. Correlation diagram for the interaction of a $^3P^o$ alkaline earth atom with a halogen molecule in the preferred C_{2v} approach geometry, illustrating the correlations between the case (e) states at large separations with the case (a) electrostatic surfaces.

The chemiluminescence channels, Eq. 4, were found to display an opposite ordering of reactivity from that of the ground state product channel, Eq. 3. The former requires charge transfer to excited ionic surfaces, which can be reached by diabatic traversal through the outermost ionic-covalent crossing [36,41,123]. Because of the multiplicity of excited $M^+(p)$-X_2^- and $M^+(d)$-X_2^- ionic surfaces, there are no symmetry constraints for charge transfer at the inner ionic-covalent crossings [36,41,123]. The larger chemiluminescence cross sections for the higher-energy reactant spin-orbit levels is thus rationalized [36,37] by the selective removal of flux at the outer crossing of the lower-energy spin-orbit levels: Since the latter react more efficiently to yield ground state products, there is less probability for these reactants to access the excited ionic surfaces.

3.5 Metastable Mercury Atoms

Mercury atoms are isoelectronic with the alkaline earths; however, the spin-orbit mixing in the 6s6p manifold is substantial (see Table 1), and the energy separations between the $^3P_J{}^o$ spin-orbit levels are comparable to electronic energy differences. Because of the large differences in reagent atomic energies, nonadiabatic mixing in the entrance channel is small, and striking differences in the reactivity of the metastable $^3P_0{}^o$ and $^3P_2{}^o$ levels are observed. Reactions of $Hg(^3P_J{}^o)$ atoms with various halogenated compounds produce electronically excited mercury halides in the $B^2\Sigma^+$ state, yielding intense HgX B-X chemiluminescence, in analogy to reaction (4). The principal difference between the alkaline earth and the mercury chemiluminescence spectra is the bound-free nature of the HgX spectra [127] because of the small dissociation energy of ground state HgX; thus, the latter are more akin to the inert gas reactions, Eq. 1.

Chemiluminescence cross sections σ_{chem} for the reaction of $Hg(^3P_{0,2}{}^o)$ with halogen molecules have been measured in both molecular beam [128,129] and flowing afterglow experiments [31,130]. A large spin-orbit effect is observed in the Cl_2 and Br_2 reactions. Here σ_{chem} is 100 $Å^2$ and greater for the J=2 level; only a very weak chemiluminescence signal is observed for J=0, and σ_{chem} is approximately 2 orders of magnitude smaller [128-131]. Somewhat smaller values of σ_{chem} have been reported for reactions of $Hg(^3P_2{}^o)$ with polyatomic halogenated molecules [128]. The branching fraction for the chemiluminescence channel for the $^3P_2{}^o$ reactant must be near unity, at least for the halogen molecule reactions, since σ_{chem} is comparable to the total cross section for collisional removal measured for the other spin-orbit levels [131,132]. This contrasts with the analogous alkaline earth reactions where the chemiluminescence branching fraction is 30% or less [36,38,123].

Dreiling and Setser [130] have given a detailed comparison of the $Hg(^3P_J{}^o)$ reactions with the corresponding alkali and metastable excited alkaline earth and inert gas reactions. The large chemiluminescence branching fraction for $^3P_2{}^o$ is consistent with negligible nonadiabatic mixing in the entrance channel: The higher energy $^3P_2{}^o$ reactant can efficiently reach excited potential energy surfaces correlating with HgX(B) product, while the lower spin-orbit states yield only non-emitting products.

4. THEORETICAL MODELLING

In order to investigate in detail theoretically the dependence on reagent spin-orbit level of cross sections involving an open-shell atom, it is necessary to use a quantum mechanical treatment of the dynamics, or at least a semiclassical method which takes into account the multiplicity of potential energy surfaces and the possible couplings between them, in the problem. The dependence of reactivity upon spin-orbit level is one of a large class of electronically nonadiabatic processes. This type of collision has evoked considerable interest. For general summaries of work in this field, we refer to review articles by Tully [133] and Rebentrost [134].

Quasiclassical surface-hopping calculations using the method of Tully and

Preston [135] have been employed to describe charge transfer collisions of alkali atoms with oxygen and halogen molecules [136]. Similar semiclassical scattering calculations have also been carried out for $F(^2P_J{}^0) + H_2$ reactive collisions [137,138]. However, such approaches to treat nonadiabatic processes, and in particular spin-orbit effects involving higher spin multiplicity atoms, may miss important dynamical features because the electronic and nuclear degrees of freedom are treated on a different footing [137]. In addition, Landau-Zener and other surface-hopping models are expected to have difficulties with nonadiabatic processes that do not necessarily involve isolated curve crossings, as is the case here. Miller and coworkers [139,140] have shown how to construct a completely classical Hamiltonian from a matrix representation of a finite-level quantum system. The formalism for a 2-level system has been applied to treat successfully nonadiabatic $F(^2P_J{}^0)$ collisions [140]. There appears to be some ambiguity in constructing such a classical model for an arbitrary number of electronic states or for a system where modelling the electronic degrees of freedom with only one classical degree of freedom is unrealistic.

Fully quantum calculations of cross sections for nonadiabatic reactions are extremely complex [141], and no actual calculations have been reported. The calculations actually carried out fall into two classes, those involving a reduced number of degrees of freedom (collinear [142], quasidiatomic [126]) and those which treat a nonreactive channel only (intramultiplet mixing [143-147]). In such treatments, it is useful to work in an electronically diabatic representation for convenience in propagating the scattering wave function out to the asymptotic region [143]. In such a basis, the electronic Hamiltonian will have off-diagonal matrix elements, as illustrated in Fig. 6 for $^3P^0$ alkaline earth atom - homonuclear diatomic interactions in C_{2v} and general C_s symmetry. We note all the matrix elements shown are in fact functions of the nuclear coordinates. These diabatic matrix elements can be obtained from electronically *adiabatic* energies and wave functions, as would be given by an *ab initio* calculation, provided matrix elements of derivatives with respect to nuclear coordinates are also computed [143,148]. In some model calculations, these matrix elements were estimated semi-empirically [126] or were calculated by an approximate electronic structure theory, such as diatomics-in-molecules [137,144] or a pseudopotential model [147,149]. For convenience in the scattering calculations, these electrostatic matrix elements, as well as the atomic spin-orbit coupling and nuclear, kinetic and rotational energy terms must all be transformed to a case (e) basis [126,141,143,147]. In this representation, the total Hamiltonian for the asymptotically separated partners is diagonal.

Several 3-dimensional studies of the nonreactive quenching of $F(^2P_{1/2}{}^0)$ in collisions with H_2 have been carried out [143-145]. This inelastic scattering process is dominated by near resonant electronic-to-rotational energy transfer,

$$F(^2P_{1/2}{}^0) + H_2(j=0) \rightarrow F(^2P_{3/2}{}^0) + H_2(j=2). \qquad (5)$$

A collinear calculation [142] of the $F + H_2$ reaction shows that the ground $^2P_{3/2}{}^0$ state is *ca.* 10 times more reactive than the excited $^2P_{1/2}{}^0$ state, as expected from the discussion in Section 3.1. Similar spin-orbit dependences were found in semiclassical scattering calculations [137,138] of this reaction.

Alexander [126] has taken a somewhat different approach to model in a fully quantum mechanical fashion spin-orbit effects in reactions of $^3P^0$ alkaline earth atoms with halogen molecules. Here the reactive encounter is simulated by a one-dimensional atom-atom collision in which the diabatic covalent curves correlating

C_{2v} symmetry

	covalent		ionic
$^3A_1$	$^3B_1$	$^3B_2$	$^3B_2$
$V_c(A_1)$	0	0	0
0	$V_c(B_1)$	0	0
0	0	$V_c(B_2)$	V_{ic}
0	0	V_{ic}	$V_i(B_2)$

C_s symmetry

	covalent		ionic
$1\,^3A'$	$^3A''$	$2\,^3A'$	$^3A'$
$V_c(1A')$	0	$V_{cc}(A')$	$V_{ic}^{(1)}$
0	$V_c(A'')$	0	0
$V_{cc}(A')$	0	$V_c(2A')$	$V_{ic}^{(2)}$
$V_{ic}^{(1)}$	0	$V_{ic}^{(2)}$	$V_i(A')$

FIGURE 6. Diabatic matrix elements of the electronic Hamiltonian for ionic-covalent $^3P^o$ alkaline earth - homonuclear diatomic interactions in C_{2v} and C_s symmetry.

with the reactant $^3P^o$ atom are crossed at long range by an attractive potential to mimic the expected ionic-covalent crossing. At smaller separations this ionic curve is crossed by and mixed with a repulsive curve which connects with a lower-energy asymptote. Reaction is then simulated by cross sections for quenching from the reactant $^3P^o$ curves to this lower asymptote. Within this pseudo-quenching model, the electrostatic and spin-orbit terms in the Hamiltonian were included without approximation.

These calculations were able to reproduce the experimentally observed [37] enhanced reactivity of the lowest reagent spin-orbit level $(^3P_0^o)$ to form ground state products. Moreover, simple correlation arguments, as illustrated in Fig. 5, were shown to have qualitative validity. However, the calculated cross sections showed an extreme sensitivity to the assumed strength of the ionic-covalent mixing and other parameters of the diabatic potentials. This suggests that 3-dimensional quantum calculations of spin-orbit effects in chemical reactions are a formidable theoretical challenge, both because of the difficulties in the scattering calculations but also because of the need for accurate diagonal and off-diagonal potential energy surfaces.

5. CONCLUSION

In this article, experimental studies of spin-orbit effects in chemical reactions and theoretical studies to model this phenomenon have been reviewed. Perhaps the most striking conclusion is that there are significant dependences of chemical reactivity on reagent spin-orbit level, as expected for atoms with large spin-orbit splittings but also for atoms with small splittings. It is common in molecular collision dynamics studies to ignore spin-orbit splittings in interpreting experiments, and such interactions are usually left out of rather sophisticated classical trajectory or quantum mechanical modelling. It now seems clear that the level of detail accessible in many experiments may warrant inclusion of the spin-orbit interaction for an accurate theoretical treatment.

The challenge to experimentalists is to devise new techniques for spin-orbit state-selective study of other classes of atoms, most notably $O(^3P)$ and other group VIA atoms. It is interesting to note that the $O(^3P)$ fine-structure splittings [12] are very close (but inverted) to those for $Ca(^3P^o)$, for which significant spin-orbit effects have already been found.

ACKNOWLEDGMENTS

I appreciate the numerous conversations with Millard Alexander on theoretical modelling of spin-orbit effects. The research carried out in our laboratory has been supported by the National Science Foundation.

REFERENCES

[1]   R.D. Levine and R.B. Bernstein, Molecular Reaction Dynamics (Oxford U.P., New York, 1974).

[2]   I.W.M. Smith, <u>Kinetics and Dynamics of Elementary Gas Reactions</u> (Butterworths, London/Boston, 1980).

[3]   R.B. Bernstein, <u>Chemical Dynamics Via Molecular Beam and Laser Techniques</u> (Oxford U.P., New York, 1982).

[4]   M.R. Levy, <u>Prog. React. Kinet.</u>, **10** (1979) 1.

[5]   I.W.M. Smith, in <u>Physical Chemistry of Fast Reactions</u>, edited by I.W.M. Smith (Plenum, New York, 1980), Vol. 2, p. 1.

[6]   B.E. Holmes and D.W. Setser, in <u>Physical Chemistry of Fast Reactions</u>, edited by I.W.M. Smith (Plenum, New York, 1980), Vol. 2, p. 83.

[7]   J.C. Polanyi and J.L. Schreiber, in <u>Physical Chemistry, an Advanced Treatise</u>, edited by W. Jost (Academic, New York, 1974), Vol. IVA, p. 383.

[8]   J.M. Farrar and Y.T. Lee, <u>Ann. Rev. Phys. Chem.</u>, **25** (1974) 357.

[9]   S.R. Leone, <u>Ann. Rev. Phys. Chem.</u>, **35** (1984) 109.

[10] D. Husain, <u>Ber. Bunsenges. Physik. Chem.</u>, **81** (1977) 168.

[11] P.J. Dagdigian and M.L. Campbell, <u>Chem. Rev.</u>, **87** (1987) 1.

[12] C.E. Moore, <u>Atomic Energy Levels</u>, NSRDS-NBS **35** (U.S. Government Printing Office, Washington, D.C., 1971).

[13] F.W. Willets, <u>Prog. React. Kinet.</u>, **6** (1971) 51.

[14] M.A.A. Clyne, in <u>Reactive Intermediates in the Gas Phase</u>, edited by D.W. Setser (Academic, New York, 1979), p. 1.

[15] K. Bergmann, S.R. Leone and C.B. Moore, <u>J. Chem. Phys.</u>, **63** (1975) 4161.

[16] P.L. Houston, <u>Chem. Phys. Lett.</u>, **47** (1977) 137.

[17] H. Hofmann and S.R. Leone, <u>Chem. Phys. Lett.</u>, **54** (1978) 314; <u>J. Chem. Phys.</u>, **69** (1978) 641.

[18] D.J. Nesbitt and S.R. Leone, <u>J. Chem. Phys.</u>, **73** (1980) 6182.

[19] H.K. Haugen, E. Weitz and S.R. Leone, <u>Chem. Phys. Lett.</u>, **119** (1985) 75.

[20] H.K. Haugen, E. Weitz and S.R. Leone, J. Chem. Phys., **83** (1985) 3041; W.P. Hess, S.J. Kohler, H.K. Haugen and S.R. Leone, *ibid.*, **84** (1986) 2143; W.P. Hess and S.R. Leone, *ibid.*, **86** (1987) 3773.

[21] R.J. Donovan and D. Husain, Chem. Rev., **70** (1970) 489; D. Husain and R.J. Donovan, Adv. Photochem., **8** (1970) 1.

[22] J.R. Wiesenfeld and G.L. Wolk, J. Chem. Phys., **69** (1978) 1797.

[23] L.G. Piper, D.W. Setser and M.A.A. Clyne, J. Chem. Phys., **63** (1975) 5018.

[24] T.D. Dreiling and N. Sadeghi, J. Phys. (Paris), **44** (1983) 1007.

[25] M.F. Golde and R.A. Poletti, Chem. Phys. Lett., **80** (1981) 18.

[26] J.E. Velasco, J.H. Kolts and D.W. Setser, J. Chem. Phys., **69** (1978) 4357.

[27] M.F. Golde and R.A. Poletti, Chem. Phys. Lett., **80** (1981) 23.

[28] M.F. Golde and R.A. Poletti, J. Chem. Phys., **82** (1985) 3160.

[29] J. Pitre, K. Hammond and L. Krause, Phys. Rev. A, **6** (1972) 2101.

[30] H. Horiguchi and S. Tsuchiya, Bull. Chem. Soc. Japan, **44** (1971) 1213.

[31] T.D. Dreiling and D.W. Setser, J. Phys. Chem., **86** (1982) 2276; F.M. Zhang, D. Oba and D.W. Setser, *ibid.*, **91** (1987) 1099.

[32] I.V. Hertel and W. Stoll, Adv. At. Mol. Phys., **13** (1978) 113.

[33] K. Bergmann, R. Engelhardt, U. Hefter and J. Witt, J. Phys. E, **12** (1979) 507.

[34] J.P.C. Kroon, H.C.W. Beijerinck, B.J. Verhaar and N.F. Verster, Chem. Phys., **90** (1984) 195.

[35] W. Happer, Rev. Mod. Phys., **44** (1972) 169.

[36] H.-J. Yuh and P.J. Dagdigian, J. Chem. Phys., **81** (1984) 2375.

[37] P.J. Dagdigian, in Gas Phase Chemiluminescence and Chemi-Ionization, edited by A. Fontijn (North-Holland, Amsterdam, 1985), p. 203.

[38] N. Furio, M.L. Campbell and P.J. Dagdigian, J. Chem. Phys., **84** (1986) 4332.

172

[39]   M.L. Campbell, N. Furio and P.J. Dagdigian, <u>Laser Chem.</u>,
        6 (1986) 391.

[40]   M.L. Campbell and P.J. Dagdigian, <u>J. Am. Chem. Soc.</u>,
        108 (1986) 4701.

[41]   M.L. Campbell and P.J. Dagdigian, <u>J. Chem. Phys.</u>, **85** (1986)
        4453.

[42]   M.L. Campbell and P.J. Dagdigian, <u>Faraday Disc. Chem. Soc.</u>,
        submitted.

[43]   C. Weiser and P.E. Siska, <u>J. Chem. Phys.</u>, **85** (1986) 4746.

[44]   M.J. Verheijen and H.C.W. Beijerinck, <u>Chem. Phys.</u>, **102**
        (1986) 255; F.T.M. van den Berg, J.H.M. Schonenberg and
        H.C.W. Beijerinck, *ibid.*, submitted.

[45]   H. Hotop, J. Lorenzen and A. Zastrow, <u>J. Electron Spectrosc.
        Relat. Phenom.</u>, **23** (1981) 347.

[46]   J. Lorenzen, H. Morgner, W. Bussert, M.-W. Ruf and H. Hotop,
        <u>Z. Phys. A</u>, **310** (1983) 141.

[47]   N. Sadeghi and D.W. Setser, <u>Chem. Phys.</u>, **95** (1985) 305.

[48]   N. Sadeghi and D.W. Setser, manuscript in preparation.

[49]   <u>American Institute of Physics Handbook</u>, 3rd ed., edited by
        D.E. Gray (McGraw-Hill, New York, 1972), pp. 8-46.

[50]   P. Grundevik, H. Lundberg, L. Nilson and G. Olsson, <u>Z. Phys. A</u>,
        306 (1982) 195.

[51]   G. Weissmann, J. Ganz, A. Siegel, H. Waibel and H. Hotop,
        <u>Opt. Commun.</u>, **49** (1984) 335.

[52]   H.-J. Yuh and P.J. Dagdigian, <u>Phys. Rev. A</u>, **28** (1983) 63.

[53]   W.A. Majewski, <u>Opt. Commun.</u>, **45** (1983) 201.

[54]   C.T. Rettner and R.N. Zare, <u>J. Chem. Phys.</u>, **77** (1982) 2416.

[55]   H. Schmidt, P.S. Weiss, J.M. Mestdagh, M.H. Covinsky and
        Y.T. Lee, <u>Chem. Phys. Lett.</u>, **118** (1985) 539; M.F. Vernon,
        H. Schmidt, P.S. Weiss, M.H. Covinsky and Y.T. Lee, *ibid.*,
        84 (1986) 5580.

[56]   J.K. Ku and D.W. Setser, <u>Appl. Phys. Lett.</u>, **48** (1986) 689.

[57]   H.C. Brashears, Jr., and D.W. Setser, J. Phys. Chem.,
       84 (1980) 224.

[58]   K. Tanaka, J. Durup, T. Kato and I. Koyano, J. Chem. Phys.,
       74 (1981) 5561.

[59]   T. Kato, K. Tanaka and I. Koyano, J. Chem. Phys., **77** (1982) 337.

[60]   F.M. Campbell, R. Browning and C.J. Latimer, J. Phys. B,
       14 (1981) 1183.

[61]   C.-L. Liao, C.-X. Liao and C.Y. Ng, J. Chem. Phys.,
       82 (1985) 5489.

[62]   C.-L. Liao, R. Xu and C.Y. Ng, J. Chem. Phys., **84** (1986) 1948;
       C.-L. Liao, J.-D. Shao, R. Xu, G.D. Flesch, Y.-G. Li and
       C.-Y. Ng, *ibid.*, **85** (1986) 3874.

[63]   J.A.R. Sampson and R.B. Cairns, Phys. Rev., **173** (1968) 80.

[64]   T. Baer, in Gas Phase Ion Chemistry, edited by M.T. Bowers
       (Academic, New York, 1979), Vol. 1, p. 153.

[65]   R.D. Smith, D.L. Smith and J.H. Futrell, Chem. Phys. Lett.,
       32 (1975) 513.

[66]   N.G. Adams, D. Smith and E. Alge, J. Phys. B, **13** (1980) 3235.

[67]   T.T.C. Jones, K. Birkinshaw, J.D.C. Jones and N.D. Twiddy,
       J. Phys. B, **15** (1982) 2439.

[68]   K.M. Ervin and P.B. Armentrout, J. Chem. Phys., **85** (1986) 6380.

[69]   P. Beadle, M.R. Dunn, N.B.H. Jonathan, J.P. Liddy and
       J.C. Naylor, J. Chem. Soc. Faraday Trans. 2, **74** (1978) 2170.

[70]   D.H. Maylotte, J.C. Polanyi and K.B. Woodall, J. Chem. Phys.,
       57 (1972) 1547.

[71]   J.C. Polanyi and W. Sklrac, J. Chem. Phys., **23** (1977) 167.

[72]   D. Brandt, L.W. Dickson, L.N.Y. Kwan and J.C. Polanyi,
       Chem. Phys., **39** (1979) 189.

[73]   J.P. Sung and D.W. Setser, Chem. Phys. Lett., **48** (1977) 413.

[74]   B.S. Agrawalla, J.P. Singh and D.W. Setser, J. Chem. Phys.,
       79 (1983) 6416.

[75]   H. Brunet, Ph. Chauvet, M. Mabru and L. Torchin, Chem. Phys.
       Lett., **117** (1985) 371.

[76] E.B. Gordon, A.I. Nadkhin, S.A. Sotnichenko and I. Boriev,
      Chem. Phys. Lett., **86** (1982) 209; I.A. Boriev, E.B. Gordon
      and A.A. Efimenko, *ibid.*, **120** (1985) 486.

[77] J.R. Wiesenfeld and G.L. Wolk, J. Chem. Phys., **69** (1978) 1805.

[78] I. Burak and M. Eyal, Chem. Phys. Lett., **52** (1977) 534.

[79] J.W. Hepburn, K. Liu, R.G. Macdonald, F.J. Northrup and
      J.C. Polanyi, J. Chem. Phys., **75** (1981) 3353.

[80] J.W. Hepburn, Israel J. Chem., **24** (1984) 273.

[81] J. Bokor, R.R. Freeman, J.C. White and R.H. Storz, Phys. Rev. A,
      24 (1981) 612.

[82] W.K. Bischel, B.E. Perry and D.R. Crosley, Chem. Phys. Lett.,
      82 (1981) 85.

[83] P. Brewer, N. van Veen and R. Bersohn, Chem. Phys. Lett., **91**
      (1982) 126.

[84] P. Das, G. Ondrey, N. van Veen and R. Bersohn, J. Chem. Phys.,
      79 (1983) 724.

[85] P. Brewer, P. Das, G. Ondrey and R. Bersohn, J. Chem. Phys.,
      79 (1983) 720.

[86] R.A. Copeland, J.B. Jeffries, A.P. Hickman and D.R. Crosley,
      J. Chem. Phys., **86** (1987) 4876.

[87] P. Das, T. Venkitachalam and R. Bersohn, J. Chem. Phys.,
      80 (1984) 4859.

[88] T.V. Venkitachalam, P. Das and R. Bersohn, J. Am. Chem. Soc.,
      105 (1983) 7452.

[89] D.M. Neumark, A.M. Wodtke, G.N. Robinson, C.C. Hayden
      and Y.T. Lee, J. Chem. Phys., **82** (1985) 3045.

[90] M.A.A. Clyne and H.W. Cruse, J. Chem. Soc. Faraday Trans. 2,
      68 (1972) 1377.

[91] M.A.A. Clyne and W.S. Nip, Int. J. Chem. Kinet., **10** (1977) 365.

[92] T. Trickl and J. Wanner, J. Chem. Phys., **78** (1983) 6091.

[93] I.W. Fletcher and J.C. Whitehead, J. Chem. Soc. Faraday
      Trans. 2, **78** (1982) 1165.

[94] D.J. Spencer and C. Wittig, Opt. Lett., **4** (1979) 1.

[95] H.V. Lilenfeld, P.D. Whitefield and G.R. Bradburn, J. Phys. Chem., **88** (1984) 6158.

[96] G.E. Hall, S. Arepalli, P.L. Houston and J.R. Wiesenfeld, J. Chem. Phys., **82** (1985) 2590.

[97] W.A. Chupka and M.E. Russell, J. Chem. Phys., **49** (1968) 5426.

[98] P.J. Kuntz and A.C. Roach, J. Chem. Soc. Faraday Trans. 2, **68** (1972) 259.

[99] S. Chapman and R.K. Preston, J. Chem. Phys., **60** (1974) 650.

[100] T.R. Govers, P.M. Guyon, T. Baer, K. Cole, H. Frohlich and M. Lavoleé, Chem. Phys., **87** (1984) 373.

[101] M.R. Spalburg and E.A. Gislason, Chem. Phys., **94** (1985) 339; G. Parlant and E.A. Gislason, J. Chem. Phys., **86** (1987) 6183.

[102] M.F. Golde, in Gas Kinetics and Energy Transfer, Specialists Periodical Reports, edited by P.G. Ashmore and R.J. Donovan (Chemical Society, London, 1976), Vol. 2, p. 121.

[103] J.H. Kolts and D.W. Setser, in Reactive Intermediates in the Gas Phase, edited by D.W. Setser (Academic, New York, 1979), p. 152.

[104] Excimer Lasers, edited by C.K. Rhodes (Springer-Verlag, Berlin, 1979).

[105] D.W. Setser, T.D. Dreiling, H.C. Brashears, Jr., and J.H. Kolts, Faraday Discuss. Chem. Soc., **67** (1979) 255.

[106] T.H. Dunning, Jr., and P.J. Hay, J. Chem. Phys., **69** (1978) 134; P.J. Hay and T.H. Dunning, Jr., *ibid.*, **69** (1978) 2209.

[107] D. Lin, Y.C. Ku and D.W. Setser, J. Chem. Phys., **81** (1984) 5830.

[108] K. Johnson, J.P. Simons, P.A. Smith, C. Washington and A. Kvaran, Mol. Phys., **57** (1986) 255.

[109] J. Derouard, T.D. Nguyen and N. Sadeghi, J. Chem. Phys., **72** (1980) 6698.

[110] T.D. Nguyen and N. Sadeghi, Chem. Phys., **79** (1983) 41.

[111] M.F. Golde and Y.-S. Ho, J. Chem. Phys., **82** (1985) 3160.

[112] D. Hausamann and H. Morgner, Mol. Phys., **54** (1985) 1085.

[113]   J. Lorenzen, H. Hotop, M.-W. Ruf and H. Morgner, Z. Phys. A,
        297 (1980) 19; W. Bussert, T. Bregel, J. Ganz, K. Harth,
        A. Siegel, M.-W. Ruf and H. Hotop, J. Phys. (Paris), **46**
        (1985) C1-199.

[114]   J. Lorenzen, H. Hotop and M.-W. Ruf, Z. Phys. D, **1** (1986) 261.

[115]   H. Morgner, J. Phys. B, **18** (1985) 251.

[116]   D.R. Herschbach, Adv. Chem. Phys., **10** (1966) 319.

[117]   N.H. Hijazi and J.C. Polanyi, J. Chem. Phys., **63** (1975) 2249;
        Chem. Phys., **11** (1975) 1.

[118]   C. Noda and R.N. Zare, J. Chem. Phys., **86** (1987) 3968.

[119]   M.H. Alexander and P.J. Dagdigian, Chem. Phys., **33** (1978) 13.

[120]   M. Menzinger, Adv. Chem. Phys., **42** (1980) 1.

[121]   J.W. Cox and P.J. Dagdigian, J. Phys. Chem., **86** (1982) 3738.

[122]   A. Kowalski and M. Menzinger, Chem. Phys. Lett., **78** (1981) 461.

[123]   M. Menzinger, in Gas Phase Chemiluminescence and Chemi-
        Ionization, edited by A. Fontijn (North-Holland, Amsterdam,
        1985), p. 25.

[124]   N. Honjou and D.R. Yarkony, J. Chem. Phys., **89** (1985) 2919,
        and unpublished results.

[125]   G. Herzberg, Molecular Spectra and Molecular Structure,
        I. Spectra of Diatomic Molecules (D. Van Nostrand, New York,
        1950).

[126]   M.H. Alexander, in Gas Phase Chemiluminescence and Chemi-
        Ionization, edited by A. Fontijn (North-Holland, Amsterdam,
        1985), p. 221.

[127]   N.-H. Cheung and T.A. Cool, J. Quant. Spectrosc. Radiat.
        Transfer, **21** (1979) 397.

[128]   H.F. Krause, S.G. Johnson, S. Datz and F.K. Schmidt-Bleek,
        Chem. Phys. Lett., **31** (1975) 577.

[129]   S. Hayashi, T.M. Mayer and R.B. Bernstein, Chem. Phys. Lett.,
        53 (1978) 419.

[130]   T.D. Dreiling and D.W. Setser, J. Chem. Phys., **79** (1983)
        5423,5439.

[131]   F.J. Wodarczyk and A.B. Harker, Chem. Phys. Lett., **62** (1979) 529.

[132]   H. Helvajian and C. Wittig, J. Chem. Phys., **76** (1982) 3505.

[133]   J.C. Tully, in State-to-State Chemistry, edited by P.R. Brooks and E.F. Hayes (American Chemical Society, Washington, D.C., 1977), p. 206.

[134]   F. Rebentrost, in Theoretical Chemistry: Advances and Prospectives, edited by D. Henderson (Academic, New York, 1981), Vol. 6B, p. 1.

[135]   J.C. Tully and R.K. Preston, J. Chem. Phys., **55** (1971) 562.

[136]   C. Evers, Chem. Phys., **21** (1977) 355; A.W. Kleyn, V.N. Kromov and J. Los, *ibid.*, **52** (1980) 65.

[137]   J.C. Tully, J. Chem. Phys., **60** (1974) 3042.

[138]   A. Komornicki, K. Morokuma and T.F. George, J. Chem. Phys., **67** (1977) 5012.

[139]   W.H. Miller and C.W. McCurdy, J. Chem. Phys., **69** (1978) 5163.

[140]   C.W. McCurdy, H.-D. Meyer and W.H. Miller, J. Chem. Phys., **70** (1979) 3177; H.-D. Meyer and W.H. Miller, *ibid.*, **70** (1979) 3214; **71** (1979) 2156.

[141]   D.L. Miller and R.E. Wyatt, J. Chem. Phys., **67** (1977) 1302.

[142]   I. Last and M. Baer, J. Chem. Phys., **82** (1985) 4954.

[143]   F. Rebentrost and W.A. Lester, Jr., J. Chem. Phys., **63** (1975) 3737; **64** (1976) 3879; **67** (1977) 3367.

[144]   R.E. Wyatt and R.B. Walker, J. Chem. Phys., **70** (1979) 1501.

[145]   D.E. Fitz and D.J. Kouri, J. Chem. Phys., **74** (1981) 3933.

[146]   D.R. Flower and J.-M. Launay, J. Phys. B, **10** (1977) 3673.

[147]   J. Pascale, F. Rossi and W.E. Baylis, Phys. Rev. A, submitted.

[148]   M.H. Alexander and G.C. Corey, J. Chem. Phys., **84** (1986) 100.

[149]   F. Rossi and J. Pascale, Phys. Rev. A, **32** (1985) 2657.

OPEN SHELL ATOMIC BEAM SCATTERING AND THE SPIN ORBIT DEPENDENCE OF POTENTIAL ENERGY SURFACES

V. Aquilanti, R. Candori, G. Liuti and F. Pirani
Dipartimento di Chimica dell'Università
06100 Perugia
Italy

ABSTRACT. Molecular beam scattering studies of open shell atoms with analysis of magnetic sublevels provide information on the dependence upon spin orbit interaction of the long range part of potential energy surfaces. Experiments on the interaction of oxygen atoms in their ground state with rare gases, molecular hydrogen and methane are reported and information is derived both on adiabatic surfaces and on nonadiabatic coupling terms.

1. INTRODUCTION

This paper reports some recent results obtained in this laboratory on the interaction of open-shell atoms with several atoms and molecules. The experimental technique exploits magnetic analysis of atomic sub-levels and provides insight on the influence of spin-orbit and electronic angular momenta on the long range part of potential energy surfaces. These surfaces are of interest not only for all kinetic processes where intramultiplet mixing and polarization phenomena are explicitly observed, but also for understanding orientation and alignment effects in reactive collisions when dominated by anisotropy at long range. Some results involving fluorine [1] and nitrogen atoms [2] have been reported previously. In this paper we focus on recent advances on the interaction of oxygen atoms in their ground states with rare gases, molecular hydrogen and methane. The experimental technique is first discussed, the theoretical framework for the analysis of results is then reviewed and results are finally reported together with the information obtained on potential energy surfaces.

2. EXPERIMENTAL

Measurements of absolute integral cross sections for atomic collisions in the thermal energy range have been carried out in this laboratory using the experimental arrangement schematized in Fig. 1. This apparatus has been used over the years [3] for the study of the interactions involving H atoms, rare gases, and simple diatomic molecules such as D_2, N_2, O_2. For many of the systems investigated, the strength of the interactions and the velocity ranges available were such that glory interference effects could be observed. A magnet for Stern-Gerlach analysis of magnetic substates

179

J. C. Whitehead (ed.), Selectivity in Chemical Reactions, 179–194.
© 1988 by Kluwer Academic Publishers.

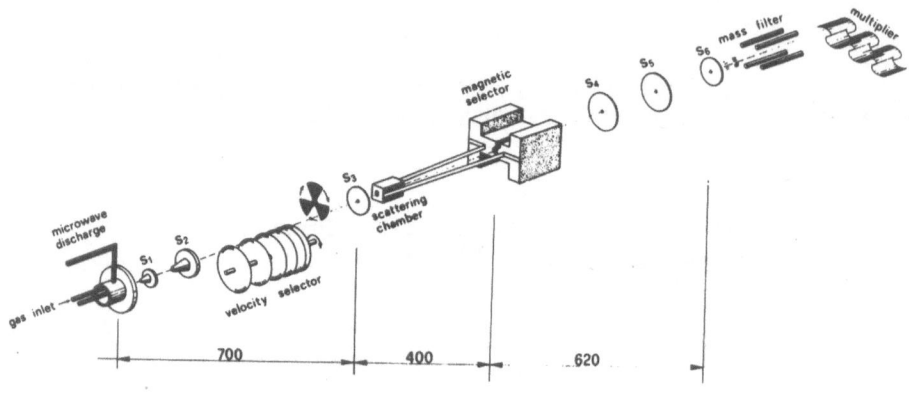

FIGURE 1. Schematic view of the apparatus for integral cross section measurements (all dimensions in mm). The diameters of the collimating slits, which separate differentially pumped chambers, are 1.5 mm for S_1, S_2, S_5 and S_6, 0.7 mm for S_3 and 1.0 mm for S_4.

(Rabi configuration, see [4]), gives information on the atomic levels involved and allows a characterization of the polarization states: this technique has been used in experiments involving nitrogen [5], oxygen [1b,6b], and fluorine [1] atoms. For oxygen atoms (Fig. 2), magnetic analysis shows that a pure beam of ground state ^3P_j atoms can be obtained by microwave discharge on O_2: the atom beam is essentially free of the possible metastable $^1D_2$. Fig. 2 also shows that by varying the magnetic field strength at fixed velocity, it is possible to vary the composition of $|jm_j\rangle$ states which are transmitted by the magnet: for this analysis, it is taken into account that $\vec{j} = \vec{L}+\vec{S}$ is a good quantum number for oxygen atoms in this range of magnetic fields, and m_j refers to the beam direction. Specifically, for $O(^3P_j)$, the electronic angular momentum L=1, and electronic spin S=1 add to j=0,1,2.

Previous experiments on the scattering of oxygen atoms by the rare gases [6] probed a velocity range where the glory interference effect could be observed for the heavier collision partners. Measurements at relatively high magnetic field strength were carried out for the target gases Ar, Kr and Xe. Anisotropy effects, as measured by differences between cross sections at zero field and at high field, were found to be negligible for Ar, but of the order of a few percent for Kr and for Xe: they showed up in the glory structure of cross sections.

From these data it was inferred that, in the region of the very shallow van der Waals wells of these rare gas oxides, Π-type electrostatic interactions are stronger than Σ ones. The analysis which led to an estimate of the interactions was based on an elastic approximation, as defined in previous papers [1b,6b]: this approximation can only be used for qualitative purposes, but the data did not allow a more quantitative approach.

The present results, obtained under much improved experimental conditions (especially concerning signal detection), supersede the old ones: they confirm previous qualitative conclusions but permit a much more refined analysis.

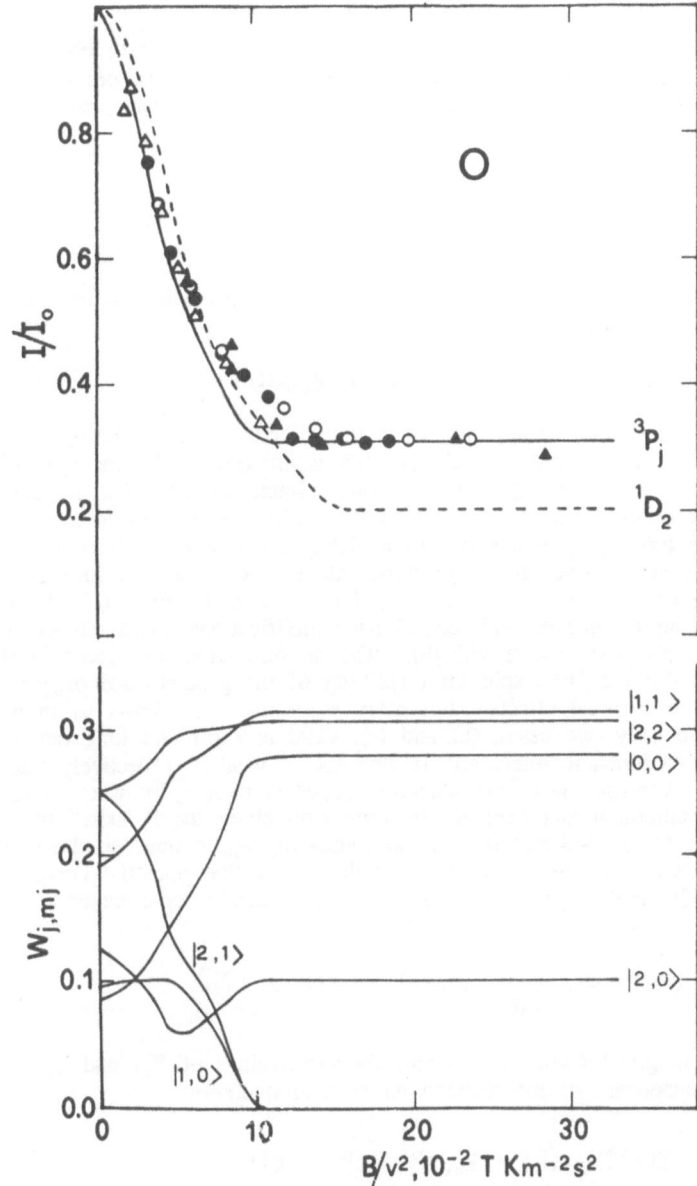

FIGURE 2. Fraction I/I_0 of the transmitted beam of O atoms as a function of the reduced parameter B/v^2, where B is the magnetic induction across the beam and v is the velocity of the atoms. (Black triangles, open dots, black dots, open triangles correspond to 1.33, 1.60, 1.80 and 2.21 km s^{-1}). The full line is calculated for the ground ^3P_j state, the dashed line for the metastable $^1D_2$ state. In the lower panel, the relative weights $W_{j,mj}$ of the $|j\ m_j\rangle$ states of O(^3P_j) are shown.

3. THEORETICAL FRAMEWORK

The theory needed to interpret the experimental results and to obtain information on the potential from them has been given and reviewed elsewhere [7-9,1b]. Basically, one has to solve a multichannel Schroedinger equation which can be written, in a matrix notation,

$$\left[- (\hbar^2/2\mu) \; \underline{1} \; \frac{d^2}{dR^2} + \underline{V}^J(R) \right] \vec{\psi} = E\vec{\psi} \tag{1}$$

where the potential energy matrix $\underline{V}^J$ is not diagonal (diabatic representation) and is the sum of three contributions:

$$\underline{V}^J(R) = \underline{V}_{SO}(R) + \underline{V}_{el}(R) + \underline{V}_{rot}(R)$$

For these low energy processes, the dimension of the basis is restricted to the fine structure components, which are split asymptotically by the spin-orbit interaction of the open shell atom, $\underline{V}_{so}$; the collision effects are described by the electrostatic interaction between atoms, $\underline{V}_{el}$, which is relatively short ranged, and by the centrifugal term $\underline{V}_{rot}$, which decays as R^{-2}. According to the relative importance of these three terms, five alternative representations are possible, corresponding to different coupling schemes for the angular momenta involved (Hund's cases). The choice of representations and recipes for simplifications based on decoupling schemes have been given elsewhere [7,8,1b]. The simplification considered in the following is justified by the large spin orbit splitting of the ground state oxygen atom with respect to centrifugal effects. For glory scattering, this allows us to restrict our attention to only two cases, (a) and (c), valid at short and long range respectively, when the electrostatic interaction is stronger or weaker respectively than spin-orbit splitting. A proper label for scattering states is then $|j\Omega\rangle$ where j is the atomic angular momentum and $\Omega=|mj|$ its projection along the R axis. In this centrifugal Sudden or Coupled States (CS) or Ω-conserving approximation, the scattering, which is considered as taking place along adiabatic effective potential curves, $\epsilon_{j\Omega}$, is described by the equivalent of Eq. 1 in the adiabatic representation:

$$[-(\hbar^2/2\mu)(\underline{1}\frac{d}{dR} + \underline{P}(R))^2 + \underline{\epsilon}(R)\vec{\psi} = E\vec{\psi} \tag{2}$$

The diagonal matrix $\underline{\epsilon}$ contains the eigenvalues of $\underline{V}_{el}$ and $\underline{V}_{so}$ (of course independent of the diabatic representation chosen):

$$\underline{T}(R)[\underline{V}_{el}(R) + \underline{V}_{so}(R)]\underline{\tilde{T}}(R) = \underline{\epsilon}(R) \tag{3}$$

These eigenvalues are effective potential energy curves in the adiabatic representation. The coupling between them is represented by the matrix $\underline{P}$, which is related to the orthogonal diagonalizing matrix $\underline{T}$ appearing in Eq. 3 by

$$\underline{P}(R) = \frac{d\underline{T}(R)}{dR} \, \tilde{\underline{T}}(R) \tag{4}$$

Its elements couple states with different j, but Ω is conserved in this approximation. Accordingly, they will be indicated as $<j\Omega \mid d/dR \mid j'\Omega>$.

In this approach, which has been used also for studying the interaction of fluorine atoms with rare gases [1], scattering of oxygen atoms is described by effective adiabatic curves, provided that the nonadiabatic coupling between them is negligible. This will be shown to be the case by explicit computation of the matrix elements of the $\underline{P}$-matrix.

Following a previous analysis [7,8,1a], for P atoms (L=1) it is convenient to introduce a parameterization of the electrostatic interaction by two functions of the internuclear distance, a spherical interaction potential, $V_0(R)$, and an anisotropic potential term, $V_2(R)$. They are simply related to the more familiar $V_\Sigma(R)$ and $V_\Pi(R)$ potential functions, representing different electronic symmetries:

$$V_0 = \frac{1}{3}(V_\Sigma + 2\ V_\Pi); \quad V_2 = \frac{5}{3}(V_\Sigma - V_\Pi)$$

Generalizations to any L are considered elsewhere [7].

In terms of V_0 and V_2, the six adiabatic effective potentials, $\epsilon_{j\Omega}$, describing the interaction of an $O(^3P_j)$ atom with a spherical target are explicitly:

$$\epsilon_{22} = V_0 - \frac{1}{5}\ V_2$$

$$\epsilon_{21} = V_0 + \frac{1}{10}\ V_2 + \frac{1}{2}\ \epsilon_1 - \frac{1}{2}(\frac{9}{25}\ V_2^2 + \epsilon_1^2)^{1/2}$$

$$\epsilon_{20} = V_0 + \frac{1}{10}\ V_2 + \frac{1}{2}\ \epsilon_0 - \frac{1}{2}(\frac{9}{25}\ V_2^2 + \epsilon_0^2 - \frac{2}{5}\ \epsilon_0 V_2)^{1/2}$$

$$\epsilon_{11} = V_0 + \frac{1}{10}\ V_2 + \frac{1}{2}\ \epsilon_1 + \frac{1}{2}(\frac{9}{25}\ V_2^2 + \epsilon_1^2)^{1/2}$$

$$\epsilon_{10} = V_0 - \frac{1}{5}\ V_2 + \epsilon_1$$

$$\epsilon_{00} = V_0 + \frac{1}{10}\ V_2 + \frac{1}{2}\ \epsilon_0 + \frac{1}{2}(\frac{9}{25}\ V_2^2 + \epsilon_0^2 - \frac{2}{5}\epsilon_0 V_2)^{1/2}$$

where $\epsilon_1 = 19.6$ meV and $\epsilon_0 = 28.1$ meV.

In this decoupling scheme nonadiabatic coupling is effective only between states having the same Ω quantum numbers and the only non zero elements of the $\underline{P}$ matrix;

$$P_{j\Omega, j'\Omega} \equiv <j\Omega \mid \frac{d}{dR} \mid j'\Omega>$$

can be shown to be given by:

$$<20|\frac{d}{dR}|00> = \frac{-10/\sqrt{2}}{9\beta_0{}^2 - 10\beta_0 + 25} \frac{d\beta_0}{dR}$$

$$<21|\frac{d}{dR}|11> = \frac{-15/2}{9\beta_1{}^2 + 25} \frac{d\beta_1}{dR}$$

where $\beta_0 \equiv V_2/\epsilon_0$ and $\beta_1 \equiv V_2/\epsilon_1$.

When the contribution of these off diagonal elements is negligible, the measured integral cross section will be given by a weighted sum of cross section $Q_{j\Omega}(v)$ for scattering by the potential $\epsilon_{j\Omega}$:

$$Q(v) = \sum_{j\Omega} W_{j\Omega} Q_{j\Omega}(v)$$

where the weights $W_{j\Omega}$, which depend on the magnetic field strength, are known (Fig. 2). This analysis has also been applied to describing interactions of $O(^3P)$ atoms with molecules: this will be justified in the following section.

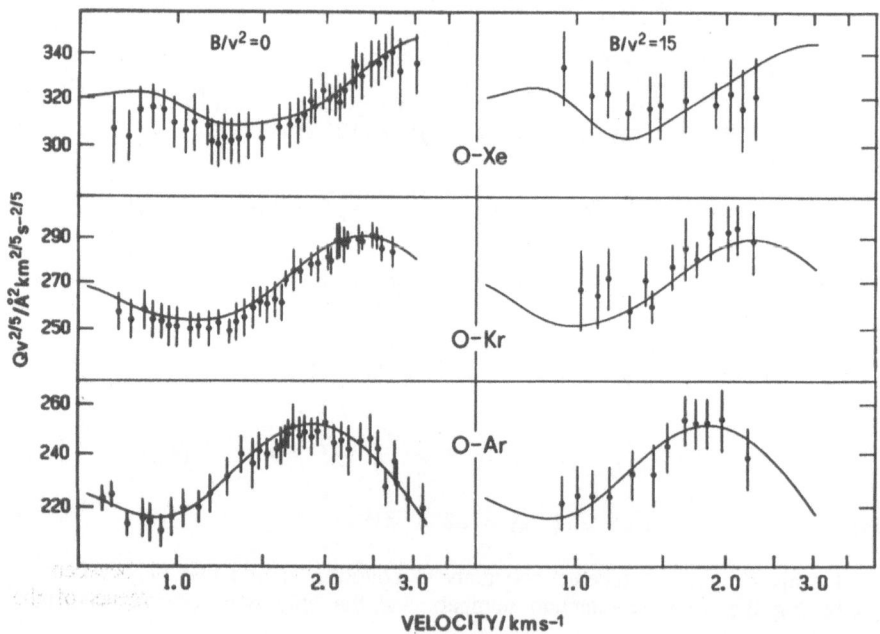

FIGURE 3. Absolute integral cross sections for the O-Ar,Kr, and Xe systems as a function of velocity v measured with (right) and without (left) magnetic field. The lines represent the cross sections calculated according to the scheme described in the text.

4. RESULTS

The experimental absolute integral cross sections for the interaction of O atoms with the heavier rare gases, methane and deuterium are reported, in Figs. 3,4 and 5 respectively, plotted as a function of the atomic beam velocity. A preliminary account of the O-He system has already been given [10]. For all the systems shown, cross section measurements have been performed also at high magnetic fields. For the different conditions, the O atom sublevel populations can be inferred from the magnetic analysis of Fig. 2: note that the atomic beam intensity is much lower at high fields.

The solid lines represent the cross sections calculated within the theoretical framework indicated in the previous section. For the fit a Morse-Spline-van der Waals model function for $V_0(R)$ and a Buckingham-Corner model function for $V_2(R)$ have been used. The six adiabatic effective potential curves obtained are shown in Figs. 6,7,8,9,10 for O-Ar,Kr,Xe,CH$_4$, and D$_2$ respectively, together with nonadiabatic coupling terms. Relevant potential parameters for the above interactions are reported in Table 1.

TABLE 1. Some derived potential parameters[a]

SYSTEM	V_0		V_2		V_Σ		V_Π	
	ϵ/meV	R_m/Å	ϵ/meV	R_m/Å	ϵ/meV	R_m/Å	ϵ/meV	R_m/Å
O-Ar	7.8	3.60	0.18	5.21	5.1	3.85	10.4	3.45
O-Kr	9.3	3.75	0.18	5.57	5.6	4.05	13.1	3.57
O-Xe	11.8	3.90	0.22	5.81	6.9	4.24	17.3	3.69
O-D$_2$	4.7	3.49	0.07	5.31	2.8	3.75	6.5	3.35
O-CH$_4$	9.3	3.75	0.17	5.61	5.5	4.06	13.4	3.56

(a) Estimated uncertainties are of the order of 10 percent for minima ϵ and 2 percent for their locations R_m.

5. DISCUSSION

As outlined above, the interaction of oxygen atoms with the atoms and molecules considered here will be represented by an effective two-body anisotropic interaction and the anisotropy, at the distances sampled by these experiments, is mainly attributed to the atomic open shell structure. For D$_2$ and CH$_4$ this amounts to assuming that the system behaves adiabatically with respect to molecular vibration while molecular rotations are averaged (a reason for using D$_2$ instead of H$_2$ is to make this last assumption more plausible).

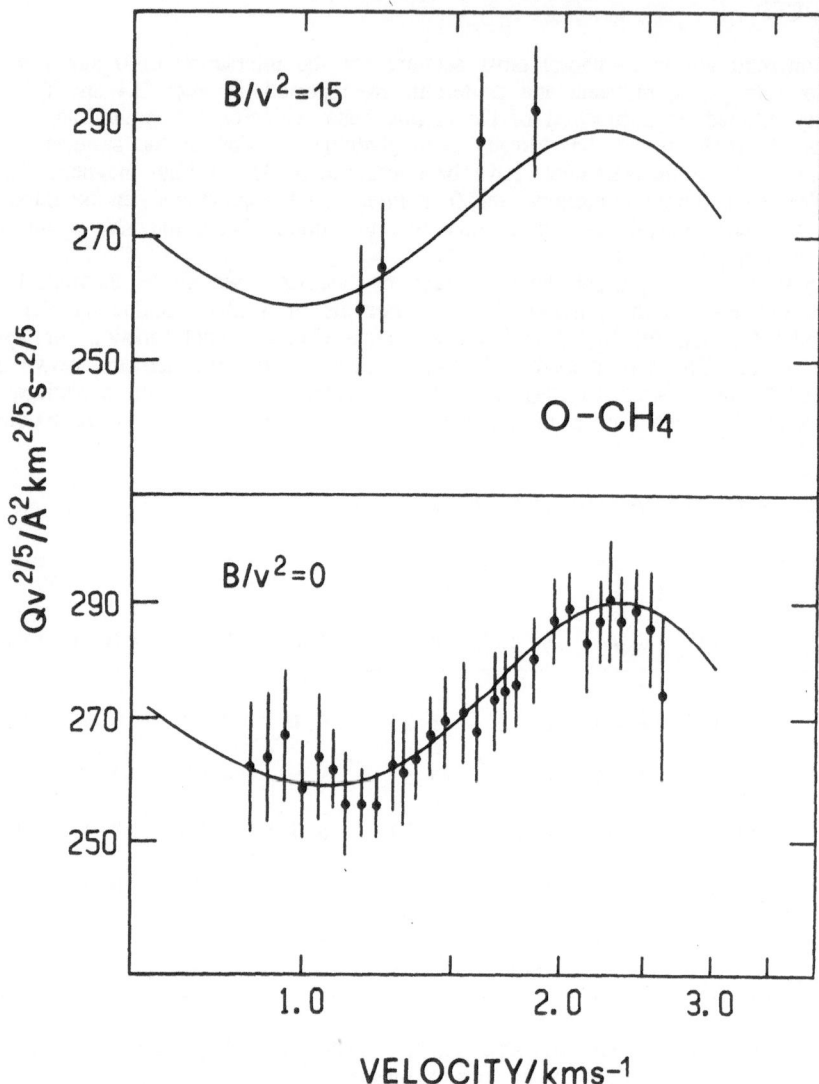

FIGURE 4. As in Fig. 3, for the O-CH$_4$ system.

For 0-rare gases it can be seen that the interaction anisotropy increases from Ar to Xe. This can be attributed to an increasing ionic contribution to the interaction due to a lower ionization potential in going from Ar to Xe.

In the 0-CH_4 and D_2 cases, the weak long range interaction obtained by the present technique is well below the energy barriers for reactive chemical processes [11]. The present cross sections are most sensitive to the van der Waals well region determined by a balance between the tail of the repulsion and the long range

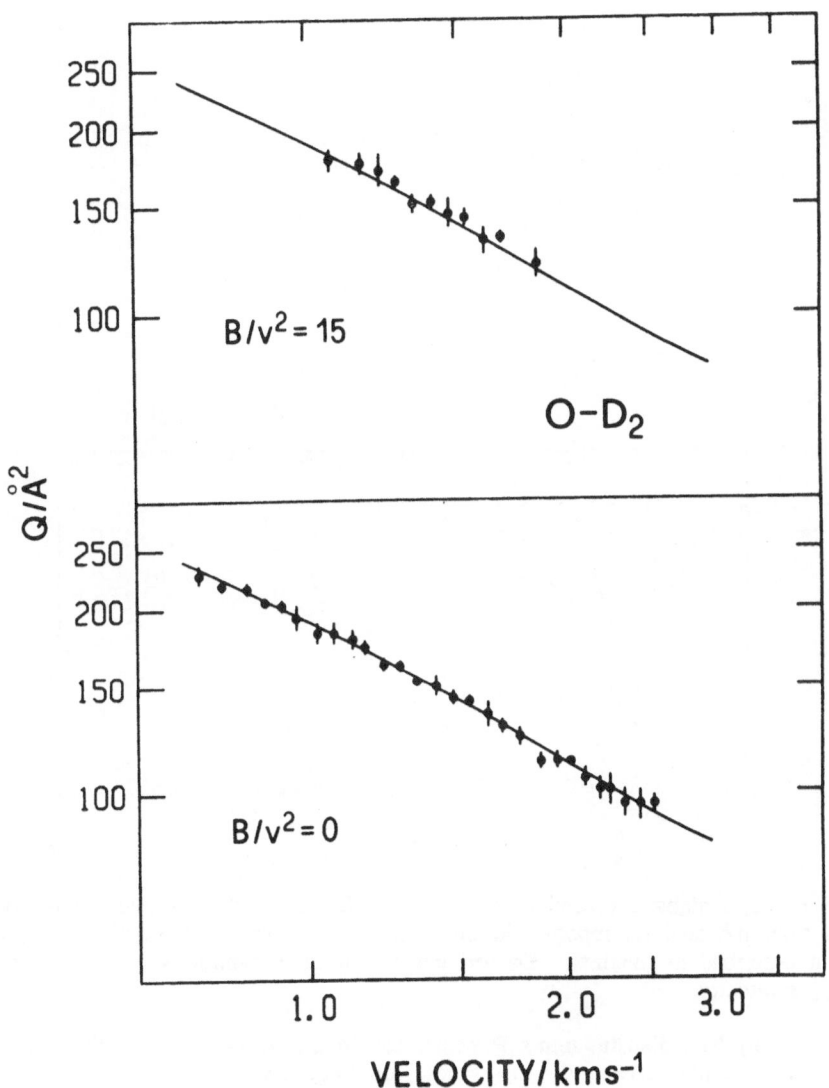

FIGURE 5. As in Fig. 3, for the 0-D_2 system.

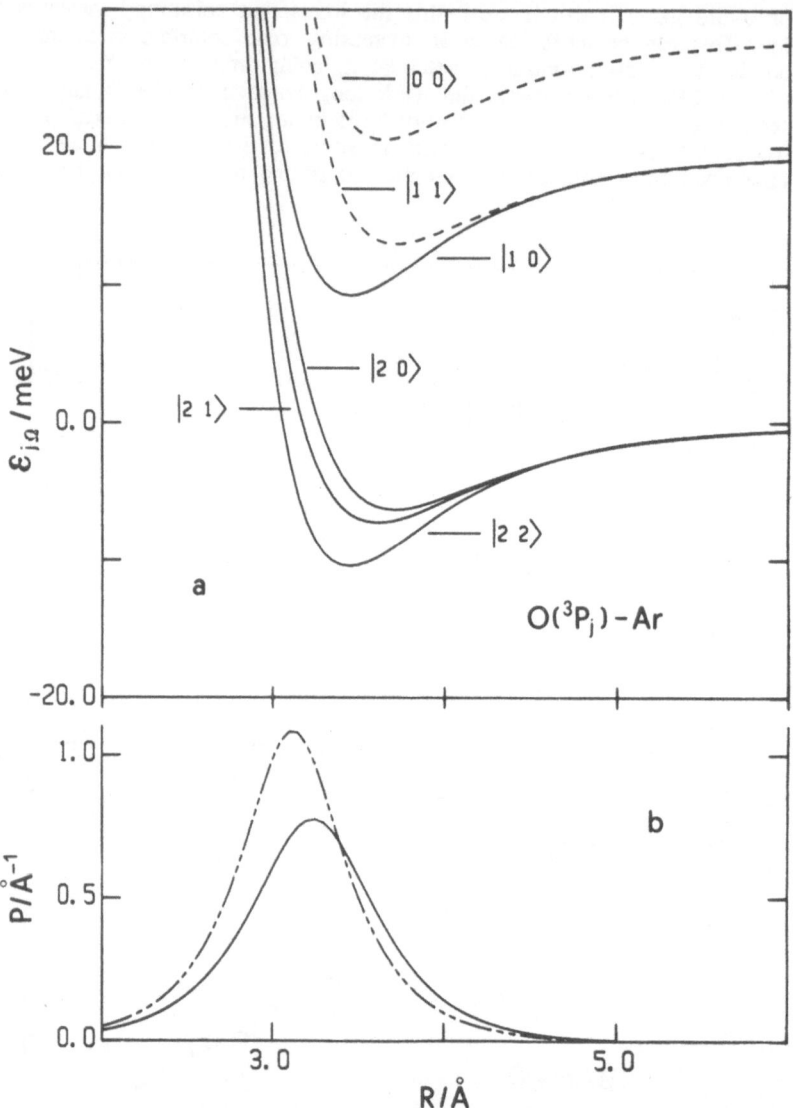

FIGURE 6. a) Adiabatic potential energy curves for the $O(^3P_j)$-Ar interactions as obtained from the analysis reported in this work. The curves are labelled using the formalism described in the text. Dashed curves are of Σ symmetry and full curves are of Π symmetry.

b) Nonadiabatic terms P which couple the $|20\rangle$ and the $|00\rangle$ states (dashed line) and the $|21\rangle$ and $|11\rangle$ states (full line). Maxima of the P functions mark the separation between coupling schemes, the molecular case (a) at short distances and the diatomic case (c) at long range.

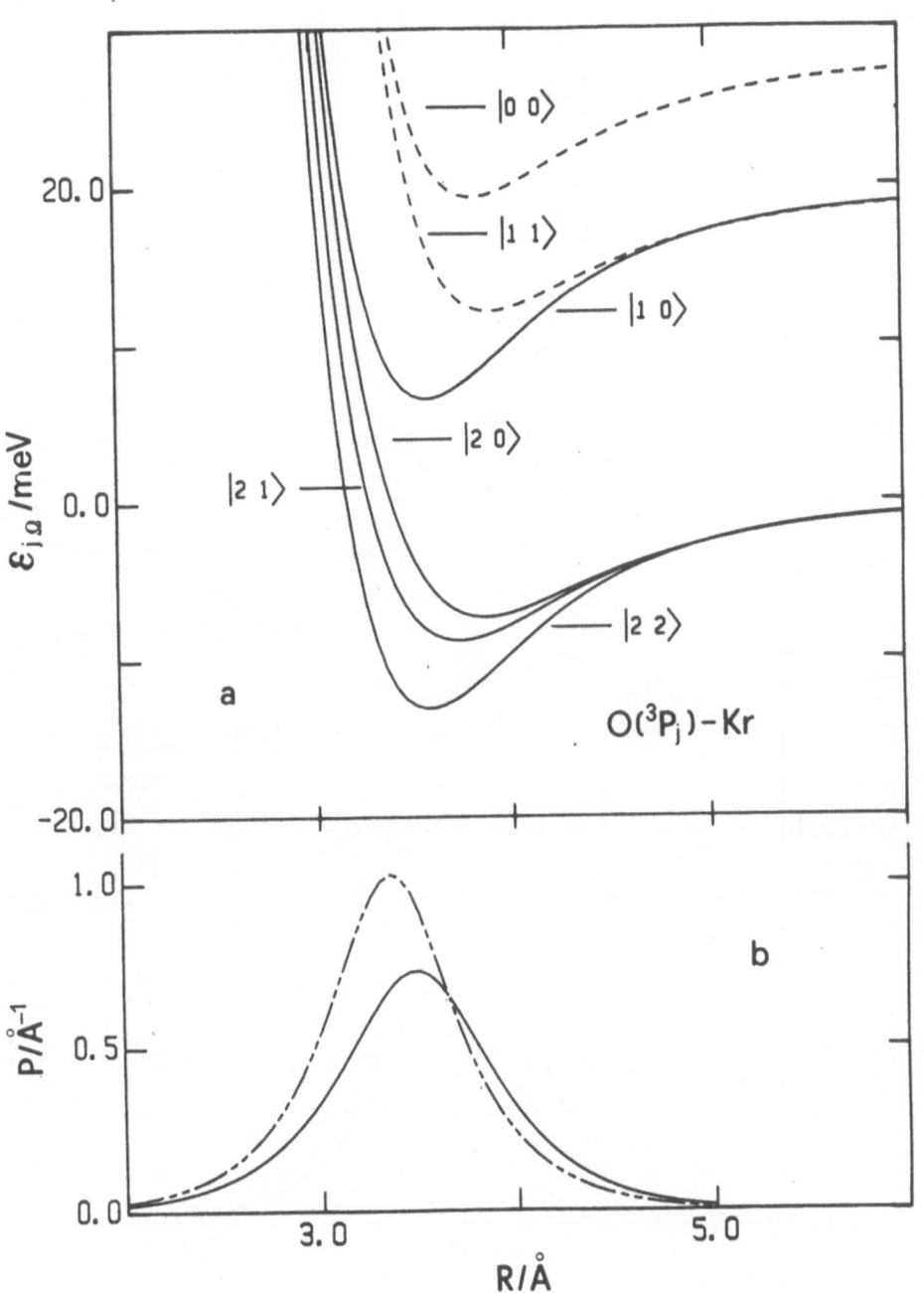

FIGURE 7.   As in Fig. 6, for 0-Kr system.

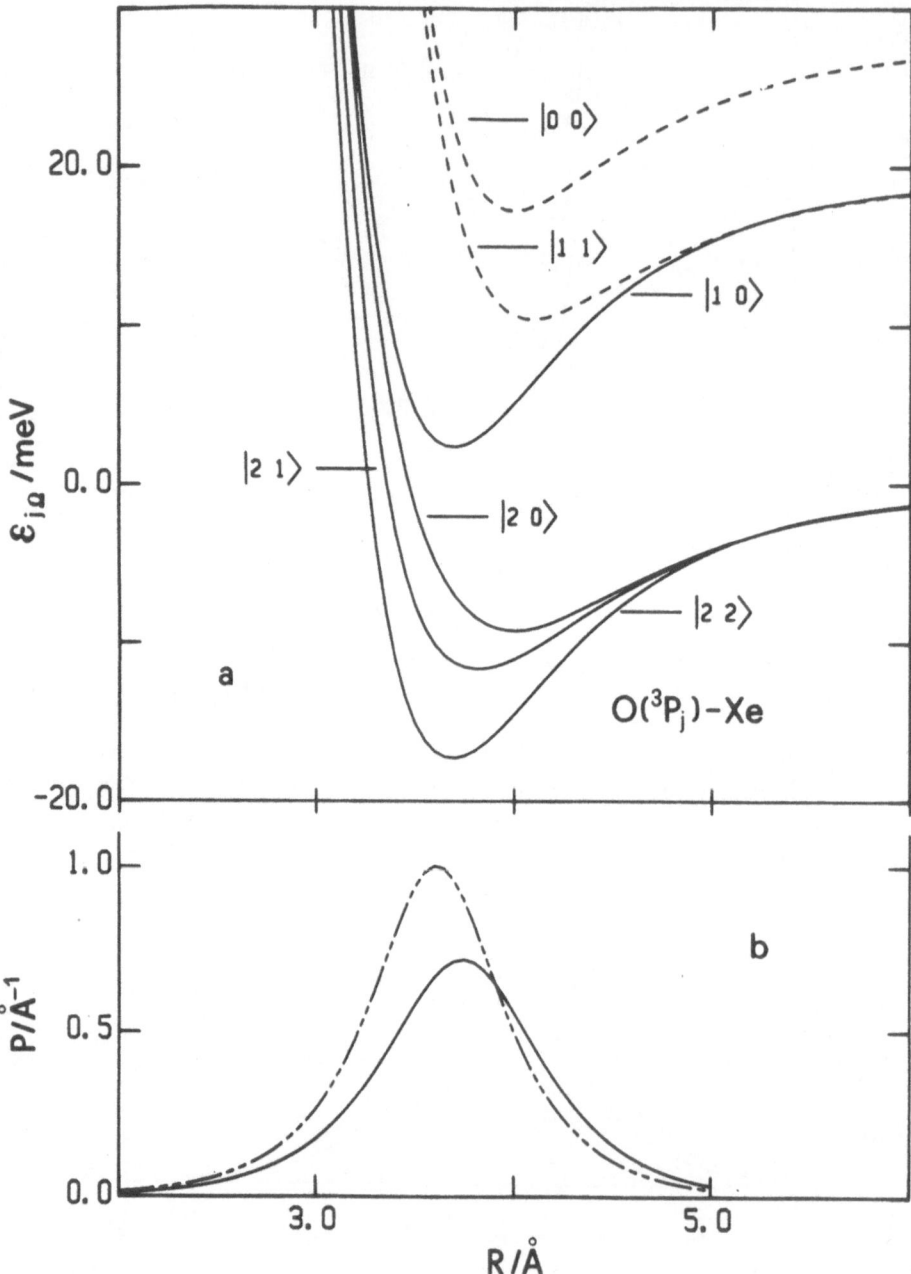

FIGURE 8. As in Fig. 6, for O-Xe system.

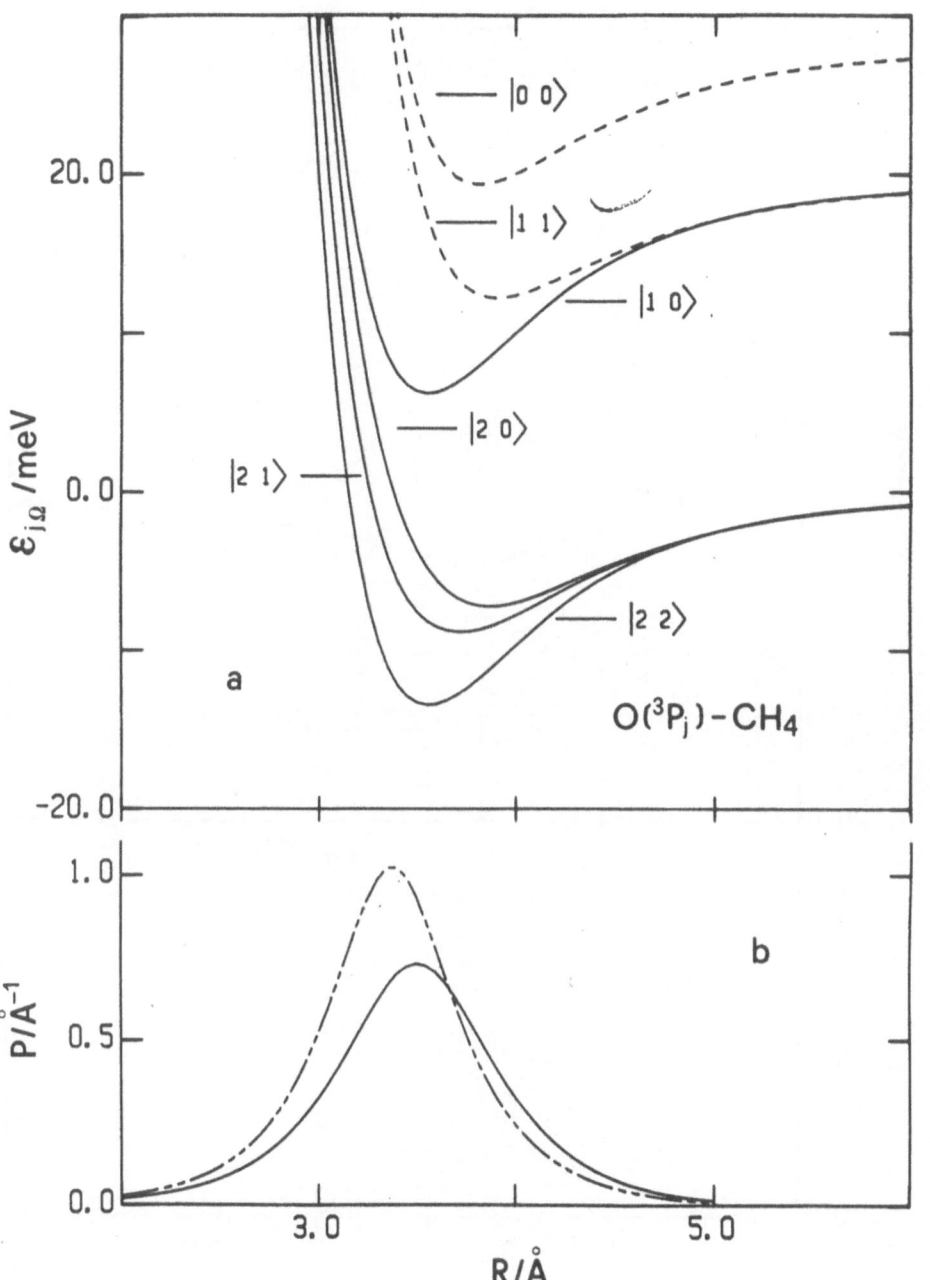

FIGURE 9.   As in Fig. 6, for 0-CH₄ system.

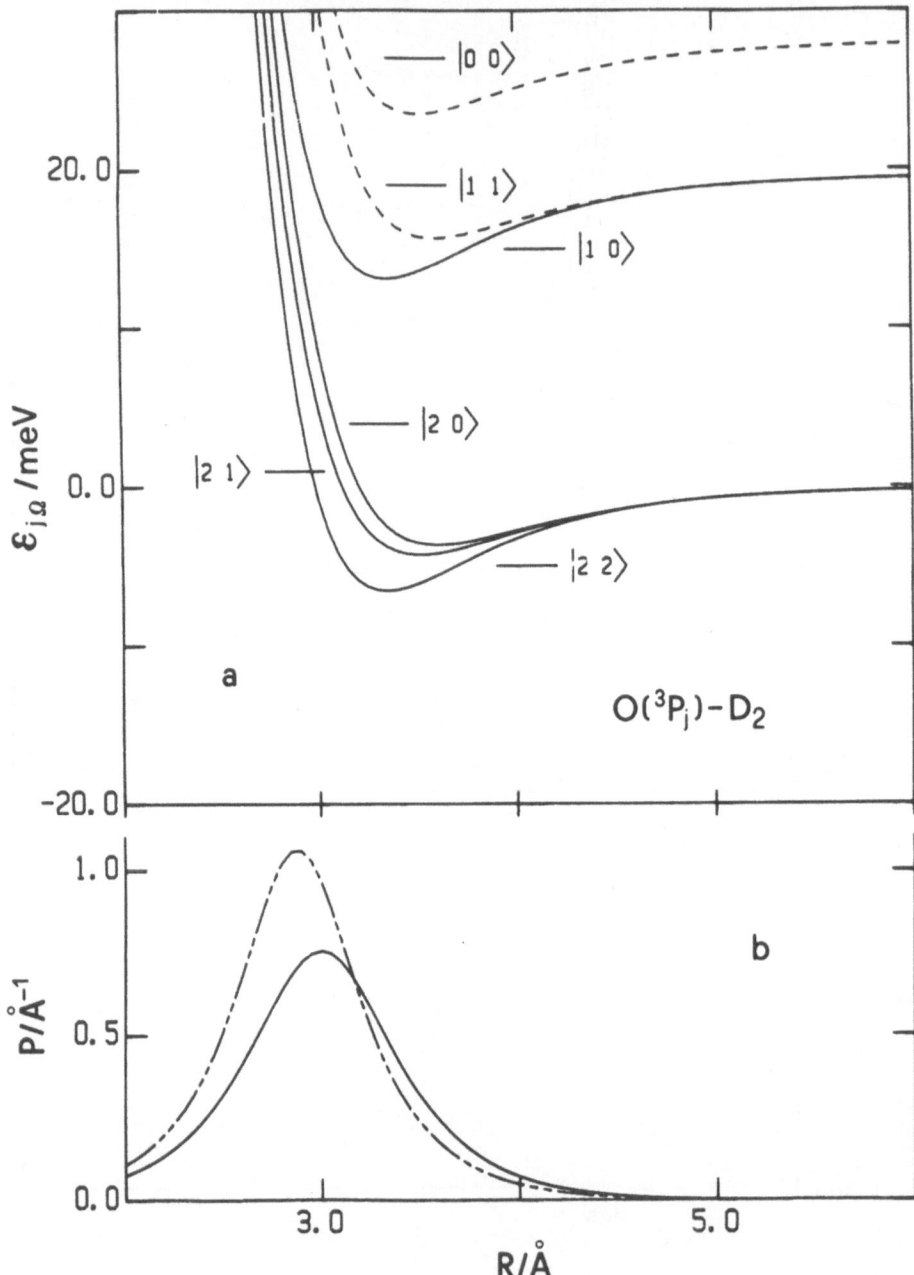

FIGURE 10.   As in Fig. 6, for O-D$_2$ system.

attraction. Furthermore, the contribution to the anisotropy interaction due to different orientations of the methane and deuterium molecules plays only a minor role under the experimental conditions of this work [12]. Hence the anisotropy effects observed can be attributed to intrinsic anisotropy of oxygen atoms.

Interestingly the potential curves obtained for O-Kr and O-CH$_4$ are very similar. This is not surprising, since the basic physical properties affecting the interaction are essentially polarizabilities and ionization potentials for the spherical part V_0 and for the anisotropic part V_2, respectively: for CH$_4$ and Kr these properties have very close values [12].

An interesting feature obtained by this analysis is given by the nonadiabatic coupling terms: they mark the transition between an atomic coupling scheme (Hund's case (c)) at long range and a molecular coupling scheme (Hund's case (a)) at short range. They are useful to estimate nonadiabatic effects and are needed to compute intramultiplet mixing, and orientation and alignment cross sections. In particular, since the first order nonadiabatic corrections to the adiabatic curves derived above is $\hbar^2 P^2/2\mu$, we can conclude that for present systems this correction is small for all systems considered here.

REFERENCES

[1] a) V. Aquilanti, E. Luzzatti, F. Pirani and G.G. Volpi, Chem. Phys. Lett., **90** (1982) 382.
b) V. Aquilanti, G. Grossi and F. Pirani, in: Electronic and Atomic Collisions, invited papers at XIII ICPEAC, eds. J. Eichler, I.V. Hertel and N. Stolterfoht (Berlin, 1983) p. 441.
c) V. Aquilanti, F. Pirani and F. Vecchiocattivi, in: Structure and Dynamics of Weakly Bound Complexes. A. Weber Ed. (Plenum, New York, 1987) p. 423.

[2] B. Brunetti, G. Liuti, E. Luzzatti, F. Pirani, and G.G. Volpi, J. Chem. Phys., **79** (1983) 273.
G. Liuti, E. Luzzatti, F. Pirani and G.G. Volpi, Chem. Phys. Lett., **121** (1985) 559.

[3] V. Aquilanti, G. Liuti, F. Vecchiocattivi, and G.G. Volpi, Chem. Phys. Lett., **15** (1972) 305.
F. Pirani and F. Vecchiocattivi, J. Chem. Phys., **66** (1977) 372.
E. Luzzatti, F. Pirani, and F. Vecchiocattivi, Molec. Phys., **34** (1977) 1239.
B. Brunetti, F. Pirani, F. Vecchiocattivi, and E. Luzzatti, Chem. Phys. Lett., **58** (1978) 504.
B. Brunetti, F. Pirani, F. Vecchiocattivi, and E. Luzzatti, Chem. Phys. Lett., **55** (1978) 565.
B. Brunetti, R. Cambi, F. Pirani, F. Vecchiocattivi, and M. Tomassini, Chem. Phys., **42** (1979) 397.
B. Brunetti, G. Liuti, E. Luzzatti, F. Pirani, and F. Vecchiocattivi, J. Chem. Phys., **74** (1981) 6734.
F. Pirani and F. Vecchiocattivi, Chem. Phys., **59** (1981) 387.

[4] N.F. Ramsey, Molecular Beams (Clarendon Press, Oxford, 1956).

194

[5] B. Brunetti, G. Liuti, F. Pirani and E. Luzzatti, Chem. Phys. Lett., **84** (1981) 201.

[6] a) V. Aquilanti, G. Liuti, F. Pirani, F. Vecchiocattivi, and G.G. Volpi, J. Chem. Phys., **65** (1976) 4751.
b) V. Aquilanti, E. Luzzatti, F. Pirani, and G.G. Volpi, J. Chem. Phys., **73** (1980) 1181.

[7] V. Aquilanti and G. Grossi, J. Chem. Phys., **73** (1980) 1165.

[8] V. Aquilanti, P. Casavecchia, G. Grossi and A. Laganá, J. Chem. Phys., **73** (1980) 1173.

[9] V. Aquilanti and G. Grossi, Lett. Nuovo Cim., **42** (1985) 157.

[10] V. Aquilanti, R. Candori, E. Luzzatti, F. Pirani and G.G. Volpi, J. Chem. Phys., **85** (1986) 5377.

[11] J.C. Whitehead in: Comprehensive Chemical Kinetics (C.H. Bamford and C.F.H. Tipper Eds., Elsevier 1983) **24,** 357 and references therein.
W. Tsang and R.F. Hampson, J. Phys. Chem. Ref. Data, **15**(1986) 1087 and references therein.

[12] G. Liuti, E. Luzzatti, F. Pirani and G.G. Volpi, Chem. Phys. Lett., **135** (1987) 387.
G. Liuti and F. Pirani, J. Chem. Phys. in press.

REACTIVE SCATTERING WITH ORIENTED MOLECULES:
SELECTIVITY IN THE Ba + $N_2O \rightarrow BaO^* + N_2$ REACTION

Henk Jalink, Maurice H.M. Janssen, Michiel Geijsberts
and Steven Stolte
Fysisch Laboratorium
Katholieke Universiteit Nijmegen
Toernooiveld
6525 ED Nijmegen
The Netherlands

David H. Parker and John Z.W. Wang
Department of Chemistry
University of California Santa Cruz
Santa Cruz
CA 95064
USA

ABSTRACT. A series of oriented reactant studies of the chemiluminescent Ba + $N_2O \rightarrow BaO^* + N_2$ reaction are described. Hexapole fields state-select and focus single rovibrational levels of N_2O for crossed beam reaction with Ba. The resulting chemiluminescence yield is measured for specified collision geometries, polarization, and coarse spectrum over a range of translational energies. Orientation dependent barriers to reaction are deduced using an angle-dependent line-of-centres model. An empirical correction for recrossing is discussed. Several features of the product channel are interpreted *via* the barrier to reaction.

1. INTRODUCTION

If one accepts that selectivity with respect to reagents leads to specificity with respect to products in a chemical reaction, then it is expected that the initial collision geometry influences many aspects of the outcome of reaction beyond the overall reactivity (i.e. the steric factor). After all, the steric effect was recognized as a dominant selective force long before vibrational or electronic state promotion was predicted. Oriented reactant scattering techniques have progressed to the point that investigation of the steric influence on several sensitive features of the product channel is possible. This paper gives an overview of the results from our own studies on selectivity in the Ba + oriented $N_2O \rightarrow BaO^* + N_2$ reaction. Here the orientation of N_2O in single selected rovibrational states is varied as we monitor the yield and other features of the product BaO* chemiluminescence. Surprisingly strong effects are seen, confirming the general expectations of the selectivity - specificity connection.

J. C. Whitehead (ed.), Selectivity in Chemical Reactions, 195–220.
© *1988 by Kluwer Academic Publishers.*

Chemiluminescence from the Ba + N_2O reaction is a convenient but not yet ideal means for probing orientation effects. In some ways its study is like looking first for a lost object under the street lamp where the light is good, rather than in the dark where the object first fell. Emission from BaO* [1,2] falls in the visible region (400 - 750 nm) and originates from the BaO $A^1\Sigma^+$ and $A'^1\Pi$ states. About 9% of the BaO chemiluminesces. The main channel is dark, vibrationally excited ground-state BaO [3]. The reaction is exceptionally exoergic (4.1 eV) and was initially considered a possible chemical laser source owing to the high light yield at intermediate reactant gas pressures. This practical interest led to a large accumulation of results from advanced chemical dynamic studies over the past two decades. A summary of the present knowledge on the Ba + N_2O system is updated in a previous paper [4] on the reaction of Ba with state-selected but unoriented N_2O. We have also reported the effects of orientation on the chemiluminescence polarization [5,6], coarse spectral features [7], and excitation function [8,9], $\sigma_{hv}(E_{tr})$, where σ_{hv} is the chemiluminescence cross section and E_{tr} the collision energy. Despite this acquired body of knowledge key features of the BaO (and N_2) internal state populations and of the actual reaction mechanism are not fully understood. Clearly, there is powerful information available from the orientation data, but partly because of the complicated nature of the reaction its full interpretation is not yet possible. At present these orientation results are most useful in uncovering general trends and more importantly in proving the effectiveness of the technique. The observations in this and so far all other oriented molecule reaction studies [10] have confirmed the selectivity - specificity connection, i.e. the experiment does control to some extent the collision geometry and the reaction carries a "memory" of this initial collision orientation into the product channels.

In the next section the main components of the experimental apparatus are illustrated. Analysis of orientation data is outlined after that, and an overview of results for Ba + oriented N_2O is then presented. We conclude by listing a few future directions of orientation selectivity in chemical reactions.

2. EXPERIMENTAL

2.1 Apparatus

Figure 1, an overview of the experimental apparatus, can be separated into three main components: the N_2O state-selection and orientation assembly, the crossed Ba beam, and the emission collection system. Since each component has been described in detail previously [5-10] only the vital features will be listed.

(a) N_2O state-selection and orientation
A seeded supersonic N_2O beam is state-selected in the $(J,\ell,M) = (1,1,1)$ or $(2,1,2)$ rotational state of the $n_2=1$ bending vibration by an electrostatic hexapole field, and focused 2.70 m downstream onto a collimator in front of the scattering chamber. E_{tr} is elevated by increasing the nozzle temperature and the fraction of He in the Ar/He seed gas mixture. The reaction zone is enclosed by "harp" orientation fields which control the collision geometry.

b) Effusive Ba beam
The important features of this simple effusive oven are the large $(2 \times 4$ mm$^2)$

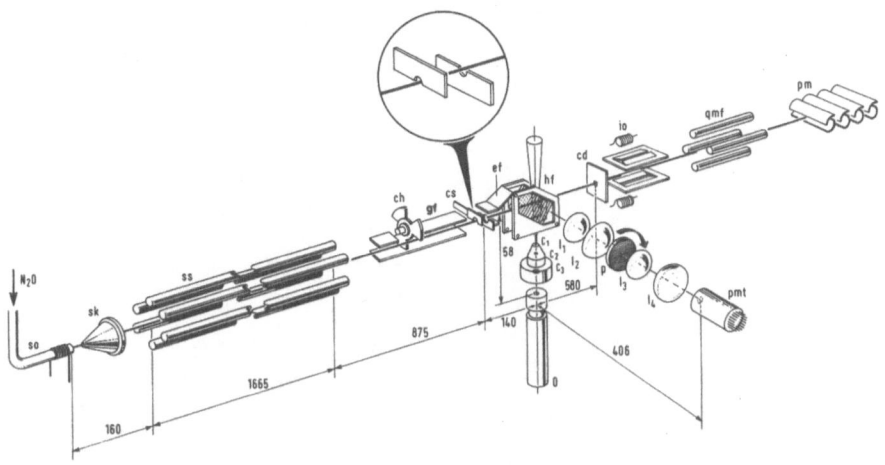

FIGURE 1. Schematic view of the Nijmegen orientation machine; sizes are in mm. The following abbreviations are used: so: heated quartz N_2O source, $\varphi=110$ μm; 2 skimmers, $\varphi_1=0.78$ mm, $\varphi_2=1.75$ mm; ss: electric hexapole state selector; ch: chopper; gf: guiding field plates; cs: collimator scattering chamber, $\varphi=2.0$ mm; ef: extra field; hf: harp orientation field plates; c: 3 circular collimators spaced 7.5 mm apart, $\varphi_3=5.0$ mm, $\varphi_1=\varphi_2=2.0$ mm; o: barium oven, T=1000 K; l_1-l_4: lens-system, f=1.5; p: polaroid sheet; pmt: photomultiplier; cd: collimator detector chamber, $\varphi=2.0$ mm; io: ionizer; qmf: quadrupole mass filter; pm: particle multiplier. The inset shows details of the scattering chamber collimator.

orifice, well-shielded heaters and extensive collimators for maximum rejection of thermal background emission from the 1000 K oven.

c)   Detection system
Figure 1 shows a rotatable polaroid sheet, p, for determination of the emission polarization with respect to $\vec{v}_r$. Bandpass and coloured glass cut off filters replace this polaroid sheet for coarse spectral analysis. Output from a bialkali cathode photomultiplier tube, pmt, is monitored by a photon counter signal averager at the hexapole modulation frequency and stored on a microcomputer that also controls the various field voltages and polaroid sheet positions.

2.2   Specification of collision geometry

Figure 2 displays the pertinent vectors defining the oriented N_2O - Ba collision. The left hand side of Fig. 2, shows the N_2O rotational vectors, $\vec{\ell}$, $\vec{N}$, $\vec{J}$, $\vec{M}$ and $\vec{\mu}_0$, the N_2O permanent dipole moment, which is directed antiparallel to the molecular axis, $\hat{r}$, pointing by convention [8] from N to O. Both precess about the applied field direction, $\hat{E}$, yielding an average value for the orientation with respect to $\hat{E}$, given by $\overline{\cos\theta} = \varrho M/(J(J+1))$ where θ is defined as the angle between $\hat{r}$ and $\hat{E}$.

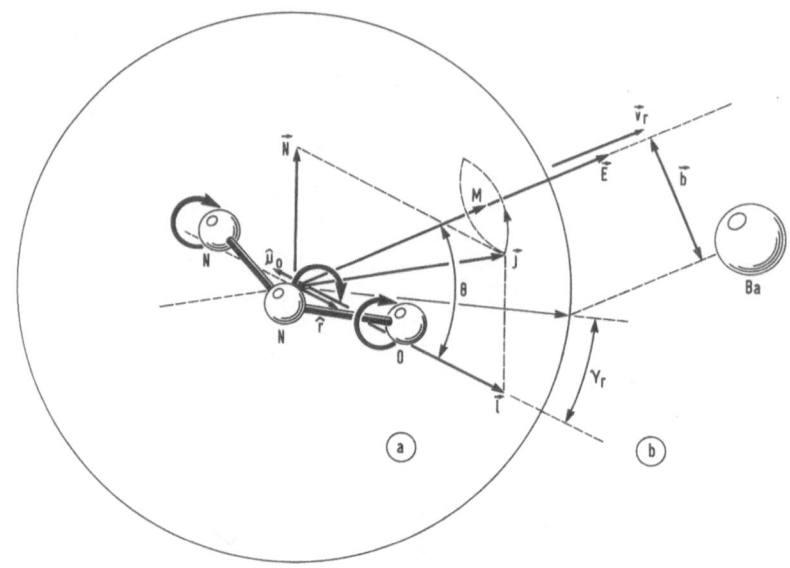

FIGURE 2. (a) Diagram of pertinent vectors defining the orientation of N_2O excited in the $n_2=1$ vibrational bending mode. Here $\vec{\ell}$, the orbital quantum number combines with $\vec{N}$, the end-over-end rotation quantum number, to give the total angular momentum $\vec{J}$, $\vec{J} = \vec{\ell} + \vec{N}$. The projection of $\vec{J}$ on an external field, $\vec{E}$, is given by $\vec{M}$. The molecular axis, $\vec{r}$, is directed opposite to the dipole moment, $\vec{\mu}_0$. Both precess around $\vec{E}$ at an angle θ, where the distributions for two (J,ℓ,M) states are plotted in Fig. 2. (b) As the Ba atom is approached at a relative velocity, $\vec{v}_r$, and impact parameter, b, a hard sphere (or ellipsoid) centred around N_2O defines the angle-of-attack at reaction, γ_r. By setting $\vec{E}$ ↑↑ or ↑↓ to $\vec{v}_r$ the value for the initial angle-of-attack at long distance, $\gamma_0 (= \gamma_r$ if b = 0), is either $\overline{\cos\theta}$ or $\overline{\cos(180^\circ-\theta)}$.

Reversing $\vec{E}$ flips the molecular axis. In this way the preferred direction of the molecular axis can be controlled spatially. Grounding the orientation field allows recoupling of the Stark-separated ℓ-doubled levels of N_2O (see next section) causing a complete loss of M, and thus orientation. A Ba atom approaches N_2O in the "favourable" orientation along $\vec{v}_r$ in Fig. 2, "unfavourable" corresponds to a reversed $\vec{E}$. By experimentally setting $\vec{E}$ ↑↑ or ↑↓ to $\vec{v}_r$ the pre-collision angle-of-attack, γ_0, of Ba on the N_2O axis at long distances is simply $\overline{\theta}$, or $180^\circ-\overline{\theta}$, respectively. The impact paramter, b, however is not controlled, thus the angle-of-attack at reaction, γ_r (= the reaction angle), is smeared somewhat [11]. A spherical (or elliptical) shell surrounding N_2O represents the position of the barrier to reaction which the Ba atom must surmount. In Fig. 2 γ_r is shown at this shell for a single b.

The molecular axis precesses at an average orientation, $\overline{\cos\theta}$, about $\vec{E}$ with a distribution [12,13], $P_{J\varrho M}(\cos\theta)$, which is quite broad for low J values. Figure 3 plots $P_{J\varrho M}(\cos\theta)$ for the (1,1,1) and (2,1,2) states. In spite of their wide range these θ distributions can be simply characterized [13-15], particularly for the (1,1,1) state, in terms of a short Legendre moment expansion:

$$P_{J\varrho M}(\cos\theta) = \sum_n C_n P_n(\cos\theta) \tag{1a}$$

where C_0 represents the isotropic contribution, C_1 represents orientation, C_2 alignment, etc. For the (1,1,1) and (2,1,2) states one obtains

$$P_{111}(\cos\theta) = \frac{1}{2}P_0(\cos\theta) + \frac{3}{4}P_1(\cos\theta) + \frac{1}{4}P_2(\cos\theta) \tag{1b}$$

and

$$P_{212}(\cos\theta) = \frac{1}{2}P_0(\cos\theta) + \frac{1}{2}P_1(\cos\theta) - \frac{5}{14}P_2(\cos\theta)$$
$$- \frac{1}{2}P_3(\cos\theta) - \frac{1}{7}P_4(\cos\theta). \tag{1c}$$

Hexapole state selection can provide a preference orientation ("heads" vs "tails") as well as alignment ("end-on" vs "side-on") information, i.e., both odd and even n > 0 values are produced at a fixed collision-geometry, while laser-photoselection methods [16] offer only the alignment moments (C_n with n even).

2.3 Representation of orientation-dependent reactivity once the reference field $\vec{E}$ is set with respect to $\vec{v}_r$

In order to extract information about the unknown $\sigma(\cos\gamma_0)$ one expands $\sigma(\cos\gamma_0)$ in a Legendre series [15]:

$$\sigma(\cos\gamma_0) = \sum_n \sigma_n P_n(\cos\gamma_0). \tag{2}$$

Note that σ_0 describes the case of "random" orientation. The probability distribution of $\cos\gamma_0$, $P_{J\varrho M}(\cos\gamma_0)_i$, for a purely selected state (J,ϱ,M) is characterized both by $P_{J\varrho M}(\cos\theta)$ and the direction of $\vec{E}$ with respect to $\vec{v}_r$ in the scattering region. Indicating with i a "favourable", "side-on", "random" or "unfavourable" orientation of $\hat{E}$ with respect to $\vec{v}_r$, one obtains [6]:

$$P_{J\varrho M}(\cos\gamma_0)_i = \sum_n C_n P_n(\cos\theta) \cdot P_n(\hat{v}_r \cdot \hat{E}) \tag{3a}$$

Substitution of Eq. 3a in Eq. 2 leads to [15]:

$$\sigma(\overline{\cos}\gamma_0, app)_i = \sum_n \sigma_n \langle P_n(\cos\theta)\rangle_{J\varrho M}^{app} \cdot \langle P_n(\hat{v}_r \cdot \hat{E})\rangle_i \tag{3b}$$

with $\langle P_n(\cos\theta)\rangle_{J\varrho M}^{app} = \langle P_n(\cos\theta)\rangle_{J\varrho M}/S$. The averaged value of $P_n(\hat{v}_r \cdot \hat{E})$ over the

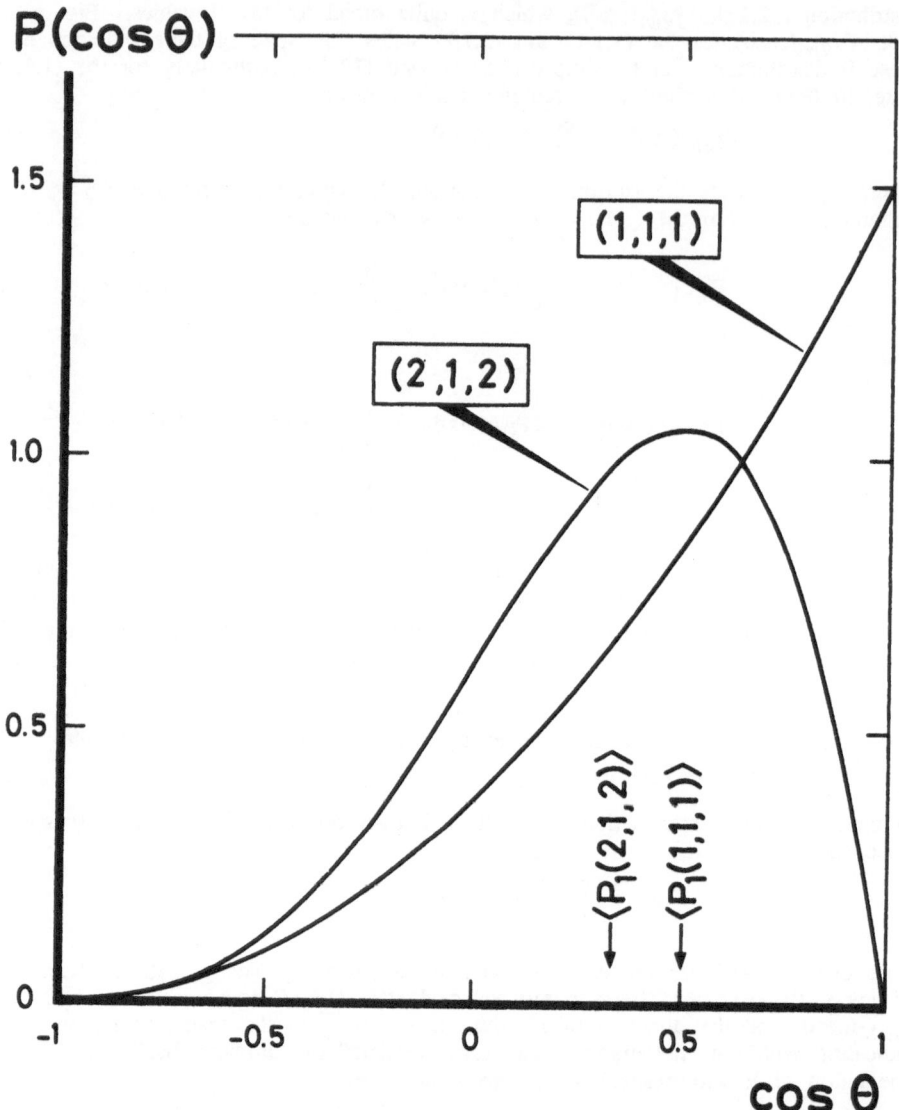

FIGURE 3. The probability function, P(cosθ), of the precession angle, θ, between the molecular axis, $\vec{r}$, and electric field, $\vec{E}$, for two N_2O rovibrational states, (J,ℓ,M) = $(1,1,1)$ and $(2,1,2)$ where ℓ is the quantum number of the ν_2-bending mode. The average angle $\overline{\cos\theta}$ = $\langle P_1(\cos\theta)\rangle_{J\ell M}$ = $\ell M/(J(J+1))$ yields 1/2 and 1/3 for the $(1,1,1)$ and $(2,1,2)$ states, respectively.

directional distribution of $\vec{v}_r$ with respect to $\vec{E}$ in the scattering region is indicated by $\langle P_n(\hat{v}_r \cdot \hat{E}) \rangle_i$ (note that $\langle P_n(\hat{v}_r \cdot \hat{E}) \rangle_{unfav} = (-1)^n \langle P_n(\hat{v}_r \cdot \hat{E}) \rangle_{fav}$). Furthermore one has:

$$\langle P_n(\cos\theta) \rangle_{J\varrho M} = \int_{-1}^{1} P_n(\cos\theta) P_{J\varrho M}(\cos\theta) d\cos\theta = \frac{2}{2n+1} C_n. \qquad (3c)$$

In the left side of Eq. 3b $\sigma(\overline{\cos\gamma_0}, app)_i$ (with $\overline{\cos\gamma_0} \equiv \int_{-1}^{+1} \cos\gamma_0 \cdot P_{J\varrho M}(\cos\gamma_0)_i d\cos\gamma_0$)

stands for the apparent reactive cross section for which the orientation can be somewhat degraded, due to imperfect strength of the orientation/guiding fields and the distribution of $\vec{v}_r$ with respect to $\vec{E}$ in the scattering region. Small corrections for kinematic averaging, spectral and polarization dependence of the chemiluminescent yield observed are already included in $\sigma(\overline{\cos\gamma_0}, app)_i$ [6]. S corrects for imperfect saturation by the orientation/guiding fields (Section 2.4.1) and $\langle P_n(\hat{v}_r \cdot \hat{E}) \rangle_i$ corrects for mismatch in direction between $\vec{E}$ and $\vec{v}_r$ and includes averaging over the velocity distribution of $\vec{v}_r$ induced by the effusive Ba beam [6]. The C_n (see Eq. 1) are calculated from Ref. 15 and the values of $\langle P_n(\cos\theta) \rangle_{J\varrho M}$ for the (1,1,1) and (2,1,2) states are given in Table 1.

TABLE 1. $\langle P_n(\cos\theta) \rangle_{J\varrho M}$ calculated from Eq. 3c. For the (1,1,1) and (2,1,2) state $\langle P_n(\cos\theta) \rangle_{J\varrho M} = 0$ for $n > 2$ and $n > 4$ respectively.

(J,ϱ,M)	$\langle P_0(\cos\theta) \rangle$	$\langle P_1(\cos\theta) \rangle$	$\langle P_2(\cos\theta) \rangle$	$\langle P_3(\cos\theta) \rangle$	$\langle P_4(\cos\theta) \rangle$
(1,1,1)	1	1/2	1/10	0	0
(2,1,2)	1	1/3	-1/7	-1/7	-2/63

The Legendre moments, σ_n, of the reactive cross section are extracted from the measured $\sigma(\overline{\cos\gamma_0}, app)_i$ at i = "favourable", "unfavourable", and "random" orientations, as:

$$\frac{a_1}{\sigma_0} = \frac{1}{2\langle P_1(\cos\theta) \rangle_{J\varrho M}} \left[\frac{\sigma_{fav}/\sigma_0 - \sigma_{unfav}/\sigma_0}{\langle P_1(\hat{v}_r \cdot \hat{E}) \rangle_{fav}} \right] \qquad (4a)$$

$$\frac{\sigma_1}{\sigma_0} = S \frac{a_1}{\sigma_0} \qquad (4b)$$

$$\frac{\sigma_2}{\sigma_0} = \frac{1}{2<P_2(\cos\theta)>_{J\varrho M}} \left[\frac{\sigma_{fav}/\sigma_0 + \sigma_{unfav}/\sigma_0 - 2}{<P_2(\hat{v}_r \cdot \hat{E})>_{fav}} \right] \qquad (4c)$$

where σ_i (i=fav or unfav) indicates $\sigma(\overline{\cos\gamma_0}, app)_i$ in short hand notation, a_1/σ_0 is defined as the anisotropy of reaction. Normally these correction factors $<P_n(\hat{v}_r \cdot \hat{E})>$ and S are all close to their ideal values of $P_n(\hat{v}_r \cdot \hat{E})$ and 1 respectively, and are described in detail elsewhere [6].

As seen from Eqs. 1b and 3 the (1,1,1) state contains only three moments, σ_0, σ_1 and σ_2, thus only three measurements, σ_{fav}, σ_{unfav} and σ_0 are needed to fully probe the basic (orientation and alignment) reaction anisotropy.

Our ability to study the orientation dependence of single rovibrational states now poses a problem. The (2,1,2) state has five moments, σ_0 - σ_4. The σ_{fav} + σ_{unfav} combination (and also $\sigma_{side-on}$) probes the even moments, σ_2 and σ_4, while σ_{fav} - σ_{unfav} probes the odd moments, σ_1 and σ_3. The five moments can not be determined independently. Several pieces of experimental evidence [6] suggest that the main differences in the J=1 and J=2 states are due to the first (and strongest) moments σ_1 and σ_2. Most significantly, the overall moment σ_0 differs by 20% between the J=1 and J=2 state reactants, suggesting that σ_1 and σ_2 also differ. We thus ignore the σ_3 and σ_4 contributions to the J=2 orientation dependence in the following analysis.

2.4 Rotational coupling effects

Molecules focused by the hexapole fields to the scattering centre start off in a single (essentially pure) (J,ϱ,M) state. Between the hexapole exit and scattering zone the local quantization axis must be steered adiabatically from the inhomogeneous electric field in the hexapole to the homogeneous orientation field along the direction of $\vec{v}_r$ in the scattering region. Whether the initial state selection remains pure during this transfer through the guiding field (gf) depends on the strength and gradients of $\vec{E}$. In N_2O the linear Stark-effect overcomes rotational coupling and hyperfine effects as long as $\vec{E}$ is large and gradients in $\vec{E}$ are small. At low field strength ϱ-doubling and hyperfine interactions substantially dilute the orientation especially of the lower J states. These types of effects are present in all systems but were ignored in most of the hexapole-oriented molecule reactions studied previously. Our apparatus and the cooperative Ba + N_2O chemiluminescence signal allows a direct gauge of the effects of hyperfine and ϱ-doubling on the quality of N_2O orientation. The good signal-to-noise ratio makes it possible to experimentally confirm that the applied field strengths are high enough to maintain state selection (practically, just M is in doubt), by using the reaction itself to probe the average precession angle of the molecular axis, i.e. by determining the laboratory anisotropy, a_1/σ_0 (see Eq. 4a), as a function of the applied voltage. Mixing, "flips" of +M and -M, transitions $M \to M - 1$, or misalignment of $\vec{E}$ with $\vec{v}_r$ drive the "favourable" and "unfavourable" reactivities towards that for unoriented N_2O molecules. Full retention of the orientation for a (J,ϱ,M) state can be checked experimentally by the appearance of (or approach to) a maximum reactive anisotropy with increasing applied voltage on the guiding or orientation fields.

2.4.1. _ℓ-doubling effects_. Mixing of $\pm\ell$ (ℓ-doubling) is the main rotational coupling effect in N_2O. Stark splitting uncouples this ℓ-doubling effect [17] to a "saturation" limit where pure (J,ℓ,M) states are recovered. Since the high field strength of the hexapole assembly ensures complete uncoupling of ℓ-doubling, the orientation field voltage must be high enough to retain this uncoupling.

"Heads" and "tails" orientations are mixed by ℓ-doubling, while at fields ⩾ 100 V/cm alignment remains unaffected. As shown in Ref. 17, for E < 100 V/cm hyperfine couplings start to dominate the Stark-effect (Fig. 4) and lead for the focused N_2O molecules, just as for NO in Ref. 18, to essentially isotropic distributions of the molecular axis (at E ~ 5 V/cm). High voltages are required to saturate the orientation moment, while alignment terms are already fully saturated. Stark-effect uncoupling of ℓ-doubling in N_2O for the orientation moment, $<\cos\theta>_{app}$, follows from calculating the expectation value of $\cos\theta$ for a two-level model [19]:

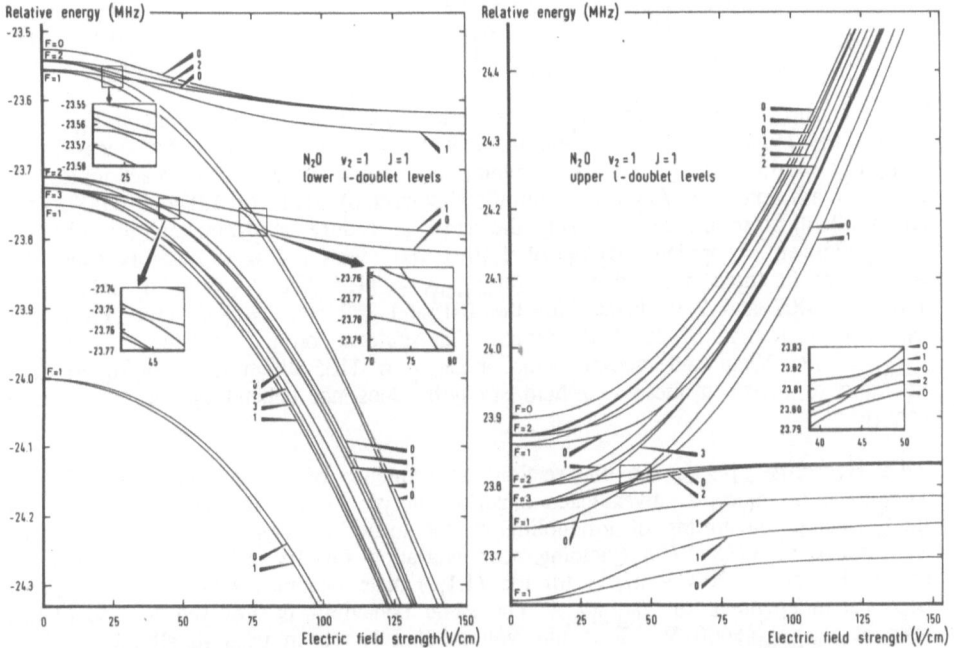

FIGURE 4. Energy level diagram of the $v_2=1, J=1$, ℓ-doublet of N_2O. The energy levels are labelled with their $|M_F|$-values. Notice that only levels with $M_F=0$ are one-fold instead of two-fold degenerate. The F quantum numbers valid at E=0 are indicated at the appropriate levels. The left panel of the figure contains levels with negative Stark-effect and connects to the lower components of the ℓ-doublet; the right panel to positive Stark-effect and the upper levels of the ℓ-doublet. Molecules in levels reaching the upper right hand corner of the figure are focused by the hexapole. In descending order from the highest upper ℓ-doublet levels at E=0, the following $N_{or}(F)$ values result in f($V_{guiding}$)=0.84, $N_{or}(F=0)=1$, $N_{or}(F=2)=5$, $N_{or}(F=1)=3$, $N_{or}(F=2)=4$, $N_{or}(F=3)=4$ and $N_{or}(F=1)=1$. In case of adiabatic following only the last three upper ℓ-doublet levels connect differently to the focusing states: $N_{or}(F=2)=5$, $N_{or}(F=3)=4$ and $N_{or}(F=1)=0$.

$$\langle\cos\theta\rangle_{app} = \frac{\langle P_1(\cos\theta)\rangle_{J\ell M}}{S} = \frac{\ell M}{J(J+1)} \left[1 + \left(\frac{\lambda_d(J)}{E_{Stark}}\right)^2\right]^{-1/2} = \frac{\ell M}{J(J+1)} \cdot \frac{1}{S}$$

(5)

where the $\lambda_d(J) = \lambda_d J(J+1)$ term is the ℓ-doubled splitting energy, $\lambda_d = 11.875$ MHz for N_2O ($n_2=1$) [17] and E_{Stark} the Stark splitting energy, $E_{Stark} = [\ell M/J(J+1)]\mu_0 E_{harp}$ ($\mu_0 = 0.173$ D [17]). Comparing at a given $\langle\cos\theta\rangle_{app}$ the $(J,\ell,M) = (1,1,1)$ and $(2,1,2)$ states, E_{harp} for a J=2 state must be 4.5 times larger than E_{harp} for the J=1 state.

Equation 5 allows a quantitative prediction of the orientation field voltage dependence of a_1/σ_0:

$$\frac{a_1}{\sigma_0} = \frac{\sigma_1}{\sigma_0} \cdot \left[1 + \left(\frac{\lambda_d(J)}{E_{Stark}}\right)^2\right]^{-1/2}$$

(6)

Figure 5 shows the measured a_1/σ_0 as a function of the electric field at $\bar{E}_{tr} = 0.10$ eV for the J=1 and J=2 states and the calculated behaviour, fitted by a least-squares computer routine through the experimental data. For the J=1 state reaction Eq. 6 gives an asymptotic σ_1/σ_0 (i.e. "saturation"-corrected) value of 1.034 ± 0.006, and an excellent fit to the data. Using the same procedure andparameters for the J=2 data yields an extrapolated σ_1/σ_0 of 1.40 ± 0.04 with a less satisfactory (but acceptable) fit through the data. S (at $E_{harp} = 1135$ V/cm) of Eq. 6 yields 1.109 ± 0.007 and 2.38 ± 0.06 for the J=1 and J=2 state respectively. From an earlier measurement of the J=1 orientation dependence on the collision energy at $E_{harp} = 694$ V/cm and measurements at $E_{harp} = 1135$ V/cm it is confirmed that the ratio of a_1/σ_0 at these two field strengths does not depend on the translational energy.

2.4.2 Hyperfine effects. In this section the influence of the guiding field (gf in Fig. 1) voltage on a_1/σ_0 is investigated experimentally. During this study the orientation field, forcing uncoupling of ℓ-doubling in the scattering region, has been left at a fixed value of 1135 V/cm (yielding near saturation for the (1,1,1) state). Figure 6 shows that above 200 V a_1/σ_0 for the (1,1,1) state becomes saturated, i.e. its value becomes independent of $V_{guiding}$. The same behaviour is observed for the (2,1,2) state at $V_{guiding} \geqslant 600$ V. It is this voltage that has been used in all other experiments of this paper. Below 200 V for (1,1,1) on the guiding field plates (~ 600 V for (2,1,2)) a significant drop in a_1/σ_0 is seen. At $V_{guiding}=0$ V there remains an appreciable orientation effect. This can be understood from Fig. 4, which shows in its right panel the upper λ-doublet levels labelled with their F and M_F quantum numbers at zero field and high field respectively. The molecules coming out of the hexapole field all populate levels with a strong positive Stark-effect. When these molecules enter a field free region (i.e. $V_{guiding} \sim 0$) their energy levels connect to the degenerate M_F manifolds at zero field. After passing the guiding fields they reach the orientation region of the machine where a high field is applied and several of the M_F states exhibit a positive Stark-effect again but some states don't. This results in a reduction of the orientation initially prepared at the hexapole.

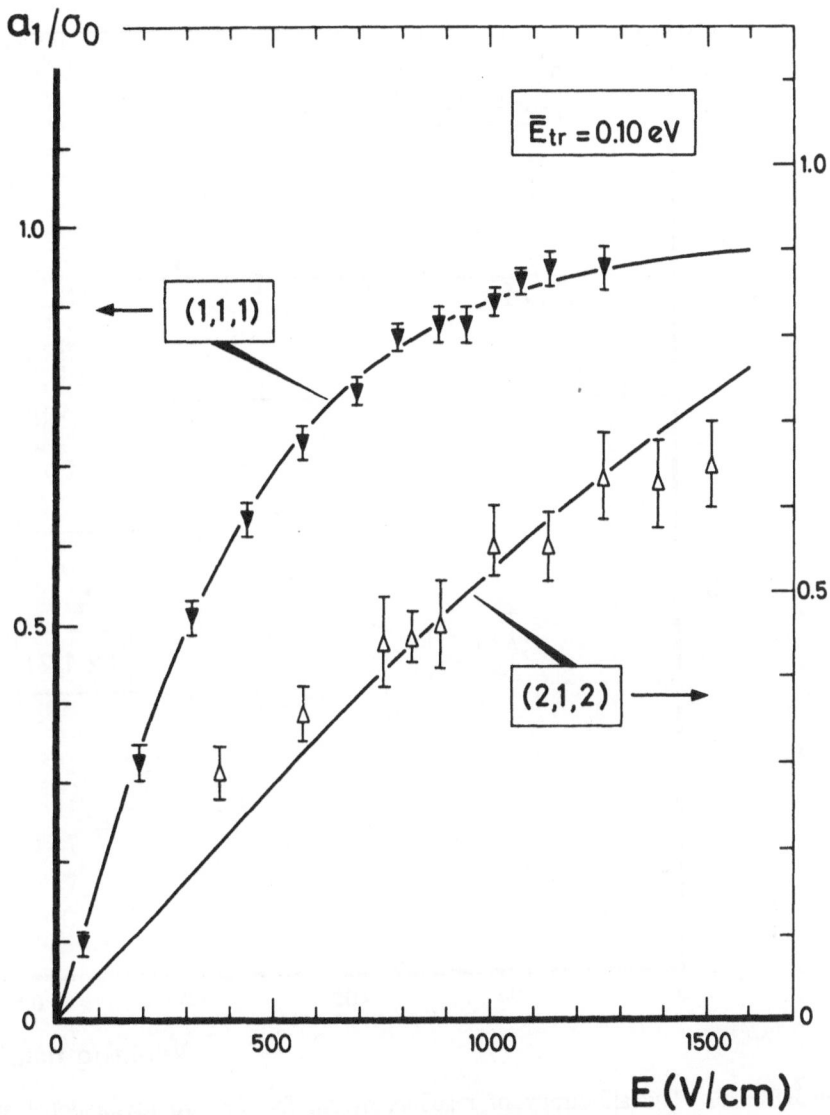

FIGURE 5. Saturation curves of the laboratory anisotropy of reaction, a_1/σ_0, as a function of the calculated electric field of the harp plates for the two rovibrational states of N_2O. The solid curves fitted through the experimental points using Eq. 6, have asymptotic values of $\sigma_1/\sigma_0 = 1.034 \pm 0.006$ for J=1 and $\sigma_1/\sigma_0 = 1.40 \pm 0.04$ for J=2 at $\bar{E}_{tr} = 0.10$ eV and $E_{guiding} = 300$ V/cm. The correction factors for saturation, S, are 1.109 ± 0.007 and 2.38 ± 0.06 for J=1 and J=2 respectively. The larger error bars in Fig. 7 are due to systematic uncertainties excluded from this fit.

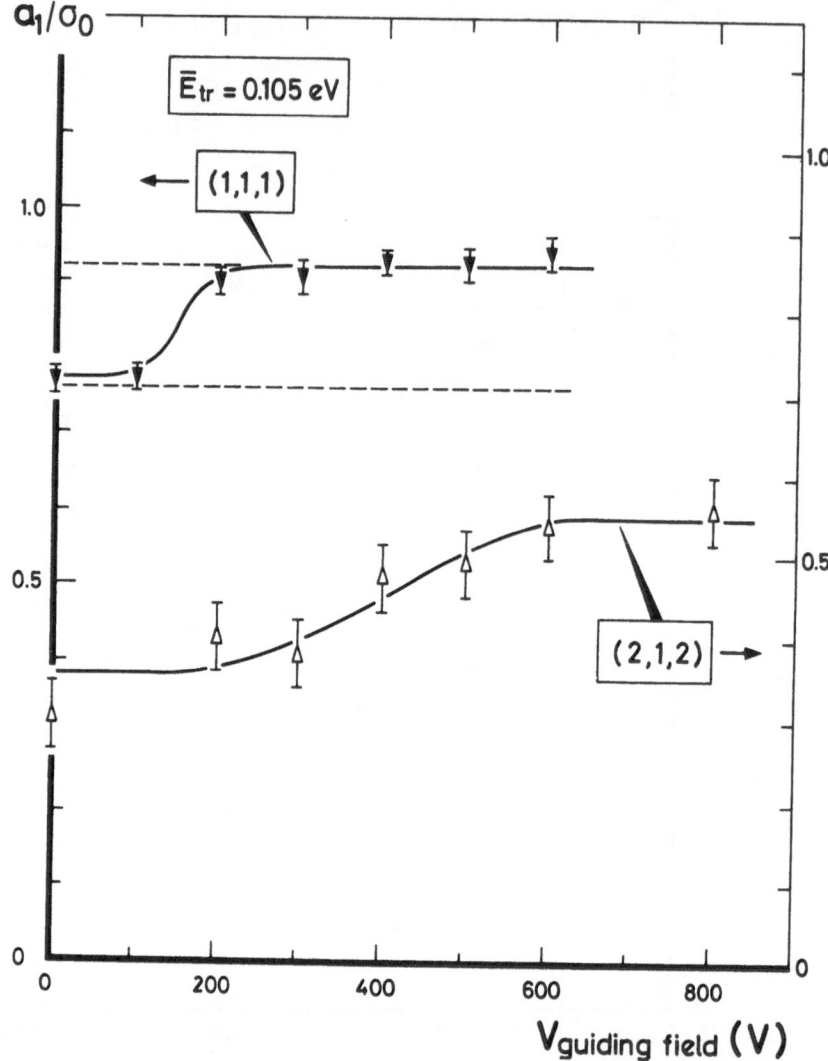

FIGURE 6. The anisotropy of reaction, a_1/σ_0, for the two rovibrational states of N_2O as a function of the voltage on the guiding plates. Smooth curves are drawn through the experimental points. The guiding plates are spaced 2 cm apart, thus $E_{guiding}=0.5\ V_{guiding}$. During this measurement the field strength on the harp plates was kept at 1135 V/cm. The upper dashed line corresponds to the value of a_1/σ_0 in Fig. 5 at this field strength. The lower dashed line corresponds to $f(V_{guiding}=0)=0.82$ as explained in Section 2.4.2.

By calculating the relative population of an F state at zero field from the filling out of the (equally populated) M_F substates with positive Stark-effect at high field and multiplying this relative population with the number of M_F states from this F manifold, $N_{or}(F)$, retaining orientation, i.e. again the M_F states with positive Stark-effect, the fraction of retained orientation at $V_{guiding}=0$, $f(V_{guiding}=0)$, can be determined. Explicitly

$$f(V_{guiding}=0) = \frac{\sum\limits_{F} (N_{or}(F))^2/(2F+1)}{\sum\limits_{F} N_{or}(F)}. \tag{7}$$

As can be seen from Fig. 4 there are a few sharp crossings, shown in the box of the right panel of Fig. 4, of levels with equal M_F from different F states. These avoided crossings result in a rather sharp bending of the curves and a non-adiabatic tunnelling from one F state to the other can be expected. Assuming this to be the case a simple counting of states yields $f(V_{guiding}=0)=0.82$ in very good agreement with the experimental result of 0.84 (see Fig. 6). Assuming an adiabatic passage through the avoided crossing one derives from Eq. 7 $f(V_{guiding}=0)=0.91$, a value not confirmed by our experiment. Qualitatively as noted in the beginning one observes a dependence of a_1/σ_0 for the (2,1,2) state upon $V_{guiding}$ resembling that for the (1,1,1) state. Similar effects are expected to play a role in both cases, unfortunately a Stark-energy level diagram as that of Fig. 4 is not available for the (2,1,2) state.

3. RESULTS

3.1 Steric Effect and Orientation-Fixed Excitation Functions

Three apparent cross sections, σ_{fav}, σ_{unfav}, and $\sigma_{random}=\sigma_0$, are measured over a range of collision energies, E_{tr}, for reactions of $(J,\ell,M) = (1,1,1)$ and $(2,1,2)$ N_2O with Ba. Figure 7 shows the data (normalized by σ_0) after deconvolution of $\vec{v}_r\cdot\vec{E}$ mismatch, saturation of ℓ-doubling and the $<P_n(\cos\theta)>$ average orientation factor. Interesting differences between J=1 and J=2 reactants are seen. The steric effect varies little with E_{tr} for the J=2 state while declining rapidly for J=1. Also, the "unfavourable" cross section for J=2 is larger than for J=1, indicating a stronger (negative) alignment contribution.

The orientation-averaged excitation function, $\sigma_0(E_{tr})$, has been measured previously for the J=1 and J=2 reactants [4]. After deconvolution of the wide velocity range (due to the Ba beam) of $\bar{E}_{tr}$, $\sigma_0(E_{tr})$ is combined (for the J=1 state) with the data of Fig. 7 and further deconvoluted for $\overline{\cos\gamma_0}$ to yield the fixed-orientation excitation functions shown in Fig. 8. Three orientations, "heads-on" ($\cos\gamma_0 = 1$), "side-on" ($\cos\gamma_0 = 0$) and "tails-on" ($\cos\gamma_0 = -1/2$), are plotted. The pure "heads-on" orientation shows a strong energy dependence whilst for a $\cos\gamma_0 = -1/2$ orientation the chemiluminescent cross section is almost independent of E_{tr}. The decline in $\sigma(E_{tr})$ at higher collision energies is often ascribed to "recrossing" and is discussed in Section 3.3.

Another striking feature of the fixed-collision geometry excitation functions is the sharp maximum seen for "favourable" collisions. The chemiluminescent cross section $\sigma_{h\nu}(\bar{E}_{tr})$ for unoriented species is calibrated [4] using the result that the total

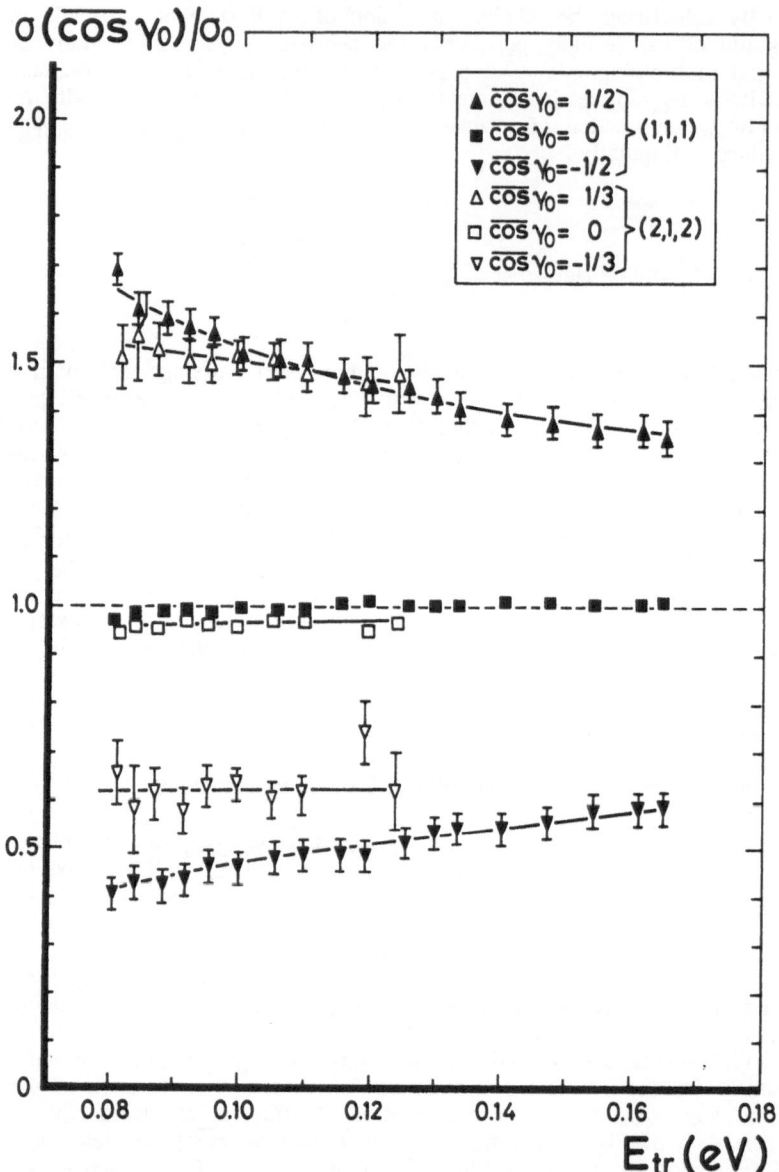

FIGURE 7. Deconvoluted steric data. The total chemiluminescent cross sections are plotted at various values of $\overline{\cos}\gamma_0$ of the selected $(1,1,1)$ or $(2,1,2)$ states. The points are corrected for velocity smearing effects due to the large spread in v_{Ba} and saturation of the σ_1 moment. The data points belonging to the $J=1$ state are also corrected for the wavelength and polarization dependence of the detection system.

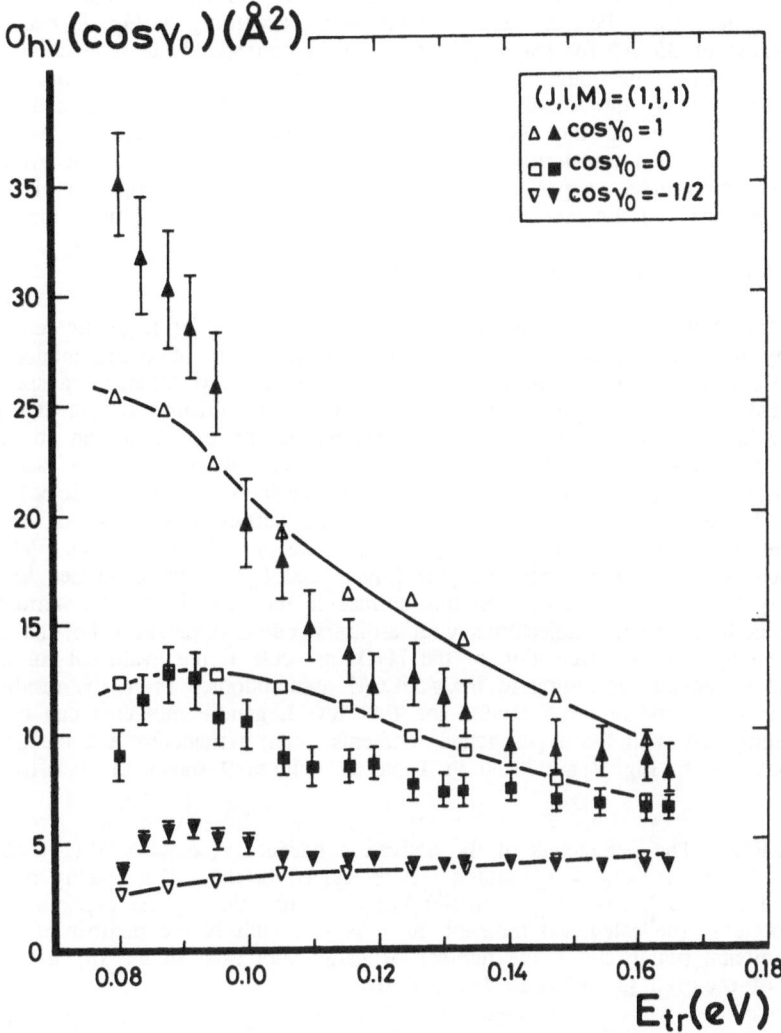

FIGURE 8. Deconvoluted steric excitation functions for "heads-on" ($\cos\gamma_0 = 1$), "side-on" ($\cos\gamma_0 = 0$) and $\cos\gamma_0 = -1/2$ for the $n_2=1, J=1$ state of N_2O. The experimental values are plotted as solid symbols, the open symbols connected by a solid curve represent the results obtained with the recrossing ADLCM model of Section 3.3.

chemiluminescence yield at 0.10 eV collision energy is 9.3% of the total (reaction) cross section, σ_R. Perfect "heads-on" collisions, according to Fig. 8, occur with a σ_{hv} value of 35 Å^2 (or σ_R = 376 Å^2). This corresponds to a reaction distance of 10.9 Å between the colliding partners, which is quite difficult to rationalize with standard electron-jump impulsive models. As the calibration of σ_{hv} and σ_R is far from perfect (e.g. σ_R has been calibrated as a total cross section for the removal of Ba out of a beam transferring through N_2O gas and could somewhat overestimate σ_R) and their values appear to be that large, a renewed investigation of the absolute cross sections and the recoil distributions of reactive products [20] is highly desirable.

3.2 Angle Dependent Line-of-Centres Model Analysis

Analysis of the results of Fig. 7 can be done by means of Legendre and empirical opacity functions, as well as *via* the angle-dependent line-of-centres model (ADLCM) [11,21,22] which will be discussed here. In the ADLCM treatment a spherical (or elliptical [6]) shell surrounds the N_2O molecule and collisions by Ba at this shell are evaluated as to whether their line-of-centres energy overcomes an empirical angle dependent barrier, $V(\cos\gamma_r)$, to reaction. For simplicity we assume a two parameter potential, employing a cut-off angle, γ_c ($V(\cos\gamma_r) = \infty$ for $\cos\gamma_r < \cos\gamma_c$) and a slope, V', describing a line that decreases to no barrier to reaction for the "favourable" collision approach ($\cos\gamma_r \to 1$), $V(\cos\gamma_r) = V'(1 - \cos\gamma_r)$. $V(\cos\gamma_r)$ is our (collision energy independent) one-dimensional (γ_r) potential surface for reaction. Equally simple, our dynamics on this surface is the line-of-centres criterium. At given collision energy trajectories with arbitrary initial orientation, i.e. the angle between $\vec{v}_r$ and the orientation of the N_2O molecule $\vec{r}$, are evaluated for all possible impact parameters according to the ADLCM assumptions. From the resulting orientation dependent cross section the first few Legendre moments can be extracted and compared with the experimental moments. The parameters and characteristics of the best fits to Fig. 7 are listed in Table 2. Figure 9 shows the best-fit

TABLE 2. The parameters of the activation barrier, $V(\cos\gamma_r) = V'(1 - \cos\gamma_r)$ with a cut-off angle, $\cos\gamma_c = x_c$, and a sphere shaped barrier. The quality of the fit is defined as $1/m \sum_n [(\sigma(\exp)_n - \sigma(\text{calc})_n)/\text{err}_n]^2$, with $\sigma(\exp)_n$ the experimental moment and $\sigma(\text{calc})_n$ the calculated moment, m = N - k with N the number of fitted experimental points and k the number of parameters used in the fit, and err_n the error of $\sigma(\exp)_n$. (a) refers to J=1 and (b) to J=2.

fitted moments	parameters	quality	V' [eV]	x_c
$\dfrac{\sigma_1}{\sigma_0}$	(a) V', x_c	0.50	0.063 ± 0.001	-0.45 ± 0.02
$\dfrac{\sigma_1}{\sigma_0}$	(b) V', x_c	0.36	0.065 ± 0.001	0.00 ± 0.02

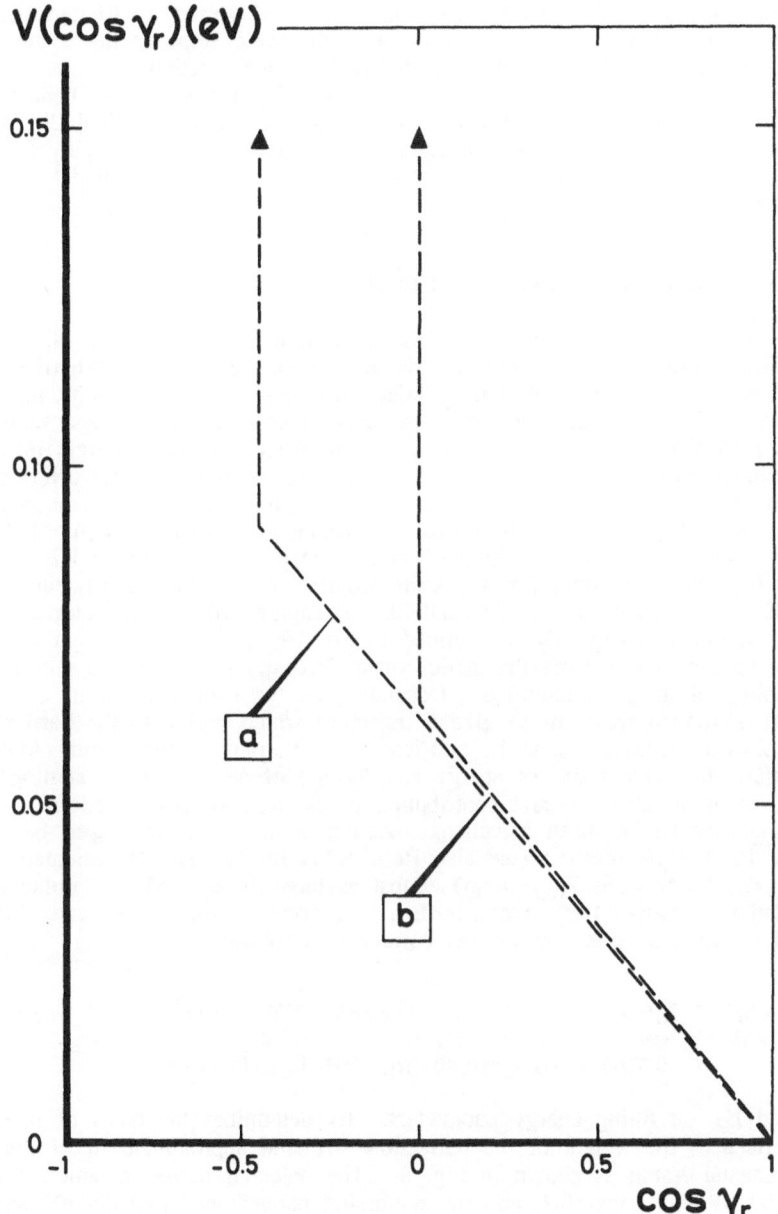

FIGURE 9. Optimal activation barriers, V(cosγ$_r$), as a function of the angle-of-attack at reaction, γ$_r$, from a least square computer fit to selected moments of the experimental data (all collision energies) using the ADLCM. The relevant parameters are listed in Table 2.

orientation-dependent barriers to reaction. Again J=2 shows a different steric behaviour than J=1, in that the barrier is of the same slope as but a much larger cut-off angle (cone-of-non-reaction) than the J=1 state reaction.

These barriers are deduced without considering the recrossing behaviour evident in Fig. 8. As discussed in the next section, adjustment of the ADLCM for recrossing modifies the barriers to reaction. These barriers are employed in the following section to rationalise the product internal energy and polarization dependence on orientation.

3.3   Recrossing and the Barrier to Reaction

A qualitative explanation of the fall-off in reactivity, evident in Fig. 8, at higher collision energies consistent with the deduced orientation dependent barrier to reaction concerns the "no-return" criterion. Trajectories passing the saddlepoint may reflect off the repulsive wall (the N_2-group in N_2O), recross the barrier and return to the reactant channel. Such behaviour is likely for early barriers (exoergic reaction) as the velocity over the barrier increases. "Heads-on" collisions (no barrier) should recross at low E_{tr} while "backside" collisions (high barriers) recross at high E_{tr} (if at all, since the latter geometries avoid "touching off" the N_2-group). Indeed, Fig. 8 shows strong recrossing for $\cos\gamma_0=+1$ and no recrossing for $\cos\gamma_0=-1/2$.

In order to account for recrossing features in the angle dependent line-of-centres model the normal ADLCM assumption, that all trajectories which have sufficient kinetic energy E_a (as defined in Eq. 10 of Ref. 11) along the line-of-centres to overcome the steric barrier $V(\cos\gamma_r)$ $(E_a>V(\cos\gamma_r))$ will react with a probability of unity, is modified. Depending on the magnitude of its excess energy, $E_{exc}\equiv E_a-V(\cos\gamma_r)$ we want to give a trajectory which will pass the barrier also a probability to recross (i.e. to be non-reactive). Employing the normal ADLCM procedure the average excess energy $\bar{E}_{exc}(\cos\gamma_0)$ of all trajectories starting with an initial angle of attack γ_0 and contributing to the reactive cross section $\sigma_{ADLCM}(\cos\gamma_0)$ has been calculated. Assuming a spherically shaped barrier relating to the Ba + N_2O reaction (see also Refs. 8,11) for "favourable" orientations $(\cos\gamma_0 \to 1)$, one obtains $\bar{E}_{exc}(\cos\gamma_0)$ almost as large as E_{tr} and for "unfavourable" orientations $(\cos\gamma_0 \to 1)$ $\bar{E}_{exc}(\cos\gamma_0)<<E_{tr}$. To account crudely for recrossing the ADLCM reactive cross section was corrected as follows:

$$\sigma(\cos\gamma_0) = \sigma_{ADLCM}(\cos\gamma_0)\cdot\exp[-(\bar{E}_{exc}(\cos\gamma_0)-E_1)/E_2] \text{ for } \bar{E}_{exc}(\cos\gamma_0) \geqslant E_1$$

$$\text{and} \qquad \sigma(\cos\gamma_0) = \sigma_{ADLCM}(\cos\gamma_0) \text{ for } \bar{E}_{exc}(\cos\gamma_0)<E_1 \qquad (8)$$

E_1 and E_2 are fitting energy parameters. E_1 determines the onset of recrossing and E_2 influences the weight of the correction. A first approximate fit of the experimental results is shown in Fig. 8. The potential barrier parameters are $V'=0.03$ eV and $\cos\gamma_c=0.2$, and the recrossing parameters $E_1=0.035$ eV and $E_2=0.04$ eV. Including this recrossing correction into ADLCM the overall features of the energy dependence of the cross sections shown in Fig. 8 agree qualitatively well. For pure "heads on" collisions $\sigma(\cos\gamma_0=1)$ decreases rapidly with increasing translational energy as the trajectories experience a relatively low barrier with a rather large line-of-centres energy and therefore yield a large excess energy.

In comparison with the results of Section 3.2 (see Table 2), the steric cone of acceptance, γ_c, had to be decreased quite strongly, from $\approx 117^0$ to 78^0. This means that trajectories with an "unfavourable" initial orientation, $\cos\gamma_0 = -0.5$, can pass the barrier at large impact parameters only. Consequently the line-of-centres energy for these reactions will be low and therefore the barrier will also be crossed with small excess energy only. Hence the resulting cross section $\sigma(\cos\gamma_0 = -0.5)$ remains essentially unaffected by the recrossing correction of Eq. 8. In Fig. 9 $\sigma(\cos\gamma_0 = 0.5)$ is found to increase slowly with E_{tr}. When E_{tr} increases also trajectories with somewhat smaller impact parameters will be able to overcome the barrier at a larger reaction angle, γ_r (see Fig. 2), i.e. at a higher potential $V(\cos\gamma_r)$ and thus contribute with a low excess energy and show absence of recrossing. Further work improving the adding of the recrossing feature into the ADLCM model is in progress.

3.4 Orientation Dependence of the Internal Energy Distribution

Figure 10 shows chemiluminescence spectra taken by observing the BaO* emission through a series of bandpass and coloured glass filters (corrected for the photomultiplier tube response) for the three experimentally prepared Ba - N_2O reaction orientations at a collision energy of 0.105 eV and 0.162 eV. The two sets of spectra are scaled to each other at 490 nm in the "randomly" oriented collisions curve. The total (undispersed) chemiluminescence cross section at 0.105 eV is 1.5 times larger than at 0.162 eV. Overall, the spectra are in rough agreement with previously reported (but non state-selected) spectra [1,3]. Much more light is produced by the "favourably" oriented reactants compared to the "unfavourable" case, especially in the 490 nm region. Interestingly, an experiment comparing spectra collected directly above and downstream from the scattering centre identifies strong bands in this region with the short-lived A state [2]. The dependence upon reactant alignment can be obtained by comparing the spectrum of unoriented N_2O and the spectrum of averaging a "favourable" and an "unfavourable" orientation [7]. In all cases (except at $\lambda = 490$ nm in panel (b) of Fig. 10) the effect induced by alignment has been found to be insignificant.

Figure 11 shows the orientation dependence of the chemiluminescence spectrum for 0.105 eV and 0.162 eV collisions. Plotted is σ_1/σ_0 versus the emission wavelength. These data are more accurate than the histogram spectra of Fig. 10 since each collision geometry was repetitively measured with the same filter. At these two collision energies the unfiltered (total) chemiluminescence data are in accord with the integrated, uncorrected spectra.

A very strong orientation dependence is apparent, especially at higher collision energy where σ_1/σ_0 ranges from nearly zero (no difference in the two orientations) for light produced at 400 nm to very strong ($\sigma_1/\sigma_0 = 1.0$) differences at the longer wavelengths. Similar trends are observed at the lower collision energy, although the difference in the blue end of the spectrum is less pronounced. Again, "favourable" reactions produce strong chemiluminescence peaked in the 490 nm region while the emission produced by reactions via the "unfavourable" geometry is much weaker and spreads to shorter wavelengths. Examining Fig. 10, the main effect of increasing the collision energy is a decrease in signal from the "favourable" orientations in the blue region of the spectrum. Overall, the "randomly" oriented spectrum shifts slightly to the red at the higher energy. We equate the higher energy emission to more highly excited BaO molecules, again, most probably in the $A^1\Sigma^+$ state. Furthermore, we assume that the higher energy emission corresponds, from energy balance, to molecules with lower recoil energy. Increasing the translational energy of the reaction then causes a relative decrease in the internal energy of the products, in accord with the observations of Parr et al. [20].

214

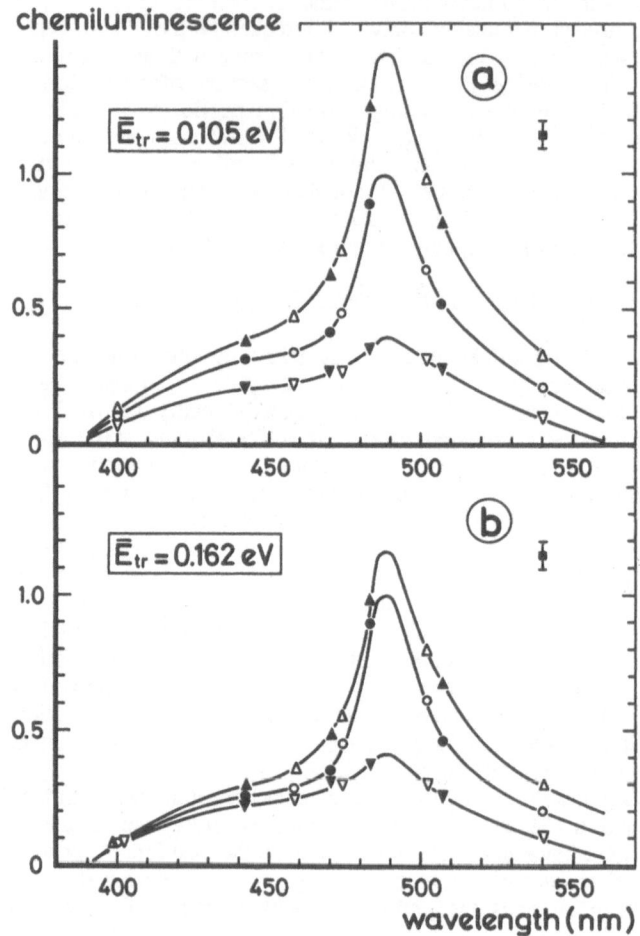

FIGURE 10. Coarse BaO* chemiluminescence spectra produced in the Ba + N_2O $(1,1,1) \rightarrow$ BaO* + N_2 reaction using selected collision geometries: (i) "favourable" (Ba approaches the "O"-end of N_2O), upper curve ($\triangle$), $\overline{\cos\gamma_0}$ = 1/2; (ii) "random", middle curve (O), $\overline{\cos\gamma_0}$ = 0; and (iii) "unfavourable" (Ba approaches the "N"-end of N_2O), lower curve ($\triangledown$), $\overline{\cos\gamma_0}$ = -1/2, at collisions of 0.105 eV and 0.162 eV translational energies. The two sets are scaled to each other at 490 nm, "random" orientations. For "randomly" oriented reactants 1.1 times more total (integrated) emission is produced at 0.105 eV than at 0.162 eV. The points at 400 nm were obtained with a 50 nm fwhm bandpass filter while the other points are differences in total signal from adjacent (closed symbols 20 nm gap) and nearby (open symbols, 40 nm gap) coloured glass cut-off filters. The spectra have been corrected for the bialkali-cathode photomultiplier spectral response. For purposes of clarity smooth curves have been drawn through the points.

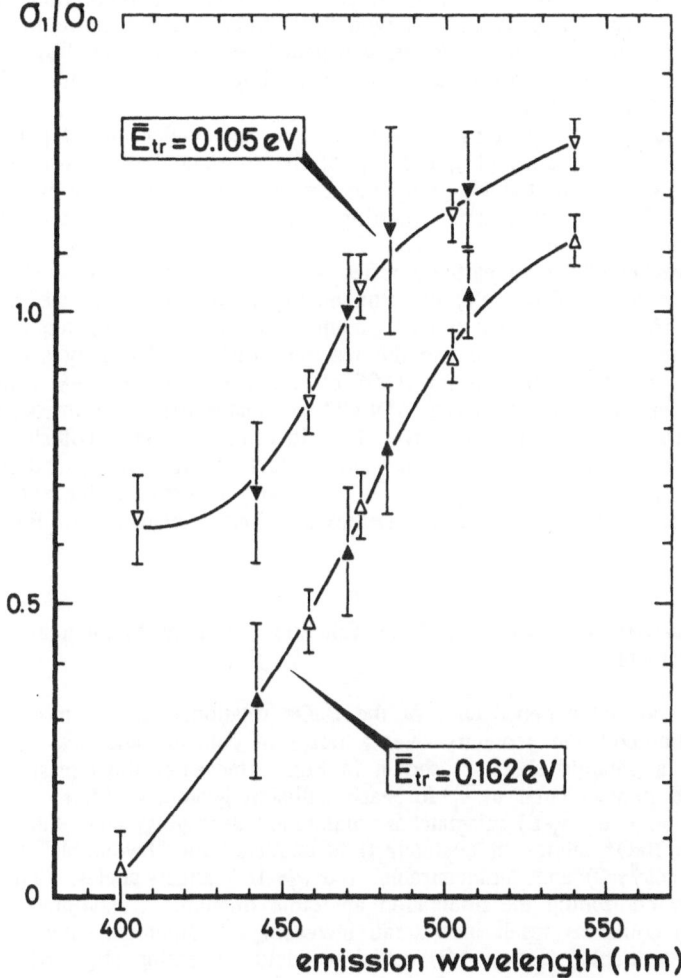

FIGURE 11. Wavelength dependence of σ_1/σ_0, the orientation moment of the reaction anisotropy, for BaO* chemiluminescence. The bandwidth at each wavelength point is described in Fig. 10. For representation smooth lines are drawn through the points.

Three features of the spectra are most distinctive: the spectral shift between "favourable" and "unfavourable" collisions, the rapidly increasing orientation dependence towards longer emission wavelengths, and finally, the changes in behaviour with increased collision energy. We offer a very simplistic explanation of these features based on the energy difference between the angle-dependent barrier to reaction (Fig. 9) and the collision energy. Collisions with (line-of-centres) energy that increasingly exceed the barrier-to-reaction may lead to proportionally higher translational energy in products. "Favourable" collisions experience lower barriers to reaction and will thus produce longer-wavelength emission while "unfavourable" reactions cross higher barriers more slowly, leading to lower recoil, and thus higher internal excitation of BaO*.

"Favourable" orientations are able to react at smaller impact parameters than "unfavourable" orientations [11], because of the substantial cone-of-no-reaction ($\cos\gamma_r < -0.45$) due to the N_2-group the Ba must avoid attacking from the "Back-end" of N_2O. Reactions occurring at small impact parameters are of course more sensitive to steric effects. The rising orientation dependence towards longer emission wavelengths may partially reflect this impact parameter sensitivity.

At the lower (0.105 eV) collision energy in the "unfavourable" case, the largest possible values of the impact parameter do not have the requisite line-of-centres energy to overcome the reaction barrier. This may account for the larger steric effect at 400 nm for 0.105 eV compared to 0.162 eV collisions. Recrossing as incorporated in our ADLCM attenuates the small impact parameter reactions first, thus lowering the overall steric effect at higher collision energies. Orientation-dependent partitioning between product internal energy and recoil are obviously highly dynamical attributes of reaction. A better understanding of the reaction mechanism itself is needed before a unique explanation of Figs. 10 and 11 is possible.

3.5 Orientation Dependence of BaO* Rotational Angular Momentum Alignment

Figure 12 shows the dependence of the BaO* rotational angular momentum alignment on the initial collision geometry over a range of collision energies. In this experiment a polaroid sheet, p, shown in Fig. 1 measures the emission polarization parallel and perpendicular to $\vec{v}_r$ for each collision geometry. After deconvolution over $(\vec{v}_r \cdot \vec{E})$ mismatches, orientation averaging, and other experimental effects, the BaO* alignment $(\bar{P}_2(\cos\gamma_0))$ is extracted for "favourable" ($\cos\gamma_0=1/2$), "side-on" ($\cos\gamma_0=0$) and "unfavourable" ($\cos\gamma_0=-1/2$) attack angles. The effects revealed by controlling the orientation are quite dramatic. "Favourable" ($\cos\gamma_0=1/2$) collisions result in a small increasing $\vec{J}'$ alignment (towards $\vec{J}' \perp \vec{v}_r$) while "unfavourable" ($\cos\gamma_0=-1/2$) collisions yield decreasing alignments (towards isotropic) as E_{tr} is elevated. No indication of these effects is provided by the (typical) "randomly" oriented signal.

The "favourable" behaviour is most interesting in that the strong alignment is not due to kinematic constraints. In a $A + BC(\vec{J}) \rightarrow AB(\vec{J}') + C$ reaction where $\vec{J}$ and $\vec{J}'$ are the reactant and product rotational angular momenta, conservation of total angular momentum, $\vec{J}_{total}$, requires that:

$$\vec{J}_{total} = \vec{L}_{A-BC} + \vec{J}_{BC} = \vec{L}'_{AB-C} + \vec{J}'_{AB} \tag{9}$$

where $\vec{L}$ and $\vec{L}'$ are the orbital angular momenta of reactants and products respectively. Since the J=1 N_2O state is selected, $|\vec{L}| >> |\vec{J}|$ thus the correlation $(\vec{v}_r, \vec{J}')$ reflects the partitioning of angular momentum between $\vec{L}'$ and $\vec{J}'$. Simons [23] has reviewed the topic of $(\vec{v}_r, \vec{J}')$ correlations. Repulsion between fragments (as in late energy release) increases $|\vec{L}'|/(|\vec{L}'| + \vec{J}'|)$, tending to diminish the $(\vec{v}_r, \vec{J}')$ correlation. In the "spectator stripping" limit, $|\vec{L}'| << |\vec{J}'|$

217

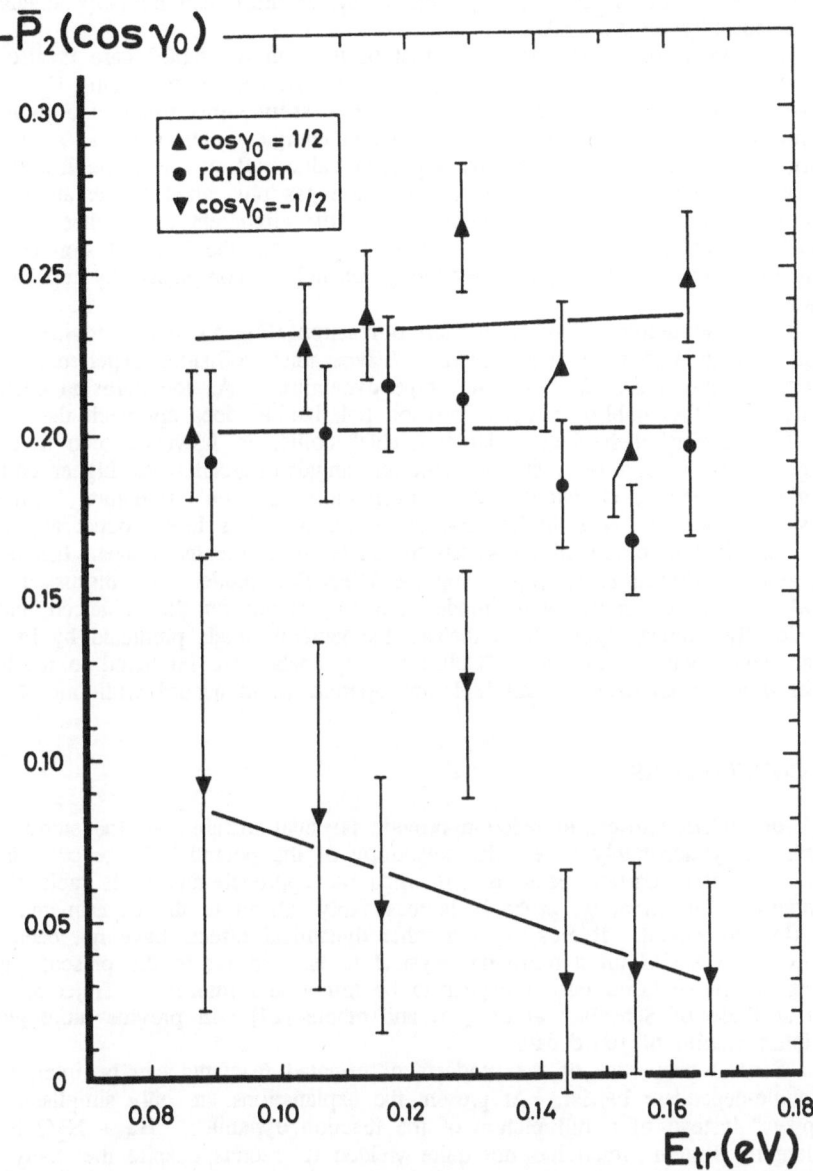

FIGURE 12. The average alignment, $\bar{P}_2(\cos\gamma_0)$, of the rotational angular momentum of BaO* is plotted as a function of the translational energy for two selected fixed orientations ($\cos\gamma_0=1/2$ and $\cos\gamma_0=-1/2$) and "random" of the (1,1,1) state of N_2O. For representation straight lines are drawn through the plotted points.

thus $\vec{L} \sim \vec{J}'$. Since $\vec{L}$ lies perpendicular to $\vec{v}_r$, $\vec{J}'$ must also lie perpendicular to $\vec{v}_r$, consequently $\bar{P}_2(\vec{v}_r.\vec{J}')=-1/2$.

Alignment behaviour similar to that of the "unfavourable" case is also uncommon. Simons and co-workers [23] have observed an increasing $\bar{P}_2$ with E_{tr} in the Ar* + N_2 excitation transfer system. They ascribe this trend to a preference for non-linear collision geometries at low collision energies. De Vries *et al.* [24] have reported isotropic, and even positive $\bar{P}_2(\vec{v}_r,\vec{J}')$ values ($\vec{J}'$ ‖ $\vec{v}_r$) at the lowest translational energies for several metastable rare gas plus alkyl halide atom transfer systems. At higher collision energies the alignment approaches the $\bar{P}_2(\vec{v}_r.\vec{J}') = -1/2$ limit, i.e. $\vec{J}' \perp \vec{v}_r$. They suggest that the low collision energy behaviour is caused by sideways scattering reactions accompanied by repulsive energy release.

The orientation dependent barrier to reaction (Fig. 9) again provides some simple insights into these observations. "Favourable" collisions experience a low barrier and may react *via* "stripping" type dynamics. As the collision energy is elevated the "favourably" induced emission polarization does approach the $\vec{L} \rightarrow \vec{J}'$ "stripping" behaviour. "Unfavourable" collisions, however, tend to experience a higher barrier and react around "side-on" angles-of-attack. At higher collision energies the reactive collisions become even more "side-on" in nature. "Side-on" collisions appear to result in late energy release and thus less product alignment.

Polarization of chemiluminescent products from oriented reagent has been addressed by Prisant *et al.* [25] using the DIPR-DIP model. As discussed in detail elsewhere [4], the electron-jump model fails to account for the enhanced reactivity of vibrationally excited N_2O. Nevertheless, the general trends predicted by Prisant *et al.* are confirmed in our data. "Collinear" approaches are predicted to result in higher product polarization than "side-on" approaches in accord with our observations.

4. CONCLUSIONS

Angle-dependent barriers to reaction provide physical insight into the steric effect. Whether they accurately reflect the anisotropy of the potential for reaction is not yet clear. For early barrier reactions and when the approach motion is rapid the orientation at reaction, γ_r, probably is reasonably related to the experimental γ_0 by the ADLCM model. Recrossing and other dynamical effects have not been included yet in the ADLCM on a profound physical basis. So far in the present study their effects on the deduced barrier appear to be minor and intuitive. Trajectory studies such as those of Schechter *et al.* [26] and others [27] will provide more guidelines for interpretation of steric data.

Several properties of the products of oriented reactants can be interpreted *via* the angle-dependent barrier. At present the explanations are only simplistic "snippets" instead of a full picture of the reaction dynamics. Ba + N_2O is a challenging system which has not quite yielded its secrets despite the many years of attention it has received. The trends uncovered by controlling the collision geometry such as the BaO* internal energy and the alignment of the rotational angular momentum will hopefully aid in finding clues to solve this puzzle.

ACKNOWLEDGMENT

The authors gratefully acknowledge the NSF (Grant INT 8619803) for support of this collaboration. We would also like to thank Prof. J. Reuss for his interest and stimulating support, J. Holtkamp, F. van Rijn and C. Sikkens for their excellent technical assistance. Constructive criticism by and discussion with Prof. R.B. Bernstein was very much appreciated.

REFERENCES

[1]   C.H. Ottinger and R.N. Zare, Chem. Phys. Lett., 5 (1970) 243.
      C.D. Jonah, R.N. Zare and C.H. Ottinger, J. Chem. Phys., 56
      (1972) 263.

[2]   J.W. Cox and P.J. Dagdigian, J. Chem. Phys., 79 (1983) 5353.

[3]   Y.S. Hsu and J.R. Pruett, J. Chem. Phys., 76 (1982) 5849.

[4]   H. Jalink, F. Harren, D. van den Ende and S. Stolte,
      Chem. Phys., 108 (1986) 391.

[5]   H. Jalink, D.H. Parker and S. Stolte, J. Phys. Chem.,
      85 (1986) 5372.

[6]   H. Jalink, S. Stolte and D.H. Parker, manuscript in preparation.

[7]   H. Jalink, S. Stolte and D.H. Parker, Chem. Phys. Lett.,
      140 (1987) 215.

[8]   H. Jalink, D.H. Parker, K.H. Meiwes-Broer and S. Stolte,
      J. Phys. Chem., 90 (1986) 552.

[9]   H. Jalink, S. Stolte and D.H. Parker, manuscript in preparation.

[10] D.H. Parker, H. Jalink and S. Stolte, J. Phys. Chem., 91 (1987)
      5427.

[11] M. Janssen and S. Stolte, J. Phys. Chem., 91 (1987) 5480.

[12] A.R. Edmonds, Angular momentum in quantum mechanics, Princeton
      University Press, Princeton (1960).

[13] S.E. Choi and R.B. Bernstein, J. Chem. Phys., 85 (1986) 150.

[14] S. Stolte, Ber. Bunsenges. Phys. Chem., 86 (1982) 413.

[15] S. Stolte, K.K. Chakravorty, R.B. Bernstein and D.H. Parker,

Chem. Phys., **71** (1982) 353.

[16] R. Altkorn, R.N. Zare and C.H. Greene, Mol. Phys., **55** (1985) 1.

[17] J.M.L.J. Reinartz, W.L. Meerts and A. Dymanus, Chem. Phys., **31** (1978) 19.

[18] D. van den Ende and S. Stolte, Chem. Phys., **89** (1984) 121.

[19] C.H. Townes and A.L. Schawlow, Microwave spectroscopy, McGraw-Hill, New York (1955).

[20] T.P. Parr, A. Freedman, R. Behrens and R. Herm, J. Chem. Phys., **67** (1977) 2181.

[21] R.D. Levine and R.B. Bernstein, Chem. Phys. Lett., **105** (1984)

467.

[22] G.T. Evans, J. Chem. Phys., **86** (1987) 3852.

[23] J.P. Simons, J. Phys. Chem., **91** (1987) 5378.
K.M. Johnson, R. Pease and J.P. Simons, unpublished work.

[24] M.S. de Vries, G.W. Tyndall, C.L. Clobb and R.M. Martin, J. Chem. Phys., **84** (1986) 3753.

[25] M.G. Prisant, C.T. Rettner and R.N. Zare, J. Chem. Phys., **75** (1981) 2222.

[26] I. Schechter, M.G. Prisant and R.D. Levine, J. Phys. Chem., **91** (1987) 5472.

[27] Special issue on dynamical stereochemistry, J. Phys. Chem., **91** (21) (1987).

PRODUCT ALIGNMENT IN REACTIVE, INELASTIC AND HALF-COLLISIONS

J.P. Simons
Department of Chemistry
University of Nottingham
University Park
Nottingham NG7 2RD
UK

1. INTRODUCTION

Traditionally, reaction kineticists have dwelt on the scalar attributes of elementary reactions - rate constants and integral cross-sections, branching ratios and quantum yields, energy utilisation and energy disposal. But reaction dynamicists have also dreamt of riding aboard the collision system itself, watching the molecular motion evolve in stereoscopic 3-D - and they have awoken to the world of vectors. Instead of energy they focus on momentum, both linear and angular; instead of distributions they focus on correlations, instead of numbers they focus on angles. They determine differential cross-sections, rotational alignments and helicities measured by the vector correlations $(\hat{k}, \hat{k}')$, $(\hat{k}, \hat{J}_j')$, $(\hat{k}', \hat{J}_j')$ where $\hat{k}, \hat{k}'$, are unit vectors directed along the reagent and product collision velocities and $\hat{J}_j'$ directed along the product rotational angular momentum. The vectorial approach has created a new branch of reaction dynamics, Dynamical Stereochemistry, whose arrival was celebrated in an earlier Workshop in Jerusalem in 1986 [1]. This article emphasises one major aspect of Dynamical Stereochemistry, the polarisation of reaction products, including those generated *via* inelastic (energy transfer) or reactive collisions, and particularly those generated through molecular photodissociation, where major progress has been made over the past year or so [2]. Conceptually, the measurement of axial angular momentum vector correlations provides an entry into the anisotropy of molecular interactions and an approach to understanding the stereo-specificity of chemical reactions. Their measurement has been facilitated by the advent of tunable, narrow line, polarised laser sources and the development of Doppler resolved detection strategies. Their analysis and interpretation have been set on rigorous foundations particularly through the efforts of Case, McClelland and Herschbach [3], Fano and Macek [4], Greene, Altkorn and Zare [5], Dixon [6], Hertel [7] and Alexander [8].

2. VOCABULARY

Any system in which the spatial distribution of angular momentum or axial vectors is not isotropic is said to be polarised. Oriented distributions are ones in which the odd moments are non-zero; aligned distributions are ones in which the even moments

J. C. Whitehead (ed.), Selectivity in Chemical Reactions, 221–244.
© *1988 by Kluwer Academic Publishers.*

are non-zero. In an oriented population the average Legendre polynomial $<P_1(\hat{J}.\hat{Z})>$ $\neq 0$; in an aligned population $<P_2(\hat{J}.\hat{Z})> \neq 0$. In a quantal system, an isotropic distribution would correspond to one in which all m_J states were equally populated: an oriented or aligned system would exhibit a unidirectional or bidirectional weighting of the populated m_J states. Angular momentum distributions are often expressed in terms of the state multipoles, expectation values of the spherical tensor operators $T_q^{(k)}$. For a monopolar distribution only the zero rank multipole, the population $<T_0^{(0)}>$ is non-vanishing. An oriented population has non-zero components $<T_{0,\pm1}^{(1)}> \equiv O_{0,\pm1}$, the orientation. An aligned population has $<T_{0,\pm1,\pm2}^{(2)}> \equiv A_{0,\pm1,\pm2}^{(2)}$, the alignment. If the vector distributions are prepared with axial symmetry, e.g. with respect to the colliding reagent relative velocity vector, or the polarisation vector of an incident photon, the multipole components with $q \neq 0$ vanish. Those that remain, $O_0^{(1)} \equiv <J_z>/(J(J+1))^{1/2}$ and $A_0^{(2)} \equiv <3J_z^2 - J^2>/J(J+1)$ reflect the first and second moments of the angular momentum distribution about the chosen symmetry axis [5a].

3. ELECTRONIC ORBITAL ALIGNMENT

When chemical bonds are made or broken there is a redistribution of the electronic charge: in a diatomic system the redistribution corresponds to a re-alignment of the electronic angular momentum about the molecular/body fixed axis to or from the mass/space fixed axis. An accompanying paper [9] describes the influence of prior "σ" or "π" electronic orbital alignment in an excited atom, on the relative probabilities of collisionally induced electronic relaxation *via* curve-crossing pathways. The initial orbital alignment, promoted by polarised laser excitations may be scrambled during the collision process, particularly when there are strong spin-orbit interactions [9,10]. On the other hand, collisional and half-collisional processes can often lead to electronic orbital alignment in the final scattered products. Where these are diatomic, or linear polyatomic molecules generated in electronic states of Π,Δ symmetry, their rotational levels are split into Λ-doublet components whose wave functions are either symmetric $\pi^+(A')$, or antisymmetric $\pi^-(A'')$ to reflection in the plane of rotation. Typical examples include NO$(...\sigma^2\pi^1$: $X^2\Pi)$, OH$(...\sigma^2\pi^3$: $X^2\Pi)$, NH*$(...\sigma^1\pi^3$: $c^1\Pi)$. In the high J limit, Hund's case (a), there is a simple relationship between the spatial distribution of the unpaired π electron density and the molecular rotation vector J: for π^+ (π^-) Λ-doublet components, the electron density tends to lie in (perpendicular to) the plane of rotation, i.e. perpendicular (parallel) to J. Unequal Λ-doublet populations therefore reflect preferential electronic orbital alignments, for example in NO or OH$(X^2\Pi_{3/2,1/2})$ or N$_2$*$(C^3\Pi_{0,2})$, as well as alignment of J with respect to the exit relative velocity vector k'. If J is reduced and the spin-orbit interaction is comparable to the rotational spacing (Hund's case (b)) the simple picture may become more clouded. The Λ-doublet levels in $^2\Pi_\Omega$ or $^3\Pi_{2,0}$ states become mixed and in the limit of J$\rightarrow$ 0, the electron density approaches cylindrical symmetry. The onset of mixing is principally determined by the ratio $\lambda = |A/B|$, where A,B are the molecular spin-orbit and rotational constants [11]. This constraint does not apply to fragments generated in $^1\Pi_1$ or $^3\Pi_1$ electronic states, e.g. NH*$(c^1\Pi_1)$ or N$_2$*$(C^3\Pi_1)$ where purity is preserved regardless of J [8].

There are many examples of unequally populated Λ-doublet state distributions, particularly among the products of bimolecular atom transfer or photodissociation reactions; rarer examples among inelastically scattered products include NO scattered from single crystal surfaces and N_2*(C) scattered from collisions with metastable $Ar(^3P_{2,0})$. Table 1 lists a representative sample, with $\pi_{\parallel}$ and $\pi_{\perp}$ symbolising preferential population of components with the electron density aligned parallel or perpendicular to the plane of rotation. The RONO systems provide an interesting contrast. Consider first the reactions

$$H + NO_2 \rightarrow \underline{HO} \ (^2\Pi)_N + NO \qquad\qquad (\pi_{\parallel} > \pi_{\perp}) \qquad (1)$$

$$HONO(^1A') + h\nu(350 \ nm) \rightarrow HONO(^1A'') \rightarrow \underline{OH}(^2\Pi)_N + NO \ \cdot (\pi_{\parallel} > \pi_{\perp}) \qquad (2)$$

where the underlined OH fragment has been probed by laser induced fluorescence excitation spectroscopy [13,14]. Parity selection rules ensure that P, R(N")↑ and Q(N")↑ features probe alternative Λ-doublet components (in both spin-orbit states). The atom transfer reaction (1) tends to proceed *via* a planar HONO transition state, since the OH fragment is generated with its π-electron density preferentially localised in the plane of rotation. A similar preference is found in the OH fragment generated by photodissociation of nitrous acid (reaction (2)) [19] but here the intermediate state is a planar photo-excited molecule of electronic symmetry A". Since the OH fragments tend to carry the symmetry A' the NO should show a preference for populating the antisymmetric π (A") component - assuming they are generated with high rotational excitation [8]. Experimental difficulties have prevented this prediction from being confirmed for HONO but the analogous molecular photodissociation

$$CH_3ONO + h\nu(364 \ nm) \rightarrow CH_3O + \underline{NO}(^2\Pi) \qquad (\pi_{\perp} > \pi_{\parallel}) \qquad (3)$$

does indeed show the expected reversal [20]. Selective population of the Λ-doublet states reflects a propensity for the retention of a plane of symmetry in the collision or half-collision process and the choice of state reflects the character of the electronic orbital symmetry with respect to that plane.

4. MOLECULAR ROTATIONAL ALIGNMENT IN INELASTIC AND REACTIVE COLLISIONS

Consider the archetypal reaction

$$A + BC(J) \rightarrow AB(J') + C \qquad\qquad (4)$$

Angular momentum conservation requires

$$J_{total} = L_{A-BC} + J_{BC} = L'_{AB-C} + J'_{AB} \qquad\qquad (5)$$

where L, L' and J, J' are the orbital and rotational angular momenta. If the reagent A were prepared as a superthermal atomic beam and if BC were rotationally cooled through supersonic jet expansion then $|L| \gg |J|$. If in addition, the atom transfer dynamics approached the "spectator stripping limit", with negligible momentum

TABLE 1. Examples of unequally populated Λ-doublet components in the products of inelastic and reactive collisions. $\pi_{||}$, $\pi_{\perp}$ indicates the alignment of electron density preferentially parallel or perpendicular to the product's rotation plane.

Excitation transfer   Ref.

$Ar^*(1^3P_{2,0}) + N_2 \rightarrow Ar + N_2^*(C^3\Pi_u)$: $\pi_{||} > \pi_{\perp}$ rotation plane   [12]
tends to contain
$\mathbf{k}'$

Atom transfer

$H + NO_2 \rightarrow \underline{OH}(^2\Pi) + NO$: $\pi_{||} > \pi_{\perp}$ [13,14]
planar transition
state
$O + HR \rightarrow \underline{OH}(^2\Pi) + R$: $\pi_{||} > \pi_{\perp}$ [15]

Photodissociation

$NH_3 + hv \xrightarrow{<135 \text{ nm}} \underline{NH}^*(c^1\Pi) + H_2$: $\pi_{||} > \pi_{\perp}$ pyramidal $\rightarrow$ planar [16,17]

$H_2O + hv \xrightarrow{157 \text{ nm}} \underline{OH}(^2\Pi) + H$: $\pi_{\perp} > \pi_{||}$ maser action?   [18]

$HONO + hv \xrightarrow{350 \text{ nm}} \underline{OH}(^2\Pi) + NO$: $\pi_{||} > \pi_{\perp}$ [19]
planar recoil from
a $^1A''$ potential
$CH_3ONO + hv \xrightarrow{364 \text{ nm}} CH_3O + \underline{NO}(^2\Pi)$: $\pi_{\perp} > \pi_{||}$ [20]

$H_2O_2 + hv \xrightarrow[266 \text{ nm}]{248 \text{ nm}} \underline{OH}(^2\Pi) + \underline{OH}(^2\Pi)$: $\pi_{\perp} > \pi_{||}$ [21,22]

torsional motion

$HONO_2 + hv \xrightarrow{280 \text{ nm}} \underline{OH}(^2\Pi) + NO_2$: $\pi_{\perp} > \pi_{||}$ [23]

Surface scattering

$NO + Ag(111) \rightarrow \underline{NO}(^2\Pi)$: $\pi_{||} > \pi_{\perp}$ scattering from two   [27]
potentials

---

transfer between the recoiling products, $|L'| \ll |J'|$ and

$$L_{A-BC} \simeq J'_{AB}. \tag{6}$$

Under such conditions, since L lies perpendicular to the reagent relative velocity vector k, so too must the product rotational angular momentum J'. The AB molecule would be fully aligned with $\langle \hat{J}'.\hat{k}\rangle^2 \to 0$ and $\langle P_2(\hat{J}'.\hat{k})\rangle \to -1/2$. Repulsion between the separating products would break down this simple picture; instead of the direct mapping $L \to J'$ there would be partitioning of the initial angular momentum between orbital and product rotation $L \to J'$, L' and the alignments $\sim \langle P_2(\hat{J}',\hat{k})\rangle$ would reflect the mean angular momentum partitioning $\langle \frac{L'}{L' + J'}\rangle$ [25].

Measurements of product rotational alignments therefore provide a measure of angular momentum disposal.

The spectator stripping limit is most readily approached when the departing atom has negligible mass, i.e. for $C \equiv H$. Three examples illustrate alternative ways in which the product rotational alignment can be manifested experimentally. Other examples are listed in Table 2.

(i) Electric deflection [25-27]:

e.g. $M + HX \to (MX)_{J'} + H$: $M = K$, Cs; $X = I$, Br $\qquad$ (7)

The first pioneering crossed-beam studies of Herschbach's group used an electric deflection analyser system to probe the product $|M_{J'}|$ distributions and hence the rotational alignment of the scattered alkali halide molecules, generated under thermal collisional energies [25-27].

(ii) Spontaneous fluorescence polarisation [28,29,30]:

e.g. $Xe^*(^3P_2) + HX \to XeX^*(B)_{J'} + H$ $\qquad$ (8a)

$Ca^*(^1D) + HCl \to CaCl^*(B)_{J'} + H$ $\qquad$ (8b)

Rotational alignment of the chemiluminescent products is manifested by the polarisation of their fluorescence, since the transition dipole μ_f is referenced to the molecular rotation axis (at the high J' limit). For "parallel" transitions, P, R$\downarrow$, $\mu_f \perp J'$ and for "perpendicular" transitions, Q$\downarrow$, $\mu_f \parallel J'$. Alignment of J' with respect to the reagent relative velocity k is reflected in an accompanying spontaneous fluorescence polarisation, R_k where

$$R_k^{P,R} \equiv \frac{I_{\parallel} - I_{\perp}}{I_{\parallel} + 2I_{\perp}} = -1/2\langle P_2(\hat{J}'.\hat{k})\rangle \tag{9a}$$

$$R_k^Q = + \langle P_2(\hat{J}'.\hat{k})\rangle \tag{9b}$$

TABLE 2. Some examples of inelastic and reactive collision systems where it has been possible to characterise scattered product rotational alignments. Detection *via* (1) spontaneous fluorescence polarisation (2) electric field deflection of scattered molecular beams (3) polarisation dependent LIF spectroscopy and sub-Doppler spectroscopy.

Excitation transfer

		Ref.
e.g. $Ar(^3P_2) + N_2 \rightarrow Ar + N_2^*(C^3\Pi_u)_{N'}$	(1)	[38]
$Xe(^3P_2) + I_2 \rightarrow Xe + I_2(D'(2g))_{N'}$	(1)	[37]

Atom transfer

e.g. $M + HX \rightarrow MX + H$; $M \equiv K,Cs$; $X \equiv I$, Br	(2)	[26,27]
$Ca^*(^1D) + HCl \rightarrow CaCl^*(A,B) + H$	(1)	[30]
$Xe^*(^3P_2) + HX \rightarrow XeX^*(B,C) + H$; $X \equiv I$, Br, Cl	(1)	[28,29]
$Cs + CH_3I \rightarrow CsI + CH_3$	(2)	[34]
$Xe(^3P_2) + CH_3X \rightarrow XeX^*(B,C) + CH_3$; $X \equiv I$, Br	(1)	[55]
$M + Br_2 \rightarrow MBr + Br$; $M \equiv K$, Cs	(2)	[26]
$Ca^{(*)} + F_2,HF \rightarrow CaF^*$ (A,B,C) + F,H	(1) (3)	[31,32,33]
$Xe^*(^3P_J) + X_2 \rightarrow XeX^*(B,C,D) + X$	(1)	[29,35,37]

Photodissociation

$H_2O + h\nu \rightarrow OH(X,A)_{N'} + H$	(1,3)	[18,42,44]
$XCN + h\nu \rightarrow CN^*(B)_{N'} + X$; $X \equiv Cl$, I	(1)	[56,57]
e.g. $ICN + h\nu \rightarrow CN(X)_{N'} + I$	(3)	[58]
$H_2O_2 + h\nu \rightarrow 2OH(X)_{N'}$	(3)	[21,22,23]
$(HCO)_2 + h\nu \rightarrow H_2 + 2CO(X)_{J'}$	(3)	[48]

Surface scattering

e.g. $NO + Ag(111) \rightarrow NO(X)_{N'}$	(3)	[59]

In both of the examples (8), the chemiluminescence is associated with a "parallel" transition: the positive polarisation ratios found experimentally reflect the expected negative alignment, i.e. a tendency for $\mathbf{J}' \perp \mathbf{k}$. However, the ultimate spectator stripping limit was only attained at superthermal collision energies, see Fig. 1 [28]. Despite the low mass of the H atom there is a significant release of orbital motion in the separating products and a clear indication of a preferred non-linear collision geometry.

(iii)   Polarisation dependent laser induced fluorescence [5b,6,31]:

$$\text{e.g. } Ca*(^3P_J) + HF \rightarrow (CaF)_{J'} + H \tag{10}$$

any polarisation of the product angular momentum distribution will be reflected by a sensitivity to the chosen alignment of the polarisation vector ϵ_a of the incident laser photon.   Since LIF is a two photon process, the spatial distribution of the fluorescence emission is the resultant of two dipolar distributions, - one for the absorbed and one for the emitted photons - allowing determination of a higher moment of the rotational alignment, namely $A_0^{(4)}$ as well as $A_0^{(2)}$.   More importantly, tunable laser excitation also offers a very selective and sensitive probe of the product quantum state distributions (just how sensitive and selective will be seen in Section 5).

Atom transfer reactions are much more sensitive to the dynamical features imposed by the potential over which the reaction proceeds when kinematic constraints are unimportant, i.e. when the masses of the atomic/molecular fragments are comparable and the collision energies are low.   Typical examples include (see Table 2):-

$$M + CH_3X \rightarrow MX + CH_3 \tag{11a}$$

$$Xe*(^3P_2) + CH_3X \rightarrow XeX*(B,C) + CH_3 \tag{11b}$$

$$M + X_2 \rightarrow MX + X \tag{12a}$$

$$Xe*(^3P_2) + X_2 \rightarrow XeX*(B,C,D) + X \tag{12b}$$

$$Ca + F_2 \rightarrow CaF*(A,B) + F \tag{12c}$$

The average rotational alignments are far lower than in the corresponding kinematically constrained reactions.

The Ca/F_2 system [32,33] provides a rare example of product rotational alignments being measured for resolved product vibrational states.   As expected, the alignment is greatest when the energy released in translational motion is least, i.e. in CaF* product molecules populating the highest accessible vibrational levels.   In the Cs/CH_3I system, the usual azimuthal averaging was broken down by simultaneous measurement of the rotational alignment (via electric deflection) and the differential cross-sections (reflecting the correlation $(\hat{\mathbf{k}},\hat{\mathbf{k}}')$) - which allowed the first determination of the triple vector correlation $(\hat{\mathbf{k}},\hat{\mathbf{k}}',\hat{\mathbf{J}}')$ [34].   There is a tendency for the product rotation to be aligned perpendicular to the $\mathbf{k},\mathbf{k}'$ collision plane.   The advent of sub-Doppler laser induced fluorescence spectroscopy is providing a powerful new route to this line of enquiry (see Section 5).   The rare gas halide chemiluminescence generated by the reactions of $Xe(^3P_2)$ with the halogens and methyl halides becomes

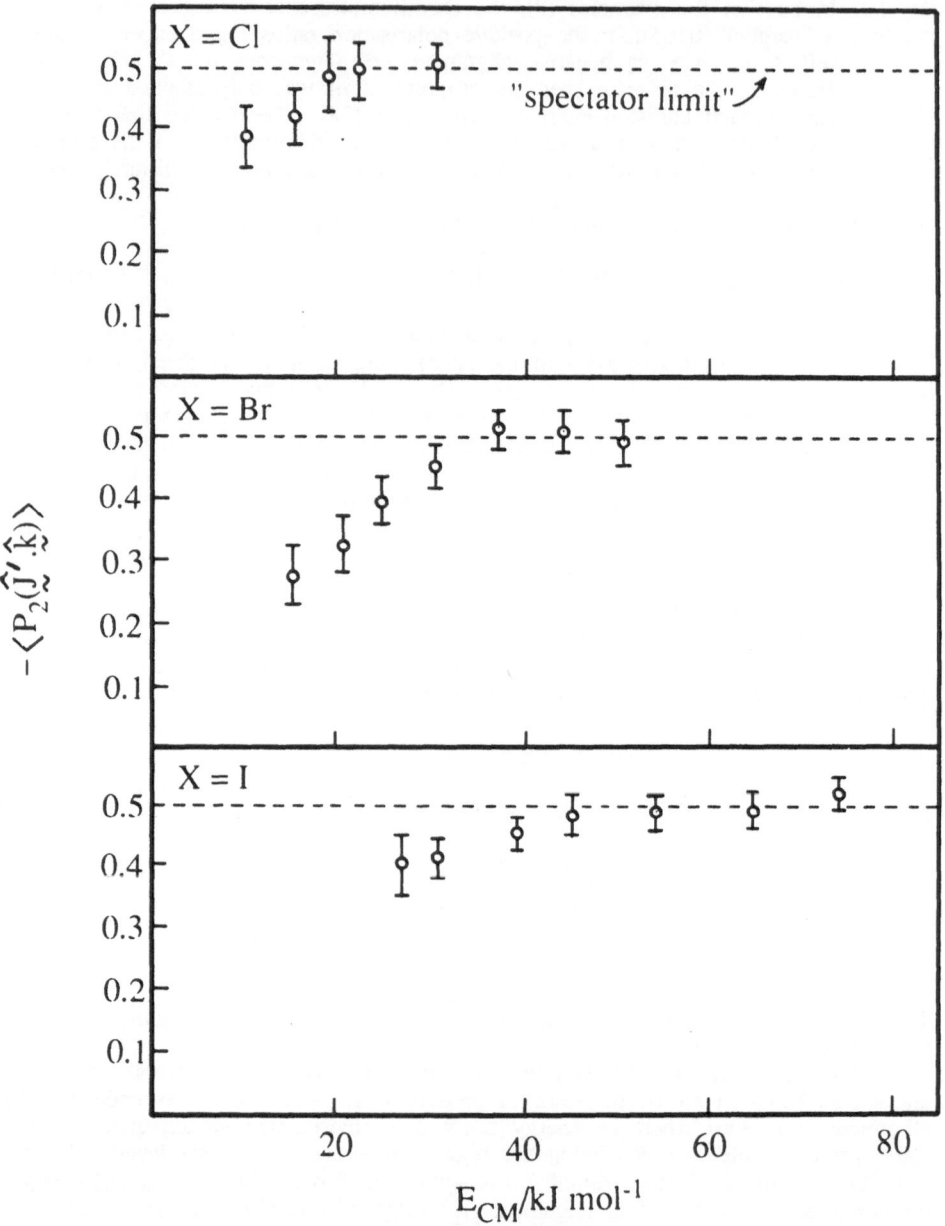

FIGURE 1. The approach to the spectator stripping limit of reaction dynamics: rotational alignment of xenon halides generated through collisions of a superthermal beam of Xe($^3P_2$) with hydrogen halides.

increasingly polarised as the collision energy increases, though it always remains well below the kinematic limit [29,35] see Fig. 2. Qualitatively at least, this behaviour as well as that found in the Ca/F_2 and Cs/CH_3I systems, follows the trends predicted by the DIPR model - an approximate 3-body classical dynamical model which assumes a Direct Interaction of the reagents, followed by Product Repulsion between the central and departing atoms [32,36]. However, there are complicating features in the metastable rare gas/halogen systems, where excitation transfer channels

$$\text{e.g. } Xe(^3P_2) + I_2 \to Xe + I_2^* \ (D'...) \tag{13}$$

become increasingly competitive in the sequence $Cl_2 < Br_2 < I_2$ [35,37]. There is a correlation between the onset of this competition and a reduction both in the mean vibrational energy disposal and the rotational alignment of the rare gas halide product. If the increased energy disposal into translation promotes an increased exit orbital angular momentum L', the reduced alignment would necessarily follow.

In contrast to reactive atom transfer collisions, relatively little attention has so far been paid to rotational alignment in the products of inelastic, excitation transfer. The $I_2^*(D'\to A')$ fluorescence excited through collision of I_2 in the superthermal atomic beam of $Xe(^3P_2)$ is found to be weakly polarised (see Fig. 2), reflecting a slight preference for its rotation to become aligned perpendicular to the incident relative velocity. Similar behaviour is found in the system [38]

$$Ar(^3P_2) + N_2 \to Ar + N_2(C^3\Pi_u)_{J'} \tag{14}$$

monitored *via* the structured $N_2(C\to B)$ fluorescence emission, where earlier measurements of the rotationally resolved fluorescence spectra (measured under thermal discharge flow conditions) had revealed unequal Λ-doublet populations and a tendency for the alignment $J' \perp k'$ [12]. Combination of the two sets of data establishes a further tendency for $J' \perp (k,k')$ plane [38], cf. Cs/CH_3I [34].

A few final remarks before moving on:

(i) In many chemiluminescence systems, there are great signal: noise advantages to be gained by conducting experiments under beam-gas, rather than crossed-beam collision conditions. Not the least of these is the ability to focus sufficient intensity on an analysing monochromator (equipped with a Soleil-Babinet compensator to allow for its varying sensitivity to polarised light). Under beam-gas conditions, the appropriate reference polarisation reflects the moment $\langle P_2(\hat{J}'.\hat{Z})\rangle$. The azimuthal symmetry allows this to be related to the true alignment through the relation [30,39].

$$\langle P_2(\hat{J}' \cdot \hat{Z})\rangle = \langle P_2(\hat{J}' . \hat{k})\rangle \langle P_2(\hat{k}.\hat{Z})\rangle \tag{15}$$

It might be thought that the spread of relative velocity directions, reflected in the "blurring factor" $\langle P_2(\hat{k}.\hat{Z})\rangle$ would very seriously reduce the sensitivity to product alignment, but as the relative velocity and/or the mass of the target molecule increases, the factor rapidly approaches unity. In practice the polarisations measured under beam-gas conditions are never reduced by more than ~20% [30,39].

230

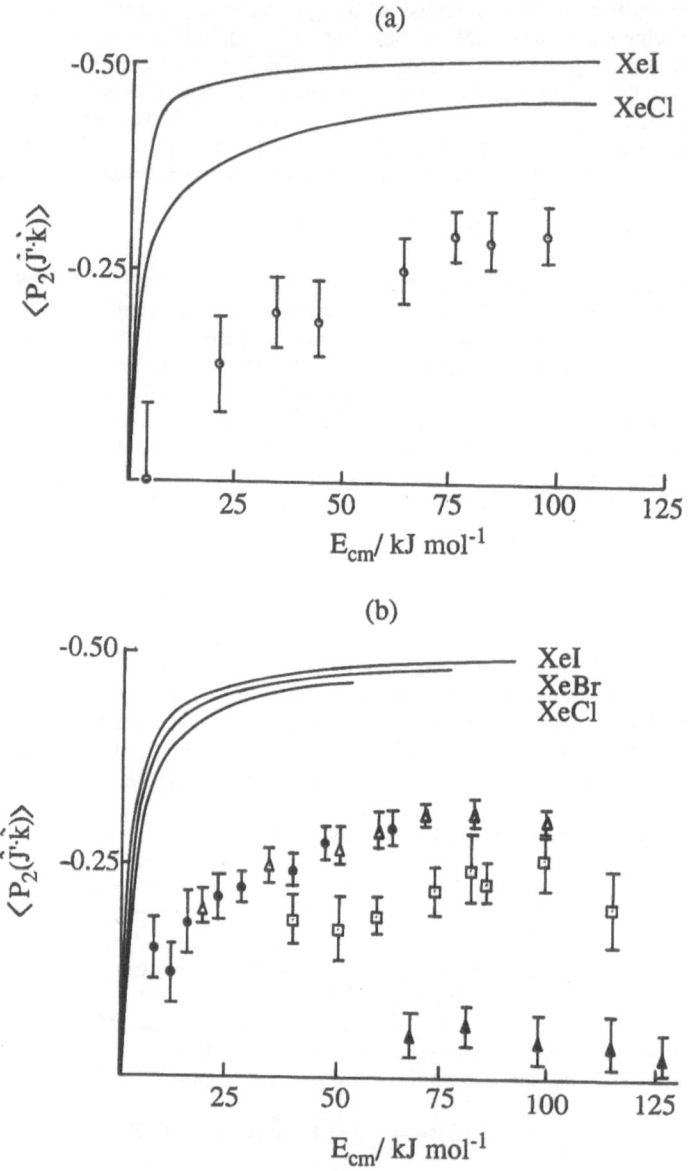

FIGURE 2. Rotational alignment in the products of atom transfer or excitation transfer collisions involving $Xe(^3P_2)$ and (a) ICl, (b) Cl_2, Br_2, I_2. Data points include (a) XeCl(B) and (b) XeCl(B) (●), XeBr(B) (△), XeI(B) (□) and I_2(D') (▲). Full curves are the predicted alignments calculated on the basis of the purely kinematic model of I.R. Elsum and R.G. Gordon, J. Chem. Phys., 76 (1982) 3009. Clearly "kinematics are not enough".

(ii) The ability to measure the polarisation for individually resolved product vibrational states or better yet, partially resolved rotational envelopes (chemiluminescence) or even better yet, fully resolved rotational features (LIF) is no longer a distant dream (see Section 5).

(iii) Different methods of monitoring rotational alignment can give different types of average, and experimental data obtained through different methods should not be compared recklessly. Spontaneous fluorescence is monitored over all scattering directions and the derived alignments are averaged over the full CM scattering angular distributions. LIF measurements sample populations/velocities along the probe beam direction. Unless the measured spectral intensities are integrated across the full Doppler width the measured "alignments" will be false. Similarly the electric beam deflection technique only samples those products whose CM scattering angles lie within the cone of acceptance of the polarisation analyser. When the leaving atom is light, e.g. in the M/HX systems, there is very little angular discrimination, but this is not the case for systems which are not kinematically constrained.

5.   PHOTOFRAGMENT ROTATIONAL ALIGNMENT

When a triatomic molecule is dissociated through absorption of light

$$ABC + h\nu \rightarrow AB^{(*)} \ (J') + C \tag{16}$$

rotational alignment of the product $AB^{(*)}$ will be manifested *via* the polarisation of its spontaneous fluorescence [5a,40] if it is produced in an electronically excited state, or the polarisation dependence of its LIF [5b], or its resonantly enhanced multiple photon ionisation, REMPI, spectrum if it is produced in its ground electronic state [41]. If the incident photolysis beam is unpolarised the polarisation/alignment is referenced to the incident wave vector k; if the beam is polarised, the reference is to the polarisation vector ϵ_p. In the latter case the spontaneous fluorescence polarisation R (J') is related to the laboratory alignment $A_0^{(2)}$ (J') by the simple equation [5a]

$$R \ (J') = 1/2 h^{(2)} \ A_0^{(2)}(J') \equiv h^{(2)} \ <P_2(\epsilon_p.\hat{J}')> \tag{17}$$

where $h^{(2)}$ is a geometric factor reflecting the anisotropy of the fluorescence transition dipole; it takes the values

$$h^{(2)} \ = \ -J/(2J + 3) \qquad\qquad \text{for P}\downarrow$$
$$+1 \qquad\qquad\qquad\qquad \text{for Q}\downarrow \tag{18}$$
$$-(J + 1)/(2J - 1) \qquad\quad \text{for R}\downarrow$$

reducing to $h^{(2)} \rightarrow$ -1/2 (PR$\downarrow$), $h^{(2)} \rightarrow$ +1 (Q$\downarrow$) in the high J limit. For unpolarised incident light,

$$A_{0,k}^{(2)} = -1/2 \; A_0^{(2)} \tag{19}$$

and for undispersed fluorescence associated with a 'perpendicular' electronic transition, i.e. where $\Delta\Lambda$ or $\Delta\Omega = 1$

$$<h(2) \; (PQR\downarrow)> \; \to \; +1/4. \tag{20}$$

The sign of the alignment reflects the symmetry of the photoexcited parent molecule, through the alignment of its transition dipole, μ_p. If μ_p lies in the molecular plane, perpendicular to the fragment rotation J', $A_0^{(2)} > -2/5$; if it lies perpendicular to the plane $J' \parallel \mu_p$ $A_0^{(2)} < +4/5$ [5a,40]. Thus for OH(A ^{2} Σ^+) generated *via* photodissociation of $H_2O(\tilde{B}^1A_1 \leftarrow \tilde{X}^1A_1)$ the maximum experimental alignment approach $A_0^{(2)} \sim -2/5$ as indeed it does in the high J limit [42]. However, this is not generally the case and the question arises "What can be learned from the absolute magnitude of the alignment?" The answer is "Quite a lot, provided due corrections have been made for the extraneous factors which obscure reductions associated with the photodissociation dynamics".

These factors include precessional effects promoted by coupling of the fragment molecular rotation to its nuclear and/or electron spin [59,42] or to external magnetic fields, (e.g. that of the earth). The latter generally presents no problems unless the photofragment has a long fluorescence lifetime, but the nuclear hyperfine precession does have a major effect at low J particularly when the fluorescence is laser induced [18]. Additional corrections are required, again at low J, when the fluorescent transition involves a degenerate electronic state, e.g. OH($A^2\Sigma^+ \leftrightarrow X^2\Pi$). The change from case (b) to case (a) coupling, which promotes mixing of the two Λ-doublet components, destroys the simple picture in which the transition moment is aligned either with $\mu_f \parallel J$ ($Q\downarrow$) or $\mu_f \perp J$ (P,R$\downarrow$) [11]. If the corrected rotational alignments still remain well below the limiting value, two potential causes remain: either a (trivial) overlap of two electronic transitions in the photodissociation continuum, associated with opposite symmetries, e.g. A' and A", and hence oppositely signed alignments, or rotational motion in the photoexcited molecule prior to dissociation. If the angular rotational frequency of the molecule, ω, is greater than its dissociation rate, τ^1, so that $<\omega\tau> \; > 1$, memory of the initial correlation between μ_p and ϵ_p may be reduced, with a corresponding loss of the alignment $2<P_2(\hat{J}'.\epsilon_p)>$ [43]. If dissociation is fast on the timescale of the molecular rotational clock, there is no time for "amnesia" to set in. Once again the photodissociation of H_2O provides an excellent case study [44]. Two photon absorption of tunable, narrow-line laser radiation near 248 nm, excites both the rotationally resolved (but strongly predissociated) Rydberg state, $H_2O(\tilde{C}^1B_1)$ and the dissociative continuum $H_2O(\tilde{B}\;^1A_1)$ upon which it is superposed (see Fig. 3). The structured state survives for about a picosecond, the latter for much less: both lead to generation of OH($A^2\Sigma^+$) (and OH($X^2\Pi$) fragments) [44,45]. Excitation into the continuum between the rotational features produces strongly polarised OH(A→X) fluorescence, reflecting a strongly aligned, rotationally excited photofragment population with $A_0^{(2)} < + 0.6$. Such a high level of alignment can only be understood if the 2-photon absorption populates a continuum of A_1 symmetry, principally *via* a virtual intermediate state of B_1 symmetry [44,46]

$$\tilde{X}^1A_1 \to B_1 \to A_1 \; : \; A_0^{(2)} < +8/7. \tag{21}$$

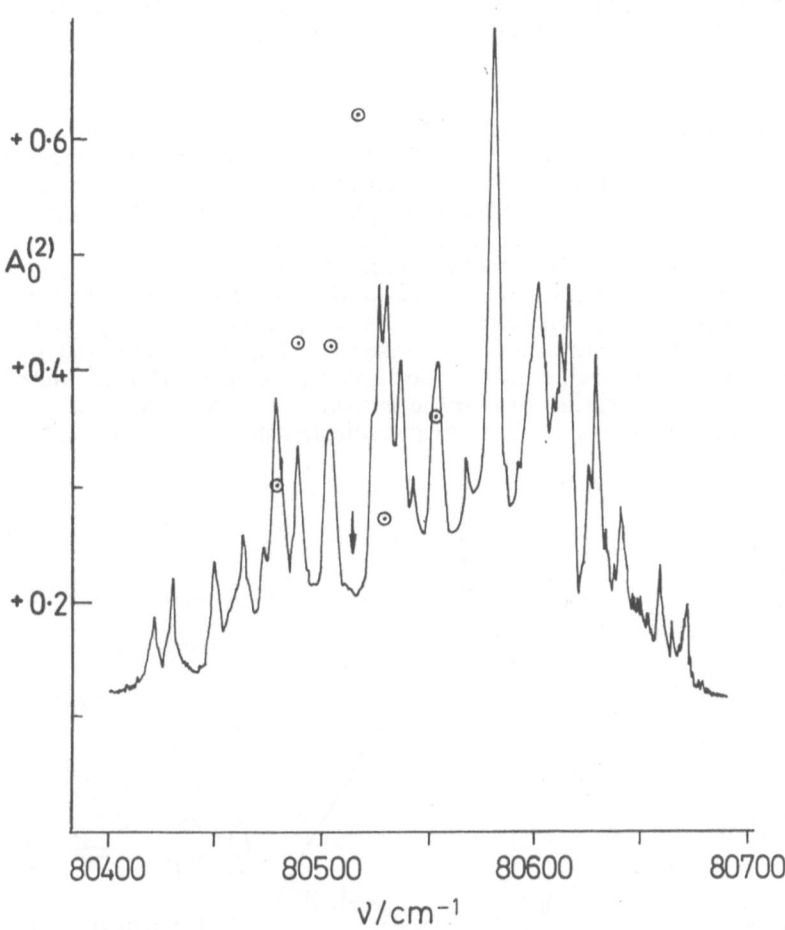

FIGURE 3. Rotational alignments of $OH(A^2\Sigma^+)$ generated through 2-photon excitation of water in the region of the short-lived $\tilde{B}^1A_1$ continuum and the longer-lived quasi-bound $\tilde{C}^1B_1$ Rydberg state.

The nearest intermediate real state is $\tilde{A}^1B_1$ and the anticipated final state is $\tilde{B}^1A_1$. On the other hand, the alignment of $OH(A)$ fragments generated through excitation into the structured features associated with $H_2O(\tilde{C}^1B_1)$ is dramatically reduced, and the reduction is in proportion to the relative contribution "structured: continuum" in the photofragment fluorescence excitation spectrum. Predissociation of $H_2O(\tilde{C}^1B_1)$ into $OH(A)$ is only possible when out-of-plane rotation about the a-inertial axis couples the structured $\tilde{C}$ state into the $\tilde{B}$ state continuum [44]. This is just the motion necessary to tilt the molecular framework away from its initial alignment with respect to ϵ_p and hence reduce the final product rotational alignment $\sim \langle P_2(\hat{\mathbf{J}}'.\epsilon_p)\rangle$.

6. PHOTOFRAGMENT VECTOR CORRELATIONS: TRANSLATIONAL ALIGNMENT AND PHOTOFRAGMENT HELICITY

The photofragment alignment $A_0^{(2)}$ is the first member of an interlocking triangle of vector correlations [6,38,47] see Fig. 4. The other two are the translational anisotropy $\beta = 2 < P_2(\hat{\mathbf{k}}' \cdot \epsilon_p) >$ which reflects the correlation between ϵ_p and the photofragment recoil velocity $\mathbf{k}'$, and the photofragment helicity, $<P_2(\hat{\mathbf{k}}' \cdot \hat{\mathbf{J}}')>$. The translational anisotropy, like the rotational alignment may be blurred by parent molecular rotation since both are referenced to the LAB frame *via* ϵ_p, but the helicity will persist independently of the previous history [6,48]. It reflects the imposition of a $(\hat{\mathbf{k}}', \hat{\mathbf{J}}')$ correlation in the recoiling fragments through exit channel interactions. A bending motion, for example, or the torque associated with rapid recoil from a bent, nuclear configuration will tend to align $\mathbf{J}' \perp \mathbf{k}'$ while a torsional motion about the breaking bond would encourage $\mathbf{J}' \parallel \mathbf{k}'$. In broader terms, the $(\hat{\mathbf{J}}', \hat{\mathbf{k}}')$ correlation is not restricted to photodissociation systems only: it certainly must

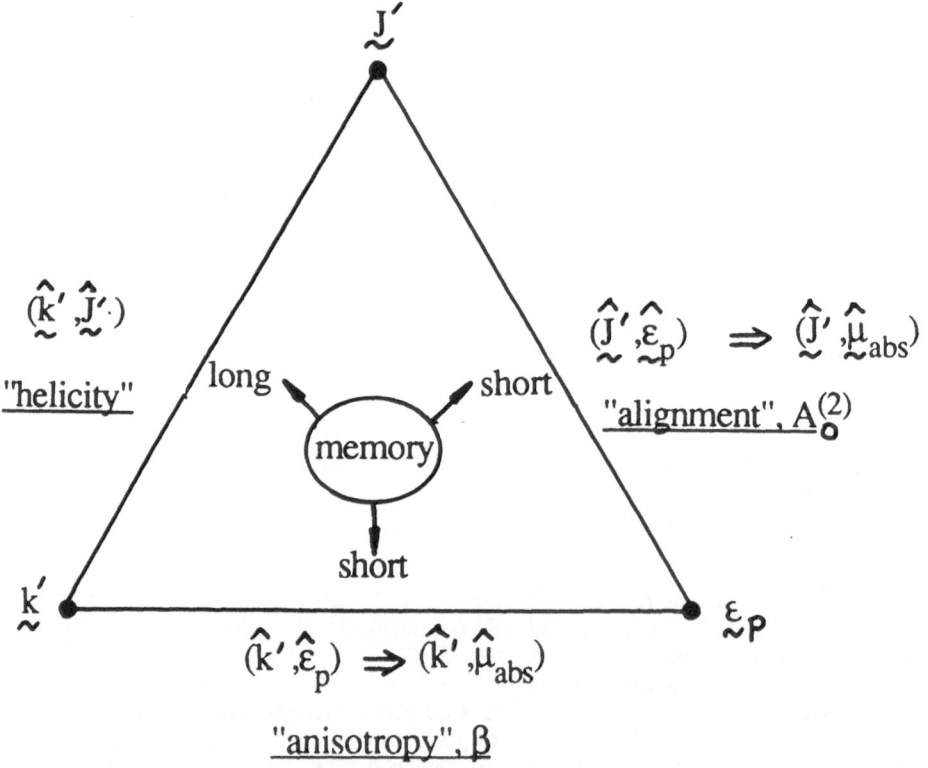

FIGURE 4. The photofragment vector triangle: $\mathbf{k}'$ is the recoil velocity vector, $\mathbf{J}'$ is the rotational angular momentum, μ_{abs} is the parent molecular transition dipole and ϵ_p is the polarisation vector of the absorbed photon.

exist in all products of exothermic bimolecular collision systems where there is a significant release of potential energy in the exit channel. Indeed we have already met it in the $N_2(C)$ molecules generated through excitation transfer from $Ar(^3P_2)$. Similarly, the translational anisotropy has its analogue in the full collisional differential cross-section and the rotational alignment $A_0^{(2)}$ is clearly common to both types of system. The product helicity, being an exit channel "dynamical indicator" can in principle, provide direct information on the preferred geometry at the transition state when the final repulsive forces are "turned on".

Rotational polarisation and translational anisotropy in the recoiling fragments have a profound influence on the intensity distributions in their Doppler resolved LIF excitation (or spontaneous fluorescence) spectra [6] or their multiple photon ionisation spectra [41]. Analysis of the Doppler resolved LIF spectra has provided a powerful method for determining each of the three 2-vector correlations as well as their mutual correlation $(\epsilon_p(\hat{\mu}_p),\hat{k}^{\cdot}\hat{J}^{\cdot})$ (cf. the triple vector correlations $(\hat{k}, \hat{k}^{\cdot}, \hat{J}^{\cdot})$ found in the Cs/CH_3I full collision system).

When the vector properties of the scattered fragments are correlated, both the integrated intensities and the resolved Doppler profiles are profoundly sensitive to changes in the relative directors and polarisations of the photolysis (k_p, ϵ_p) and probe/analysis (k_a, ϵ_a) laser beams and to the nature of the probed transition, P, R↑ or Q↑. Alternative quantal and classical procedures for relating the observed changes to the vector correlations have been developed by Dixon [6] and by Houston and his co-workers [47,49] and these have already been used to analyse the photodissociation dynamics in H_2O_2 [21,22,23], SCO [48,50], $(HCO)_2$ [48], HNO_3 [23], $(CH_3)_2$ NNO [51] and $(CH_3)_3CONO$ [52]. Dixon in particular [6], has shown how the correlations can be expressed as a set of bipolar moments $\beta_0^K(k_1,k_2)$ which are contained in the expansion coefficients g_{k_1} of the Doppler lineshape function

$$g(\nu) = \frac{1}{2\Delta\nu_D} \left\{ g_0 + g_2P_2(\chi_D) + g_4P_4(\chi_D) + g_6P_6(\chi_D) \right\} \qquad (22)$$

$\Delta\nu_D = \nu_0(v/c)$ is the Doppler width for a fragment recoil velocity component v ($\equiv k^{\cdot}$) along the probe laser axis k_a, and $\chi_D = (v - v_0)/\Delta\nu_D$ is the fractional Doppler shift from the line centre v_0. The two leading terms in the expansion are the most significant, and their coefficients

$$g_0 = b_0 + b_1\beta_0^2(02) \qquad (23a)$$

and
$$g_2 = b_2\beta_0^2(20) + b_3\beta_0^0(22) + b_4\beta_0^2(22) + b_5\beta_0^2(24) \qquad (23b)$$

contain

$$\beta_0^2(02) \equiv (5/4)A_0^{(2)} \quad \rightarrow \text{the rotational alignment} \qquad (24a)$$

$$\beta_0^2(20) \equiv (1/2)\ \beta \quad \rightarrow \text{the translational anisotropy} \qquad (24b)$$

$$\beta_0^0(22) \equiv <P_2(\hat{k}^{\cdot}.\hat{J}^{\cdot})> \rightarrow \text{the helicity} \qquad (24c)$$

and
$$\beta_0^2(22) \equiv (\epsilon_p(\hat{\mu}_p),\ \hat{k}^{\cdot},\hat{J}^{\cdot}) \rightarrow \text{the mutual correlation.} \qquad (24d)$$

The final term $\beta_0^2(24)$ is generally neglected since its accompanying multiplier b_5 is small. The remaining multipliers $b_0 \to b_4$, which have all been tabulated [6], are dependent on the photolysis-analysis laser geometry and the nature of the spectral transition P, R↑ and Q↑. Provided there is a single recoil velocity or only a very narrow velocity spread, Eq. 22 can be used quantitatively to determine the listed set of vector correlations (23). In practice this is effected by recasting the equation in the form

$$g(\nu) = \frac{1}{2\Delta\nu_D} \left\{ 1 + \beta_{eff} \, P_2(\hat{k}_a \cdot \epsilon_p) P_2(\chi_D) \right\} \tag{25}$$

and using non-linear least squares procedures to return values of $\Delta\nu_D$ and $\beta_{eff} \equiv g_0/g_2$ for the alternative experimental conditions. The term g_0, which only contains the alignment, is generally determined separately, from relative integrated intensity measurements. Two examples will illustrate the approach:

(a) the photodissociation of H_2O_2, where the assumption of a very narrow photofragment recoil velocity distribution is fully justified, and

(b) the photodissociation of $HONO_2$ where it certainly is not.

6.1 H_2O_2 [21-23]

The near u.-v. absorption of H_2O_2 is a broad, structureless continuum. Excitation into the continuum generates rotationally excited (N"<15 at 248 nm) but vibrationally cold $OH(X^2\Pi)$ fragments. The two spin-orbit states $^2\Pi(3/2)$ and $^2\Pi(1/2)$ are near equally populated but in high rotational levels the relative Λ-doublet populations, $\pi^-(A")/\pi^+(A') \to 2$. The Doppler profile of the $R_2(1)$ feature in the OH(A-X) LIF spectrum is broad and displays a characteristic "forward-backward" symmetry under coaxial probe conditions, $k_a \, || \, k_p$, with a strong central dip. The $R_2(1)$ transition is necessarily free of complications associated with rotational vector correlations, and its Doppler profile can be analysed to give a translational anisotropy $\beta \approx -0.9$, close to the limiting value $\beta = -1$, and a recoil velocity of 4.06 km s^{-1}. This implies fast axial recoil perpendicular to the parent molecular transition dipole μ_p, consistent with μ_p directed along the C_2 symmetry axis, and the electronic electronic assignment $4b(n_0) \to 5b(\sigma^*_{0-0})$, $\tilde{A}^1A \leftarrow \tilde{X}^1A$ to the near u.-v. absorption continuum.

The translational anisotropy in the rotationally excited fragments retains the value $\beta \sim -0.9$, but the striking changes in their Doppler resolved LIF spectral contours with changes in the detection geometry and rotational branch (see Fig. 5) betray a marked polarisation of the photofragment rotational vector distributions The alignment, $A_0^{(2)} = 0.10 \pm 0.05$ which is also insensitive to increasing rotation, indicates a small correlation between J' and μ_p (C_2) with an average angle $\widehat{J'C_2} \sim 50°$; the helicity $\langle P_2(\hat{k}'.\hat{J}')\rangle$ is much more pronounced and increases steadily with the photofragment rotation to a final value which corresponds to an average angle $\widehat{J'k'} \sim 40°$. This reflects an important contribution from torsional motion about the O-O axis as the two OH fragments separate on the repulsive upper state potential energy surface. In addition, the sum of the average alignment and helicity angles approaches 90°, restricting J' to the plane containing the recoil axis

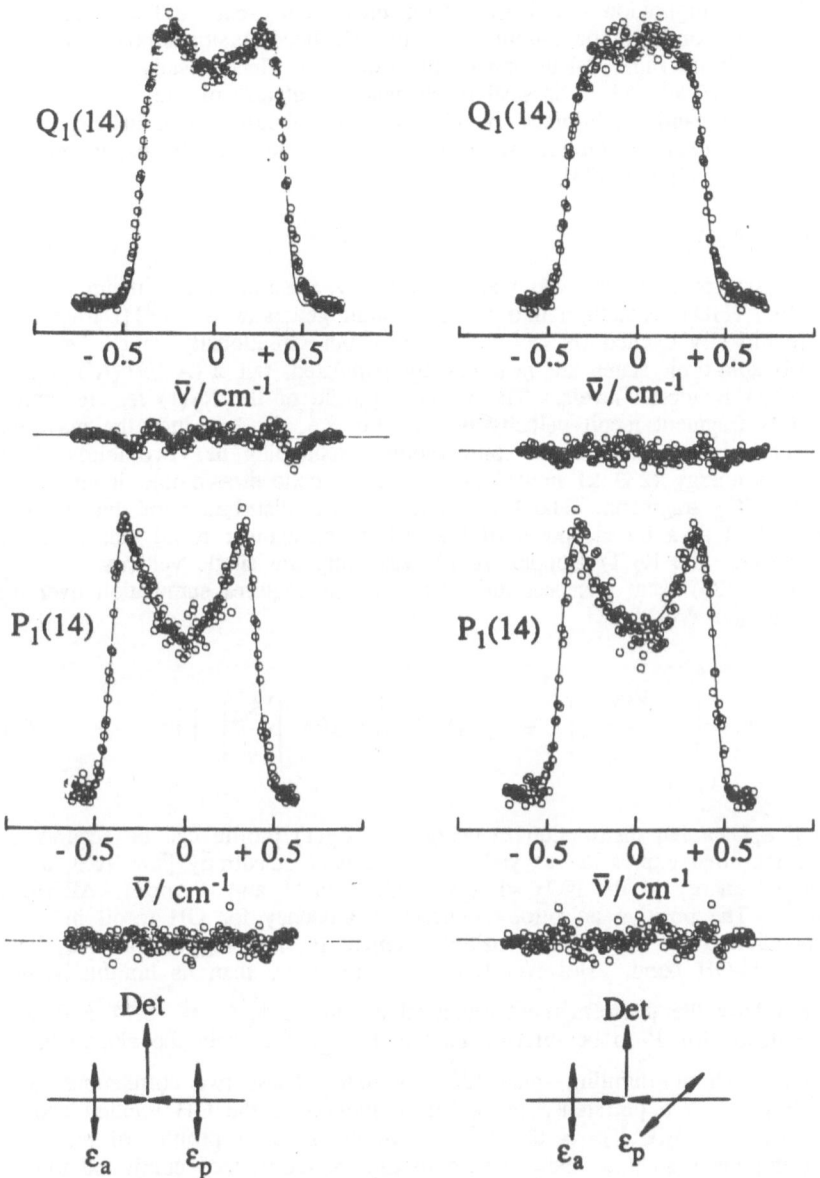

FIGURE 5. Doppler resolved profiles of the $P_1(14)$ and $Q_1(14)$ features in the LIF excitation spectrum of $OH(X)_{v=0}$ generated by photo-dissociation of H_2O_2 at 248 nm. Curves correspond to best fits to Eq. 25 for $A_0^{(2)} = +0.12$; $\beta^2 = -0.9$; $\langle P_2(\hat{k}'\hat{J}')\rangle = +0.38$;

and the perpendicular two-fold rotation axis and implying dissociation from a near-trans-planar configuration - at least for the most highly excited fragments. This configuration corresponds to the minimum in the calculated torsional potential energy determined through *ab initio* calculation of the low-lying, electronically excited $\tilde{A}^1A$, potential energy surface [53]. Classical mechanical calculation of the photodissociation dynamics reinforces the view that the photofragment helicity arises primarily from the torsional motion in the excited state promoted by excitation into the $\tilde{A}^1A \leftarrow \tilde{X}^1A$ continuum [23].

6.2 HONO$_2$ [23]

Nitric acid vapour presents an even weaker near u.-v. absorption continuum than H_2O_2 and, like H_2O_2, excitation into the continuum generates $OH(X^2\Pi)$ fragments which are rotationally excited (N"<14 at 280 nm), but vibrationally cold. Once again the two spin-orbit states are near equally populated, but $\pi^-(A")/\pi^+(A') \rightarrow 2$ in the most excited rotational levels. The Doppler profile of the $R_2(1)$ feature indicates an average OH fragment recoil velocity of ~2.4 km s^{-1}, far less than the maximum dictated solely by energy/momentum conservation. Something like two thirds of the available excess energy (233 kJ mol^{-1}) at 280 nm is concentrated into internal motion in the NO$_2$ fragment. The broad spread in the distribution of this internal energy is reflected in a broad spread of the OH photofragment recoil velocities and attempts to analyse the $R_2(1)$ Doppler profile assuming the single velocity expressions (22), (25) were unsuccessful. The analysis requires summation over the velocity distribution $W(v)dv$

$$
g(v) \sim \int_{|v_{k_a}|}^{\infty} \frac{W(v)}{2v} \left\{ 1 + \beta_{eff} P_2(\hat{k}_p \cdot \epsilon_a) P_2 \left[\frac{v_{k_a}}{v} \right] \right\} dv \qquad (26)
$$

By assuming a Gaussian distribution in energy the $R_2(1)$ profile can be accurately fitted to the modified expression to yield a translational anisotropy $\beta = +0.3$, an average internal energy in the NO$_2$ $<E_{int}> \sim 13000$ cm^{-1} and a spread $<\Delta E>_{FWHM} \sim 6800$ cm^{-1}. The positive anisotropy indicates a tendency for OH recoil in directions parallel to the molecular transition moment μ_p and suggests that μ_p lies parallel to the N-OH bond. However β is very much less than its limiting value $\beta \rightarrow +2$. Similarly the photofragment rotational alignment $A_0^{(2)} = +0.14 \pm 0.05$, reflects a tendency for J' to be directed parallel to μ_p, but again the alignment lies far below the maximum limiting value $A_0^{(2)} \rightarrow +0.8$. These two correlations taken together both point to a preference for a helical motion of the OH rotation about the recoil axis, i.e. $<P_2(\hat{k} \cdot \hat{J}')> > 0$. Analysis of the Doppler profiles of rotationally excited OH fragments such as those shown in Fig. 6, has indeed confirmed this (initially surprising) result, which is attributed to a torsional motion associated with movement of the H atom out of the molecular plane in the initially photoexcited state. However, that is not the only molecular distortion promoted by excitation in the u.-v. continuum; the strong reduction of the alignment, $A_0^{(2)}$ and the anisotropy, β both point to a planar $\rightarrow$ pyramidal configurational change, which confers a strong recoil velocity component perpendicular to the initial parent molecular plane, and hence to the radial component directed along the original N-OH bond axis, and

239

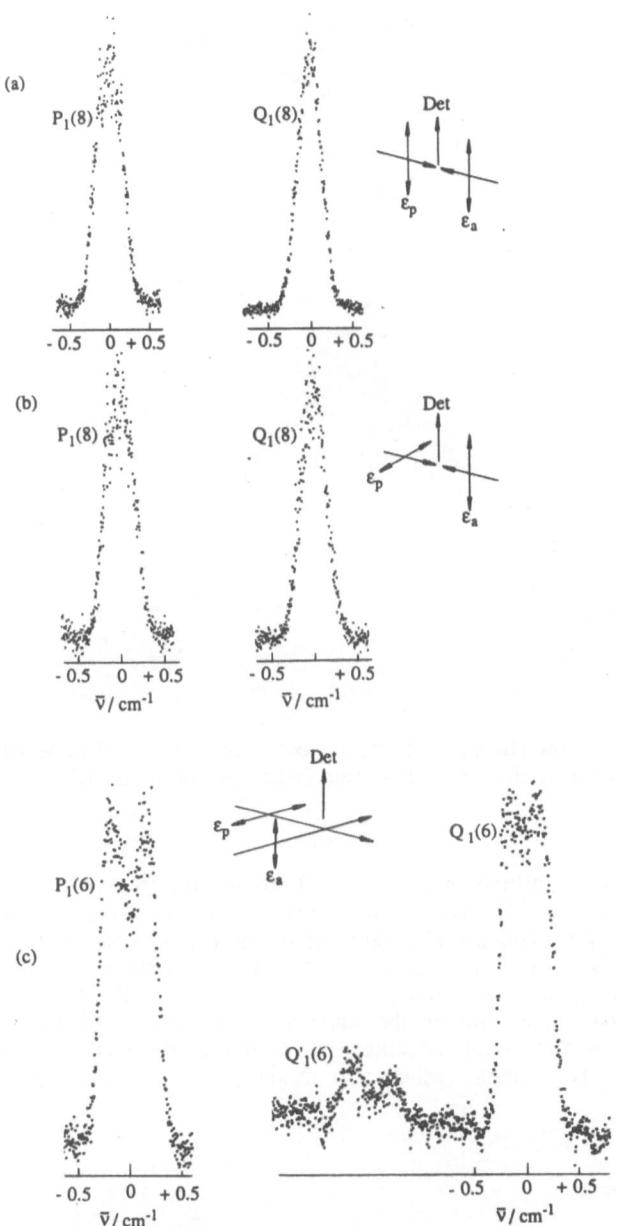

FIGURE 6. Doppler resolved profiles of $P_1(N'')$ and $Q_1(N'')$ features in OH(X) generated by photodissociation of $HONO_2$ at 280 nm. Note the change in spectral contours when the excitation-detection geometry changes from coaxial to perpendicular and the differences in contrast in the double peaked $P_1(6)$ and $Q_1(6)$ lines.

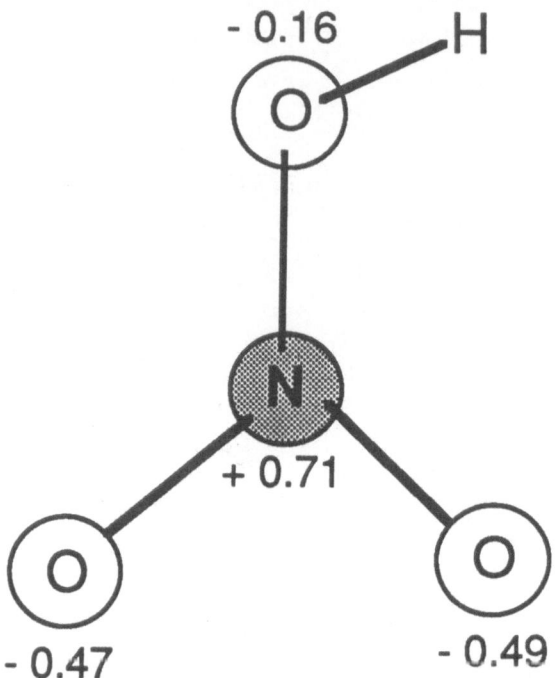

FIGURE 7. The nodal structure of the lowest unoccupied molecular orbital in $HONO_2$ - the orbital coefficients were calculated assuming the CNDO approximation [54].

perpendicular to the transition moment μ_p. Both the departures from planarity and the torsion about the N-OH axis can be understood if the photo-excited state is populated through the predicted [54] electron promotion into the $4a''(3b_1)$ π-antibonding orbital, shown in Fig. 7. Furthermore, the strong π-antibonding character in the NO_2 moiety would promote an increase in the $N-O_2$ bond lengths (which in $HONO_2(\tilde{X})$ are virtually the same as in $NO_2(\tilde{X})$). Simple Franck Condon arguments would predict strong vibrational excitation in the NO_2 fragment, as would any change in the bond angle (which also changes little between $HONO_2(\tilde{X})$ and $NO_2(\tilde{X})$).

7. CONCLUSIONS

After a long gestation period, the measurement of vector (momentum) correlations among the scattered products of bimolecular collisions and unimolecular fragmentations has finally come of age. They often reflect far more selectivity in the final product state distributions than would have been gauged from measurements of scalar (energy) distributions alone and provide a very sensitive diagnostic for the

detailed reaction dynamics which determine that selectivity. Coupled with the
pioneering work of Stolte, Bernstein, Zare and Leone (see Ref. 1) on stereoselectivity
in the initial reagents, the "grand synthesis" of full state-to-state steric characterisation
cannot be far away.

ACKNOWLEDGMENTS

I am most grateful to Professor P.L. Houston for sending manuscripts prior to
publication, to the many colleagues who have contributed to some of the work
described herein, particularly Drs Andrew Hodgson, Mark Brouard, Michael Docker,
Julie August, Keith Johnson and Peter Smith, and to Mrs Margaret Krause for typing
the manuscript in record time.

REFERENCES

[1] "Dynamical Stereochemistry", Fritz Haber Institute, Jerusalem,
     1986: proceedings published in J. Phys. Chem., **91** (21) (1987).

[2] "Dynamics of Molecular Photofragmentation", Faraday Discuss.
     Chem. Soc., **82** (1986).

[3] D.A. Case, G.R. McClelland and D.R. Herschbach, Mol. Phys.,
     35 (1978) 541.

[4] U. Fano and J.H. Macek, Rev. Mod. Phys., **45** (1973) 553.

[5a] C.H. Greene and R.N. Zare, Ann. Rev. Phys. Chem., **33** (1982) 119.

[5b] R. Altkorn and R.N. Zare, Ann. Rev. Phys. Chem., **35** (1984) 265;
     C.H. Greene and R.N. Zare, J. Chem. Phys., **78** (1983) 6741.

[6] R.N. Dixon, J. Chem. Phys., **85** (1986) 1866.

[7] I.V. Hertel, H. Schmidt, A. Bahring and F. Mayer, Rep. Progr.
     Phys., **48** (1985) 375.

[8] M.H. Alexander and P. Dagdigian, J. Chem. Phys., **80** (1984) 4325.

[9] S.R. Leone, "Alignment Effects in Electronic Energy Transfer
     and Reactive Events", paper in this volume.

[10] B. Pouilly and M.H. Alexander, J. Chem. Phys., **86** (1987) 4790.

[11] P. Andresen and E.W. Rothe, J. Chem. Phys., **82** (1985) 3634.

[12] J. Derouard, T.N. Nguyen and N. Sadeghi, J. Chem. Phys.,
     72 (1980) 6698.

[13] R.P. Mariella and A.C. Luntz, J. Chem. Phys., **67** (1977) 5398;
     R.P. Mariella, B. Lantzeh, V.T. Maxson and A.C. Luntz,

242

[14] E.J. Murphy, J.H. Brophy, G.C. Arnold, W.F. Dimpole and J.L. Kinsey, J. Chem. Phys., 74 (1981) 324.

[15] P. Andresen and A.C. Luntz, J. Chem. Phys., 72 (1980) 5842.

[16] F. Alberti and A.E. Douglas, Chem. Phys., 34 (1978) 399.

[17] A.M. Quinton and J.P. Simons, Chem. Phys. Lett., 81 (1981) 214.

[18] P. Andresen, G.S. Ondrey, B. Titze and E.W. Rothe, J. Chem. Phys., 80 (1984) 2548.

[19] R. Vasudev, R.N. Dixon and R.N. Zare, J. Chem. Phys., 80 (1984) 4863.

[20] U. Bruhlmann, M. Dubs and J.R. Huber, J. Chem. Phys., 86 (1987) 1249; B.A. Keller, P. Felder and J.R. Huber, J. Phys. Chem., 91 (1987) 1114; and Z. Phys. D, in press.

[21] K.H. Gericke, S. Klee, F.J. Comes and R.N. Dixon, J. Chem. Phys., 85 (1986) 4463.

[22] M.P. Docker, A. Hodgson and J.P. Simons, Chem. Phys. Lett., 128 (1986) 264; Faraday Discuss. Chem. Soc., 82 (1986) 25.

[23] J. August, M. Brouard, M.P. Docker, A. Hodgson, C.J. Milne and J.P. Simons, Ber. Bunsenges., 1988, in press.

[24] A.C. Luntz, A.W. Kleyn and D.J. Auerbach, J. Chem. Phys., 76 (1982) 727.

[25] D.R. Herschbach, Faraday Discuss. Chem. Soc., 55 (1973) 233.

[26] C. Maltz, N.D. Weinstein and D.R. Herschbach, Mol. Phys., 24 (1972) 133.

[27] D.S.Y. Hsu, N.D. Weinstein and D.R. Herschbach, Mol. Phys., 29 (1975) 257.

[28] See K. Johnson, J.P. Simons, P.A. Smith, C. Washington and A. Kvaran, Mol. Phys., 57 (1986) 255 and references therein.

[29] M.S. deVries, G.W. Tyndall, C.L. Cobb and R.M. Martin, J. Chem. Phys., 84 (1986) 3753.

[30] M.G. Prisant, C.T. Rettner and R.N. Zare, J. Chem. Phys., 75 (1981) 2222.

[31] F. Engelke and K.H. Meiwes-Broer, Chem. Phys. Lett., 108 (1984) 32.

[32] M.G. Prisant, C.T. Rettner and R.N. Zare, J. Chem. Phys., **81** (1984) 2699.

[33] F. Engelke and K.H. Meiwes-Broer, Z. Phys. A, **320** (1985) 39.

[34] D.S.Y. Hsu, G.M. McClelland and D.R. Herschbach, J. Chem. Phys., **61** (1974) 4927.

[35] K.M. Johnson, R. Pease, J.P. Simons, P.A. Smith and A. Kvaran, J. Chem. Soc. Faraday Trans. 2, **82** (1986) 1281; K.M. Johnson, J.P. Simons, P.A. Smith and A. Kvaran, J. Chim. Phys., **84** (1987) 371.

[36] G.M. McClelland and D.R. Herschbach, J. Phys. Chem., **91** (1987) 5509: P.J. Kuntz, E.M. Nemeth and J.C. Polanyi, J. Chem. Phys., **50** (1969) 4607.

[37] R.J. Donovan, P. Greenhill, M.A. MacDonald and A.J. Yencha, W.S. Hartree, K. Johnson, C. Jouvet, A. Kvaran and J.P. Simons, Faraday Discuss. Chem. Soc., **84** in press.

[38] J.P. Simons, J. Phys. Chem., **91** (1987) 5378.

[39] K. Johnson, R. Pease and J.P. Simons, Mol. Phys., **52** (1984) 955.

[40] G.A. Chamberlain and J.P. Simons, Chem. Phys. Lett., **32** (1975) 355; M.T. Macpherson, J.P. Simons and R.N. Zare, Mol. Phys., **38** (1979) 2049.

[41] M. Mons and I. Dimicoli, personal communication.

[42] J.P. Simons, A.J. Smith and R.N. Dixon, J. Chem. Soc. Faraday Trans 2, **80** (1984) 1489.

[43] T. Nagata, T. Kondrow, K. Kuchitsu, G.W. Loge and R.N. Zare, Mol. Phys., **50** (1983) 49.

[44] A. Hodgson, J.P. Simons, M.N.R. Ashfold, J.M. Bayley and R.N. Dixon, Mol. Phys., **54** (1985) 351.

[45] H.J. Krautwald, L. Schneider, K.H. Welge and M.N.R. Ashfold, Faraday Discuss. Chem. Soc., **82** (1986) 99.

[46] G.W. Loge and J.R. Wiesenfeld, J. Chem. Phys., **75** (1981) 2795,

[47] P.L. Houston, J. Phys. Chem., **91** (1987) 5388.

244

[48] G.E. Hall, N. Sivakumar, R. Ogorzalek, G. Chawla, H.P. Haerri,
      P.L. Houston, I. Burak and J.W. Hepburn, Faraday Discuss. Chem.
      Soc., 82 (1986) 13; I. Burak, J.W. Hepburn, N. Sivakumar,
      G.E. Hall, G.Chawla and P.L. Houston, J. Chem. Phys., 86 (1987)
      1258.

[49] G.E. Hall, N. Sivakumar, G. Chawla, P.L. Houston and I. Burak,
      personal communication.

[50] P.L. Houston, Ber. Bunsenges, (1988) in press: N. Sivakumar,
      G.E. Hall, P.L. Houston, J.W. Hepburn and I. Burak, personal
      communication.

[51] M. Dubs, U. Bruhlmann and J.R. Huber, J. Chem. Phys., 84 (1986)
      3106.

[52] J. August, M. Brouard, R. Lavi, D. Schwartz-Lavi, S. Rosenwaks
      and J.P. Simons, unpublished work.

[53] V. Staemmler, reported in Ref. 21.

[54] L.E. Harris, J. Chem. Phys., 58 (1973) 5615.

[55] R.J. Hennessy and J.P. Simons, Mol. Phys., 44 (1981) 1027;
      R.J. Hennessy, Y. Ono and J.P. Simons, Mol. Phys., 43 (1981) 181.

[56] J.A. Guest, M.A. O'Halloran and R.N. Zare, Chem. Phys. Lett.,
      103 (1984) 261.

[57] M.A. O'Halloran, H. Joswig and R.N. Zare, J. Chem. Phys.,
      87 (1987) 303.

[58] G.E. Hall, N. Sivakumar and P.L. Houston, J. Chem. Phys.,
      84 (1986) 2120.

[59] A.W. Kleyn, A.C. Luntz and D.J. Auerbach, Surface Science,
      113 (1982) 33.

ALIGNMENT EFFECTS IN ELECTRONIC ENERGY TRANSFER AND REACTIVE EVENTS

Stephen R. Leone[†]
Joint Institute for Laboratory Astrophysics
National Bureau of Standards and University of Colorado
and Department of Chemistry and Biochemistry
University of Colorado
Boulder
Colorado 80309-0440
USA

ABSTRACT. The rates of electronic curve crossing processes depend critically on the alignment of atomic orbitals, which determine the symmetries of the electronic potentials participating in the reaction or energy transfer event. Recent work from our laboratory is presented on the effect of orbital alignment in near resonant energy transfer processes of electronically excited Ca and Sr atoms. Several energy transfer events are carried out on aligned p-states in collisions with rare gases. The simplicity of the rare gas systems in terms of their symmetry and nonreactive nature is advantageous for comparison to accurate theoretical treatment. In the context of understanding chemical phenomena, collisions of these atoms with molecular partners are also investigated. This opens the possibility to study the correlation of alignment dependent effects in competing reactive and energy transfer pathways. Remarkably state-specific alignment effects are also observed when two or more independent energy transfer pathways are accessible.

1. INTRODUCTION

The simplest mechanistic pictures of organic reactions incorporate considerable sophistication in recognizing the orbital alignment and hybridization effects that participate in chemical transformations. It is only relatively recently, however, that chemical dynamicists are exploring experimental methods to measure directly the effects of orbital alignment on energy transfer and reactive events. Studies of the chemical effects of "alignment" and "orientation" are rapidly becoming well-established fields of investigation [1-3]. In this article we consider primarily orbital alignment effects, although there is considerable work on molecular geometrical orientation effects as well [1,2].

Perhaps some of the earliest types of experiments in this field are the

---

† Staff member, Quantum Physics Division, National Bureau of Standards.

J. C. Whitehead (ed.), Selectivity in Chemical Reactions, 245–263.

investigations of depolarization of fluorescence by collisions [4]. A dramatic demonstration of the differences in potential energy curves caused by orbital alignments comes from the different rainbow angles observed in crossed beam differential scattering of optically aligned excited states with other atoms, e.g. Na*($^2P_{3/2}$) + Hg [5]. Alignment effects are also evident from polarized far wing absorption and emission lineshape profiles, which are studied for many collision systems [6].

More recently, there have been several studies of electronic energy transfer using laser-aligned atomic excited states [7,8]. These include detailed crossed beam investigations of the transfer from aligned Na 3p atoms to 3d states upon collisions with Na$^+$ [9,10] and the intramultiplet mixing in excited Ne states as a function of orbital alignment [11]. In our own recent work, we studied the effects of orbital alignment on the cross sections for near-resonant electronic energy transfer between Ca (or Sr) excited states with rare gas and molecular collision partners [12-16].

There have also been recent investigations of chemiionization processes and chemical reaction dynamics as a function of orbital alignment. Substantial orbital alignment effects are observed for the ionization of Ne metastable states by collisions with Ar [17]. There are also many detailed investigations on the associative ionization of two excited Na $^2P_{3/2}$ states, also as a function of all the possible combinations of alignment of both atoms [18-21]. In a pioneering series of experiments, Rettner and Zare observed remarkably selective dependencies on orbital alignment of the Ca($4\ ^1P_1$) state in reactions with Cl_2, HCl, and CCl_4 [22,23].

Whilst the first studies of chemical reactions demonstrate that there are many exciting alignment effects to be found in chemical transformations, there is also something appealingly simple about the study of energy transfer systems. Energy transfer effects may be tractable theoretically, and the information which is learned can readily be applied to reactive systems, since there have been many correlations between energy transfer and reactive events. For example, there has been considerable controversy concerning whether or not there will be "retention" of orbital alignment in a strong collision. Energy transfer events offer a fertile testing ground to probe the limitations of models of orbital following and "locking" [3,24-27], especially as the strength of the interaction is varied from weak and nonreactive to strong and potentially reactive. This will be one important consideration of the present work.

In the studies of orbital alignment effects considered in this article, a linearly polarized pulsed laser is used to prepare aligned p-states of either electronically excited Ca or Sr. In a crossed beam apparatus, the excited Ca or Sr collides with a pulsed jet of a target gas. Near resonant energy transfer events occur which are highly dependent on the initial p-orbital alignment. The relative cross sections are obtained as a function of the orbital alignment by measuring fluorescence emission from both the initially excited state and the state which receives the transfer. Systems are explored which vary from weakly perturbing collisions to strong, potentially reactive ones. We are able to investigate some of the energy transfer processes in both the "forward" and "reverse" directions, thus allowing for the first time an experimental determination of both electronic potential curves that participate in the curve crossing of the energy transfer. We also investigate systems which have competing pathways and find remarkably state-specific alignment dependencies in cases where there are a multitude of intertwined electronic states.

2. GENERAL PICTURE OF ALIGNMENT-DEPENDENT ENERGY TRANSFER

The physical basis for an alignment-dependent energy transfer process, or reactive event, comes about through the differences in the electronic states that are formed as the asymptotically aligned atomic orbitals couple into the reference frame of the transient molecules during a collision. Figure 1 shows a set of approximate van der Waals potential curves which are applicable to the $Ca(4s5p\ ^1P_1)$ + He rare gas system. This Ca state transfers its excitation readily to the near resonant $Ca(4s5p\ ^3P_J)$ states because of the mixed character of their wave functions ($\overline{\Delta E} \approx 177\ cm^{-1}$) [12,14]. The laser is used to excite the $Ca(5\ ^1P_1)$ state or the $Ca(5\ ^3P_1)$ state. In the case of the $5\ ^1P_1$ state the linearly polarized light forms a resulting p-orbital which is parallel to the direction of polarization. Rotation of the laser polarization allows the initial p-orbital to be aligned asymptotically either parallel (Σ state) or perpendicular (Π state) to the relative velocity of approach in the Ca*-rare gas collision.

Ca and Sr are relatively simple atoms for such studies, because in the case of excitation of a $^1S_0$ - $^1P_1$ transition with zero nuclear spin, the resulting p-orbital electron density lies rigorously along the direction of the laser polarization. There is no precession to deteriorate the alignment, except for the essentially negligible effect of precession due to the Earth's magnetic field.

If the splitting between the Π and Σ states becomes large enough at long range, then the p-orbital will "lock" into the reference frame of the van der Waals Ca*-rare gas molecule and remain on either the asymptotically-prepared Π or Σ

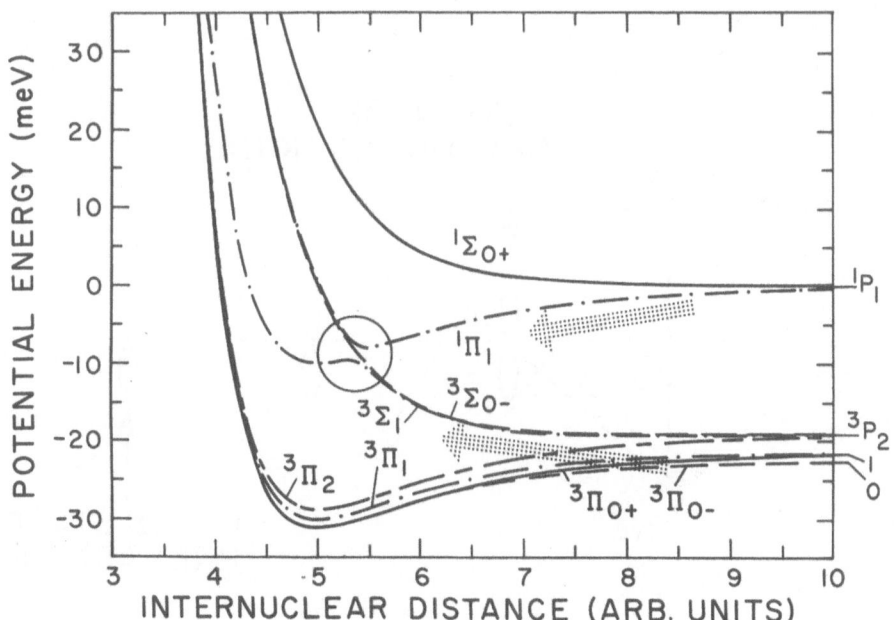

FIGURE 1. Schematic of the approximate van der Waals potential energy curves for $Ca(5\ ^1P_1)$ and $Ca(5\ ^3P_J)$ + He.

curve throughout the collision [3]. It is this condition which must be met in order to observe any alignment effects at all. If the orbital direction becomes scrambled in the collision, then there is no opportunity to observe a dependence of the cross section for the energy transfer process or reactive event on the initial direction of the orbital.

If the orbital following and locking occurs only at very close range, then non-zero impact parameters can cause a shift of the initial orbital alignment by as much as 90^0, since the orbital is transformed into the molecular reference frame only after the initial orbital angle has changed as the atoms approach. Figure 2 displays this effect, in which an incident Π orbital alignment locks into the molecular frame at non-zero impact parameter with substantial Σ character.

As depicted in Fig. 1, if the electronic energy transfer takes place at a well-defined curve crossing between an initial $^1\Pi$ van der Waals state and a final $^3\Sigma$ state, then initial $^1\Pi$-state preparation is expected to show an enhanced cross section over an initial $^1\Sigma$ state. This is the case observed for the transfer from Ca(5 $^1P_1$) to Ca(5 ^3P_J) when induced by collisions with the light rare gases. Observation of a strong enhancement (50-60%) of the cross section for the perpendicular (Π) orbital alignment is interpreted to mean both that orbital locking occurs reliably at long range and that the specific curve crossing which is responsible for the energy transfer involves the $^1\Pi$ state in the entrance channel. In the sections that follow we will investigate in more detail the meanings of the magnitudes and directions of the alignment effects observed in a variety of energy transfer events, the current theoretical interpretations, and the relationship to chemical reaction dynamics.

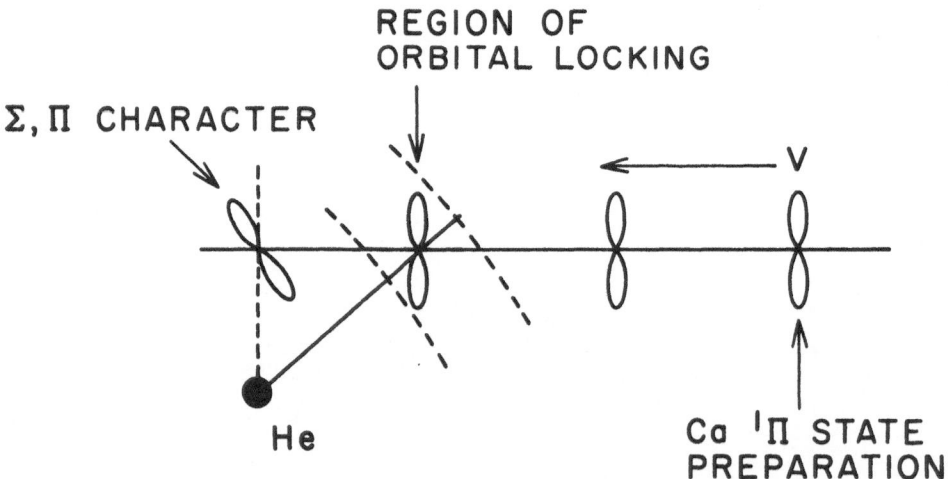

FIGURE 2. An example of "late" orbital locking, in which an orbital of initial Π character locks into the molecular frame with substantial Σ character.

3. EXPERIMENTAL METHODOLOGY

The experimental apparatus is depicted in Fig. 3 and is described in detail elsewhere [14]. It consists of a molecular beam scattering chamber, pulsed frequency-doubled dye laser, double fresnel rhomb polarization rotator, fibre bundles to collect fluorescence from excited states, and time-gated photon counting and boxcar averaging equipment.

The Ca or Sr beam is a skimmed effusive beam emanating from an oven which contains the metal. The target gas beam approaches at right angles to the Ca beam and is an unskimmed, pulsed supersonic expansion. In the interaction region, the laser enters at the third mutually perpendicular direction. The Ca atoms are excited with the laser, and a small fraction of the excited atoms undergo collisions with the target gas and change state (e.g. from $5\ ^1P_1$ to $5\ ^3P_J$, Fig. 4). Fluorescence emissions from both the initially excited state and the states which receive the transfer of excitation (Fig. 4) are collected by fibre bundles and brought out through interference filters to photomultipliers. The output signals of the photomultipliers are recorded with either boxcar integrators or photon counting equipment as a function of the angle of polarization of the laser.

Figure 5 shows the geometries of the crossed beam arrangement and the laser polarization ($\vec{E}_L$) and gives the definitions of the angles used in the formulae for the alignment effects [14]. For each scattering system, the velocity vectors are calculated by standard molecular beam formulae, in order to determine the expected

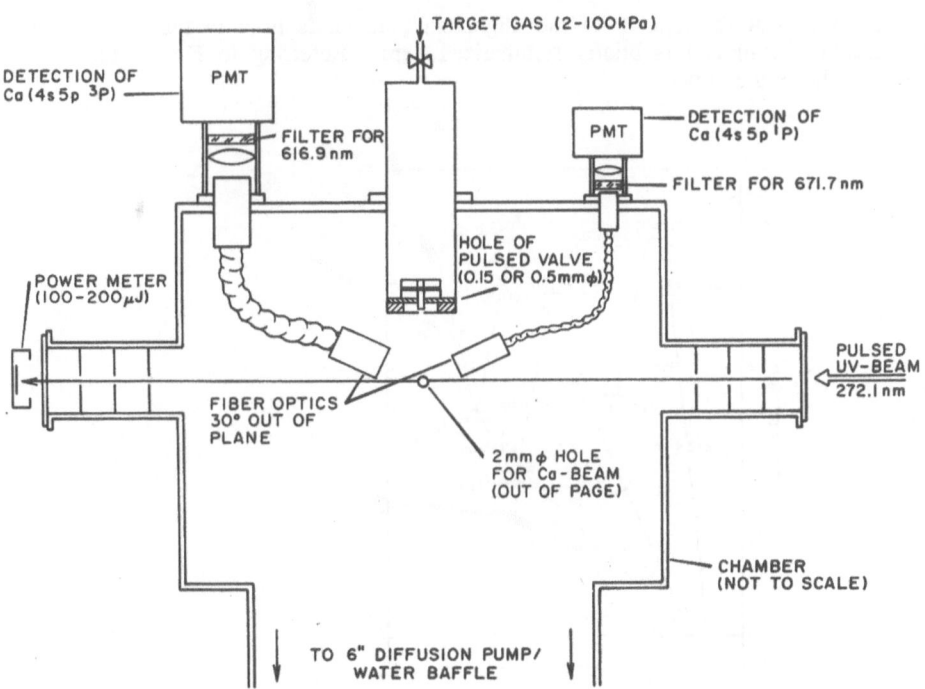

FIGURE 3. Sketch of the experimental apparatus for orbital alignment studies.

polarization angles for parallel and perpendicular alignments. In practice, the observed angles of maximum or minimum signals are obtained by an empirical fit of a cosine function to the data, and the phase angle is compared to the predicted value. Due to the cylindrical symmetry of the impact parameter, angles other than $0^{\circ}(\Sigma)$ and $90^{\circ}(\Pi)$ for the maximum signal are not expected. Because of possible uncertainties in the beam angle and the spread of angles and velocities, the measurements are sometimes as much as 15° in error from 0° and 90° in some systems [14].

Signals are acquired with the laser both on and off, with the target gas on and off, and from both the initially excited state and the final state. A detailed series of corrections [14] are applied to account for background scattered light, leakage of emission through the filters, and to normalize for the extent of initial excitation. In favourable cases, the alignment effect is immediately evident in the raw data of just the final state (e.g. Ca 5 $^{3}P_J$ when exciting 5 $^{1}P_1$, Fig. 6). In more difficult cases when there are very small transfer cross sections, the careful subtractions and normalization are essential to detect the alignment effect. Typical signals in the best cases can be acquired with only several minutes of integration at each angle. In unfavourable cases, such as in the studies of the reverse transfer process Ca(5 $^{3}P_J$) → Ca(5 $^{1}P_1$), data integration times of 30-60 minutes per angle are required.

4. ALIGNMENT PARAMETERS

The mathematical formulation of the alignment parameters used in these studies is discussed in detail and is briefly summarized here. Referring to Fig. 4, the final data are fit to the function

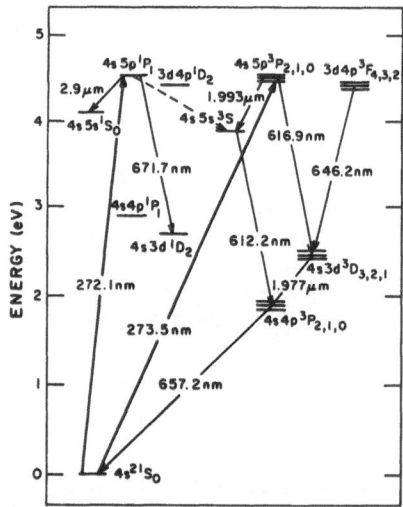

FIGURE 4. Energy level diagram for the Ca atom showing the levels and transitions of relevance to the 5 $^{1}P_1 \leftrightarrow$ 5 $^{3}P_J$ transfer.

$$I(\varphi) = C_0 + C_2 \cos[2(\varphi - \varphi_{max})]$$

to obtain the angle of maximum signal in the laboratory frame, φ_{max}, and the magnitudes of the alignment effect:

$$I_{max} = I(\varphi_{max}) = C_0 + C_2$$

$$I_{min} = I(\varphi_{max}\text{-}90^0) = C_0 - C_2.$$

In the centre-of-mass frame, the angle of maximum signal is obtained as:

$$\varphi_{max}^{cm} = \varphi_{max} - \varphi_{rel},$$

where φ_{rel} is computed from the velocity vectors of the two crossed beams, i.e. the average direction of the relative velocity of the colliding atoms in the lab frame.

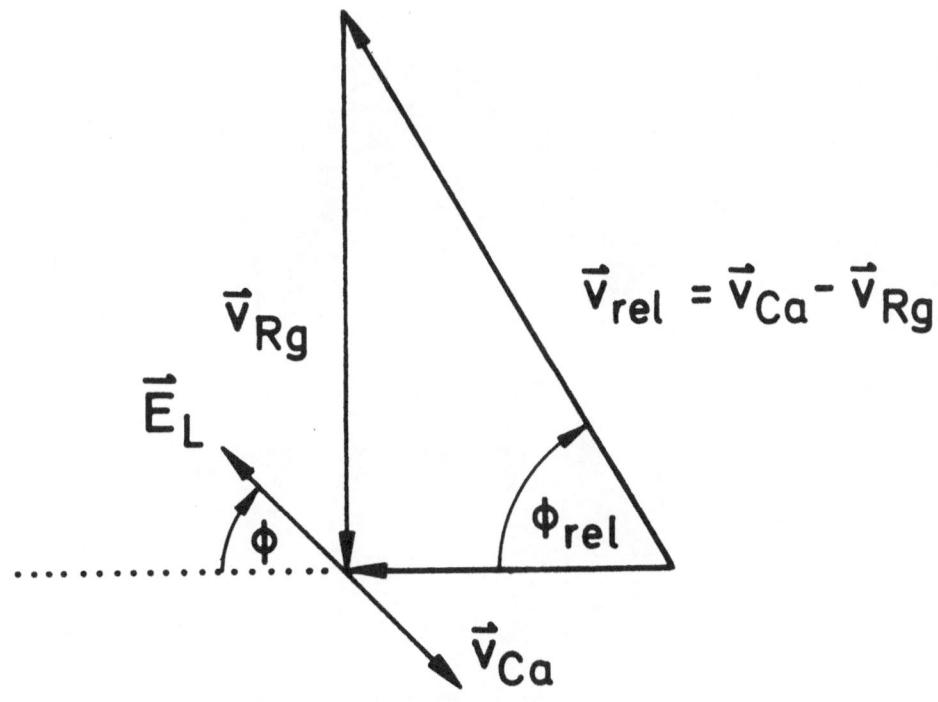

FIGURE 5. Vector diagram of the Ca-rare gas collision system. $\bar{E}_L$ is the laser polarization.

252

The size and direction of the alignment effect can be described by the angle φ_{max}^{cm} and the ratio I_{max}/I_{min}, or by the polarization quantity

$$P = [I(\perp) - I(\| \|)][I(\perp) + I(\| \|)].$$

The values of P range from 100% to -100%. When $I(\perp)$ has the higher cross section, P is positive, and when $I(\| \|)$ is larger, P is negative. The absolute value of the polarization is directly related to the fit constants:

$$|P| = |C_2/C_0|.$$

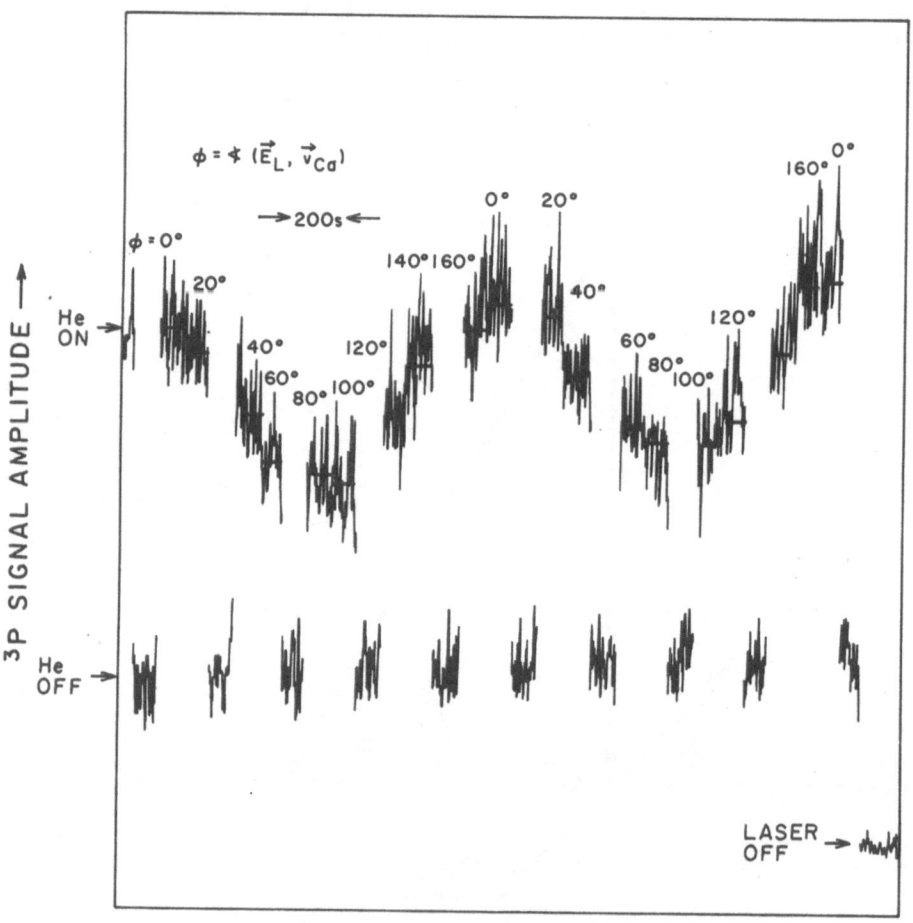

FIGURE 6. Raw data for the Ca(5 ^3P_J) emission as a function of laser polarization used to excite 5 $^1P_1$.

5. SPECIFIC SYSTEMS

5.1 $Ca(4s5p\ ^1P_1)$ + rare gas $\rightarrow$ $Ca(4s5p\ ^3P_J)$ + rare gas

5.1.1   Forward transfer.   Optical excitation of the $Ca(5\ ^1P_1)$ state with linearly polarized light gives a straightforward interpretation of the resulting orbital alignment.   Since the ^{40}Ca isotope has an abundance of 97% and no hyperfine structure, the selection rule $\Delta M_J = 0$ prepares specifically $M_L = 0$ for a spin S = 0 state.   Thus the p-orbital electron density is aligned parallel to the electric vector of the laser $\bar{E}_L$.   When $\bar{E}_L$ is aligned parallel to the relative velocity vector v_{rel}, an asymptotic Σ state is prepared.   When $\bar{E}_L$ is perpendicular to v_{rel}, an asymptotic Π state results, and the p-orbital is perpendicular to the quasi-molecular axis.   From the angle of maximum transfer probability, it can be determined whether the asymptotic Σ or Π state gives the larger cross section.

The measured data are shown in Figs. 7a-f for the collision partners ^3He, He, Ne, Ar, Kr, and Xe.   This series of rare gases have maximum angles φ_{max}^{cm} of 105^0, 103^0, 109^0, 107^0, (-20^0), and 7^0.   The parentheses indicate that the data for Kr may show no effect.   It is evident from the data that the lighter rare gases prefer the Π state, while Kr shows little or no effect, and Xe prefers the molecular Σ state.   The magnitudes of the effects vary, with I_{max}/I_{min} ratios of 1.60(He), 1.55(Ne), 1.5(Ar), <1.2(Kr), and 1.4(Xe) [14].   While an excellent interpretation can be made for the enhancement of the cross section by the perpendicular orbital alignment for He, Ne, and Ar, there is not yet a definitive explanation for the change in direction of the effect for Xe and the absence of an effect for Kr.

Referring to Fig. 1, it is evident, from these approximate potentials for $Ca(5\ ^1P_1)$ and $Ca(5\ ^3P_J)$ with He, that a primary curve crossing can occur between the $5\ ^1P_1$ Π state and the $5\ ^3P_J$ Σ state.   For Kr and Xe, several facts will alter these potentials.   The well depths will be much greater and the turning point of the repulsive wall will be displaced to larger internuclear separation.   In addition, the spin-orbit coupling interaction will be greater for Kr and Xe.   This could give rise to long range transition matrix elements which can cause a change of the state of the system without a curve crossing.   Either the ordering of the Π and Σ states is changed for Kr and Xe, or the long range spin-orbit interaction mixes the states, yielding for Xe the larger cross section for asymptotic preparation of the Σ state and no preference for Kr.

5.1.2.   Reverse direction.   Optical excitation of the $Ca(5\ ^3P_1)$ state results in the reverse transfer to the $5\ ^1P_1$ by utilizing the kinetic energy of the collision, since the process is ~200 cm^{-1} endothermic.   The initial Ca state now has a spin S = 1, and the initial state preparation again selects $M_J = 0$.   However the interpretation of this state in terms of the orbital electron density requires a more complicated assessment of the coupling into the molecular frame.   Our analysis [14] suggests that preparation with perpendicular laser polarization gives equal parts of the $^3\Pi_1$ and $^3\Sigma_1$ curves at closer range.   Preparation with the parallel laser polarization results in the system remaining on the $^3\Pi_{0^+}$ curve throughout the collision.

The data in the reverse direction show that the parallel laser polarization is strongly preferred for collisions with $Xe(I_{max}/I_{min} = 1.9)$, and the perpendicular laser preparation gives the larger cross section for He collisions $(I_{max}/I_{min} = 1.6)$.   We interpret these results to indicate that the Π molecular state is active for Xe in the reverse process.   For He, either the Π or Σ state is responsible for the reverse process, although as shown in Fig. 1, we strongly prefer the assignment of the Σ

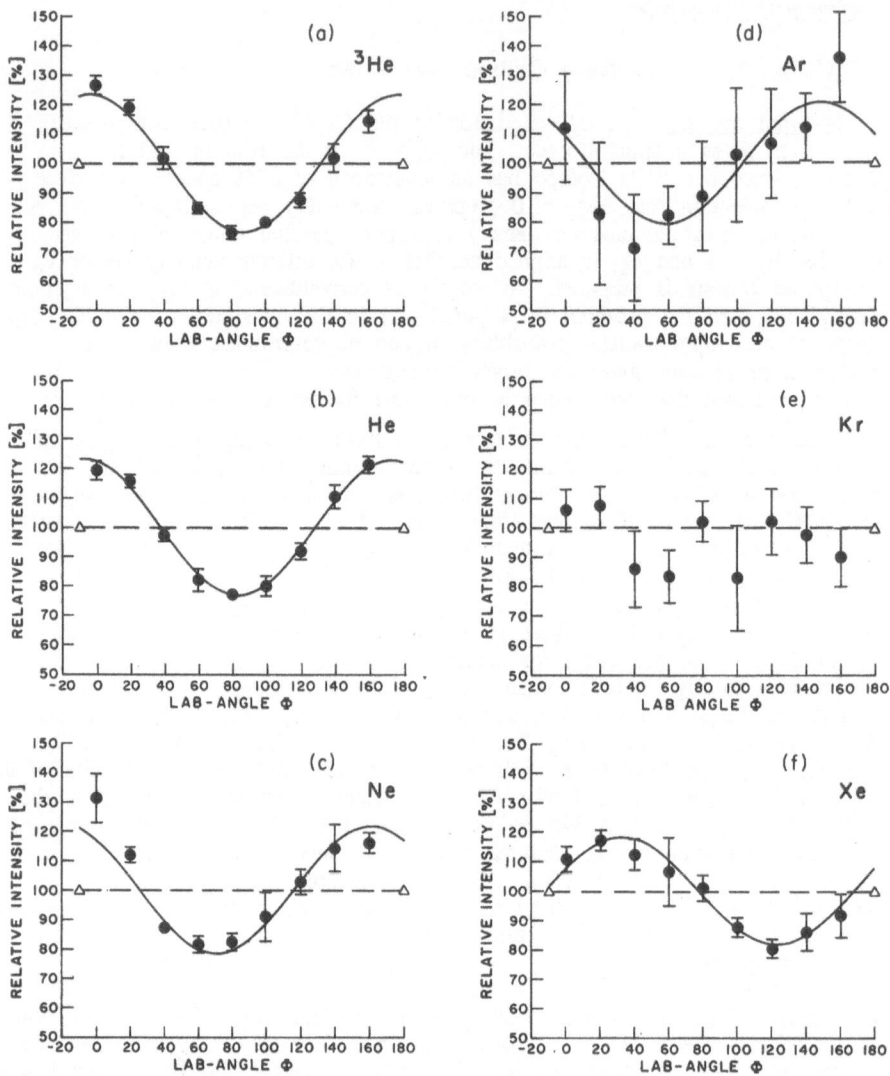

FIGURE 7. Normalized alignment effects for the Ca($5\ ^1P_1 \leftrightarrow 5\ ^3P_J$) energy transfer with rare gases.

state in the reverse direction. These results are the first dynamical experiments to extract the symmetries of the two electronic potentials involved in the curve crossing, and they demonstrate a powerful new way to investigate electronic curve crossing phenomena in general.

The larger size of the alignment effect for Xe in the reverse direction is highly suggestive that a smaller range of impact parameters may be responsible for the reverse transfer. This makes sense when one considers that the endothermicity of the process will require smaller impact parameters to utilize the available kinetic energy.

5.1.3. Theory. From a parameterization of the magnitudes of the alignment effects [12], information can be deduced on the radius for orbital locking, i.e. the region where the energy of the molecular orbitals becomes sufficient enough to "lock" the alignment of the p-orbital to the molecular axis. Our estimates [12] show that this occurs at a radius about twice the radius of the curve crossing for He. From the measured cross sections for the rare gases of 25, 5, 3, 13, and 31 x 10^{-16} cm^2 for He, Ne, Ar, Kr, and Xe [28], reasonable estimates of the locking and crossing radii for He are 6 and 3 x 10^{-8} cm, respectively.

Published theoretical descriptions of the Ca(5 $^1P_1$) - Ca(5 ^3P_J) alignment system have considered the formal Landau Zener curve crossing probability [29] and have used full quantum mechanical descriptions [30]. Unfortunately, all the theoretical descriptions are limited by the lack of accurate potential surfaces for the van der Waals states of the electronic levels. However, in the future, accurate information may become available from recent experiments to investigate metal atom + rare gas van der Waals potentials using supersonic jet spectroscopy [31-34]. Thus there is an excellent chance that it will also be possible to obtain more accurate theoretical descriptions, which will elucidate important subtleties of alignment effects in energy transfer and reactions.

Thus far, the classic Landau Zener curve crossing calculation [29] obtains a magnitude for the alignment effect for Ca(5 $^1P_1$) + He which is considerably larger than observed experimentally. Although the simple representations of orbital locking and curve crossing are appealing, there is reason to suspect that there may be weaknesses in this type of interpretation [30]. The full quantum mechanical calculations of Pouilly and Alexander find that the experimental data can be reproduced with some accuracy, given trial model potentials which reproduce the total cross section data. In addition, however, the Pouilly and Alexander interpretation does not find a physical basis as yet for the orbital following and locking concept. They suggest that the memory of the initial orbital direction may be lost in the collision. These provocative comments invite further theoretical and experimental investigations to elucidate the dynamics of alignment effects.

5.2   Ca(4s5p $^1P_1$) + Molecules

With molecular collision partners, the rotational structure of the incoming particle may enhance the scrambling of the initial orbital alignment. It is not a priori obvious that orbital alignment effects would be observed in either energy transfer or chemical reactions with molecules. The pioneering work of Rettner and Zare confronted this complexity headlong and definitively showed that in a variety of Cl transfer reactions with Ca(4 $^1P_1$) that orbital alignment effects are observed. Their results showed substantial alignment selectivities in some product channels as well as a lack of a dependence on alignment for others.

256

Our results in this section confirm that relatively complex molecular systems can display a memory of the initial orbital alignment without scrambling in the collision. The systems exhibit an alignment dependence even in light of numerous curve crossings, possible reactive channels, and the highly anisotropic potentials for the p-orbital interacting with the structured particle. The concepts of orbital following and locking, and the correlation between isolated atomic orbitals and specific molecular orbitals still appear to be valid in some cases. The results offer encouraging prospects for further studies of reactive events with aligned states of atoms.

Figure 8 shows four successful sets of alignment data for the Ca 5 $^1P_1 \rightarrow 5$ ^3P_J transfer system with molecules H_2, D_2, and CO_2 and for H_2 in the reverse direction [15]. As was observed for Xe, the alignment effect for H_2 in the reverse direction is larger than in the forward direction, suggesting that a smaller range of impact parameters participates in the endothermic direction. While H_2 and D_2 exhibit a preference for the initial perpendicular orbital alignment, similar to the lighter rare gases, CO_2 shows a strong preference for the parallel orbital alignment, similar to Xe.

A large number of other molecules exhibit no alignment effects (N_2, O_2, CO, and C_2H_6) or an insufficient cross section for the energy transfer pathway (CH_4 and SF_6). The results of all the Ca energy transfer alignment dependencies are summarized in Fig. 9. An assessment of the data for the "negative" results shows that some energy transfer does occur with N_2, O_2, CO, and C_2H_6, but that there is complete scrambling of the alignment effect. For CH_4 and SF_6, the loss of the Ca $^1P_1$ excited state is complete, but without concomitant formation of the Ca ^3P_J states.

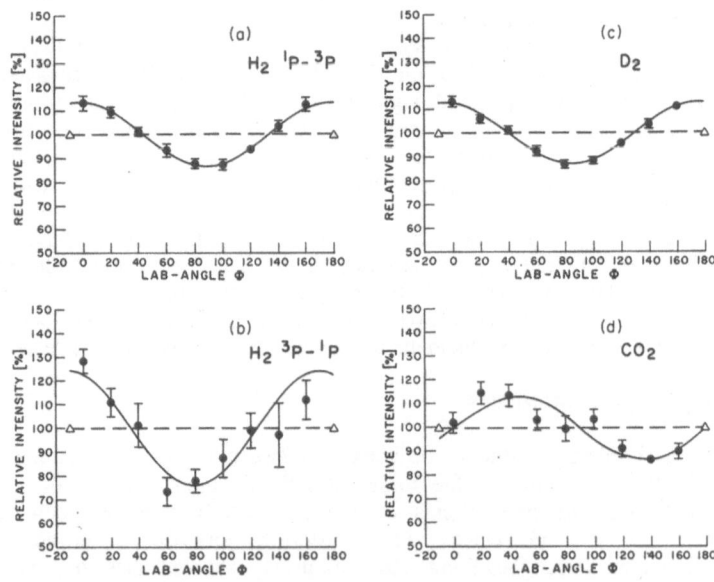

FIGURE 8. Normalized alignment effects for the Ca(5 $^1P_1 \leftrightarrow 5$ ^3P_J) energy transfer with molecules.

Reactive channels are available and are exothermic for all of the molecular collision partners tested, a fact which is expected to play a significant role in the alignment-dependence of the energy transfer process. The lack of an alignment effect for many of the molecules may be due to the competition between the desired energy transfer process and reactive events. A strong collision that involves a deeply attractive, reactive potential surface is much more likely to cause a loss of the initial orbital alignment, even if reaction does not occur.

The magnitudes of the "forward" effects for H_2, D_2 and CO_2 are somewhat lower than those observed for most of the rare gases ($I_{max}/I_{min} \approx 1.3$, compared to 1.6). It is tempting to speculate that the rotational structure of the molecular partner deteriorates the alignment effect by some partial scrambling of the orbital directionality. However, it should be noted that the reverse process for H_2 shows an effect of $I_{max}/I_{min} \approx 1.6$. Thus, it does not appear that the rotational structure alone can be responsible for the smaller alignment effects with molecules.

The effects of molecular structure on the rates of energy transfer are manifested in many ways. Additional pathways are available in collisions with molecules, such as the influence of ionic potential surfaces [35], the availability of near resonant electronic-to-vibrational and rotational energy transfer pathways [36], and the introduction of nonadiabatic transitions due to the breaking of the molecular orbital symmetry [37]. For the studies considered here, we might also add the competition between reaction and the desired energy transfer process, the possibility of energy transfer processes in the entrance or exit channels, selective changes in

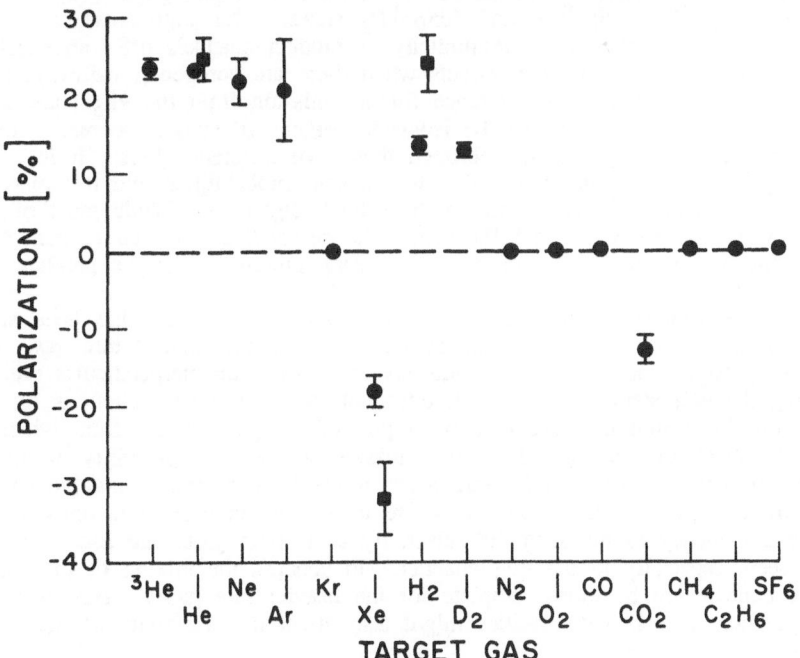

FIGURE 9. Summary of polarization values for the Ca($5\,^1P_1 \leftrightarrow 5\,^3P_J$) energy transfer system; o forward direction and ■ reverse direction.

fine structure states [38], and the influence of differing spin-orbit reactivities [39]. These might all affect the alignment dependence observed for the $^1P_1$ - ^3P_J transfer which is of primary interest. At the present time we have no data on the occurrence of energy transfer to populate other energy transfer states or to participate in reactions, but these aspects will be the subjects of future intriguing investigations.

The energy transfer with CO_2 is unique among all these systems because its cross section for energy transfer is equally as small as for O_2, N_2, CO, and C_2H_6, but a significant alignment effect is observed. One possible explanation may be found in a consideration of the accessibility of ionic states, in which the formation of a transient negative ion of CO_2 must be bent in order to accept the electron. Thus, the same types of ionic states that may be involved for collisions with O_2 may not be accessible for CO_2. However, participation of ionic states with N_2 are not expected to be significant either. Possibly CO_2 can induce the energy transfer by long range impact parameters, and thus a small fraction of the collisions behave like the highly polarizable Xe, but without ever entering the reactive zone.

5.3   Sr(5s6p $^1P_1$): A Case for Competing Channels

Most of the energy transfer processes studied thus far can be approximated as two-state systems. The Ca(5 $^1P_1$) collisions with molecules may have competing non-resonant energy transfer pathways and reactions. However these have not yet been investigated. A recent study of Sr(5s6p $^1P_1$) excitation transfer showed that several near-resonant states are populated in collisions with rare gases [40]. These include the 5s6p ^3P_J, 4d5p ^3F_J, and 4d5p $^1D_2$ states. The large cross sections observed for this system and the multiplicity of product channels offer an excellent system for the study of alignment effects when there are competing pathways [16].

With two narrow band interference filters, emission from the $^3F_3$ state and a mixture of ^3P_J and $^3F_4$ states can be isolated. Figure 10 shows the results for the alignment-dependent energy transfer through these two different filters. It is immediately evident from the figure that the transfer probabilities into different states depend in remarkably different ways on the orbital alignment. While most of the collisions with the rare gases and H_2 favour the perpendicular orbital alignment, transfer with He into the $^3F_3$ state shows a strong preference for the parallel alignment.

The $^3F_3$ state is the furthest level away from resonance that has been studied, and the special selectivity in this state is observed with the lightest rare gas. Molecular hydrogen shows a similar tendency away from the perpendicular preference in forming the $^3F_3$ state, but the result does not quite change direction as for He.

Figure 11 summarizes this data on a plot of the polarization, from which it is evident that there are some trends with increasing size and polarizability of the rare gas collision partner. The heavier rare gases might be expected to involve stronger interactions, and in a regime where there are many electronic potential curves, there might be a tendency to diminish the selectivity as heavier gases are used. Such a trend is seen, especially in the $^3F_3$ channel. In general, the mixing of the states might be expected to be more complete for the heavier rare gases. However, the large differences in relative velocities might also affect the selectivity of the curve crossings.

What is most intriguing is that there are dramatic differences in the alignment effects for the two channels, especially for the lighter gases. In spite of the high density of states embedded in this region of the Sr energy levels, there is a

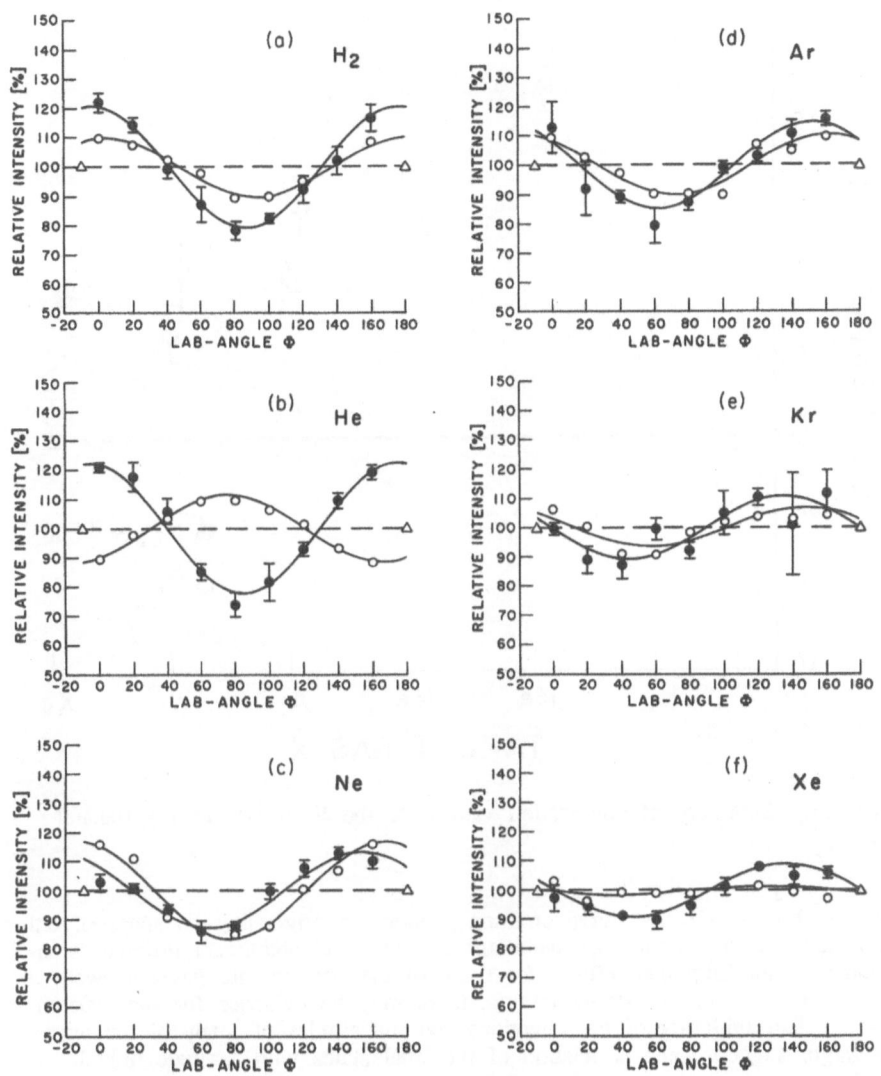

FIGURE 10. Normalized alignment effects for the Sr energy transfer system, o 6 $^1P_1 \rightarrow {}^3P_J + {}^3F_4$ states, ● 6 $^1P_1 \rightarrow {}^3F_3$ state.

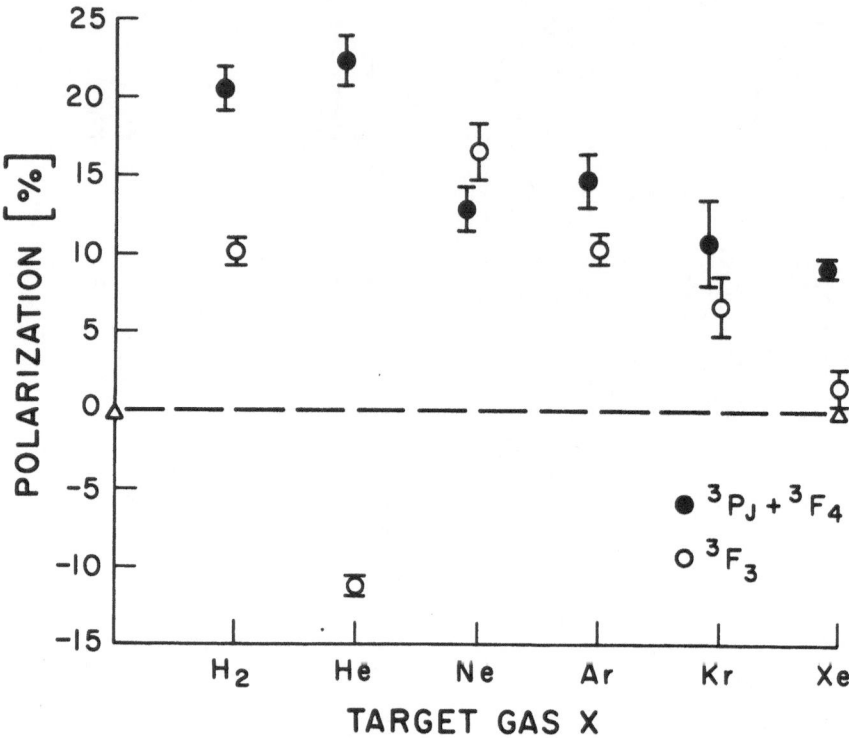

FIGURE 11. Summary of polarization values for the Sr 6 $^1P_1$ energy transfer system.

retention of highly selective curve crossing phenomena originating on states of either perpendicular (Π) or parallel (Σ) symmetries. There are significant differences in the magnitudes of the alignment effects for some of the heavier rare gases as well. Thus, even in the limit of strong mixing, there may be evidence for state specificity here also. The similarity of the directions and magnitudes of many of the alignment effects might suggest some correlation of the final states with a single, initial potential curve which accesses most of the product channels.

6. CONCLUSION

A high degree of specificity of orbital alignment dependence is observed for numerous near-resonant energy transfer processes. Alignment effects are found for collisions with rare gases as well as with several molecules. Many of the molecular systems show little or no alignment dependence, an effect which is likely to be due to traversal on deeply attractive, reactive surfaces. A study of the competition between several accessible energy transfer channels shows remarkable degrees of selectivity, in which, for example, one state prefers a parallel alignment and another the perpendicular. The results are successfully interpreted in terms of orbital following, locking, and curve crossing pictures, although recent quantum mechanical calculations call into question the pictorial representations of the orbital following and locking description. Many challenging new opportunities for the study of orbital alignment effects in chemical reaction dynamics will be opened up by the use of similar pulsed laser, state preparation experiments.

ACKNOWLEDGMENTS

The author gratefully acknowledges the exceptional work of Wolfgang Bussert, Dieter Neuschäfer, and Michael Hale, and numerous invigorating conversations with Ingolf Hertel, Alan Gallagher, and Millard Alexander. The Air Force Office of Scientific Research and the National Science Foundation generously support this research.

REFERENCES

[1] R.N. Zare, Ber. Bunsenges. Phys. Chem., 86 (1982) 422; S. Stolte, *ibid*, p. 413.

[2] S.R. Leone, Ann. Rev. Phys. Chem., 35 (1984) 109.

[3] E.E.B. Campbell, H. Schmidt and I.V. Hertel, Adv. Chem. Phys., (in press).

[4] For a recent reference, see A. Sieradzan, Opt. Comm., 56 (1985) 243, and references therein.

[5] L. Hüwel, J. Maier, H. Pauly, J. Chem. Phys., 76 (1982) 4961.

[6] W.J. Alford, N. Anderson, K. Burnett and J. Cooper, Phys. Rev. A, 30 (1984) 2366.

[7] J.M. Mestdagh, J. Berlande, P. dePujo, J. Cuvellier and A. Binet, Z. Phys. A, 304 (1982) 3.

[8] W. Reiland, G. Jamieson, U. Tittes and I.V. Hertel, Z. Phys. A, 307 (1982) 51.

[9] A. Bahring, I.V. Hertel, E. Meyer, N. Spies and H. Schmidt, J. Phys. B, 17 (1984) 2859.

[10] A. Bahring, E. Meyer, I.V. Hertel and H. Schmidt, Z. Phys. A, 320 (1985) 141.

[11] M.P.I. Manders, J.P.J. Driessen, H.C.W. Beijerinck and B.J. Verhaar, Phys. Rev. Lett., 57 (1986) 1577.

[12] M.O. Hale, I.V. Hertel and S.R. Leone, Phys. Rev. Lett., 53 (1984) 2296.

[13] D. Neuschäfer, M.O. Hale, I.V. Hertel and S.R. Leone, in Electronic and Atomic Collisions XIV ICPEAC, edited by D.C. Lorents, W.E. Meyerhof and J.R. Peterson (Amsterdam, North Holland, 1986), p. 585.

[14] W. Bussert, D. Neuschäfer and S.R. Leone, J. Chem. Phys., 87 (1987) 3833..

[15] W. Bussert and S.R. Leone, Chem. Phys. Lett., (in press).

[16] W. Bussert and S.R. Leone, Chem. Phys. Lett., (in press).

[17] W. Bussert, T. Bregel, R.J. Allan, M.W. Ruf and H. Hotop, Z. Phys. A, 320 (1985) 105.

[18] J.G. Kirez, R. Morgenstern and G. Nienhuis, Phys. Rev. Lett., 48 (1982) 610.

[19] M.-X. Wang, M.S. DeVries and J. Weiner, Phys. Rev. A, 33 (1986) 765.

[20] M.-X. Wang, M.S. DeVries and J. Weiner, Phys. Rev. A, 33 (1986) 1612.

[21] M.-X. Wang, J. Keller, J. Boulmer and J. Weiner, Phys. Rev. A, 34 (1986) 4497.

[22] C.T. Rettner and R.N. Zare, J. Chem. Phys., 74 (1981) 3630.

[23] C.T. Rettner and R.N. Zare, J. Chem. Phys., 77 (1982) 2417.

[24] H.W. Hermann and I.V. Hertel, Comments At. Mol. Phys., 12 (1982) 61, 127.

[25] I.V. Hertel, H. Schmidt, A. Bahring and E. Meyer, Rep. Prog. Phys., 48 (1985) 375.

[26] R. Witte, E.E.B. Campbell, C. Richter, H. Schmidt and I.V. Hertel, Z. Phys. D, 5 (1987) 101.

[27] J. Cooper, in Spectral Line Shapes, Vol. 2, edited by K. Burnett (de Gruyter, Berlin, 1983) p. 737.

[28] M.O. Hale and S.R. Leone, J. Chem. Phys., 79 (1983) 3352.

[29] A.Z. Devdariani and A.L. Zagrebin, Chem. Phys. Lett., 131 (1986) 197.

[30] B. Pouilly and M.H. Alexander, J. Chem. Phys., 86 (1987) 4790.

[31] K. Fuke, T. Saito and K. Kaya, J. Chem. Phys., 81 (1984) 2591.

[32]   M.-C. Duval, O. Benoist D'Azy, W.H. Breckenridge, C. Jouvet
       and B. Soep, J. Chem. Phys., **85** (1986) 6324.

[33]   K. Yamanouchi, J. Fukuyama, H. Horiguchi, S. Tsuchiya,
       K. Fuke, T. Saito and K. Kaya, J. Chem. Phys., **85** (1986) 1806.

[34]   A. Kowalski, D.J. Funk and W.H. Breckenridge, Chem. Phys. Lett.,
       132 (1986) 263.

[35]   E.A. Andreev and A.I. Voronin, Chem. Phys. Lett., **3** (1969) 488.

[36]   F. Rebentrost and W.A. Lester, J. Chem. Phys., **67** (1977) 3367.

[37]   A.P. Hickman, Phys. Rev. Lett., **47** (1981) 1585.

[38]   H.-J. Yuh and P.J. Dagdigian, J. Phys. B, **17** (1984) 4351.

[39]   H.-J. Yuh and P.J. Dagdigian, J. Chem. Phys., **81** (1984) 2375.

[40]   R.W. Schwenz and S.R. Leone, Chem. Phys. Lett., **133** (1987) 433.

FUNDAMENTAL CONSIDERATIONS IN THE STUDY OF INELASTIC AND REACTIVE COLLISIONS INVOLVING ATOMS IN ^{1}P ELECTRONIC STATES

Millard H. Alexander and Brigitte Pouilly*
Department of Chemistry
University of Maryland
College Park, Maryland 20742
USA

1. INTRODUCTION

There has been considerable recent interest, both experimental [1-7] and theoretical [8,9] in inelastic as well as reactive collisions involving atoms in ^{1}P electronic states. Leone and co-workers [2-5] have carried out a series of experimental studies of collision-induced, spin-changing transitions between the ^{1}P and ^{3}P electronic states of the 4s5p Rydberg levels of Ca in collisions with noble gases and closed-shell molecules. The electronically excited Ca atoms (4s5p ^{1}P) were prepared by laser excitation out of the $4s^2$ ^{1}S ground state. More recently, similar studies have been carried out with electronically excited Sr [6,7]. The primary motivation for this type of experiment is that the orientation of the excited p orbital can be selected by varying the direction of the polarization vector of the laser. It is thus possible to study the dependence of spin-changing transitions or another collisional process on orbital alignment - orbital stereochemistry. Ultimately, this type of study can be extended to reactive collisions. An example is the work of Rettner and Zare [1] who used laser excitation to investigate the effect of orbital alignment on cross sections for reactions of Ca(4s4p ^{1}P) with several halogen-containing reactants.

The mechanism for this type of process, even the simpler inelastic events, is quite complex, involving collisional coupling between the electronic angular momentum $\mathbf{L}$ of the electronically excited atom and the orbital angular momentum ℓ of the collision partners. In processes involving singlet→triplet transitions, the spin $\mathbf{S}$ of the atom must also be included. In addition, the electrostatic interaction between the open-shell atom and either an atomic or molecular collision partner will depend on the orientation of the initially excited p orbital. Thus the description of the collision will involve more than one potential energy curve (or surface). Another complication is that the orientation of the initially excited orbital is selected in a *laboratory frame* coordinate system which may or may not coincide with the coordinate system used to describe the collision.

In the initial interpretation of their experiments Leone and Zare and their co-workers used qualitative concepts developed by Hertel and co-workers [10-12] to

---

* NATO science fellow, 1987; Permanent address: Laboratoire de Spectroscopie des Molécules Diatomiques, UA 779, Université de Lille Flandres-Artois, Bâtiment P5, 59655 Villeneuve d'Ascq Cedex, France.

J. C. Whitehead (ed.), Selectivity in Chemical Reactions, 265–305.
© 1988 by Kluwer Academic Publishers.

interpret earlier experimental work on collisions involving electronically excited Na atoms in ^{2}P states. Recently, we presented [9] the first fully-quantum description of the inelastic ^{1}P $\leftrightarrow$ ^{3}P processes under investigation in Leone's group.

Our intent in the present article is to provide the reader with the theoretical background necessary for a fuller understanding of the dynamics of these spin-changing ^{1}P $\leftrightarrow$ ^{3}P transitions and, ultimately, of the dynamics of reactive collisions involving laser-excited atoms in ^{1}P electronic states. Rather than present the results of calculations for individual systems we shall concentrate on a general and thorough discussion of the underlying methodology. On the one hand we shall review the fully-quantum formulation of the collision dynamics which is necessary for *quantitative* calculations of cross sections. We shall also present and apply two techniques which can be used to gain *qualitative* insight into the dynamics. These involve the use of Hund's case coupling schemes [13-22], well known to spectroscopists [13-15], and the use of a fully adiabatic description of the collision [9,16,20-29]. The former, discussed in more detail in section 2, can be a useful guide in constructing the proper basis to describe the collision both at large and small values of the interatomic separation [16-22]. The use of a fully adiabatic description will allow us to isolate those regions of coordinate space where the coupling between given initial and final states of the open-shell atom is largest. The differences and transformations between diabatic and adiabatic descriptions is the subject of section 3.

Section 4 contains an analysis of the proper description of the state of the system which is initially prepared in a collision experiment involving a laser-excited atom. Here an adiabatic analysis will be used to point out several inadequacies in simple semiclassical treatment of these spin-changing transitions. Finally, section 5 is devoted to a presentation of the orbital-locking models of Hertel and co-workers [10-12]. The insights gained in our more exact quantum treatment will be used to examine critically the validity of these models. A brief conclusion follows.

2. SYSTEM WAVEFUNCTIONS, HUND'S CASE COUPLING SCHEMES, AND SYMMETRY DESIGNATION

Consider the collision between an open-shell atom and a closed-shell atomic partner. Let **R** designate the interatomic distance (the bond axis of the diatomic) and let **r** designate the coordinates of the electrons. The total Hamiltonian of the diatomic system can be written as

$$H(\mathbf{R},\mathbf{r}) = \frac{\hbar^2}{2\mu R^2} \frac{d}{dR} R^2 \frac{d}{dR} + H_{rot}(\hat{R}) + V(\mathbf{R},\mathbf{r}) + V_{fs}(\mathbf{R},\mathbf{r}), \quad (1)$$

where μ is the collision reduced mass and $V_{fs}(\mathbf{R},\mathbf{r})$ is the fine- and/or hyperfine-structure Hamiltonian for the open-shall atom. Here $H_{rot}(\hat{R})$ is the Hamiltonian for the orbital motion of the two nuclei, namely

$$H_{rot}(\hat{R}) = \frac{\hbar^2}{2\mu R^2} \varrho^2 = \frac{\hbar^2}{2\mu R^2} (\mathbf{J} \text{-} \mathbf{L} \text{-} \mathbf{S})^2, \quad (2)$$

where **L**, **S**, ϱ and **J** denote, respectively, the electronic orbital and spin angular momenta of the open-shell atom, the orbital angular momentum of the two nuclei, and the total angular momentum. Although a diatomic spectroscopist would use $\mathbf{R}$ [14,15], rather than ℓ to designate the orbital angular momentum of the two nuclei, we will not adopt this notation in order to avoid confusion with the interparticle distance $\mathbf{R}$. In Eq. 1 V(R,r) designates the total electronic Hamiltonian, independent of spin-orbit, spin-spin, or hyperfine coupling, namely

$$V(R,r) = H_0(r) + W(R,r), \tag{3}$$

where $H_0(r)$ represents the *electronic* Hamiltonian (exclusive of spin-orbit, spin-spin, or hyperfine coupling) of the isolated atoms. The term W(R,r) is the electrostatic interaction which arises as the two atomic collision partners approach. We assume that W(R,r) vanishes as $R \to \infty$.

In general it is impossible to find a basis in which the fine-structure Hamiltonian V_{fs}, the rotational Hamiltonian H_{rot}, and the electrostatic Hamiltonian W are all three diagonal. Since H_{rot} and W vanish as $R \to \infty$, in the asymptotic region it is appropriate to use a basis in which V_{fs} is diagonal. As the two atoms approach one another the electrostatic and rotational Hamiltonians increase in magnitude and eventually dominate the fine-structure Hamiltonian. In this region it is more appropriate to describe the wavefunction in a basis in which either H_{rot} or W is diagonal.

In molecular spectroscopy the various bases in which one can describe the electronic wavefunction correspond to one of the so-called Hund's coupling cases [13-15]. For the collision of two atomic partners the most suitable Hund's coupling case changes as R decreases from ∞ to the classical turning point. Much of the observed dynamics can be understood in terms of transitions induced as the two atoms pass from a region in which one coupling scheme is appropriate into a region where another scheme is more suited [16-22].

To illustrate much of this general discussion it is useful to focus on a particular system: the collision of an atom with a singly-filled p orbital with a closed-shell atom. The total term symbol of the open-shell atom (^1P, ^2P, ^3P etc.) will depend on the spin coupling between the singly-filled p-orbital and the inner shells. As the closed-shell partner approaches, the 3-fold spatial degeneracy of the p orbital is lifted. If the molecular axis is taken to denote the body frame z-axis, then at short range the electrostatic interaction will be more repulsive if the p orbital is oriented along $\hat{z}$ (p_0, Σ in $C_{\infty v}$ symmetry) and less repulsive if the p orbital is oriented perpendicular to $\hat{z}$ ($p_{\pm 1}$ or p_x and p_y, Π in $C_{\infty v}$ symmetry). At large internuclear separation, the situation is *reversed*, since the attractive dispersion interaction is stronger if the p orbital is oriented along $\hat{z}$. This is because the polarizability (parallel to the z-axis) of the P-state atom is larger if the p orbital lies along $\hat{z}$. For clarity we shall denote the molecule-frame coordinate system, with its origin fixed at the center of mass of the molecule, by lower-case letters, and the space- or laboratory-fixed coordinate system, by upper-case letters.

Asymptotically, when the separation between the collision partners is so large that the electrostatic interaction potential is negligibly small, the wavefunction of the combined system is most conveniently expanded in an *uncoupled* basis, namely a product of the wavefunction of the open-shell atom, $|LSjm_j\rangle$, multiplied by the wavefunction which describes the relative orbital motion of the two collision partners, $|\ell m_\rho\rangle$. Here j=L+S is the total angular momentum of the open-shell atom (e.g. j=1 for ^1P atoms, j=0, 1, or 2 for ^3P atoms) and m_j and m_ρ denote the projections of $\mathbf{j}$ and ℓ, respectively, along the space-fixed Z-axis. For collisions involving atoms in P states, L=1. The coordinate representation of the orbital wavefunction is just [30]

$$|\ell m_\varrho> = Y_{\ell m_\varrho}(\theta,\varphi),\qquad\qquad(4)$$

where $Y_{\ell m_\varrho}$ is a spherical harmonic and θ and φ are the polar and azimuthal angles which describe the orientation of the molecular axis $\mathbf{R}$ with respect to the space-fixed coordinate system. In the case of a partially filled p-shell, the wavefunction describing the open-shell atom $|LSjm_j>$ can be written as a sum of electronic wavefunctions in an uncoupled basis, namely

$$|LSjm_j> = \sum_{m_L m_S} (Lm_L Sm_S | jm_j) |Lm_L Sm_S>,\qquad\qquad(5)$$

where $(....|..)$ is a Clebsch-Gordan coefficient [30]. Here the electronic wavefunction which appears in the right hand side of Eq. 5 can be written as one, or a linear combination of, Slater determinants, depending on the exact number of electrons assigned to the unfilled s- and p-atomic orbitals in the atom - the electron occupancy. For sp and p^2 electron occupancies the $|Lm_L Sm_S>$ wavefunctions for several values of m_L and m_S are given in Table 1. For more detail the reader should consult the classic book by Condon and Shortley [31].

TABLE 1. Electronic wavefunctions $|Lm_L Sm_S>$ for a number of different ^1P and ^3P state electron occupancies.

		$m_L=1,$ $m_S=1$	$m_L=0,$ $m_S=1$	$m_L=1, m_S=0$	$m_L=0, m_S=0$
sp	S=1	$\lvert sp_1 \rvert$	$\lvert sp_0 \rvert$	$\{\lvert s\bar{p}_1\rvert+\lvert \bar{s}p_1\rvert\}/\sqrt{2}$	$\{\lvert s\bar{p}_0\rvert+\lvert \bar{s}p_0\rvert\}/\sqrt{2}$
	S=0			$\{\lvert s\bar{p}_1\rvert-\lvert \bar{s}p_1\rvert\}/\sqrt{2}$	$\{\lvert s\bar{p}_0\rvert-\lvert \bar{s}p_0\rvert\}/\sqrt{2}$
p^2	L=1,S=1	$\lvert p_1 p_0\rvert$	$\lvert p_1 p_{-1}\rvert$	$\{\lvert p_1\bar{p}_0\rvert+\lvert \bar{p}_1 p_0\rvert\}/\sqrt{2}$	$\{\lvert p_1\bar{p}_{-1}\rvert+\lvert \bar{p}_1 p_{-1}\rvert\}/\sqrt{2}$

In an actual scattering calculation it is more convenient to work with states which are eigenfunctions of the total angular moment $\mathbf{J}$ [9,20,21,32,33]. These states, which correspond to Hund's case (e) coupling [13,19,33], can be obtained by vector coupling the $|LSjm_j>|\ell m_\varrho>$ states, namely [30]

$$|LSj\ell JM> = \sum_{m_j m_\varrho} (jm_j \ell m_\varrho | JM) |LSjm_j>|\ell m_\varrho>\qquad\qquad(6)$$

where M is the projection of $\mathbf{J}$ along the space-fixed Z-axis. In this case (e) basis the matrix elements of *both* the rotational and spin-orbit Hamiltonians are diagonal and given by [9]

$$\langle L'S'j'\ell'JM|H_{rot}|LSj\ell JM\rangle = \delta_{LL'}\cdot\delta_{SS'}\cdot\delta_{\ell\ell'}\cdot\delta_{jj'}\cdot\hbar^2\ell(\ell+1)/(2\mu R^2) \quad (7)$$

and

$$\langle L'S'j'\ell'JM|V_{fs}|LSj\ell JM\rangle = \delta_{\ell\ell'}\cdot\{\delta_{jj'}\cdot\delta_{S1}\delta_{S'1}\frac{1}{2}\,a_3(R)j(j+1) +$$

$$\delta_{j1}\delta_{j'1}(1-\delta_{SS'})a_{13}(R)\}. \quad (8)$$

Here $a_3(R)$ denotes the spin-orbit coupling within the ^3P state and $a_{13}(R)$ denotes the off-diagonal (in S) spin-orbit coupling between the ^1P and ^3P states.

In the case (e) basis the electronic Hamiltonian is *not* diagonal. The matrix elements are [9]

$$\langle L'S'j'\ell'JM|W(R,r)|LSj\ell JM\rangle = \delta_{SS'}\cdot\delta_{\ell\ell'}\cdot\delta_{jj'}\cdot[^{2S+1}W_{\Pi}(R)+\delta_{S1}\Delta E_{13}] +$$

$$[\ell\ell'jj']^{1/2}$$

$$\times\,[^{2S+1}W_{\Sigma}(R)-{}^{2S+1}W_{\Pi}(R)]\,\sum_{\Omega}\begin{pmatrix}\ell & j & J\\0 & \Omega & -\Omega\end{pmatrix}\begin{pmatrix}\ell' & j' & J\\0 & \Omega & -\Omega\end{pmatrix}\begin{pmatrix}L & S & j\\0 & \Omega & -\Omega\end{pmatrix}\begin{pmatrix}L & S & j'\\0 & \Omega & -\Omega\end{pmatrix}, \quad (9)$$

where $(\vdots)$ is a 3j symbol [30] and $[x_1x_2...x_n] = (2x_1+1)(2x_2+1)...(2x_n+1)$ denotes a product of rotational degeneracy factors. Here $^{2S+1}W_{\Pi}(R)$ and $^{2S+1}W_{\Sigma}(R)$ designate the electronically adiabatic potential curves of Π and Σ symmetry which arise when, as discussed above, the spatial degeneracy of the unfilled p orbitals is lifted by the approach of the closed-shell partner. It is these curves which would result from an *ab initio* or pseudopotential calculation of the electronic potential curves of the diatomic system [34-36]. The quantity ΔE_{13} represents the splitting between the ^1P and $^3P_{j=0}$ levels of the isolated atom. For simplicity, in the discussion below, we shall suppress the electronic orbital angular momentum L in the designation of the case (e) wavefunctions, except where it is explicitly needed.

As discussed above, for interactions involving an atom in a ^{2S+1}P state with a closed shell atom we anticipate that the Σ curve will be more repulsive at short range, so that $^{2S+1}W_{\Sigma}(R)$ will be larger at short range than $^{2S+1}W_{\Pi}(R)$. The *qualitative* behaviour of these Σ and Π potential curves is illustrated in Fig. 1. Here we plot two sets of Σ and Π curves, originating from singlet and triplet asymptotes which arise from the same electron occupancy of the open-shell atom (as, for example, the 4s5p ^1P and 4s5p ^3P states of the Ca atom).

The case (e) description is most appropriate at long-range, since the electrostatic interaction between an excited atom and a neutral, closed-shell partner will go to zero as R^{-6}, which becomes negligibly small compared to either the centrifugal potential, which goes to zero only as R^{-2}, or the spin-orbit coupling, which becomes constant asymptotically. Thus, in the case (e) basis the Hamiltonian becomes asymptotically *diagonal* except for the residual mixing between the ^1P and $^3P_{j=1}$

270

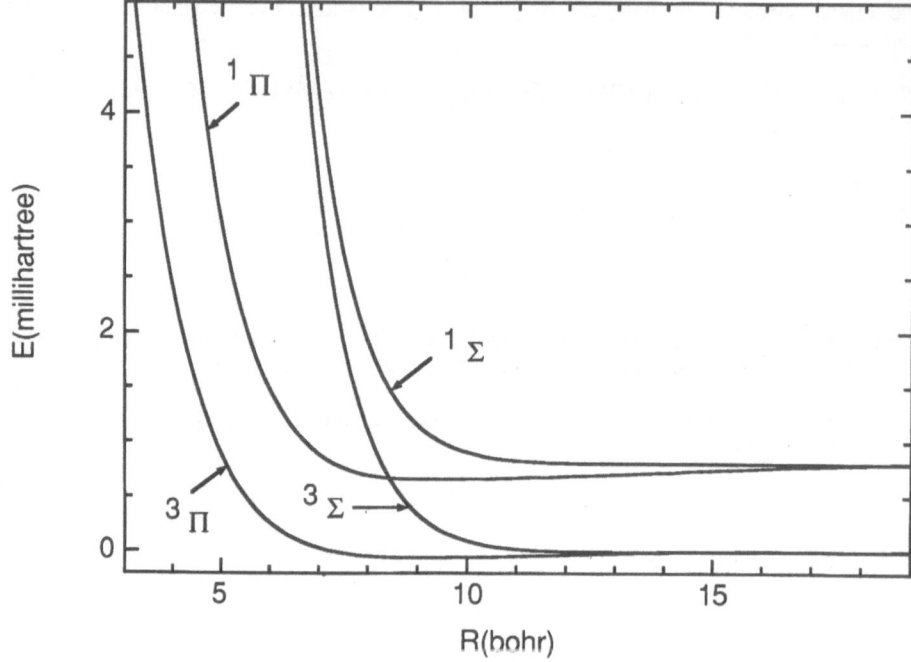

FIGURE 1. Typical potential energy curves of $^1\Sigma$, $^3\Sigma$, $^1\Pi$ and $^3\Pi$ symmetry which would arise from the interaction of a closed-shell atom with an atom in ($...nsn'p$ ^1P) and ($...nsn'p$ ^3P) electronic states. The actual potential energy curves shown are appropriate to the interaction of Ca(4s5p) with He and are identical to those defined as "set III" in Table 2 of Ref. 9.

states, due to the off-diagonal spin-orbit coupling $a_{13}(R=\infty)$. One can fully diagonalize the Hamiltonian by working with a linear combination of these two j=1 states, namely

$$| F_0 j \ell JM \rangle = \cos\theta \, | S=0, j \ell JM \rangle + \sin\theta \, | S=1, j \ell JM \rangle \qquad (10a)$$

and

$$| F_1 j \ell JM \rangle = -\sin\theta \, | S=0, j \ell JM \rangle + \cos\theta \, | S=1, j \ell JM \rangle. \qquad (10b)$$

Here we have introduced a rotation angle defined by

$$\theta = \frac{1}{2}\tan^{-1}\{2a_{13}(\infty)/[E(^1P_1) - E(^3P_1)]\}. \qquad (11)$$

In the limit of small mixing, θ is small so that the $|F_0\rangle$ state is nominally singlet and the $|F_1\rangle$ state nominally triplet. Since the spin-orbit mixing is independent of ℓ, so will be the rotation angle θ. In collisions involving a closed-shell partner, where the interaction involves no bond formation, it is reasonable to assume that the off-diagonal spin-orbit coupling in the open-shell atom $[a_{13}(R)]$ varies little from its asymptotic value, at least at moderate to large value of R. Thus, the rotation angle θ will be virtually independent of R. The F_S labelling can be extended to include the two triplet states which are not mixed with the $^1P_1$ state by defining, for $j \neq 1$

$$|F_1 j \neq 1, \ell JM\rangle = |S=1, j \neq 1, \ell JM\rangle. \tag{12}$$

For the interaction between a closed-shell atom and an atom in a ^{2S+1}P electronic state the parity of the case (e) wavefunctions (i.e. the symmetry with respect to reflection of the space and spin coordinates of all the electrons) is *independent* of j and given by

$$i \, |Sj\ell JM\rangle = (-1)^{n+\ell} |Sj\ell JM\rangle, \tag{13}$$

where n denotes the number of p electrons. This result follows from the fact that the parity of the electronic wavefunction of an atom is given by (-1) raised to the power $\sum \ell_i$, [31] where the sum runs over all the orbitals. This must be multiplied by the parity of the orbital wavefunction in Eqs. 4 and 6, namely $(-1)^\ell$ [37]. As is clear in Eq. 8, the off-diagonal spin-orbit coupling is diagonal in ℓ and so will not mix wavefunctions of different total parity.

In modern spectroscopic notation states of *odd* multiplicity with parity $(-1)^J$ are labelled e and states with parity $-(-1)^J$ are labelled f [38]. States of *even* multiplicity with parity $(-1)^{J-1/2}$ are labelled e and states with parity $-(-1)^{J-1/2}$ are labelled f. The correlation between the j, ℓ, and J quantum numbers and the e/f label is clarified by Table 2. For completeness, we have included the case (e) levels with S=1/2, appropriate to collisions involving atoms in ^2P electronic states [22], although these will not be the focus of our attention here. We observe that for systems with an *odd* number of p electrons [as, for example Ca(4snp ^3P)], the unique triplet state corresponding to j=0 is f labelled. This implies, in particular, that spin-conserving (e.g. intramultiplet) or spin-changing transitions into or out of the S=1, j=0 fine-structure level can occur only by coupling within the manifold of f levels. An entirely analogous table can be constructed for interactions involving a P-state atom with an *even* number of p-electrons.

At close interparticle separation, the splitting between the Σ and Π potential curves will, in general, increase exponentially, and become larger than the rotational or spin-orbit terms in the Hamiltonian. In this close-in, molecular region it is quite natural to use a basis in which the electrostatic Hamiltonian is diagonal. This is the Hund's case (a) basis, with wavefunctions designated $^{2S+1}\Lambda_\Omega$ and given by [9,15,39]

$$|\Lambda S\Sigma\Omega \in JM\rangle = 2^{-1/2}(|\Lambda S\Sigma\rangle|JM\Omega\rangle + \in |-\Lambda, S, -\Sigma\rangle|JM, -\Omega\rangle). \tag{14}$$

Here Λ and Σ denote the projections of L and S along the molecular axis (R), and $\Omega = \Lambda + \Sigma$. The symmetry index $\in$ can take on the values ± 1 except for the $^3\Sigma_0^+$ and $^1\Sigma^+$ states in which case $\Lambda = \Sigma = \Omega = 0$ and the wavefunctions are

$$|\Lambda=0, S, \Sigma=\Omega=0, JM\rangle = |\Lambda=0, S, \Sigma=0\rangle|JM0\rangle. \tag{15}$$

TABLE 2. Association of quantum numbers and e/f labels for states arising from the interaction of a closed-shell atom with an atom with an *odd* number of p-electrons in a ^{2S+1}P electronic state.

e levels	f levels

case (e) wavefunctions
(parity and e/f label independent of j)

	e levels	f levels
S=0	j=1, ℓ=J-1, J+1	j=1, ℓ=J
S=1/2	j=1/2, ℓ=J+1/2 j=3/2, ℓ=J-3/2, J+1/2	j=1/2, ℓ=J-1/2 j=3/2, ℓ=J-1/2, J+3/2
S=1	... j=1, ℓ=J-1, J+1 j=2, ℓ=J-1, J+1	j=0, ℓ=J j=1, ℓ=J j=2, ℓ=J-2, J, J+2

case (a) wavefunctions

	e levels	f levels
S=0	$^1\Sigma_0^+$, $^1\Pi_1$ (ϵ=+1)	$^1\Pi_1(\epsilon$=-1)
S=1/2	$^2\Sigma_{1/2}^+(\epsilon$=+1), $^2\Pi_{1/2,3/2}(\epsilon$=+1)	$^2\Sigma_{1/2}^+(\epsilon$=-1), $^2\Pi_{1/2,3/2}(\epsilon$=-1)
S=1	$^3\Sigma_1^+(\epsilon$=-1), $^3\Pi_{0,1,2}(\epsilon$=-1)	$^3\Sigma_0^+,^3\Sigma_1^+(\epsilon$=+1), $^3\Pi_{0,1,2}(\epsilon$=+1)

In a case (a) basis L, j, and ℓ are no longer good quantum numbers [14,15]; they are replaced by Λ, Σ, Ω, and ϵ. As anticipated, the electrostatic Hamiltonian is diagonal with matrix elements [9]

$$\langle\Lambda'S'\Sigma'\Omega'\epsilon'JM\,|\,W(R,r)\,|\,\Lambda S\Sigma\Omega\epsilon JM\rangle \;=\; \delta_{\Lambda\Lambda'}\delta_{SS'}\delta_{\Sigma\Sigma'}\delta_{\Omega\Omega'}\delta_{\epsilon\epsilon'}\; ^{2S+1}W_\Lambda(R),$$

$$(16)$$

where the $^{2S+1}W_\Lambda(R)$ functions represent the usual electrostatic potential energy curves for the two atoms system, depicted qualitatively in Fig. 1.

One can show that for states arising from the interaction between a closed-shell atom and an atom with an *odd* number of p electrons the parity of the case (a) states is ϵ $(-1)^{J-S}$, except for the $^{1,3}\Sigma_0^+$ states, whose parity is $(-1)^{J-S}$ [40]. Thus for S=0 and S=1/2 the ϵ=+1 $^{2S+1}\Lambda_\Omega$ states are labelled e and the ϵ=-1 $^{2S+1}\Lambda_\Omega$ states, f, *except for* the unique $^1\Sigma_0^+$ state [Eq. 15] which is labelled e. By contrast, for S=1 the ϵ=-1 $^{2S+1}\Lambda_\Omega$ states are labelled e and the ϵ=+1 $^{2S+1}\Lambda_\Omega$ states, f, *except for* the unique $^3\Sigma_0^+$ state [Eq. 15] which is labelled f.

The relation between the case (e) and case (a) wavefunctions is [20,21]

$$|LSj\ell JM> = (2[j][\ell])^{1/2}(-1)^{J+S-L} \; [1\pm\epsilon(-1)^{J+S+\ell+L}]$$

$$\times \sum_{\Lambda\Sigma\Omega} \begin{bmatrix} \ell & j & J \\ 0 & \Omega & -\Omega \end{bmatrix} \begin{bmatrix} L & S & j \\ \Lambda & \Sigma & -\Omega \end{bmatrix} \; |\Lambda S\Sigma\Omega\epsilon\, JM>, \tag{17}$$

where the sum is restricted to $\Omega \geqslant 0$. Here the plus sign in the term in square brackets refers to singlet states and the minus sign to triplets. In principle L no longer remains a good quantum number in the molecular region (small R). In other words, atomic p(L=1), d(L=2), f(L=3), etc. orbitals can all contribute to a molecular orbital of π (Λ=1) symmetry. Strictly speaking, then, the case (a) states can no longer be indexed in L [21,33]. One consequence is that correction terms must be added to the expressions for the matrix elements of the Hamiltonian in the case (e) basis [Eqs. 7-9], which were derived within a "pure precession" limit [14] in which L is assumed to remain equal to its asymptotic value [21,33]. In practice, all previous quantum treatments of collisions involving atoms in P electronic states have always assumed a pure precession limit [9,20,21,31,33]. Thus we assume that the index L which appears in the phase factors and in the second 3j symbol in Eq. 17 is equal to the asymptotic value of the orbital angular momentum for a P state atom (L=1).

3. QUANTUM FORMULATION OF THE COLLISION DYNAMICS: DIABATIC AND ADIABATIC BASES

To treat the collision dynamics we expand the total wavefunctions in terms of the case (e) wavefunctions, namely

$$Y(R,r) = \sum_{JMSj\ell} \frac{1}{R} C_{Sj\ell}^{JM}(R) \, |F_Sj\ell JM>. \tag{18}$$

Here we assume that proper linear combinations of the j=1, S=1 and j=1, S=0 states have been taken in order to diagonalize the off-diagonal spin-orbit coupling term $a_{13}(R=\infty)$ [see Eq. 10]. The expansion coefficients in Eq. 18 satisfy the usual close-coupled (CC) equations, namely [20,27,32-34,36,41-44]

$$\sum_{S'j'\ell'} \left[-\delta_{SS'}\delta_{jj'}\delta_{\ell\ell'} \left[\frac{\hbar^2}{2\mu R^2}\frac{d^2}{dR^2} + E \right] + U_{S'j'\ell',Sj\ell}^J(R) \right] C_{S'j'\ell'}^{JM}(R)=0, \tag{19}$$

with E denoting the total energy and

$$U_{S'j'\ell',Sj\ell}^J(R) = <F_{S'}j'\ell'JM|H_{rot}+V_{fs}+W(R,r)|F_Sj\ell JM>. \tag{20}$$

Here the matrix elements of H_{rot}, V_{fs}, and $W(R,r)$ in the F_S basis correspond to the matrix elements in Eqs. 7-9, with the exception that the matrix elements involving states with j=1 have been further transformed in accordance with Eq. 10.

Explicitly, for $W(R,r)$

$$\langle F_1 j=1,\ell' \mid W \mid F_0 j=1,\ell \rangle = \sin\theta\cos\theta(-\langle S=0,j=1,\ell' \mid W \mid S=0,j=1,\ell \rangle$$

$$+\langle S=1,j=1,\ell' \mid W \mid S=1,j=1,\ell \rangle), \tag{21a}$$

and

$$\langle F_1 j\neq1,\ell' \mid W \mid F_0 j=1,\ell \rangle = \sin\theta\langle S=1,j\neq1,\ell' \mid W \mid S=1,j=1,\ell \rangle, \tag{21b}$$

and similarly for the other $\langle F_S \mid W \mid F_S \rangle$ matrix elements with $j=1$ and/or $j'=1$. For simplicity we have suppressed the quantum numbers J and M in the bras and kets in these last two expressions as well as suppressing the dependence on R and r of the W operator.

Since H_{rot} is diagonal in the case (e) basis, and since the transformation to the F_S basis is diagonal in ℓ, this operator remains diagonal with matrix elements given by Eq. 7. Also, the transformation to the F_S basis is chosen specifically to diagonalize the spin-orbit Hamiltonian; the diagonal elements of V_{fs} are given by

$$\langle F_S \cdot j' \ell' \mid V_{fs} \mid F_S j \ell \rangle = \delta_{\ell\ell'} \Big[\delta_{j1}\delta_{j'1} \{ \delta_{S0}\delta_{S'0}$$

$$[\tfrac{1}{2}a_3(R)j(j+1)\sin^2\theta + 2a_{13}(R)\sin\theta\cos\theta]$$

$$+ \delta_{S1}\delta_{S'1}[\tfrac{1}{2}a_3(R)j(j+1)\cos^2\theta - 2a_{13}(R)\sin\theta\cos\theta]\}$$

$$+ \delta_{j2}\delta_{j'2}\delta_{S1}\delta_{S'1}\tfrac{1}{2}a_3(R)j(j+1) \Big]. \tag{22}$$

In the original case (e) basis the electrostatic Hamiltonian is diagonal in S. However, the transformation to the F_S basis will give rise to matrix elements of the electrostatic potential which couple states which correlate to the ^1P and ^3P asymptotes. It is these terms which give rise to the singlet $\rightarrow$ triplet transitions observed experimentally by Leone and co-workers [2-5] in collisions of Ca(4s5p ^{1,3}P).

As discussed in the preceding section, at moderate to large values of R the rotation angle can be assumed to be constant. Thus the coupling between ^1P-like and ^3P-like states will be a *constant* function of the $^{2S+1}W_\Pi(R)$ and $^{2S+1}W_\Sigma(R)$ potential curves. This is illustrated in Fig. 2, where we show the $\langle F_0 j=1,\ell=J-1 \mid H \mid F_1\ j=1,\ell=J+1 \rangle$ and $\langle F_0 j=1,\ell=J-1 \mid H \mid F_1 j=2,\ell=J+1 \rangle$ coupling matrix elements for a total angular momentum of $J=2$ determined using the potential curves shown in Fig. 1 and assuming a mixing angle of $\theta=18.4^o$ ($\sin^2\theta=0.1$). These two matrix elements are representative of those which couple case (e) states which are nominally singlet with those which are nominally triplet. On the same figure we have also plotted the difference between the $^3W_\Sigma(R)$ and $^3W_\Pi(R)$ potential curves. It is this difference which appears in the expression for the off-diagonal matrix elements in a case (e) basis [Eq. 9]. Both of the plotted matrix elements behave qualitatively like $W_\Sigma(R) - W_\Pi(R)$; in fact Eqs. 9 and 21 imply that $\langle F_0 j=1,\ell=J-1 \mid H \mid F_1 j=2,\ell=J+1 \rangle$ is directly proportional to $^3W_\Sigma(R)-^3W_\Pi(R)$.

A number of algorithms and computer codes have been developed for solution of the close-coupled equations [45-48]. For the problem under discussion here, which involves only 6 channels, the calculations can be done very quickly, on any

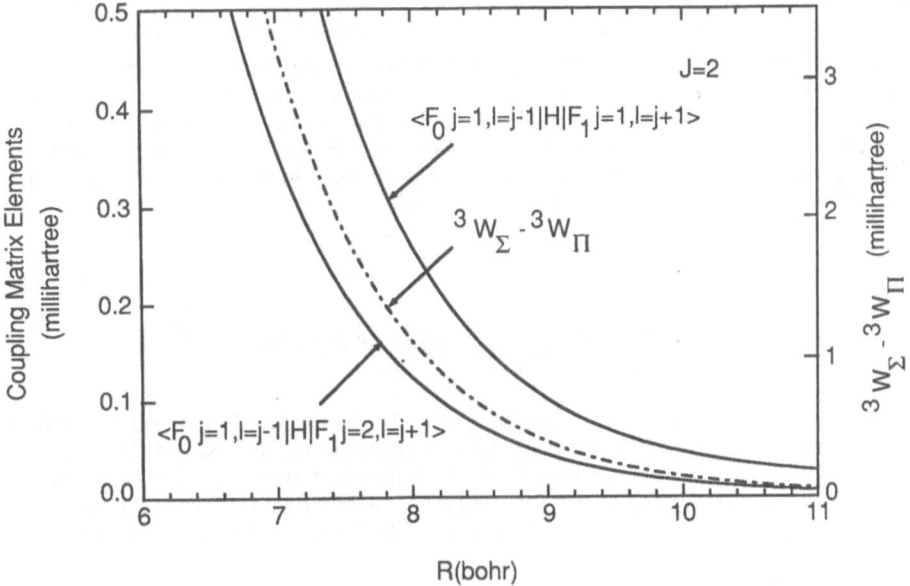

FIGURE 2. Coupling matrix elements in the case (e) (asymptotic) basis: $<F_0 j=1,\ell=J-1|H|F_1 j=1,\ell=J+1>$ and $<F_0 j=1,\ell=J-1|H|F_1 j=2,\ell=J+1>$. Curves calculated using the potential curves shown in Fig. 1 for a total angular momentum of J=2 and assuming a mixing angle of $\theta=18.4^0$ ($\sin^2\theta=0.1$). The dashed curve corresponds to the difference between the $^3W_\Sigma(R)$ and $^3W_\Pi(R)$ potential curves.

current minicomputer or mainframe, or, indeed, a microcomputer based on 32 bit architecture. Solution of the close-coupled equations results in a set of T-matrix elements for transitions between all pairs of $|F_S j \ell>$ states which are diagonal in total angular momentum J. Integral cross sections for a particular transition between an initial $(F_S j)$ and a final $(F_{S'} j')$ state, averaged over the m_j projection quantum number of the initial state and summed over the projection quantum number of the final state, can be written as [20,32,33,41]

$$\sigma_{F_S j \rightarrow F_{S'} \cdot j'} = \frac{\pi}{(2j+1)k_j^2} \sum_{J\ell\ell'} [J] \left| T^J_{F_S j \ell, F_{S'} \cdot j' \ell'} \right|^2 , \quad (23)$$

where k_j is the wavevector in the initial channel. The sum in Eq. 23 extends over all values of the total angular momentum for which the $T_{F_Sj\ell,F_{S'}j'\ell'}$ matrix element is non-zero. Once J increases to the point where the centrifugal barriers $[\ell(\ell+1)/2\mu R^2]$ in the initial and final channels are so large that the matrix elements of the coupling potential are negligible at the classical turning point, then the T-matrix elements vanish. The weighting by [J]=(2J+1) in Eq. 23 implies that large values of the total angular momentum (glancing collisions) can make a dominant contribution to the total cross section. Eq. 23 is equivalent to the semi-classical expression

$$\sigma_{F_Sj \to F_{S'}j'} = \frac{1}{2j+1} \int_0^\infty 2\pi b \left[\sum_{m_j, m_{j'}} P_{F_Sjm_j \to F_{S'}j'm_{j'}}(b) \right] db, \quad (24)$$

where $P_{i \to f}(b)$ is the inelastic transition probability as a function of impact parameter.

In beam experiments involving laser preparation of the excited [1]P state the relevant cross sections are not equal to the degeneracy-averaged cross sections defined by Eq. 23. Imagine an experiment involving two perpendicular atomic beams and a pump laser at right angles to both beams. This is illustrated

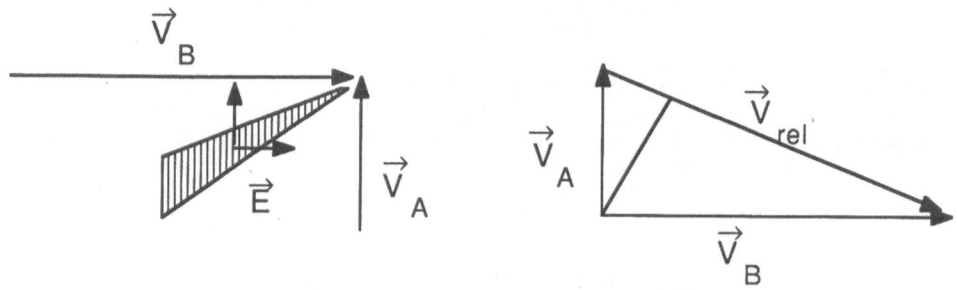

FIGURE 3. (Left panel) Schematic representation of two atomic beams crossing at right angles intersected at right angles by a linearly polarized laser beam. The electric field vector of the laser can be rotated in the plane defined by the two atomic beams. (Right panel) Newton velocity diagram corresponding to the left panel. Note that the initial relative velocity vector of the two beams is perpendicular to the direction of propagation of the laser.

schematically in the left panel of Fig. 3. The relevant Newton diagram [49] is shown in the right panel. We assume that the excited ^1P state is prepared by excitation out of a ^1S state. If the laser is *linearly* polarized, we can take the electric field vector of the laser to define the laboratory-fixed Z-axis. The usual optical selection rules [13-15,31] then imply that only the $m_j=0$ projection level will be prepared. [Here we consider only a $^1P \leftarrow {}^1S$ transition; for a $^{2S+1}P \leftarrow {}^{2S+1}S$ transition, where the total angular momentum in the ground state can range from $-(2S+1)$ to $+(2S+1)$ all excited states with $|m_j| < (2S+1)$ will be prepared in the initial excitation process]. The general quantum analysis of the collision refers to a coordinate system with z-axis defined by the initial relative velocity vector - the so-called collision frame [41]. It is important to distinguish the collision frame, which is *fixed* in space, from the molecular frame, whose origin is fixed at the center of mass of the molecule and whose z-axis is defined by $\mathbf{R}$, the relative separation vector. (In fact, the molecular frame is identical to the so-called R-helicity frame [50]). In the case of a collision between two atoms, the collision frame and the molecular frame coincide, in general, only prior to the collision.

Thus in the quantum analysis of a beam experiment it is necessary to obtain an expression for the scattering amplitude out of the $^1P_{m_j=0}$ state, under the constraint that the axis to which the projection quantum number refers may differ from the collision frame z-axis. The appropriate expression is [51-53]

$$f_{F_S j m_j \rightarrow F_{S'} j' m_{j'}}(\hat{k},\hat{R}) = 2\pi \sum_{J,M,\ell,m_\ell,\ell',m_\ell'} i^{\ell-\ell'} (-1)^{\ell+\ell'+j+j'} [J]$$

$$\times \begin{pmatrix} j & \ell & J \\ m_j & m_\ell & -M \end{pmatrix} \begin{pmatrix} j' & \ell' & J \\ m_{j'} & m_{\ell'} & -M \end{pmatrix} Y^*_{\ell m_\ell}(\hat{k}) Y_{\ell' , m_{\ell'}}(\hat{R}) T^J_{F_S j \ell, F_{S'} j' \ell'} \quad (25)$$

where $m_j=0$. Here $\hat{R}=(\theta,\phi)$ designates the scattering angle in the collision frame. The orientation of the collision frame with respect to the laboratory-fixed Z-axis, is designated $\hat{k}=(\theta_k,\phi_k)$. Note that the projection quantum numbers in Eq. 25 all refer to the same laboratory system. In the case of excitation by *circularly* polarized light, only the $m_j=+1$ or $m_j=-1$ projection states are prepared, where the Z-axis refers now to the direction of propagation of the laser. Again, the appropriate expression for the scattering amplitude is given by Eq. 25 with now $m_j=+1$ or $m_j=-1$.

In many cases, the final state is detected by fluorescence emission. If the region subtended by the detector is much larger than the volume defined by the intersection of the two beams, then an integral, rather than differential, cross section is measured. The expression for the cross section is obtained by taking the absolute value squared of the scattering amplitude [Eq. 25] and integrating over all values of the scattering angle in the collision frame $\hat{R}$. One obtains, after some algebra [9]

$$\sigma F_{Sjm_j \to F_{S'}j'm_{j'}}(\hat{k}) = (4\pi^2/k^2) \sum_{J,J',M,\ell,\ell_1,\ell',m_\ell} i^{\ell-\ell_1}(-1)^{\ell+\ell_1}[JJ']$$

$$\times \begin{bmatrix} j & \ell & J \\ m_j & m_\ell & -M \end{bmatrix} \begin{bmatrix} j & \ell_1 & J' \\ m_j & m_\ell & -M \end{bmatrix} \begin{bmatrix} j' & \ell' & J \\ m_{j'} & m_{\ell'} & -M \end{bmatrix} \begin{bmatrix} j' & \ell' & J' \\ m_{j'} & m_{\ell'} & -M \end{bmatrix}$$

$$\times Y^*_{\ell m_\ell}(\hat{k}) Y_{\ell_1,m_\ell}(\hat{k}) T^J_{F_{S}j\ell,F_{S'}j'\ell'} T^{J'\,*}_{F_{S}j\ell_1,F_{S'}j'\ell''} \qquad (26)$$

The dependence on the azimuthal angle φ in the spherical harmonic $Y_{\ell m_\ell}$ is $\exp(im_\ell\varphi)$ [37]. Consequently, the presence of the two spherical harmonics in Eq. 26, one conjugated, but both of the same order m_ℓ, implies that the integral cross sections will be independent of the azimuthal orientation of the molecular axis R with respect to the space- or laboratory-fixed Z-axes.

If, furthermore, the final projection quantum number of the atom is *not* resolved, then the effective cross section is given by

$$\sigma_{Sj,m_j \to S'j'}(\hat{k}) = \sum_{m_{j'}} \int d\hat{R} \left| f_{F_{S}jm_j \to F_{S'}j'm_{j'}}(\hat{k},\hat{R}) \right|^2. \qquad (27)$$

A dependence on $(\hat{k})$ remains. This is, of course, responsible for the dependence on direction of laser polarization of the $Ca(4s5p\ ^1P) \to Ca(4s5p\ ^3P)$ cross sections observed by Leone and co-workers [2-5].

Within the close-coupled formulation given above, the Schrödinger equation is solved by expansion of the wavefunction in the asymptotic [case (e)] basis [9,20,33]. This basis can be designated a *diabatic* basis, since it is *independent* of the interparticle separation. Other diabatic bases are possible, for example a case (a) basis. The advantage of the case (e) basis is that the Hamiltonian is diagonal asymptotically, so that the close-coupled equations become uncoupled as $R \to \infty$.

Alternatively, one can expand the total wavefunctions in an *adiabatic* basis [16,20,21,23-26]. This basis is defined by an orthogonal transformation of the diabatic basis, namely

$$|nJM\rangle = \sum_{Sj\ell} A^{JM}_{n,Sj\ell}(R) |F_{S}j\ell JM\rangle, \qquad (28)$$

where the index n ranges over all the states in the coupled ($|F_{S}j\ell JM\rangle$) case (e) basis. The transformation matrix is chosen to diagonalize the matrix of the total Hamiltonian

$$A(R)U^J(R)A^T(R) = \lambda^J(R), \qquad (29)$$

where $U^J(R) \equiv U^J_{F_{S'}j'\ell',F_{S}j\ell}(R)$ is defined in Eq. 20. Since the matrix of the Hamiltonian in the case(e) basis is real and symmetric, the transformation defined by Eq. 28 is an orthogonal transformation and the eigenvalues $\lambda^J(R)$ are real. The adiabatic basis is *unique*; each choice of diabatic basis will result in a different transformation matrix **A**, but the resulting adiabatic states $|nJM\rangle$ will be the same. Note also that the transformation is *independent* of the total energy.

In the adiabatic basis the total system wavefunction can be expanded similarly to Eq. 18, namely [9,20,21]

$$Z(R,r) = \sum_{nJM} \frac{1}{R} B_n^{JM}(R) |nJM\rangle \qquad (30)$$

The structure of the close-coupled equations is now

$$\sum_n \left[\delta_{nn'} \left| - \frac{\hbar^2}{2\mu} \frac{d^2}{dR^2} - E + \lambda_{n'}^J \right| B_{n'}^{JM}(R) - \frac{\hbar^2}{2\mu} \right.$$

$$\left. \left[2G_{n'n}^J(R) \frac{d}{dR} B_n^{JM}(R) + F_{n'n}^J(R) B_n^{JM}(R) \right] \right] = 0, \qquad (31)$$

where the $G^J_{n'n}$ and $F^J_{n'n}$ coupling matrices are defined by

$$G^J(R) = A^T(R) \frac{d}{dR} A(R) \qquad (32)$$

and

$$F^J(R) = A^T(R) \frac{d^2}{dR^2} A(R). \qquad (33)$$

In practice these matrices can be obtained by numerical differentiation of the matrix of eigenvectors, as follows:

$$A^T(R)\frac{d}{dR}A(R) = \lim_{\Delta R \to 0} \frac{1}{\Delta R}\{A^T(R)[A(R+\Delta R)-A(R)]\} \approx \frac{1}{\Delta R}[A^T(R)A(R+\Delta R)-1] \qquad (34)$$

and similarly for the second derivative.

The obvious difference between the close-coupled equations in an adiabatic, as compared to diabatic, basis, is that the coupling occurs through the derivative terms rather than through the potential. Qualitatively, however, the derivative terms will be large only when a significant change is taking place in the relative contribution of the individual case (e) diabatic states to some (or all) of the adiabatic states, in other words when the eigenvectors in the diabatic basis [the columns of the $A(R)$ matrix] are changing rapidly. This can be illustrated with the set of Ca(4s5p)+He potentials displayed in Fig. 1. For a value of the total angular momentum J=2, the adiabatic energies - the eigenvalues $\lambda(R)$ in Eq. 29 - of the 6 e labelled states

TABLE 3. Correlation of adiabatic and diabatic [case (e) or case (a)] states for Figs. 2-7.

n	case(e)			case(a)	n	case(e)			case(a)
	S	j	ℓ			S	j	ℓ	
1	1	1	J-1	$^3\Pi_0$	5	0	1	J-1	$^3\Sigma_1{}^+$
2	1	1	J+1	$^3\Pi_1$	6	0	1	J+1	$^1\Sigma_0{}^+$
3	1	2	J-1	$^3\Pi_2$					
4	1	2	J+1	$^1\Pi_1$					

[parity $(-1)^J$], are plotted in Fig. 4. The asymptotic correlation between these adiabatic states and the appropriate case (e) and case (a) states is given in Table 3. It is worthwhile observing that the $^3\Sigma_1{}^+$ case (a) state with $\Omega=1$, correlates asymptotically with the *singlet* $^1P_1$ case (e) state with $\ell=$J-1 and, conversely the $^1\Pi_1$ case(a) state correlates with the *triplet* $^3P_2$ case (e) with $\ell=$J+1. The strong avoided crossing between the $^3\Sigma$ and the $^1\Pi$ electrostatic potential curves (Fig. 1) is clearly apparent; remember that the $^1P_1$ and $^3P_1$ atomic states are mixed slightly by the spin-orbit Hamiltonian (see section 2).

In order to focus clearly on the Ca(4s5p ^{1}P) → Ca(4s5p ^{3}P) transition studied by Leone and co-workers [2-5], it is helpful to examine the coupling between the two adiabatic states which undergo the avoided crossing, namely $|n=4\rangle$, nominally $^3P_2$, $\ell=$J+1 [case (e)] or $^1\Pi_1$ [case (a)], and $|n=5\rangle$, nominally $^1P_1$, $\ell=$J-1 [case(e)] or $^3\Sigma_1{}^+$ [case (a)]. Also, we should look at the coupling between the two closest adiabatic states which correlate with case (e) states which are directly mixed by the off-diagonal (in S) spin-orbit Hamiltonian. This corresponds to the mixing between the $|n=5\rangle$ state, mentioned just above, and the $|n=2\rangle$ state, nominally $^3P_1$, $\ell=$J+1 [case (e)] or $^3\Pi_1$ [case (a)]. The $\langle 4|H|5\rangle$ ($\langle S=1,j=2,\ell=$J+1,$J|H|S=0,j=1,\ell=$J-1,$J\rangle$) and $\langle 2|H|5\rangle$ ($\langle S=1,j=1,\ell=$J+1,$J|H|S=0,j=1,\ell=$J-1,$J\rangle$) matrix elements in the asymptotic [case (e)] basis are displayed in Fig. 2, and the *same* matrix elements, but in the *adiabatic* basis are displayed in the upper panel of Fig. 5.

We observe that in the asymptotic basis, the off-diagonal coupling matrix elements extend smoothly over all the range of values of R sampled by the collision and, furthermore, as discussed previously, have the same qualitative form as the difference between the Σ and Π electrostatic potentials. By contrast, in the adiabatic basis, the coupling matrix elements are localized in space, here reaching their maxima in the region of avoided crossing between the $^3\Sigma$ and the $^1\Pi$ potential curves. We see, also, that the strongest off-diagonal coupling is now between the n=4 and n=5 adiabatic states, whereas in the case (e) basis the strongest off-diagonal coupling, at least between initial and final states which are relevant to the ^{1}P→^{3}P transition, is the coupling between the S=1,j=1,$\ell=$J+1 and S=0,j=1,$\ell=$J-1 states.

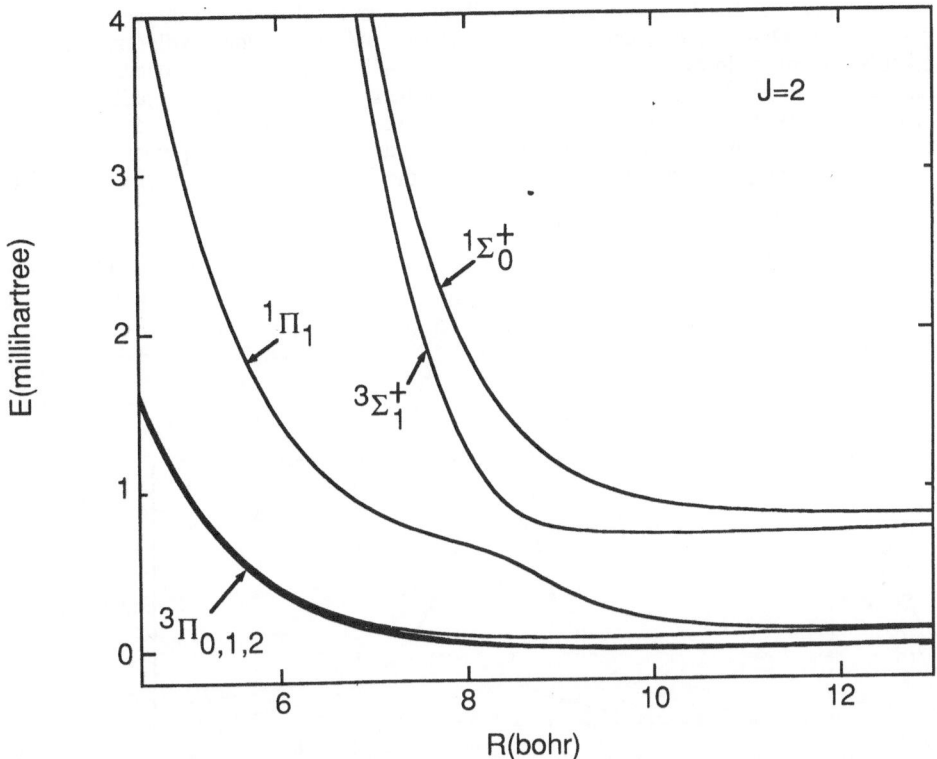

FIGURE 4. Adiabatic energies [λ^J, Eq. 29] of the 6 e-labelled states which correlate asymptotically to the diabatic [case (e)] states corresponding to the interaction of an atom in ^1P and ^3P electronic states with a closed-shell atom. The curves were calculated using the $^1\Sigma$, $^3\Sigma$, $^1\Pi$ and $^3\Pi$ potential curves shown in Fig. 1 for a total angular momentum of J=2 and assuming a mixing angle of $\theta=18.4^o$ ($\sin^2\theta=0.1$). The correlation between the 6 adiabatic states and the case(e) and case(a) states is given in Table 3. The curves for the lowest two adiabatic states are virtually indistinguishable.

This discrepancy can be understood as follows: In a case (e) basis, the strongest off-diagonal coupling is that between states which are directly connected by the spin-orbit Hamiltonian, namely the $^3P_1$(n=2) and the $^1P_1$(n=5) states. By contrast, in the case (a) basis both the $^1\Pi_1$ (n=4) state and the $^3\Pi_1$ (n=2) state are coupled to the $^3\Sigma_1^+$ (n=5) state by the spin-orbit Hamiltonian. However, because of the strong mixing between the $^1\Pi_1$ and $^3\Sigma_1^+$ states in the vicinity of the avoided crossing, the strongest off-diagonal coupling in the *adiabatic* basis is that between the n=4 and n=5 states.

The above discussion was illustrated by Figs. 2 and 5, in which the displayed curves were obtained using a value of the total angular momentum of J=2. However, the major contribution to both the degeneracy-averaged and m-state resolved integral $^1P \to {}^3P$ cross sections will arise from collisions at larger impact parameters. One can estimate the largest impact parameter which will make a contribution as follows: We first assume that $^1P \to {}^3P$ transitions will occur only if the classical turning point of the two collision partners lies inside (at a smaller value of R than) the point of avoided crossing between the $^3\Sigma$ and the $^1\Pi$ potential curves. For potentials with weak long-range tails the distance of closest approach is just the classical impact parameter b. Semiclassically, the relation between ℓ and b is given by

$$\ell = kb = \mu vb = (2\mu E)^{1/2}b/\hbar. \tag{35}$$

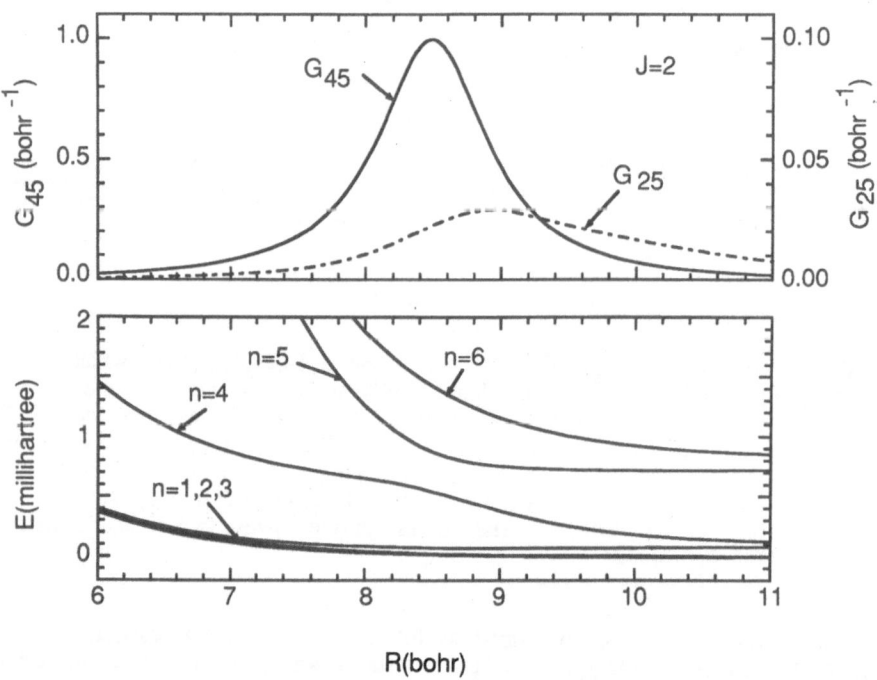

FIGURE 5. (Lower panel) Enlargement of the adiabatic energies shown in Fig. 4. The curves for the lowest two adiabatic states are virtually indistinguishable. (Upper panel) Absolute values of the first derivative coupling terms [Eq. 32] G_{45} (solid curve) and G_{25} (dashed curve). The G_{45} matrix element corresponds to the coupling between the adiabatic states n=4 {nominally $^3P_2, \ell=J+1$ [case(e)] or $^1\Pi_1$ [case(a)]} and n=5 {nominally $^1P_1, \ell=J-1$ [case(e)] or $^3\Sigma_1{}^+$ [case (a)]}. The G_{25} matrix element corresponds to the coupling between the adiabatic states n=5, described above and n=2 {nominally $^3P_1, \ell=J+1$ [case(e)] or $^3\Pi_1$ [case(a)]}.

Thus if R_C designates the point of avoided crossing, then the maximum orbital angular momentum, which for large ℓ is nearly equal to the total angular momentum J, is just $(2\mu E)^{1/2}R_C$. Consider collisions of He with Ca, in which case μ=3.64 amu. Introduction of the correct conversion factors gives

$$J_{max}\approx41.7\ R_C(\text{Å})\sqrt{E(\text{eV})}=22.1\ R_C(\text{bohr})\sqrt{E(\text{eV})}. \tag{36}$$

Thus, as an example, for E=0.1 eV and R_C=8.5 bohr (see Fig. 1), we find J_{max}=56.
We have determined adiabatic potentials and coupling elements analogous to those shown in Figs. 2 and 5 but with J=55. These are shown in Figs. 6 and 7. The presence of the large centrifugal barrier at J=55 obscures both the overall difference between the nominally Σ and the nominally Π adiabatic potential curves at moderate to large values of R as well as the avoided crossing between the nominally $^3\Sigma$ and the nominally $^1\Pi$ curves. Nevertheless, the coupling matrix elements in both the asymptotic [case (e)] and adiabatic bases are qualitatively similar to those shown in Figs. 2 and 5, with one exception: The n=5 $\leftrightarrow$ n=2 matrix

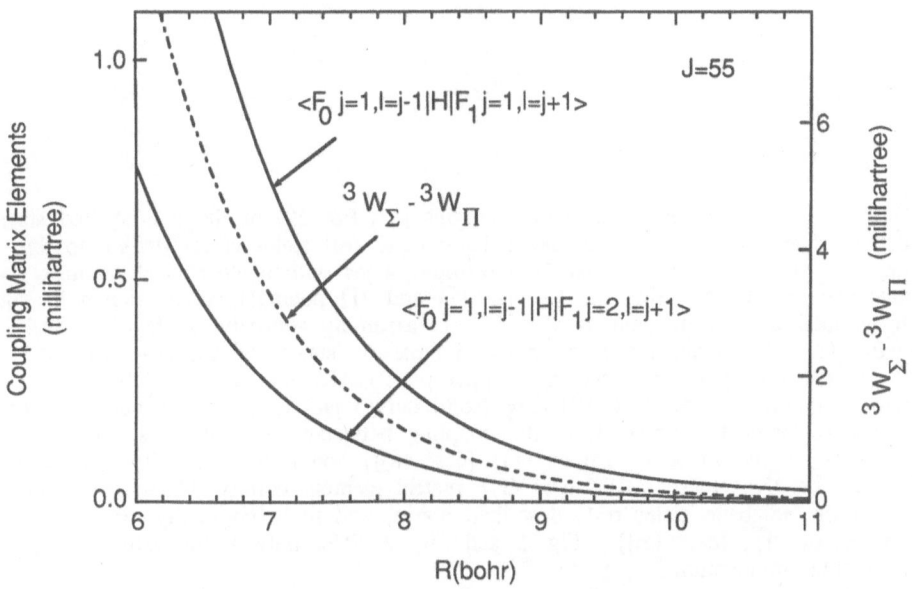

FIGURE 6. Coupling matrix elements in the case (e) (asymptotic) basis: $\langle F_0j=1,\ell=J-1\,|\,H\,|\,F_1j=1,\ell=J+1\rangle$ and $\langle F_0j=1,\ell=J-1\,|\,H\,|\,F_1j=2,\ \ell=J+1\rangle$. Curves calculated using the potential curves shown in Fig. 1 for a total angular momentum of J=55 and assuming a mixing angle of θ=18.4^0 (sin$^2\theta$=0.1). The dashed curve corresponds to the difference between the $^3W_\Sigma(R)$ and $^3W_\Pi(R)$ potential curves. Fig. 2 and Fig. 6 differ only in the value of the total angular momentum.

284

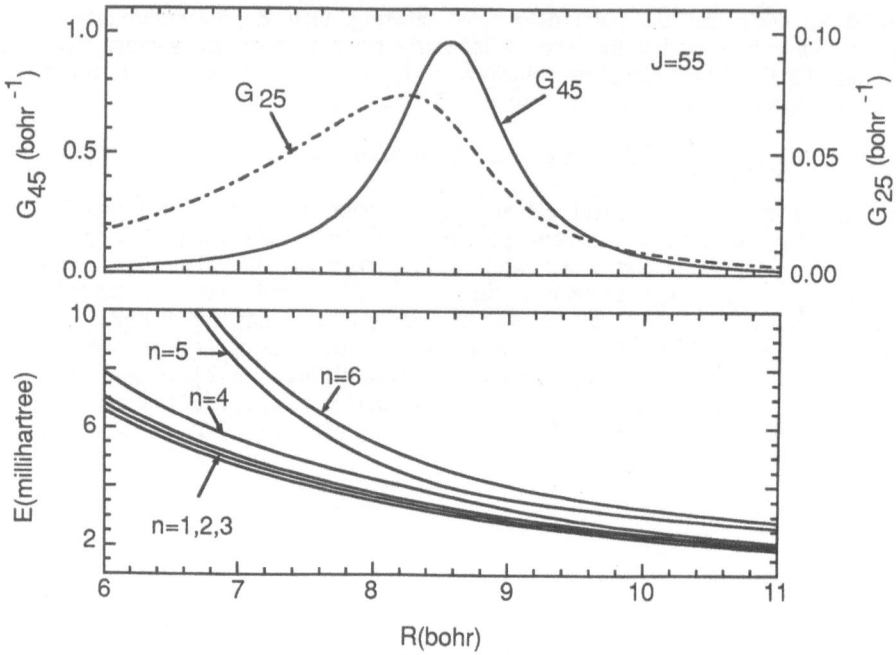

FIGURE 7. (Lower panel) Adiabatic energies [λ^J, Eq. 29] of the 6 e-labelled states which correlate asymptotically to the diabatic [case (e)] states corresponding to the interaction of an atom in ^1P and ^3P electronic states with a closed-shell atom. The curves were calculated using the $^1\Sigma$, $^3\Sigma$, $^1\Pi$ and $^3\Pi$ potential curves shown in Fig. 1 for a total angular momentum of J=55 and assuming a mixing angle of θ=18.4° ($\sin^2\theta$=0.1). The correlation between the 6 adiabatic states and the case (e) and case (a) states is given in Table 3. (Upper panel) Absolute values of the first derivative coupling terms [Eq. 32] G_{45} (solid curve) and G_{25} (dashed curve). The G_{45} matrix element corresponds to the coupling between the adiabatic states n=4 {nominally $^3P_2,\ell$=J+1 [case (e)] or $^1\Pi_1$ [case (a)]} and n=5 {nominally $^1P_1,\ell$=J-1 [case (e)] or $^3\Sigma_1^+$ [case (a)]}. The G_{25} matrix element corresponds to the coupling between the adiabatic states n=5, described above, and n=2 {nominally $^3P_1,\ell$=J+1 [case (e)] or $^3\Pi_1$ [case (a)]}. Fig. 5 and Fig. 7 differ only in the value of the total angular momentum.

element in the adiabatic basis is considerably larger at small R, relative to the n=5 ↔ n=4 matrix element, than was seen for J=2. This can be understood within the context of the case (a) matrix elements. At small values of R, $^3\Pi_1$ and $^3\Sigma_1$ case (a) states will be mixed not only by the spin-orbit Hamiltonian but also by the J.L term in the expansion of H_{rot} [Eq. 2]. The strength of this mixing is roughly proportional to J/R^2, so that it becomes increasingly large at small R and large J.

The advantage of formulating the scattering problem in an adiabatic basis lies chiefly in the insight one can gain from inspection of the coupling matrix elements in this basis, which are well localized in space. This localization permits one to identify the range (or ranges) of interparticle distances over which collision-induced transitions can be expected to occur. Consequently, it is the form and strength of the underlying couplings (electrostatic, spin-orbit) in these regions, to which the kinetic processes in question will be most sensitive. Formulation of the scattering in a diabatic basis, in which the coupling matrix elements are by nature *delocalized*, does not allow this type of insight. By contrast, solution of the close-coupled equations is simpler and faster in a diabatic basis. This is because (a) a time-consuming diagonalization of the coupling Hamiltonian [Eq. 31] at each value of R is not necessary and (b) algorithms for solution of the CC equations in a diabatic basis are faster and more widely available [45-48,54].

4. INITIAL STATE PREPARATION

As discussed in section 3 the analysis of beam experiments involving laser preparation of an atom in an excited ^1P electronic state is considerably more complex than that of an experiment in which one measures only degeneracy averaged cross sections. This is particularly true if one is interested not only in the calculation of a series of cross sections, but also in a mechanistic interpretation of the dependence of the cross section on various parameters (directions of laser polarization, form of potential curves, singlet-triplet mixing, etc.). As we have seen in the preceding section, analysis within an adiabatic basis provides the most direct way to obtain this kind of mechanistic interpretation, at least within a time-independent formulation of the dynamics. In this section we will discuss how one carries out an adiabatic analysis of a beam experiment involving laser excitation of one of the atoms.

The primary difficulty arises from the fact that although it is most convenient to carry out the scattering calculation in a *coupled* basis [section 2] in which the total angular momentum is a good quantum number, the wavefunction of two atoms at infinite separation can best be expressed in an *uncoupled* basis. To illustrate this point, consider the initial state of the diatomic (atom+laser-excited atom) prior to collision. Prior to the collision the relative orbital angular momentum ℓ is always oriented perpendicular to the collision plane, in other words ℓ is always perpendicular to the collision-frame z-axis, which, as discussed in section 3, is coincident with v_{rel}, the initial relative velocity vector. If the electric field vector of the pump laser, which defines the laboratory-fixed Z axis, is chosen to lie parallel to v_{rel}(Fig. 3), and if we consider a $^1P \leftarrow ^1S$ excitation process, then, as discussed in section 3, only the $^1P_{m=0}(|j=1,m_j=0\rangle)$ atomic state is prepared [13-15,31]. Since $\hat{Z}$ and v_{rel} are coincident prior to the collision, the collision-frame and laboratory-frame z-axes are *identical*. This we shall refer to as *parallel* excitation.

Thus, for parallel excitation, prior to the collision only the $|j=1,m_j=0>|\ell m_\varrho>$ *uncoupled* state is prepared. Remember that the collision-frame and laboratory-frame z-axes coincide. Since the prepared p orbital has $m_L=0$, it lies along v_{rel} and thus is always perpendicular to ℓ. Yet, although $\hat{Z}$ and v_{rel} are coincident prior to the collision, the *azimuthal* orientation of ℓ is not restricted - a unique collision plane is not selected. This is shown clearly in Fig. 8. At first thought, and after a cursory inspection of Fig. 3, it might seem surprising that the orientation of the collision plane is not unique. In fact if we assign a definite value to the orbital angular momentum ℓ and its projection m_ϱ, then the conjugate angle - the azimuthal orientation of the collision plane - must be totally undetermined.

This unique uncoupled state, which we shall designate $|1_{||}>$, can be expressed as a linear combination of *two coupled* [case (e)] wavefunctions corresponding to *two* values of the total angular momentum J. This can be done by reversing the angular

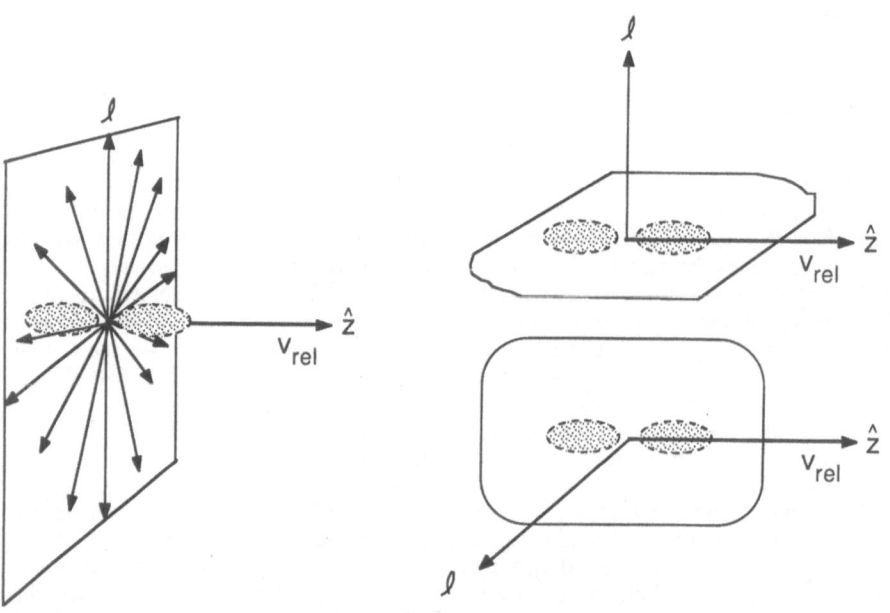

FIGURE 8. Schematic representation of the initial relative velocity vector v_{rel}, the initial relative orbital angular momentum ℓ, and the initially excited p orbital in the case of parallel excitation. Since the prepared p orbital has $m_L=0$, it lies along v_{rel} and thus is always perpendicular to ℓ. Yet, although $\hat{Z}$ and v_{rel} are coincident prior to the collision, the azimuthal orientation of ℓ is not restricted - a unique collision plane is not selected.

momentum coupling [30,37] contained in Eq. 6. In general we have

$$|LSjm_j\rangle\,|\ell m_\varrho\rangle = \sum_{JM} (jm_j\ell m_\varrho\,|\,JM)\,|LSj\ell JM\rangle, \qquad (37)$$

or, in the specific case of L=1, m_L=0, S=0, m_ϱ=0,

$$|1_{||}\rangle = |S=0,j=1,m_j=0\rangle\,|\ell m_\varrho\rangle$$

$$= (1\ell 0 0\,|\,\ell+1,0)\,|S=0,j=1,\ell,J=\ell+1,M=0\rangle$$

$$+ (1\ell 0 0\,|\,\ell-1,0)\,|S=0,j=1,\ell,J=\ell-1,M=0\rangle. \qquad (38a)$$

We observe (see Table 2) that the two prepared states are both e levels. An alternative analysis can be made in terms of case (a) wavefunctions. Since the prepared orbital has m_L=0 (with respect to $\hat{Z}$) and since $\hat{Z}$ and v_{rel} coincide prior to the collision, only a $^1\Sigma$ molecular state is prepared, which is e labelled.

The correct mechanistic interpretation of the collision dynamics in this case of parallel excitation will then involve the study of the evolution of this initially prepared state as the collision partners approach. Since the total angular momentum J and its projection M are conserved during the collision, and since the Hamiltonian will not mix e and f levels, for a given value of ℓ only e levels with J=$\ell\pm$1 will mix with the $|1_{||}\rangle$ state. There are only *three* other *singlet* case (e) states which meet these criteria, namely

$$|2\rangle = -(1\ell 0 0\,|\,\ell+1,0)\,|S=0,j=1,\ell,J=\ell+1,M=0\rangle$$
$$+(1\ell 0 0\,|\,\ell-1,0)\,|S=0,j=1,\ell,J=\ell-1,M=0\rangle \qquad (38b)$$

$$|3\rangle = |S=0,j=1,\ell+2,J=\ell+1,M=0\rangle \qquad (38c)$$

$$|4\rangle = |S=0,j=1,\ell-2,J=\ell-1,M=0\rangle. \qquad (38d)$$

as well as *eight triplet* case(e) states, namely

$$|5\rangle = |S=1,j=1,\ell,J=\ell+1,M=0\rangle \qquad (39a)$$

$$|6\rangle = |S=1,j=1,\ell+2,J=\ell+1,M=0\rangle \qquad (39b)$$

$$|7\rangle = |S=1,j=1,\ell,J=\ell-1,M=0\rangle \qquad (39c)$$

$$|8\rangle = |S=1,j=1,\ell-2,J=\ell-1,M=0\rangle \qquad (39d)$$

$$|9\rangle = |S=1,j=2,\ell,J=\ell+1,M=0\rangle \qquad (39e)$$

$$|10\rangle = |S=1,j=2,\ell+2,J=\ell+1,M=0\rangle \qquad (39f)$$

$$|11\rangle = |S=1,j=2,\ell,J=\ell-1,M=0\rangle \qquad (39g)$$

$$|12\rangle = |S=1,j=2,\ell-2,J=\ell-1,M=0\rangle. \qquad (39h)$$

For the system under study here it will be instructive to examine the R dependence of the adiabatic state which is described at large distances by the $|p_z> |\ell 0>[|1_{||}>]$ state defined in Eq. 38a, as well as the R dependence of the first derivative coupling terms between this and the other three *singlet* wavefunctions and the other eight *triplet* wavefunctions. At large R, the primary coupling will involve the four singlet functions, since the triplet states will be well separated asymptotically. It is important to notice that states $|1_{||}>$ and $|2>$ have the *same* value of ℓ and thus will become degenerate asymptotically, when the difference between $^1W_\Sigma(R)$ and $^1W_\Pi(R)$ becomes negligibly small compared to the centrifugal barrier. In fact, the wavefunction for state $|2>$, expressed in an *uncoupled* basis, is

$$|2> = 2^{-1/2}[|j,m_j=+1>|\ell,m_\varrho=-1>+|j,m_j=-1>|\ell,m_\varrho=+1>]. \tag{40}$$

This corresponds to a state in which the p orbital is no longer aligned parallel to v_{rel}, in other words the $^1\Pi_{1e}(\in=+1)$ molecular state.

Since states $|1_{||}>,|2>,|3>$, and $|4>$ are defined by Eq. 38 in terms of the case (e) basis, the matrix elements of the Hamiltonian connecting these states can be obtained from Eq. 9. The off-diagonal coupling will arise only from the electrostatic Hamiltonian. The matrix of the electrostatic Hamiltonian in the basis defined by the states $|1_{||}>,|2>,|3>$, and $|4>$ is given in Table 4. Here the quantity Δ denotes $^1W_\Sigma(R) - ^1W_\Pi(R)$. We observe that the off-diagonal matrix elements are *all* proportional to $^1W_\Sigma(R) - ^1W_\Pi(R)$. In addition, with the exception of the

TABLE 4. Matrix elements of the electrostatic Hamiltonian in the singlet basis defined by Eq. 38.

$|1_{||}>$	$|2>$	$|3>$	$|4>$
$W_\Pi + \Delta \dfrac{2\ell^2+2\ell+1}{(2\ell-1)(2\ell+3)}$	$-\Delta\dfrac{\sqrt{\ell(\ell+1)}}{(2\ell-1)(2\ell+3)}$	$-\Delta\dfrac{(\ell+1)\sqrt{\ell+2}}{(2\ell+3)\sqrt{(2\ell+1)}}$	$\Delta\dfrac{\ell\sqrt{\ell-1}}{(2\ell-1)\sqrt{2\ell+1}}$
	$W_\Pi + \Delta\dfrac{2\ell(\ell+1)}{(2\ell-1)(2\ell+3)}$	$-\Delta\dfrac{\sqrt{\ell(\ell+1)(\ell+2)}}{(2\ell+3)\sqrt{2\ell+1}}$	$-\Delta\dfrac{\sqrt{\ell(\ell-1)(\ell+1)}}{(2\ell-1)\sqrt{2\ell+1}}$
		$W_\Pi + \Delta\dfrac{(\ell+2)}{(2\ell+3)}$	0
			$W_\Pi + \Delta\dfrac{(\ell-1)}{(2\ell-3)}$

$<1|W|2>$ matrix element, all the matrix elements become independent of ℓ as ℓ becomes large. The $<1|W|2>$ matrix element decreases as $1/\ell$ in the large-ℓ limit. The behaviour of the $<1|H|2>$ and $<1|H|4>$ matrix elements as a function of R is shown in Figs. 9 and 10 for $\ell=2$ and $\ell=55$. Here we have taken the electrostatic potential energy curves $^1W_\Sigma(R)$ and $^1W_\Pi(R)$ from Fig. 1. We observe, as discussed previously, that in a diabatic basis the off-diagonal coupling matrix elements are not well localized in space. Since states $|1_{||}>$ and $|2>$ become degenerate asymptotically, the non-zero electrostatic mixing will cause them to mix strongly. This will be true even at large values of ℓ (large impact parameter). Although the $<1|H|2>$ matrix element decreases as $1/\ell$, two virtually degenerate states will be thoroughly mixed even by the smallest coupling.

Compared to state $|1_{||}>$, state $|2>$ represents a reorientation of both the p orbital as well as the orbital angular momentum. This reorientation is accomplished by rotation of L and ℓ in an opposite sense, so that the total projection quantum number $(M=m_j+m_\ell)$ remains 0. Physically the mixing of states $|1_{||}>$ and $|2>$ implies that as soon as R becomes small enough that the Σ and Π electrostatic potentials begin to differ significantly, the initially prepared orientation of the p-orbital will be scrambled. There is another implication: in many semiclassical treatments of atomic and molecular collisions both the orientation and the magnitude of ℓ are assumed to remain *fixed* during the collision [55,56]. The discussion in

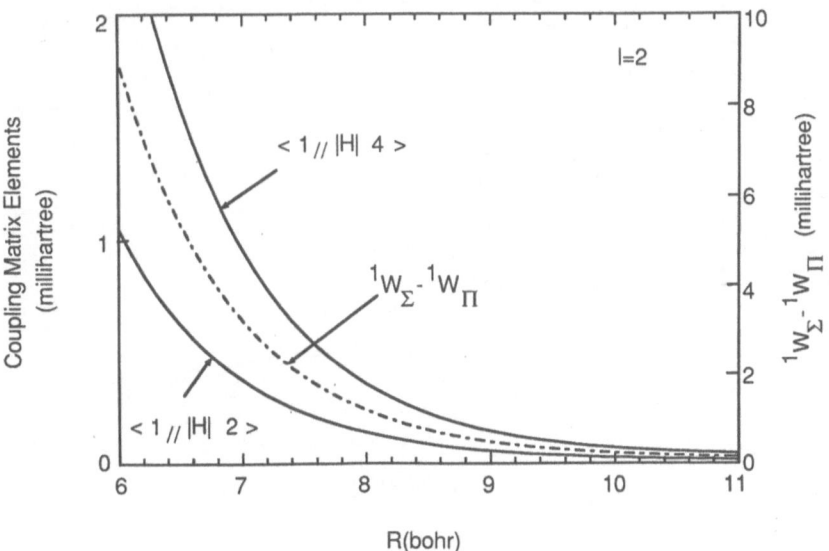

FIGURE 9. Absolute value of the $<1_{||}|H|2>$ and $<1_{||}|H|4>$ coupling matrix elements in the asymptotic uncoupled basis defined by Eq. 38. The curves were calculated using the $^1\Sigma$, $^3\Sigma$, $^1\Pi$ and $^3\Pi$ potential curves shown in Fig. 1 for an orbital angular momentum of $\ell=2$ and assuming a mixing angle of $\theta=18.4^0$ $(\sin^2\theta=0.1)$. The dashed curve corresponds to the difference between the $^1W_\Sigma(R)$ and $^1W_\Pi(R)$ potential curves.

the beginning of this paragraph casts serious doubts on the validity of this approximation (and, consequently, on the accuracy of cross sections calculated by this type of semiclassical method), since the mixing between states $|1_{||}>$ and $|2>$ can not occur if the orientation of ℓ is assumed to remain fixed. As the collision energy increases, the orbital angular momentum corresponding to a given value of the impact parameter will increase [Eq. 35]. Since the coupling between states $|1_{||}>$ and $|2>$ decreases as $1/\ell$, one might expect that a model in which the orientation of ℓ remains fixed might become more accurate in the high-energy limit.

An alternative picture involves a case (a) basis. As discussed above, the initially prepared state corresponds to a $^1\Sigma$ molecular state, which is e labelled. At large R, this state becomes degenerate (Fig. 1) with a $^1\Pi$ molecular state. Because of this degeneracy the J.L term in the rotational Hamiltonian [Eq. 2] will cause a complete mixing between the $^1\Sigma$ state and the e labelled ($\epsilon=+1$) component of the $^1\Pi$ state.

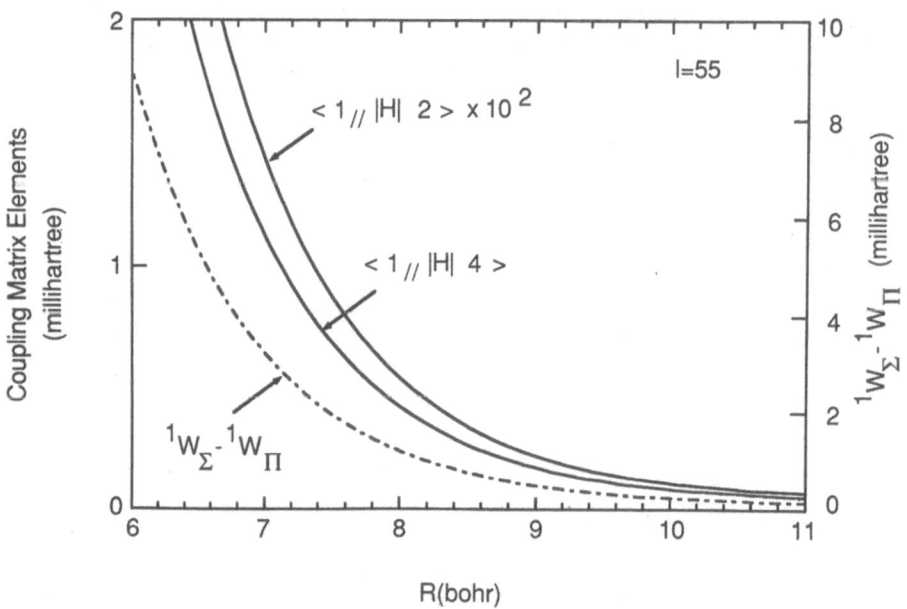

FIGURE 10. Absolute value of the $<1_{||}|H|2>$ and $<1_{||}|H|4>$ coupling matrix elements in the asymptotic uncoupled basis defined by Eq. 38. The curves were calculated using the $^1\Sigma$, $^3\Sigma$, $^1\Pi$ and $^3\Pi$ potential curves shown in Fig. 1 for an orbital angular momentum of $\ell=55$ and assuming a mixing angle of $\theta=18.4^0$ ($\sin^2\theta=0.1$). The dashed curve corresponds to the difference between the $^1W_\Sigma(R)$ and $^1W_\Pi(R)$ potential curves. Fig. 9 and Fig. 10 differ only in the value of the orbital angular momentum. Note that the curve for the $<1_{||}|H|2>$ matrix element has been multiplied by 100 here.

Additional insight can be gained by diagonalizing the full Hamiltonian in the basis formed by states $|1_{||}>, |2>, |3>$, and $|4>$, or, eventually, in the basis formed by states $|1_{||}>-|12>$, and looking at the resulting adiabatic energies and off-diagonal derivative couplings. The four singlet adiabatic energies (obtained by diagonalizing within the basis formed by states $|1_{||}>-|4>$) are shown in the lower panel of Fig. 11. We shall denote the corresponding adiabatic states by ψ_1-ψ_4. As in Fig. 10 we have taken the electrostatic potential energy curves $^1W_\Sigma(R)$ and $^1W_\Pi(R)$ from Fig. 1 and used $\ell=55$. The reader might wonder why there are *four* adiabatic singlet e-labelled states here whereas there appear only *two* adiabatic states which correlate with the ^1P asymptote in Fig. 4 which depicts the e-labelled adiabatic states in a coupled angular momentum representation (see also Table 2). The reason is that here we are concentrating on the evolution of a system prepared

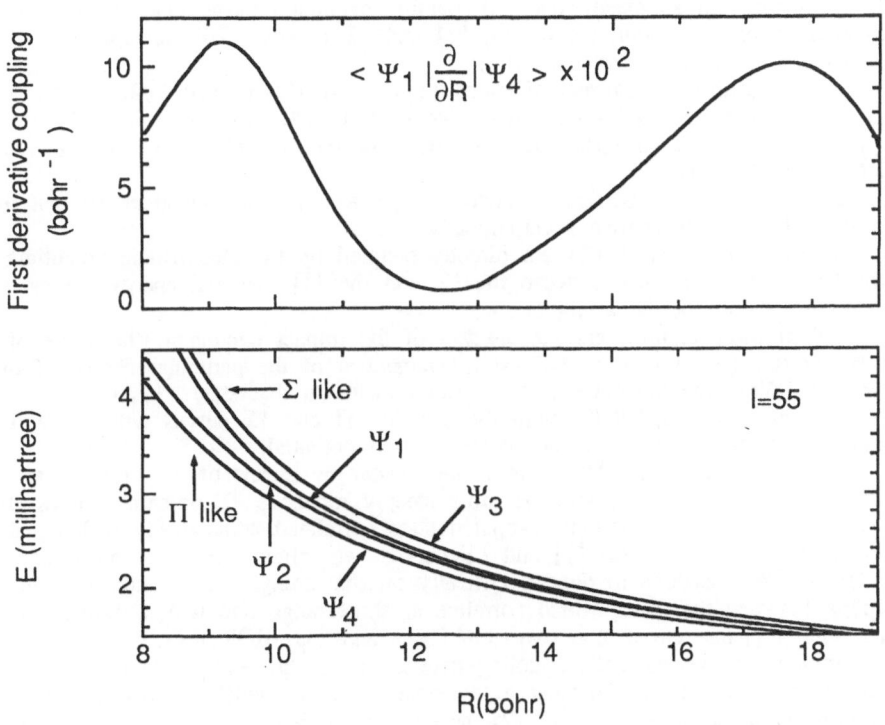

FIGURE 11. (Lower panel) Adiabatic energies of the four e-labelled states (denoted ψ_i) which correlate asymptotically to the four uncoupled basis states defined by Eq. 38 [$|1_{||}>, |2>, |3>$, and $|4>$] for an orbital angular momentum of $\ell=55$. (Upper panel) Absolute value of the first derivative coupling matrix element $<\psi_1|\partial/\partial R|\psi_4>$ for an orbital angular momentum of $\ell=55$.

in a unique state of the relative orbital angular momentum ℓ, rather than, as was the case in section 3, the total angular momentum J. Although J is conserved during the collision, it is not a good quantum number in an uncoupled basis. As discussed earlier in this section, *two* values of J are associated with a given value of $\ell (J=\ell \pm 1)$.

As R decreases, the curves separate into two pairs, which correlate at short range with, respectively, the $^1W_\Sigma(R)$ and $^1W_\Pi(R)$ curves in Fig. 1. For a given value of J, there is one Σ-like and one Π-like adiabatic e-labelled curve which correlate with the ^1P asymptote. Thus in the present case where states with both $J=\ell+1$ and $J=\ell-1$ must be included, there will be two Σ-like and two Π-like adiabatic e-labelled curves to consider. We notice, in particular, that the two adiabatic states which correlate asymptotically with the degenerate states $|1_{||}>$ and $|2>$ split apart at shorter distance, with one state acquiring an increasing degree of $^1\Sigma$ character and the other an increasing degree of $^1\Pi$ character. Since the curves labelled ψ_1 and ψ_2 in Fig. 11 become degenerate asymptotically, and since the states $|1_{||}>$ and $|2>$ are directly coupled by the electrostatic Hamiltonian, the state initially prepared in an experiment with parallel excitation (state $|1_{||}>$) will correlate with *equal probability* with the case(a) $^1\Pi$ and $^1\Sigma$ curves. This unexpected conclusion is a result of three features:

(1) The centrifugal barriers of states $|1_{||}>$ and $|2>$ $[\hbar^2 \ell(\ell+1)/2\mu R^2]$ are less than that of state $|3>$ and greater than that of state $|4>$, so, adiabatically, states $|1_{||}>$ and $|2>$ must correlate with the *higher* of the two Π-like and the *lower* of the two Σ-like curves.

(2) The centrifugal barriers of states $|1_{||}>$ and $|2>$ are identical, so that these two states become degenerate asymptotically.

(3) States $|1_{||}>$ and $|2>$ are directly coupled by the electrostatic potential, in particular by the *difference* between the $^1\Sigma$ and the $^1\Pi$ potential energy curves which correlate to the ^1P asymptote.

All of these features are *independent* of the impact parameter (the value of ℓ), except for the special case of b=0, and *independent* of the particular choice of the $^1W_\Sigma$ and $^1W_\Pi$ potential curves. Thus our conclusion that state $|1_{||}>$ will correlate with *equal probability* with the case(a) $^1\Pi$ and $^1\Sigma$ curves will be valid *irrespective* of the particular choice of interaction potential.

As discussed before, $^1P \rightarrow ^3P$ transitions occur by means of the crossing at short-range between the $^1\Pi$ and the more steeply repulsive $^3\Sigma$ potential curves (Fig. 1) [2-6,9,57]. Since the initially prepared state correlated adiabatically with equal probability with the case (a) $^1\Pi$ and $^1\Sigma$ curves, we might expect the magnitude of the $^1P \rightarrow ^3P$ cross sections to depend critically on the strength of the nonadiabatic coupling between the states which correlate at short range one with a Σ-like and the other with a Π-like curve, e.g. $<\psi_1 | \partial/\partial R | \psi_4>$ and $<\psi_1 | \partial/\partial R | \psi_2>$.

In fact, the nonadiabatic coupling between states ψ_1 and ψ_2 is null. This can be understood as follows: Consider a two-state model in which we restrict our attention just to states $|1_{||}>$ and $|2>$ as defined in Eqs. 38a and 38b. Similarly to Eq. 10 the adiabatic states can then be written as

$$|\psi_1> = -\sin\theta(R)|1_{||}> + \cos\theta(R)|2> \tag{41a}$$

and

$$|\psi_2> = \cos\theta(R)|1_{||}> + \sin\theta(R)|2>. \tag{41b}$$

Here the rotation angle is given by

$$\theta(R) = \frac{1}{2} \tan^{-1}\{2W_{12}(R)/[W_{11}(R)-W_{22}(R)]\}, \tag{42}$$

where $W_{ij} \equiv \langle i|W|j\rangle$. Using the expressions for these matrix elements given in Table 4 we find

$$\theta(R) = \frac{\pi}{4} + \frac{1}{4\ell} - \frac{1}{8\ell^2} + O(\ell^{-3}), \tag{43}$$

which is *independent* of R. Since the mixing angle is close to $\pi/4$, increasingly so as ℓ increases, the mixing between states $|1_{||}\rangle$ and $|2\rangle$ is almost complete. Despite this strong mixing the nonadiabatic coupling between these states vanishes, since, in a two-state model the nonadiabatic first-derivative coupling matrix element is given by [58]

$$\langle \psi_1|\frac{\partial}{\partial R}|\psi_2\rangle = -\frac{\partial \theta}{\partial R}, \tag{44}$$

which is zero here because θ is independent of R. We thus see that the absence of nonadiabatic coupling between states ψ_1 and ψ_2 is not a reflection of the absence of mixing between states $|1_{||}\rangle$ and $|2\rangle$ but rather a consequence of the fact that the mixing is independent of R.

The nonadiabatic coupling between states ψ_1 and ψ_4 is displayed in the upper panel of Fig. 11, again for $\ell=55$. We observe that there exist two regions of strong-coupling, well separated, both extending over a considerable distance.

Let us turn now to an examination of the situation where the polarization vector of the excitation laser is *perpendicular* to v_{rel}. Without loss of generality we can take $E||\hat{x}$. In this case the ^1P_x atomic state will be prepared, with wavefunction $2^{-1/2}$ $(-|m_L=1\rangle + |m_L=-1\rangle)$. Again, the wavefunction of two atoms at infinite separation can best be expressed in an *uncoupled* basis, namely

$$|1_\perp\rangle = 2^{-1/2}[-|j=1,m_j=1\rangle|\ell,m_\ell=0\rangle + |j=1,m_j=-1\rangle|\ell,m_\ell=0\rangle], \tag{45}$$

or in a coupled [case (e)] basis as

$$|1_\perp\rangle = 2^{-1/2}[-(11\ell0|\ell+1,1)|j\ell,J=\ell+1,1\rangle - (11\ell0|\ell,1)|j\ell,J=\ell,1\rangle - $$

$$(11\ell0|\ell-1,1)|j\ell,J=\ell-1,1\rangle + (1-1\ell0|\ell+1,-1)|j\ell,J=\ell+1,-1\rangle + $$

$$(1-1\ell0|\ell,-1)|j\ell,J=\ell,-1\rangle + (1-1\ell0|\ell-1,-1)|j\ell,J=\ell-1,-1\rangle]. \tag{46a}$$

Prior to the collision the relative orbital angular momentum ℓ is still oriented perpendicular to the collision plane and perpendicular to the collision-frame z-axis, which, as discussed in section 3, is coincident with v_{rel}, the initial relative velocity vector. In this case of perpendicular excitation the electric field vector of the pump laser E, which defines the laboratory-fixed Z axis, is also perpendicular to v_{rel} (Fig. 3). However, as shown clearly in Fig. 12, although both ℓ and E must lie perpendicular to v_{rel}, the angle between ℓ and E is *unrestricted*. Consequently, the orientation of the initially excited p orbital with respect to the collision plane is also *unrestricted*. As in the case of parallel excitation, dicussed above, at first thought, and after a cursory inspection of Fig. 3, it might seem surprising that the orientation of the collision plane with respect to the initially excited orbital is not unique. Again this is a consequence of the fact that if we assign a definite value to the orbital angular momentum ℓ and its projection m_ρ, then the conjugate angle - the azimuthal orientation of the collision plane - must be totally undetermined.

It is worthwhile observing that in the case of perpendicular excitation both e-labelled ($\ell = J\pm1$) as well as f-labelled ($\ell = J$) states are prepared, whereas in the case of parallel excitation, only e-labelled states are prepared. In section 2 (Table 1) we discussed how production of the $^3P_{j=0}$ level by collisional transfer out of a ^1P level can occur only by coupling within the manifold of f levels. One consequence is that $^1P\rightarrow^3P_0$ transitions will be *forbidden* in the case of parallel excitation, where only e-labelled ^1P states are populated. The simultaneous preparation of both e- and f-labelled states can be understood easily within a case (a) description. For a molecule in a $^1\Pi$ electronic state, the electronic wavefunction is *symmetric* with respect to reflection in the plane of rotation of the molecule for

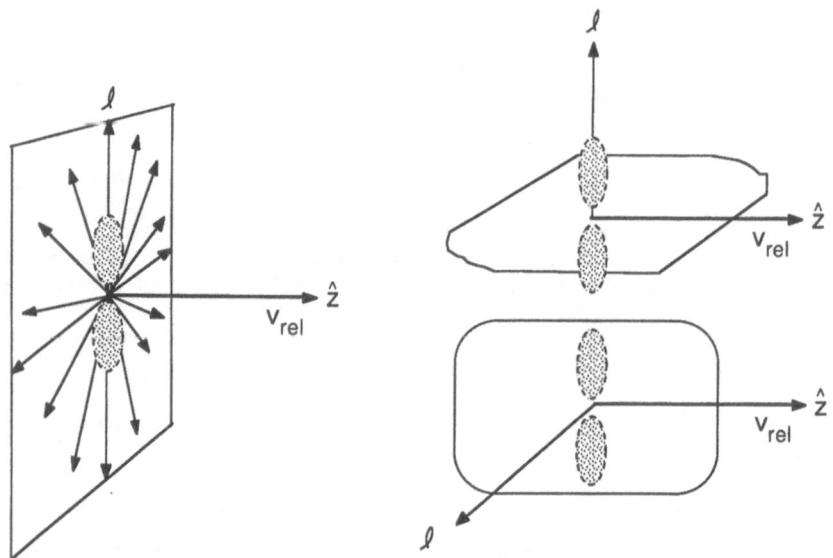

FIGURE 12. Schematic representation of the initial relative velocity vector v_{rel}, the initial relative orbital angular momentum ℓ, and the initially excited p orbital in the case of perpendicular excitation. Since the prepared p orbital has $m_L = \pm1$, it lies perpendicular to v_{rel} but is not necessarily perpendicular to ℓ. Yet, although $\hat{Z}$ and v_{rel} are coincident prior to the collision, the *azimuthal* orientation of ℓ is not restricted - a unique collision plane is not selected. Also, the initially prepared p orbital can be either perpendicular to or in the plane of collision. Thus the initially excited electronic wavefunction can be either symmetric or antisymmetric with respect to reflection in the initial plane of rotation of the two atom system - both e and f Λ-doublet states are excited.

the e-labelled ($\epsilon=+1$) component of the Π-state Λ-doublet and *antisymmetric* for the f-labelled ($\epsilon=-1$) Λ-doublet [59]. The plane of rotation is here just the collision plane. As we see clearly in Fig. 12, for perpendicular excitation an initially excited p orbital can be both symmetric or antisymmetric with respect to reflection of the wavefunction in the collision plane. Hence both the e and f components of the $^1\Pi$ state are prepared.

In the case of perpendicular excitation, an adiabatic analysis will involve *seven* singlet case (e) states, instead of the four states for parallel excitation [9]. The initially prepared state is now degenerate at long range with *two* of the other states (instead of one in the case of parallel excitation). These three states can be expressed in an uncoupled basis as

$$|2>=|j,m_j=0>|\ell,m_\varrho=1> \tag{46b}$$

$$|3>=|j,m_j=0>|\ell,m_\varrho=-1> \tag{46c}$$

$$|4>=2^{-1/2}[|j=1,m_j=1>|\ell=1,m_\varrho=0> + |j=1,m_j=-1>|\ell,m_\varrho=0>] \tag{46d}$$

Compared to state $|1_\perp>$, states $|2>$ and $|3>$ correspond to a reorientation of *both* the p orbital as well as the orbital angular momentum. As in the case of parallel excitation, the presence of states with both $m_j=0$ and $m_j=\pm1$ as members of this degenerate triplet implies that as soon as R becomes small enough that the Σ and Π electrostatic potentials begin to differ significantly, the initially prepared orientation of the p orbital will be scrambled. Again, this casts serious doubts on the validity of semiclassical approximations in which both the orientation and the magnitude of ℓ are assumed to remain *fixed* during the collision.

At short range, these four degenerate levels ($|1_\perp>$, $|2>$, $|3>$ and $|4>$) correlate adiabatically with two Σ-like (steeply rising) and two Π-like (less steeply rising) curves. Thus, exactly as in the case of parallel excitation, the initially prepared state correlates with *equal probability* to the case (a) $^1\Pi$ and $^1\Sigma$ curves *regardless of the choice of interaction potential*.

5. ORBITAL LOCKING MODELS

The initial experiments of Leone and co-workers on $^1P \to ^3P$ transitions in collisions of electronically excited alkaline earth atoms [2-5] and those of Rettner and Zare on reactive collisions of Ca(4s4p ^1P) [1] were interpreted using the concept of "orbital following" [10-12]. This model was developed by Hertel and co-workers [10-12] to interpret earlier experimental work on collisions involving electronically excited Na atoms in ^2P states. This model is illustrated in Figs. 13-15. If the pump laser initially excites a p orbital either lying along, or perpendicular to, R, the orientation of the orbital with respect to R during the collision can remain fixed or "frozen" with respect to a space- or laboratory-frame coordinate system (Fig. 13). Alternatively, as shown in Fig. 14, the orientation of the orbital can "follow" R. Indeed as shown in Fig. 15, the orbital can remain frozen at large distances but then "follow" R once the atoms have approached within a certain distance, designated the "locking radius". If the initially excited p orbital follows or locks, then perpendicular excitation will result in a π orientation of the p orbital at close range (p orbital perpendicular to R), while parallel excitation will result in a σ orientation (p orbital parallel to R). The reverse will occur if the initially excited p orbital remains frozen: perpendicular excitation will lead to σ orientation and parallel excitation, to π orientation.

This analysis is conceptually and pictorially appealing. It has to be used to provide a qualitatively satisfactory interpretation of the experimental results of Leone, Zare, and their co-workers. Despite this apparent success, it is necessary to ask whether this model is justified on the basis of the exact quantum analysis of the collision dynamics presented here.

At first glance the concept of a locking radius seems consistent with the analysis in terms of Hund's case coupling schemes, presented in section 2. At long range a case (e) representation is appropriate, where, for an atom in a ^1P state, the electronic orbital angular momentum L precesses about a space-fixed axis and so remains "frozen" as the atoms approach. At shorter range, however, a case (a) representation is more appropriate, wherein the orbital is "locked" to the molecular axis, R, so that Λ, the projection of L along R, remains fixed. Thus the concept of a "locking radius" could be equated with the point at which a transition from a case (a) to a case (e) representation becomes more appropriate. As discussed earlier, this transition occurs *roughly* at the point where the splitting between the $^{2S+1}W_\Sigma(R)$ and $^{2S+1}W_\Pi(R)$ potentials becomes equal to the centrifugal barrier. In an adiabatic treatment of the dynamics, the transition between the regions where case(e) and case(a) representations are appropriate is marked by a dramatic change in the adiabatic states, so the first-derivative coupling matrix elements [Eq. 32] should be large. Hence one might associate the "locking radius" with a point of large nonadiabatic coupling. Unfortunately, as we have seen in the preceding section, the region of strong nonadiabatic coupling may be delocalized and may indeed display several maxima.

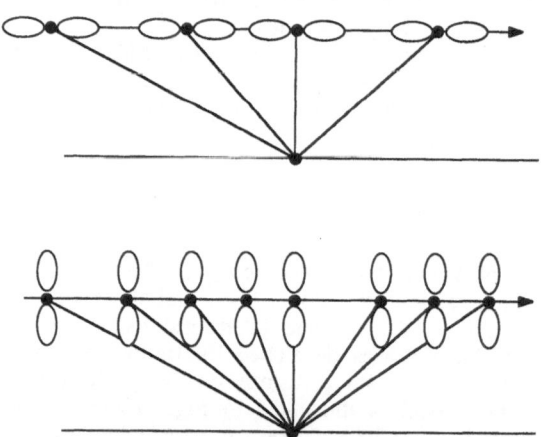

FIGURE 13. Schematic representation of an initially excited p orbital which remains fixed in space - "frozen" - during the collision.

Further examination points to several flaws in orbital locking models. Implicit in the pictorial description of orbital locking contained in Figs. 13-15 is a collinear trajectory for the two colliding atoms. This is obviously unrealistic, but could be easily altered to allow curvature, either attractive or repulsive, produced by the

$^{2S+1}W_\Sigma(R)$ and $^{2S+1}W_\Pi(R)$ potentials. A more subtle point concerns the restriction of the trajectory to a plane, implicit in many semiclassical models [55,56], which is based on the assumption that if ℓ is large, then its projection will not be altered. As we have seen in the preceding section, at long range the initially prepared state, both for parallel as well as perpendicular excitation, is degenerate with one (or more) states which differ only by reorienting both m_L and m_ϱ by ± 1 unit of angular momentum, so that the total projection quantum number remains constant. The result was that the initially prepared state, in which the space- or laboratory-frame orientation of the atomic p orbital was well-defined, became an equal mixture of the degenerate levels, with a consequent loss of information about the space-frame orientation of **L**. This scrambling was seen to be induced by the splitting between the $^{2S+1}W_\Sigma(R)$ and $^{2S+1}W_\Pi(R)$ potentials.

If m_ϱ is assumed to be constant, so that the collision plane remains fixed, then the asymptotic mixing between states with $m_\varrho=0$ and $m_\varrho=\pm 1$ is no longer permitted. Hence we see that only if m_ϱ can be treated as fixed, can we consider the initially prepared p orbital to remain frozen as the atoms approach. Dynamical models in which ℓ is treated as fixed have been proposed under a number of names [60-62]. It is fair to say that if these models can be shown to reproduce the results of exact quantum scattering calculations for a given choice of collision mass, energy, and interaction Hamiltonian, then it would be justifiable to neglect the asymptotic scrambling of the initially excited p orbital.

Another subtle defect inherent in the simple pictorial description of Figs. 13-15 is the assumption that in the case of perpendicular excitation, the initially prepared p orbital lies in the plane of the collision.. As discussed in section 4, in the case of perpendicular excitation it is not possible to specify the orientation of the collision

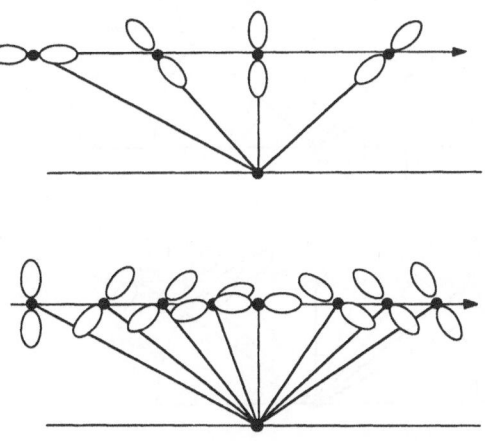

FIGURE 14. Schematic representation of an initially excited p orbital which "follows" the internuclear axis **R** during the collision.

plane (or, alternatively, the orientation of ϱ) with respect to the electric field vector of the laser. Thus in the case of perpendicular excitation the simple orbital locking pictures (Figs. 13-15) must be extended to include a situation where the initially prepared orbital lies perpendicular to **R** as well as perpendicular to the plane of collision (as shown in Fig. 16). In this case, since the initially prepared wavefunction is *antisymmetric* with respect to reflection in the plane of the collision (A" in C_S symmetry), and since the model assumes the plane of collision to remain fixed, the wavefunction must remain antisymmetric. Thus the initially prepared p orbital must always remain frozen. Hence we see that a locking model, properly applied to the case of perpendicular excitation, must allow for the fact that for one-half of the prepared atoms the p orbital can never "lock" onto the internuclear axis, even at distances shorter than the "locking radius". In terms of a case (a) analysis, perpendicular excitation produces equal populations in both the e and f Λ doublets. The wavefunction for the e Λ doublet level is symmetric with respect to reflection in the plane of rotation [59] and so can mix with the $^1\Sigma$ state, itself e-labelled. However, the wavefunction for the f Λ doublet level is *antisymmetric* with respect to reflection in the plane of rotation and so *cannot* mix with the $^1\Sigma$ state.

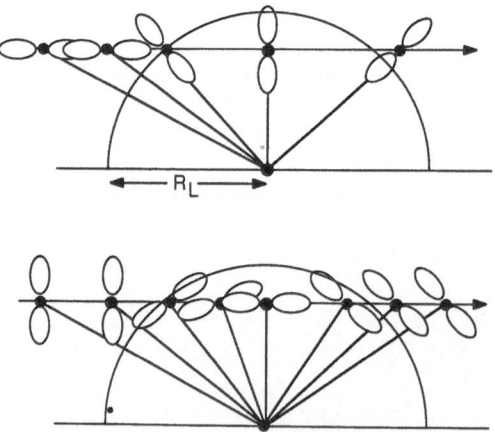

FIGURE 15. Schematic representation of an initially excited p orbital which remains fixed in space initially but then "follows" the internuclear axis **R** once the two collision partners have approached to within the "locking radius".

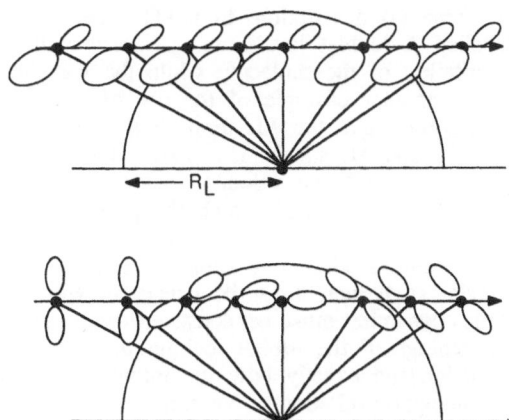

FIGURE 16. Schematic representation of orbital following in the case of perpendicular excitation. If the initially excited p orbital is in the plane of the collision, it can "follow" **R** by means of a rotation in this plane. If, however, the initially excited p orbital is *perpendicular* to the plane of the collision, it will remain in this orientation.

6. CONCLUSION

We have presented here a detailed review of the fully-quantum description of inelastic collisions involving atoms in ^{1}P electronic states. Our focus has been threefold, stressing

(1) The way in which various Hund's case coupling schemes can help to clarify and understand the interactions and couplings which are present.

(2) The use of a fully adiabatic description to circumscribe the range of interparticle separations over which significant mixing will occur between given initial and final states.

(3) The care which must be used in describing correctly the initial state in beam experiments involving laser excitation.

As an example of the insight gained we cite the case of perpendicular excitation, where, contrary to what one might guess, the initially excited orbital is not constrained to be perpendicular to the collision plane. This is understandable from the viewpoint of a Hund's case (a) description: in a $^1\Pi$ state, the electronic wavefunctions of the e-labelled Λ-doublet levels are *symmetric* with respect to reflection in the plane of rotation of the molecule while the wavefunctions of the f levels are *antisymmetric*. As another example of the use of a case (a) description, it is well known to spectroscopists that the J.L term in the rotational Hamiltonian will mix nearly degenerate $^1\Pi$ and $^1\Sigma$ states [15]. This implies that a p orbital initially oriented parallel (Σ) to the relative velocity vector will rapidly scramble and become mixed with the e-labelled Π state in which the p orbital lies perpendicular to **R**.

As stressed in the preceding section, any simple, interpretative models of the spin-changing transitions under study here must be consistent with the full quantum description. In other words, dynamics must be consistent with spectroscopy. In spite of the apparent shortcomings of the orbital locking models [10-12], it would be intellectually satisfying to understand exactly how the initially prepared state evolves, both for parallel and perpendicular excitation. We might then be able to ascertain whether the dependence of the ^{1}P-^{3}P cross section on the direction of laser polarization observed by Leone and co-workers [2-5] arises because in one or the other case the initially prepared state accesses preferentially at small R the $^1\Pi$ potential curve, which ultimately crosses the $^3\Sigma$ curve (Fig. 1). This, and similar, questions might be answered by determining the scattering wavefunction corresponding to a properly defined initial state and projecting it onto the adiabatic basis as a function of the internuclear distance.

The ideas and methods presented here should certainly be extended to deal with encounters between an electronically excited atom and a molecule. The situation will be more complicated in that the doubly degenerate Π potential curves of the two atom system will be replaced by two potential energy *surfaces*, of A' and A" symmetry in C_S geometry. Similarly the Σ potential curve of the two atom system will become a potential energy surface, of A' symmetry, if the atomic partner is replaced by a closed-shell molecule. A formally equivalent situation, collisions of F(^{2}P) with H_2, has been treated by Rebentrost and Lester [63]. More recently, Alexander and Corey [64] have presented the exact quantum theory of collisions of CN(A$^2\Pi$,$^2\Sigma^+$) with a closed-shell atom. Here too, it is necessary to consider two potential surfaces of A' and one of A" symmetry.

However, despite the additional complexity, it is clear that the same scrambling of the initially prepared orbital orientation will occur in atom-molecule collisions. For example, in the case of parallel excitation one of the A' wavefunctions will be prepared which can subsequently mix with the other, degenerate, A' wavefunction. In the case of a molecular partner this mixing will be induced not only by the splitting between the two A' potential energy surfaces as a function of the atom-molecule separation (as in the case of atom-atom collisions) but also by the variation in this splitting as the orientation of the molecule with respect to **R** changes.

The major impediment to calculations of cross sections for spin-changing transitions in collisions of, for example, Ca(4s5p^{1}P) with a molecule is the total lack of information on the necessary *six* potential energy surfaces (two A' and one A" surfaces of both singlet and triplet multiplicity) plus the additional off-diagonal coupling matrix elements in a diabatic basis [63,64] which arise in collisions involving a molecular partner. In the case of atomic collisions it is easily possible to carry out a series of calculations based on qualitatively reasonable model potentials [9]. By varying the parameters on which these potentials depend one can explore [9,65] which features of the potential energy curves are probed experimentally. This type of study will be much more difficult, if not impossible, in the case of a molecular partner. First, the scattering calculations will be far more costly. Secondly, it is not clear what functional form can provide an accurate and easily varied description of the necessary molecular potential energy surfaces.

In the case of *reactive* collisions, such as the halogen reaction

$$Ca(4s4p\ ^1P)+Cl_2 \rightarrow CaCl+Cl$$

studied by Rettner and Zare [1], we would hope to be able to explore the dependence on orbital alignment while avoiding the complexity of a full atom-molecule quantum scattering calculation. A possibly promising approach is the pseudoquenching model [29], developed to study the dependence on initial spin-orbit state of cross sections for the related *triplet* reaction, namely

$$Ca(4s4p\ ^3P_j)+Cl_2 \rightarrow CaCl+Cl,$$

which has been studied by Dagdigian and co-workers [66]. In this model all entrance channel effects, that is the distribution of population out of the initially prepared state, are treated by considering the atom-molecule system to be a diatomic. The effect of the exoergic reaction is then simulated by adding on an additional, basically repulsive, channel which is coupled to the Ca*+M channels at small values of R and which correlates to an energetically lower asymptote.

ACKNOWLEDGMENTS

The authors would like to acknowledge their debt to the U.S. Army Research Office, whose support under grants DAAG29-81-K-0102, DAAG29-84-G-0078, and DAAG29-85-K-0018 has permitted their long-standing collaboration. During the past five years they have benefited immensely from the encouragement, help, and criticism of many colleagues on numerous subtle points concerning collisions of atoms in P electronic states. They wish to acknowledge in particular V. Aquilanti, P. Dagdigian, M. Hale, I. Hertel, S. Leone, D. Neuschäfer, E. Nikitin, and J.-M. Robbe.

REFERENCES

1. C.T. Rettner and R.N. Zare, J. Chem. Phys., **75** (1981) 3636; **77** (1982) 2416.

2. M.O. Hale and S.R. Leone, J. Chem. Phys., **79** (1983) 3352.

3. M.O. Hale, I.V. Hertel and S.R. Leone, Phys. Rev. Lett., **53** (1984) 2296.

4. M.O. Hale and S.R. Leone, Phys. Rev. A, **31** (1985) 103.

5. D. Neuschäfer, M.O. Hale, I.V. Hertel and S.R. Leone, in Electronic and Atomic Collisions, XIV ICPEAC, edited by D.C. Lorents, W.E. Meyerhof, and J.R. Peterson (North Holland, Amsterdam, 1986), p. 585.

6. W. Bussert, D. Neuschäfer and S.R. Leone, in press.

302

7. R.W. Schwenz and S.R. Leone, in press; W. Bussert and S.R. Leone, in press.

8. A.Z. Devdariani and A.L. Zagrebin, Chem. Phys. Lett., **131** (1986) 197.

9. B. Pouilly and M.H. Alexander, J. Chem. Phys., **86** (1987) 4790.

10. H.W. Hermann and I.V. Hertel, Comments At. Mol. Phys., **12** (1982) 61, 127.

11. I.V. Hertel, H. Schmidt, A. Bähring, and E. Meyer, Rep. Prog. Phys., **48** (1985) 375.

12. R. Witte, E.E.B. Campbell, C. Richter, H. Schmidt, and I.V. Hertel, to be published (as cited in Ref. 6).

13. G. Herzberg, Molecular Spectra and Molecular Structure. I. Spectra of Diatomic Molecules (D. Van Nostrand, Princeton, 1950).

14. J.T. Hougen, Natl. Bur. Stand. (U.S.) Monogr. **115** (1970).

15. H. Lefebvre-Brion and R.W. Field, Perturbations in the Spectra of Diatomic Molecules (Academic, New York, 1986).

16. E.E. Nikitin, J. Chem. Phys., **43** (1965) 744; Adv. Chem. Phys., **28** (1975) 317; E.E. Nikitin and S. Ya. Umanskii, Theory of Slow Atomic Collisions (Springer-Verlag, Berlin, 1984).

17. E.I. Dashevskaya, E.E. Nikitin and A.I. Reznikov, J. Chem. Phys., **53** (1970) 1175.

18. F. Masnou-Seeuws and R. MacCarroll, J. Phys. B, **7** (1974) 2230.

19. V. Aquilanti and G. Grossi, J. Chem. Phys., **73** (1980) 1165; V. Aquilanti, P. Casavecchia, G. Grossi and A. Laganá, ibid, **73** (1980) 1173. V. Aquilanti, G. Grossi and A. Laganá, Nuovo Cimento **63B** (1981) 7.

20. M.H. Alexander, T. Orlikowski and J.E. Straub, Phys. Rev. A, **28** (1983) 73.

21. B. Pouilly, T. Orlikowski and M.H. Alexander, J. Phys. B, **18** (1985) 1953.

22. D. Lemoine, J.-M. Robbe and B. Pouilly, J. Phys. B, submitted.

23. A.I. Voronin and V.A. Kvilividze, Theor. Chim. Acta, **8** (1967) 334.

24. M.S. Child, Molecular Collision Theory (Academic, New York, 1974).

25. R.K. Preston, C. Sloane and W.H. Miller, J. Chem. Phys., **60** (1974) 4961.

26. F.H. Mies, Mol. Phys., **41** (1980) 973.

27. R.W. Anderson, J. Chem. Phys., **77** (1982) 5426.

28. J. Pascale, J.-M. Mestdagh, J. Cuvelier, and P. de Pujo, J. Phys. B , **17** (1984) 2627.

29. M.H. Alexander, in Gas-Phase Chemiluminescence and Chemi-Ionization, edited by A. Fontijn (Elsevier, Amsterdam, 1985), p. 221.

30. D.M. Brink and G.R. Satchler, Angular Momentum, 2nd ed. (Clarendon, Oxford, 1975).

31. E.U. Condon and G.H. Shortley, The Theory of Atomic Spectra (Cambridge, 1951).

32. R.H.G. Reid, J. Phys. B, **6** (1973) 2018.

33. F.H. Mies, Phys. Rev. A, **7** (1973) 942.

34. R.P. Saxon, R.E. Olson, and B. Liu, J. Chem. Phys., **67** (1977) 2692.

35. B. Pouilly, B.H. Lengsfield and D.R. Yarkony, J. Chem. Phys., **80** (1984) 5089.

36. J. Pascale and M.Y. Perrin, J. Phys. B, **13** (1980) 1839, J. Pascale, Phys. Rev. A, **28** (1983) 632.

37. A.R. Edmonds, Angular Momentum in Quantum Mechanics (Princeton University, Princeton, 1960).

38. J.M. Brown, J.T. Hougen, K.-P. Huber, J.W.C. Johns, I. Kopp, H. Lefebvre-Brion, A.J. Merer, D.A. Ramsay, J. Rostas and R.N. Zare, J. Mol. Spectrosc., **55** (1975) 500.

39. R.N. Zare, A.L. Schmeltekopf, W.J. Harrop, and D.L. Albritton, J. Mol. Spectrosc., **46** (1973) 37.

40. M. Larsson, Phys. Scr., **23** (1981) 835.

41. A.M. Arthurs and A. Dalgarno, Proc. R. Soc. London A, **256** (1960) 540.

42. R.H.G. Reid and A. Dalgarno, Phys. Rev. Lett., 22 (1969) 1029;
    Chem. Phys. Lett., 6 (1970) 85.

43. A.D. Wilson and Y. Shimoni, J. Phys. B, 7 (1974) 1543; 8
    (1975) 1392, 2393.

44. R.E. Olson, Chem. Phys. Lett., 33 (1975) 250.

45. D. Secrest, in Atom-Molecule Collision Theory. A Guide for the
    Experimentalist, edited by R.B. Bernstein (Plenum, New York,
    1979) p. 265.

46. L.D. Thomas, M.H. Alexander, B.R. Johnson, W.A. Lester, Jr.,
    J.C. Light, K.D. MacLenithan, G.A. Parker, M.J. Redmon,
    T. Schmalz, D. Secrest, and R.B. Walker, J. Comput. Phys.,
    41 (1981) 407.

47. M.H. Alexander, J. Chem. Phys., 81 (1984) 4510; M.H. Alexander
    and D.E. Manolopoulos, ibid, 86 (1987) 2044.

48. D.E. Manolopoulos, J. Chem. Phys., 85 (1986) 6625.

49. R.D. Levine and R.B. Bernstein, Molecular Reaction Dynamics
    (Clarendon, Oxford, 1974). Chap. 3.

50. V. Khare, D.J. Kouri, and R.T. Pack, J. Chem. Phys., 69 (1978)
    4419.

51. M.D. Rowe and A.J. McCaffery, Chem. Phys., 43 (1979) 35; see
    also footnote 47 of Ref. 53.

52. R.T. Pack and J.O. Hirschfelder, J. Chem. Phys., 73 (1980) 3833.

53. M.H. Alexander and S.L. Davis, J. Chem. Phys., 78 (1983) 6754.

54. S.A. Evans, J.S. Cohen and N.F. Lane, Phys. Rev. A, 4 (1971)
    2235; F. Mrugala and D. Secrest, J. Chem. Phys., 78 (1983)
    5954; 79 (1983) 5960.

55. S. Geltman, Topics in Atomic Collision Theory (Academic,
    New York, 1969).

56. G.G. Balint-Kurti, in Theoretical Chemistry, MTP International
    Review of Science, Physical Chemistry, Series 2 (Butterworths,
    London, 1975) Vol. 1.

57. W.H. Breckenridge and O.K. Malmin, J. Chem. Phys., 76 (1982)
    1812.

58. F. Rebentrost, in Theoretical Chemistry: Advances and

Perspectives, edited by D. Henderson (Academic, New York, 1981).

59. M.H. Alexander and P.J. Dagdigian, J. Chem. Phys., **80** (1984) 4325.

60. J.-M. Launay and E. Roueff, J. Phys. B, **10** (1977) 879.

61. J.S. Cohen, L.A. Collins, and N.F. Lane, Phys. Rev. A, **17** (1978) 1343.

62. D.E. Fitz and D.J. Kouri, J. Chem. Phys., **73** (1980) 5115.

63. F. Rebentrost and W.A. Lester, Jr., J. Chem. Phys., **64** (1976) 2879; **67** (1977) 3367.

64. M.H. Alexander and G.C. Corey, J. Chem. Phys., **84** (1986) 100.

65. B. Pouilly, J.-M. Robbe and M.H. Alexander, Abstracts, Conference on the Dynamics of Molecular Collisions, Wheeling, WV, USA (1987) and to be published.

66. P.J. Dagdigian, in Gas-Phase Chemiluminescence and Chemi-Ionization, edited by A. Fontijn (Elsevier, Amsterdam, 1985), p. 203; P.J. Dagdigian and M.L. Campbell, Chem. Rev., **87** (1987) 1.

IONIZING COLLISIONS OF FAST K ATOMS WITH ORIENTED SYMMETRIC TOP MOLECULES: PRELIMINARY RESULTS

Howard S. Carman, Jr[1], Leon F. Phillips[2] and
Philip R. Brooks*
Chemistry Department and Rice Quantum Institute
Rice University
Houston
Texas 77251
USA

1. INTRODUCTION

A harpoon-type mechanism has been proposed [1] to explain the angular distribution of the scattering of low-energy K atoms from oriented CF_3I. This proposes that an electron from the K atom is transferred to the CF_3I at long range giving rise to a CF_3I^- ion in a repulsive state. The CF_3I^- then dissociates impulsively with the CF_3 radical and I^- ion separating along the instantaneous direction of the C-I bond and the KI product recoiling with a momentum given by the resultant of the K^+ and I^- momenta. This predicts that KI will be backward scattered for "heads" orientation (K incident on the I end of the molecule) and forward scattered for "tails" orientation. The model is in good agreement with experiment, and moreover, correctly accounts for the experimental angular distribution resulting from "sideways" oriented molecules.

This model presumes that there is no great difference in reactivity between the two ends of the CF_3I molecule, the dynamics being dominated by processes associated with the long-range electron transfer under conditions where the internal structure of the top does not play a significant role. This is suggestive, in some ways, of results obtained in an earlier study of the scattering of K by oriented $CHCl_3$ [2], where the scattering was found to be independent of orientation. In contrast, for the scattering of K by CH_3I [3], scattering has been found to be very dependent on orientation, with a substantial cone of no reaction at the CH_3 end of the molecule. An intermediate case is provided by the reaction of K atoms with CF_3Br [4], where the results are consistent with the long-range harpoon-type model mentioned above. For CF_3Br there is a difference in reactivity by about a factor of 3 between the two ends of the molecule, the Br end being the more reactive. In CF_3I, the two ends are roughly equally reactive.

The question now arises as to whether the differences in reactivity for the different ends of CF_3I and CF_3Br arise from simple steric effects, or whether they are a reflection of differences in the efficiency of the initial electron-transfer process.

---

[1] Present address, Oak Ridge National Laboratory, Tennessee
[2] On leave from the Chemistry Department, University of Canterbury, Christchurch, New Zealand.

J. C. Whitehead (ed.), Selectivity in Chemical Reactions, 307–310.
© 1988 by Kluwer Academic Publishers.

Differences in electron-transfer efficiency might arise either from an orientation dependence of the H_{12} matrix element, which controls the probability of non-adiabatic transition in Landau-Zener and related theories [5], *or* from an orientation dependence of the crossing-radius itself. The latter could be regarded as due to either an orientation dependence of the vertical electron affinity of the symmetric top or an orientation dependence of the difference between the position of the centre of mass of the symmetric top and the centre of charge of its lowest unoccupied molecular orbital. The aim of the work reported here was to search for an orientation dependence of the electron transfer process by measuring relative cross-sections for ionization in collisions of energetic (15 eV) potassium atoms with oriented CF_3I, CF_3Br, $CHCl_3$ and CH_3I.

2. EXPERIMENTAL

The experimental system was similar to that described previously [4] except that improved pumping of the seeded source for symmetric-top species (in argon carrier gas) gave rise to lower rotational temperatures and narrower angular distributions, and the effusive potassium beam was replaced by a charge-exchange source of the type described by Helbing and Rothe [6]. Positive ions produced in the collision region were detected by a pair of channeltron particle multipliers mounted on the orienting-field conductors, only one channeltron - that on the negative side - being used at a time. The channeltrons were operated with the same cone voltage, differences in operating characteristics being largely compensated by applying positive bias to the anode of the channeltron having the lower operating voltage. Relative sensitivities of the two channeltrons were determined by counting the ions produced by collision of the 15 eV K atoms with a seeded beam of SF_6 in argon, for which no orientation dependence is possible.

3. RESULTS AND DISCUSSION

Results, in the form of the ratio of count rates for 0^0 and 180^0 orientations, are given in Table 1. The 0^0 orientation corresponds to the channeltron and orienting plate nearest the K source being positive, so the K is incident on the positive end of the molecule. For CH_3I the 0^0 orientation corresponds to collision with the CH_3 end of the molecule; for the other CX_3Y species the 0^0 orientation corresponds to collision of K with the Y end of the molecule. Limits of error shown correspond to two standard deviations, as calculated from count rates assuming a Poisson distribution such that the standard deviation of a total count is equal to the square root of the number of counts. These error limits necessarily incorporate the standard errors of the relatively small numbers of counts obtained with SF_6, so that the relative count rates are more accurate than the limits of error would indicate. The results show that there is no significant orientation dependence of the ionization cross-section for any of the symmetric tops except CH_3I, and that for this species ionization is 40% more likely for collisions with the CH_3 end of the molecule than for collisions with the iodine end.

TABLE 1

Reaction (0° orientation)	Ratio 0°/180°	Conclusion
K + HCCl$_3$	0.95 ± 0.16	No effect
K + BrCF$_3$	0.96 ± 0.16	No effect
K + ICF$_3$	1.04 ± 0.10	No effect
K + H$_3$CI	1.39 ± 0.14	Steric effect (see text)

The absence of an orientation dependence of the ionization rates for the first three entries in Table 1 implies that the differences in forward and backward scattering found for CF$_3$Br are indeed most likely due to steric effects and not to differences in the probability of electron transfer. A remarkable degree of compensation between the effect of orientation on the relative yields of neutral products and ions and the effect of orientation on electron transfer would otherwise be needed, *for all three species simultaneously*, in order to account for these results. For CH$_3$I, on the other hand, there is the interesting and counter-intuitive result that ionization is more likely to occur during collision with the CH$_3$ end of the molecule. The simplest interpretation of this observation is that ionization is here occurring in a fairly even competition with reaction to form neutral products, a conclusion which is interesting in view of the relatively low reaction efficiency expected at energies of the order of 15 eV. At the CH$_3$ end of the molecule the efficiency of forming neutral products is very low. Further work is intended to provide more insight into the CH$_3$I + K system by measuring the orientation dependence of ionization efficiency over a range of energies.

We gratefully acknowledge support of this work by the National Science Foundation under grant CHE 8210438, by the Robert A. Welch Foundation, and by the National Aeronautics & Space Administration.

REFERENCES

[1] P.R. Brooks, J. McKillop and H.G. Pippen, Chem. Phys. Lett., 66 (1979) 144.

[2] G. Marcelin and P.R. Brooks, J. Am. Chem. Soc., 95 (1973) 7885, *ibid* 97 (1975) 1710.

[3] D.H. Parker, K.K. Chakravorty and R.B. Bernstein, Chem. Phys. Lett., 86 (1982) 113, and D.H. Parker, K.K. Chakravorty and R.B. Bernstein, J. Phys. Chem., 85 (1981) 466.

[4]   H.S. Carman, P.W. Harland and P.R. Brooks, J. Phys. Chem., 90 (1986) 944.

[5]   E.E. Nikitin, Theory of Elementary Atomic and Molecular Processes in Gases, Oxford: Clarendon Press 1974.

[6]   R.K.B. Helbing and E.W. Rothe, Rev. Sci. Instrum., 39 (1968) 1948, *ibid* 33 (1962) 841.

SATURATION IN LASER-INDUCED FLUORESCENCE: DETECTION OF NASCENT PRODUCT ROTATIONAL ANGULAR MOMENTUM ALIGNMENT

M.H.M. Janssen*, M.A. Quesada, D.H. Parker
Department of Chemistry
U.C. Santa Cruz
Santa Cruz
CA 95064
USA

S. Stolte
Molecular and Laser Physics,
Department of Physics
Catholic University of Nijmegen
Toernooiveld
6525 ED Nijmegen
The Netherlands

ABSTRACT. Laser-induced-fluorescence (LIF) with polarization analysis is a powerful means of analysing angular momentum disposal in a chemical reaction. In many cases saturation effects can strongly perturb the information provided by the LIF signal on state populations and rotational angular momentum alignment. In this paper we extend a previous rate equation based treatment of LIF alignment analysis to cover most experimental configurations probing cylindrically symmetric anisotropic distributions produced by photodissociation and inelastic or reactive scattering processes.

1. INTRODUCTION

Laser-induced-fluorescence is a simple, yet powerful means of probing the outcome of scattering events, particularly the polarization of rotational angular momentum J, in nascent product molecules. Ideally, by varying the polarization plane of the probe laser and measuring the fluorescence signal through another rotatable polarizer, the spatial alignment of the nascent product distribution is directly obtained. The high fluences of typical lasers, however, can greatly complicate this analysis and must be carefully accounted for.

In a recent paper [1] we described an analytical method of extracting alignment information on nascent product rotational angular momentum distributions using laser-induced-fluorescence under arbitrary saturation conditions. A variety of

---

* visiting scientist, permanent address: Molecular and Laser Physics, Catholic University of Nijmegen, The Netherlands.

J. C. Whitehead (ed.), Selectivity in Chemical Reactions, 311–321.
© 1988 by Kluwer Academic Publishers.

processes including photofragmentation, inelastic or reactive scattering and gas surface collisions produce anisotropic distributions of molecular fragments in space. To probe such a distribution it is convenient to use laser-induced-fluorescence where both the laser and the fluorescence polarization are varied, allowing the determination of four in general independent fluorescence intensities. Linear combinations of these experimental signals yield the moments describing the anisotropy. Saturation, i.e. a considerable depletion of the lower state population, by intense fields strongly affects the accurate determination of relative populations and anisotropy with polarized laser excitation [2]. Case *et al.* [3] developed the general mathematical formalism for analysis of LIF polarization in terms of rotational angular momentum alignment. Their treatment was applied by Greene and Zare [4] to derive explicit expressions for axially symmetric distributions and a perpendicular excitation-detection geometry. Saturation effects in the so called high-J limit (which allow a classical treatment of the molecule-laser beam interaction) were discussed for a few selected anisotropic distributions [5]. The rate-equational approach was applied in a subsequent study [6] to treat saturation in the resonance enhanced two-photon ionization process focusing on a comparison of the classical high-J and the quantal treatment. It was found that such a classical modelling is already applicable for rotational quantum numbers $J > 4$ for which the results deviate by less than 5% from the exact quantal calculation.

In a recent paper by Vigué and co-workers [7] coherent saturation effects in laser-induced-fluorescence are discussed. Coherent effects should be considered in situations where the laser coherence time is comparable to the time the molecule interacts with the laserfield. In the saturation regime where the Rabi frequency ω_R is much larger than the total decay rate out of the excited state the fluorescence intensity turns out to be proportional to ω_R, i.e. proportional to the square root of the laser power.

Many experiments employing LIF to extract the population and or the anisotropy of the nascent angular momentum distribution are carried out with pulsed lasers exhibiting rather limited coherence properties. Under these conditions coherence between the lower and excited state may be neglected and the more simple rate equational approach to model the laser-molecule interaction is appropriate.

In this paper we extend our previous [1] rate-equation-based treatment to cover other possible experimental configurations. Analytic expressions for extracting the alignment moments of an arbitrary anisotropic distribution are derived for the complete range of saturation conditions. In each configuration fluorescence is collected perpendicular to the probe laser beam and four signals are measured corresponding to orthogonal positions of two (excitation and detection) linear polarizers. Besides the rate equation approach and high J limitation we further assume that the signals are integrated over any Doppler profiles and that the chosen absorption and detection lines can be characterized as Q or PR transitions, i.e. that the spectroscopy is understood. Linear combinations of the results for PR and Q branches can be used to analyse mixed transitions.

Figure 1 shows three common experimental configurations and the corresponding four measured signals, I_{ad}, where a refers to the axis of the probe laser polarization and d the direction of the detector polarizer. Owing to cylindrical symmetry about the $\hat{z}$ axis, $I_{xx} \equiv I_{yy}$, $I_{zx} \equiv I_{zy}$, $I_{xz} \equiv I_{yz}$, and $I_{yx} \equiv I_{xy}$. Case (a) of Fig. 1 is most appropriate for beam gas scattering and case (c) for photofragmentation studies, each yielding four independent signals. Case (b) is less informative since two of the signals (I_{zx} and I_{zy}) are geometrically equivalent. In a photodissociation experiment where the dissociation and probe lasers propagate

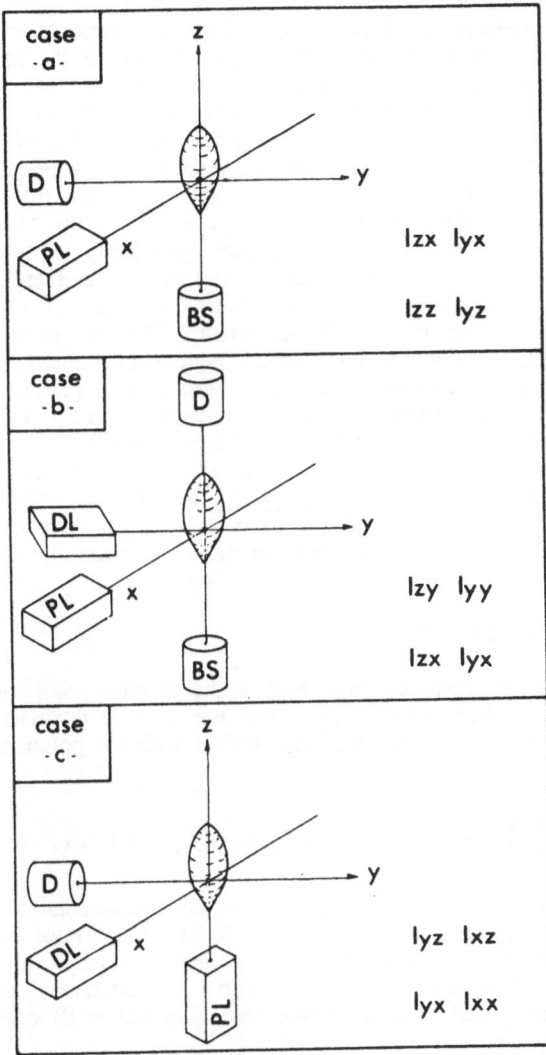

FIGURE 1. Experimental configurations for laser-induced-fluorescence probing of cylindrically symmetric nascent product anisotropic rotational angular momentum distributions. The shaded area represents an arbitrary anisotropic distribution defined around the z-axis. PL is the probe laser, D the fluorescence detector, BS the beam source in a beam gas experiment and DL the dissociation laser in a photofragmentation experiment. Also indicated for each configuration are the four measured intensities I_{ad} where a is the polarization direction of the probe laser and d the detected fluorescence polarization direction. Case (a) is the typical configuration for a beam-gas experiment, case (c) for a photofragmentation experiment whereas case (b) is appropriate for both. In photofragmentation studies the polarization of the dissociation laser is selected such that the transition moment lies along the z-axis.

collinearly the dissociation transition moment is set perpendicular to the detector (case (a)) or along the detector axis (case (b)). When the same laser both photodissociates the parent molecule and induces fluorescence in the probed fragment only two signals, I_{zz} and I_{zx} (= I_{zy}) are measured, meaning that just two moments can be determined uniquely (in the high fluence limit) from the two intensities. We have treated case (a) previously, deriving the saturation dependence of the contribution of the anisotropic moments to the fluorescence intensities I_{zz}, I_{zx}, I_{yz} and I_{yx}. Only one further moment (given in this text) I_{yy} (case (b)) or I_{xx} (case (c)) is needed to completely describe the possible geometries. Five independent signals, in principle, can be measured. As discussed previously [1] and in the following section at the high fluence limit only two moments and at the low fluence limit only three moments of the anisotropic distribution are accessible. Measuring all five moments in these extremes is certainly redundant. The system is not overdetermined, however, since in the intermediate regions of saturation higher moments become accessible. Conversion between configuration (a) and (b) is trivial for probes of photodissociation fragments, but not recommended for beam-gas experiments (especially using heated ovens or discharge beams!).

Another possible configuration, not shown in Fig. 1, directs the laser along the $\hat{x}$ axis and into the fluorescence detector, I_{yy}, I_{yz}, I_{zz} and I_{zy} (all independent) are measured in this case. McCaffery et al. [8] have shown this geometry can be used to probe product helicity when right and left circularly polarized light is employed.

2. RESULTS AND DISCUSSION

Following the treatment and notations from Ref. 1 the starting point for the calculation of the detected fluorescence I_{ad}, where the laser is polarized along $\hat{e}_a$ and the detector is positioned on the symmetry axis $\hat{z}$ with its polarization set along $\hat{e}_d$, is

$$I_{ad} = K \frac{3}{4\pi} \int d\Omega \frac{1}{2} \left\{ 1 - \exp[-6\bar{r}t f_a(\hat{e}_a.\mu)] \right\} f_d(\hat{e}_d.\mu) n(\hat{J}.\hat{z}), \qquad (1)$$

where K is a proportionality constant accounting for all instrumental factors, $\bar{r}$ is a directionally averaged pumping rate (see Eq. 4 of Ref.1), t the laser interaction time, μ the direction of the transition dipole moment and $n(\hat{J}.\hat{z})$ the molecular angular momentum distribution. The function $f(\hat{e}.\mu)$ describes the directional property of the excitation and fluorescence step. For a Q-type transition ($\Delta J = 0$) one has:

$$f(\mu.\hat{e}) = |\hat{J}.\hat{e}|^2, \qquad (2)$$

and for a P- or R-type transition ($\Delta J = \pm 1$)

$$f(\mu.\hat{e}) = \frac{1}{2}(1 - |\hat{J}.\hat{e}|^2). \qquad (3)$$

Because of the assumed cylindrical symmetry of the distribution $n(\hat{J}.\hat{z})$ can be expanded in Legendre polynomials, $P_{2l}(\hat{J}.\hat{z})$,

$$n(\hat{J}.\hat{z}) = \sum_{l=0}^{\infty} a_{2l} P_{2l}(\hat{J}.\hat{z}).\tag{4}$$

The moments a_{2l} describe the anisotropy of the angular momentum distribution (only the even moments can be measured with linearly polarized light) and they are the physical parameters of interest to be extracted from the measured intensity I_{ad}. For a totally isotropic distribution $a_0 = 1$ and $a_{2l} = 0$ for $l > 1$. As was discussed in the introduction in this experimental set-up (see case (b) of Fig. 1) one additional intensity, I_{yy}, as compared to case (a) of Fig. 1 can be measured. I_{yx} and I_{zx} ($= I_{zy}$) are the same in both cases and the analytical expressions and results for the saturation effects on the alignment parameters can be taken from Table 3, Table 4 and the corresponding graphs of Ref. 1.

A brief outline of the derivation of the analytical expressions for I_{yy} will be given here for a $Q\uparrow Q\downarrow$ transition. Defining $cos\chi \equiv \hat{J}.\hat{e}_a$ as the angle between $\hat{J}$ and the polarization of the laser and φ the azimuthal angle we have $\hat{J}.\hat{e}_d = cos\chi$ and $\hat{J}.\hat{z} = sin\chi cos\varphi$. Employing the polynomial expansion $P_{2l}(\hat{J}.\hat{z}) = \Sigma_{k=0} g_{l,k}(\hat{J}.\hat{z})^{2k}$ (see Eq. 2.5.13 of Ref. 9) and performing the integration over φ one obtains from Eq. 1

$$I_{ad} = 3K \sum_{l=0}^{\infty} a_{2l} \sum_{k=0}^{l} g_{l,k} \begin{bmatrix} 2k \\ k \end{bmatrix} 2^{-2k} \sum_{s=0}^{k} \begin{bmatrix} k \\ s \end{bmatrix} (-)^s \frac{1}{2} h_{s+1}(X_0),\tag{5}$$

with $X_0 = \sqrt{6\bar{r}t}$ and the function $h_s(X_0)$ is defined by the recursion relation

$$h_{k+1}(X_0) = \frac{1}{2k+3} + (e^{-X_0^2} - 1)\frac{1}{2X_0^2} + [(2k+1)\frac{1}{2X_0^2}]h_k(X_0),\tag{6a}$$

$$h_0(X_0) = 1 - \frac{1}{2}\sqrt{\pi}\,erf(X_0)\frac{1}{X_0}.\tag{6b}$$

Defining $I_{AV} = \frac{1}{2}K(1 - exp(-2\bar{r}t))$ and $I_{ad}^*(\bar{r}t) = I_{ad}(\bar{r}t)/I_{AV}(\bar{r}t)$ one obtains

$$I_{ad}^* = \sum_{l=0}^{\infty} a_{2l} C_{2l}^{ad}(\bar{r}t).\tag{7}$$

By employing the computer algebra routine REDUCE explicit analytical expressions for the coefficients $C_{2l}^{ad}(\bar{r}t)$ can be derived from Eqs. 5, 6 and 7. For a $Q\uparrow$ type excitation

$$C_{2l}^{yy}(\bar{r}t) = \frac{s_{2l}}{(1-e^{-2\bar{r}t})} \sum_{n=0}^{l+1}$$

$$\left[p_{2l,n} + q_{2l,n}\left[1 - \frac{\sqrt{\pi}\,erf(\sqrt{6\bar{r}t})}{2\sqrt{6\bar{r}t}} \right] + r_{2l,n}e^{-6\bar{r}t} \right](12\bar{r}t)^{n-l-1}\tag{8}$$

TABLE 1. Analytical expansion coefficients of $C_{2l}^{yy}(\bar{r}t)$ for PR↑ excitation and $l = 0,1,2$ or 3 (see Eq. 9). As an example for Q↓ fluorescence one obtains:

$$C_0^{yy}(\bar{r}t) = \frac{1}{1-e^{-2\bar{r}t}} \left[1 - \frac{1}{2\bar{r}t} \left(1 - \frac{\text{Daw}(\sqrt{3\bar{r}t})}{\sqrt{3\bar{r}t}}\right)\right]$$

	PR↑Q↓	PR↑PR↓
s_0	1	1/2
$p_{0,0}$	0	0
$q_{0,0}$	-3	3
$p_{0,1}$	1	-1
$q_{0,1}$	0	3
s_2	-1/20	1/40
$p_{2,0}$	0	0
$q_{2,0}$	135	135
$p_{2,1}$	-45	-45
$q_{2,1}$	15	60
$p_{2,2}$	4	-11
$q_{2,2}$	0	15
s_4	-9/64	9/128
$p_{4,0}$	0	0
$q_{4,0}$	525	525
$p_{4,1}$	-175	-175
$q_{4,1}$	90	195
$p_{4,2}$	5	-30
$q_{4,2}$	3	33
$p_{4,3}$	0	-3
$q_{4,3}$	0	3
s_6	-15/256	15/512
$p_{6,0}$	0	0
$q_{6,0}$	24255	24255
$p_{6,1}$	-8085	-8085
$q_{6,1}$	4725	8190
$p_{6,2}$	42	-1113
$q_{6,2}$	315	1260
$p_{6,3}$	-21	-105
$q_{6,3}$	5	110
$p_{6,4}$	0	-5
$q_{6,4}$	0	5

TABLE 2. Analytical expansion coefficients of $C_{2l}^{yy}(\bar{r}t)$ for $Q\uparrow$ excitation and $l=0,1,2$ and 3 (see Eq. 8). As an example, for $PR\downarrow$ fluorescence one obtains:

$$C_0^{yy}(\bar{r}t) = \frac{1}{1-e^{-2\bar{r}t}} \left[\frac{1}{12\bar{r}t} \left(3e^{-6\bar{r}t} + 3 \left(1 - \frac{\sqrt{\pi}\, \mathrm{erf}(\sqrt{6\bar{r}t})}{2\sqrt{6\bar{r}t}}\right) - 3\right) + 1 \right]$$

	Q↑Q↓	Q↑PR↓		Q↑Q↓	Q↑PR↓
s_0	1	1/2	s_6	15/256	15/512
$P_{0,0}$	-3	3	$P_{6,0}$	24255	-24255
$q_{0,0}$	3	-3	$q_{6,0}$	-24255	24255
$r_{0,0}$	3	-3	$r_{6,0}$	-24255	24255
$P_{0,1}$	1	-1	$P_{6,1}$	-4725	8190
$q_{0,1}$	0	3	$q_{6,1}$	4725	-8190
$r_{0,1}$	0	0	$r_{6,1}$	-3360	-105
			$P_{6,2}$	315	-1260
s_2	1/20	1/40	$q_{6,2}$	-315	1260
$P_{2,0}$	135	-135	$r_{6,2}$	-357	147
$q_{2,0}$	-135	135	$P_{6,3}$	-5	110
$r_{2,0}$	-135	135	$q_{6,3}$	5	-110
$P_{2,1}$	-15	60	$r_{6,3}$	-16	-5
$q_{2,1}$	15	-60	$P_{6,4}$	0	-5
$r_{2,1}$	-30	-15	$q_{6,4}$	0	5
$P_{2,2}$	-4	-11	$r_{6,4}$	0	0
$q_{2,2}$	0	15			
$r_{2,2}$	0	0			
s_4	9/64	9/128			
$P_{4,0}$	-525	525			
$q_{4,0}$	525	-525			
$r_{4,0}$	525	-525			
$P_{4,1}$	90	-195			
$q_{4,1}$	-90	195			
$r_{4,1}$	85	20			
$P_{4,2}$	-3	33			
$q_{4,2}$	3	-33			
$r_{4,2}$	8	-3			
$P_{4,3}$	0	-3			
$q_{4,3}$	0	3			
$r_{4,3}$	0	0			

and a PR↑ type excitation

$$C_{2l}^{yy}(\bar{r}t) = \frac{s_{2l}}{(1 - e^{-2\bar{r}t})} \sum_{n=0}^{l+1}$$

$$\left[P_{2l,n} + q_{2l,n}\left[1 - \frac{Daw(\sqrt{3\bar{r}t})}{\sqrt{3\bar{r}t}}\right]\right] (6\bar{r}t)^{n-l-1}, \tag{9}$$

where *erf* is the error function and *Daw* is the Dawson function (e.g. Ref. 10). The coefficients $P_{2l,n}, q_{2l,n}, \bar{r}_{2l,n}$ and s_{2l} turn out to be simple integers or fractions and are listed in Table 1 and Table 2 for $l=0,1,2$ and 3. To gain an impression about the saturation dependence of $C_{2l}^{yy}(\bar{r}t)$ in Fig. 2 C_{2l}^{yy} is plotted in the case of a Q↑ PR↓ transition and $\bar{r}t$ ranging from 10^{-4} to 10^{4}. Note that C_4 and C_6 have been magnified so the contribution to the LIF signal from these higher order moments is orders of magnitude smaller than the isotropic and the alignment contribution. As can be seen from Fig. 2 the saturation behaviour for all coefficients is quite smooth. In the low saturation regime, $\bar{r}t \to 0$, it can be shown [5] that there are only contributions to the fluorescence intensity from a_0, a_2 and a_4.

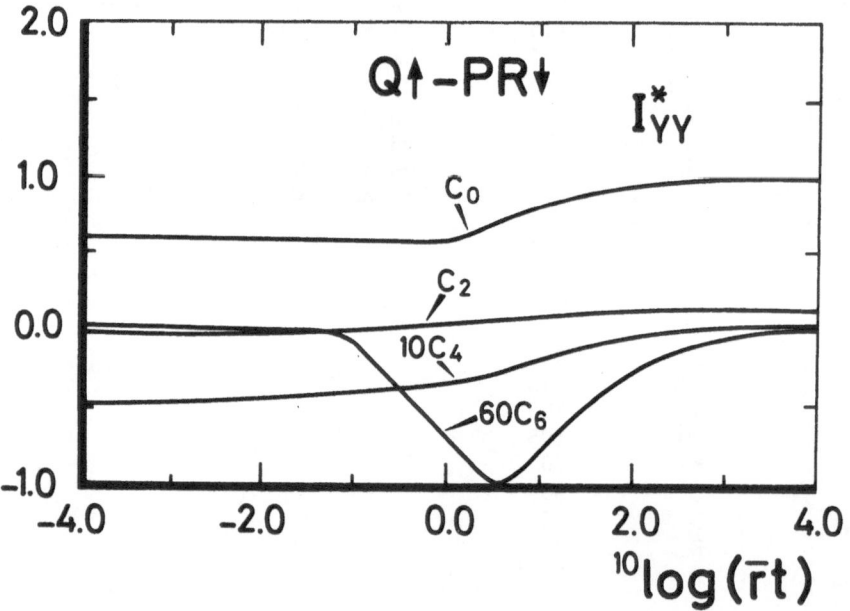

FIGURE 2. LIF saturation dependence of $C_{2l}^{yy}(\bar{r}t)$ for a Q↑ PR↓ transition (see Eq. 8 and Table 2) and $\bar{r}t$ ranging between 10^{-4} and 10^{4}. Notice that C_4^{yy} and C_6^{yy} have been multiplied by 10 and 60 respectively so the real contribution of a_4 and a_6 to the fluorescence is order(s) of magnitude smaller compared to a_0 and a_2.

All higher order contributions vanish for low saturation. In Table 3 the low saturation limit values of C_{2l}^{yy} are given. As was remarked earlier [1] care must be taken to use these low fluence limit values outside this region as for some cases deviations from the low fluence limit start to occur for saturation parameters $\bar{r}t \geqslant 0.05$. In the high fluence region there are only contributions from a_0 and a_2 and this is illustrated in Fig. 2 by the vanishing of C_4 and C_6 for $\bar{r}t \to \infty$. Again the high fluence limits of C_0 and C_4 can be calculated straightforwardly from Eq. 1 and are given in Table 3. In between the low- and high saturation regime there are contributions from higher order moments $a_{2l}(l=3,4,...)$ but due to the oscillatory character of the higher order Legendre polynomials the magnitude of the contributions to the fluorescence decrease rapidly with increasing order. As was discussed before in this set-up (case (b)) there are three independent intensities which means that from one data set $(I_{yx}, I_{zx} = I_{zy}, I_{xy})$ at a fixed saturation parameter one can at most extract the moments up to a_4. The fact that I_{zx} should be equal to I_{zy} might serve as an experimental verification of the reliability of the polarization set-up. Assuming contributions from higher moments than a_4 may be neglected a simple inversion of the set of Eqs. 5 results directly into a_0, a_2 and a_4. Because the saturation dependence of C_{2l} turns out to be quite smooth a very precise knowledge of the saturation parameter is not required. It can be estimated from Eq. 5 or if the polarization measurements have been done at several (widely varying) laser intensities the variation of the fluorescence intensity with laser intensity can be used to eliminate the unknown saturation conditions from the analysis.

TABLE 3. Low fluence limit value of C_{2l}^{yy}. In the limit $\bar{r}t \to 0$ all coefficients vanish except C_0^{yy}, C_2^{yy}, and C_4^{yy}.

	low fluence			
	c_0^{yy}	c_2^{yy}	c_4^{yy}	c_6^{yy}
Q↑Q↓	$\dfrac{9}{5}$	$-\dfrac{18}{35}$	$\dfrac{3}{35}$	0
Q↑PR↓	$\dfrac{3}{5}$	$-\dfrac{3}{70}$	$-\dfrac{3}{70}$	0
PR↑Q↓	$\dfrac{3}{5}$	$-\dfrac{3}{70}$	$-\dfrac{3}{70}$	0
PR↑PR↓	$\dfrac{6}{5}$	$\dfrac{6}{35}$	$\dfrac{3}{140}$	0

TABLE 4. High fluence limit value of C_{2l}^{yy}. In the limit $\bar{r}t \to \infty$ all coefficients vanish except C_0^{yy} and C_2^{yy}.

	high fluence	
	C_0^{yy}	C_2^{yy}
PR↑,Q↑-Q↓	1	$-\dfrac{1}{5}$
PR↑,Q↑-PR↓	1	$\dfrac{1}{10}$

In summary, the weight C_{2l}^{yy} of the Legendre moments a_{2l} to the fluorescence intensity $I_{yy}(= I_{xx})$ from any LIF study of an arbitrary anisotropic distribution which is cylindrically symmetric (the symmetry axis is defined here as the $\hat{z}$ axis) has been derived for the complete range of laser fluence. When combined with the results for the $I_{zx}(= I_{zy}), I_{zz}, I_{yx}(=I_{xy})$ and $I_{yz}(= I_{xz})$ moments presented previously [1], any experimental geometry employing linear polarization that conforms to the stated assumptions (rate equations, high-J limit,..) can be treated. Once the saturation parameter is estimated (or extracted from a series of measurements at different laser fluences) the signals can be directly converted to the moments of the anisotropic distribution.

This paper most readily extends our analytical treatment to photodissociation experiments. Applications to such systems including the analysis of NH rotational angular momentum polarization following the photodissociation of HNCO [11] are now in progress [12].

ACKNOWLEDGMENTS

The authors gratefully acknowledge the National Science Foundation, Grant INT 8619803 for support of this collaboration.

REFERENCES

[1] M.H.M. Janssen, D.H. Parker and S. Stolte, Chem. Phys., 113 (1987) 357.

[2] J. Allison, M.A. Johnson and R.N. Zare, Faraday Discuss. Chem. Soc., 67 (1979) 124.

[3] D.A. Case, G.M. McClelland and D.R. Herschbach, Mol. Phys., 35 (1978) 541.

[4] C.H. Greene and R.N. Zare, J. Chem. Phys., 78 (1983) 6741.

[5] R. Altkorn and R.N. Zare, Ann. Rev. Phys. Chem., 35 (1984) 265.

[6] D.C. Jacobs and R.N. Zare, J. Chem. Phys., 85 (1986) 5457.

[7] N. Billy, B. Girard, G. Gouedard and J. Vigué, Mol. Phys., 61 (1987) 65.

[8] A.J. Bain and A.J. McCaffery, J. Chem. Phys., 83 (1985) 2627; ibid., 83 (1985) 2632; ibid., 83 (1985) 2641.

[9] A.R. Edmonds, Angular momentum in quantum mechanics (Princeton Univ. Press, Princeton (1960)).

[10] M. Abramowitz and I.A. Stegun, Handbook of mathematical functions (Dover, New York (1965)).

[11] T.A. Spiglanin and D.W. Chandler, J. Chem. Phys., 87 (1987) 1577.

[12] M.H.M. Janssen, D.H. Parker and D.W. Chandler, private communication.

INVESTIGATION OF THE ORIENTATION DEPENDENCE OF THE F + H$_2$
REACTION USING SCALED BARRIER FUNCTIONS IN THE
ANGLE-DEPENDENT LINE-OF-CENTRES MODEL

J.N.L. Connor
Department of Chemistry
University of Manchester
Manchester M13 9PL
U.K.

W. Jakubetz
Institut für Theoretische Chemie und Strahlenchemie
Universität Wien
A-1090 Wien
Austria

ABSTRACT. The use of scaled barrier functions in the angle-dependent
line-of-centres (ADLOC) model has been investigated. Application has been made to
the F + H$_2$ → HF + H reaction on the Muckerman 5 potential energy surface in
two and three dimensions. Three different scaled barrier functions have been used,
together with an unscaled one obtained by inversion of zero impact-parameter
quasiclassical trajectory (QCT) data. None of the barrier functions studied is able to
reproduce consistently within ±2 standard deviations the QCT excitation functions
over an extended energy range. The calculations complement earlier work by the
authors and Whitehead in Ref. 10.

1. INTRODUCTION

There is a growing interest in the role of steric effects on the dynamics of
elementary chemical reactions [1-11,16-18]. In particular, the steric requirements of a
reaction can provide information about the anisotropy and associated cone of
acceptance of the potential energy surface.
    A simple model for reaction cross sections which incorporates steric effects is
the angle-dependent line-of-centres (ADLOC) model. This model was introduced by
Smith [1,2], and has been extended by Levine and Bernstein [3]. There are two
parameters in the ADLOC model: the collision diameter D and the angle-dependent
barrier height $E_0(\varphi)$.
    In an earlier paper with Whitehead [10], we have investigated the validity of
the ADLOC model by testing it against the results of quasiclassical trajectory (QCT)
calculations. The reaction considered was

$$F + H_2(v=0,j=0) \rightarrow HF + H,$$

J. C. Whitehead (ed.), Selectivity in Chemical Reactions, 323–339.
© 1988 by Kluwer Academic Publishers.

on the Muckerman 5 (M5) London-Eyring-Polanyi-Sato (LEPS) potential surface [12], and QCT computations were performed in one (collinear, 1D), two (coplanar, 2D) and three dimensions (3D) [10,13].

Two approaches were adopted for the ADLOC calculations of Ref. 10. In the first approach, the theoretically best-justified choices for $E_0(\varphi)$ and D were employed in the calculations. In particular for $E_0(\varphi)$, we used the ground-state classical vibrationally adiabatic (VA) barrier height, calculated along cuts through the potential surface at fixed collision angles φ. For D, the physically motivated value $D \equiv 1.96$ Å was employed, which was obtained by extrapolation of the cut-off impact-parameters for the opacity functions in the 2D and 3D QCT computations [10]. Comparison of the ADLOC excitation functions with the 2D and 3D QCT results revealed some significant discrepancies [10], which however provide insight into the assumptions and limitations of the ADLOC model.

In the second approach, D was allowed to vary, being used as a least-squares fitting parameter to the 2D and 3D QCT excitation functions. Three further $E_0(\varphi)$ were employed in these calculations [10], in addition to the VA barrier function. In this second approach, the errors in the ADLOC assumptions are evidently (in part) absorbed in to $E_0(\varphi)$ and D, which therefore become effective quantities.

The purpose of the present Chapter is to further pursue this second approach by using linearly scaled barrier functions in the ADLOC theory. The scaling parameter is then used in a least-squares fitting procedure, whilst keeping D fixed at a physically meaningful value (e.g. 1.96 Å). We also present some additional results based on our earlier paper [10]; in particular we include in the ADLOC calculations a barrier function obtained by inversion of the QCT opacity functions at zero impact-parameter [10].

This Chapter is arranged as follows. In Section 2 we outline the 3D and 2D ADLOC theory. The various barrier functions we use are defined and discussed in Section 3. Results from the 3D ADLOC theory are presented in Section 4, whilst the 2D results are in Section 5. Our concluding remarks are in Section 6.

2.   ADLOC THEORY

In this section, we summarize some of the key equations of the ADLOC model [2,3,10]. The 3D and 2D cases will be considered separately. It should also be noted that ADLOC incorporates concepts similar to sudden transition state theory [16].

2.1   Three Dimensions

For a fixed value of φ, we can use the notion of a fixed-angle partial cross section $\sigma_3(E_T|\varphi)$, where E_T is the translational (collision) energy [3,16]. Averaging $\sigma_3(E_T|\varphi)$ over φ then gives the total cross section

$$\sigma_3(E_T) = \int_0^\pi \sigma_3(E_T|\varphi)\sin\varphi d\varphi \Big/ \int_0^\pi \sin\varphi d\varphi,$$

$$= \frac{1}{2} \int_{0}^{\pi} \sigma_3(E_T|\varphi)\sin\varphi d\varphi. \qquad (1)$$

For a homonuclear target molecule, we can evidently restrict φ to the interval $0 < \varphi < \pi/2$ in Eq. 1, provided we also multiply by 2 to allow for double-ended attack. This gives

$$\sigma_3(E_T) = \int_{0}^{\pi/2} \sigma_3(E_T|\varphi)\sin\varphi d\varphi. \qquad (2)$$

Next we assume that the line-of-centres (LOC) expression for the partial cross section [14,15] is valid at each φ:

$$\sigma_3(E_T|\varphi) = \pi D^2[1 - E_0(\varphi)/E_T]\theta(E_T-E_0(\varphi)), \qquad (3)$$

where $\theta(x)$ is a unit step function: $\theta(x) = 1$ for $x > 0$, $\theta(x) = 0$ for $x < 0$. Combining Eqs. 2 and 3 gives

$$\sigma_3(E_T) = \pi D^2 \int_{0}^{\pi/2} [1-E_0(\varphi)/E_T]\theta(E_T-E_0(\varphi))\sin\varphi d\varphi, \qquad (4)$$

$$= \pi D^2 \int_{0}^{\varphi_{max}} [1-E_0(\varphi)/E_T]\sin\varphi d\varphi. \qquad (5)$$

In going from Eq. 4 to Eq. 5, we have assumed a collinearly dominant reaction with $E_T > E_0(\varphi=0)$ (otherwise we would have $\sigma_3(E_T) \equiv 0$). The angle $\varphi_{max}(E_T)$ in Eq. 5 is the maximum angle of attack at which reaction can occur, and is obtained from

$$E_0(\varphi_{max}) = E_T \quad \text{if } E_T < E_0(\pi/2), \qquad (6)$$

or

$$\varphi_{max} = \pi/2 \quad \text{if } E_T > E_0(\pi/2). \qquad (7)$$

In our calculations for $F + H_2$, we are always in an energy range to which Eq. 6 applies. Finally, the orientation-averaged opacity function $P_3(b|E_T)$ for the ADLOC model is given by [2,3,10]

$$P_3(b|E_T) = 1 - \cos\varphi. \qquad (8)$$

The accessible angle φ for a given E_T and impact-parameter b is obtained from the equation for the energy along the line-of-centres [2,3,10]:

2.2 Two Dimensions

In 2D, the expression for the cross length corresponding to Eq. 1 is [10]

$$\sigma_2(E_T) = \int_0^{2\pi} \sigma_2(E_T|\varphi)d\varphi \Big/ \int_0^{2\pi} d\varphi,$$

$$= (1/2\pi) \int_0^{2\pi} \sigma_2(E_T|\varphi)d\varphi,$$

$$= (1/\pi) \int_0^{\pi} \sigma_2(E_T|\varphi)d\varphi, \qquad (10)$$

where the last line follows from the fact that the angular ranges $0 < \varphi < \pi$ and $\pi < \varphi < 2\pi$ are clearly equivalent. If we allow for double-ended attack, Eq. 10 becomes

$$\sigma_2(E_T) = (2/\pi) \int_0^{\pi/2} \sigma_2(E_T|\varphi)d\varphi. \qquad (11)$$

The 2D LOC expression for the partial cross length at a fixed value of φ is [10]

$$\sigma_2(E_T|\varphi) = 2D[1-E_0(\varphi)/E_T]^{1/2} \; \theta(E_T-E_0(\varphi)), \qquad (12)$$

so that Eq. 11 becomes

$$\sigma_2(E_T) = (4D/\pi) \int_0^{\varphi_{max}} [1-E_0(\varphi)/E_T]^{1/2}d\varphi, \qquad (13)$$

where $\varphi_{max}(E_T)$ is again obtained from Eq. 6. Finally the 2D orientation-averaged opacity function is [10]

$$P_2(b|E_T) = 2\varphi/\pi, \qquad (14)$$

where φ is given by Eq. 9.

3. ANGLE-DEPENDENT BARRIER FUNCTIONS

We have used the same barrier functions $E_0(\varphi)$ as in our earlier
work [10], namely

(a) The ground-state classical vibrationally adiabatic barrier height, evaluated for b=0
along cuts through the potential surface at fixed values of φ. The angle φ is
the angle between the relative velocity vector and a vector along the axis of
the H_2 molecule. These VA barriers have already been computed by Jellinek
and Pollak [16], and we fitted their data to a quartic polynomial in $\kappa \equiv 1-\cos\varphi$
[10] [this barrier function will be denoted $E_0^{VA}(\varphi)$].

(b) The potential barrier obtained directly from the M5 surface for b=0, fixed φ
collisions [denoted $E_0^{CM}(\varphi)$], shifted globally so that $E_0^{CM}(\varphi=0) = B = 2.469$
kJ mol^{-1}, where B is the collinear VA barrier height. The shifting is done so
as to ensure that $E_0^{CM}(\varphi)$ has the correct QCT threshold for reaction in
1D, 2D and 3D [10]. This barrier function can be generated directly from the
M5 surface at any value of φ. We have also fitted the CM data to a cubic
polynomial in κ [10]. However the ADLOC results using it are virtually
identical to those from $E_0^{CM}(\varphi)$, so this fitted CM barrier function will not be
considered separately in what follows.

(c) The barrier height for fixed φ, when the F atom points towards one of the H
atoms [denoted $E_0^{H}(\varphi)$]. Again this barrier function is shifted globally so that
$E_0^{H}(\varphi=0) = B$. Note that for this barrier function, φ is defined to be the
angle <(FHH).

(d) The ADLOC model allows the b=0 QCT data to be inverted to yield a barrier
function, which will be denoted $E_0^{INV}(\varphi)$ [10]. The 3D inversion follows
from Eqs. 6, 8 and 9 on setting b=0, i.e.

$$P_3(b=0 \,|\, E_T) = 1 - \cos[\varphi_{max}(E_T)]. \qquad (15)$$

An iterative solution of Eq. 15 then yields $E_0^{INV}(\varphi_{max})$. It proved difficult to
accurately fit $E_0^{INV}(\varphi)$ to a polynomial in κ; however an accurate (smoothed)
fit was possible using cubic splines. This fitted barrier function will be
denoted $E_0^{INV}sf(\varphi)$
    In 2D, the inversion uses Eqs. 6, 9 and 14 for b=0, namely

$$P_2(b=0 \,|\, E_T) = 2\varphi_{max}(E_T)/\pi, \qquad (16)$$

together with a correction to allow for non-reactive back-reflection from the
strong interaction region of the M5 surface [10]. We found the resulting 2D
inverted barrier data could also be accurately fitted by $E_0^{INV}sf(\varphi)$, and so for
the INVsf barrier function, we do not distinguish between 2D and 3D.
    Figure 1 (upper panel) shows $E_0^{VA}(\varphi)$, $E_0^{CM}(\varphi)$, $E_0^{H}(\varphi)$, $E_0^{INV}(\varphi)$
and $E_0^{INV}sf(\varphi)$ plotted against $\kappa \equiv 1 - \cos\varphi$. The $E_0^{INV}(\varphi)$ barrier function
has a discrete representation, in which the horizontal bars corresponds to ±1
standard deviation of the 3D, b=0, QCT data. It is evident from Fig. 1 that
the VA, CM, H and INV barrier functions are all different from one another.

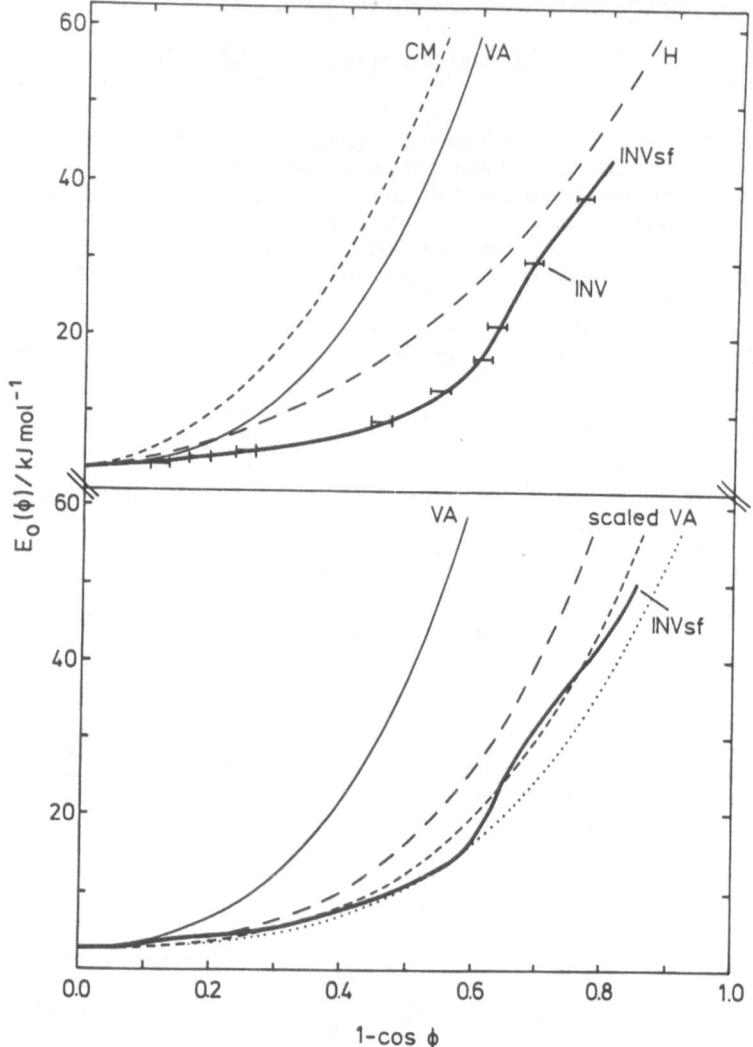

FIGURE 1. Plots of barrier functions versus 1 - cosφ. Upper panel: The curves represent the unscaled barrier functions $E_0^{VA}(\varphi)$, $E_0^{CM}(\varphi)$, $E_0^H(\varphi)$ and $E_0^{INVsf}(\varphi)$. The horizontal bars are a pointwise representation of $E_0^{INV}(\varphi)$ obtained by inversion of the 3D, b=0, QCT data. The widths of the bars correspond to ±1 standard deviation. Lower panel: Comparison of $E_0^{INVsf}(\varphi)$ with $E_0^{VA}(\varphi)$ and the scaled functions $^sE_0^{VA}(\varphi)$. Short dashed line (------): s = 0.2855 [direct fit of $^sE_0^{VA}(\varphi)$ to $E_0^{INV}(\varphi)$]. Long dashed line (— — —): s = 0.390, D ≡ 1.96 Å (s fitted to 3D QCT cross sections). Dotted line (······): s = 0.230, D = 1.78 Å (simultaneous fit of s and D to 3D QCT cross sections).

An interesting question is whether a scaling of the VA, CM and H barrier functions can map them onto INV. We have used the following linear scaling

$$^sE_0^X(\varphi) = B + s[E_0^X(\varphi) - B], \qquad X = VA, CM, H, \qquad (17)$$

where s is the scaling parameter. This question is investigated in Table 1 and Fig. 1 (lower panel), in which s is used as a least-squares fitting parameter. Note that the eight $\kappa_i \equiv 1-\cos\varphi_i$ values in Table 1 correspond to an energy range of 3.35 $\langle$ E_T $\langle$ 38.5 kJ mol^{-1}. To assess the accuracy of the fit, we also report the root mean square (rms) or standard error

$$\Delta = \left\{(1/N) \sum_{i=1}^{N} [E_0^{INV}(\kappa_i) - {}^sE_0^X(\kappa_i)]^2\right\}^{1/2}, \quad X = VA, CM, H, \quad (18)$$

where N = 8 is the number of κ_i values.

The magnitude of s for the fits in Table 1 (columns 4,7,9) follows the order CM < VA < H. This is the order expected from the upper panel in Fig. 1. The barrier function most accurately mapped onto INV is VA (for which Δ = 1.17 kJ mol^{-1}) followed by CM (Δ = 2.11 kJ mol^{-1}) and H (Δ = 3.05 kJ mol^{-1}). However if a smaller E_T range is considered, 3.35 $\langle$ E_T $\langle$ 13.4 kJ mol^{-1}, we found that the order changes to CM (Δ = 0.35 kJ mol^{-1}), H (Δ = 0.40 kJ mol^{-1}) and VA (Δ = 0.66 kJ mol^{-1}). The other entries in Table 1 are discussed in Section 4.1.

4.   3D ADLOC CROSS SECTIONS

In this Section, we investigate the use of the scaled barrier functions $^sE_0^X(\varphi)$, X = VA, CM, H (see Eq. 17) in the ADLOC model. In particular, the scaling parameter s is used to least-squares fit, for a range of E_T, the ADLOC cross sections (Eq. 5) to the QCT results. The collision diameter D is kept fixed in these calculations.

We also compare with the method used in our earlier work [10], in which $E_0^Y(\varphi)$, Y = VA, CM, H, INVsf are left unscaled and D is the least-squares fitting parameter. For the VA barrier function, we have also allowed for the simultaneous fitting of s and D. It is also interesting to compare the ADLOC results with classical reactive infinite-order-sudden (CRIOS) cross sections [17], because the assumption of fixed-angle collisions is common to both approaches.

4.1   Comparison with 3D QCT Cross Sections

Table 2 and Fig. 2 compare eight different ADLOC excitation functions with the QCT results for the energy range 3.35 $\langle$ E_T $\langle$ 38.5 kJ mol^{-1}. Also included is the rms error associated with each ADLOC excitation function.

First it should be noted that none of the ADLOC excitation functions agrees with the QCT data to within ±2 standard deviations for more than four of the eight E_T values in the Table. However the Δ values in Table 2 do show that the VA, H, CM cross sections obtained by scaling $E_0^X(\varphi)$ for D ≡ 1.96 Å are more accurate than those calculated by fitting D with s ≡ 1. For both these cases the VA cross sections are more accurate than the H or CM ones. Even better VA

TABLE 1. INV and INVsf barrier functions together with scaled versions of VA, H and CM. All energies are in kJ mol⁻¹. Entries for D in round brackets denote fixed values. Entries that are starred agree with $E_0^{INV}(\kappa)$ to within ±2 standard deviations. Note that $\kappa \equiv 1 - \cos\varphi$.

quantity	$E_0^{INV}(\kappa)$	$E_0^{INVsf}(\kappa)$	$sE_0^{VA}(\kappa)$ fitted to: $E_0^{INV}(\kappa)$	3D QCT	3D QCT	$sE_0^{H}(\kappa)$ fitted to: $E_0^{INV}(\kappa)$	3D QCT	$sE_0^{CM}(\kappa)$ fitted to: $E_0^{INV}(\kappa)$	3D QCT
s	-	-	0.2855	0.390	0.230	0.773	0.720	0.277	0.285
D/Å	-	-	-	(1.96)	1.78	-	(1.96)	-	(1.96)
$\kappa_1 = 0.118$	3.35 ± 0.17^a	3.35*	2.85	3.01*	2.80	3.97	3.89	3.51*	3.51*
$\kappa_2 = 0.180$	4.18 ± 0.21	4.18*	3.39	3.72	3.18	5.10	4.90	4.31*	4.35*
$\kappa_3 = 0.249$	5.02 ± 0.25	5.02*	4.27	4.94*	3.93	6.65	6.36	5.65	5.77
$\kappa_4 = 0.456$	9.20 ± 0.59	9.22*	10.3*	13.1	8.79*	13.8	13.1	12.9	13.2
$\kappa_5 = 0.546$	13.4 ± 0.9	13.2*	15.4	20.2	12.9*	18.4	17.3	17.8	18.3
$\kappa_6 = 0.629$	21.8 ± 2.8	21.3*	22.2*	29.5	18.4*	23.7*	22.2*	23.5*	24.1*
$\kappa_7 = 0.685$	30.1 ± 2.0	30.0*	28.2*	37.6	23.2	27.9*	26.1*	28.1*	28.9*
$\kappa_8 = 0.763$	38.5 ± 1.4	38.6*	38.7*	52.0	31.7	34.6	32.4	35.7*	36.7*
Δ/kJ mol⁻¹	-	0.18	1.17	6.71	3.69	3.05	3.41	2.11	2.53

ᵃ Uncertainties denote ±1 standard deviation and are estimated from the horizontal bars in Fig.1.

TABLE 2. 3D QCT and ADLOC cross sections. All energies are in kJ mol^{-1} and cross sections in Å^2. Entries for s and D in round brackets denote fixed values. ADLOC cross sections that are starred agree with the QCT results to within ±2 standard deviations.

quantity	3D QCT	ADLOC $\sigma_3(E_T)$ with barrier function:							
		INVsf	VA	H	CM	VA	H	CM	VA
s	-	(1)	(1)	(1)	(1)	0.390	0.720	0.285	0.230
D/Å	-	1.85	2.36	2.12	2.69	(1.96)	(1.96)	(1.96)	1.78
E_T = 3.35	0.22 ± 0.03ᵃ	0.19*	0.27*	0.12	0.12	0.32	0.13	0.19*	0.34
E_T = 4.18	0.57 ± 0.06	0.47*	0.61*	0.33	0.34	0.69*	0.37	0.49*	0.74
E_T = 5.02	0.94 ± 0.05	0.78	0.94*	0.57	0.58	1.05	0.64	0.79	1.10
E_T = 9.20	2.45 ± 0.06	2.25	2.21	1.74	1.76	2.35*	1.87	2.06	2.43*
E_T = 13.4	3.47 ± 0.06	3.24	3.09	2.70	2.72	3.21	2.83	2.98	3.27
E_T = 21.8	4.50 ± 0.06	4.48*	4.33	4.17	4.18	4.37	4.29	4.31	4.38*
E_T = 30.1	5.16 ± 0.06	5.20*	5.21*	5.30	5.29	5.16*	5.38	5.29	5.14*
E_T = 38.5	5.54 ± 0.06	5.76	5.90	6.22	6.22	5.78	6.27	6.08	5.72
Δ/Å^2	-	0.15	0.21	0.48	0.47	0.15	0.43	0.31	0.14

ᵃ Uncertainties denote ±1 standard deviation.

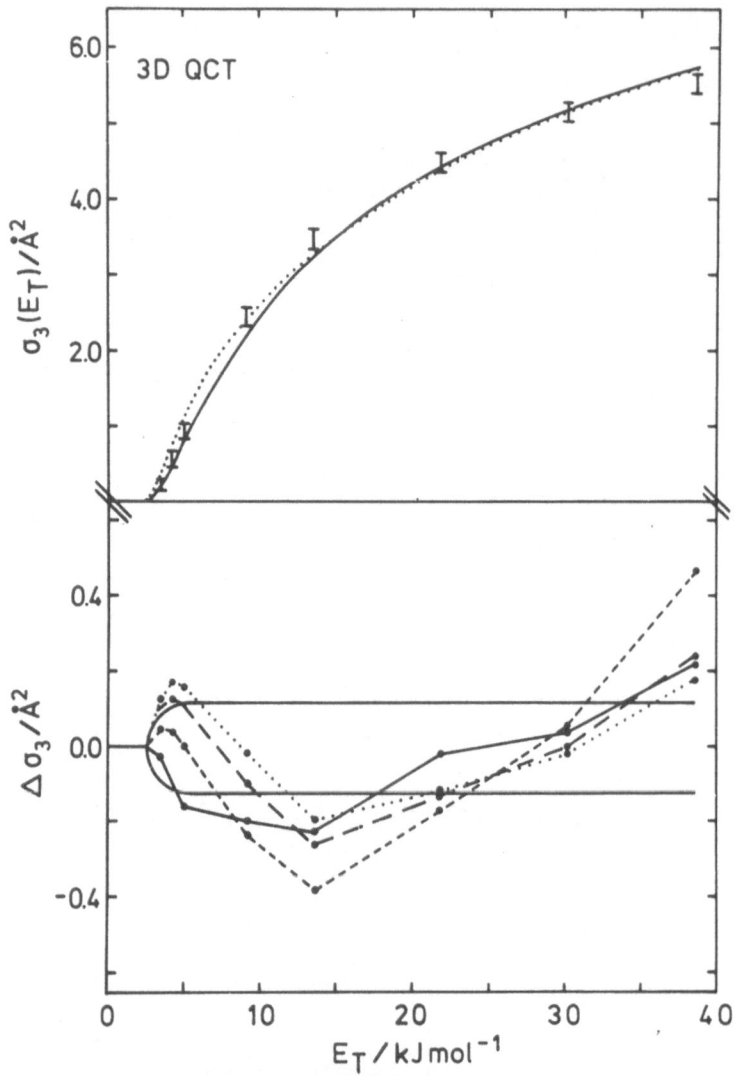

FIGURE 2. Comparison of ADLOC and 3D QCT cross sections. <u>Upper panel:</u> The vertical bars are QCT cross sections ±1 standard deviation. Solid line (————): INVsf with $s \equiv 1$, $D = 1.85$ Å. Dotted line (······): scaled VA barrier function with $s = 0.230$, $D = 1.78$ Å. <u>Lower panel:</u> Deviations of ADLOC cross sections from 3D QCT results, $\Delta\sigma_3(E_T) \equiv \sigma_3^{ADLOC}(E_T) - \sigma_3^{QCT}(E_T)$. The full line near ±0.12 Å^2 spans ±2 standard deviations in the QCT results. Solid line (————): INVsf with $s \equiv 1$, $D = 1.85$ Å. Long dashed line (— — —): scaled VA, with $s = 0.390$, $D \equiv 1.96$ Å. Short dashed line (------): scaled VA, with $s \equiv 1$, $D = 2.36$ Å. Dotted line (······): scaled VA, with $s = 0.230$, $D = 1.78$ Å. The calculated points have been joined by straight lines as an aid to visualization.

results can be obtained if both s and D are allowed to vary (see column 10).

If a smaller E_T range is considered than in Table 2, then much closer agreement between ADLOC and QCT can be achieved. For example, we found the ADLOC excitation function for the scaled CM barrier function even agreed to within ±1 standard deviation of the QCT results at all five E_T values in the range $3.35 < E_T < 13.4$ kJ mol^{-1}.

Table 1 gives the explicit values of ${}^sE_0{}^X(\varphi)$ corresponding to the last four columns of Table 2. It also shows (as expected) that the s value for fixed $D \equiv 1.96$ Å, which is obtained by mapping ${}^sE_0{}^X(\varphi)$ onto $E_0{}^{INV}(\varphi)$, differs from the s value obtained by fitting ${}^sE_0{}^X(\varphi)$ to the QCT cross sections via the ADLOC σ_3. However the differences are only marked for the VA case. The lower panel of Fig. 1 shows the unscaled and scaled VA barrier functions together with $E_0{}^{INV}sf(\varphi)$.

4.2 Comparison with 3D CRIOS Cross Sections

The ADLOC and CRIOS cross sections are compared in Table 3 and Fig. 3. We present results at two fixed values of D. The first of these, $D \equiv 1.96$ Å is the same as in our earlier work [10] and is also used in Tables 1 and 2 for the ADLOC/QCT comparison. The second value, $D \equiv 1.86$ Å, was obtained by extrapolating to high E_T the CRIOS opacity functions in Fig. 3 of Ref. 16. The 3D QCT limiting impact parameter of $D \equiv 1.96$ Å is expected to be greater than the CRIOS one, because the target diatomic molecule can adopt a more favourable orientation during the approach of the attacking atom (this phenomenon is called "funnelling" in Ref. 18).

Table 3 shows that no ADLOC excitation function is able to consistently reproduce to within ±5% the CRIOS results [although $E_0{}^{VA}(\varphi)$ comes close]. The scaled barrier results are comparable to (for VA), or slightly worse (for H and CM), than those which set $s \equiv 1$ and fit D. Apart from this difference, the trends in Table 3 are similar to those in Table 2. In particular, the VA cross sections are more accurate than the corresponding H or CM ones. Although the rms errors in Table 3 are about half those in Table 2, it should be noted that the CRIOS cross sections are themselves about half the QCT ones, so the relative errors are approximately the same.

5. 2D ADLOC CROSS LENGTHS

The ADLOC cross lengths are compared with the 2D QCT results in Table 4 and Fig. 4. All the ADLOC cross lengths include a correction for non-reactive back-reflection from the strong interaction region of the M5 surface [10]. This correction arises because the 2D QCT opacity functions computed in Ref. 10 show evidence of 3-5% back reflection at low impact-parameters, an effect which is absent from the corresponding 3D QCT results. Notice also that the 2D QCT cross length for $E_T = 2.51$ kJ mol^{-1} is not used in the ADLOC least-squares fitting of s or D. This is because there are large relative errors in the ADLOC cross lengths at this near-threshold energy, which are associated with uncertainties in the choice of B.

Table 4 shows that the ADLOC results for ${}^sE_0{}^X(\varphi)$ with fixed $D \equiv 1.96$ Å are of equal or better accuracy than those which fit D for $s \equiv 1$. Again the VA excitation functions are more accurate than those for H or CM. It is interesting to compare the 3D and 2D ADLOC results in Tables 2 and 4

TABLE 3. 3D CRIOS and ADLOC cross sections. All energies are in kJ mol⁻¹ and cross sections in Å². Entries for s and D in round brackets denote fixed values. ADLOC cross sections that are starred agree with the CRIOS results to within ±5%.

quantity	CRIOS	ADLOC $\sigma_3(E_T)$ with barrier function:									
		VA	H	CM	VA	H	CM	VA	H	CM	VA
s	–	(1)	(1)	(1)	1.42	1.53	0.730	1.78	1.80	0.870	0.540
D/Å	–	1.71	1.61	2.04	(1.86)	(1.86)	(1.86)	(1.96)	(1.96)	(1.96)	1.50
E_T = 4.14	0.26	0.31	0.18	0.19	0.30	0.17	0.20	0.29	0.16	0.19	0.33
E_T = 7.03	0.89	0.85*	0.67	0.68	0.84	0.64	0.71	0.83	0.62	0.70	0.88*
E_T = 8.95	1.20	1.13	0.97	0.98	1.12	0.94	1.01	1.12	0.92	1.00	1.16*
E_T = 15.1	1.80	1.79*	1.75*	1.76*	1.79*	1.76*	1.77*	1.79*	1.75*	1.77*	1.79*
E_T = 22.5	2.28	2.33*	2.47	2.46	2.36*	2.54	2.46	2.36*	2.54	2.47	2.30*
Δ/Å²	–	0.05	0.17	0.17	0.06	0.20	0.14	0.06	0.22	0.15	0.04

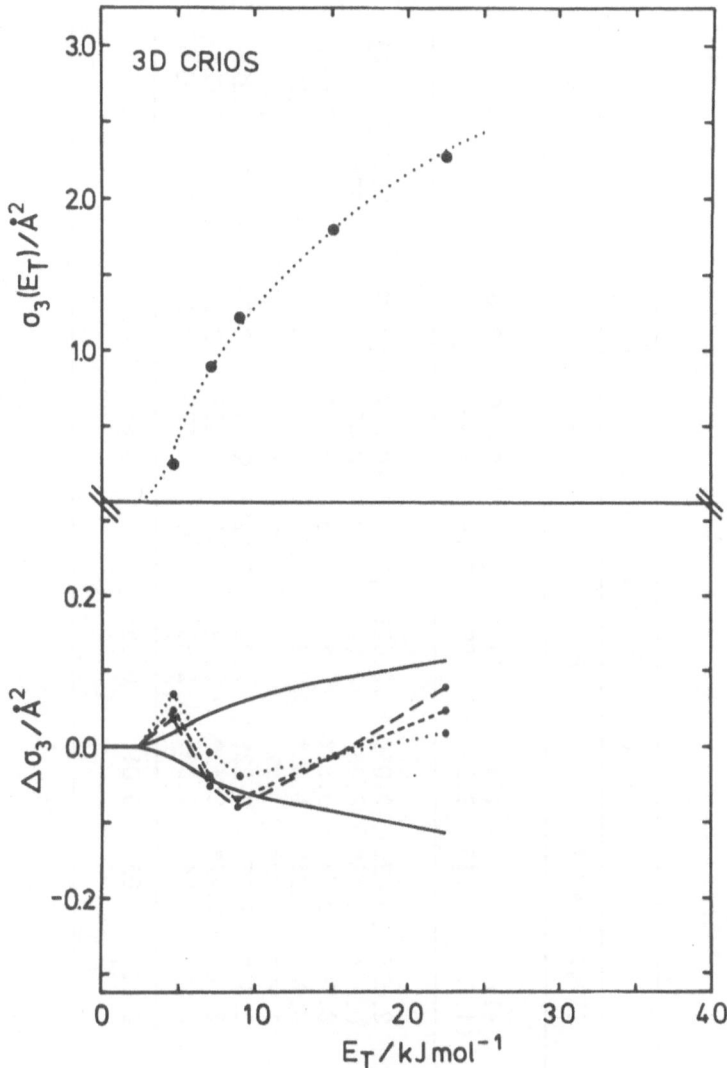

FIGURE 3. Comparison of ADLOC and 3D CRIOS cross sections. The notation is the same as in Fig. 2. The diverging full lines in the lower panel span ±5% of the CRIOS cross sections. Note that no INVsf barrier function is available for the CRIOS data.

336

TABLE 4. 2D QCT and ADLOC cross lengths. The ADLOC results all include a correction for non-reactive back-reflection from the strong interaction region of the M5 surface. All energies are in kJ mol⁻¹ and cross lengths in Å. Entries for s and D in round brackets denote fixed values. Entries in square brackets denote fixed values for the near threshold energy, E_T = 2.51 kJ mol⁻¹, are not used in the least-squares fitting of s or D. ADLOC cross lengths that are starred agree with the QCT results to within ±2 standard deviations.

quantity	2D QCT	ADLOC $\sigma_2(E_T)$ with barrier function:								
		INVsf	INVsf	VA	H	CM	VA	H	CM	VA
s	-	(1)	(1)	(1)	(1)	(1)	0.450	0.800	0.315	0.903
D/Å	-	(1.85)	1.815	2.34	2.09	2.69	(1.96)	(1.96)	(1.96)	2.28
E_T = 2.51	0.055 ± 0.011ᵃ	0.028	[0.026]	[0.050]*	[0.021]	[0.021]	[0.057]*	[0.022]	[0.027]	[0.052]*
E_T = 3.14	0.38 ± 0.02	0.39*	0.38*	0.46	0.29	0.30	0.48	0.30	0.37*	0.46
E_T = 4.18	0.83 ± 0.03	0.74	0.73	0.83*	0.62	0.63	0.86*	0.64	0.74	0.83*
E_T = 5.02	0.99 ± 0.03	0.96*	0.95*	1.02*	0.81	0.83	1.06	0.83	0.94*	1.03*
E_T = 13.4	1.97 ± 0.03	1.98*	1.94*	1.87	1.77	1.79	1.88	1.80	1.83	1.87
E_T = 21.8	2.27 ± 0.03	2.31*	2.27*	2.22*	2.22*	2.22*	2.21*	2.23*	2.23*	2.22*
E_T = 30.1	2.41 ± 0.03	2.46*	2.42*	2.42*	2.49	2.48	2.39*	2.50	2.46*	2.42*
E_T = 38.5	2.48 ± 0.03	2.59	2.54*	2.56	2.68	2.67	2.52*	2.68	2.63	2.55
Δ/Å	-	0.06	0.05	0.06	0.16	0.14	0.06	0.14	0.09	0.06

ᵃ Uncertainties denote ±1 standard deviation.

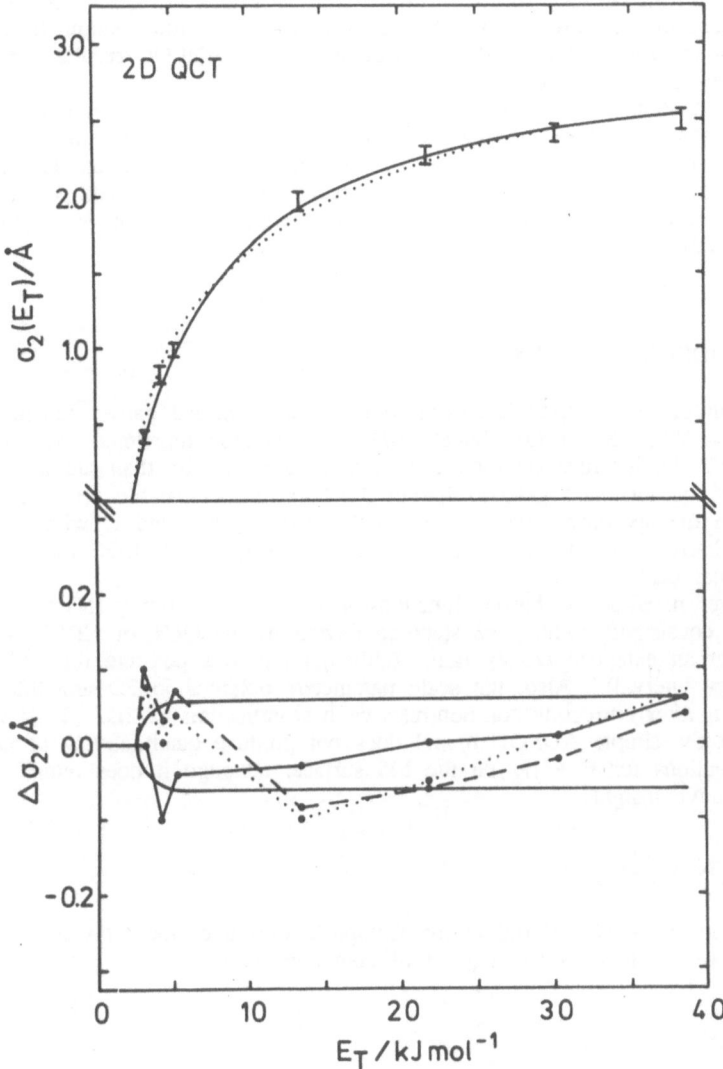

FIGURE 4. Comparison of ADLOC and 2D QCT cross lengths. The notation is the same as in Fig. 2. In the lower panel, the results for scaled VA with $s \equiv 1$, $D = 2.34$ Å are almost coincident with those for scaled VA, with $s = 0.903$, $D = 2.28$ Å [dotted line (·····)] and have been omitted for clarity of presentation. The solid line (———) represents INVsf with $s \equiv 1$, $D \equiv 1.85$ Å. The long dashed line (– – – –) represents scaled VA with $s = 0.450$, $D \equiv 1.96$ Å.

respectively. It can be seen that the fitted s and D values are quite similar in 3D and 2D (except for the VA case when both parameters are fitted simultaneously). If the correction for back-reflection is not made to the 2D ADLOC results, then the agreement between 3D and 2D gets worse.

Finally it should be noted that as in 3D (see Tables 2 and 3), none of the ADLOC excitation functions in Table 4 is able to reproduce consistently the QCT results to within ±2 standard deviations, although the overall agreement is clearly better in 2D than in 3D. In particular, the results for the INVsf and VA barrier functions are of acceptable overall accuracy. This suggests that the assumptions of the ADLOC model are more closely obeyed in 2D than in 3D.

6. CONCLUDING REMARKS

The main aim of this Chapter has been to use linearly scaled barrier functions, $^sE_0{}^X(\varphi)$, X = VA, CM, H to calculate ADLOC excitation functions. We found that scaling the VA barrier function produced more accurate results than did a scaling of the H or CM barriers, and gave excitation functions of comparable (or greater) accuracy than the approach employed originally [10], which fitted D with s ≡ 1. Thus scaled barrier functions are a useful way of fitting the ADLOC model to QCT or experimental data.

However none of the barrier functions investigated (including INVsf) was able to reproduce consistently within ±2 standard deviations the QCT or CRIOS excitation functions over an extended energy range (although this was possible for CM over a smaller energy interval). Also, the scale parameters obtained in 2D and 3D are different, even after correcting for non-reactive back-reflection in 2D. Thus we conclude that the simple ADLOC model does not produce quantitatively accurate excitation functions for F + H_2 on the M5 surface, although it does remain useful for its qualitative insights.

ACKNOWLEDGMENTS

The calculations were carried out at the Computer Centre of the University of Vienna and we are grateful for a grant of computer time.

REFERENCES

[1] I.W.M. Smith, Kinetics and Dynamics of Elementary Gas Reactions (Butterworths, London, 1980) pp. 81-84.

[2] I.W.M. Smith, J. Chem. Educ., 59 (1982) 9.

[3] R.D. Levine and R.B. Bernstein, Chem. Phys. Lett., 105 (1984) 467.

[4] N.C. Blais, R.B. Bernstein and R.D. Levine, J. Phys. Chem., 89 (1985) 10.

[5]   R.B. Bernstein, J. Chem. Phys., **82** (1985) 3656.
      S.E. Choi and R.B. Bernstein, J. Chem. Phys., **83** (1985) 4463.
      N.C. Blais and R.B. Bernstein, J. Chem. Phys., **85** (1986) 7030.

[6]   I. Schechter, R. Kosloff and R.D. Levine, Chem. Phys. Lett.,
      121 (1985) 297; J. Phys. Chem., **90** (1986) 1006.
      I. Schechter and R.D. Levine, Int. J. Chem. Kinet., **18** (1986)
      1023.

[7]   G.T. Evans, R.S.C. She and R.B. Bernstein, J. Chem. Phys., **82**
      (1985) 2258.
      R.S.C. She, G.T. Evans and R.B. Bernstein, J. Chem. Phys., **84**
      (1986) 2204.
      G.T. Evans, J. Chem. Phys., **86** (1987) 3852; **87** (1987) 3865.

[8]   R.D. Levine and R.B. Bernstein, Chem. Phys. Lett., **132** (1986)
      11.

[9]   W.C. Gardiner, Jr. and R.D. Levine, Chem. Phys., **111** (1987) 1.

[10] J.N.L. Connor, J.C. Whitehead and W. Jakubetz, J. Chem. Soc.
      Faraday Trans. 2, **83** (1987) 1703.

[11] M.H.M. Janssen and S. Stolte, J. Phys. Chem., (in the press).

[12] J.T. Muckerman, Theor. Chem. (Theory of Scattering: Papers in
      Honor of Henry Eyring), **6A** (1981), 1.

[13] J.N.L. Connor, W. Jakubetz, J. Manz and J.C. Whitehead, Chem.
      Phys., **39** (1979) 395.

[14] R.D. Present, Proc. Natl. Acad. Sci. USA, **41** (1955) 415;
      Kinetic Theory of Gases, (McGraw-Hill, New York, 1958), Section
      8.2.

[15] M.A. Eliason and J.O. Hirschfelder, J. Chem. Phys., **30** (1959)
      1426.

[16] J. Jellinek and E. Pollak, J. Chem. Phys., **78** (1983) 3014.

[17] J. Jellinek and M. Baer, Chem. Phys. Lett., **82** (1981) 162.

[18] R.M. Whitnell and J.C. Light, J. Chem. Phys., **86** (1987) 2007.

CHEMISTRY AND VAN DER WAALS COMPLEXES

C. Jouvet and M. Boivineau
Laboratoire de Photophysique Moléculaire
Université de Paris-Sud
91405 Orsay Cedex
France

1. INTRODUCTION

Van der Waals complexes have become, during the last few years, very useful in the study of reactive collision processes. They have been used for two main reasons.
   The selection of the initial geometry that can be achieved in the entrance channel.
   The solvent effect of the reaction in a single collision regime.
   The control of the geometry of the collision is readily achieved when the chemical reaction is studied in a van der Waals complex. Here the cold complex formed by the collision partners, in a supersonic expansion, is optically excited to begin the reactive process. The starting point on the reactive surface is defined by the geometry of the ground state complex. It is worthwhile noting that the optical excitation also defines the starting time of the reaction.
   For the reaction B+CD → BCD occurring in the ground state, an AB partner in an AB-CD complex is photodissociated forcing the B fragment into collision with the CD partner within the defined geometry of the complex leading to a B-CD → BCD aligned reaction [1]. In this kind of experiment a precise control of the initial geometry and of the kinetic energy of the colliding partners is obtained. Here the kinetic energy can be changed by varying the energy of the photon, and the influence of the A partner upon the studied reaction has to be considered.
   The method had been developed earlier [2] to study chemical reactions occurring when one of the partners is in an excited state: $A^*+BC \rightarrow AB+C$. Here the optical excitation of the A-BC complex is no longer localised on one of the components of the complex; instead the intermediate state of the reaction is directly excited. In addition to the well defined geometry of the starting point on the reactive surface one obtains direct spectroscopic information on the surface itself. As opposed to the preceding case the kinetic energy is zero and can not be varied. We present in the following part of the paper the main results obtained by this technique on the Xenon halogen systems.
   There are ionic reactions where site selective chemistry has been studied by this technique using a one colour resonant two photon ionization [3]. For example in the fluorobenzene methanol (d_3) complex the first photon selects one isomer in the $S_0 \rightarrow S_1$ transition. The second photon ionizes the complex, and depending which isomer is selected the ionic complex or the anisole ion is observed.

J. C. Whitehead (ed.), Selectivity in Chemical Reactions, 341–352.
© *1988 by Kluwer Academic Publishers.*

Van der Waals complexes can also be used to study the influence of an extra partner on the reactivity.

The first type of these studies concerns the photodissociation of neutral or ionic molecules. For example in the I_2-Ar system when I_2 is excited above the dissociation limit of the B state, the dissociation is inhibited by a cage-like effect and the fragment is found in the high vibrational levels of the B state [4]. This effect is not observed in the ionic Ar_n-X^+ cluster [5].

The complexes can also be used as a colliding partner, and the extra partner can change the reactivity of the bare monomer. As an example in the reaction Ba + $CO_2 \rightarrow$ BaO + CO it was found that the reactive cross section for a CO_2 dimer is between four and eight times larger than for the monomer [6].

The studies of the reactivity of the van der Waals complex are still in their early stage, and promise, in our opinion, significant developments.

We present in the following sections some results on the Xe halogen systems that we have recently obtained.

2. THE RARE GAS(Rg)+HALOGEN(X_2) REACTIONS WITHIN THE Rg-X_2 VAN DER WAALS COMPLEX

During the last ten years, reactive collisions Rg*+X_2 or Rg+X_2* leading to the RgX* excimer through the harpoon mechanism have been widely studied [7]. These systems (mainly Xe+Cl_2 or Br_2) have been the first ones where the collision complex was excited in the gas phase by one photon [8] or two photon excitation [9]. Here the same two-photon excitation technique has been used for the excitation of the Xe-X_2 Van der Waals complexes (X_2 = Cl_2, Br_2, I_2). In this experiment only one frequency doubled dye laser was used, and visible light (from 584 to 500 nm) or the second harmonic or both were focused into the supersonic jet where the complexes are formed. The resulting fluorescence was detected with 10 nm resolution. Our results will then be compared with the collisional ones.

In the following we shall consider that the excited states of the complex are correlated to the electronic states of the components (Xe or X_2), this assumption being incorrect in some cases since the localised character of the excitation may be lost. Nevertheless it is useful to characterise the excitation process. In this approximation when 2 UV photons (second harmonic) are used we can consider that the Rydberg states of the halogen are excited. Alternatively when one first harmonic and one second harmonic photon are used in order to excite the complex we can consider the valence states of the halogen, the first photon (visible) brings the complex into a state which correlates to the $B(^3\Pi_u)$ state, and the second UV photon induces the transition between the B state and the valence states (Fig. 1).

The main reactive features obtained for these systems are summarized as follows:

For all systems the fluorescence of the first excited state Xe-$X_2(B^3\Pi_u)$ has not been observed as it has been for lighter rare gas [10]. Here the excitation does not bring enough energy into the system to allow a chemical reaction. The absence of fluorescence arises from the electronic predissociation of the halogen induced by the presence of the heavy atom in the complex.

2.1 The Xe-I_2 Complex

For the Xe-I_2 system no fluorescence of the excimer XeI* or of I_2* resulting from

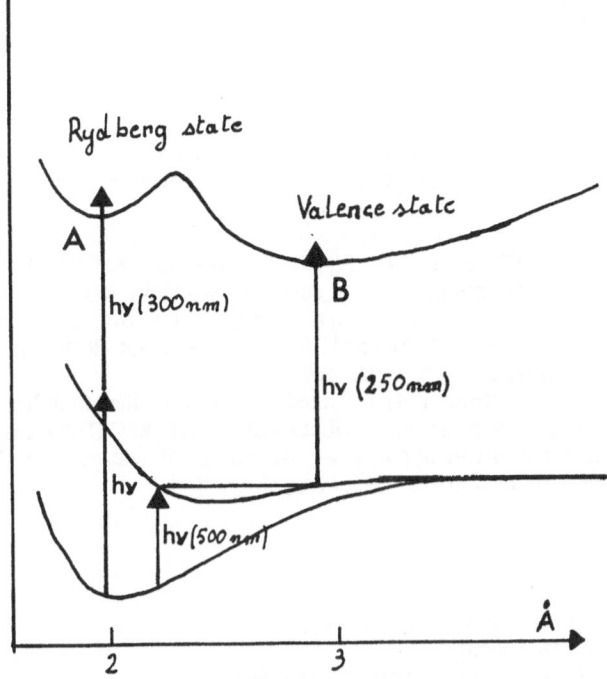

FIGURE 1. Schematic of the excitation process.
a) direct two photon (UV) excitation of the Rydberg states
b) excitation of the valence states; 1 photon visible on the X→B transition, one photon UV from the B state to the valence state.

the dissociation of the Xe-I_2 excited complex have been observed whatever state of I_2 (valence or Rydberg) is excited; yet the existence of the complex was proven by extra lines occurring in the photoionisation spectrum [11]. It seems that when the complex is excited the major channel is a black channel, presumably Xe+I+I in contrast with the full collision where XeI* or I_2* is obtained.

2.2 The Xe-Br_2 Complex

The excimer XeBr(B,C) fluorescence was obtained by optical-optical double excitation of the complex Xe-Br_2 (excitation of the B state of Br_2 by the first harmonic of the laser and followed by the excitation of the valence state by the second harmonic [12]). The action spectrum, excitation of the complex and observation of the XeBr(B→X) fluorescence, is structureless and starts at the energetical threshold. The fluorescence spectrum obtained with an excess of energy of 3800 cm^{-1} is characteristic of a cold XeBr(B,C), and presents a small emission assigned to the Br_2(D′→A′) transition. As the reaction occurs at the energetical threshold, then no barrier to the reaction was found. However, when the Rydberg states of the Br_2 are excited no fluorescence is observed (direct two UV photon absorption). Probably the excitation of the Rydberg states leads to the dissociation into Xe+Br+Br. This is similar to the collisional case where no reactions starting from the Rydberg states of Br_2 have been observed.

2.3 The Xe-Cl_2 Complex

For the Xe-Cl_2 complex the XeCl(B) excimer is obtained by two or three-photon absorption from 480 up to 270 nm (Fig. 2), but the signal is only strong in the 310 to 285 nm region.

In this spectral region two action spectra have been recorded depending on which XeCl state (B or C) is monitored [13]. In one the XeCl(B→X) transition at 308 nm is observed as the pump laser is scanning while in the second the XeCl(C→A) emission is collected at 345 nm. Only in the first spectrum is a vibrational progression observed (380 cm^{-1} for one photon at 300 nm), and these bands seem to be intrinsically broad (50 cm^{-1}).

The fluorescence spectrum [14] obtained by exciting the complex at 290 nm was characteristic of the emission of a vibrationally cold XeCl(B,C) excimer although the excitation process has given about 2 eV of excess of energy into the system, and this is not likely to be due to the cluster formation.

3.   DISCUSSION

We shall now discuss the following points:
- The XeCl(B,C) action spectra.
- The differences in the reactivity between the halogens.

3.1 The Xe-Cl_2 Action Spectra

Surprisingly no increase of the absorption cross section is observed in the 440 nm region where the three-photon atomic transition $Xe(^1S_0 \rightarrow ^3P_1)$ is expected (Fig. 2). One interpretation is that we are in a situation distinct from the Hg-Cl_2 case (Ref. 2a) where the absorption is correlated to the atomic transition since the crossing radius between the ionic and covalent potentials coincides with the ground state equilibrium distance. In the Xe-Cl_2 case the Franck-Condon accessible region from the ground state equilibrium distance (3.5 Å) is much shorter than the crossing radius (≈ 5 Å). The excitation leads to well inside the intermediate state where some other crossings occur. The approximation of considering a localised excitation of the xenon atom is no more a good approximation. Another interpretation can be that any signal coming from the direct Xe excitation is weaker than the excitation processes via Cl_2 since in the first case it is a three photon non resonant absorption process while in the second the first step is quasiresonant (X→B absorption).

The threshold in the spectra at 460 nm does not correspond to the energetical threshold of the reaction (3 photons of 570 nm) but seems to correspond to the three photon absorption of the first Rydberg state of Cl_2 [15]. Moreover, using the same optical-optical double resonance technique as for Br_2 to access the valence states (one photon in the 500 nm region leading to the B state and its second harmonic leading to the valence $F(O_g^+)$ state) we were not able to detect any signal from the XeCl* excimer. This was also very surprising since in the collisional conditions only these valence states seem to lead to reaction when the Cl_2 is the excited partner.

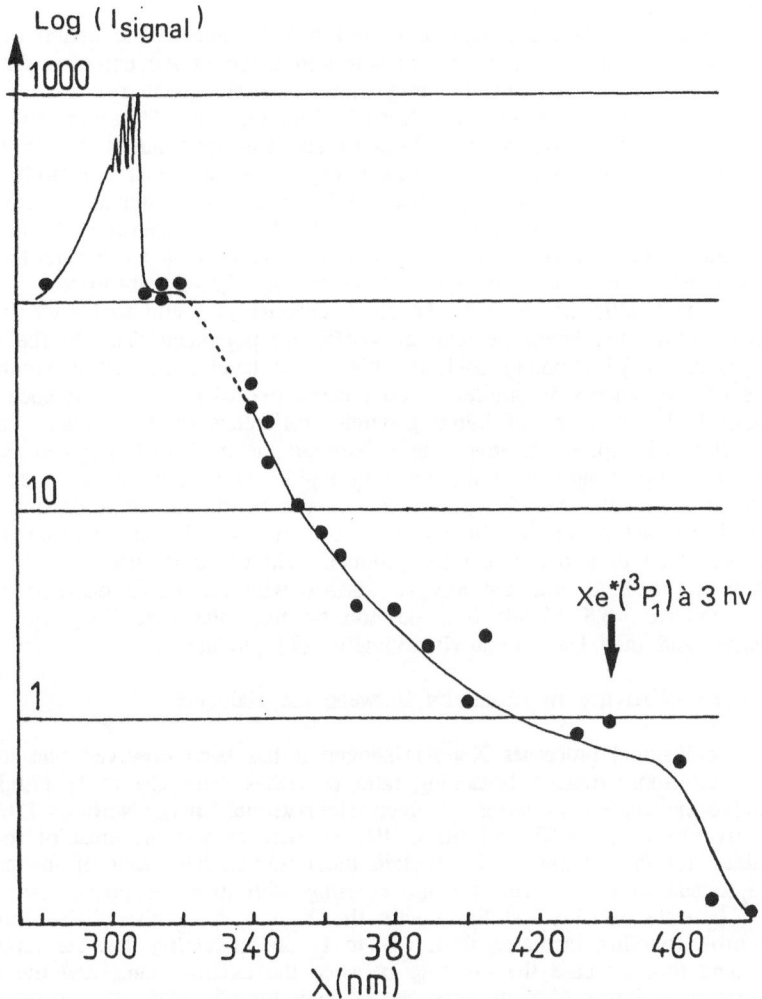

FIGURE 2. Action spectrum of the Xe-Cl_2 complex. Excitation of the complex *via* a two or three photons observation of the XeCl(B→X) transition. Experimental conditions: P_0=20 atm, X/D=100, D=0.2 mm, mixture: He(99%), Cl_2(0.5%), Xe(0.5%).

One possible explanation can be found in an experimental artefact. The UV power in this spectral region was roughly ten times smaller than that used in the experiment with Br_2, due to the relative inefficiency of the doubling crystal in this region. We can think that the predissociation of the B state of Cl_2 induced by the presence of Xe is then quicker than the absorption (B→F) process at this low UV power.

As we will see, this high threshold may be due to the nature of the potential energy surface and the very specific selection made in the entrance channel when the complex is excited (no kinetic energy and a specific geometry).

The difference between the XeCl(B) and the XeCl(C) action spectra in the 300 nm region has been discussed in Ref. 13 and was rationalised by the fact that the initial geometry is fixed in the complex experiment and that a two-photon transition is used. The interpretation was that the excimer can be obtained either by a two-photon excitation, enhanced by the $Cl_2(X^1\Sigma_g \to {}^1\Pi_u)$ transition leading to an unstructured intermediate state, or by a sequential process leading to the vibrational structure which has the same structure as the XeCl(X$\to$B) absorption.

In the collisional process the XeCl excimer is found to be vibrationally excited; vibrational levels as high as v=100 are populated [7]. In the complex case the product is vibrationally cold; the highest v' level populated is around 10, though the excess of energy is similar in both cases ($\approx$2 eV). The difference was explained [14] in terms of initial geometry difference in the entrance channel. In the collisional experiment involving a harpoon mechanism, the system reaches the triatomic ion-pair surface in the crossing region, here about 5 Å. Assuming a spectator model the Xe$^+$-Cl$^-$ pair which is the asymptotic limit of the XeCl(B,C) states is formed at $\approx$5 Å which is far from their equilibrium distance ($\approx$2.9 Å). This will lead to a highly excited product. On the other hand, in the complex, the system is promoted onto the ion-pair surface with the Xe-Cl distance of the ground state complex ($\approx$3.5 Å) which is not too far from the XeCl(B,C) equilibrium distance, and then leads to a vibrationally cold product.

3.2 The Difference in Reactivity Between the Halogens

In the collisional processes Xe(^{3}P)+halogen it has been observed that the reactive versus electronic transfer branching ratio decreases from Cl_2 to I_2 [7,h,i]. A tentative explanation in terms of electronic potential energy surfaces [16] has been recently given, which fits all the collisional data as well as most of the ones obtained for the complex. To explain the observed behaviour of the complexes we need to add to these surfaces some crossing with lower repulsive states (for example, the $^1\Pi_u$ state of Cl_2) and those with the $X_{1/2} + X_{1/2}$ dissociative limit. Since the spin-orbit coupling increases from Cl_2 to I_2 and assuming that the repulsive shape is the same in each case the crossing between the excited states and the repulsive state will occur at larger X-X distance for I_2 than for Cl_2 (Fig. 3). Moreover in the complex case a strong electronic relaxation in the halogen is induced by the rare gas, since the presence of the rare gas breaks the $D_{\infty h}$ symmetry of the halogen. This is shown by the absence of the fluorescence from the Xe-X$_2$ B($^3\Pi_u$) state and by the structureless action spectrum obtained by double optical resonance in the Xe-Br$_2$ case [12].

The observed reaction in a collision and its absence in the Xe-I$_2$ complex may be then due to the entrance channel geometry difference. In the collision process there are still some trajectories which avoid the crossing, leading to the black channel Xe+X*+X*. In the complex case, the system is promoted on the reactive surface at a short distance where these crossings are seen in the exit channel for I$_2$ when the I-I distance stretches. This is not so in the case of Cl$_2$ since the crossing occurs at shorter Cl-Cl distance (Fig. 4).

These surfaces may also explain the difference in reactivity between the Br$_2$ and Cl$_2$ when the valence states are excited in the complex. Assuming that the

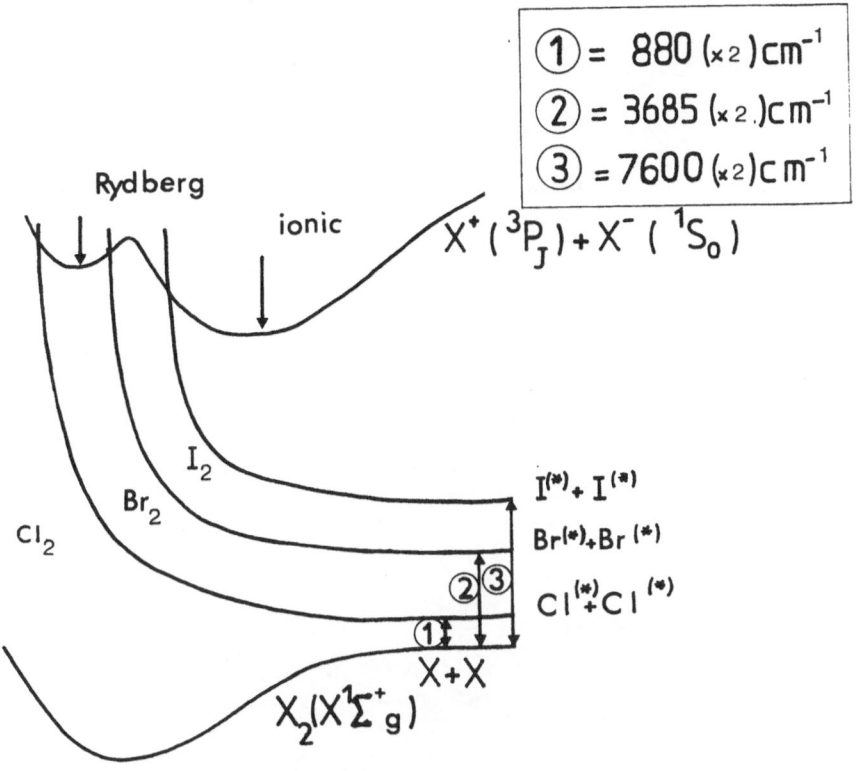

FIGURE 3. Schematic of the intersections between the repulsive curves and the Rydberg and valence states in Cl_2, Br_2, I_2.

reaction occurs at the crossing between the XeX* and the X_2* surfaces it can be seen from Fig. 4 that the reaction with Br_2 can lead to XeBr* with no vibrational energy thus at the energetical threshold. On the other hand in the case of Cl_2 the reaction will lead to an excess of vibrational energy (crossing occurring at a larger Xe-X distance), and then products will not occur at the energetical threshold but at higher energy. This explanation is obviously only tentative and will need further theoretical and experimental work to be confirmed.

4. EXCITATION OF THE Xe-Cl$_2$ COMPLEX BY AN ELECTRICAL DISCHARGE

In the preceding Sections we have reported some experiments on the reactivity of Xe-X$_2$ systems in which we had some selectivity in the geometry in the entrance channel through the excitation of the van der Waals complex, but also in the excitation process by the use of an optical excitation. We report here some preliminary results of the excitation of the van der Waals complex by an electrical

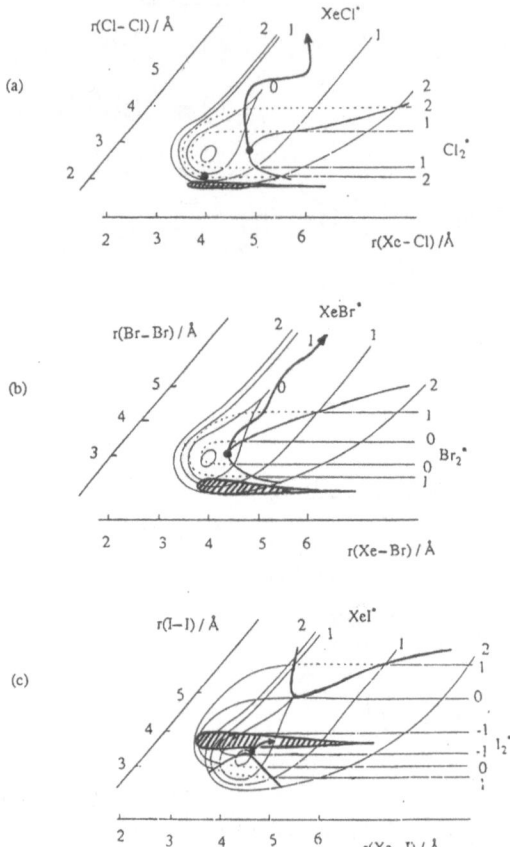

FIGURE 4. Qualitative potential energy surface for Xe-Cl$_2$(a), Xe-Br$_2$(b), Xe-I$_2$(c) systems.

The ionic surfaces Xe$^+$+X$_2^-$ leading to XeX(B)+X have been obtained by combining the asymptotic diatomic potential XeX(B) and X$_2^-$($^2\Sigma_u$). The excited molecular halogen surfaces have been taken from spectroscopic or *ab initio* data for the X-X coordinate. The short range Xe-X$_2$ repulsions have been approximated by those of the XeX(X) ground state. The zero energy corresponds to the XeX(B,v=0) energy.

The dark line corresponds to the intersection of the two potential surfaces.

The striped area corresponds to the intersection (crudely estimated) of the reactive surface with the dissociative channel leading to Xe+X+X. This area is larger at short Xe-X distance to take into account the electronic couplings induced by the presence of the Xe atom.

a) trajectory for the reaction Cl$_2$(D')+Xe leading to a vibrationally excited product.
b) trajectory for the reaction Br$_2$(D')+Xe leading to a vibrationally cold product.
c) trajectory for the excitation of the Xe-I$_2$ complex leading to Xe+I+I.

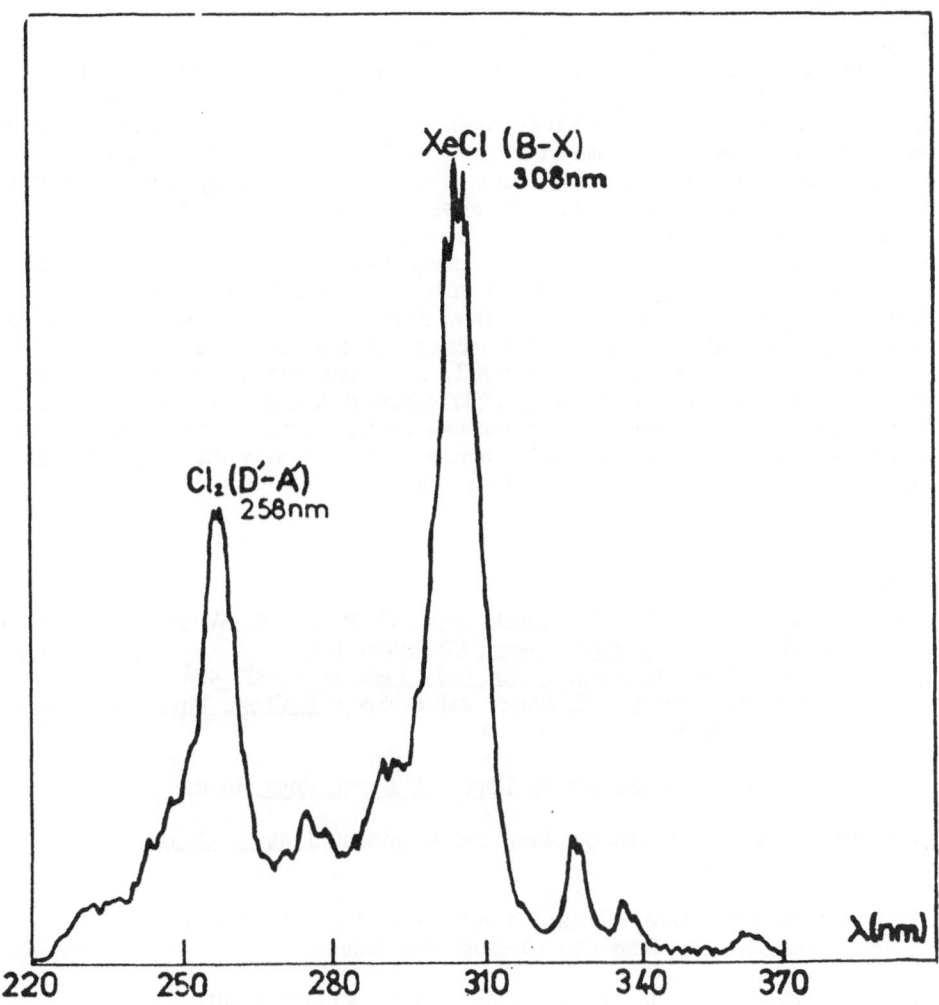

FIGURE 5. Fluorescence spectrum obtained by exciting the complex by an electrical discharge.

discharge. Here only the geometrical selectivity will be conserved but there will be no selectivity in the excitation process. Briefly two small electrodes are placed perpendicular to the jet at a distance of 4 mm. A 6kV discharge takes place within the jet, the experimental condition of the expansion being the same as the one used for the optical excitation. The resulting fluorescence is collected and dispersed through a monochromator.

The fluorescence spectrum obtained is presented in Fig. 5. The two main peaks can be assigned to the uncomplexed $Cl_2(D' \rightarrow A')$ emission and to the XeCl(B→X) emission. The XeCl(B→X) peak is fairly narrow which indicates that the product is not highly vibrationally excited. No emission from the C state is recorded (C→A emission around 345 nm). From this spectrum we can deduce that the reaction leading to the excimer still exists when the complex is excited with a non-selective process, and that some selectivity upon the electronic and vibrational energy distributions still persists. Let us mention that this preliminary result suggests the possibility of making a CW excimer laser based upon the excitation of the Xe-Cl_2 complex due to the very rapid flow obtained in the jet, since it is very easy to make electrical discharge at the low pressure obtained in the jet.

This work on Xe-X_2 is still not fully completed, and will need some two independent colour experiments for example. Nevertheless it shows that the van der Waals technique brings some new information on the reactive surface, and that the selectivity upon the entrance channel geometry leads to a very drastic effect upon the vibrational energy distribution of the products.

REFERENCES

[1] S. Buelow, M. Noble, G. Radhakrishnan, H. Reisler, C. Wittig
     and G. Hancock, J. Phys. Chem., 90 (1986) 1015.
[2] a: C. Jouvet and B. Soep, Chem. Phys. Lett., 96 (1983) 426.
     b: W.H. Breckenridge, C. Jouvet and B. Soep, J. Chem. Phys.,
     84 (1986) 1443.

[3] B. Brutschy, Cl. Janes and H. Eggert, J. Chem. Phys., in the
     press.
[4] J.M. Philippoz, H. van der Berg and R. Monot, J. Phys. Chem.,
     91 (1987) 2545.

[5] A.J. Stace and D.M. Bernard, 'Reactions of Molecular Ions in
     Association with Inert Gas Clusters', this volume.

[6] a: J. Nieman and R. Naaman, Chem. Phys., 90 (1984), 407,
     b: J. Nieman and R. Naaman, J. Chem. Phys., 84 (1986) 3825

[7] a: Y.C. Yu, D.W. Setser and H. Horiguchi, J. Phys. Chem.,
     87 (1983) 2199 and Refs. therein.
     b: K. Tamagake, J.H. Kolts, D.W. Setser, J. Chem. Phys.,
     71 (1979) 1264.
     c: J. Le Calvé, M.C. Castex, B. Jordan, G. Zimmerer, T. Moller
     and D. Haaks, Photophysics and Photochemistry above 6 eV,
     Elsevier Science Publishers, Amsterdam (1985) p. 639.
     d: T. Ishiwata, A. Tokunaga and I. Tanaka, Chem. Phys. Lett.,
     112 (1984) 356.
     e: B.V. O'Grady and R.J. Donovan, Chem. Phys. Lett., 122 (1985)
     503.

f: J.P.T. Wilkinson, E.A. Kerr, K.P. Lawley, R.J. Donovan, D. Shaw, A. Hopkirk and I. Munro, Chem. Phys. Lett., 130 (1986) 213.
g: K. Johnson, R. Pease, J.P. Simons, P.A. Smith and A. Kvaran, J. Chem. Soc., Faraday Trans. 2, 82 (1986) 1281.
h: K. Johnson, J.P. Simons, P.A. Smith, C. Washington and A. Kvaran, Mol. Phys., 57 (1986) 255.
i: K. Johnson, J.P. Simons, P.A. Smith and A. Kvaran, J. de chimie-physique, 84 (1987) 371.
j: A.C. Vikis and D.J. Le Roy, Can. J. Chem., 51 (1973) 1207; and Refs. therein.

[8] a: V.S. Dubov, Y.E. Lapsker, A.N. Samoilova and L.V. Gurvich, Chem. Phys. Lett., 83 (1981) 518.
b: H.P. Grieneisen, Hu Xue-Jing and K.L. Kompa, Chem. Phys. Lett., 82 (1981) 421.
c: B.E. Wilcomb and R. Burnham, J. Chem. Phys., 74 (1981) 6784.

[9] a: J.K. Ku, G. Inoue, D.W. Setser, J. Phys. Chem., 87 (1983) 2989.
b: D.W. Setser and J.K. Ku, in Photophysics and Photochemistry above 6 eV, Elsevier Science Publishers, Amsterdam (1985), p. 621.

[10] a: R.E. Smalley, D.H. Levy and L. Wharton, J. Chem. Phys., 64 (1976) 3266.
b: K.E. Johnson, L. Wharton and D.H. Levy, J. Chem. Phys., 69 (1978) 2719.
c: W. Sharfin, K.E. Johnson, L. Wharton and D.H. Levy, J. Chem. Phys., 71 (1979) 1292.

d: J.A. Blazy, B.M. Dekoven, T.D. Russel and D.H. Levy, J. Chem. Phys., 72 (1980) 2439.
e: K.E. Johnson, W. Sharfin and D.H. Levy, J. Chem. Phys., 74 (1981) 163.
f: D.E. Brinza, C.M. Western, D.D. Evard, F. Thommen, B.A. Swartz and K.C. Janda, J. Phys. Chem., 88 (1984) 2004.
g: B.A. Swartz, D.E. Brinza, C.M. Western and K.C. Janda, J. Phys. Chem., 88 (1984) 6272.
h: F. Thommen, D.D. Evard and K.C. Janda, J. Chem. Phys., 82 (1985) 5295.
i: L.J. van de Burgt, J.P. Nicolai, M.C. Heaven, J. Chem. Phys., 81 (1984) 5514.

[11] M. Boivineau, Thèse Université Paris-Sud (1987).

[12] M. Boivineau, J. Le Calvé, M.C. Castex and C. Jouvet, J. Chem. Phys., 84 (1986) 4712.

[13] M. Boivineau, J. Le Calvé, M.C. Castex and C. Jouvet, Chem. Phys. Lett., 130 (1986) 208.

[14] M. Boivineau, J. Le Calvé, M.C. Castex and C. Jouvet,
      Chem. Phys. Lett., **128** (1986) 528.

[15] S.D. Peyerimhoff, R.J. Buenker, Chem. Phys., **57** (1981) 279.

[16] R.J. Donovan, P. Greenhill, M.A. MacDonald, A.J. Yencha,
      W.S. Hartree, K. Johnson, C. Jouvet, A. Kvaran and J.P. Simons,
      Faraday Discuss. Chem. Soc., **84** in the press.

A THEORETICAL STUDY OF THE ORIENTED REACTION
HBr·CO$_2$ + hv → OH + CO + Br

George C. Schatz and Michael S. Fitzcharles
Department of Chemistry
Northwestern University
Evanston
IL 60201 USA

ABSTRACT. This paper presents a quasiclassical trajectory study of the photoinduced intracluster reaction HBr·CO$_2$ + hv → OH + CO + Br. In this study the potential energy surface for the photoexcited state is approximated by summing together an excited HBr potential curve, a previously developed HOCO surface, and a set of two body interactions between Br and CO$_2$. Trajectory simulations start on this surface with either 2.6 eV or 1.9 eV of excess energy and with the Br-H--- O-C-O geometry sampled from a distribution of HBr orientations relative to a linear equilibrium geometry. We find that only orientations within 15° of collinear lead to OH + CO formation, and that within this range two reaction mechanisms occur: a complex mechanism (which accounts for 70% of the reactivity) in which HOCO moves away from Br before dissociating, and a direct mechanism (30% of the reactivity) in which CO moves away from Br·OH and the OH suffers several secondary encounters with Br which cause rotational relaxation before separating from it. Both mechanisms give final state distributions which are different from the bulk reaction at 1.9 eV, but at 2.6 eV only the direct mechanism is different. Neither mechanism gives OH rotational distributions which are significantly colder than in the bulk. This result disagrees with the observations of Wittig and coworkers, and it is likely that errors in the assumed potential surface are responsible for this disagreement. Estimates of HOCO lifetimes in the complex mechanism are 0.3 ps at 2.6 eV and 0.4 ps at 1.9 eV.

1. INTRODUCTION

Recently, Radhakrishnan, Buelow and Wittig (RBW) [1] have introduced a new method for observing orientation effects in chemical reactions through the use of van der Waals complexes between the two reagent molecules as precursors to a photoinduced reaction. The specific reaction they studied is H + CO$_2$ → OH + CO, and in this case the precursor used was HBr·CO$_2$. Photolysis of this at 193 nm produces 2.6 eV H atoms which undergo intracluster reaction to produce OH + CO + Br. RBW found that the OH produced is rotationally colder (with peak quantum numbers of 4-6) than in the corresponding bulk reaction (where the peak is at 7-11). The corresponding ratio of vibrational populations v"=1/v"=0 was not, however, noticeably different.

The reaction H + CO$_2$ has been studied by other groups in the past [2,3]

J. C. Whitehead (ed.), Selectivity in Chemical Reactions, 353–364.
© *1988 by Kluwer Academic Publishers.*

(and there also has been a great deal of work on nonreactive collisions which produce vibrational and rotational excitation [4]), and recently Harding and we [5] have presented a theoretical study of this system based on trajectory calculations and an *ab initio*-based potential surface. In this study the calculated reactive cross sections and OH vibrational and rotational distributions were found to be in excellent agreement with the measured results. An analysis of the trajectories from this calculation indicated that the reaction proceeds almost exclusively through the formation of short lived HOCO intermediate complexes, and as a result the OH + CO product state distributions were found to be reasonably close to statistical. Since the RBW results are quite different from statistical, the H + CO_2 trajectory results suggest that the effect of the van der Waals precursor is to change the reaction mechanism in an important way such as from a complex to a direct mechanism or by changing the angular momentum and energy distribution in the complex significantly. Very recently Scherer *et al.* (SKBZ [6]) have presented time resolved measurements on the $HI \cdot CO_2$ analog of $HBr \cdot CO_2$ which show that at a somewhat different energy of 2.1 eV, the oriented reaction proceeds through an HOCO complex with a complex lifetime of 5-15 ps. At this point it is not known what the OH vibrational and rotational distribution is in the Scherer *et al.* measurement and thus it is not clear whether the apparently nonstatistical behaviour seen at 2.6 eV is at odds with the fairly long lived complexes seen at 2.1 eV.

In this paper we present the results of a trajectory study of $HBr \cdot CO_2$ using a potential surface which is an extension of our earlier HCO_2 surface. This surface incorporates the current best estimates of the saddle points and minima associated with the $HBr \cdot CO_2$ system, but since these stationary points are for the most part not well known, it is likely that the dynamical picture provided by our study is rather crude. Thus we will emphasize mainly the qualitative results from our calculations. Most of our analysis will concentrate on the 2.6 eV energy associated with the RBW experiment, but a few calculations done at 1.9 eV will also be presented.

2. THEORY

2.1 Van der Waals Structure and Potential Energy Surface

In our trajectory simulations, the $HBr \cdot CO_2$ system is started on the excited surface that is prepared by excitation from the ground state of the van der Waals cluster. The structure of the ground state cluster has not been measured, but can be inferred based on the known structures of $HF \cdot CO_2$ and $HCl \cdot CO_2$ [7]. Fig. 1 shows the relevant variables to define this structure and Table 1 gives the values of several geometrical parameters. In all calculations, CO_2 is assumed to have zero point displacements (sampled by Monte Carlo) relative to its equilibrium structure. The H-Br separation r is determined by the diatomic excited potential curve (taken from Ref. 8) and the desired excitation energy. The value given in Table 1 is that appropriate for excitation 2.6 eV above HBr dissociation. At equilibrium, it is presumed [1] that ground state $HBr \cdot CO_2$ has a linear Br-H---0-C-0 geometry such that the HBr rotor angle θ and the impact parameter b in Fig. 1 are both zero. In addition, the H---0 distance has been estimated to be 4.16 a_0, which makes the distance R between the HBr and CO_2 centers of mass equal to 9.11 a_0 for the r value given in Table 1. However it is important in the simulations that b and θ be sampled away from equilibrium since these variables undergo rather large amplitude motions. Indeed, $<\theta>$ is estimated by RBW to be 25^0 and b has most probable values of 3-4 a_0. Details of the

TABLE 1. Properties of HBrCO$_2$ and HCO$_2$ Potential Surfaces

	Quantity	Value
HBrCO$_2$	b (a$_0$)	0
	θ (deg)	varied from 0-13^0
	r (a$_0$)	2.8[a]
	R (a$_0$)	9.11[b]
HCO$_2$[c]	V$_{min}$(H + CO$_2$)	0 (reference point)
	V$_{min}$ (HOCO-trans)	-0.63 eV
	V$_{min}$ (HOCO-cis)	-0.54 eV
	V$_{min}$ (HCO$_2$)	-0.22 eV
	V$_{min}$ (OH + CO)	+0.97 eV
	V$^+$ (H--OCO)	1.10 eV
	V$^+$ (H--CO$_2$)	0.49 eV
	V$^+$ (O $\overset{H}{}$ CO)	1.25 eV
	V$^+$ (HO---CO)	0.93 eV
	V$^+$ (cis $\leftrightarrow$ trans)	-0.31 eV

a. Value determined by excited state HBr potential curve at 2.6 eV above dissociation.

b. Value determined from the estimate given in Ref. 1 that the OH distance in HBr·CO$_2$ is 2.2 Å.

c. HCO$_2$ surface properties are for the fitted surface of Ref. 5. These values are similar to the best estimates given in that paper.

sampling of b and θ will be discussed in section 3. One additional variable that needs to be defined is the azimuthal angle φ associated with rotation of the HBr about the R axis at fixed θ. The distribution of φ values is determined by b and θ. To develop a potential surface for the van der Waals cluster in Fig. 1, we

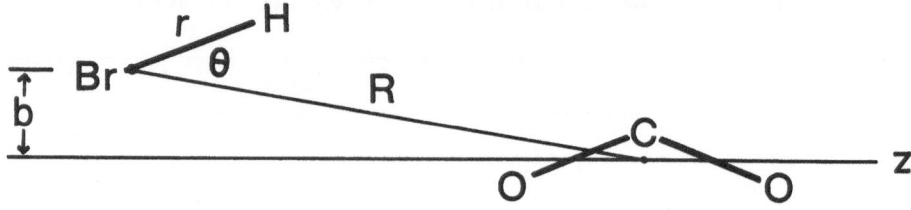

FIGURE 1. Coordinates used to define the HBr·CO₂ van der Waals complex. R is the distance from the HBr center of mass to the CO₂ center of mass, θ is the angle between R and the HBr axis, and b is the impact parameter. The z axis is defined as the equilibrium axis of the CO₂ molecule.

assume that the HCO₂ part can be taken from the ground state surface of Ref. 5 and that only two-body forces are needed in describing the additional Br. For HBr we take the excited state potential from Ref. 8 while for OBr and CBr we use simple 6-12 potentials which are approximated as NeKr potentials from Ref. 9. Thus the overall potential is given by

$$V = V_{HBr}(r) + V_{HCO_2}(r_{CO}, r_{CO'}, r_{OO'}, r_{HO}, r_{HO'}, r_{HC})$$
$$+ V_{6-12}(r_{BrO}) + V_{6-12}(r_{BrO'}) + V_{6-12}(r_{BrC}) \quad (1)$$

It should be apparent that this potential has a number of defects, although it probably describes at least the initial HBr dissociation and collision with CO₂ correctly and it has all the correct energetics with respect to HOCO and OH + CO formation (as summarized in Table 1). What it does not describe is the possibility of hopping to the ground HBr curve after the initial HBr dissociation. As a result, secondary encounters of H with Br take place on the excited surface. Also once the OH fragment is formed, subsequent interaction of this with the Br will involve repulsive two body terms in Eq. 1 that are certainly not correct for these two open shell species. Finally it should be noted that the HCO₂ surface itself is known to be deficient in that inelastic collisions of H with CO₂ on it give too much vibrational and too little rotational excitation [5]. This problem has recently been traced [10] to a defect in the repulsive interaction outside the H---OCO addition barrier. It does not appear that this defect has an important effect on the reactive dynamics involving H + CO₂ so we will assume that it is not important here.

Fig. 2 shows contours of the HBr·CO₂ surface as a function of the H atom location, with the Br to CO₂ center of mass distance fixed and the CO₂ bend and stretch coordinates chosen so as to minimize the potential. In this figure the highest contour is at 2.6 eV and the contour spacing is 0.4 eV. Initially the H atom is located on the innermost circular contour of the hill around the Br. At this point, when the H atom is between the Br and CO₂, there is already 0.4 eV of repulsion between HBr and CO₂. As the HBr dissociates, the H atom "falls down" the hill around the Br, but note that only about 1.0 eV of the repulsive H-Br interaction is decayed when the H atom gets to the H---OCO addition barrier. This indicates that

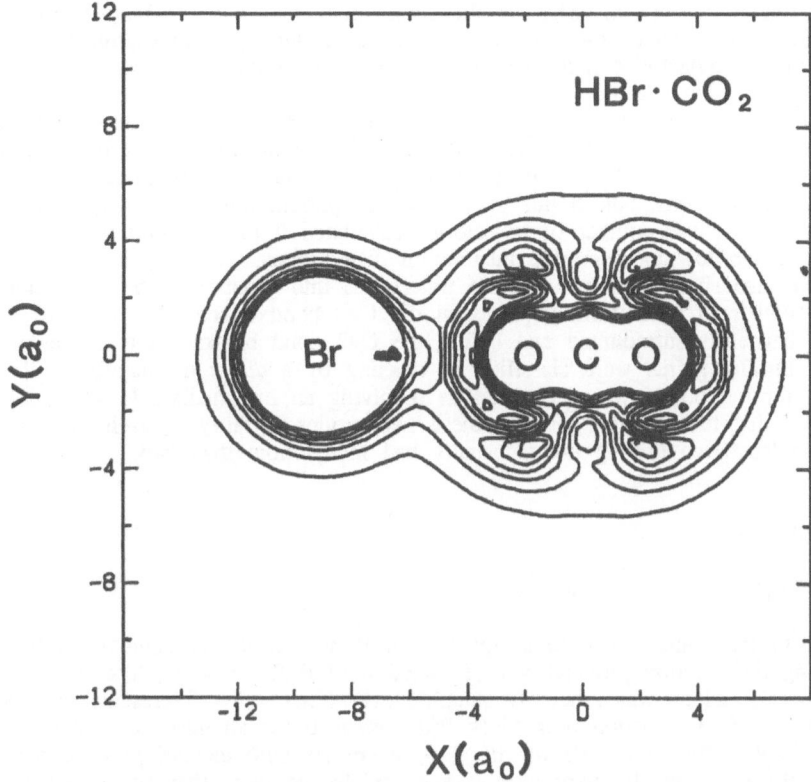

FIGURE 2. Contours of the HBr·CO$_2$ potential surface as a function of the H atom location, with the Br atom fixed on the x axis at 9.15 a$_0$ from the CO$_2$ center of mass. For each H atom location, the energy is minimized with respect to the CO$_2$ bend and stretch coordinates with the constraint that the O-O axis be parallel to x.

most of the energy release takes place when there are significant "three-body" interactions between Br, H and CO$_2$. The figure also shows that only a very small range of θ values cause the H atom to be directed towards the HOCO well.

2.2 Trajectory Calculations

A standard quasiclassical trajectory method similar to that used in Ref. 5 was used to determine the motion of the atoms starting from initial conditions that were sampled by Monte Carlo methods. Trajectories were terminated when one of the following two events took place: (1) the separation between H and the center of mass of CO$_2$ and also between H and Br became greater than 15 a$_0$ while neither C-O distance was greater than twice its CO$_2$ equilibrium value (corresponding to nonreactive scattering), (2) the separation between the OH and CO centers of mass became greater than 15 a_0 while at the same time one of the C-O bonds in CO_2 was at least twice its equilibrium value (reactive scattering). An examination of movies of our trajectories indicated that no significant secondary encounters would occur after the above criteria were satisfied.

Final state analysis of the OH and CO involved calculating vibrational and rotational actions for each diatomic and then rounding off the actions (divided by $\hbar$) to the nearest integer. This enabled us to determine detailed vibrational and rotational distributions, but in this paper we will present only average quantum numbers because the number of trajectories calculated (typically 1000 with ~50 reactive) is too low to determine meaningful distributions.

We also subdivided our reactive trajectories into groups which were determined by the number of outer turning points in the C-O bond which take place after the addition barrier is surmounted and before the C-O bond breaks. Trajectories which had 0-1 turning points were classified as reacting by a *direct* mechanism while those with ⟩2 turning points were classified as involving an intermediate HOCO *complex*. We found that the division between these two groups is fairly clean in that those having at least two turning points usually had many more than two.

3. RESULTS

3.1 θ,b Dependence of Reaction Probability

In order to learn about how to sample b and θ, we began our simulations by calculating the reaction probability as a function of θ for b = 0. The angle φ in all these calculations was randomly sampled over (0,π). Fig. 3 presents the results in the form of the reaction probability P(θ) versus θ for an initial excitation energy E of 2.6 eV. This figure shows that P(θ) decreases with increasing θ, becoming zero by θ=15°. This θ distribution is substantially narrower than the distribution which naturally occurs in the van der Waals complex (where <θ> = 25°), so further results were obtained by sampling from the narrower distribution and then normalizing the results to be consistent with the broader distribution. For either distribution it is necessary to weight the results by sinθ. The product P(θ)sinθ, shown in Fig. 3, is seen to have a broad peak between 5° and 10°.

Note that for b = 0 the angle θ = 15° leads to trajectories which strike the H---OCO addition barrier at a sufficiently oblique angle that they are deflected from it. For smaller θ the H atom often gets caught in the HOCO minimum (the mechanism for this is discussed in the next section), but the largest reaction probability in Fig. 3 is still much less than unity. The angle integrated reaction probability in Fig. 3 (weighted by sinθ over the angular range where P(θ) ≠ 0) is about 0.06, but the properly normalized probability is smaller. If we assume that θ has constant probability for 0 < θ < 34° (and zero otherwise), then <θ> = 25° (in agreement with estimates given earlier) and the properly normalized reaction probability is 0.01. This is only slightly larger than the bulk reaction probability which is about 0.007 (based on a reactive cross section of 1.3 a_0^2 [3] and a gas kinetic cross section of 179 a_0^2 [11]).

A set of runs which sampled b ≠ 0 was considered, and in all cases we found results which were similar to those obtained for b = 0 (except for the obvious changes in the ranges of θ and φ which give rise to reaction and for the usual reduction in reaction probability for large enough b). Most important, we found that the OH + CO product state distributions were insensitive to changes in

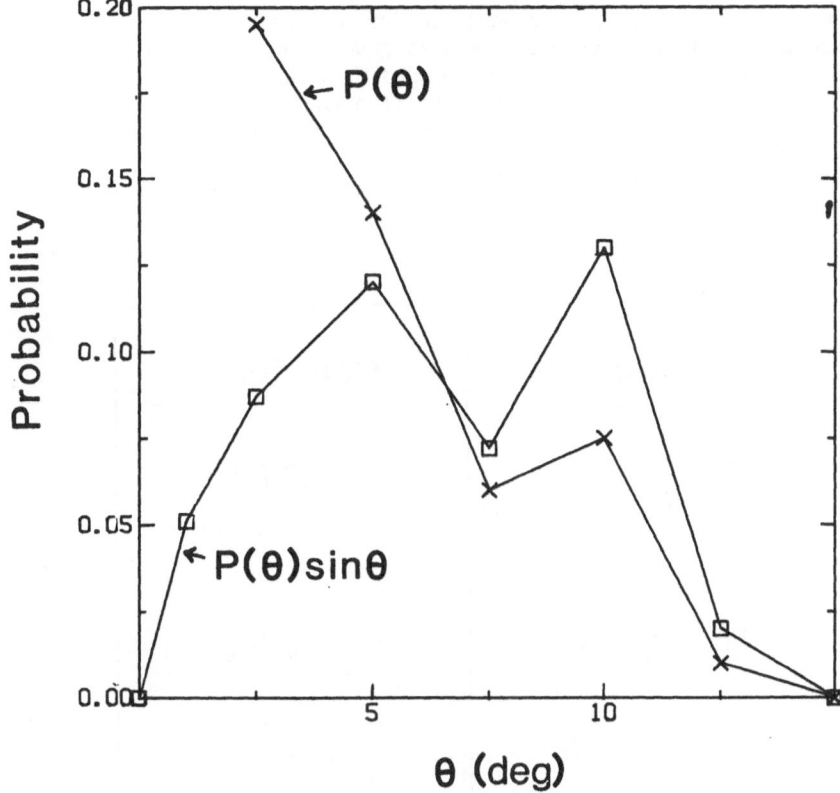

FIGURE 3. $P(\theta)$ (x's) and $P(\theta)\sin\theta$ (squares) versus θ for $b = 0$. Note that typical uncertainties on both curves are $\pm$ 20 - 30%.

b. To make our initial condition sampling as efficient as possible we chose b to be zero in calculating the ensemble averaged results which we present in section 3.3.

3.2 Reaction Mechanisms

By examining movies of the trajectories and by considering the C-O outer turning point analysis, we found that there are two mechanisms by which reaction occurs, complex and direct, as defined in section 2.2. Snapshots of trajectories which show these two mechanisms are plotted in Fig. 4, and they reveal that the initial stages of these two mechanisms (picture (a) of each trajectory) are the same in that they both involve shooting the H atom across the HOCO well toward the isomerization saddle point region (picture (b) of each trajectory). In the complex mechanism, the H atom then rebounds and starts vibrating as part of a distinct HOCO complex (see picture (c) of the complex trajectory). The complex then moves away from the Br, and after several vibrational periods breaks up into OH and CO (picture (d)). In the direct mechanism, the C-O bond begins to break during the initial encounter so that before the OH rotates to the point indicated in picture (c) of the direct trajectory, CO has already moved well away. The OH is still next to the Br at this point, and before they separate, the H atom hits the Br several times (as in picture (d)). An examination of the rotational angular momentum J_{OH} of the OH reveals that these secondary collisions of OH with Br cause substantial changes in J_{OH} (usually relaxation by a few quanta).

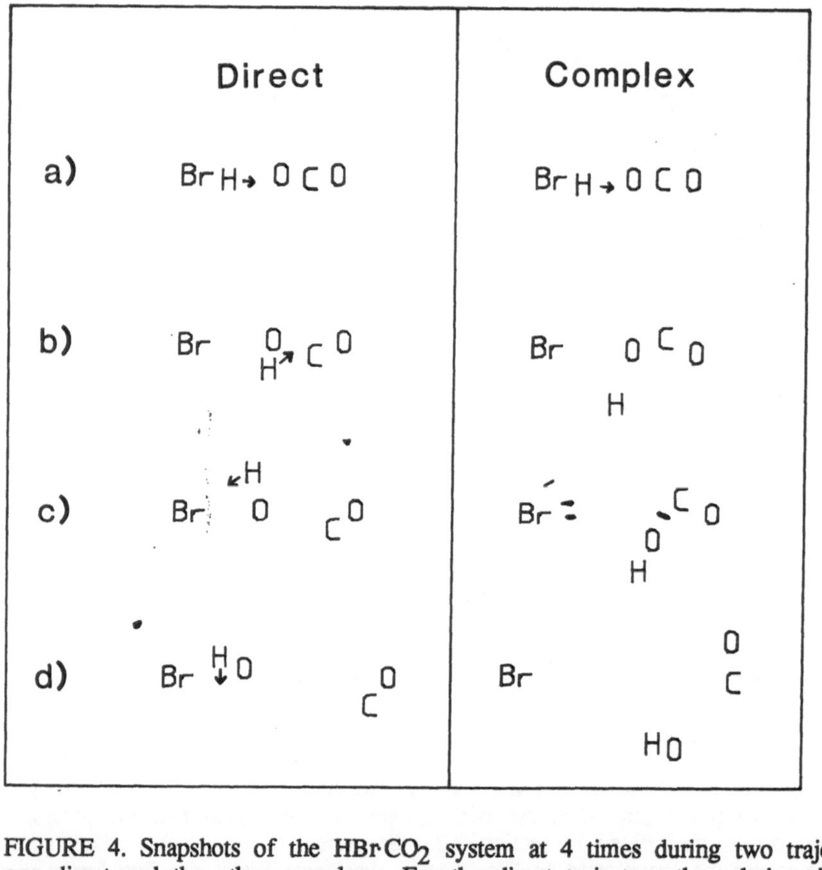

FIGURE 4. Snapshots of the HBr·CO_2 system at 4 times during two trajectories, one direct and the other complex. For the direct trajectory the relative times of each snapshot are (a) 0 ps (reference), (b) 0.02 ps (c) 0.06 ps and (d) 0.13 ps while for the complex trajectory the times are (a) 0 ps (b) 0.02 ps (c) 0.12 ps and (d) 0.28 ps. Note that the atom location is determined by the center of the first symbol (H, C, O, B) associated with it. The Cartesian coordinate system has been chosen so that most of each trajectory being viewed is near to the plane of observation. Arrows in some of the snapshots indicate the instantaneous direction of motion of the H atom.

An examination of energy flow in the trajectories pictured in Fig. 4 reveals that in both mechanisms, significant energy is initially deposited in the OH stretch and bending motions of HOCO. In the complex mechanism this energy is able to flow to the CO before the HOCO comes apart, but in the direct mechanism the CO is vibrationally and rotationally cold compared to that in the complex mechanism, while the OH is vibrationally and rotationally hot. The secondary collisions of OH with Br are therefore the only events by which OH can be cooled off in the direct mechanism.

3.3 Final State Information

Table 2 presents the ensemble averaged final state information for E = 1.9 and 2.6 eV, including separate results for the direct and complex mechanisms, and the results from bulk simulations on the same HCO_2 potential surface at the same energy. Included in Table 2 are

TABLE 2. OH + CO Final State Information

Quantity	Direct	Complex	Total van der Waals	Bulk[a]
(A) E = 1.9 eV				
fraction of total	0.29	0.71	1.00	
$<v_{OH}>$	0.5	0.6	0.6	-0.2
$<J_{OH}>$	8	9	9	8
$<v_{CO}>$	0.2	0.6	0.5	0.1
$<J_{CO}>$	17	17	17	28
(B) E = 2.6 eV				
fraction of total	0.30	0.70	1.00	
$<v_{OH}>$	0.9	0.6	0.7	0.5
$<J_{OH}>$	11	10	11	11
$<v_{CO}>$	0.5	1.3	1.1	1.2
$<J_{CO}>$	18	21	20	25

[a]. Ref. 5.

the OH and CO vibrational and rotational quantum numbers (v_{OH}, J_{OH}, v_{CO} and J_{CO}). This table shows that the direct mechanism makes up about 30% of the van der Waals reaction probability at both energies which is to be contrasted with the bulk reaction where we previously found that nearly 100% of the reaction arises from the complex mechanism. The reason for this difference seems to be due to the fact that in the van der Waals reaction the H atom can shoot across the HOCO well into the isomerization barrier region with high probability while in the bulk reaction, the H atom most frequently approaches from large impact parameters where it undergoes one or more oscillations in the HOCO well before getting to the isomerization barrier.

Table 2 indicates that the direct and complex mechanisms produce OH and CO's that are for the most part similar. The biggest difference occurs for CO vibration which is colder in the direct mechanism. OH rotation is about the same in the two mechanisms, and we assume that this is due to an accidental cancellation between the larger rotational excitation that is initially produced by the direct mechanism and subsequent relaxation of this excitation by the secondary OH + Br collisions.

Since the complex mechanism makes up 70% of the total reaction probability, the vibrational and rotational excitation seen in the column labelled "Total" are closest to that labelled "Complex". More important, many of the entries in the "Bulk" column are different from those in the "Complex" column, particularly at 1.9 eV. Since both columns represent reaction that proceeds through an intermediate complex, this result suggests that the complexes are different enough to give different final state distributions. This is sensible given that much of the H atom motion in the van der Waals complex involves simultaneous Br-H-CO_2 interactions that are absent in the bulk.

3.4 Comparison with Experiment

Comparison of our results in Table 2 at 2.6 eV with RBW indicates agreement with respect to the OH vibrational distributions but disagreement with respect to the rotational distributions. From Ref. 1 one can infer that $<v_{OH}> \geqslant 0.3$ for both van der Waals and bulk, which agrees with the values in Table 2. However, RBW find a major decrease in $<J_{OH}>$ between bulk and van der Waals with $<J_{OH}>$ being approximately 6 in the van der Waals complex (ignoring the difference between J_{OH} and K) and 10 in the bulk. Clearly we do not see any significant difference between J_{OH}(total) and J_{OH}(bulk) at either energy in Table 2. The fact that they see any change at all suggests that either a change in reaction mechanism or a change in the statistical properties of the HOCO complex must be occurring. Our results show evidence for both possibilities but neither is significant enough to change the OH rotational distributions. Presumably defects in our potential surface are responsible for this disagreement. Probably the most obvious defect is the absence of an accurate Br·OH interaction. This would cause us to underestimate the amount of OH rotational relaxation during secondary collisions with Br in the direct mechanism. The relative fraction of direct versus complex trajectories is also subject to large errors in our results, since the region of the potential surface near the isomerization saddle point that is probably responsible for the branching between these two mechanisms is not accurately fit by our potential surface [5]. Finally, quantum zero point effects could play an important role in determining which OH and CO final states can be populated via the complex mechanism.

Now let us consider the SKBZ experiment. Since this was done using

HI·CO_2 at 2.1 eV, it is not possible to make a direct comparison of our HBr·CO_2 results with theirs, but our simulations at 1.9 eV should be somewhat relevant. At this energy Table 2 indicates that the complex mechanism contributes 70% to the reactivity, while SKBZ find 100% complex formation in their results. An analysis of the lifetime distribution associated with the complex mechanism indicates lifetimes of 0.4 ps at 1.9 eV and 0.3 ps at 2.6 eV. These values are substantially smaller than the 5-15 ps lifetime estimated by SKBZ, but the difference is not surprising given the substantial uncertainties in our potential surface. In addition, the use of classical mechanics to estimate lifetimes can cause errors due to the absence of zero point constraints as mentioned above.

4. CONCLUSION

In this initial attempt at understanding the HBr·CO_2 reaction dynamics we have found that both direct and complex reaction mechanisms are possible in the van der Waals cluster and that the relative amounts of these mechanisms and the product state distributions obtained from them can be different from that in the bulk reaction. At this point we are not able to quantitatively reproduce the observed change in the OH rotational distribution in going from bulk to van der Waals complex, but reasons for this discrepancy are understandable based on known errors in the potential surface and due to quantum effects.

In addition to refining the theoretical description of HBr·CO_2, progress in understanding this system would be aided by the following two measurements: (1) repeating the OH final state distribution measurements at lower energies (as the results in Table 2 suggest that the different mechanisms have different dependence of $<J_{OH}>$ on energy), and (2) measuring the CO product state distributions (as these do not depend on the secondary Br---OH collisions that we found were crucial to relaxing the OH in the direct mechanism, but they do depend on the branching between direct and complex mechanisms).

5. ACKNOWLEDGMENTS

We thank Curt Wittig for many valuable discussions, and R.B. Bernstein, N.F. Scherer and A.H. Zewail for helpful correspondence. This research was supported by NSF Grant CHE-8416026.
REFERENCES

1. G. Radhakrishnan, S. Buelow and C. Wittig, J. Chem. Phys., 84 (1986) 727.

2. C.R. Quick and J.J. Tiee, Chem. Phys. Lett., 100 (1983) 223.

3. K. Kleinermanns and J. Wolfrum, Chem. Phys. Lett., 104 (1984) 157; K. Kleinermanns, E. Linnebach and J. Wolfrum, J. Phys. Chem., 89 (1985) 2525.

4. J.A. O'Neill, C.X. Wang, J.Y. Cai, G.W. Flynn and R.E. Weston, Jr., J. Chem. Phys., 85 (1986) 4195 and references therein.

5. G.C. Schatz, M.S. Fitzcharles and L.B. Harding, Far. Disc. Chem. Soc., (1987) in press.

6. N.F. Scherer, L.R. Khundkar, R.B. Bernstein and A.H. Zewail, J. Chem. Phys., **87** (1987) 1451.

7. F.A. Baiocchi, T.A. Dixon, C.H. Joyner and W. Klemperer, J. Chem. Phys., **74** (1981) 6544; R.S. Altman, M.D. Marshall and W. Klemperer, J. Chem. Phys., **77** (1982) 4344.

8. C.F. Goodeve and A.W.C. Taylor, Proc. Roy. Soc. London, **152** (1935) 221.

9. J.M. Parson, T.P. Schaefer, F.P. Tully, P.E. Siska, Y.C. Wong, and Y.T. Lee, J. Chem. Phys., **53** (1970) 2123.

10. G.K. Chawla, P.L. Houston and G.C. Schatz, to be published.

11. J.O. Chu, C.F. Wood, G.W. Flynn and R.E. Weston, Jr., J. Chem. Phys., **81** (1984) 5533.

REACTIONS OF MOLECULAR IONS IN ASSOCIATION WITH INERT GAS CLUSTERS

A.J. Stace and D.M. Bernard
School of Molecular Sciences
University of Sussex
Falmer
Brighton BN1 9QJ
UK

1. INTRODUCTION

For the most part, the forces responsible for bonding in clusters are weak van der Waals interactions. It is, therefore, somewhat surprising that even the most fragile clusters, e.g. Ar_n, appear capable of forming stable ions when ionized by electrons with impact energies of the order of 70 eV. To account for such behaviour, it has been proposed that the formation of Ar_n^+ clusters proceeds through the generation of high-lying Rydberg states which then autoionize, followed by a self-trapping of the positive charge. The latter process is particularly efficient in both liquid and solid argon, and arises from the comparative stability of the Ar_2^+ ($^2\Sigma_u^+$, $D_0 = 1.3$ eV) dimer. Obviously, the trapping mechanism leads to an increase in internal temperature which the ion cluster reduces through evaporation. Just how many argon atoms are lost in this cooling process is difficult to estimate; however, there is good evidence that evaporation on quite a large scale continues even 10^{-5} s after ion formation.

This picture of comparative stability on the part of inert gas ion clusters is further enhanced when one considers the behaviour of mixed clusters of the type $Ar_n.X$, where X is a molecule which in the present experiments ranges from I_2 to $C_3H_7COCH_3$ [1-6]. It is observed that upon ionization, X undergoes unimolecular decomposition without significantly disrupting the inert gas component. In some cases the degree of fragmentation is quite extensive, and so far all the examples studied have exhibited very selective decomposition routes. For example, all the following product ions arise from the electron impact ionization of isolated diethyl ether. Shown in **bold** are the product ions that are observed when the ether is clustered with argon [1]:

J. C. Whitehead (ed.), Selectivity in Chemical Reactions, 365–372.
© *1988 by Kluwer Academic Publishers.*

$$\rightarrow CH_3CH_2O=CHCH_3^+ \rightarrow CH_3CH=OH^+ \tag{1}$$

$$(C_2H_5)_2O^+$$

$$\rightarrow CH_2=OH^+ \tag{2}$$

$$\rightarrow C_3H_5^+ \tag{3}$$

$$\rightarrow CH_3CH_2OCH_2^+$$

$$\rightarrow C_2H_5^+ \tag{4}$$

$$\rightarrow CHO^+ \tag{5}$$

2. EXCITATION MECHANISM

To explain the selective nature of the Ar_nX^+ decomposition routes, we need to begin by accounting for the fact that in these mixed ion clusters, covalent bonds appear to break in preference to very much weaker van der Waals bonds. The mechanism proposed to explain such behaviour envisages a two-step route [1-3,6] leading to the appearance of fragment ions: (1) in large clusters (n > 15) the molecular ion is excited via a charge transfer process from the inert gas component to X; (2) the molecular ion decomposes because its lifetime with respect to fragmentation ($\approx 10^{-12}$ s) is very much shorter than the time necessary for energy randomization to the inert gas component. Evidence for the charge transfer mechanism has come from two sources. The initial proposal was based on a consideration of reaction energetics; no fragmentation processes have been observed for which $\epsilon_0 > IP(cluster) - IP(X)$, where ϵ_0 is the critical energy of reaction. However, a series of recent experiments [7-9] on the photoionization of van der Waals molecules have provided very strong spectroscopic support for the charge transfer mechanism. In particular, Kamke *et al.* [8] have proposed the term "intramolecular Penning ionization" as a classification for such processes.

The conclusion above regarding the time scale for fragmentation is based on RRKM calculations of the unimolecular reaction rates of the ions concerned. However, a simple classical RRK calculation can be used to illustrate that there is no energy randomization to the inert gas component. Consider, for example, the fragmentation of argon-acetone clusters, $Ar_n.(CD_3)_2CO^+$. The relative intensity of the product ion, $Ar_n.CD_3CO$ for n=2 is not much different from that observed for n=35; however, in terms of the number of vibrational degrees of freedom added in going from 2 to 35, there is a difference of over 100. The effect these extra degrees of freedom could have in terms of possible energy randomization can easily be calculated. The probability that, at a given internal energy E, an energy > ϵ_0 will be found in any one of the vibration coordinates is given by the classical expression

$$P(\epsilon_0,E) = (1 - \epsilon_0/E)^{N-1} \tag{6}$$

where N is the number of vibrational degrees of freedom. Assuming charge transfer excitation, then E for the acetone-argon combination is $\approx$ 4 eV. There are three possibilities to consider:

(1) ϵ_0 = 0.04 eV (approximately the energy necessary to remove a single argon atom) and N = 129, this gives $P(\epsilon_0,E) \simeq 0.27$;

(2) ϵ_0 = 1 eV (approximate C-C covalent bond strength), this gives $P(\epsilon_0,E) \simeq 10^{-16}$;

(3) ϵ_0 = 1 eV and N = 24 (vibrational degrees of freedom for acetone alone), this gives $P(\epsilon_0,E) \simeq 10^{-3}$. This simple calculation demonstrates that if the internal energy available to the reactant ion were randomized throughout the inert gas component, then, compared with the breaking of a covalent bond, the loss of one argon atom (or more) offers a far more facile fragmentation route.

Based on the above considerations, the sequence of events considered to be taking place inside the ion source of the mass spectrometer could be as follows:

$$Ar_n.X + e \rightarrow Ar_n^*.X + e \qquad (7)$$

where $Ar_n^*.X$ denotes the excitation of an electron to a Rydberg state. Following the formation of such a state, there are two possible routes to ionization [6]: (a) self-trapping of the positive charge followed by charge transfer, i.e.

$$Ar_n^*.X \rightarrow Ar_2^+Ar_{n-2}.X + e \qquad (8)$$

$$Ar_2^+.Ar_{n-2}.X \rightarrow Ar_n.X^+ \qquad (9)$$

or (b) direct ionization as observed by Kamke et al. [8]

$$Ar_n^*.X \rightarrow Ar_n.X^+ + e. \qquad (10)$$

Obviously, there is going to be a difference in the energy available to X^+ depending upon whether it is generated either via steps (8) and (9) or via step (10). The main difference comes from the binding energy of the stable Ar_2^+ dimer.

3. SELECTIVE FRAGMENTATION

Although it appears not to participate in energy randomization, the role of the inert gas component is by no means passive. A variation in argon cluster size can change the relative intensity of a product ion; in some examples, by a factor of three in going from Ar_n to Ar_{n+1}. In general, product ion intensity fluctuations are less pronounced for reactions on CO_2 clusters. However, more important than variations in product ion intensity is the fact that the presence of a cluster, either Ar_n or $(CO_2)_n$, can lead to the total suppression of certain types of reaction pathway. An indication of such behaviour is to be seen in the example given above. The fragment $CH_2=OH^+$ is the most intense ion in the mass spectrum of isolated diethyl ether; however, when the latter is clustered with argon, $Ar_n.CH_2=OH^+$ is not present. Similarly, the product ion $Ar_n.CH_3CH=OH^+$ is also absent, as are $Ar_n.C_2H_5^+$ and $Ar_n.C_3H_7^+$. Two further examples [5] which illustrate the selectivity imposed by the inert gas component (both argon and carbon dioxide), are butan-2-one and pentan-2-one.

$$\rightarrow CH_3CO^+ \tag{11}$$

$$CH_3COC_2H_5^+$$

$$\rightarrow C_2H_5CO^+ \rightarrow C_2H_5^+ \rightarrow C_2H_3^+ \tag{12}$$

$$\rightarrow CH_3CO^+ \tag{13}$$

$$CH_3COC_3H_7^+ \rightarrow C_3H_7CO^+ \rightarrow C_2H_5^+ \rightarrow C_2H_3^+ \tag{14}$$

$$\rightarrow CH_3C(OH)=CH_2^+. \tag{15}$$

Once again, the reaction products that are observed when each of the ketones is clustered with an inert gas are shown in **bold**.

The pattern of reactions that has emerged from the systems studied so far suggests that, in an ion cluster between argon and an oxygen-containing molecule, such as one of the ketones given above, the oxygen atom forms an integral part of the cluster's structure. Any hydrocarbon component present is assumed to protrude from the main body of the cluster. Such a structure can explain most of the reaction features given above. In all the clustered product ions, the oxygen atom is retained; no product ions are observed from reactions which would involve the oxygen atom being displaced. In addition, all rearrangement reactions involving the oxygen atom are suppressed. This explains the absence of the ion $CH_2=OH^+$ from the reactions of argon-diethyl ether clusters, and also the absence of the rearrangement product $CH_3C(OH)=CH_2^+$ given by reaction (11) above.

A very recent experiment, involving butanoic acid clustered with argon, has provided one of the most spectacular examples of reaction selectivity we have so far encountered. In the ion source of a mass spectrometer, an isolated butanoic acid ion will undergo a McLafferty rearrangement reaction

$$CH_3CH_2CH_2COOH^+ \rightarrow CH_2C(OH)_2^+ + C_2H_4 \tag{16}$$

to give a very intense fragment ion peak at m/z = 60 amu. When pure butanoic clusters are generated and ionized, they also are observed to undergo exactly the same rearrangement reaction [10]. The only difference being that the stable parent ion cluster takes the protonated form, $(CH_3CH_2CH_2COOH)_nH^+$. When butanoic acid is clustered with argon the following ions are observed: $Ar_n.CH_3CH_2CH_2COOH^+$, $Ar_n.(CH_3CH_2CH_2COOH)_2^+$, and $Ar_n.CH_3CH_2CH_2COOH.COOH^+$. Fig. 1 shows a section of the recorded mass spectrum. The composition of the clusters has been confirmed through the use of isotopes. Unlike all the examples studied previously, the combination of an argon cluster and a single butanoic acid molecule appears to be totally unreactive. This is despite the fact that the energy criterion given above favours fragmentation either via step (16) or, under the circumstances, via the fission reaction

$$CH_3CH_2CH_2COOH^+ \rightarrow COOH^+ + CH_3CH_2CH_2. \tag{17}$$

Where fragmentation is observed, it involves the butanoic acid dimer clustered with argon, and the product ion comes from a fission reaction similar to step (17)

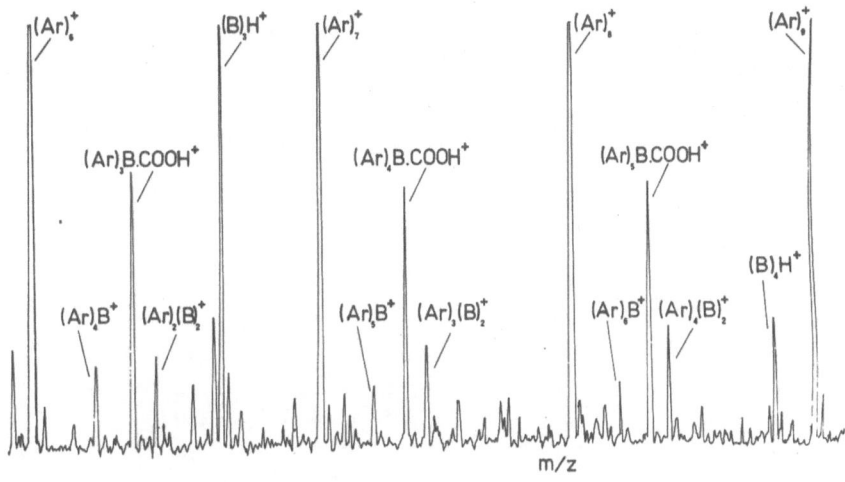

FIGURE 1. Section of the mass spectrum recorded for mixed argon-butanoic acid clusters. Ar_nB^+ and $Ar_n(B)_2^+$ denote clusters containing molecular ions, and $Ar_nB.COOH^+$ denotes fragment ions.

above. There is no evidence of a McLafferty rearrangement product, even though we know that such a reaction can proceed in ion clusters of the pure acid. The reaction product $CH_3CH_2CH_2COOH.COOH^+$ accounts for approximately 10^{-4} % of the fragments in the mass spectrum of pure butanoic acid clusters [11]. However, it can be seen in Fig. 1 that upon association with argon, the ion $Ar_n.CH_3CH_2CH_2COOH.COOH^+$ becomes the only reaction product. The effect of clustering with argon is two-fold; first the presence of the argon appears to stabilize the unprotonated form of the dimer ion and secondly, the very dominant rearrangement reaction observed in isolated butanoic acid ions is completely suppressed. Shinohara *et al.* [12] recently reported the observation of unprotonated water ion clusters in association with argon atoms. It has been suggested [13] that ion-induced dipole interactions may be responsible for stabilizing such species, those same interactions may also account for our observation of the ion $Ar_n.(CH_3CH_2CH_2COOH)_2^+$.

4. EXPERIMENTAL ASPECTS

Neutral clusters are generated by the adiabatic expansion of a gas mixture through a 100-μm diameter pulsed nozzle operating at approximately 20 Hz. Following collimation through a 1.0 mm diameter skimmer positioned 2 cm from the nozzle, the modulated cluster beam is ionized by electron impact and mass analysed on a modified AEI MS 12 mass spectrometer. The ion signal is monitored via a lock-in amplifier (Stanford Research SR510) which takes its reference from the unit responsible for driving the pulsed nozzle. An electron impact energy of 70 eV has been used in all the experiments reported here.

In order to study the reactions it has been necessary to generate mixed

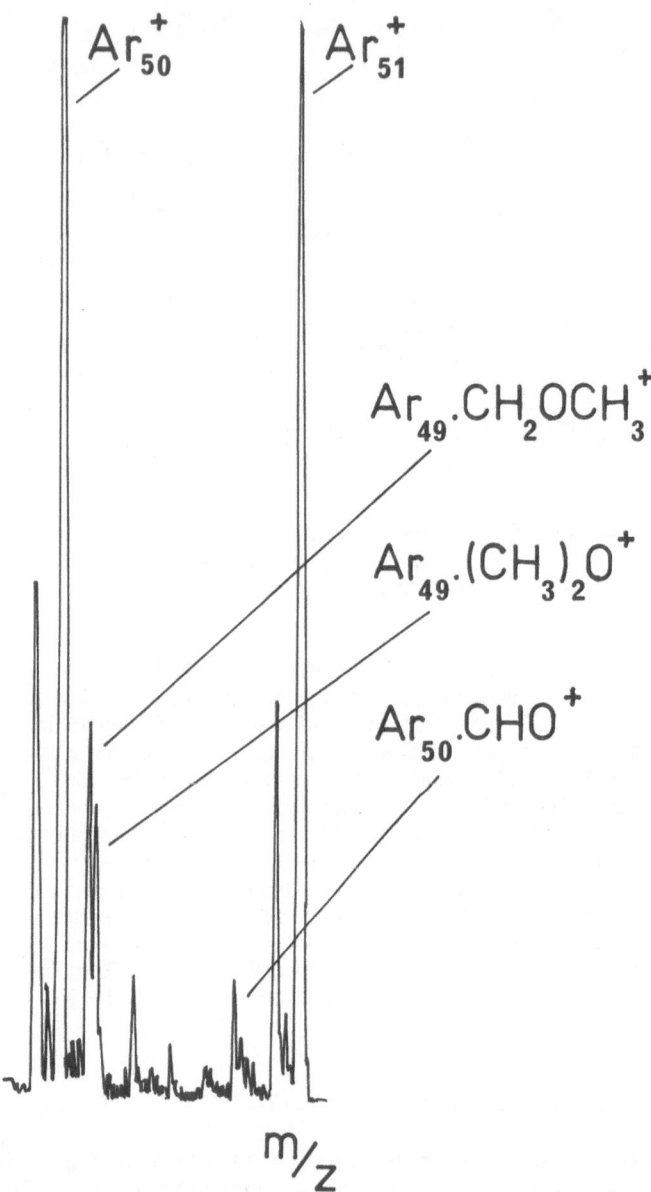

FIGURE 2. Section of the mass spectrum for mixed argon-dimethyl ether ion clusters, recorded on a VG ZAB-E mass spectrometer.

clusters of a very specific nature. In most cases, the appropriate argon/X ratio has been determined by trial and error; however, the necessary concentration of the polyatomic in most experiments does appear to be of the order of 100 ppm. If the concentration of X is too high, then only X clusters are observed. At intermediate argon/X ratios, no clusters of any type are observed; but as X is progressively diluted in argon, clusters of the type $Ar_n.X_3$, $Ar_n.X_2$ and $Ar_n.X$ appear in the mass spectrum. The intensities of the product ions from $Ar_n.X^+$ clusters have been observed to exhibit a pressure dependence, which is not present when the same molecules are clustered with carbon dioxide. It has been suggested that such behaviour arises as the result of a phase transition in the argon component [2].

The presence of impurities and possible mass coincidences with isotopes place limitations on the accuracy with which measurements can be made. Thus the ^{36}Ar peak is quite intense for clusters containing more than 10 argon atoms and, in addition to interference from ^{13}C peaks in CO_2 clusters, there are also the fragment ions $(CO_2)_n.O^+$, $(CO_2)_n.O_2^+$, and $(CO_2)_n.CO^+$ which can overlap with either $(CO_2)_n.X^+$ ions or any reaction products resulting from the fragmentation of X^+. Trace amounts of O_2 and N_2 can also present problems.

Very recently we have developed a new apparatus for studying the reactions and properties of ion clusters. The detection system is a modified VG ZAB-E double focusing, reverse geometry mass spectrometer. With the ion source operating at a potential of 8 kV the instrument has a mass range of 1-10,000 amu, and its ultimate mass range is 55,000 amu. In addition, the mass spectrometer provides a resolution which is far superior to that available on our earlier apparatus. As an example of the instruments capability, Fig. 2 presents the first ion cluster reaction to be recorded on it. In an earlier study of the reactions of the dimethyl ether ion in association with argon clusters [3], difficulty was experienced in resolving the loss of H from the molecular ion for clusters containing more than 20 argon atoms. As can be seen in Fig. 2, the new apparatus has no difficulty in resolving the products from the reaction

$$Ar_n.(CH_3)_2O^+ \rightarrow Ar_n.CH_3OCH_2^+ + H \qquad (18)$$

for n=49. It is anticipated that this equipment will provide many new exciting results on the reactions of molecular ions in association with inert gas clusters.

ACKNOWLEDGEMENTS

The authors would like to thank the SERC for an equipment grant and for the award of a studentship to DMB.

REFERENCES

[1]   A.J. Stace, J. Phys. Chem., **87** (1983) 2286.

[2]   A.J. Stace, Chem. Phys. Lett., **99** (1983) 470.

[3]   A.J. Stace, J. Am. Chem. Soc., **106** (1984) 4380.

[4]   A.J. Stace, J. Am. Chem. Soc., **107** (1985) 755.

372

[5] D.M. Bernard and A.J. Stace, J. Chem. Soc. Faraday Trans. 2,
     83 (1987) 29.

[6] A.J. Stace, J. Phys. Chem., **91** (1987) 1505.

[7] A. Ding, J.H. Futrell, R.A. Cassidy, L. Cordis and J. Hesslich,
     Surf. Sci., **156** (1985) 282.

[8] W. Kamke, B. Kamke, H.U. Kiefl and I.V. Hertel, Chem. Phys.
     Lett., **122** (1985) 356.

[9] P.D. Dao and A.W. Castleman Jr., J. Chem. Phys., **84** (1986) 1435.

[10] D.M. Bernard and A.J. Stace, submitted for publication.

[11] Y. Mori, T. Kitagawa, T. Yamamoto, K. Yanda and S. Nagahara,
     Bull. Chem. Soc. Japan., **53** (1980) 3492.

[12] H. Shinohara, N. Nishi and N. Washida, J. Chem. Phys., **84** (1986)
     5561.

[13] A.J. Stace, Chemical Society Specialist Periodical Reports:
     'Mass Spectrometry', in print.

ACCURACY AND AVAILABILITY OF POTENTIAL ENERGY SURFACES

J.N. Murrell
School of Chemistry and Molecular Sciences
University of Sussex
Falmer
Brighton BN1 9QJ
Sussex
U.K.

1. INTRODUCTION

We judge the accuracy of a potential energy surface by how well we reproduce experimental data when it is used in calculations of nuclear dynamics. Dynamical calculations are not perfect; they may assume the Born-Oppenheimer approximation, they may be classical rather than quantum, they may be approximate or unconverged quantum calculations. However, in most cases gross disagreements between theory and experiment can be attributed to inaccuracies in the potential energy surfaces.

We judge availability by the existence of analytical functions for the potential energy surface which can readily be used in dynamical programs. Tables of *ab initio* points are in this definition not generally "available" because it is rare that they can be used directly. Very complicated or non-standard functions are also "not very available".

Broadly speaking both accuracy and availability decrease as the size of the system increases. Size is mainly a reflection of the dimensions of the surface, which in turn is related to the number of atoms. However, when potential energy functions are based on *ab initio* calculations, accuracy and availability will be less for heavy atoms than for light atoms.

2. DIATOMIC MOLECULES

Diatomic molecules are a special case. Firstly, the dynamics of atomic collisions can usually be calculated accurately, and there are some inversion techniques that allow one to predict the potential from the experimental data; the RKR method of analysing spectroscopic data is the most well known of these [1]. Secondly, the potential energy functions are one dimensional and even "complicated" functions are simple compared with those of polyatomic molecules.

3. POLYATOMIC MOLECULES

In recent years there has been a considerable improvement in the accuracy and availability of small polyatomic surfaces, particularly for triatomics. Most important

373

J. C. Whitehead (ed.), Selectivity in Chemical Reactions, 373–377.
© *1988 by Kluwer Academic Publishers.*

has been the development of general strategies so that we no longer have to consider individual examples from scratch. The generalization of LEPS functions for the interaction of doublet state atoms is a very important example [2]. The application of the many-body expansion with the standard use of polynomial-exponential functions is another [3]. The important point about the use of standard functions is that it becomes worthwhile to put the effort into program packages, both for potential energy function fitting and for the dynamics of molecular collisions. The many-body expansion method has, for example, been built into function fitting programs for three and four atom systems (POT3 and POT4) [4] in which the input data may be spectroscopic frequencies, force constants or *ab initio* points.

For systems having more than four atoms one would rarely need to know details of the whole 3N-6 dimensional surface because the full functional representation of the surface, even if available, could not be tested by experiment. In this case one usually seeks functions which represent sections of the surface of lower dimensions, or optimum paths, or paths with dynamical averaging over other dimensions. Another strategy is to use the full dimensionality but to recognise the fact that the potential functions apply only to local regions of space. The popular Molecular Mechanics potentials are of this type [5]. They have been constructed to give the energies and geometries (sometimes force constants) of minima on the surface which represent traditional chemical structures but they will not generally give a good representation of the paths between these minima and they notably fail in their representation of the bond breaking process.

I now wish to return to the matter of using quantum mechanical methods (normally *ab initio* methods) to calculate points on a potential energy surface, and from these to deduce the potential function. There are two important questions of strategy. The first is how to choose the points for calculation and the second is how to allow for different levels of accuracy at different points. I am assuming that a choice has to be made; there is really no system for which an unlimited number of points can be calculated with unlimited accuracy.

It is my understanding that there is no solely mathematical solution to the problem of how best to distribute K points over specified ranges of 3N-6 variables to optimize the fit to an unspecified function. It is not until the functional form is specified that one can define an optimization routine. There are, for example, rules for the distribution of points with a cubic spline fitting [6]. The message here is that one should not underestimate the importance of physical models for the task of converting *ab initio* points into a potential energy function.

I give one example of this. For inelastic scattering of an atom A and a diatomic molecule BC it is convenient to take an expansion of the potential in the form

$$V(r,R,\theta) = \Sigma_l V_l(r,R) P_l(\cos\theta) \tag{1}$$

where r is the diatomic bond length, R the distance between the atom and the centre of mass of the diatomic, and θ the angle between $\underline{r}$ and $\underline{R}$. Now it is well established that although this expansion is very convenient for evaluating the matrix elements that occur in the coupled channel equations [7] it is very slowly convergent in the repulsive region of the potential, and it is this region that is important for rotational inelastic scattering.

If one wants to evaluate K terms in the expansion (1) then one would expect to have to evaluate sets of points in the (r,R) plane for K different values of θ.

However, suppose that it can be shown that the potential is atom pair additive, then we write

$$V(r,R,\theta) = V_{BC}(r) + V_{AB}(R_{AB}) + V_{AC}(R_{AC}) \qquad (2)$$

and we need points along only two different angles to evaluate V_{AB} and V_{AC}. One can then obtain the expansion (1) to infinite order by multiplying (2) by $P_l(\cos\theta)$ and integrating over θ.

The additive atom pair model has been widely used for intermolecular potentials. We have examined it for a very simple case, HeH_2, for which it might be expected to work, but it is not very good. However, it can be considerably improved by generalizing to a site-site model, in which the centres of the pair potentials are allowed to float away from the hydrogen nuclei. It is further improved by treating the long and short range regions of the potential separately [8]. The important point is that even with these more flexible models one only needs to make *ab initio* calculations for two angles θ in order to obtain all orders in the expansion (1).

The problem of how to account for different levels of accuracy in calculations at different points on the potential energy surface goes back a long way and there have been many partial solutions. The atom-in-molecules method of Moffitt [9], the diatomics-in-molecules method of Ellison [10], and the counterpoise model of Boys [11] are all successful in some respects. Unfortunately none of them are helpful for the most commonly met problem by those who are trying to fit potential functions to data, which is how to join up data produced by different methods on different regions of the surface. I believe there is no answer to this except scientific judgement.

We are still in the situation where it is not possible to cover a polyatomic surface with *ab initio* points of sufficient accuracy that inaccuracies can be ignored; the exceptions are sufficiently rare to prove the rule. For this reason potential functions which are based in part on empirical data will always be superior to those that are fully *ab initio*. However, it would be exceptional to derive a potential solely from empirical data. Spectroscopic and thermodynamic data generally only give information about minima on the surface; often only the lowest minimum. Kinetic data generally only indicate barrier heights on reaction paths. Molecular beam scattering and equilibrium or transport gas phase data may provide sensitive tests of potentials but we cannot get the potential directly from the data. It is from the fusion of empirical and *ab initio* data that one obtains the best potential functions. This is, I believe, the reason for the success of LEPS functions [2], the DIM method [10] and the many-body expansion with empirical one and two-body terms [3].

The message is this: if one gets the asymptotic limits of a surface right, and one gets important stationary points right, and the permutational symmetry of the potential function right; and if the potential function has no unphysical behaviour in other regions of the surface, then one is well on the way to a good function. Further optimization can then be imposed to fit *ab initio* data. This is not to underate the importance of *ab initio* data. It may commonly provide the only evidence for other minima (isomers) on a surface, and for excited electronic states empirical data is frequently more sparse than for the ground state.

4. POTENTIAL ENERGY FUNCTION FOR HCO

From the numerous studies we have made on potential energy functions using the many body expansion method I pick out the ground state surface of HCO [12] to show the most successful use of empirical and *ab initio* data. The asymptotic limits (two-body terms) to this surface are provided by the diatomics $CH(X, {}^2\Pi)$, $CO(X, {}^1\Sigma^+)$ and $OH(X, {}^2\Pi)$ and all are represented by empirical potentials which are quite accurate. The three-body term was fitted to the empirical energy, geometry and force constants of HCO, the *ab initio* energies and geometries of the isomer HOC, the isomerization saddle point, and the saddle points for taking the hydrogen atom away from HCO or HOC. The whole three-body term was expressed through 25 parameters; a mixture of reference bond lengths, polynomial coefficients and exponents. The resulting potential function has been used for classical trajectory studies of the ro-vibrational excitation of CO by H and the cross sections of the reactions

$$O(^3P) + CH(^2\Pi) \rightarrow \begin{cases} H(^2S) + CO(^1\Sigma^+) \\ C(^3P) + OH(^2\Pi) \end{cases}$$

There is currently no evidence to doubt the reliability of the surface.

5. CONCLUSION

Potential energy surfaces are more accurate and much more available than they were ten years ago but there is plenty of room for improvement. To look on the bright side I would say that the subject is advancing at about the same rate as the techniques for performing dynamical calculations.

REFERENCES

[1] J.T. Vanderslice, E.A. Mason, W.G. Maisch and E.R. Lippincott, J. Mol. Spectrosc., 3 (1959) 17; 5 (1960) 83.

[2] P.J. Kuntz, E.M. Nemeth, J.C. Polanyi, S.D. Rosner and C.E. Young, J. Chem. Phys., 44 (1966) 1168.

[3] Molecular Potential Energy Functions, J.N. Murrell, S. Carter, S.C. Farantos, P. Huxley and A.J.C. Varandas, J. Wiley & Sons, 1984.

[4] S. Carter, Department of Chemistry, The University of Reading, Reading RG6 2AD, UK.

[5] N.L. Allinger, Adv. Phys. Org. Chem., 13 (1976) 1.

[6] A.R. Curtis in Numerical Approximations to Functions and Data, Ch. 4, Ed. J.G. Hayes, Athlone Press, 1970.

[7] A.M. Arthurs and A. Dalgarno, Proc. Roy. Soc. A, **256** (1960) 50.

[8] J.N. Murrell and B. Hudson, unpublished work.

[9] W. Moffitt, Proc. Roy. Soc. A, **210** (1951) 245.

[10] F.O. Ellison, J. Amer. Chem. Soc., **85** (1963) 3540.

[11] S.F. Boys and F. Bernardi, Molec. Phys., **19** (1970) 553.

[12] J.N. Murrell and J.A. Rodriguez, J. Molec. Struct. (Theochem), **139** (1986) 267.

THE CALCULATION OF POTENTIAL ENERGY SURFACES FOR REACTIVE SYSTEMS

David M. Hirst
Department of Chemistry
University of Warwick
Coventry
CV4 7AL
UK

ABSTRACT. The criteria to be satisfied in the calculation of potential energy surfaces for reactive systems are outlined and accounts are given of the molecular orbital, multi-configuration SCF and configuration interaction methods. Some recent calculations from the literature are discussed briefly. Selectivity is of importance in ion-molecule reactions for which different electronic states of the reactant ion are possible or for which there are several competing product channels. Potential energy surfaces for the reactions $Si^+ + H_2$ and $Al^+ + H_2$ are discussed.

1. INTRODUCTION

The potential energy surface plays a key role in the understanding of the dynamics and kinetics of chemical reactions [1]. An increasing amount of information about molecular potential energy surfaces is being obtained from *ab initio* calculations and in this chapter we give a general outline of the criteria that must be satisfied, the methods that can be used in such calculations together with some examples.

A system of electrons and nuclei is described by the total wavefunction $\Psi(R,r)$ which is obtained by solving the Schrödinger equation

$$[T_{nuc}(R) + T_{el}(r) + V_{nn}(R) + V_{ne}(R,r) + V_{ee}(r)]\Psi(R,r) = E\Psi(R,r)$$

$$(1)$$

In this equation $T_{nuc}(R)$ is the kinetic energy operator for the nuclei, at positions R, $T_{el}(r)$ is the kinetic energy operator for the electrons, at positions r, and the electrostatic potentials arising from nuclear-nuclear, nuclear-electron and electron-electron interactions are given by the terms $V_{nn}(R)$, $V_{ne}(R,r)$ and $V_{ee}(r)$ respectively. It is customary to make the Born-Oppenheimer separation in which it is assumed that the interaction between electronic motion and nuclear motion is negligible. For dynamical calculations for reactions which occur on a single potential energy surface this is a reasonable assumption but for non-adiabatic processes the interaction terms have to be taken into account. With the assumption of the separability of electronic and nuclear motion, the total wavefunction $\Psi(R,r)$ is written as the product of an electronic wavefunction $\Phi(R,r)$, which depends

J. C. Whitehead (ed.), Selectivity in Chemical Reactions, 379–391.
© 1988 by Kluwer Academic Publishers.

parametrically on the positions of the nuclei, and a nuclear wavefunction $\chi(R)$

$$\Psi(R,r) = \Phi(R,r)\chi(R) \tag{2}$$

The electronic wavefunction $\Phi(R,r)$ is obtained by solving the electronic Schrödinger equation

$$[T_{el}(r) + V_{ne}(R,r) + V_{ee}(r)]\Phi(R,r) = W(R)\Phi(R,r) \tag{3}$$

and nuclear motion is described by the nuclear Schrödinger equation

$$[T_{nuc}(R) + V_{nn}(R) + W(R)]\chi(R) = E\chi(R) \tag{4}$$

The sum $U(R)$ of the terms $V_{nn}(R)$ and $W(R)$ represents the potential energy to which the nuclei are subject and the function $U(R)$ is known as the potential energy function or potential energy surface. Any dynamical calculation requires knowledge of the potential energy function $U(R)$. In order to obtain $U(R)$ it is necessary to solve the electronic Schrödinger equation to obtain $W(R)$ which represents the contribution of the electronic distribution to the potential experienced by the nuclei. It is, of course, impossible to solve the electronic Schrödinger equation exactly for systems containing more than one electron and in this chapter we shall be concerned with *ab initio* methods which can be used to give a reasonably accurate description of the electronic contribution $W(R)$ to $U(R)$.

2. COMPUTATIONAL METHODS

For a reactive system, in which bonds are being broken or formed, the potential energy function must give comparable descriptions of the reactant and product asymptotes as well as describing realistically the surface connecting the two asymptotes. If we take the simple example of the dissociation of methane to a methyl radical CH_3 and a hydrogen atom, it is essential that the potential energy function reproduces the dissociation energy reasonably accurately and also has the correct shape as the hydrogen atom is pulled away [2].

The electronic structures of molecules are most commonly discussed in terms of molecular orbital theory whereby the electronic wavefunction $\Phi(R,r)$ for a closed shell system containing an even number of electrons is represented by a Slater determinant of the form

$$\Phi(R,r) = \frac{1}{\sqrt{(n!)}} \begin{bmatrix} \psi_1(r_1)\alpha & \psi_1(r_1)\beta & \cdots & \psi_{n/2}(r_1)\beta \\ \cdots & \cdots & \cdots & \cdots \\ \psi_1(r_n)\alpha & \psi_1(r_n)\beta & \cdots & \psi_{n/2}(r_n)\beta \end{bmatrix} \tag{5}$$

where ψ_i are the molecular orbitals. Writing the wavefunction as a determinant in this way ensures that the antisymmetry requirement of the Pauli principle is satisfied. This form of wavefunction is appropriate to an independent particle model in which the interactions between electrons (electron correlation) are ignored. For open shell systems the wavefunction is often more complicated and can be expressed as a linear combination of a relatively small number of Slater determinants in order to ensure that the wavefunction is an eigenfunction of the spin angular momentum operator S^2. This form of wavefunction is capable of giving a reasonably good description of a molecule in its equilibrium configuration but is not capable of describing bond breaking processes such as

$$CH_4 \rightarrow CH_3 + H \tag{6}$$

because of the constraint of doubly occupied orbitals. At the dissociation asymptote there are single electrons on each of the fragments CH_3 and H and a single Slater determinant cannot transform smoothly from the closed shell molecular wavefunction to the open shell configuration required to describe CH_3 + H.

In order to describe correctly the dissociation process, it is necessary to use the method of configuration interaction (CI). A result which is qualitatively correct for the dissociation of methane can be obtained by writing the electronic wavefunction Φ as a linear combination of two terms

$$\Phi = C_0\Phi_0 + C_1\Phi_1 \tag{7}$$

where Φ_0 and Φ_1 represent the electronic configurations

$$\Phi_0 = 1a_1^2 2a_1^2 1e^4 3a_1^2 \tag{8}$$

$$\Phi_1 = 1a_1^2 2a_1^2 1e^4 4a_1^2 \tag{9}$$

and C_0 and C_1 are coefficients whose values are obtained in a variational calculation. The equilibrium configuration of CH_4 is described reasonably well by Φ_0 but at the dissociation asymptote the wavefunction has the form $0.69(\Phi_0 - \Phi_1)$ [3].

Thus in the calculation of potential energy surfaces for reactive systems there are, in general, two stages. Firstly it is necessary to obtain a set of molecular orbitals $\{\psi_i\}$ and then, in order to correct for the inadequacies of the independent particle model, it is necessary to make allowance for electron correlation. For the calculation of potential energy surfaces this is best done by the method of configuration interaction.

2.1 The Molecular Orbital Method

The best single determinantal wavefunction of the form of Eq. 5 is obtained by minimizing the energy W_{approx} given by

$$W_{approx} = \frac{\int \Phi^*(R,r)H_{e1}\Phi(R,r)d\tau}{\int \Phi^*(R,r)\Phi(R,r)d\tau} \tag{10}$$

and by the variation theorem the optimized energy is an upper bound to the electronic energy $W(R)$. The volume element $d\tau$ represents integration with respect to all of the electronic coordinates. The condition of minimum energy leads to the Hartree-Fock equations which can be solved numerically for atoms but cannot be solved for molecular systems. However, solutions can be obtained by the Roothaan procedure in which the molecular orbitals ψ_i are expanded in terms of a set of basis functions $\{\chi_r\}$

$$\psi_i = \sum_r c_{ir}\chi_r. \tag{11}$$

In order to obtain a satisfactory set of molecular orbitals it is essential to choose an adequate basis set. The choice of basis sets has been discussed extensively in the literature [4-6] and we limit ourselves here to stating that in general one needs a basis set of at least double zeta plus polarization (DZP) quality. By this we mean that we include two functions for each of the atomic orbitals included in a qualitative discussion of the bonding (a double zeta basis) and in addition, to represent distortions of the orbitals occurring in bond formation, we include orbitals of higher angular momentum (polarization functions), namely p functions for hydrogen atoms and d functions for first and second row atoms. Thus for methane one would have two s functions and a set of p functions for each hydrogen atom and four s functions, two sets of p functions and one set of d functions for the carbon atom.

2.2. The Multi-configuration SCF Method

However, as mentioned above, a molecular orbital function will not, in general, be a satisfactory zeroth order wavefunction for a molecular potential energy surface because it is incapable of describing dissociation correctly. Qualitatively correct asymptotic behaviour can usually be obtained by using a relatively simple CI wavefunction as illustrated in Eq. 7. It is necessary to include the configurations Φ_i appropriate to the dissociation asymptotes and to the equilibrium configuration. The accuracy of the description will be increased if the calculation simultaneously optimises the linear variational coefficients C_0, C_1 in Eq. 7 and the coefficients c_{ir} used in the expansion of the molecular orbitals ψ_i in terms of the basis set $\{\chi_r\}$ in Eq. 11. This is a multi-configuration SCF (MC-SCF) calculation and if the configurations Φ_i are carefully chosen, the method is capable of giving very reasonable potential energy surfaces [7].

The wavefunction of Eq. 7 for the dissociation of methane can be cast in a different form by writing it in terms of determinants in which the condition of double occupancy is relaxed i.e.

$$\Phi = 1a_1^2 2a_1^2 1e^4 3a_1 \alpha 3a_1'\beta \tag{12}$$

where the orbitals $3a_1$ and $3a_1'$ are different. In the region of the equilibrium configuration, the orbitals $3a_1$ and $3a_1'$ will be very similar but as the dissociation limit is approached one of them will transform into an a_1 orbital on CH_3 and the other into a 1s orbital on the hydrogen atom. This is an example of a generalized valence bond (GVB) wavefunction [8]. For bond breaking processes such as the methane dissociation, the GVB method provides very useful zeroth order wavefunctions.

The MC-SCF method as outlined above involves an element of choice in the selection of the configurations to be included. This may lead to the omission of a configuration which may be quite important. This problem can be overcome in the complete active space SCF (CASSCF) version of the MC-SCF method [9]. In this method the molecular orbitals are partitioned into three subspaces, namely the inactive orbitals, the active orbitals and the virtual orbitals. The inactive space includes those orbitals which can reasonably be regarded as always being doubly occupied. In the case of methane this would include the $1a_1$ orbital. The active space included the occupied valence orbitals and those orbitals to which excitations are included in the construction of the configurations Φ_i. In the case of methane the simplest set of orbitals would be $2a_1$, $3a_1$, $4a_1$ and $1e$. The remaining orbitals constitute the virtual space. The configurations included in the CASSCF calculation consist of all of the configurations that can be generated by arranging the active electrons in the orbitals of the active space. For a small molecule the active space will usually consist of the occupied valence orbitals and the corresponding anti-bonding orbitals. For a more complete treatment of electron correlation a larger active space can be chosen. However, as the size of the active space is increased, the number of configurations generated increases very rapidly. The development of efficient algorithms for CASSCF calculations is an active area of research and programs capable of handling very large numbers of configurations are available [10,11].

2.3 Configuration Interaction

Use of an MC-SCF, GVB or CASSCF method results in a potential energy surface which describes correctly bond breaking and dissociation processes. However, unless a very large CASSCF calculation has been performed, one has not taken sufficient account of the neglect of electron correlation in using a wavefunction derived from the independent particle model. In order to approach quantitative accuracy it is necessary to take into account the fact that the contribution to the energy from electron correlation will vary over the potential energy surface. To do this one has to include more extensive CI than that arising from the fairly limited number of configurations included in a MC-SCF calculation. A full CI calculation for methane would consist of writing the wavefunction as

$$\Phi = \sum_{i=0}^{N} C_i \Phi_i \tag{13}$$

where

$$\Phi_0 = 1a_1^2 2a_1^2 1e^4 3a_1^2 \tag{14}$$

and the remaining N configurations Φ_i are all of those functions of the same spin and symmetry as Φ_0 which can be obtained by promoting electrons from the occupied orbitals of Φ_0 to the unoccupied orbitals. The coefficients C_i are optimized in a variational calculation but, in contrast to the MC-SCF or CASSCF methods, there is no optimization of the molecular orbitals ψ_i in the functions Φ_i. A full CI calculation represents the best calculation that can be made with a given basis set. A number of full CI calculations have been reported recently [12] and provide benchmarks for other work. A full CI calculation is only feasible for small molecules or for rather restricted basis sets. The major contribution to the electron correlation can be taken into account by limiting the configuration list to single and double excitations from those configurations included in the MC-SCF or CASSCF calculations, provided that the CASSCF calculation is based on a modest active space. The direct CI method developed by Roos and Siegbahn [13] makes it possible to perform CI calculations with over a million configurations. An alternative approach is to truncate the configuration list by selecting the most important configurations by a perturbation technique [14].

An alternative approach to the treatment of electron correlation is to use many-body perturbation theory [15,16]. However, these methods are usually only valid for cases where there is one dominant configuration in the CI wavefunction. Thus in the case of the dissociation of methane one can use many-body perturbation theory for the equilibrium configuration and for the dissociation asymptote but the bond breaking process is not described correctly [2,3,17] (Fig. 1). Work is in progress on the development of many-body perturbation and coupled cluster techniques which are applicable to potential energy surface calculations and which will give a correct description of the bond breaking process [18].

3. APPLICATIONS

3.1. Some Examples from the Recent Literature

The calculation, by *ab initio* methods, of potential energy surfaces for reactive systems has been reviewed recently by Dunning *et al.* [19,20]. These articles present surveys of the results of a number of

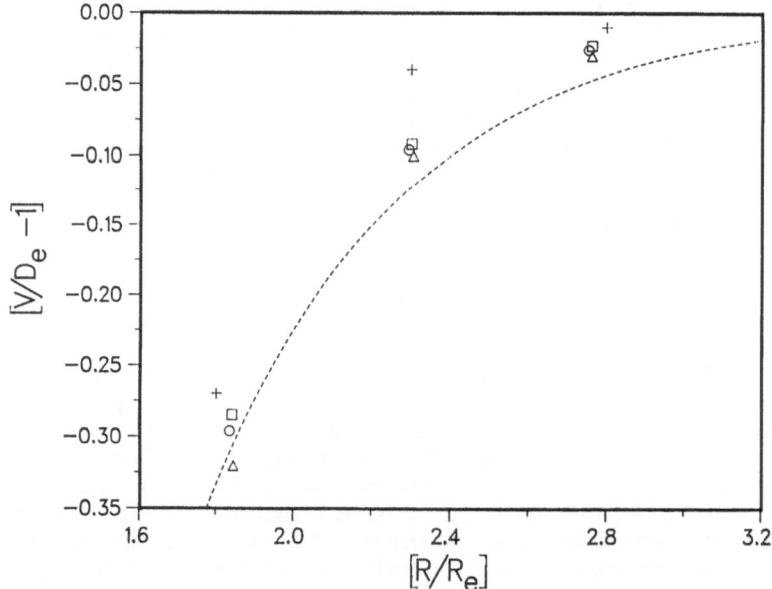

FIGURE 1. Reduced potential energy curves $[V(R)/D_e-1]$ versus R/R_e for CH stretching potential in $CH_4 \rightarrow CH_3 + H$. + MP4,
□ spin-projected MP4, O CI (Ref. 3), Δ CI (Ref. 17). The dashed line is the Morse curve derived from the CI calculations in Ref. 3. (Reproduced, with permission, from Ref. 2].

recent investigations for neutral systems such as $O(^3P) + H_2, H + HX(X = \text{halogen})$, $CH + H_2$, $OH + H_2$. A broad review of potential energy surfaces for reaction dynamics has been given by Truhlar *et al.* [21]. We mention briefly two recent calculations which indicate the sort of information which can currently be obtained from *ab initio* calculations.

The reactions

$$N + O_2 \rightarrow NO + O \qquad (15)$$

and

$$O + N_2 \rightarrow NO + N \qquad (16)$$

have been considered by Walch and Jaffe [22]. The calculations were restricted to the $^2A'$ and $^4A'$ surfaces for reaction (15) and to the $^3A'$ and $^3A''$ surfaces for reaction (16). For these surfaces ground state reactants are correlated with ground state products. Large CASSCF calculations were made with between 1404 and 2652 configurations and for selected portions of the potential energy surfaces very extensive CI was included. For reaction (15) the calculations yielded a value for the exothermicity which was in reasonable agreement with experiment. The $^2A'$ and $^4A'$ surfaces were qualitatively similar with early barriers of 42.7 and 75.3 kJ mol^{-1} respectively. The calculated barrier height for the $^2A'$ surface is 8-13 kJ mol^{-1} higher than the experimental activation energy. However, the accurate calculation of barrier heights on potential energy surfaces is very difficult and we will discuss this further below. Reaction (16) is endothermic and the calculated endothermicity was comparable with the experimental value. The calculations indicated an additional barrier in the N + NO region of the $^3A''$ surface of about 4 kJ mol^{-1} but this was thought to be a consequence of limitations in the calculations. There is a much larger barrier (of about 62 kJ mol^{-1}) in the product region of the $^3A'$ surface. These differences were interpreted in terms of the different strengths of the initial N-NO bond.

A very comprehensive treatment of the Ca + HF system has been made by Jaffe *et al.* [23] in which 175 points on the potential energy surface were obtained in CI calculations based on CASSCF orbitals using an active space consisting of Ca 4s and 4p, F 2p and H 1s orbitals. This surface was then fitted to an analytical function which was subsequently used in quasi-classical trajectory calculations. The error in the calculated endothermicity of the reaction of Ca with HF was less than 4 kJ mol^{-1}. The potential surface has a deep well corresponding to a stable H-Ca-F intermediate and consequently many trajectories involved long-lived intermediate complexes. From the trajectory calculations it was deduced that vibrational excitation of HF to $v=1$ is much more effective than relative translational energy in promoting reaction. The reaction cross-sections were found to increase with increasing rotational energy of the HF for fixed translational energy. Escape from the H-Ca-F well to products is facilitated by increasing HF rotation whereas increasing translational energy favours return to reactants.

The calculation of a sufficient number of points to generate a potential energy surface which can be used for dynamical calculations is computationally very expensive and in many investigations attention has been devoted to the calculation of the energy and geometry of the saddle point. Transition states and equilibrium structures can be readily located in molecular orbital or MC-SCF calculations by the use of gradient techniques [24,25]. However, gradient techniques are not routinely applicable to CI calculations. The accurate calculation of barrier heights on potential surfaces is very difficult, even for relatively simple systems such as

$$F + H_2 \rightarrow HF + H. \tag{17}$$

Schaefer [26] has reviewed many of the calculations for this system. Extensive CI with large basis sets yielded a theoretical barrier height of 13.6 kJ mol^{-1} which is considerably larger than the experimental activation energy which is of the order of 4.2 kJ mol^{-1}. It has been suggested that part of this discrepancy may be attributed to differences in the zero-point vibrational energies of the reactants and the transition state. Full CI calculations with limited basis sets [27] have indicated the importance of higher excitations than those included in single and double excitations from the configurations considered in a CASSCF calculation. Schwenke *et al.* [28] have devised a technique for scaling the energy to compensate for deficiencies in the basis set and the neglect of higher excitations. Using their technique, they obtain a value of 6.7 kJ mol^{-1} which is much closer to the experimental activation energy.

3.2 Selectivity in Ion-molecule Reactions

Many chemical reactions occur on a single potential energy surface. For neutral systems the reactants are usually in their ground electronic states and the nature of the products will be defined by the usual correlation rules. Selectivity is important in cases where it is possible to form one of the reactants in several electronic states or when molecular potential energy surfaces are in close proximity and non-adiabatic transitions are important. These effects are particularly important in ion-molecule reactions. Reactant ions are often produced by electron impact or in plasmas and a mixture of electronic states may be present. By varying the conditions of the source it is often possible to produce reactant ions which are predominantly in one electron state. The asymptotes for $X^+ + AB$ and $X + AB^+$ are often close to each other and there may be a considerable range of geometries for which the two potential energy surfaces are in close proximity. In such cases non-adiabatic effects may be important. It is possible in some cases to envisage the formation, by a non-adiabatic transition, of products which would not be expected to correlate with reactants. If, in a particular point group, two surfaces are of different symmetry they will intersect. However, if the symmetry is lowered and the two surfaces transform to the same irreducible representation in the point group of lower symmetry then, for these geometries, non-adiabatic transitions can provide a route from one surface to another.

The reaction cross-section for the reaction

$$Si^+ + H_2 \rightarrow SiH^+ + H \tag{18}$$

has been measured as a function of relative kinetic energy and has been observed to proceed without any activation energy other than the endothermicity of the reaction [29]. We have recently made calculations for potential energy surfaces of SiH$_2^+$ [30]. The saddle point on the collinear $^2\Sigma^+$ surface, which correlates with Si$^+$(^{2}P) + H$_2$ and with SiH$^+$($^1\Sigma^+$) + H, lies at about 0.14 eV above the product asymptote and thus for collinear geometries a small activation energy would be expected. The C_{2v} surface of $^2A_1$ symmetry, which correlates with the above asymptotes, has a large barrier (*ca.* 4.2 eV) for the perpendicular approach of Si$^+$ to H$_2$ and reaction on this surface would require considerable activation energy. However, the $^2B_2$ surface is initially very flat and has a shallow well relative to the reactants. In C_{2v} symmetry, the $^2A_1$ and $^2B_2$ surfaces will intersect but on distortion to C_s symmetry both surfaces transform to $^2A'$ and the intersection will be avoided. Calculations [31] show that there is a range of geometries for which the intersection of the $^2A_1$ and $^2B_2$ surfaces occurs at energies comparable with that of the product asymptote $SiH^+(^1\Sigma^+)$ + H. Thus motion on the lower $^2A'$ surface can provide a low energy route from $Si^+(^2P)$ + H_2 to $SiH^+(^1\Sigma^+)$ + H.

As an example of an ion-molecule reaction in which selectivity is important, we consider the reactions of Al^+ with H_2. The ground state of Al^+ is ^1S but the ^3P state lies at 4.56 eV. In the experimental investigation by Müller and Ottinger [32] it was possible to vary the proportion of the metastable ion by varying the operating conditions of the Coultron hot cathode discharge. They observed chemiluminescence from the reactions

$$Al^+(^3P) + H_2 \rightarrow AlH^+(A\ ^2\Pi) + H \qquad (-2.59\ eV) \qquad (19)$$

$$Al^+(^1S) + H_2 \rightarrow AlH^+(A\ ^2\Pi) + H \qquad (-7.25\ eV) \qquad (20)$$

$$Al^+(^1S) + H_2 \rightarrow AlH^+(B\ ^2\Sigma^+) + H \qquad (-7.57\ eV). \qquad (21)$$

Ground state $AlH^+(X^2\Sigma^+)$ (which cannot be observed in chemiluminescence experiments) can also be formed in the reactions

$$Al^+(^1S) + H_2 \rightarrow AlH^+(X\ ^2\Sigma^+) + H \qquad (-3.82\ eV) \qquad (22)$$

and

$$Al^+(^3P) + H_2 \rightarrow AlH^+(X\ ^2\Sigma^+) + H \qquad (0.84\ eV). \qquad (23)$$

Thus, in order to interpret the data from these experiments one requires information about several potential energy surfaces for AlH_2^+. We have recently completed a set of CI calculations [33] for the surfaces relevant to reactions (19,20,22 and 23). The formation of the second excited state $B^2\Sigma^+$ of AlH^+ in reaction (21) is much more complicated because the products correlate with the third state of $^1\Sigma^+$ - $^1A_1$ symmetry. The basic principles outlined in Section 2 apply to these calculations but one has to ensure that asymptotic separations such as $Al^+(^1S)$ + H_2, $Al^+(^3P)$ + H_2 and $AlH^+(X\ ^2\Sigma)$ + H, $AlH^+(A\ ^2\Pi)$ + H are in reasonable agreement with experiment. It is, of course, necessary to include in the CI those reference configurations required to give correct descriptions of the reactant and product asymptotes. For cases where more than one surface of a given symmetry is required, the reference configurations should be chosen to give a balanced treatment of all of the surfaces.

Consideration of the correlation diagram (Fig. 2) for the AlH_2^+ system indicates that reactions (19,22 and 23) can occur adiabatically on the potential energy surfaces $^3\Pi$ ($^3A'$ or $^3A''$), $^1\Sigma^+(^1A')$ and $^3\Sigma^+(^3A')$ respectively. For the endothermic reactions (19) and (22), the collinear surfaces have no barrier other than that due to the endothermicity. The $^3\Sigma^+$ surface relevant to reaction (23) is attractive and there is no activation barrier for collinear geometries.

Müller and Ottinger deduced that reaction (19) has an activation energy of about 3.5 eV. As mentioned above, the collinear $^3\Pi$ surface correlating $Al^+(^3P)$ + H_2 with $AlH^+(A\ ^2\Pi)$ + H does not have an activation barrier and Müller and Ottinger interpreted their findings in terms of intersections between the $^3A_1$ and 2

388

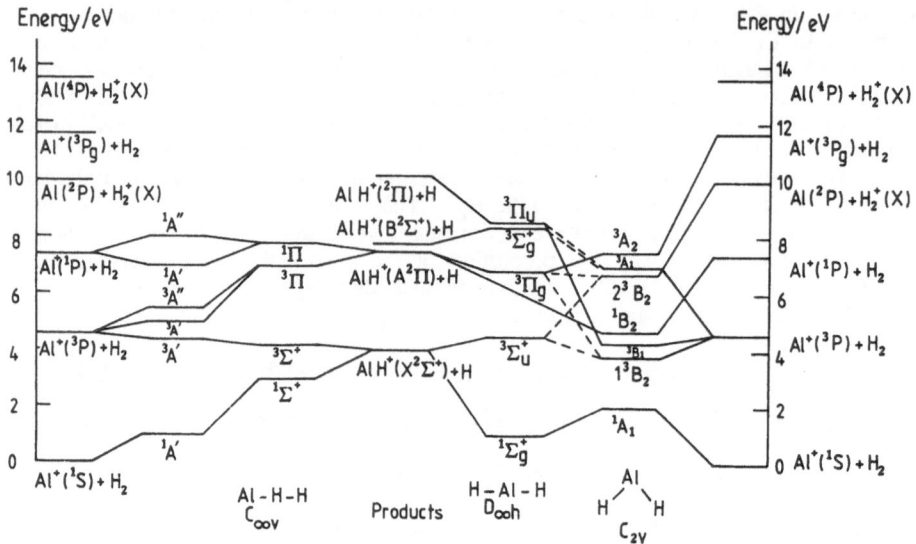

FIGURE 2. Correlation diagram for the AlH$_2^+$ system. (Reproduced, with permission, from Ref. 33).

$^3B_2$ surfaces and between the $^3B_1$ and $^3A_2$ surfaces. The $^3A_1$ and $^3B_1$ surfaces correlate with the reactants and the 2 $^3B_2$ and $^3A_2$ surfaces correlate with the $^3\Pi_g$ state for symmetric H-Al-H$^+$. This in turn correlates with AlH$^+$($^2\Pi$) + H as one hydrogen atom is pulled away. On distortion from C$_{2v}$ symmetry the intersections become avoided and provide routes from reactants to the $^3\Pi_g$ surface. We have examined the regions of intersection of the appropriate surfaces and cannot suggest a simple interpretation for the activation barrier deduced by Müller and Ottinger. Dynamical calculations will be required to provide an understanding of these observations.

Reaction (20) cannot occur on adiabatic surfaces of C$_{\infty v}$ or C$_{2v}$ symmetry because the reactants correlate with $^1\Sigma^+$ or $^1A_1$ and the products with $^1\Pi$ or $^1B_2$. However, the $^1A_1$ and $^1B_2$ surfaces (Figs. 3,4) intersect and the seam of intersection is shown in Fig. 3. On distortion to C$_s$ symmetry both surfaces transform to $^1A'$ and the intersection will be avoided. Thus a trajectory which starts on the surface correlating with $^1A_1$ in C$_{2v}$ can readily reach the surface correlating with $^1B_2$ resulting in the formation of AlH$^+$(A $^2\Pi$) + H.

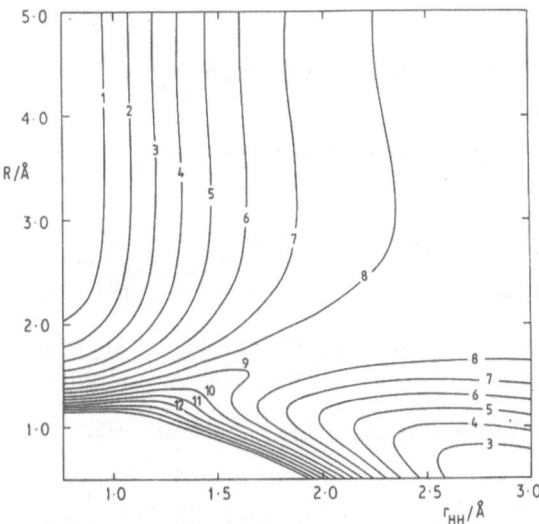

FIGURE 3. Potential energy surface for the $^1A_1$ state of AlH$_2^+$. Contour 1 = -242.84 E_h, contours drawn at intervals of 0.02 E_h. (Reproduced, with permission, from Ref. 33).

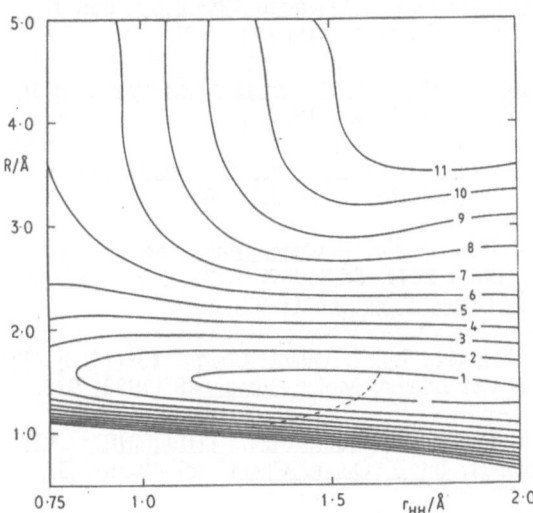

FIGURE 4. Potential energy surface for the $^1B_2$ state of AlH$_2^+$. Contour 1 = -242.68 E_h, contours drawn at intervals of 0.02 E_h. Intersection with $^1A_1$ surface is indicated by the dashed line. (Reproduced, with permission, from Ref. 33).

REFERENCES

[1]   D.M. Hirst, Potential Energy Surfaces: Molecular Structure and
      Reaction Dynamics, Taylor and Francis, London, 1985.

[2]   W.L. Hase, S.L. Mondro, R.J. Duchovic and D.M. Hirst, J. Amer.
      Chem. Soc., 109 (1987) 2916.

[3]   D.M. Hirst, Chem. Phys. Lett., 122 (1985) 225.
[4]   T.H. Dunning and P.J. Hay in Methods of Electronic Structure
      Theory, (Ed. H.F. Schaefer). Plenum Press, New York, 1977, p. 1.

[5]   W.J. Hehre, L. Radom, P.v R. Schleyer and J.A. Pople, Ab Initio
      Molecular Orbital Theory, Wiley, New York, 1986.

[6]   R. Ahlrichs and P.R. Taylor, J. Chem. Phys., 78 (1981) 315.

[7]   A.C. Wahl and G. Das in Methods of Electronic Structure Theory,
      (Ed. H.F. Schaefer), Plenum Press, New York, 1977, p. 51.

[8]   F.W. Bobrowicz and W.A. Goddard in Methods of Electronic
      Structure Theory, (Ed. H.F. Schaefer), Plenum Press, New York,
      1977, p. 79.

[9]   B.O. Roos in Ab Initio Methods in Quantum Chemistry, Part II,
      (Ed. K.P. Lawley), Wiley, Chichester, 1987, p. 399.

[10] H.-J. Werner in Ab Initio Methods in Quantum Chemistry, Part II,
      (Ed. K.P. Lawley), Wiley, Chichester, 1987, p. 1.

[11] R. Shepard in Ab Initio Methods in Quantum Chemistry. Part II,
      (Ed. K.P. Lawley), Wiley, Chichester, 1987, p. 63.

[12] C.W. Bauschlicher and P.R. Taylor, J. Chem. Phys., 86 (1987) 858,
      86 (1987) 1420, 86 (1987) 2844, 86 (1987) 5600; C.W. Bauschlicher
      and S.R. Langhoff, J. Chem. Phys., 86 (1987) 5595.

[13] B. Roos and P.E.M. Siegbahn, Int. J. Quant. Chem., 17 (1980) 485;
      V.R. Saunders and J. van Lenthe, Molec. Phys., 48 (1983) 923.

[14] R.J. Buenker and R.A. Phillips, J. Mol. Struct., THEOCHEM, 123
      (1985) 291; R.J. Buenker, Int. J. Quant. Chem., 29 (1986) 435.

[15] J.A. Pople, J.S. Binkley and R. Seeger, Int. J. Quant. Chem.
      Symp., 10 (1976) 1; J.A. Pople, R. Krishnan, H.B. Schlegel and
      J.S. Binkley, Int. J. Quant. Chem., 14 (1978) 545.

[16] S. Wilson, Electron Correlation in Molecules, Clarendon Press,
      Oxford, 1984.

[17] F.B. Brown and D.G. Truhlar, Chem. Phys. Lett., **113** (1985) 441.

[18] W.D. Laidig, P. Saxe and R.J. Bartlett, J. Chem. Phys., **86** (1987) 887.

[19] T.H. Dunning and L.B. Harding in Theory of Chemical Reaction Dynamics. Volume 1, CRC Press, Boca Raton, Florida, 1985, p. 1.

[20] T.H. Dunning, L.B. Harding, R.A. Bair, R.A. Eades and R.L. Shepard, J. Phys. Chem., **90** (1986) 344.

[21] D.G. Truhlar, R. Steckler and M.S. Gordon, Chem. Rev., **87** (1987) 217.

[22] S.P. Walch and R.L. Jaffe, J. Chem. Phys., **86** (1987) 6946.

[23] R.L. Jaffe, M.D. Pattengill, F. Mascarello and R.N. Zare, J. Chem. Phys., **86** (1987) 6150.

[24] F. Bernardi and M.A. Robb in Ab Initio Methods in Quantum Chemistry. Part I, (Ed. K.P. Lawley), Wiley, Chichester, 1987, p. 155.

[25] H.B. Schlegel in Ab Initio Methods in Quantum Chemistry. Part I, (Ed. K.P. Lawley), Wiley, Chichester, 1987, p. 249.

[26] H.F. Schaefer, J. Chem. Phys., **89** (1985) 5336.

[27] C.W. Bauschlicher and P.R. Taylor, J. Chem. Phys., **86** (1987) 858.

[28] R. Steckler, D.W. Schwenke, F.B. Brown and D.G. Truhlar, Chem. Phys. Lett., **121** (1985) 475.

[29] J.L. Elkind and P.B. Armentrout, J. Phys. Chem., **88** (1984) 5454.

[30] D.M. Hirst and M.F. Guest, Molec. Phys., **59** (1986) 141.

[31] D.M. Hirst, J. Chem. Soc. Farad. II, **83** (1987) 61.

[32] B. Müller and Ch. Ottinger, J. Chem. Phys., **85** (1986) 232.

[33] D.M. Hirst, J. Chem. Soc. Farad. II, **83** (1987) 1615.

ELECTRONIC DEEXCITATION OF OH(A$^2\Sigma^+$) WITH CO(X$^1\Sigma^+$): AN *AB INITIO* STUDY

A. Vegiri[a], S.C. Farantos[b], P. Papagiannakopoulos[b]
and C. Fotakis[a]
Institute of Electronic Structure and
Laser Research Centre of Crete
Iraklion 711 10
Crete
Greece

ABSTRACT. *Ab initio* electronic structure calculations are employed for studying the electronic deexcitation of OH(A$^2\Sigma^+$) in collisions with ground state CO. An attractive interaction potential between the two diatomics is found when carbon approaches oxygen. A conical intersection for linear geometries and an avoided crossing between the X^{2}A' and 2^{2}A' potential surfaces for C$_s$ configurations guarantee the efficiency of electronic quenching when a collision complex is formed. The topography of the potential energy surface reveals that the complex formation depends on the orientations of the two molecules and this can explain the relatively strong relation of the deexcitation rate constants to the rotational quantum number of OH.

1. INTRODUCTION

The hydroxyl radical is an important species in combustion and atmospheric chemistry [1-3]. Therefore it is not surprising that the quenching of the excited state A$^2\Sigma^+$ of OH has attracted considerable investigation over the last years. McDermid *et al.* [4], and Crosley and his co-workers [5-7] have studied the quenching rate of OH(A$^2\Sigma^+$) in collisions with several partners such as He, Ar, CO, N$_2$, O$_2$, H$_2$, H$_2$O and others. It is found that except for the rare gases the quenching cross sections are larger than the gas kinetic. There is also a dependence of the electronic relaxation on the rotational level of OH(A$^2\Sigma^+$). This is manifested with the decrease of the quenching rate of OH(A$^2\Sigma^+$, v'=0) by increasing the rotational excitation [4-8]. Copeland and Crosley [6,7] measured state specific rate constants over the range of rotational levels N'=0-7 and for several partners. Recently Papagiannakopoulos and Fotakis [8] extended these studies to N'=13-17 and to collisions with CO, N$_2$ and H$_2$O and found the same trend in the quenching rate constant. Series of similar experiments were performed with hydride molecules

---

(a) Also Dept. of Physics, University of Crete, Iraklion, Crete, Greece
(b) Also Dept. of Chemistry, University of Crete, Iraklion, Crete, Greece.

J. C. Whitehead (ed.), Selectivity in Chemical Reactions, 393–402.
© *1988 by Kluwer Academic Publishers.*

such as $NH(A^3\Pi)$ [9] and $CH(A^2\Delta)$ [10].

Fairchild *et al.* [5], in order to explain the large cross sections, invented a simple model which is based on the assumption of a complex formation due to long range multipole attractive interactions between the collision partners. In spite of the reasonable correlation found for most of the quenchers, it turns out that an *ab initio* calculation is needed to justify, or not, the assumption of an attractive potential and to clarify the mechanism responsible for the quenching. Working towards this direction we have recently computed the interaction potentials of OH with He. The excited $OH(A^2\Sigma^+)$ + He and the ground $OH(X^2\Pi)$ + He surfaces are repulsive and a $(\Sigma\text{-}\Pi)$ conical intersection at 2 eV above the entrance channel explains the inefficiency of the rare gases to the quenching process [11]. However, the anisotropy of the repulsive potential leads to an efficient rotational energy transfer and this has been studied in the close coupling formalism [12].

In this article we report the results of the MRD-CI *ab initio* calculations for the ground and the first two excited interaction potentials of OH with ground state $CO(X^1\Sigma^+)$. Experimentally it was found [13] that the removal of $OH(A^2\Sigma^+)$ by CO is three orders of magnitude faster than the reaction of $OH(X^2\Pi)$ with CO. The efficiency of deactivation of $OH(A^2\Sigma^+)$ from CO, also decreases with increasing the rotational quantum number of OH, N'. The calculations which we have carried out indeed reveal attractive forces between the two diatomics, the magnitude of which depend on their mutual orientation.

Peyerimhoff *et al.* [14] have calculated the three lowest lying electronic states of the formyloxyl radical, HCO_2 $(^2B_2, {}^2A_1, {}^2A_2)$ for C_{2v} geometries. They find that all three states have potential minima with an energy range of about 0.36 eV from one another and were primarily distinguished from the different behaviour of the surfaces with respect to OCO angle. We are interested in describing that part of the potential energy surface (pes), which corresponds to the approach of OH to CO; this is a dissociation channel of the formyloxyl radical. MCSCF and CI calculations have also been carried out by Feller *et al.* [15] predicting a σ ground state and a low π excited state. The agreement with the results of Peyerimhoff *et al.* [14] is at the CI level.

2. NUMERICAL TECHNIQUES

2.1 Dissociation Limits

By symmetry considerations, we distinguish the following possible dissociation channels of the potential surfaces which are involved in the quenching process of $OH(A^2\Sigma^+)$.

$$HOCO(X^2A') \longrightarrow \begin{array}{l} CO(X^1\Sigma^+) + OH(X^2\Pi) \\ O_2(X^3\Sigma_g) + CH(X^2\Pi) \\ HCO(X^2A') + O(^3P_g) \\ HO_2(X^2A'') + C(^3P_g) \\ CO_2(X^1\Sigma_g) + H(^2S) \end{array}$$

$$HOCO(1^2A'') \longrightarrow \begin{array}{l} CO(X^1\Sigma^+) + OH(X^2\Pi) \\ O_2\ (X^3\Sigma_g) + CH(X^2\Pi) \\ HCO(X^2A') + O(^3P_g) \\ HO_2(X^2A'') + C(^3P_g) \\ CO_2(^3A_2) + H(^2S) \end{array}$$

395

$$HOCO(2^2A') \longrightarrow \begin{array}{l} CO(X^1\Sigma^+) + OH(A^2\Sigma^+) \\ O_2(a^1\Delta_g) + CH(X^2\Pi) \\ HCO(X^2A') + O(^1D) \\ HO_2(X^2A'') + C(^3P_g) \\ CO_2(^3B_2) + H(^2S) \end{array}$$

The energy ordering of the different reaction channels of $OH(A^2\Sigma^+) + CO(X^1\Sigma^+)$ is presented in Fig. 1. We can see that there are only three channels energetically

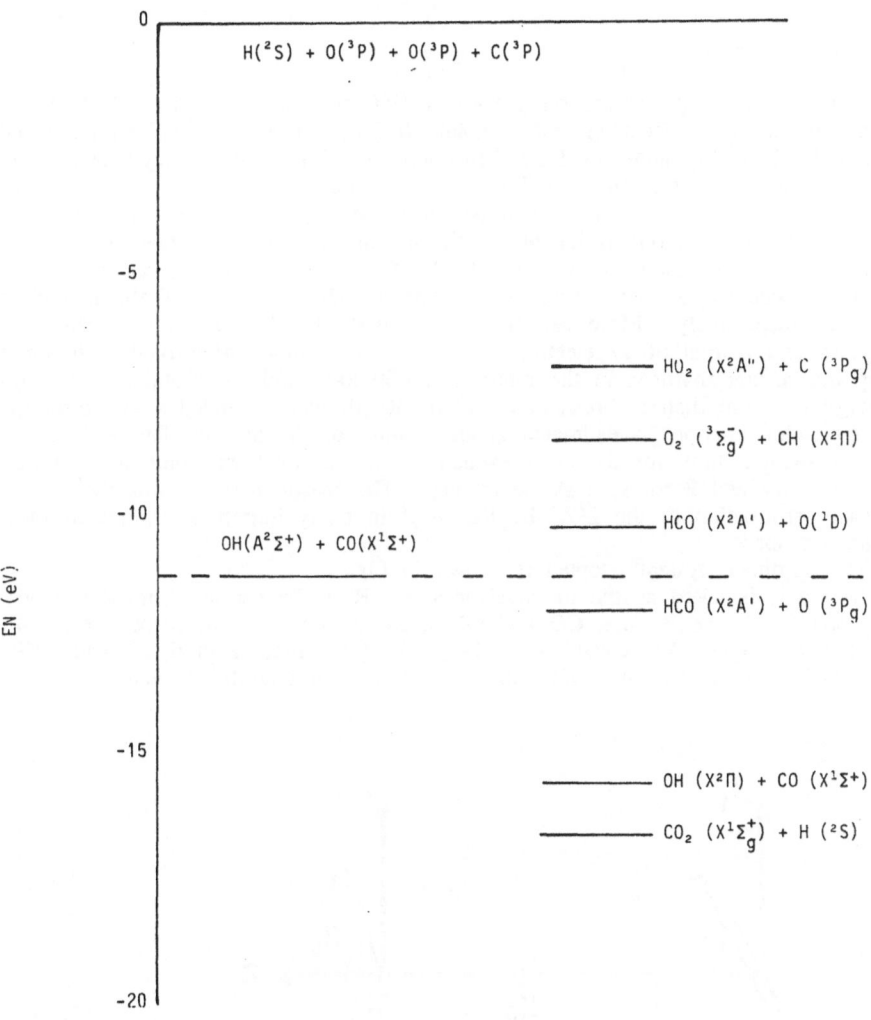

FIGURE 1. Energy diagram for several reaction channels of $CO(X^1\Sigma^+)$ + $OH(A^2\Sigma^+)$.

accessible. $CO_2(X^1\Sigma_g) + H(^2S)$ is accessible only through the (X^2A') state, whereas the $OH(X^2\Pi) + CO(X^1\Sigma^+)$ and $HCO(X^2A') + O(^3P_g)$ are accessible both from (X^2A') and $(1^2A'')$ states.   However, as we shall discuss later on, the approach of H to the carbon end of CO, makes all three surfaces highly repulsive and we do not expect $HCO(X^2A')$ to be formed at least through direct type collisions at room temperature experiments.   On the other hand the production of $HCO(X^2A')$ through complex formation, where the statistical laws are valid, is less probable because of the small exoergicity of this channel.   Therefore we expect ground state OH, CO and CO_2 to be the main products of the quenching of $OH(A^2\Sigma^+)$.

2.2   Numerical Details

The calculations were performed on a VAX 11/750 computer with the MRD-CI programs developed by Buenker and Peyerimhoff [16].   The AO basis set employed is of double $-\zeta$ quality augmented by d-functions on each of the heavy atoms.   The basis was taken from Ref. 14 with the s-bond functions at the centre of each of the CO and CH bonds omitted.   Hydrogen was described by 2s functions in a 4/1 contraction scheme, and one p function with an exponent $\alpha=1.0$.   Thus the calculation for HCO_2 comprises a total of 54 AOs.   Throughout the calculations we kept doubly occupied the inner three shells, corresponding to the K shells in carbon and oxygen respectively.   Moreover the three highest virtual MOs were discarded entirely, so that a total of 17 electrons was distributed among 48 orbitals.   In order to keep the secular matrices in the range of [7000-8000] only configurations lowering the energy by more than a threshold of 30 or 40 μh were included.   The error of the extrapolation energies is estimated about $\pm$ 4mh for the ground state and the $1^2A''$ and about $\pm$ 8mh for the other excited states.   In total we computed 3 roots of A' symmetry and 2 roots of A'' symmetry.   The reason for that was the significant perturbation of the $2^2A'$ by the $3^2A'$ in many important regions of the configuration space.

The coordinate system adopted is shown in Fig. 2.

The z-axis is taken as the internuclear axis.   R is the centre of mass distance and θ_1 and θ_2 the angles that CO and OH make with the z-axis, respectively. Φ is the dihedral angle.   We considered only planar geometries, with $\Phi=0^0$ and 180^0 degrees, and with the CO and OH bonds frozen at their equilibrium values.

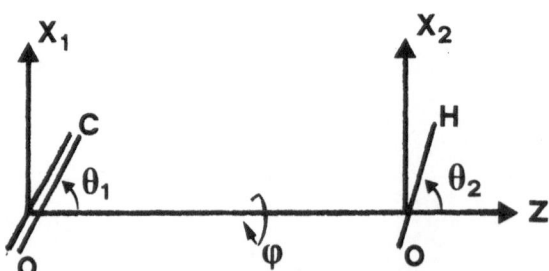

FIGURE 2.   Coordinate system for two colliding diatomic molecules.

3. RESULTS AND DISCUSSION

About 100 energies were computed for R=6.0, 5.0, 4.5, 4.0 and 3.5 a_0, Θ_1=0°, 30°, 60°, 90°, Θ_2=0°, 60°, 90°, 130° and Φ=0°, 180°. From the energies above, we draw the following characteristics of the potential energy surfaces.

(1) As CO approaches OH(A$^2\Sigma^+$) from long distances and with carbon pointing towards oxygen atom, an attractive potential is developed with the minimum energy path along linear geometries.

(2) A Σ-Π conical intersection occurs between the CO + OH(A$^2\Sigma^+$) and CO + OH(X$^2\Pi$) potential surfaces at about R=5.0 a_0, and at an energy 1.4 eV below the entrance channel of the excited state 2^2A'. This conical intersection could be anticipated by using similar arguments as those for He + OH [11].

In the A$^2\Sigma^+$ state of OH there is an empty MO across the line of approach of carbon monoxide. In contrast the X$^2\Pi$ state has a MO occupied by one electron. Therefore it is expected that CO can approach OH closer in the excited state than in the ground state before experiencing the electronic cloud. Thus the OH(X$^2\Pi$) + CO(X$^1\Sigma^+$) potential increases faster and intersects the excited surface. Figure 3 shows the computed energies for the ground and the first excited states at the aforementioned collinear approach (carbon points towards oxygen of OH). Notice the repulsive character of the ground state.

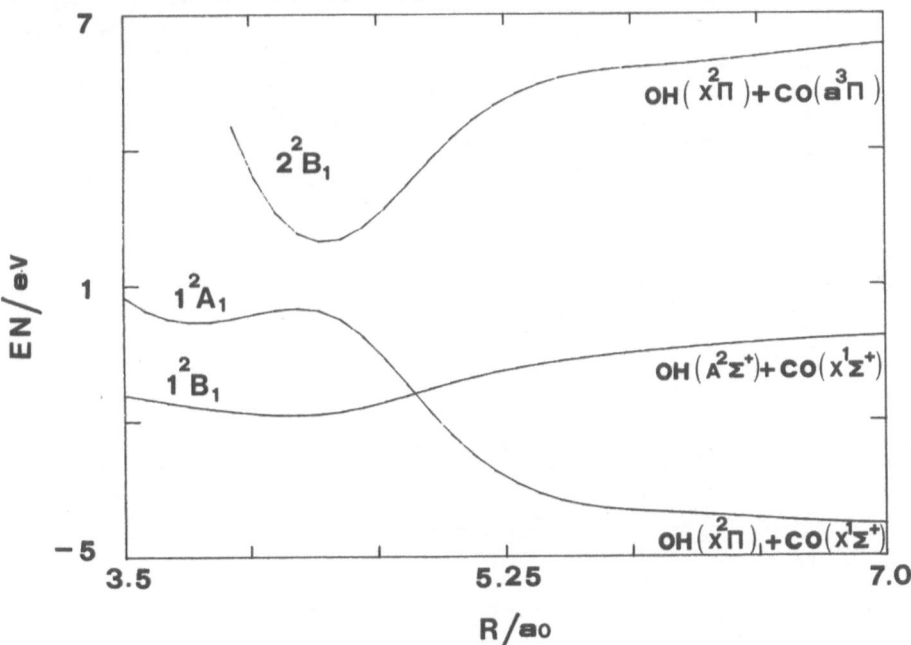

FIGURE 3. Potential energy curves for a collinear approach of CO(X$^1\Sigma^+$) to OH, R is the centre of mass distance.

Similar repulsive behaviour is found when OH approaches collinearly, with H atom pointing towards carbon. In this case however, since both states are repulsive they intersect at energies high enough to be taken into account.

(3) A minimum was located in the $2^2A'$ state at about R=4.0 a_0, $\Theta_1=0^0$ and $\Theta_2=60^0$. This is 1.8 eV below the entrance channel. In order to confirm that this is a relative minimum with respect to the absolute minimum at C_{2v} geometries of the formyloxy radical, we have carried out a few calculations by relaxing the OH bond length around the relative minimum position. It is found that while the energy of the $2^2A'$ and $1^2A''$ states increases, in contrast the X^2A' state quickly drops to its C_{2v} minimum, found previously [14,15]. Although we have not done an extensive minimization search around the minimum of the $2^2A'$ surface, our calculations do not show any route with a low barrier for the system to pass to the absolute C_{2v} minimum of $2^2A'$.

Potential energy contours for $\Theta_1=0^0$ as functions of R and Θ_2 are shown in Fig. 4. These contours are constructed by interpolating between the computed points by cubic splines. Notice the position of the conical intersection and the presence of an energy barrier of about 0.4 eV above the entrance channel, that separates the region of the conical intersection from the relative minimum region, found at

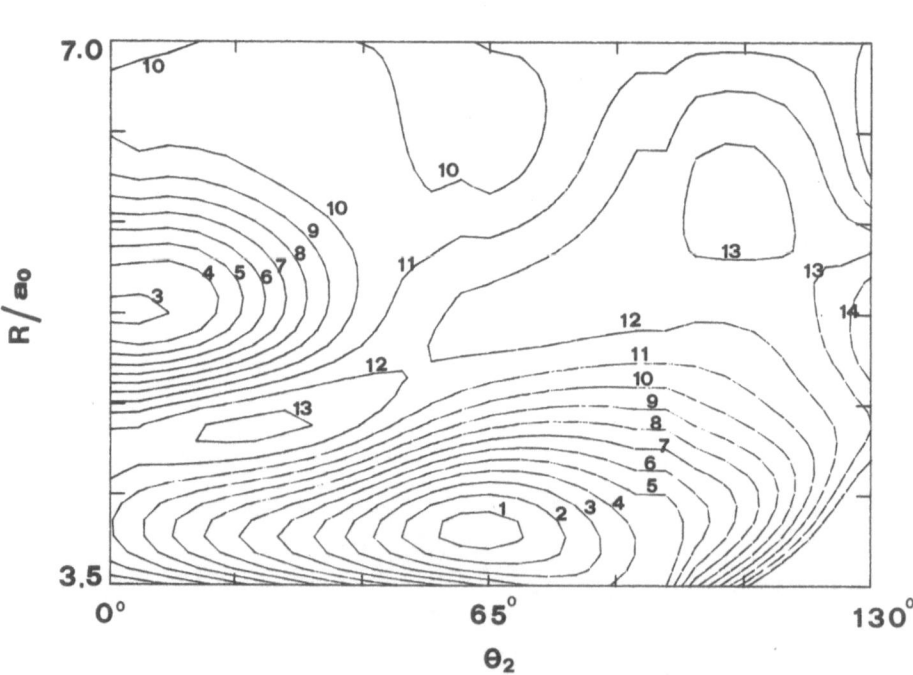

FIGURE 4. Potential energy contours for the $2^2A'$ state of CO + OH with $\Theta_1=0^0$ contour 1 is -1.8 eV and the spacing between the contours is 0.2 eV.

$\Theta_2=60^0$. The formation of a barrier is a consequence of the interaction with the $3^2A'$ state as can be seen in Fig. 3.

This barrier can be surmounted when the angle Θ_1 is relaxed (Fig. 5). Plot 5b shows that for R=4.5 a_0 the system passes from the minimum at the conical intersection to the Cs minimum by overcoming a barrier which lies 0.4 eV below the entrance channel.

(4) In the area around the minimum the two A' symmetry states avoid each other and this is an indication of a non-adiabatic coupling. The avoidance is strongly localized at $\Theta_1=0^0$ (Fig. 6). For R greater than 4.0 a_0, the interaction diminishes, in order to vanish at R=4.5 a_0 (Fig. 6b).

An estimate of the coupling between the two surfaces is obtained by calculating the non-adiabatic matrix elements numerically from the CI wavefunctions [17]. Coupling is mainly induced by the CO rotation and it is an order of magnitude larger than the OH rotation and the radial approach. Numerical values are presented in Table 1.

TABLE 1. Values of the non-adiabatic matrix elements of $\partial_1=0^0$ and $\partial_2=60^0$.

	R/a_0		
	3.5	4.0	4.5
$< X^2A' \mid \frac{\partial}{\partial R} \mid 2^2A' >$	4.8	-17.3	0.2
$< X^2A' \mid \frac{\partial}{\partial \theta_1} \mid 2^2A' >$	12.6	-133.3	3.3
$< X^2A' \mid \frac{\partial}{\partial \theta_2} \mid 2^2A' >$	1.5	-16.2	0.2

(5) The $2^2A'$ and $1^2A''$ surfaces remain approximately parallel and very close to each other around $\Theta_1=0^0$ and for every value of Θ_2 when R varies in the range between 3.5 and 4.5 a_0. This is shown in Fig. 6.

Although for planar geometries these two states, being of different symmetry, do not intersect (since the nonadiabatic matrix elements are zero for out of plane vibrational modes) the planar Cs symmetry is destroyed, and the non adiabaticity, which now becomes non zero, can lead to the deexcitation of $OH(A^2\Sigma^+)$. For triatomics the planar symmetry is conserved and the mixing of different spatial symmetry states can occur through the Coriolis interaction part of the Hamiltonian. Such a behaviour was found in the N_2 + Na system [18], and can also be present for CO + OH.

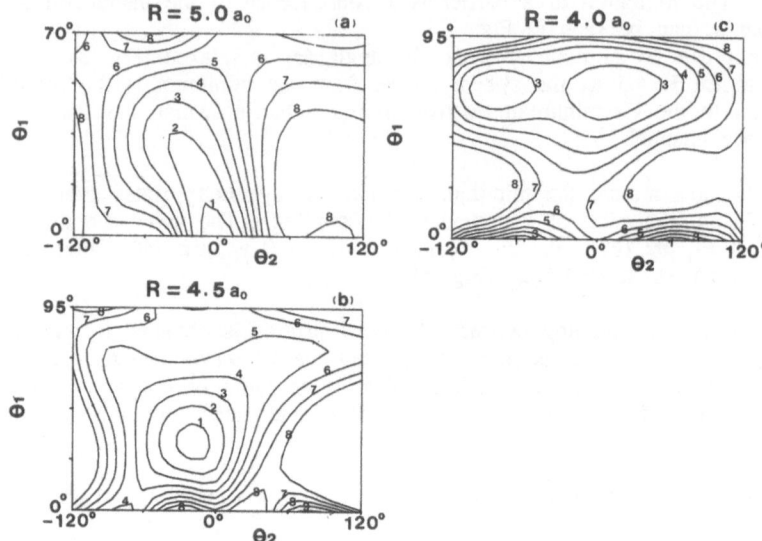

FIGURE 5. Potential energy contours of the $2^2A'$ surface. Contour 1 in Figs. a and b is -1.0 eV and the contour spacing is 0.2 eV. In Fig. c, contour 1 is -1.8 eV and the contour spacing is 0.4 eV.

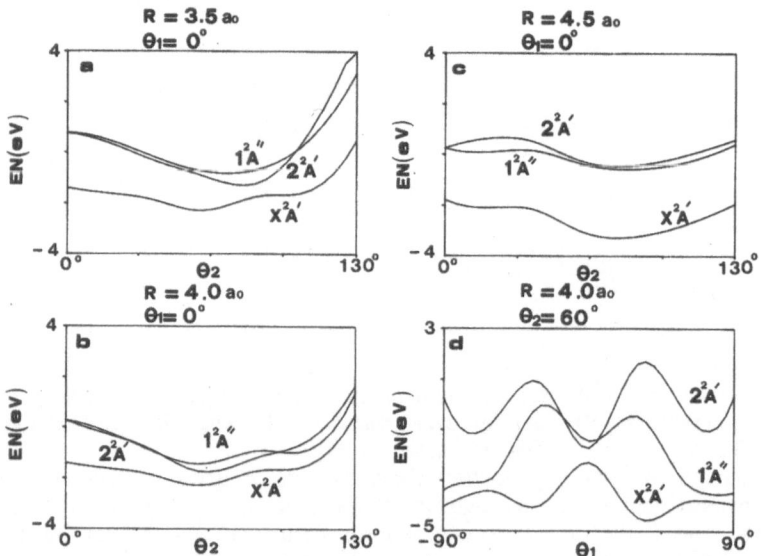

FIGURE 6. Potential interaction curves for CO + OH. Plot d shows the localization of the avoided crossing between X^2A' and $2^2A'$ at the region of the minimum.

4. CONCLUSIONS

The above topographical characteristics of the potential energy surfaces not only explain why the quenching of the $A^2\Sigma^+$ state of OH is so efficient but also they elucidate the detailed mechanisms of the process. CO initially approaches OH collinearly because of the attractive interaction potential. The Σ-Π conical intersection can lead to a radiationless deexcitation at a distance of about 5.0 a_0. This picture partly justifies the model of complex formation introduced by Crosley [5]. However, now the strict orientational dependence of the complex formation is clearly shown. The minimum of the potential is reached when $\Theta_2=60^0$. Around the minimum there is a strong coupling of the $2^2A'$ with the X^2A' state, whereas for out of plane vibrational excitations coupling of the $2^2A'$ with the $1^2A''$ state is also expected to be significant. It was said before that no route was found to connect this minimum with the true C_{2v} minimum of the $2^2A'$ surface of CO + OH($A^2\Sigma^+$). Nevertheless even if the system can pass to the deep C_{2v} minimum, the quenching is still expected to be efficient since, as shown by Peyerimhoff et al. [14], the surfaces approach very close to each other at this region.

It has been demonstrated that a complex is formed only for specific orientations of the molecules. Therefore the rotational excitation of OH will affect the rate of electronic quenching. Generally by increasing the rotational excitation of the diatomic, the probability of complex formation decreases and so does the rate of quenching. Such behaviour has been found for He + $H_2(B^1\Sigma_u)$ the potential energy surface of which has a minimum when the distance of He from the centre of mass of H_2 forms an angle of 45^0 with the H_2 bond length [19]. Classical trajectories have shown the dependence of the quenching cross sections on the rotational level of H_2 [20].

For a quantitative description of CO + OH($A^2\Sigma^+$) deexcitation a dynamical theory is needed and this will be presented in a future publication.

ACKNOWLEDGMENT

We are grateful to Professor Buenker for making the MRD CI programs available to us.

REFERENCES

[1] M.O. Rodgers, J.D. Bradshaw, S.T. Sandholm, S. Kesheng and D.D. Davis, J. Geophys. Res., 90 (1985) 12819.

[2] D.M. Bakalyar, L.I. Davis, Jr., C. Guo, J.V. James, S. Kakos, P.T. Morris and C.C. Wang, Appl. Opt., 23 (1984) 4076.

[3] D.H. Campbell, Appl. Opt., 23 (1984) 689.

[4] S. McDermid and James B. Laudenslager, J. Chem. Phys., 76 (1982) 1824.

[5] P.W. Fairchild, G.P. Smith and D.R. Crosley, J. Chem. Phys., 79 (1983) 1795.

[6]   R.A. Copeland and D.R. Crosley, Chem. Phys. Lett., **107**
      (1984) 295.

[7]   R.A. Copeland, M.J. Dyer and D.R. Crosley, J. Chem. Phys.,
      82 (1985) 4022.

[8]   P. Papagiannakopoulos and C. Fotakis, J. Phys. Chem., **89**
      (1985) 3439.

[9]   A. Holzumahaus and F. Stuhl, J. Chem. Phys., **82** (1985) 3152.

[10] H. Hontzopoulos, Y. Vlahogiannis, C. Fotakis (in preparation).

[11] A. Vegiri, S.C. Farantos, J. Phys. Chem. (submitted).

[12] S.C. Farantos, A. Vegiri, J. Phys. Chem. (submitted).

[13] M. Kaneko, Y. Mori, I. Tanaka, J. Chem. Phys., **48** (1968) 4468.

[14] S.D. Peyerimhoff, P.S. Skell, D.D. May, R.J. Buenker,
      J. Am. Chem. Soc., **104** (1982) 4515.

[15] D. Feller, E.S. Huyser, W.T. Borden, E.R. Davidson,
      J. Am. Chem. Soc., **105** (1983) 1459.

[16] R.J. Buenker, S.D. Peyerimhoff, Theor. Chim. Acta, **35** (1974) 33;
      R.J. Buenker, in Proceedings of Workshop on Quantum Chemistry and
      Molecular Physics, Wollongong, Australia 1970;
      R.J. Buenker in Studies in Physical and Theoretical Chemistry,
      Vol. 21 (Current Aspects of Quantum Chemistry 1981) and
      R. Carbo, Elsevier Scientific Publ. Co., Amsterdam 1982,
      pp. 17-34; R.J. Buenker and R.A. Phillips, J. Mol. Struct. Theo.
      Chem., **123** (1985) 291.

[17] G. Hirsch, P.J. Bruna, R.J. Buenker and S.D. Peyerimhoff,
      Chem. Phys., **45** (1980) 335.

[18] P. Archirel and P. Habitz, Chem. Phys., **78** (1983) 213.

[19] S.C. Farantos, G. Theodorakopoulos and C.A. Nicolaides,
      Chem. Phys. Lett., **100** (1983) 263.

[20] S.C. Farantos, Mol. Phys., **54** (1985) 835.

DIRECT USE OF THE DIATOMICS-IN-MOLECULES METHOD IN LARGE SYSTEMS AND IN DYNAMICAL CALCULATIONS

P.J. Kuntz
Hahn-Meitner Institut GmbH
Glienicker Strasse 100
D-1000 Berlin 39
W. Germany

ABSTRACT.   The rapid calculation of potential energy surfaces is a pre-requisite for dynamical calculations as well as for the investigation of systems having a large number of degrees-of- freedom.   Such calculations can be effected readily by the method of diatomics-in-molecules (DIM), which is capable of treating ground and excited states on a nearly equal footing.   The development of a new computer program enabling application of the method to an arbitrary molecule is described. The utility of the new program is illustrated by reference to three physical systems with widely differing requirements: (1) electronic structure calculations of ionised rare-gas clusters (many degrees-of-freedom); (2) The interaction of excited and ground state alkali atoms in the neighbourhood of a metal surface (excited states, large number of degrees-of-freedom); and (3) the reaction dynamics of $O(^1D) + H_2 \rightarrow OH(X^2\Pi) + H$ (excited states, trajectory calculations).   The DIM program is so constructed as to be directly usable in a trajectory program.

1.   INTRODUCTION

This paper deals with the application of the method of diatomics-in-molecules (DIM) [1-4] to the calculation of potential energy surfaces (PES) for use in the theoretical study of reaction dynamics.   Model potential surfaces for dynamical studies must meet several severe requirements: in regions where the PES is known from other theoretical computations, the model must represent the PES in such a way as to satisfy the criteria of accuracy and availability, already mentioned by John Murrell; in addition, in regions where no other theoretical information is at hand, the model must provide a realistic estimate of the PES.   The DIM method offers a good practical approach to this problem, scoring high on availability and global representation, although it cannot simultaneously reproduce local properties to spectroscopic accuracy.   The computer program to be described here increases the availability of DIM, since it is able to produce the expressions for the DIM hamiltonian matrix elements (neglecting spin-orbit coupling) for an arbitrary polyatomic molecule in an arbitrary state; these expressions are written to a disc-file which can be used directly as input to a dynamics program.   This paper describes the structure of the new program and presents applications to three systems with widely differing requirements.

J. C. Whitehead (ed.), Selectivity in Chemical Reactions, 403–416.
© 1988 by Kluwer Academic Publishers.

2. THE DIM METHOD

DIM distinguishes itself from other approximate quantum mechanical methods in that no calculation of integrals is necessary in the evaluation of the hamiltonian matrix elements of the polyatomic molecule, for these are expressed entirely in terms of the matrix elements of diatomic and atomic hamiltonians in a corresponding basis. This is possible because the polyatomic (and diatomic) basis is constructed from composites which are products of atomic functions:

$$X_M = \prod_{i=1}^{N} \varphi_{k_i}^{\alpha_i}, \tag{1}$$

for $M = 1, N_X$. Here i labels the N atomic centres, M the composite basis functions, k_i the atomic functions, and α_i the chemical species on the centre i (e.g. H, O^+, etc). Note that α_i and k_i depend on the index of the composite, M, and that the superscript α_i implicitly fixes the number of electrons associated with centre i (for composite function M), n_{α_i}. The hamiltonian matrix in the composite basis, B_X, can be expressed in terms of diatomic and atomic fragment matrices in the related composite basis by utilising the (exact) partitioning of the polyatomic hamiltonian operator:

$$\hat{H} = \sum_{i=1}^{N-1} \sum_{j=i+1}^{N} \hat{H}_{ij} - (N - 2) \sum_{i=1}^{N} \hat{H}_i. \tag{2}$$

The MM' matrix element of B_X is expressed as a sum of fragment terms by applying the partitioning of the polyatomic hamiltonian (Eq. 2) appropriate to the electron distribution implied by the index M, (which specifies the superscripts α_i and thereby the number of electrons associated with centre i, n_{α_i}):

$$<X_{M'} | \hat{H} | X_M> = <X_{M'} | \sum_{i=1}^{N-1} \sum_{j=i+1}^{N} \hat{H}_{ij} - (N - 2) \sum_{i=1}^{N} \hat{H}_i | X_M>. \tag{3}$$

The resulting fragment terms are of two kinds: diatomic,

$$<X_{M'}^{ij} | X_M^{ij}> <\varphi_{k_i'}^{\alpha_i'} \varphi_{k_j'}^{\alpha_j'} | \hat{H}_{ij} | \varphi_{k_i}^{\alpha_i} \varphi_{k_j}^{\alpha_j}>,$$

and atomic,

$$<X_{M'}^{i} | X_M^{i}> <\varphi_{k_i'}^{\alpha_i'} | \hat{H}_i | \varphi_{k_i}^{\alpha_i}>,$$

where we have used the notation

$$X_M^{ij} \equiv \prod_{\substack{l=1 \\ l \neq i,j}}^{N} \varphi_{k_l}^{\alpha_1}(M) \tag{4}$$

$$X_M^{i} \equiv \prod_{\substack{l=1 \\ l \neq i}}^{N} \varphi_{k_l}^{\alpha_1}(M) \tag{5}$$

The functions φ are chosen to be orthogonal when all centres are infinitely separated from each other:

$$\langle \varphi_{k_i}^{\alpha_i} | \varphi_{k_j}^{\alpha_j} \rangle = \delta_{\alpha_i \alpha_j} \delta_{k_i k_j} . \tag{6}$$

It is customary to assume that this relation also holds for finite distances between the centres; this is called the no-overlap approximation. The atomic terms in Eq. 3 are therefore zero unless $\alpha_i' = \alpha_i$ for all i; hence, the non-zero terms may be identified with the hamiltonian matrix elements of the atomic species α_i in the basis φ^{α_i}: $b_\varphi^{\alpha_i}$. The diatomic terms in Eq. 3 are zero unless $(n_{\alpha_i'} + n_{\alpha_j'})$ $= (n_{\alpha_i} + n_{\alpha_j})$; i.e. unless the diatomic species associated with centres i and j is the same for functions X_M and $X_{M'}$. The diatomic terms fulfilling this condition have the value

$$b_\chi^{\alpha ij}(m'm) \prod_{\substack{l=1 \\ l \neq i,j}}^{N} \delta_{\alpha_1 \alpha_1'} \delta_{k_1 k_1'} ,$$

where

$$b_\chi^{\alpha ij}(m'm) = \langle \chi_m^{ij} | \hat{H}_{ij} | \chi_m^{ij} \rangle, \tag{7}$$

and the functions $\{\chi_m^{ij}\}$ are the composite basis functions for the diatomic species $\alpha_i \alpha_j$:

$$\chi_m^{ij} \equiv \varphi_{k_i}^{\alpha_i} \varphi_{k_j}^{\alpha_j} . \tag{8}$$

From Eq. 3 it therefore follows that a knowledge of the atomic fragment matrices, $b_\varphi^{\alpha_i}$, and diatomic fragment matrices, $b_\chi^{\alpha_{ij}}$, suffices to construct the polyatomic hamiltonian matrix:

$$B_\chi = \sum_{i=1}^{N-1} \sum_{j=i+1}^{N} B_\chi^{ij} - (N-2) \sum_{i=1}^{N} B_\varphi^{i} . \tag{9}$$

This is the working equation for DIM. It is analogous to the partitioning of the hamiltonian, Eq. 2, but is not exact. A DIM model is constructed by specifying the basis functions φ^{α}_i for each centre i and then constructing the diatomic and polyatomic bases as direct products of these. From Eq. 9 the DIM hamiltonian matrix is then constructed.

There are two considerations which make the DIM procedure somewhat difficult to program: 1) the bases χ and X are not very convenient - one would like to exploit the symmetry (spin and spatial) of the polyatomic and diatomic molecules; 2) the polyatomic basis is defined relative to a fixed coordinate system, so that the internuclear distance of the diatomic fragments does not lie along the z-axis but rather takes up a direction specified by the angles ω, the rotations required to bring the z-axis of the polyatomic coordinate system into coincidence with the internuclear distance vector, r_{ij}. This means that the fragment matrices $b^{\alpha}_{\chi}{}^{ij}$ need to be calculated so as to correspond to the geometrical configuration of the polyatomic molecule. Normally, the diatomic hamiltonian matrices, $b_{\chi'}$, are defined in a basis χ' in which the internuclear axis of the molecule lies along the z-axis ($\omega = 0$). The two bases are related by a rotation:

$$\chi' = \chi R(\omega),$$ (10)

and the relation between the two types of diatomic fragment matrix is

$$b_{\chi} = Rb_{\chi'} \cdot \tilde{R}.$$ (11)

In the present program, the functions $\{\varphi^{\alpha}\}$ are chosen such that the rotation matrices R have real elements.

As mentioned above, it is convenient to transform the polyatomic basis, X, so that the individual functions have a particular spin and spatial symmetry (it is assumed that the direct product functions $\{\varphi^{\alpha} \otimes \varphi^{\beta} \otimes...\}$ are complete in this sense). Denoting these symmetry functions by Ψ, we have

$$\Psi = XT$$ (12)

$$B_{\Psi} = \tilde{T}B_{X}T.$$ (13)

Similarly, the diatomic basis can be transformed into symmetry functions:

$$\psi^{\alpha}{}_{ij} = \chi^{\alpha}{}_{ij} t^{\alpha}{}_{ij}$$ (14)

$$b^{\alpha}_{\chi}{}_{ij} = t^{\alpha}{}_{ij} b^{\alpha}_{\psi}{}_{ij} \tilde{t}^{\alpha}{}_{ij}.$$ (15)

The equations presented here suffice to construct the polyatomic hamiltonian in the symmetry-adapted basis, B_{Ψ}, from the diatomic and atomic fragment matrices, $b^{\alpha}_{\psi}{}^{ij}$ and $b^{\alpha}_{\varphi}{}^i$ respectively. The latter constitute the numerical input to the DIM method which, in addition to the mappings:

$$B_{\chi}^{ij} \leftarrow b_{\psi}^{\alpha ij} \qquad\qquad (16)$$

$$B_{\varphi}^{i} \leftarrow b_{\varphi}^{\alpha i}, \qquad\qquad (17)$$

specifies the DIM polyatomic hamiltonian matrix. The latter is diagonalised at each desired geometrical configuration to obtain the potential energy functions of the polyatomic molecule.

3. THE DIM PROGRAMS

The DIM program package consists of two separate programs: the master program and the potential energy subroutines. The latter program can be appended to some main program (e.g. a trajectory program) which requires information about the PES (e.g. potential energy, derivatives, non-adiabatic coupling, etc.) for a given geometrical configuration. These subroutines require two data files in order to function: the B-file, containing the diatomic and atomic fragment matrices, $b_{\psi}^{\alpha ij}$ and $b_{\varphi}^{\alpha i}$, respectively, and the D-file, containing the information for the construction of B_{ψ} from the fragment matrices. The B-file consists of spline knots for the individual matrix elements of the fragments at selected values of the internuclear distance, r, in addition, information about the long-range behaviour of these elements (for $r \rightarrow \infty$) and about the energy of the various atomic states is present. The D-file contains the mappings (16) and (17) and the transformation matrix T so that B_{ψ} can be constructed from the information in the B-file. The B-file must be constructed laboriously from valence bond (or MO) calculations and semi-empirical data; the D-file is constructed in a few seconds from the master program.

The master program, which constitutes the major advance here, is made up of two principal sections: the model designer and the D-file generator. The model designer is an interactive program which produces the list of polyatomic composite functions, X, which define the DIM model structure. The basic input to this program is kept as short as possible and consists of the attributes of the polyatomic molecule (no. of centres, chemical element on each centre, total charge, and spin symmetry) and the specification of the spin and spatial symetries of the atomic functions, $\varphi^{\alpha i}$, which define the DIM basis. If no restrictions are given, the designer proceeds to construct the list of atomic symmetries on each centre from which the DIM polyatomic basis will be constructed; from this list of "state-groups", the direct product composite basis functions, X, for the polyatomic molecule are generated (these are very near to VB structures, but without spin-coupling).

As an example, consider the molecule H_2O in the $^1A'$ state. Allowing the atomic symmetries $O(^1D_g)$, $O(^3P_g)$, $O^-(^2P_u)$, $O^{--}(^1S_g)$, $H(^2S_g)$, and $H^+(^1S_g)$, one obtains the following state-groups:

1. $O(^3P_g)H(^2S_g)H(^2S_g)$

2. $O(^1D_g)H(^2S_g)H(^2S_g)$

3. $O^-(^2P_u)H(^2S_g)H^+(^1S_g)$

4. $O^-(^2P_u)H^+(^1S_g)H(^2S_g)$

5. $O^{--}(^1S_g)H^+(^1S_g)H^+(^1S_g)$

These state-groups yield 1, 3, 2, 2, and 1 (i.e. a total of 9) polyatomic composite functions respectively.

Since the DIM model is ultimately destined for use in a dynamical calculation, where the subprograms will be called many times, it is best to restrict the size of the basis set as much as possible. The designer allows for three basic restrictions: 1) in the distribution of electrons among the centres, 2) in the number of spin-coupled functions, and 3) in the number of functions in the state-group list. Implementation of these restrictions requires some previous knowledge of the polyatomic wave function for the states of interest. For example, it is known from valence bond calculations [5] that only a small subset of spin-couplings need be taken into account for many molecules; this reduces the size of the DIM model enormously.

The program to generate the D-file starts from the list of polyatomic composite functions. First, it generates the list of diatomic fragments for which information is needed. Then, for each such fragment, it makes a list of diatomic composites and the functions ψ^{α}_{ij} which correspond to them, putting this information into a skeleton B-file, which must later be fleshed out (with the addition of the spline knots referred to above) by the user. The D-file itself is then written out; it consists of the specific algebraic formulae for the elements of the polyatomic hamiltonian, $B\psi$, in terms of the diatomic fragment elements and rotation matrix elements. The generating program removes much of the labour associated with a DIM calculation. The master program performs a great deal of matrix algebra to arrive at its results, but it should be borne in mind that this is done only once; the formula implicit in the D-file is very direct, ensuring that the DIM hamiltonian matrix can later be computed with a minimum of operations.

4. APPLICATIONS

4.1 N-atom Rare-gas Ionic Clusters

As an application of the above program, we present some recent results [6] for rare-gas ionic clusters, which are of current interest [7-13]. The simplest DIM model is obtained by admitting only one electronic state for each atomic centre: neutral atoms are assumed to be in the ^1S_g state and the positive ions are assumed to be in the ^2P_u state. The positive charge is allowed to reside on any of the centres so that there are N state-groups possible for a singly ionized N-atom cluster (Rg stands for rare gas):

1. $Rg(^1S)Rg(^1S) \ldots Rg(^1S)Rg^+(^2P)$

2. $Rg(^1S)Rg(^1S) \ldots Rg^+(^2P)Rg(^1S)$

  .

  .

  .

N-1 $Rg(^1S)Rg^+(^2P) \ldots Rg(^1S)Rg(^1S)$

N $Rg^+(^2P)Rg(^1S) \ldots Rg(^1S)Rg(^1S).$

Corresponding to each of the state-groups there are 3 polyatomic basis functions, since the ^{2}P state is 3-fold degenerate. Denoting the three ^{2}P-functions as x,y,z, and the Rg(^{1}S) function by s, the 3 basis functions for a particular state group k are

kx: ss ... x ... ss

ky: ss ... y ... ss

kz: ss ... z ... ss.

In all, there are 3N polyatomic basis functions for a singly-ionized N-atom cluster.

The DIM hamiltonian matrix **B** is written as a sum of pair contributions, $\mathbf{B}^{ij}$, and single-centre contributions, $\mathbf{B}^i$, where i and j label the atomic centres: The atomic contributions, $\mathbf{B}^i$, are diagonal and the diagonal elements B^i_{kk} are either E(Rg) or E(Rg$^+$), so that adopting the convention that E(Rg) $\equiv$ 0, the sum over i simply yields the ionization potential of Rg, I(Rg); i.e. the atomic terms yield a contribution $-(N - 2) \cdot I(Rg)$ to each diagonal element of the DIM hamiltonian matrix.

The symmetric pair-matrices, $\mathbf{B}^{(ij)}$, consist mostly of zeros except for elements between those basis functions having the positive charge on atom i or j, (there are exactly 6 such functions: ix, iy, iz, jx, jy and jz) and diagonal elements for all the remaining functions. The value of the elements is expressed in terms of the diatomic fragment energies:

$$Rg_2(^1\Sigma_g^+) \equiv S$$

$$Rg_2^+(^2\Sigma_g^+) \equiv G$$

$$Rg_2^+(X^2\Sigma_u^+) \equiv U$$

$$Rg_2^+(^2\Pi_g) \equiv \bar{G}$$

$$Rg_2^+(^2\Pi_u) \equiv \bar{U}.$$

It is convenient to define $Q = \frac{1}{2}(U + G)$ and $J = \frac{1}{2}(U - G)$ with similar definitions for $\bar{Q}$ and $\bar{J}$ in terms of $\bar{U}$ and $\bar{G}$. If atoms i and j lie on the z-axis, the 6 x 6 block corresponding to the functions iy, iz, ix, jy, jz, jx is

$$
\mathbf{B}^{ij}(0,0) =
\begin{array}{c|cccccc}
 & iy & iz & ix & jy & jz & jx \\
\hline
iy & \bar{Q}_{ij} & & & & & \\
iz & 0 & Q_{ij} & & & & \\
ix & 0 & 0 & \bar{Q}_{ij} & & & \\
jy & \bar{J}_{ij} & 0 & 0 & \bar{Q}_{ij} & & \\
jz & 0 & J_{ij} & 0 & 0 & Q_{ij} & \\
jx & 0 & 0 & \bar{J}_{ij} & 0 & 0 & \bar{Q}_{ij}
\end{array}
,
$$

where Q_{ij} and J_{ij} are obtained from the diatomic fragment potential curves at the internuclear distance r_{ij}, and $(0,0)$ denotes the dependence of the matrix on rotational angles (*vide infra*). The remaining diagonal elements of B^{ij} have the value S.

For the case that the vector r_{ij} is not parallel to the z-axis, the above matrix must be constructed from the rotated diatomic fragment matrices as defined in Eq. 11; this is equivalent to transforming $B^{ij}(0,0)$ by a rotation matrix:

$$B^{ij}(\alpha,\beta) = R(\alpha,\beta) \cdot B^{ij}(0,0) \cdot \tilde{R}(\alpha,\beta)$$

where here we have defined

$$R(\alpha,\beta) = 1_{3N-6} \otimes R^{(i)}(\alpha,\beta) \otimes R^{(j)}(\alpha,\beta)$$

and $R^{(i)}(\alpha,\beta)$ is a 3 x 3 matrix connecting the functions iy, iz and ix:

$$
\tilde{R}^{(i)}(\alpha,\beta) \equiv
\begin{array}{c}
iy' \\
iz' \\
ix'
\end{array}
\begin{bmatrix}
\cos\alpha & \sin\alpha\sin\beta & \sin\alpha\cos\beta \\
0 & \cos\beta & -\sin\beta \\
-\sin\alpha & \cos\alpha\sin\beta & \cos\alpha\cos\beta
\end{bmatrix}
\begin{array}{ccc}
iy & iz & ix
\end{array}
.
$$

The matrix $R^{(j)}(\alpha,\beta)$ is a similar 3 x 3 block connecting jy, jz and jx. Carrying out the matrix multiplication yields a symmetric matrix which is diagonal except for the 6 x 6 block already mentioned. This has the structure

$$
B^{ij}(\alpha,\beta) =
\begin{bmatrix}
M & N \\
N & M
\end{bmatrix}
$$

where M and N are both 3 x 3 symmetric matrices which are conveniently written in terms of a matrix $W(a,b)$

$$W_{11} \equiv W_{yy} = a(\cos^2\alpha + \sin^2\alpha\cos^2\beta) + b\sin^2\alpha\sin^2\beta$$

$$W_{21} \equiv W_{zy} = (b - a)\sin\alpha\sin\beta\cos\beta$$

$$W_{31} \equiv W_{xy} = (b - a)\sin\alpha\cos\alpha\sin^2\beta$$

$$W_{22} \equiv W_{zz} = a\sin^2\beta + b\cos^2\beta$$

$$W_{32} \equiv W_{xz} = (b - a)\cos\alpha\sin\beta\cos\beta$$

$$W_{33} \equiv W_{xx} = a(\sin^2\alpha + \cos^2\alpha\cos^2\beta) + b\cos^2\alpha\sin^2\beta.$$

In terms of W, the matrices M and N are

$$M = W(\bar{Q},Q)$$

$$N = W(\bar{J}, J).$$

The angles α and β defining the transformation are the angles of the polar coordinates of the relative vector r_{ij}:

$$r_{ij} = x\hat{i} + y\hat{j} + z\hat{k}$$

$$x = x_j - x_i$$

$$y = y_j - y_i$$

$$z = z_j - z_i.$$

In terms of polar coordinates (R,α,β),

$$x = R\sin\beta\sin\alpha$$

$$y = R\sin\beta\cos\alpha$$

$$z = R\cos\beta.$$

Hence $\cos\beta = z/R$ and $\alpha = \tan^{-1}(y/x)$.

The DIM hamiltonian matrix, $B_{3N \times 3N}$ is now completely determined. The eigenvalues and eigenvectors are determined by solving the usual secular equation:

$$BC = CE$$

i.e. the diagonal eigenvalue matrix, E, is determined by diagonalising B:

$$E = \tilde{C}BC$$

$$\tilde{C} = C^{-1}.$$

As an example, the above procedure is applied to Ar and Xe clusters. The DIM models for all clusters of a particular noble gas are fully defined only when the atomic and diatomic fragment matrices are specified. The former were fixed by taking I(Ar) to be 15.76 eV and I(Xe) to be 12.13 eV [14]; the latter were defined by taking the points on the curves U,G,Ū and Ḡ for Ar_2^+ from the *ab initio* computations of Böhmer and Peyerimhoff [15]. The corresponding points for Xe_2^+ were taken from Wadt [16]. The $Ar_2(^1\Sigma_g^+)$ and $Xe_2(^1\Sigma_g^+)$ interactions were taken from Watts [17]. For each diatomic curve, the points were fitted to a cubic spline function; in the asymptotic region, the interaction was represented by A/R^n, where A and n were chosen to match the spline function at its largest knot. Some relevant input data is collected in Table 1.

In agreement with earlier theoretical work [18], the trimer ions are found to be most stable in the $^2\Sigma_u^+$ state of the linear symmetric molecule. The ground $(^2\Sigma_u^+)$ state is bound by 1.376 eV at R = 4.94 bohr (cf. Böhmer and Peyerimhoff who get 1.35 eV at R = 4.95 bohr); the $^2\Pi_u$ state is very weakly bound at long range and the other states are essentially repulsive in character. The energies of the

TABLE 1

Position of the minima for Ar_3^+ and Xe_3^+ in C_{2v} symmetry and for Ar_2^+ and Xe_2^+. β is the ABC angle, where A,B, and C denote the individual atoms.

		r_{A-BC}[bohr]	r_{BC}[bohr]	$\beta[^\circ]$	E[eV]
Ar_3^+	1	0.00000	9.88311	0.00	-1.37621
	2	4.22669	6.26997	53.44	-0.84734
	3	6.76105	4.62946	71.10	-1.21917
Ar_2^+	-	-	4.6296	-	-1.179
Ar_2	-	-	7.11	-	-0.0122
Xe_3^+	1	0.00000	12.68220	0.00	-1.42412
	2	5.37750	7.70670	54.38	-0.94048
	3	7.34729	6.16493	67.24	-1.22093
Xe_2^+	-	-	6.1639	-	-1.1300
Xe_2	-	-	8.253	-	-0.02432

excited states relative to the ground state agree very well with the earlier results of Wadt [18]. The results for the Xe_3^+ ion are very similar to those for Ar_3^+.

For the larger clusters, our major finding is that the noble gas ionic clusters consist of a small ionic molecule (trimer, tetramer or dimer) surrounded by shells of nearly neutral atoms. The trimer ion is particularly stable and forms the core of clusters up to size 13, where the first shell of neutrals is complete. With the filling of the second shell, there is a tendency for the central ion to change from a trimer to a tetramer and for a small fraction of the positive charge to spread out over the rest of the atoms in the cluster. The clusters are very floppy, there being many geometrical configurations with an energy near the lowest energy found.

Improvements to the DIM model can be effected in two ways: increasing the basis, or incorporation of (small) 3-body effects into the polyatomic matrix elements. The latter procedure has been carried out [19], where account was taken of the induced-dipole induced-dipole forces between the pairs of neutrals under the influence of a neighbouring ion. Strictly speaking, this is no longer a DIM procedure; however, it may point the way to a systematic way of representing the PES: instead of expanding the PES itself in a many-body expansion [20-21], one should first obtain a model hamiltonian for the system and then introduce the many-body effects into the matrix elements [21].

4.2 Alkali Atoms near Surfaces

Inclusion of many-body effects as outlined above is absolutely essential in some systems. An example is the interaction of excited sodium atoms with a tungsten surface covered with sodium ions [22-23]

$$Na^*(^2P_u) + Na | W \rightarrow Na^+$$

$$Na^*(^2P_u) + Na | W \rightarrow Na_2^+.$$

On a molecular scale, a surface is of infinite extent; hence, a model must in principle consider a large number of atoms, especially if long-range forces are important, as they are here. A tractable and yet realistic model must treat a small number of adsorbed surface atoms explicitly, taking the influence of the remaining atoms into account in an approximate way. This is done by assuming that the sodium is adsorbed on the surface in the form of ions, each having an associated image charge; hence, the approaching excited sodium atom sees an infinite dipole layer. This layer determines the position of ionic potential energy surfaces relative to neutral ones and as such influences the position of "avoided crossings" between the adiabatic potentials. A DIM model can be constructed by considering the entire tungsten surface to be an "atom" and then adding (N-1) Na^+ ions and one sodium atom. In addition to the Na_2^+ diatomic fragment interactions (in ground and excited states), it is important to incorporate the interaction between a Na^+ ion and the entire dipole layer into the Na^+ - W "diatomic fragment".

Such a model yields a physically realistic description of the processes occurring in the neighbourhood of the surface. In particular, examination of the DIM polyatomic wave functions leads to a simple harpooning [24] picture for the production of ions: if the electron from $Na^*(^2P_u)$ jumps into the metal surface, Na^+ ions are formed. If, however, the electron jumps onto one of the adsorbed ions, the latter is repelled from the surface and joins the incident sodium atom (now an ion) to form Na_2^+. This process is illustrated graphically in Fig. 1.

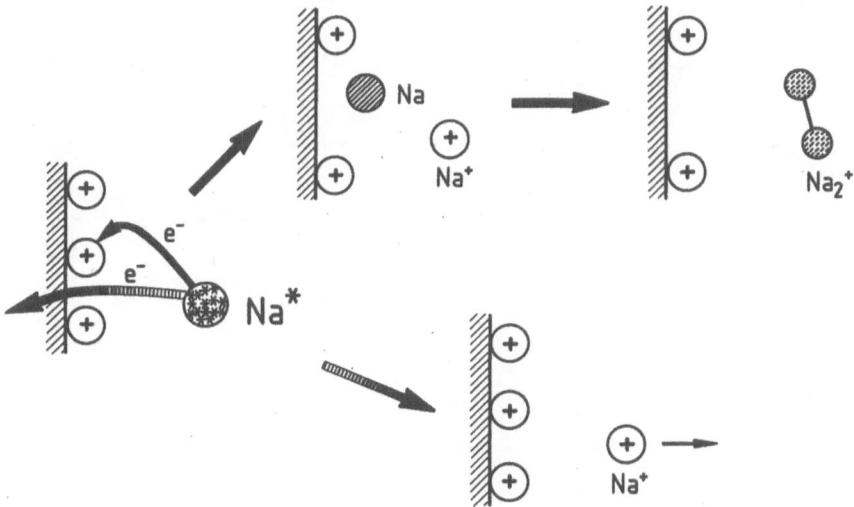

FIGURE 1. Schematic diagram showing the harpooning mechanism for production of Na^+ and Na_2^+ from $Na^*(^2P_u)$ in the neighbourhood of a partially covered metal surface.

4.3 Reaction of $O(^1D_g)$ with H_2

As a last example, we consider the reaction of $O(^1D_g)$ with hydrogen molecules to form OH in its ground state [4]. The state-groups leading to a 9 x 9 hamiltonian matrix have already been listed. The important feature here is that there are 2 $^1A'$ and one $^1A''$ PES which connect the reagents with the products, as can be seen from Fig. 2. Also, it is seen from the figure that, in the collinear abstraction pathway, the Σ and Π surfaces lie very close to each other and in fact cross each other. This crossing is explicitly shown in Fig. 3. Thus, the possibility of surface hopping [25] must be taken into account in the dynamics. Surface hopping trajectory calculations have been carried out using the DIM model outlined here directly in the trajectory program [26]. The entire program was remarkably efficient, considering that the 9 x 9 hamiltonian matrix needed to be diagonalised every time the potential energy was needed. These trajectories showed that, in fact, surface hopping is not very important, largely because the collinear configuration is not preferred over others, and in other configurations there is a direct single-surface pathway connecting the reagents with the products. On the other hand, trajectories starting on the second surface correlate directly with the ground-state OH molecule, and this surface resembles the well-known LEPS surfaces in that linear configurations are favoured. The two surfaces lead to OH products having very different vibrational distributions: trajectories starting on the ground-state surface lead to products with low vibrational energy, whereas the second surface produces highly

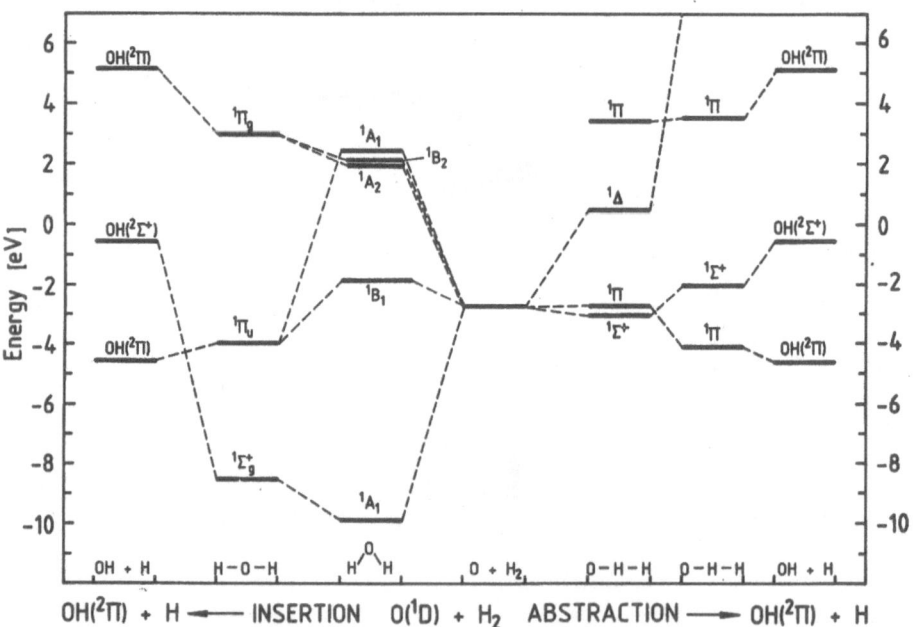

FIGURE 2. State correlation diagram for H_2O based on the energy eigenvalues of the 9 x 9 DIM model.

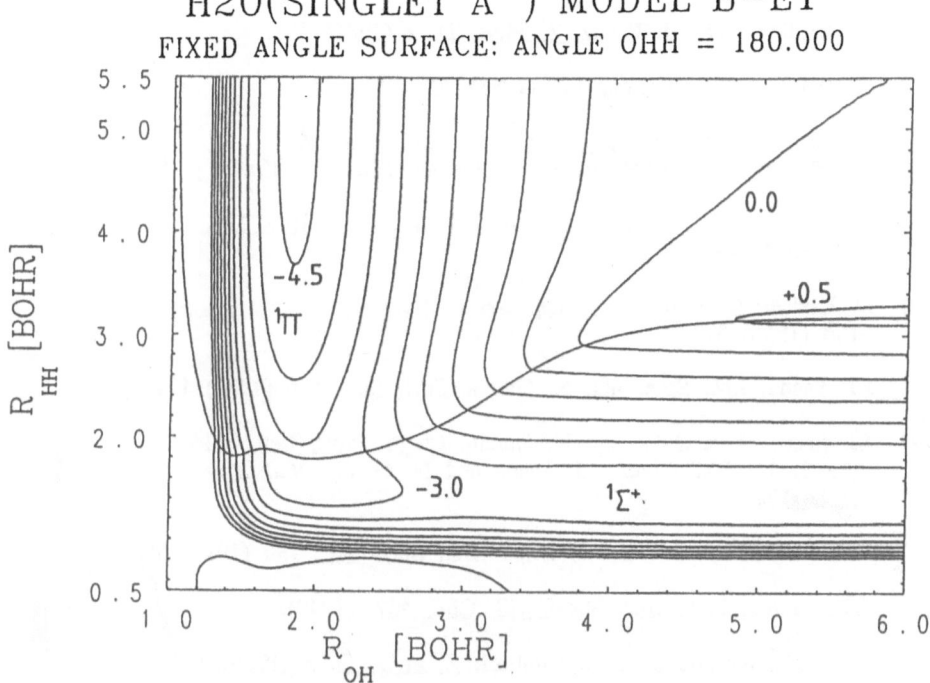

FIGURE 3. The lowest potential energy surface for collinear O-H-H showing the crossing between the Σ(entrance channel) and Π(exit channel) surfaces. Contour energies in eV.

vibrationally excited OH. This result suggests an experimental test: increasing the reagent translational energy of O(^{1}D$_g$) should lead to a sharp increase in vibrational excitation of product OH as soon as the reagent translational energy becomes substantially greater than the barrier on the Π surface.

REFERENCES

[1] F.O. Ellison, J. Am. Chem. Soc., 85 (1963) 3540.

[2] J.C. Tully, Adv. Chem. Phys., 42 (1980) 63.

[3] P.J. Kuntz, in Atom-Molecule Collision Theory, ed. R.B. Bernstein, (Plenum, New York, 1979), chap 3.

[4] R. Polak, I. Paidarova and P.J. Kuntz, J. Chem. Phys., 82 (1985) 2352.

[5]   J. Gerratt, Chem. in Britain, **23** (1987) 327.

[6]   J. Hesslich and P.J. Kuntz, Z. Phys. D, **2** (1986) 251.

[7]   H. Haberland, Surf. Sci., **156** (1985) 305.

[8]   H. Haberland, in Electronic and Ionic Collisions ed.
      J. Eichler, I.V. Hertel and N. Stolterfoht, (Elsevier, 1984), p. 597.

[9]   I.A. Harris, R.S. Kidwell and J.A. Northby, Phys. Rev. Lett.,
      53 (1984) 2390.

[10] D. Kreisel, O. Echt, M. Knapp and E. Recknagel, Surf. Sci.,
      156 (1985) 321.

[11] J.J. Saenz, J.M. Soler and N. Garcia, Surf. Sci., **156** (1985) 121.

[12] J.J. Saenz, J.M. Soler and N. Garcia, Chem. Phys. Lett., **114**
      (1985) 15; J.M. Soler, J.J. Saenz and N. Garcia, *ibid.* **109**
      (1984) 71.

[13] E.E. Polymeropoulos and J. Brickmann, Surf. Sci., **156** (1985) 563.

[14] C.E. Moore, Natl. Bur. Stand. US. Circ. 467, (1949).

[15] H.U. Böhmer and S.D. Peyerimhoff, Z. Phys. D, **3** (1986) 195.

[16] W.R. Wadt, J. Chem. Phys., **68** (1978) 402.

[17] R.O. Watts and I.J. McGee, in Liquid State Chemical Physics,
      Chap. 7, (Wiley, 1976).

[18] W.R. Wadt, Appl. Phys. Lett., **38** (1981) 1030.

[19] M. Amarouche, G. Durand and J.P. Malrieux (private
      communication).

[20] K.S. Sorbie and J.N. Murrell, Mol. Phys., **29** (1975) 1387; *ibid.*
      31 (1976) 905.

[21] J.N. Murrell, S. Carter, I.M. Mills and M.F. Guest, Mol. Phys., **42** (1981) 605.

[22] P.J. Kuntz, Int. J. Quantum Chem., **29** (1986) 1105.

[23] B. Auschwitz and *K.* Lacmann, Chem. Phys. Lett., **113** (1985) 230.

[24] J.W. Gadzuk, Comments At. Mol. Phys., **16** (1985) 219.

[25] J.C. Tully, J. Chem. Phys., **65** (1976) 1002.

[26] P.J. Kuntz, B.I. Niefer and J.J. Sloan, J. Chem. Phys., **88** (1988) 3629.

SPIN-ORBIT INTERACTION ON POTENTIAL ENERGY SURFACES INVOLVING HALOGEN ATOMS

Neil C. Firth and Roger Grice
Chemistry Department
University of Manchester
Manchester
M13 9PL
UK

1. INTRODUCTION

The role of spin-orbit interaction in reaction dynamics so far has been little studied. The current state of experimental work has recently been reviewed by Dagdigian and Campbell [1]. Many experiments have focussed on the reactivity of spin-multiplet states of metastable electronically excited inert gas, alkali and alkaline earth atoms. The product spin-multiplet states of halogen atoms have also been investigated for a number of halogen atom-hydrogen halide [2,3] and halogen atom- halogen molecule [4-7] reactions. Such measurements give information on the form of the low lying potential energy surfaces and their mutual interaction arising from spin-orbit interaction. The strength of spin-orbit interaction varies strongly along the homologous series of halogen atoms. Hence reactions involving halogen atoms may be expected to exhibit a variety of effects arising from spin-orbit interaction [1]. However there have been very few calculations of reaction potential energy surfaces which have included spin-orbit interaction, a notable exception being the diatomics-in-molecules (DIM) calculation of the F + H$_2$ potential energy surface by Tully [8]. DIM calculations [9-11] of potential energy surfaces involving more than a single halogen atom have recently been extended to include spin-orbit interaction [12-14], in order to investigate the forms of interaction which may be encountered in such halogen systems.

2. RESULTS

Potential energy surfaces have been calculated by the DIM method for the H + F$_2$, Cl$_2$ reactions [12] including spin-orbit interaction in the collinear configuration. The two lowest $\Omega = \frac{1}{2}$ potential energy profiles along the minimum energy pathway for H + F$_2$ are shown in Fig. 1. The lowest $^2\Sigma^+$ surface correlates ground state reactants H(^{1}S) + F$_2$($^1\Sigma_g^+$) with ground state HF($^1\Sigma$) + F(^{2}P$_{3/2}$) products. The first $^2\Pi$ excited state surface correlates H(^{1}S) + F$_2$($^3\Pi$) reactants with HF($^1\Sigma$) + F(^{2}P$_{1/2}$) products. The non-adiabatic coupling strength [8] between these $\Omega = \frac{1}{2}$ surfaces is given by

$$Z_{13} = \langle\Psi_1|\partial\Psi_3|\partial R\rangle/|E_1 - E_3| \qquad (1)$$

417

J. C. Whitehead (ed.), Selectivity in Chemical Reactions, 417–426.
© *1988 by Kluwer Academic Publishers.*

418

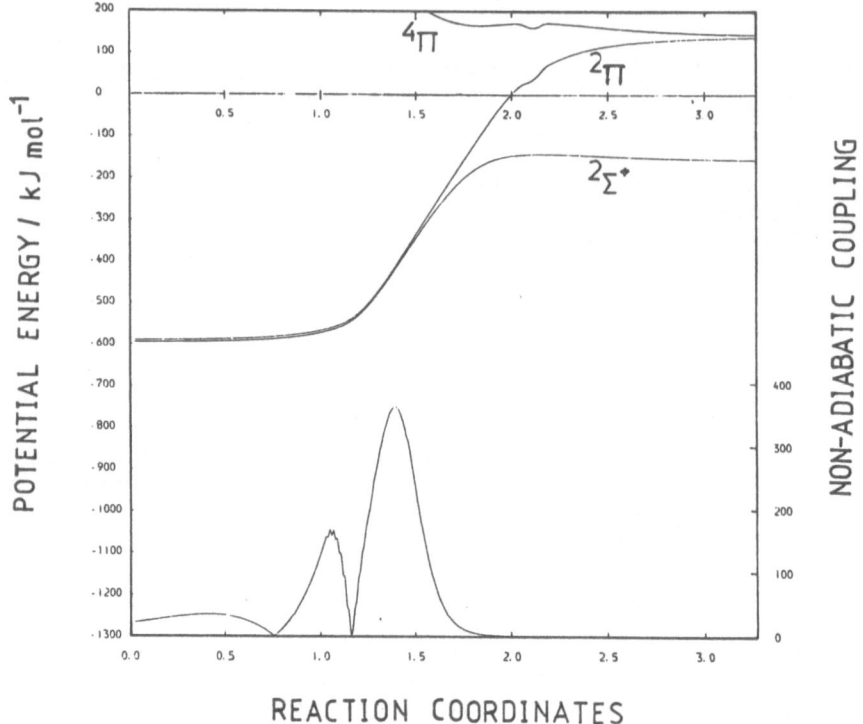

REACTION COORDINATES

FIGURE 1. Potential energy profiles for the two lowest $\Omega = \frac{1}{2}$ potential energy surfaces for the collinear $H + F_2$. The non-adiabatic coupling strength Z_{13} for motion along the minimum energy pathway is shown below.

where Ψ_1, Ψ_3 denote the wavefunctions for the surfaces with potential energies E_1, E_3 and R is the distance along the minimum energy pathway. The magnitude of the coupling strength Z_{13} shown by a curve at the bottom of Fig. 1, exhibits two principal maxima corresponding to the ends of the region of closest approach between the two $\Omega = \frac{1}{2}$ surfaces. Due to the strong exoergicity of the $H + F_2$ reaction these surfaces lie almost parallel over a significant distance along the steeply descending part of the minimum energy pathway. The dependence of the non-adiabatic coupling strength on the nuclear coordinates is shown in more detail by the contour maps of Figs. 2 and 3 where the direction of nuclear motion is indicated by the vector $\tilde{R}$ and the minimum energy pathway by a broken curve. Fig. 2 demonstrates that H-F bond stretching contributes to the first maximum in the non-adiabatic coupling strength which occurs late in the entrance valley of the potential energy surface, while Fig. 3 demonstrates that F-F bond stretching contributes both to the first maximum and the second maximum in the non-adiabatic coupling strength which occurs late in the exit valley. The situation is similar for the $H + Cl_2$ potential energy surfaces but a model calculation for $H + Br_2$ [12]

indicates that the increased spin-orbit interaction maintains sufficient separation between the $\Omega = \frac{1}{2}$ potential energy surfaces that the non-adiabatic coupling strength will be very small.

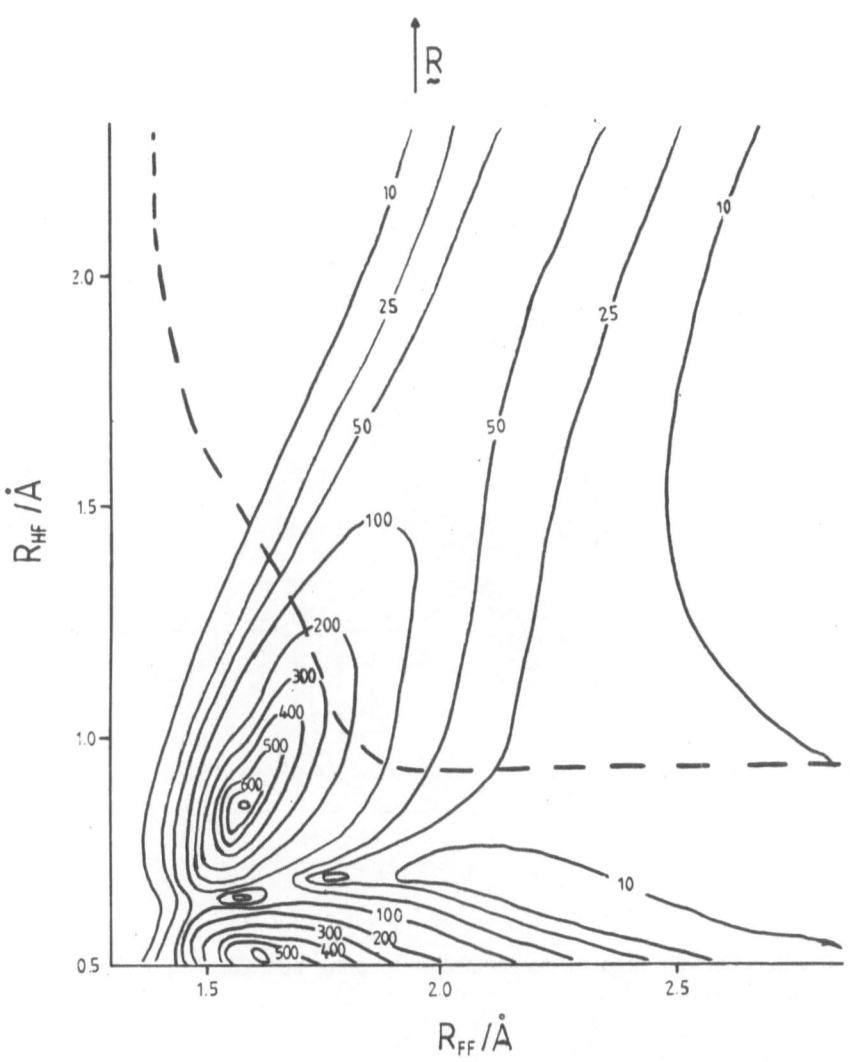

FIGURE 2. Contour map of non-adiabatic coupling strength Z_{13} for extension of the H-F bond as a function of H-F and F–F bond lengths for $H + F_2$. Minimum energy pathway is shown by a broken curve.

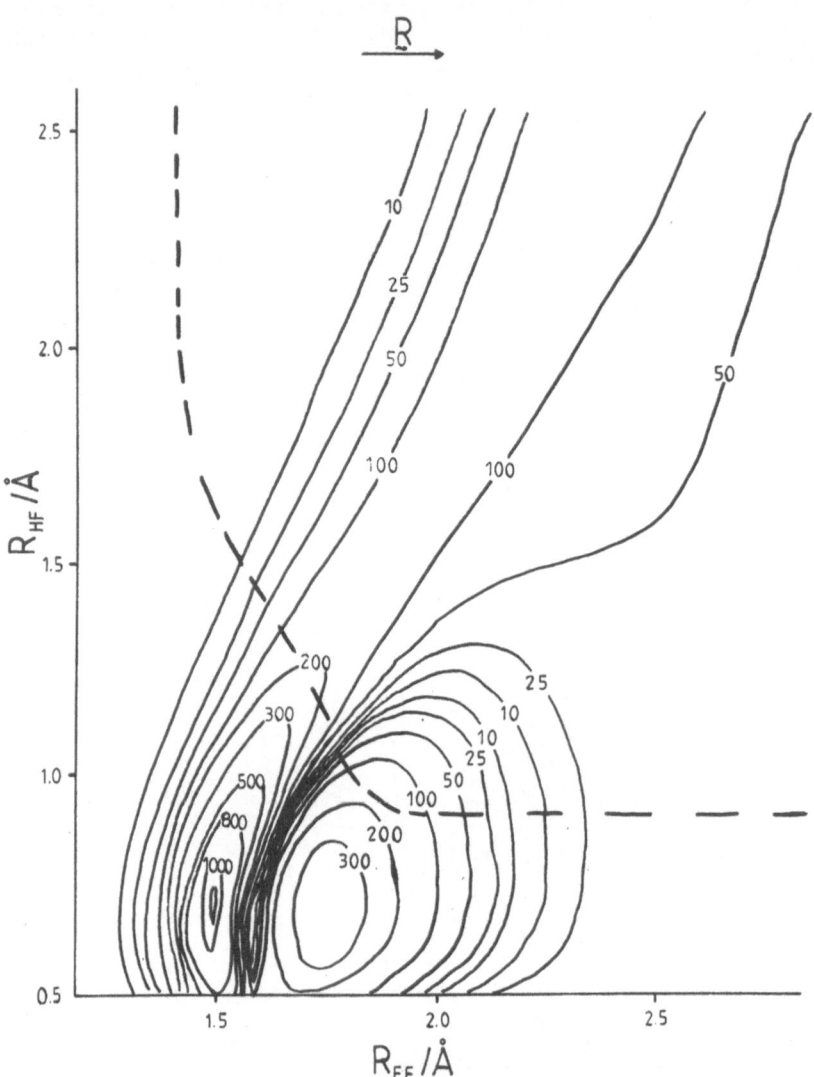

FIGURE 3. Contour map of non-adiabatic coupling strength Z_{13} for extension of the F-F bond for $H + F_2$.

Potential energy surfaces have also been calculated for the collinear configuration of the F + HF and Cl + HCl reactions [13] including spin-orbit interaction. The lowest $\Omega = \frac{1}{2}$ potential energy profiles along the minimum energy pathway for F + HF are shown in Fig. 4. The lowest $^2\Sigma^+$ and $^2\Pi$ surfaces both correlate ground

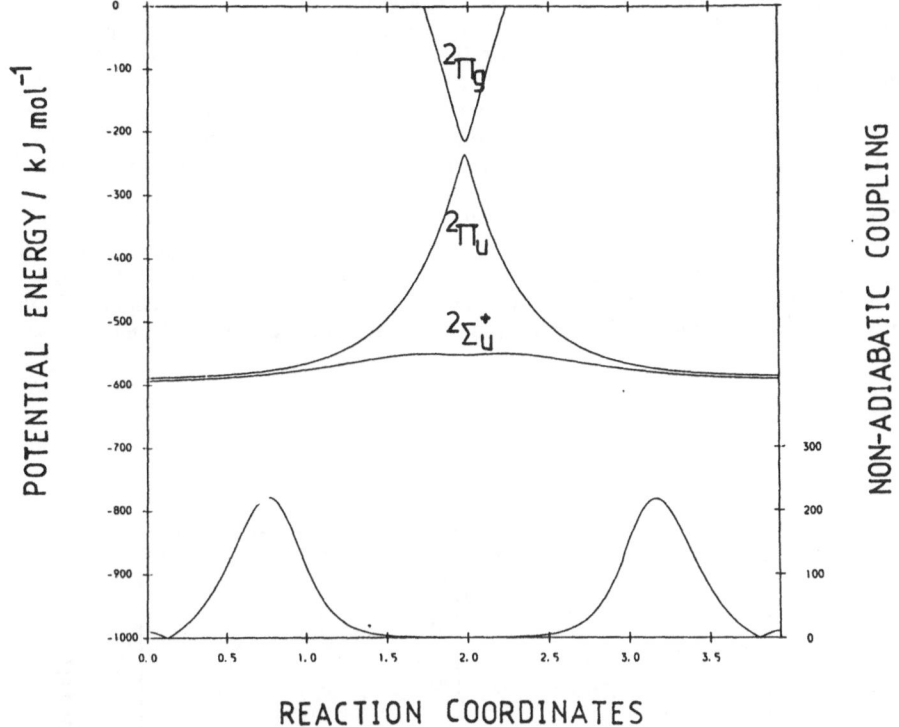

FIGURE 4. Potential energy profiles for the two lowest $\Omega = \frac{1}{2}$ potential energy surfaces for collinear F + HF.

state reactants and products but the $^2\Pi$ state exhibits an avoided intersection with the excited $^2\Pi$ surface which correlates excited HF($^3\Pi$) + F($^2P_{3/2}$) reactants and products. The non-adiabatic coupling strength $Z_{1,3}$ between the two lowest $\Omega = \frac{1}{2}$ surfaces, shown by a curve at the bottom of Fig. 4, exhibits a single maximum far out in the reactant and the product valley. The maximum non-adiabatic coupling strength for F + HF is similar in magnitude and location to the smaller maximum for H + F$_2$ which is also located far out in the product valley. Indeed the non-adiabatic coupling strength depends on the variation of the internuclear distance between the F atom and the HF molecule but not on stretching of the H-F bond. In both these cases the magnitude of the non-adiabatic coupling strength is largely independent of the orientation of the hydrogen halide molecule. However the first maximum of the non-adiabatic coupling strength on the H + F$_2$ potential energy surface decreases on bending from the collinear H + F$_2$ configuration, since the near degeneracy of the $^2\Sigma^+$ and $^2\Pi$ states is lost and the separation of the resulting $^2A'$ and $^2A''$ surfaces becomes greater than the spin-orbit interaction. The non-adiabatic coupling strengths for the Cl + HCl and Br + HBr reactions [13] are found to be much smaller than that for F + HF.

Potential energy surfaces calculated in the collinear configuration for the F +

422

F_2 and $Cl + Cl_2$ reactions [14] including spin-orbit interaction are similar in form to those for F + HF and Cl + HCl. The lowest $\Omega = \frac{1}{2}$ potential energy profiles along the minimum energy pathway for $F + F_2$ are shown in Fig. 5. The non-adiabatic coupling strength $Z_{1,3}$ between the two lowest $\Omega = \frac{1}{2}$ surfaces, shown by a curve at the bottom of Fig. 5, represents an intermediate case. The single maxima in the reactant and product valleys lie closer to the symmetrical F-F-F transition state than is the case for F + HF but are similar in magnitude. The magnitude of the non-adiabatic coupling strength decreases on bending from the collinear $F + F_2$ configuration in a similar manner to the first maximum on the $H + F_2$ potential energy surface. The non-adiabatic coupling strength for the Cl + Cl_2 reaction is similar in form and magnitude to that for $F + F_2$ but a model calculation for Br + Br_2 indicates that the non-adiabatic coupling strength will be very much smaller due to the large spin-orbit interaction of the Br atom.

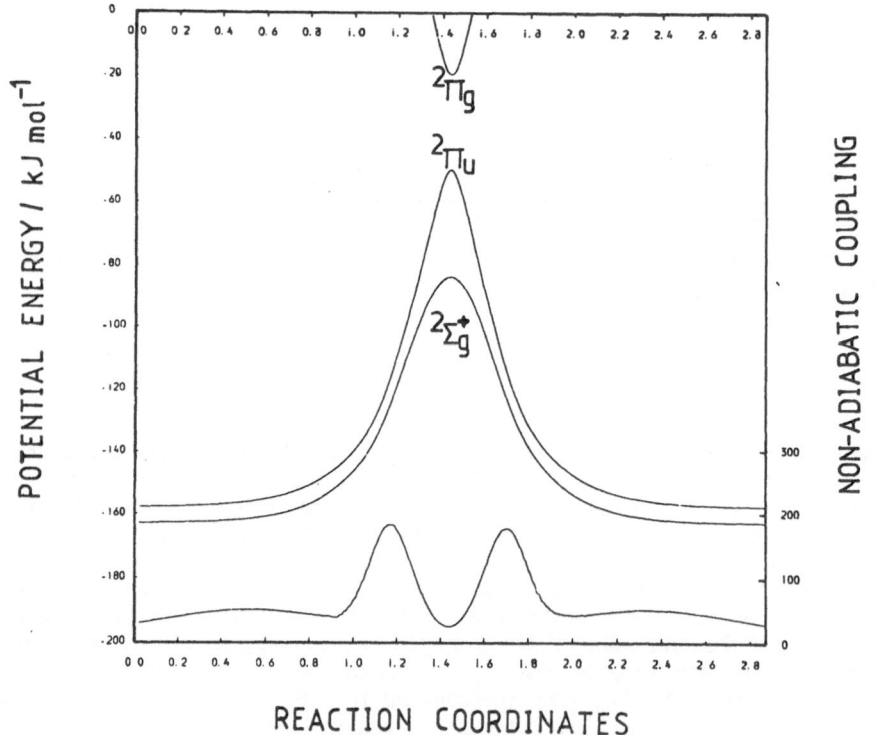

FIGURE 5. Potential energy profiles for the two lowest $\Omega = \frac{1}{2}$ potential energy surfaces for collinear $F + F_2$.

For all the surfaces considered the $\Omega = 3/2$ component of the $^2\Pi$ surface correlates with the $X(^2P_{3/2})$ halogen atom spin multiplet state. However it is coupled with the $\Omega = \frac{1}{2}$ surfaces only by Coriolis interaction [15]. The non-adiabatic coupling strength may be significant for $H + F_2$ trajectories where a large component of angular momentum $K > 10$ due to the H atom motion lies along the F-F bond. The $\Omega = 3/2$ surfaces have been omitted from Figs. 1-5 in order to maintain clarity of presentation.

3. DISCUSSION

The probability of non-adiabatic transition depends not only on the non-adiabatic coupling strength but also on the velocity with which the trajectory crosses the region of interaction. The magnitude of the transition probability may be estimated from the Massey parameter [16,17]

$$\zeta_{13} = |Z_{13}dR/dt|. \tag{2}$$

The probability of non-adiabatic transition is large for $\zeta_{13} \gg 1$ and for $\zeta_{13} \ll 1$ the behaviour is adiabatic. Estimates calculated [12] under typical experimental collisions indicate that significant non-adiabatic transition probability on the collinear $H + F_2$, Cl_2 potential energy surfaces may be associated with H atom motion as the trajectory turns the corner and descends the exit valley due to the rapid H atom motion. The transition probability associated with motion of the halogen-halogen bond is estimated to be rather lower. The estimated transition probabilities decrease for the bent configurations which will certainly be explored by reaction trajectories even when the near collinear geometry is energetically preferred. Any trajectory following a transition from the reactant potential energy surface to the upper surface which correlates with the excited halogen atom spin multiplet state, can lead to reaction products without being impeded by a potential energy barrier. Estimates of non-adiabatic transition probabilities on the F + HF, Cl + HCl and $F + F_2$, Cl + Cl_2 collinear potential energy surfaces are lower than those for $H + F_2$, Cl_2. Moreover any trajectory arising from a halogen atom in the excited spin multiplet state or following a non-adiabatic transition to the upper potential energy surface in the entrance valley will be impeded from reaching the exit valley by an energy barrier on the upper surface. Consequently product halogen atoms in the excited spin multiplet state may be formed only by trajectories which follow the ground state potential energy surface and experience a non-adiabatic transition to the upper surface in the exit valley. This is in accord with the molecular beam measurements of Polanyi and coworkers [2] on the F + HBr reaction where only a small proportion ~ 6% of $Br(^2P_{\frac{1}{2}})$ atoms is observed. The non-adiabatic transition probability estimated [13] for a single passage through the region of interaction is low. However the dynamics of the F + HBr reaction are characterised by rapid oscillation of the light H atom between the slowly separating heavy halogen atoms [18,19]. Hence the H atom vibrational motion may accumulate many traversals of the region of interaction in the exit valley and thus accumulate a significant transition probability. Indeed this conclusion is supported by the observation of a lower yield ~ 1% of $Br(^2P_{\frac{1}{2}})$ atoms in the F + DBr reaction with its slower D atom motion and the higher yield of $Br(^2P_{\frac{1}{2}})$ at lower collision energy which is associated with slower separation of the halogen atoms.

The DIM calculations outlined above [12-14] are concerned with systems with

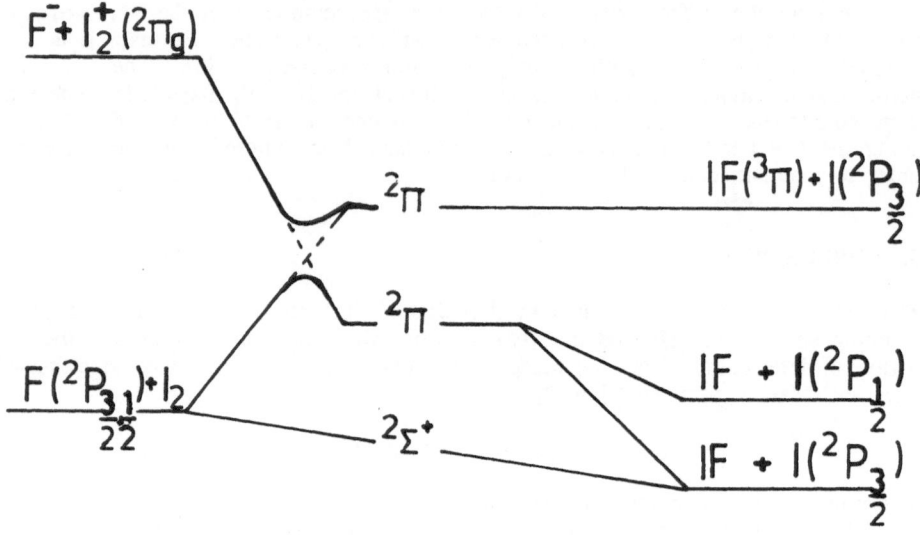

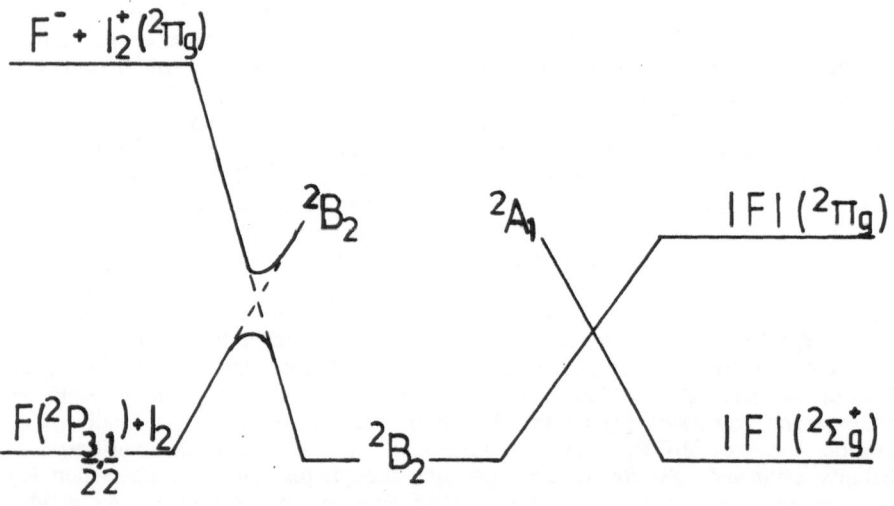

FIGURE 6. Schematic correlation diagram for covalent and charge transfer states in the F + I₂ reaction in the collinear and broadside configurations.

light F,Cl halogen atoms in near collinear configurations, without the inclusion of charge transfer states. However systems with both light and heavy halogen atoms may involve strongly bent configurations [20] which are stabilised by charge transfer interaction [21]. A correlation diagram for the F + I_2 reaction is shown in Fig. 6 which includes the F^- + $I_2^+(^2\Pi)$ charge transfer state. In the collinear configuration the charge transfer state interacts directly with the excited $^2\Pi$ covalent state but it interacts with the $^2\Sigma^+$ ground state only through spin orbit inter- action. In the broadside configuration there is direct interaction with the covalent $^2B_2$ state which lies lowest at large internuclear distance but interaction with the covalent $^2A_1$ state which lies lowest in the I-F-I collinear configuration occurs only through spin-orbit interaction. In intermediate bent configurations, the F^- + I_2^+ charge transfer state and the F + I_2 covalent state both have $^2A'$ symmetry and interact directly. Consequently the long-range potential energy surface for F + I_2 is lowered by charge transfer interaction [21] over a wide range of bent configurations and the resulting attractive potential energy surface gives rise to migratory reaction dynamics [22, 23]. The large spin-orbit interaction of the I atom compared with that of the F atom implies that non-adiabatic transitions to the excited $^2\Pi$ potential energy surface will be confined to the entrance valley. Such trajectories will be impeded from reaching the exit valley by an energy barrier on the $^2\Pi$ surface. Consequently reactive trajectories are expected to be confined to the lowest potential energy surface. This is confirmed by several experiments [5-7] which find little or no yield of spin multiplet excited $I(^2P_{\frac{1}{2}})$ atoms in the F + I_2 reaction.

REFERENCES

[1] P.J. Dagdigian and M.L. Campbell, Chem. Rev., 87 (1987) 1.

[2] J.W. Hepburn, K. Liu, R.G. Macdonald, F.J. Northrup and J.C. Polanyi, J. Chem. Phys., 75 (1981) 3353.

[3] K. Bergmann, S.R. Leone and C. Bradley Moore, J. Chem. Phys., 63 (1975) 4161.

[4] H.K. Haugen, E. Weitz and S.R. Leone, Chem. Phys. Lett., 119 (1985) 75.

[5] H. Brunet, P. Chauvet, M. Mabru and L. Torchin, Chem. Phys. Lett., 117 (1985) 371.

[6] P. Das, T. Venkitachalam and R. Bersohn, J. Chem. Phys., 80 (1984) 4859.

[7] B.S. Agrawalla, J.P. Singh and D.W. Setser, J. Chem. Phys., 79 (1983) 6416.

[8] J.C. Tully, J. Chem. Phys., 59 (1973) 5122.

[9] J.J. Duggan and R. Grice, J. Chem. Phys., 78 (1983) 3842.

[10] J.J. Duggan and R. Grice, J. Chem. Soc. Faraday Trans. II,
      80 (1984) 729, 739.

[11] J.J. Duggan and R. Grice, J. Chem. Soc. Faraday Trans. II,
      80 (1984) 795, 809.

[12] N.C. Firth and R. Grice, J. Chem. Soc. Faraday Trans. II,
      83 (1987) 1011.

[13] N.C. Firth and R. Grice, J. Chem. Soc. Faraday Trans. II,
      83 (1987) 1023.

[14] N.C. Firth and R. Grice, J. Chem. Soc. Faraday Trans. II,
      83 (1987) 1029.

[15] R.N. Dixon, Molec. Phys., 54 (1985) 333.

[16] E.E. Nikitin, in Chemische Elementarprozesse, ed. H. Hartmann
      (Springer, Berlin, 1968).

[17] J.C. Tully, Adv. Chem. Phys., 42 (1980) 63.

[18] J.C. Polanyi, Acc. Chem. Research, 5 (1972) 161.

[19] C.A. Parr, J.C. Polanyi and W.H. Wong, J. Chem. Phys.,
      58 (1973) 5.

[20] N.C. Firth, D.J. Smith and R. Grice, Molec. Phys., 61 (1987) 859.

[21] L.M. Loewenstein and J.G. Anderson, J. Phys. Chem.,
      91 (1987) 2993.

[22] I.W. Fletcher and J.C. Whitehead, J. Chem. Soc.
      Faraday Trans. II, 80 (1984) 985.

[23] N.C. Firth, N.W. Keane, D.J. Smith and R. Grice,
      Faraday Disc. Chem. Soc., 84 (1987), in press.

BRANCHING RATIOS IN CHEMICAL REACTIONS

Ch. Ottinger
Max-Planck Institut für Strömungsforschung
Bunsenstraße 10
3400 Göttingen
Federal Republic of Germany

ABSTRACT. In this review branching between various exit channels of a chemical reaction is considered. Exit channels are here defined by chemical species, electronic state and charge state, while the finer branching between "internal" states is generally excluded. Classic as well as recent examples are discussed, drawn from many different areas: Neutral and ionic bimolecular reactions, chemiluminescence, chemiionization, charge transfer, Penning and associative ionization, excimer formation, electronic energy transfer, photodissociation. Only those examples have been included for which an explanation, however tentative, has been offered in the literature, and this theoretical background is sketched briefly in each case. The determining factors considered by various authors include effects of mass, orientation, geometry of the heavy particles; electronic symmetries, Woodward-Hoffmann rules, alignment, non-adiabatic transition requirements; energetics; time scales; shapes of potential curves or surfaces; statistics.

1. INTRODUCTION

We speak of branching in an elementary chemical process if a given set of reactants yields different sets of products in a certain proportion, called the branching ratio, which usually depends on the reaction conditions:

$$A + B \rightarrow C + D \quad x\%$$

$$\rightarrow E + F \quad y\% \quad \text{branching ratio} \quad x{:}y{:}z$$

$$\rightarrow G + H \quad z\%.$$

Branching, although very common, is nevertheless a surprising phenomenon: It seems to contradict the fundamental expectation that given initial conditions lead, invariably, to predetermined final states (within, of course, the limits set by the quantum mechanical uncertainty principle). Obviously finer unobservable details of the reaction mechanism are at work which control the outcome of a particular reactive encounter. Studies of branching are intriguing, because they offer insight into the true, microscopic nature of the process concerned. Having achieved such an

J. C. Whitehead (ed.), Selectivity in Chemical Reactions, 427–455.
© 1988 by Kluwer Academic Publishers.

advanced level of understanding one can then hope to actually control a reaction, by manipulating its "hidden variables", and to steer it towards a particular product channel.

This review attempts to give a kaleidoscopic view of many different examples of branching, a total of about 35 systems, taken from the literature of the last decade or so. Of necessity in each case the experimental method, the results and the theoretical explanation are only briefly summarized. However, it is hoped that highlighting the salient features of both experiment and theory will help to stimulate fresh thinking about these fascinating effects and point the way to further studies.

2. EXCHANGE REACTIONS BETWEEN NEUTRAL SPECIES

2.1 K_2 + Br_2: Harpooning Reactions

At the origin of modern reaction dynamics are the scattering experiments on alkali-halogen systems which Herschbach and his school pioneered. Very rich chemical branching was observed in reactions of alkali dimer molecules with halogens [1,2], e.g.

$$K_2 + Br_2 \rightarrow K + KBr + Br \qquad\qquad (1a)$$

$$\rightarrow 2KBr. \qquad\qquad (1b)$$

The angular distribution of the KBr product was found to be strongly forward-peaked (see Fig. 7 of Ref. 1). This indicates immediately a substantial contribution from (1a) since (1b), from symmetry, can only give a forward-backward symmetric product distribution. However, measuring the K atom product separately it was found that it matched the KBr only in the forward direction, but fell short at 90°. Thus there must also be a contribution from (1b) which gives KBr predominantly at wider angles. A branching ratio (1a):(1b) = 3:1 was estimated.

These reactions are certainly, like the triatomic alkali-halogen systems (e.g. K + Br_2), initiated by an electron jump ("harpooning"):

$$K_2 + Br_2 \rightarrow K_2^+ + Br_2^-,$$

followed by dissociation of the negative ion. The Br⁻ then attaches itself to the K_2^+ to give forward-scattered K_2Br. Thus the mechanism is completely analogous to the K + Br_2 case, which gives forward-scattered KBr. However, in the present case K_2Br is short-lived and quickly dissociates, which explains the observation of forward-scattered K + KBr. Reaction (1b) requires a second electron jump, since two product molecules with ionic structure are formed. Intuitively this is less likely, because a more concerted motion of all four atoms is necessary to orient the K_2 and Br_2 molecules roughly parallel to each other and thus allow "simultaneous" formation of two KBr molecules (see Fig. 10 of Ref. 1). The observed 3:1 branching appears therefore reasonable. The latter argument can, in fact, be substantiated by electronic symmetry considerations [1]: The first electron jump is, as a result of the conical intersection between the homopolar and ionic surfaces of the K_2Br_2 system, only allowed for orientations with the Br_2 pointing towards the K_2 centre (just like in the case of K instead of K_2). The K_2 itself can be oriented randomly for K_2Br formation and reaction (1a) to follow. The more or less parallel arrangement of K_2 and Br_2, however, which would be conducive to (1b), is certainly discriminated against at the instant of the first electron jump. Moreover it can be shown [3] that the second electron jump necessary for (1b) is inhibited when the K_2 is oriented broadside to the Br_2. Thus the low yield of (1b) is entirely consistent with the expectation based on the electronic structure.

Several further reaction paths were observed in this system. Production of ions occurs ("chemiionization"), the possible channels being

$$K_2 + Br_2 \rightarrow K^+ + KBr + Br^- \qquad (1c)$$

$$\rightarrow K_2Br^+ + Br^- \qquad (1d)$$

$$\rightarrow K^+ + KBr_2^-. \qquad (1e)$$

While no mass analysis has yet been performed to distinguish between these paths, all of them demonstrate directly the occurrence of electron jumps to form charged intermediates. These paths can be viewed as events during the course of which the ions somehow missed the chance of recombining into neutral products. Also, if a recombination does occur, it will not always lead to the electronic ground state of the products. Thus formation of electronically excited alkali atoms or halide molecules can be expected and has, in fact, been observed [1] ("chemiluminescence").

Generally speaking, branching in these systems has been discussed in a largely intuitive way in terms of the relative importance of sequential vs. concerted motions. The controlling factor is the development of the system with time. This concept could be followed up quantitatively by trajectory calculations, but none appear to have been done on these four-atomic systems.

2.2 H + HO_2: Energy Barriers

A system of great importance in hydrocarbon combustion chemistry is H + HO_2 (the radical being formed by the three-body addition of initiating H atoms to O_2). Three strongly exothermic reaction paths are

$$H + HO_2 \rightarrow OH + OH \qquad (2a)$$

$$\rightarrow H_2 + O_2 \qquad (2b)$$

$$\rightarrow H_2O + O. \qquad (2c)$$

For the overall chemistry particularly relevant is the branching between (2a), giving two OH radicals which are very aggressive towards hydrocarbons, and (2b), giving two stable molecules. Measurement in a flow system [4] yielded branching fractions of (2a) : (2b) : (2c) = 69% : 29% ζ 2%. Noteworthy is especially the low yield for (2c), the reaction with the highest exothermicity, 63.5 kcal. On the other hand the branch (2a) has the lowest exothermicity, 43.4 kcal, and is the most abundant. This was rationalized by *ab initio* calculations [5] which showed that (2c) requires surmounting a high (17 kcal) potential barrier, (2b) involves a very low barrier (6.2 kcal), while (2a) proceeds without barriers smoothly downhill to form a strongly bound (89.1 kcal) H_2O_2 intermediate which then breaks up into 2 OH. Thus the branching is here explained, in a global way, simply from the energetics along the various reaction paths.

2.3 O + CN: Branching to Electronically Excited Atomic Product

A very interesting system is the reaction

$$O(^3P) + CN \rightarrow CO + N(^2D) \qquad\qquad (3a)$$

$$\rightarrow CO + N(^4S). \qquad\qquad (3b)$$

An electronic state correlation diagram shows that (3a) can proceed *via* a strongly bound NCO intermediate (complex formation path), while (3b) occurs on a quartet surface directly towards the products, bypassing the low-lying NCO states (all doublets). The branching is potentially important because the electronically excited, metastable $N(^2D)$ atoms carry 2.38 eV and could be responsible for chemiionization in flames. Also a visible chemical laser could perhaps be based on $N(^2D)$ formed via (3a). The measurement was made [6] in a flow reaction using O atoms (from an O_2 MW discharge) and CN (from C_2N_2 UV photolysis). The $N(^2D)/N(^4S)$ ratio was derived from atomic VUV absorption signals [7] and indicated a strong inversion. This was beautifully corroborated and made more quantitative by a detailed measurement of the CO vibrational distribution, using the absorption of a continuous CO chemical laser [8]. The result (see Fig. 4 of Ref. 8) was bimodal. In addition to a dominant quasi-statistical monotonically decreasing distribution of v_{CO}, an inverted component, peaking at $v_{CO} = 9$, was found. These two very different distributions indicated clearly two different mechanisms and could be associated uniquely with the complex and the direct reaction path, respectively. This was done on the basis of trajectories, calculated on LEPS surfaces designed to model the potentials for (3a) and (3b). The calculated v_{CO} distributions showed the expected features, monotonically decreasing and inverted, respectively (see Fig. 10 of Ref. 8). A breakdown of the measured distribution according to the theoretical results finally gave a branching ratio (3a):(3b) = 85%:15% [7]. In these studies, no explanation on the basis of first principles was given for the branching ratio. It is obviously determined by the details of the long-range portions of the two competing potentials. Also, non-adiabatic transitions between them may take place en route, which were explicitly excluded from the trajectory calculations [8].

The calculated v_{CO} distributions did not match the experiment perfectly, in particular for the direct path the peak occurred at too high v_{CO}. This is clearly due to deficiencies of the LEPS surface which is not very important for the determination of the branching ratio from the method described above. However, in a different approach taken earlier by the same group the potential imperfections proved very critical. A measurement was made of the dependence of the overall rate of the reaction on the reactant vibrational excitation, v_{CN}, which varied during the photolysis flash due to Franck-Condon pumping. The measured rate constant k increased slightly from $v_{CN} = 0$ to 6 (see Fig. 5 of Ref. 6). The corresponding trajectory calculations gave an increasing trend for the "direct" potential, and a decreasing trend for the "complex" potential (see Figs. 5 and 6 of Ref. 9). This would have indicated a predominance of the former, in disagreement with the above conclusion. Obviously the calculated rate constant is much more sensitive to potential imperfections than is the calculated product vibration, making a branching measurement based on the $k(v_{CN})$ dependence very unreliable.

The experiments even yielded the dependence of v_{CO} on v_{CN} (see Fig. 9 of

Ref. 8). As expected, and also confirmed by trajectory calculations, the (dominant) fraction from the complex mechanism showed little ν_{CN} dependence (vibrational scrambling in the NCO complex), while along the direct path ν_{CN} and ν_{CO} show a strong positive correlation (so-called vibrational adiabaticity).

2.4   H + Dihalogen: Macroscopic and Microscopic Branching

J.C. Polanyi and his co-workers have applied their IR chemiluminescence method to several reactions of H atoms with asymptotic dihalogen molecules, e.g.

$$H + ClF \rightarrow HF + Cl + 75 \text{ kcal} \tag{4a}$$

$$\rightarrow HCl + F + 42 \text{ kcal} \tag{4b}$$

$$H + BrCl \rightarrow HCl + Br + 51 \text{ kcal} \tag{5a}$$

$$\rightarrow HBr + Cl + 29 \text{ kcal.} \tag{5b}$$

The macroscopic branching ratios, derived from the integrated IR intensities, were for reaction (4) (a):(b) = 13%:87% [10], for reaction (5) (a):(b) = 29%:71% [11]. Thus the less exothermic channel is, surprisingly, strongly favoured. The explanation given is, briefly: The attacking H atom interacts most with that end of the target molecule at which both the highest occupied and the lowest unoccupied molecular orbitals are concentrated. This is the less electronegative of the two atoms (Cl in (4), Br in (5)), e.g. H-Cl-F is more stable than H-F-Cl. The attack at the more electronegative end is disfavoured, though not completely inhibited, by a potential barrier, such as is shown for the case of H + ICl in Fig. 13a of Ref. 12. However, there exists a second pathway to bond formation with the more electronegative atom, e.g. Cl in ICl: The H atom initially interacts with the I end in the "regular" way. However, this interaction loosens the I-Cl bond. Now it can be shown that the barrier around the Cl end disappears rapidly as the I-Cl bond stretches (see Fig. 13b of Ref. 12). Then the H atom lingering around the I can, at an appropriate vibrational phase, insert itself into the I-Cl bond and finally end up in the energetically most favourable position, i.e. in the HCl product. This mechanism is called "migration" and has been confirmed by trajectory calculations (see Fig. 15 of Ref. 12). The branching between the two paths for the (macroscopically less likely) reaction with the more electronegative atom, namely the direct and the migratory path, is called microscopic branching. It is clearly reflected in both calculated and observed rotational product distributions: The direct path gives rotationally rather "cold", the migratory path highly excited products. Thus for H + ICl $\rightarrow$ HCl + I the HCl IR spectrum showed a highly bimodal rotational structure in all but the highest vibrational levels (see Fig. 4 of Ref. 11), quite unlike HCl from H + Cl$_2$. The microscopic branching ratio was here about 1:1. The HI product, which is not directly observable, is expected to show no bimodality, because it is only formed via the "regular", direct route. This conclusion could be experimentally confirmed by analogy in the case of reactions (4) and (5), where both products are observable. For example, in reaction (5) HCl, the less abundant product, was rotationally bimodal, HBr was not.

2.5   Mg (^{1}P) + H$_2$: Insertion

A very pretty example of microscopic branching between different modes of attack on a target molecule is the reaction [13] of laser-excited Mg atoms:

$$Mg(^1P_1) + H_2 \rightarrow MgH + H. \tag{6}$$

The MgH rotational distribution was monitored with a second laser. It was found to be very bimodal. The high-J component is, like in Polanyi's experiments, clearly due to a side-on, insertive mechanism. *Ab initio* potential calculations show a smooth downhill pathway towards a bent H-Mg-H complex. When it separates into products, the H-H repulsion "spins the light H atom of MgH around the heavy Mg atom" [13], and this accounts for the high J observed. The low-J component, on the other hand, is probably due to nearly linear Mg–H-H configurations (end-on attack). In this system insertion is by far the dominant microscopic channel (~ 90%), which is in contrast to the H + ICl, ClF, BrCl reactions.

2.6 $Br_2 + (Cl_2)_2$: Van der Waals Reactions

A four-centre reaction $Br_2 + Cl_2 \rightarrow 2BrCl$ is expected to have a large activation barrier, a consequence of the Woodward-Hoffmann symmetry rules. Experimentally an upper limit to the cross section could be set at 0.01 Å² [14]. However, with six participating atoms the rules do allow a low-energy path. Reaction of Br_2 with van der Waals $(Cl_2)_2$ dimers revealed three facile parallel reactions [15]:

$$Br_2 + Cl_4 \rightarrow 2\ BrCl + Cl_2 \tag{7a}$$

$$\rightarrow BrCl_3 + BrCl \tag{7b}$$

$$\rightarrow Br_2Cl_2 + Cl_2 \tag{7c}$$

(a) and (b) involve breaking and making three strong (chemical) bonds in a cyclic six-membered transition state [16]. By contrast, in (c) only van der Waals bonds are exchanged, the Br_2 and the two Cl_2 molecules remaining bound like "quasi-atoms". The energy dependence of the branching ratio (in the 3-25 kcal range) shows strong preference for (7c) at low energies. This is as expected from the above, especially since this path does not require a cyclic intermediate. At 10-15 kcal this branch disappears in favour of (7a). (7b) is minor at all energies. Thus the ratio (c):(a), in particular, demonstrates nicely how the Woodward-Hoffmann symmetry requirements enforced on (a) control the branching.

3. ISOTOPE EFFECTS

3.1 Neutral Systems

Reactions of atoms X with HD molecules provide an especially obvious example of a branching, into XH + D *vs* XD + H products. If the relative yield of the two channels differs from unity, one speaks of an isotope effect. Studies of isotope effects offer the unique possibility of isolating the effects of mass and of the potential on the overall dynamics, since the potential is, at least in the adiabatic approximation, identical for the two branches.

An early example of a precise determination of an isotope effect is the reaction

$$F + HD \rightarrow HF + D \qquad\qquad (8a)$$

$$\rightarrow DF + H. \qquad\qquad (8b)$$

Two very different techniques (a flow system with mass spectrometric detection [17] and a measurement of relative gain on HF vs. DF lines in a chemical laser [18]) gave very similar results: (a):(b) 1.45 ± 0.03 [17] and 1.42 ± 0.1 [18]. The "prior" statistical expectation would be (a):(b) = 0.88. The large difference could be explained [19] on the basis of the information theory of Bernstein and Levine. Their prediction is that the product with the least surprisal is formed most. The rotational surprisals for both HF and DF were measured by IR chemiluminescence [19] and were both found to be unequal to zero, but much smaller for HF than for DF. This tends, therefore, to increase the HF yield, and the quantitative application of the information theory gave a "theoretical" branching ratio of 1.41. However, this treatment only reduces the branching ratio to the empirical rotational surprisals, (although these, in turn, could be modelled by trajectory calculations); it does therefore not really provide an "explanation" of the effect. An alternative mechanistic rationalization of the branching is also given in Ref. 19. It simply hinges on the fact that the skewing angle of the collinear potential energy surfaces is greater for (8a) than for (8b), and this makes the HF exit channel on the skewed surface 1.39 times as wide as the DF exit channel. A correspondingly larger portion of reagent vibrational phases will then lead to reaction, and this could also explain the observed isotope effect (for non-rotating, in particular collinearly oriented target molecules).

Using an entirely different experimental approach, the F + HD reaction has recently been reinvestigated and contrasted with two further branching reactions [20]

$$O(^1D) + HD \rightarrow OH + D \qquad\qquad (9a)$$

$$\rightarrow OD + H \qquad\qquad (9b)$$

$$S(^1D) + HD \rightarrow SH + D \qquad\qquad (10a)$$

$$SD + H. \qquad\qquad (10b)$$

The reagent atoms were produced by laser photolysis (F by CO_2 laser IR-MPD of SF_6, O and S by excimer laser UV photodissociation from O_3 and CS_2, respectively), and the H and D product atoms were detected by tunable UV LIF on the respective 82259 cm^{-1} and 82282 cm^{-1} L_α transitions. The following branching ratios (a):(b) were obtained: 1.52 for reaction (8), 0.88 for reaction (9), and 0.52 for reaction (10). The first of these is in reasonable agreement with the other measurements [17, 18]. It was here explained on the basis that the reaction (8) proceeds by abstraction in largely collinear conformations, as has been shown by scattering experiments in Lee's laboratory. An incoming F atom interacting with a rotating HD molecule will then preferentially be intercepted by the H rather than the D atom, because during the rotation around the molecular centre of mass the H atom sweeps through a larger circle. This very simple picture would not apply to reactions (9) and (10), because ^1D atoms are commonly believed to proceed via insertion. Qualitatively one would then expect that in the decomposition of an excited complex "the same forces act on both H and D and the lighter atom should have a higher probability of escape" [20], thus explaining a branching ratio (a):(b) < 1.

For the case of reaction (9), this question has recently been thoroughly explored [21]. Trajectory calculations were performed on two surfaces, due to Murrell-Carter (MC) and Schinke-Lester (SL). The branching ratios obtained were OH/OD = 0.91 (MC) vs. 0.56 (SL). More important than the actual numbers is, however, the way they come about. The two surfaces imply totally different reaction mechanisms: From MC, the branching stems from a long-range orientation effect, favouring a collinear (!) loose complex with a slight preference for bond formation with the D end. According to SL, a long-range orientation into perpendicular (!) geometry takes place, followed by formation of a tight complex, out of which the H atom is knocked out preferentially, and OD is formed. The principal lesson to be learned from this careful study is the great sensitivity of the reaction mechanism and the ensuing isotope effect on the details of the potential surface. The better agreement with experiment for the MC calculation must not be valued too highly, since the collinear mechanism in this system appears to be at variance with the general chemical expectation [20]. On the other hand, the SL result is also unsatisfactory. Obviously both surfaces have their shortcomings, and improved *ab initio* calculations are needed.

3.2   Ionic Systems

Two interesting prototype systems are the reactions:

$$Ar^+ + HD \rightarrow ArH^+ + D \qquad\qquad (11a)$$

$$\rightarrow ArD^+ + H \qquad\qquad (11b)$$

$$Kr^+ + HD \rightarrow KrH^+ + D \qquad\qquad (12a)$$

$$\rightarrow KrD^+ + H \qquad\qquad (12b)$$

Here the branching ratio (a):(b) is clearly < 1 at very low energy (0.02 eV_{CM} and 0.01 eV_{CM}, respectively), but increasing with energy, becoming unity at ~ 0.15 eV_{CM} for both reactions [22]. At higher energies, an almost energy independent branching ratio (11a):(11b) ~ 1.2 was measured between 0.2 and 2 eV_{CM} [23]. The Kr^+ reaction, by contrast, shows in this energy range a dramatic energy dependence of the (12a):(12b) ratio, which rises from ~ 1.2 to a maximum of ~ 2.5 at 0.8 eV_{CM}, followed by a steep decline to ~ 0.3 at 3 eV_{CM} [24]. The angular distributions of KrH^+ (KrD^+) are also very interesting [24]. At 1.4 eV_{CM} they are quite similar, both peaking in the forward direction. At higher energies, KrH^+ still peaks forward of the C.M., close to the stripping energy, while KrD^+ switches to peaking in the backward direction.

The (a):(b) < 1 ratio could be explained, as was done in [22], on the basis of a statistical partitioning, based on the assumption that a long-lived intermediate complex is formed. However, product angular distribution measurements showed that this is not the case even at very low energies [25], so that an alternative explanation had to be sought.

Hierl [26] developed a model which can (qualitatively) account for the (a):(b)

< 1 ratio at very low energy as well as (quantitatively) for the peaking of the branching ratio for reaction (12), against a constant branching ratio for reaction (11), at higher energies. The model considers the torque exerted by the polarization of the ion on the HD molecule. At very low energy, this leads to an HD rotation at uneven speed, the molecule spending the majority of the time with its D end towards the ion and hence favouring XD^+ formation. At moderate collision energies, however, the H end becomes locked in towards the ion, and the XH^+ product becomes favoured. Specifically, in this energy regime the orientation is assumed to terminate at the instant a) the centrifugal barrier in the effective potential is reached or b) charge transfer takes place, whichever occurs first. The isotope effect is taken to be the fraction of target molecules oriented with the H atom towards X^+ at this instant. The calculated isotope effect is in remarkable agreement with the experimental result for both systems, (11) and (12) (see Figs. 4 and 5 of Ref. 26), the different behaviour being due to the different ionization potentials, which determine the end of the orientation process.

This model cannot explain the renewed reversal of the Kr^+ + HD isotope effect at high energies - the polarisation effect will then diminish, and the branching ratio will tend to unity. For this regime, however, the observation of a shift in the KrD^+ angular distribution sheds light onto the mechanism:

There are generally two classes of collision geometries:

a) large-impact parameter encounters, which lead to the stripping-type forward scattering, as is observed in both channels of reaction (12) [24] and also in Ar^+ + H_2 reactions [25]. This mechanism usually has a cross section which decreases rapidly with increasing energy.

b) In addition collinear collisions can occur, with a small cross section which, however, extends to high energies and then starts to dominate the overall behaviour. It can then be shown by simple hard-sphere kinematics [24] that this type of collision will only give a bound product if the ion approaches from the D end, and this leads to backward-scattered KrD^+, as is in fact observed. High-energy collinear collisions with the H end, on the other hand, do not give a bound product, due to chattering of the H atom between its two heavier neighbours. The reverse isotope effect, (12a):(12b) < 1, can then be understood provided the central collision mechanism does indeed overwhelm the glancing reaction type at high energies. For Ar^+ + HD this does not appear to be the case up to 2 eV_{CM}.

In Hierl's model the very different isotopic branching of Ar^+ and Kr^+ is essentially attributable to the different electronic properties (ionization potentials) of these two species. Recently it has been demonstrated very directly how changing the electronic structure of one and the same ion affects its isotopic branching behaviour [27]. Vanadium ions were prepared using electron of different energy. The branching ratio of

$$V^+ + HD \rightarrow VH^+ + D \tag{13a}$$

$$\rightarrow VD^+ + H \tag{13b}$$

depends in a complicated way on the kinetic energy (threshold to 10 eV_{CM}), twice crossing the 1.00 mark, provided the V^+ ions are in their electronic ground state ^5D, $3d^4$. If, however, metastable V^+ states (^3F, $3d^34s$, at 1.1 eV) participate, then the (a):(b) ratio is on the order of 3-4 through most of the energy range. In terms of an MO correlation diagram an explanation is given of why V^+ ions in the $3d^3 4s$ configuration might mainly react collinearly, while $3d^4$ ions tend to insert. The latter gives (statistically) an (a):(b) $\approx$ 1 ratio, the former, however, should prefer (a) over (b).

An especially clear-cut example of an almost purely kinematic, very pronounced isotope effect is given in Ref. 28. This concerns the reactions

$$Ne^+ + HD \rightarrow NeH^+ + D^* \qquad (14a)$$

$$\rightarrow NeD^+ + H^* \qquad (14b)$$

which are endothermic by 4.6 eV because the atomic products are (presumably) formed in the n=2 state. The corresponding cross sections rise from a common threshold at $\sim$ 9 eV_{CM} to peaks at 13.5 eV_{CM} for (14b) and 27 eV_{CM} for (14a), (the maximum value being only 0.01 Å^2). Beyond the respective peaks both cross section curves decline rapidly, intersecting each other at $\sim$ 17 eV. Thus the branching ratio changes from very strongly in favour of (14b) (between 9 and $\sim$ 16 eV) to just the reverse (above 18 eV).

The difference in the position of the cross section maxima, which causes this very striking isotope effect pattern, is explained kinematically: The interaction of Ne^+ is "pairwise", either only with H or only with D (which is also one of the sumptions of the stripping model). The relevant kinetic energies are therefore the relative energies within these pairs of atoms. They are, from the reduced masses, approximately $E(Ne^+\text{-}H) = 1/3 \ E_{CM}$, $E(Ne^+ - D) = 2/3 \ E_{CM}$, thus differing by a factor of two as observed. In terms of these pairwise energies, the peaks occur both at 9 eV (1/3 of 27 eV_{CM} or 2/3 of 13.5 eV_{CM}). The cross section fall-off beyond the peak may be due, again as in the stripping model, to the limit set by the product dissociation energy.

A very large isotope effect was also recently observed from the product luminescence in the reaction [29]

$$P^+ \ (^1D) + HD \rightarrow PH^+ \ (A \ ^2\Delta) + D \qquad (15a)$$

$$\rightarrow PD^+ \ (A \ ^2\Delta) + H \qquad (15b)$$

At 6 eV_{CM}, the (a):(b) ratio in this non-adiabatic reaction was measured to be at least 4.0. Pending a measurement of the complete energy dependence of both cross sections, a tentative interpretation would be similar to that for reaction (14).

4. PRODUCT CHANNEL BRANCHING IN SIMPLE IONIC REACTIONS

4.1 $H^+ + D_2$: The Prototype System

H_3^+ is of great fundamental importance. Possessing only two electrons, it is the simplest of all triatomics. Nevertheless, reactions in this system offer a rich variety of branching channels. Using isotopic labelling, precise absolute integral cross sections have been measured by the Freiburg group, by means of the guided ion beam technique developed there [30]. Branching between three channels of the H^+ + D_2 reaction was studied between 0.2 and 12 eV_{CM}:

$$H^+ + D_2 \rightarrow D^+ + HD \qquad (16a)$$

$$\rightarrow HD^+ + D \qquad (16b)$$

$$\rightarrow D_2^+ + H. \qquad (16c)$$

(16a) is about thermoneutral (to within the HD-D_2 zero point energy difference), (16b,c) are both (within the HD^+ and D_2^+ zero point energy difference) endothermic by ~ 1.85 eV ($\approx$ IP (H_2) - IP(H)). Starting from the lowest energies, the cross section for (16a) was found to decrease strongly (see Fig. 1 of Ref. 30), but dominates the branching up to well above the onset of the competing reactions (16b,c). These two are equally probable at first, but from about 4 eV_{CM} on (16c) becomes by far the most important branch, while (16b) decreases sharply from 5 eV_{CM} on, and (16a) gains again in relative importance.

Theory has been remarkably successful in explaining not only this highly structured pattern of energy dependent branching ratios, but even in predicting quantitatively the absolute cross sections for the branching channels individually. At low energies, complex formation is to be expected due to the existence of a strongly bound (4.92 eV) H_3^+ ion. The partial cross sections are then the product of the complex formation cross section, times the complex decay branching ratio.

The concept of complex formation was quantified by means of trajectory calculations which were performed on a DIM H_3^+ surface [31]. For each event the number of times, M_w, was determined that the shortest of the three interatomic distances switched from one pair of atoms to another. Thus, a large M_w corresponds to a long-lived complex. Carrying the calculations through up to separation into products, (which may require a great deal of computer time), the branching ratio could, in these exploratory studies, be determined directly. It was found to attain a nearly constant (of course, energy-dependent) value once $M_w = 4$ was exceeded, and this value was therefore adopted as the definition of statistical conditions having been reached (see p. 113 of Ref. 32). (In recent work, an "induction period", i.e. an M_w range in which the branching still changes somewhat with increasing M_w, of up to $M_w = 20$ was found [33]). Given this definition for complex formation, the trajectory calculations then yield directly the absolute capture cross section at each energy.

The (energy dependent) branching ratios of the complex decomposition into product channels was determined according to Light's phase space theory. Comparison of the results with the "exact" ones, i.e. as obtained directly from the limited sample of trajectory calculations on the decomposing complex, confirmed that the phase space theory is correct to within 30% or so [34]. This is further supported by the success of this theory in explaining the rotational structure in the energy distribution of D^+ from (16a), measured in a differential scattering apparatus (see Fig. 5.9 of Ref. 32).

The absolute energy-dependent cross sections for reactions (16 a-c), calculated as described above, agree very well with the experimental results up to an energy of about 2.5 eV_{CM} (see Fig. 4.31 of Ref. 32). The breakdown of the statistical approach at higher energies is not surprising since above 3 eV_{CM} the fraction of trajectories reaching the necessary minimum number of $M_w = 4$ becomes less than 20% [32] (i.e. the majority of trajectories correspond to "direct collisions" [35]).

In the energy region beyond 2.5 eV_{CM} absolute cross sections for the reactions (16a,b,c) have been calculated using the trajectory surface hopping (TSH) method: (16a) can (but need not) proceed only on the lowest adiabatic HD_2^+

surface, whereas (16b,c) require a jump to the next higher surface somewhere along the seam between the two. Early TSH results [36], at only three energies, gave very good agreement with experiment [20]. They have recently been confirmed and extended over a wide energy range [37]. Again the agreement (on an absolute scale!) is excellent up to the highest measured energy (12 eV$_{CM}$), some discrepancies for (16c) being attributed to deficiencies of the DIM surface.

4.2 H_2^+ + Ar: Exchange Reaction vs. Charge Transfer

Another well-studied ionic system is H_2^+ + Ar where branching between exchange reaction and charge transfer occurs:

$$H_2^+ + Ar \rightarrow ArH^+ + H + 1.30 \text{ eV} \qquad (17a)$$

$$\rightarrow Ar^+ + H_2 - 0.31 \text{ eV}. \qquad (17b)$$

The branching has here been measured as a function of H_2^+ vibrational level, v, both by "TESICO" [38] and by a guided ion beam total cross section measurement, using photoionization to produce known H_2^+(v) distributions [39]. Both experiments agree in the main results: The (17a):(17b) branching ratio is generally greater than unity at low collision energy (although this is reversed at high energy [39]). (17b) shows, however, a dramatic dependence on v, being a very minor channel at v=0, but passing through a "resonance enhancement" at v=2 which makes it competitive with (17a). The branch (17a), on the other hand, exhibits only a minor "vibrational" enhancement. These features, which are qualitatively plausible considering the location of the seam between the two interacting potential surfaces of this system, were reproduced quite accurately by a TSH calculation [40]. Very recently, three-dimensional quantum scattering calculations [41] have been made on the same DIM surface as in Ref. 40, which allow interesting comparisons with the quasi-classical TSH results. The overall agreement is very reasonable, for example as regards the opacity functions for the two branches of the reaction inverse to (17):

$$Ar^+ + H_2 \rightarrow ArH^+ + H \qquad (18a)$$

$$\rightarrow Ar + H_2^+ \qquad (18b)$$

(see Fig. 12 of Ref. 41). Both reactions (17) and (18) have opacity functions which are quite different for the two branches: Atom transfer (a) is more probable in near-central, charge transfer in more glancing collisions. This demonstrates nicely (and in agreement with intuition) how the branching in a given encounter is largely predetermined by the initial orbital angular momentum (i.e. impact parameter).

4.3 N^+ + CO: A "Complex" Triatomic System

Being composed of atoms from the middle of the periodic table and therefore having "very open shells" [42], this system exhibits a very rich chemistry:

$$N^+ + CO \rightarrow N + CO^+ + 0.52 \text{ eV} \qquad (19a)$$

$$\rightarrow C + NO^+ + 0.65 \text{ eV} \qquad (19b)$$

$$\rightarrow C^+ + NO \quad - 1.35 \ eV \qquad\qquad (19c)$$

$$\rightarrow O^+ + CN \quad - 2.61 \ eV \qquad\qquad (19d)$$
$$\rightarrow O \ + CN^+ \ - 2.90 \ eV. \qquad\qquad (19e)$$

Absolute integral cross sections σ for all five channels have been measured with the guided beam technique in the energy range 0.15-13 eV_{CM} [42]. Charge transfer (19a) dominates at all energies (σ between 2 $Å^2$ and 20 $Å^2$), (19d) is the least important channel ($\sigma \approx 0.01 - 0.1 \ Å^2$), and there is clear evidence of intricate competition between these and the other channels, including further dissociative ones. Some of the branching ratios, e.g. (19b):(19c) and (19d):(19e), could be explained with reasonable success on the basis of statistical models, either phase space theory or the Bernstein-Levine information theoretic approach. This can be rationalized on the grounds that NCO^+ forms a linear stable complex (the least electronegative atom being, characteristically, in the central position).

Further details of the reaction mechanism were revealed by measurements of the product kinetic energy distribution in the forward direction [43]. For example, the product ions from (19b) and (19e), NO^+ and CN^+, at low reactant energy both show clearly their origin from the postulated collision complex. They appear at just the right translational energy expected for this mechanism. With increasing energy, however, the peak of the distribution switches distinctly to that energy which corresponds to a stripping mechanism. At still higher primary energies ($\sim$ 20 eV_{CM}) the peak moves beyond the stripping position and finally disappears, quite in accordance with the stripping model: The necessary high internal excitation can then no longer be accommodated by the product, and dissociative channels take over. Interestingly, though, NO^+ and CN^+ can still be observed at projectile energies as high as 100 eV_{lab}. The product energy is then concentrated in a sharp peak at practically zero eV_{lab}. The appearance of the "slow" peak at high collision energy is clear indication of a very special "knock-on" mechanism: For a collinear arrangement of three (nearly) equal masses A + BC a hard-sphere collision of fast A with BC at rest enables C to take over the momentum, with AB staying behind as a bound product at rest. Thus we have here, for each product separately, a nice example of mechanism branching and its energy dependence.

Measurements of this type were also made for the charge transfer channel (19a) [44]. Here, as expected, most of the product CO^+ appears at $E_{lab} \approx 0$, corresponding to glancing collisions. There is, however, a minor CO^+ component appearing at high translational energy and obviously resulting from head-on charge transfer. This fast peak lies, interestingly, always a little below the maximum energy allowed on simple kinematic grounds, indicating that this mechanism always involves substantial electronic excitation of the products (groups of states around 9-12 and 21-24 eV). This can be discussed in terms of *ab initio* potential energy curves which are available for this system in collinear arrangement, such as is appropriate here (see Fig. 6 of Ref. 45): The entrance channel, $N^+(^3P) + CO$, is in fact the first excited state of the system. It lies asymptotically $\sim$ 0.5 eV above the N + CO^+ ground state and is repulsive. The attractive ground state, on the other hand, leads adiabatically to the bound NCO^+. However, diabatically the N^+ + CO channel does connect with the $^3\Sigma^-$ ground state of NCO^+ [42]. The large cross section for (19a) then indicates that even at low impact energy the non-adiabatic transition into the NCO^+ potential well is easily made, followed by adiabatic dissociation to ground state N + CO^+. Also the complex formation stage of the other reaction channels is obviously preceded by this charge transfer surface hop

440

occurring at low energy. Charge transfer at high impact energy is then all the more expected to go via this initial non-adiabatic transition and through the NCO^+ well, but then followed by further non-adiabatic transitions high up on the inner repulsive wall. This will ultimately lead to electronically excited products such as were observed in Ref. 44. In fact, Fig. 6 in Ref. 45 shows one of the low electronically excited product channels explicitly. The alternative, "directly uphill" route along the repulsive adiabatic entrance channel and towards high-lying curve crossings, although suggestive at first sight, is much less likely, considering the ease with which the NCO^+ ground state is reached even at low energies.

4.4 CH_4^+ + CH_4: Mechanism Branching

The N^+ + CO system provides many examples of formation of a given product via different mechanisms, depending on the collision energy. The CH_4^+ + CH_4 system now considered is a case where the product is formed via different mechanisms even at one and the same reactant energy:

$$CH_4^+ + CH_4 \rightarrow CH_5^+ + CH_3 + 0.17 \text{ eV}. \tag{20}$$

Low-energy (0.7 eV_{CM}) differential scattering experiments with product energy and mass analysis [46] show, in the form of CM system intensity contours, three distinct peaks: a) CH_5^+ appearing at the CM velocity originates through formation of an intermediate $C_2H_8^+$ complex; b) Another CH_5^+ intensity peak is located at "forward" velocities (with respect to the CH_4^+ CM direction of flight), at the position expected for a stripping mechanism: These CH_5^+ ions are formed by H atom transfer from the target to the projectile; c) A third, and in fact the largest peak, is found at a "backward" CM velocity (again relative to the CH_4^+ CM direction): It corresponds to proton transfer to the target molecule which, from the measured peak position, also proceeds in stripping-like fashion. The relative intensities ("mechanism branching ratios") can be nicely rationalised: Although the $C_2H_8^+$ complex lifetime is estimated to be fairly long ($> 10^{-12}$ s) and its dissociation energy fairly large (> 0.5 eV), *ab initio* calculations show that it is structurally not very accessible from CH_4^+ + CH_4. Consequently the complex formation is only a minor branch ($< 10\%$). Of the rest, atom transfer (20b) is less likely than proton transfer (20c), because (20b) must involve a non-adiabatic transition, while (20c) does not: Transferring a bare proton does not affect the electronic state correlations.

This reaction was further studied by TESICO at controlled internal reactant ion energy and using isotopic substitution, but without angular energy product analysis [47]; e.g.

$$CH_4^+ + CD_4 \rightarrow CH_3D_2^+ + CHD_2 \tag{21a}$$

$$\rightarrow CH_4D^+ + CD_3 \tag{21b}$$

$$\rightarrow CD_4H^+ + CH_3 \tag{21c}$$

$$\rightarrow CD_4^+ + CH_4. \tag{21d}$$

(21a) is, from the isotopic scrambling, ascribed to the complex mechanism, whereas (21b,c) obviously are (D-) atom and proton transfer, respectively. The measured branching ratios are in good agreement with those from Ref. 46, although some modification due to the isotopic substitution is to be expected. In fact, this has been directly observed in the corresponding reactions [47]

$$CD_4^+ + CH_4 \rightarrow CH_3D_2^+ + CHD_2 \qquad (22a)$$

$$\rightarrow CD_4H^+ + CH_3 \qquad (22b)$$

$$\rightarrow CH_4D^+ + CD_3 \qquad (22c)$$

$$\rightarrow CH_4^+ + CD_4. \qquad (22d)$$

The relative ordering (a) < (b) < (c) is the same as in (21), but the branching ratios are changed by an isotope effect of 1.3 - 2.0: (21c) takes a larger share (65-80%) from the total of (21) than does (22c) from (22) (50-70%), while conversely (21b) amounts to only 20%, compared to 30-40% for (22b). All this is as expected: H^+ and H are more readily transferred than D^+ and D. The effect of the internal and the translational energy on all branching ratios is similar: Channels (b) are decreased, channels (c) are increased slightly. The greatest overall effect of reagent energy is on channel (a), which decreases rather strongly in relative importance, as expected. The pure charge transfer channels (d), surprisingly, have a cross section which is smaller than the cross sections for chemical reaction (a)-(c) combined.

Many more examples of the rich branching chemistry of ion-molecule reactions are given in Ref. 48, e.g.

$$CH^+ + H_2CO \rightarrow CH_3^+ \qquad 30\%$$

$$\rightarrow H_3CO^+ \qquad 30\%$$

$$\rightarrow HCO^+ \qquad 30\% \qquad (23)$$

$$\rightarrow H_2C_2O^+ \qquad 10\%.$$

Most of these, particularly at the low collision energies considered specifically in Ref. 48, will result from the statistical break-up of a long-lived complex. A calculation of the various parallel, competing fragmentation paths would follow similar lines as the statistical theory of mass spectra, except that in the case of very low energy ion-molecule reactions the initial energy of the complex is rather better defined than that of an organic ion produced by electron impact in a conventional mass spectrometer.

5. CHEMILUMINESCENCE AND CHEMIIONIZATION

The observation of optical emission from electronically excited reaction products (chemiluminescence, CL) and the formation of ions in encounters between neutrals (chemiionization, CI) provide many beautiful examples of branching among the multitude of accessible states. Beam-gas or crossed beam arrangements have been used, the reactants usually being atoms or atomic ions (in their ground or metastable states) plus diatomic target molecules. CL is highly product specific thanks to the spectroscopic detection, in CI mass analysis is sometimes employed.

5.1 Ca + F$_2$: A Typical Case

The highly exothermic group II atom and halogen molecule reactions are typical of systems capable, on energetic grounds, of populating electronically excited states including ionization. The CL cross sections usually comprise only a small fraction of the overall reactive cross section, the remainder leading to ground or metastable states. Examples are

$$Ca(^1S) + F_2 \rightarrow CaF(A^2\Pi) + F \qquad 1.2 \ \text{Å}^2 \qquad (24a)$$

$$\rightarrow CaF(B^2\Sigma) + F \qquad 0.3 \ \text{Å}^2 \qquad (24b)$$

$$\rightarrow CaF^+ + F^- \qquad 0.1 \ \text{Å}^2 \qquad (24c)$$

$$\rightarrow dark \ channels \qquad 58 \ \text{Å}^2 \qquad (24d)$$

and the corresponding reactions of metastable Ca(^{3}P,^{1}D), whose cross sections are 2-4 times greater. The low quantum yield, which is unfortunate as regards the prospects for designing a chemical electronic-transition laser, is basically a consequence of a common characteristic of nearly all CL/CI systems: These reactions are highly non-adiabatic, involving many curve crossings between reactants and luminescent products.

In more detail, reactions of the type (24) are thought to be initiated by harpooning [49,50], both for CL and for CI. The gradual decrease of the CL cross section with increasing energy, compared to the more abrupt drop, at quite low energy, for CI (e.g. for Ba + Cl$_2$, see Fig. 9 of Ref. 50) suggests that different curve crossings are responsible for these two channels: CI appears to originate from a transition between two weakly interacting curves, such as is typical of a crossing at a fairly large internuclear distance R$_1$ ("outer harpooning"), while the CL behaviour corresponds to two strongly interacting curves, crossing closer in, at R$_2$ ("inner harpooning"). The two ionic curves crossing the covalent entrance potential at R$_1$ and R$_2$ correlate, respectively, with ground state and excited ions on the reactant side, e.g. Ca$^+$(^{2}S) + F$_2^-$ and Ca^{+*}(^{2}D) + F$_2^-$. The first of these leads primarily to ground state products. If it is subsequently crossed again by a curve correlating with Ca^{++} + F$_2^{--}$ on the reactant and with CaF$^+$ + F$^-$ on the product side, it leads to CI. The ionic curve which crosses at R$_2$, on the other hand, correlates with CaF* + F (CL). The CI/CL branching ratio will be determined, in the first place, by the relative transition probabilities at R$_1$ and R$_2$ (note that, in order to reach R$_2$, the system has to traverse R$_1$ diabatically!). Other transition probabilities, such as that to the Ca^{++} + F$_2^{--}$ curve, also play a role.

As for the branching ratio between the different CL channels, e.g. (24 a,b) the explanation is advanced in Ref. 50 that the crossing at R$_2$ in reality consists of three subcrossings, deriving from the ^{2}D symmetry of Ca^{+*}. Among these, the curve correlating with the CL product in the A$^2\Pi$ state is reached before the curve correlating with B$^2\Sigma$ product. This, plus different symmetries at the two R$_2$ subcrossings, could cause the generally observed predominance of the A over the B state channel.

The effect of reagent electronic symmetry on the A/B branching has been demonstrated in a beautifully direct way by observation of CL with laser excited Ca(^{1}P) reacting with HCl and Cl$_2$ [51]. This atomic state behaves analogously to the much-studied metastable Ca(^{3}P) [49,50], but can be electronically aligned relative to the collision velocity vector v_{rel}, utilizing polarized laser light. The laser prepares the Ca 4s4p configuration with the 4p orbital either parallel or perpendicular to v_{rel}. At the "inner crossing", the electron jump removes the s electron, leaving the system on an ionic surface correlating with Ca$^+$(^{2}P) + X$_2^-$ (similar to the correlation with Ca$^+$(^{2}D) postulated in Refs. 49,50). Rotating the polarizer it could be shown that the yield of CaCl(A$^2\Pi$) from HCl goes through a maximum when the Ca p electron is in π alignment relative to the CaCl bond, while the CaCl(B$^2\Sigma$) yield is maximized for a σ alignment. With Ca(^{1}P) + Cl$_2$ the same effect should be operative, but the observation is obscured by an additional contribution to CL from the outer crossing. The persistence of the asymptotically prepared alignment into the products could indicate that these reactions occur only at low impact parameter, as quasi head-on collisions. More likely, however, off-axis collisions do contribute, but in such a way that "orbital following", i.e. adiabatic reorientation of the molecular orbitals during the collision, takes place.

5.2 C$^+$ + H$_2$: A Chemiluminescent Ion-Molecule Reaction

The essence of the current rationalization of chemiluminescence branching ratios is that a particular product channel is the more disfavoured the more non-adiabatic surface hops are required to reach it. This can be checked especially conveniently using ions as projectiles, which allow a very wide energy range to be studied. A prototype system is [52]

$$C^+(^2P) + H_2 \rightarrow CH^+(^1\Delta) + H \qquad -6.9 \text{ eV} \qquad (25a)$$

$$\rightarrow CH(^2\Delta) + H^+ \qquad -6.3 \text{ eV} \qquad (25b)$$

$$\rightarrow CH^+(^1\Pi) + H \qquad -3.4 \text{ eV} \qquad (25c)$$

$$\rightarrow CH^+(^3\Pi) + H \qquad -1.5 \text{ eV} \qquad (25d)$$

$$\rightarrow CH^+(^1\Sigma^+) + H \qquad -0.4 \text{ eV} \qquad (25e)$$

$$C^+(^4P) + H_2 \rightarrow CH^+(^3\Sigma^-) + H \qquad +0.36 \text{ eV}. \qquad (25f)$$

The cross sections are ~ 0.001 Å^2 for (25a) and (25b) and 0.04 Å^2 for (25c), compared to ~ 2Å^2 (from mass spectrometry) for the dark channels (25d,e) combined. These two latter ones can proceed adiabatically, while (25a-c) require non-adiabatic transitions. (25f), a CL reaction of metastable C$^+$ ions, also requires a non-adiabatic transition between two adjacent surfaces, but these are presumably close to each other, because of the small overall exothermicity of only 0.36 eV. As a result, the cross section is here quite large, ~ 5 Å^2.

In the case of ionic CL exchange reactions, there will generally be two types according to whether the ionic or the neutral product is the emitter:

$$A^+ + BC \rightarrow AB^{+*} + C \qquad (26a)$$

$$\rightarrow AB^* + C^+ \qquad (26b)$$

The branching between these two classes was found to be governed by the relative ionization potentials of AB and C, at least in the case of B = hydrogen (the excitation energies of hydrides and hydride ions are usually similar and therefore cancel roughly in the energy balance between (26a) and (26b)), AH^{+*} was observed in preference over AH^* (see Fig. 15 of Ref. 52). This is really just a variant of the above rule: The energy level of AH^{+*} + B will then lie below that of AH^* + B^+ so that, for these generally endothermic reactions, the former can be reached with fewer non-adiabatic surface hops.

5.3   La + O_2: A Very Complex Chemiluminescent System

For systems with very many potential surfaces there appears to be, irrespective of the symmetry of the product state, always such a large number of possible adiabatic/non-adiabatic pathways that it is no longer possible, or indeed necessary, to account for them individually. Instead, statistical considerations take over. Reactions of La with O_2 are a case in point [53]. The low-lying $La(^2D_{3/2,5/2})$ and $La(^4F_{3/2,5/2,7/2,9/2})$ levels give rise, with O_2, to a large number of adiabatic surfaces. Three emitting product states were observed:

$$La + O_2 \rightarrow LaO(A^2\Pi_{1/2,3/2}) + O \quad + 35 \text{ kcal} \quad (27a)$$

$$\rightarrow LaO(B^2\Sigma) \quad + O \quad + 21 \text{ kcal} \quad (27b)$$

$$\rightarrow LaO(C^2\Pi_{1/2}) \quad + O \quad + 7 \text{ kcal}. \quad (27c)$$

The branching ratio between these exothermic reactions varied considerably over the energy range 130-900 meV$_{CM}$, favouring (27c). Interestingly, this trend followed very well that predicted by the "prior probability model", i.e. the branching occurred in the ratio of the density of product translational states, summed over the rotational states. The success of the statistical treatment demonstrates that non-adiabatic transitions are no longer an impediment to a reaction once the density of the electronic states is sufficiently high. The model does not work quite so well for some endothermic CL reactions of Sc and Y with O_2 [53].

5.4 B^+ + H_2: Comparison of CL with Scattering Results

As a rule, CL experiments suffer from the fact that the branching between the radiating and the (usually dominant) ground state product channels cannot be determined directly. This measurement generally requires a knowledge of the absolute light detection efficiency plus a determination of the cross section for total reactant consumption. (Only in the case of spontaneously emitting, long-lived electronically excited reactants can the first of these requirements be replaced by a relative measurement of the CL and the reactant emission intensities, provided the reactant's radiative lifetime is known [54]). For chemiluminescent ion reactions the possibility exists of obtaining the CL/ground state branching ratio from scattering experiments with product energy analysis. The B^+ + D_2 system was originally studied in this way [55]. BD^+ was observed in the forward direction in two well-separated groups on the energy scale. The fast component was identified as BD^+ formed in the ground state $X^2\Sigma^+$, the slow component as BD^+ which, at formation, originated in the $A^2\Pi$ state:

$$B^+(^3P) + D_2 \rightarrow BD^+(X^2\Sigma^+) + D \quad +2.09 \text{ eV} \quad (28a)$$

$$\to \text{BD}^+(\text{A}^2\Pi) + \text{D} \qquad -1.19 \text{ eV}. \qquad (28\text{b})$$

This assignment was nicely verified by direct optical observation of A → X emission [56]. The spectroscopic analysis revealed the rotational-vibrational A-state distribution with high precision [57] but it gave of course no information on the ground state. The vibrational distribution derived from the spectra accounts very well for the (unresolved) band contour of the scattering experiment (see Fig. 17 of Ref. 58). The A/X branching ratio is obtained from the translational spectrum (see Fig. 1 of Ref. 55) after lab → CM transformation [58]. The result is (28a):(28b) = 1:0.25, a surprisingly high CL yield. Apparently formation of A-state product is facilitated by the existence of an adiabatic pathway for (28b) [56], a rather exceptional circumstance. Details of this path were elucidated by another scattering experiment on $\text{B}^+ + \text{H}_2$, where even very slow product ions, corresponding to CM backwards scattering, could be observed [59]. From the resolved velocity groups it was found that, at E_{CM} = 1.6 eV, the A-state product was symmetrically distributed between forward and backward hemispheres, while X state ions exhibited some forward asymmetry. Apparently both reaction paths involve a bound complex. In fact, *ab initio* calculations of the $^3\text{A}''$ surface for (28b) have shown a deep (several eV) well. The adiabatic surface for (28a) is much more complicated [60].

6. REACTIONS OF METASTABLE ATOMS

6.1 He*: High-energy Metastables

Reactions of helium metastables, e.g. He(2 $^1\text{S}_0$) with an excitation energy of 20.6 eV, will nearly always involve ionization. For example

$$\text{He}^* + \text{Ar} \to \text{HeAr}^+ + \text{e}^- \qquad (29\text{a})$$

$$\to \text{He} + \text{Ar}^+ + \text{e}^-. \qquad (29\text{b})$$

(29a) is called associative ionization (AI), (29b) Penning ionization (PI).

The branching (a):(b) is well understood and can be discussed in terms of the entrance and exit channel diatomic potential energy curves [61]: If the energy of the ejected electron, E_{el}, is greater than the asymptotically available energy $E_0 + E_{kin}$ (the sum of the metastable excitation and the collision energies), then the product will be stabilized into a bound state (AI), otherwise not (PI). Assuming vertical autoionizing transitions between the potential curves, AI can then be associated with electron emission at close internuclear distance, PI with long-range transitions (see Fig. 46 of Ref. 61 for the typical case of a strongly interacting incoming channel and a moderately strongly interacting outgoing channel). The partitioning AI/PI can be read from the E_{el}-spectrum, the dividing line being $E_0 + E_{kin}$ (see Fig. 48 of Ref. 61). Alternatively the branching can be obtained directly from the product mass spectrum. Figure 11 in Ref. 61 gives a nice example for reaction (29), where, in addition, the dependence on E_{kin} is shown (increasing E_{kin} strongly favours (29b) over (29a), as expected from the stabilization condition). The angular distribution of Ar^+ and HeAr^+ is also given in this figure: HeAr^+ is scattered along the CM direction, as it must be, Ar^+ is strongly forward scattered, indicating a small translational exoergicity Q.

For molecular targets further branching in the PI channel occurs:

$$He^* + X_2 \rightarrow X_2^+ + He + e^- \tag{30a}$$

$$\rightarrow X_2^{+*} + He + e^- \tag{30b}$$

$$\rightarrow X^+ + X + He + e^- \tag{30c}$$

i.e. the product can be electronically excited or dissociate. If X is a halogen, then the incoming covalent He^* + X_2 potential curve will be crossed by an ionic He^+ + X_2 curve. Two mechanisms of electron emission can then be distinguished [62]:

a) PI during the approach along the covalent curve; He^* then acts simply as an energy donor or quasi-photon, and the product distribution resembles that observed in photoionization with He(I) light (hv = 21.21 eV).

b) PI out of the ionic complex. This is the chemically more interesting case, and the measured E_{el} spectra help to map out the transition state. Synthetic E_{el} spectra for He(2 ^{3}S) + Br_2 were calculated [62] from trajectories in a two-dimensional model (intramolecular distance r and molecule-atom distance R as variables), and the potential parameters were adjusted to achieve an optimum fit with the experimental spectrum. This corresponds to the energy distribution in the nascent Br_2^+ products, irrespective of whether or not they fragment subsequently into Br + Br^+. Important additional experimental input for the potential adjustment was in this work an E_{el} spectrum measured in coincidence with Br_2^+ ions, i.e. corresponding to only those ionizing transitions which lead to stable Br^+. This spectrum is related to the above, total E_{el} spectrum by the fact that the fate of a nascent Br_2^+ (fragmentation or not) can be predicted, at each excitation energy, from the known Br^+ potential curves.

The coincidence E_{el} spectrum was found to be particularly sensitive to the potential parameters and the autoionization widths. A consistent fit of both the total and the coincidence E_{el} spectrum was achieved. The dominant mechanism proved to be autoionization from the ionic potential. Structure in the latter could be assigned to formation of excited Br_2^+, e.g. Br_2^+ ($A^2\Pi_u$) formed in the ionic well. In addition $Br_2^+(A)$ is formed in the covalent entrance channel and is trivially identifiable, like in photoelectron spectroscopy, from the corresponding E_{el}.

For He^* + Cl_2, optical emission from the analogous Cl_2^{+*} has been directly observed [63]. The areas under the corresponding peaks in the covalent-dominated part of the E_{el}-spectrum account only for 1/3 of the observed fluorescence intensity (calibrated against photoionization-induced fluorescence as measured in the same apparatus). Two thirds of the fluorescence comes from (highly vibrationally) excited Cl_2^+ formed in the ionic intermediate. This, then, is another type of microscopic branching, identifiable by careful quantitative analysis of Penning electron spectra.

Very recently an interesting correlation with the electron angular distribution has been observed [64]: the covalently emitted electrons appear preferentially in the background direction (relative to the He^* beam), the electrons emitted out of the ionic complex are more isotropically distributed. The first is in keeping with a well-known model for PI [65], the second is also plausible, considering the highly attractive Coulomb potential. Thus the observation of angular distributions may help to distinguish between the two microscopic channels without recourse to the difficult coincidence measurements.

In purely covalent systems like He^* + N_2 a correlation between the angular distribution and the electronic state of the N_2^+ product was found: Formation of $N_2^+(X^2\Sigma_g^+)$ gives backward peaking, $N_2^+(B^2\Sigma_u^+)$ an isotropic distribution. This, too, can be rationalized in terms of orbital overlap, [64], in the spirit of Ref. 65

6.2 Xe^*: Low-energy Metastables

Xe has two metastable states, $^3P_2$ at 8.3 eV and $^3P_0$ at 9.4 eV. This energy is insufficient to ionize most species. Instead, rich chemiluminescence spectra have been observed, for example with halogen molecules [66]. Two types of reaction occur, excitation transfer or atom transfer, forming an excimer molecule:

$$Xe^* + X_2 \rightarrow Xe + X_2^* \tag{31a}$$

$$\rightarrow XeX^* + X. \tag{31b}$$

A monotonic decrease of the cross section was measured in both channels for collision energies from ~ 0.1 to ~ 1 eV (attained by high-speed rotor acceleration of the atoms). This behaviour is typical of an ionic/covalent curve crossing mechanism and suggests that the reactions are initiated by an electron jump $Xe^* + X_2 \rightarrow Xe^+ + X_2^-$. The excited X_2^* is then formed by a secondary recrossing into the neutral potential curve manifold. Analysis of the X_2^* spectra shows that the various observed emitting states can all be populated by a one-electron jump from the first (repulsive) X_2^- state, $^2\Pi_g$, rather than from the X $^2\Sigma_u^+$ ground state. Atom transfer may, nevertheless, proceed also via X_2^- $(^2\Sigma_u^+)$. The branching between channels (31a,b) will therefore, in the first place, depend on the relative probabilities of X_2^- $^2\Sigma_u^+$ vs. $^2\Pi_g$ formation. Given the second possibility, it can further be viewed as a competition between the neutralization and the dissociation of $X_2^- (^2\Pi_g)$, the first path yielding X_2^* (excitation transfer), the second $X + X^-$ and, by recombination of $X^- + Xe^+$ within the ionic complex, $XeX^* + X$ (atom transfer, for example, as specified in the limiting case of the DIPR model [67]). The measured fractions of X_2^* formation (31a), obtained by subtraction of computer simulated XeX^* spectra, were $\sim 40\%$, $\sim 20\%$, $< 2\%$ for $X_2 = I_2$, Br_2, Cl_2, respectively [66]. These branching ratios reflect, in a way which is not yet clear in detail, the electronic and kinematic effects on the two reaction paths.

The excited halogen negative ion states also play an important role in controlling the branching between the two different atom transfer channels which are possible in the case of a mixed halogen molecule, e.g.

$$Xe^* + IBr \rightarrow XeI^* + Br \tag{32a}$$

$$\rightarrow XeBr^* + I. \tag{32b}$$

Resolved crossed-beam chemiluminescence spectra [68] have yielded a 0.4:1 branching ratio between channels (32a) and (32b) at 40 meV collision energy. Additional time-of-flight measurements, using coarser filtering of the spectral features, showed that the fraction of (32b) increased by a factor of about 1.5 between 15 and 230 meV_{CM}.

These results were explained in Ref. 68 considering the crossings of the incoming $Xe^* + IBr$ covalent curve with the ionic curves (I), $Xe^+ + IBr^-$ (X $^2\Sigma^+$), and (II), $Xe^+ + IBr^- (^2\Pi_{1/2})$. The former (outer) crossing will lead to $XeBr^*$ excimer formation, reaction (32b), because the IBr^- (X $^2\Sigma^+$) ground state dissociates adiabatically to Br^- + I. Conversely the (inner) transition to potential (II) will give XeI^* via (32a), because $IBr^-(^2\Pi_{1/2})$ dissociates to I^- + Br. The possibility was also considered that initially all trajectories switch to the ionic potential (I), with subsequent coupling between (I) and (II) at closer range. (This coupling might be induced by the presence of the Xe^+ ions which "fixes" the negative charge on the I atom where it was initially deposited, thus preventing the curve (I) from dissociating to its proper limit Br^- + I, and giving it some of the character of curve (II)). In either case the product from (II), i.e. XeI^* + Br, will be disfavoured, in agreement with the observation.

The velocity dependence of the branching ratio can also be explained in either case: If (32a) arises from direct harpooning to curve (II), then the competition with the outer harpooning, to curve (I), will be important. At low velocities, "prestretching" of the IBr by the approaching Xe^* will occur, increasing its electron affinity and thereby moving the curve crossing radius $R_c(I)$ to larger values, where an electron jump becomes less likely. This will then favour a jump at $R_c(II)$, and hence channel (32a). On the other hand, if (32a) arises from close-range coupling of curve (I) to curve (II), then the vibrational phase of the IBr $(X ^2\Sigma^+)$ will be decisive, at the instant the Xe^+ arrives at the short distance necessary for the atom transfer. At low velocities about one full vibrational period elapses between the electron jump at $R_c(I)$ and the "arrival" of Xe^+. The Xe^+ will then meet the IBr^- with the charge again localized at the I end, as it was at the instant of harpooning at $R_c(I)$, and this will favour XeI^* formation, (32a). At higher impact velocity, Xe^+ will arrive at a vibrational phase such that the IBr is halfway to its dissociation limit, and thus is more in a I + Br^- configuration, favouring (32b), again in agreement with the observation. The experiments were inconclusive in deciding which of the two proposed mechanisms, population of surface (II) directly from the covalent surface or via surface (I), is dominant in XeI* formation.

Reactions of Xe^* with CO and N_2 are qualitatively different from the foregoing in that ionic curves play no role here, because these molecules have negative electron affinity. Only electronic energy transfer, but no excimer formation occurs. In either target molecule at least two radiating states are formed:

$$Xe(^3P_2) + CO \rightarrow CO(a'^3\Sigma^+) + Xe \tag{33a}$$

$$\rightarrow CO(d\ ^3\Delta) + Xe \tag{33b}$$

$$Xe(^3P_2) + N_2 \rightarrow N_2(B\ ^3\Pi_g) + Xe \tag{34a}$$

$$\rightarrow N_2(W\ ^3\Delta_u) + Xe. \tag{34b}$$

Very striking is the highly selective vibrational level population: In each case the closest-to-resonance endothermic levels (CO (a', v=11), CO(d, v=6), N_2 (B, v=5), N_2 (W, v=6)) are strongly favoured over the exothermic levels [69,70]. This can be explained by a crossing of the (semiempirically calculated) entrance and exit covalent curves, the former being long-range repulsive. The branching ratio (33a):(33b) was found to be about 1:1 and within 10% independent of the collision energy from 50 to 350 meV$_{CM}$ [70]. This is not surprising, because both CO(a') and CO(d) derive from the same electronic configuration, $\pi^3\sigma\pi^*$. The exit channel interaction potential and the transition probabilities are therefore expected to be similar.

The branching between (34a) and (34b) is in the ratio 1:(0.2 ± 0.1) at 100 meV$_{CM}$ [70]. It is difficult to measure, because $N_2(W)$ emits to $N_2(B)$ in the IR and was in this experiment detected indirectly, via the subsequent $N_2(B \rightarrow A)$ emission. A selectivity towards the slightly endothermic level $N_2(W, v=6)$ was, however, clearly established from the spectrum. The energy transfer mechanism is therefore basically the same as in (33a,b) and (34a), although in (34) the electron configurations of the two products are different: $N_2(W, \pi_u^3\sigma_g^2\pi_g)$ vs. $N_2(B, \pi_u^4\sigma_g\pi_g)$.

7. PHOTODISSOCIATION ("HALF-COLLISIONS")

7.1 Small Molecules

Photodissociation very commonly produces molecular fragments in a variety of electronically excited states. This fact alone does not constitute branching in the sense implied throughout this review: Usually each asymptotic exit channel is related uniquely to some repulsive or predissociating state prepared in the photon absorption act, and the product distribution simply reflects the accessibility of the various excited states from the ground state (Franck-Condon factors and transition moments).

One would speak of true dynamical branching only if different products result from the initial preparation of one and the same state. An example amenable to quantum calculations is the photodissociation of H_2:

$$H_2 + h\nu \rightarrow H \ (2p) + H \tag{35a}$$

$$\rightarrow H \ (2s) + H. \tag{35b}$$

Using synchrotron radiation the branching was studied at 839 Å, corresponding to the resonant excitation of bound H_2 (D $^1\Pi_u$, v'=3, J=1 and 2) [71]. These levels lie close to (900 cm^{-1} above) the H(n=2) dissociation limit. H(2p) was detected by its spontaneous L_α radiation, H(2s) by the additional L_α photons emitted in the presence of a 20 V/cm quench field. A branching ratio (35a):(35b) = 0.75 was measured.

Recent theoretical work has been devoted to this experiment, as well as future ones using different wavelengths [72]. The initial state predissociates primarily to $H_2(B' \ ^1\Sigma_u^+)$ which connects adiabatically to H(2s) + H(1s). However, at very large internuclear distance (~ 10 a.u.) the $H_2(B')$ state comes close to $H_2(B \ ^1\Sigma_u^+)$. Via radial coupling between the two states, transitions to $H_2(B)$ can take place. This state leads adiabatically to H(2p) + H(1s) and is the exit channel for (35a). (Coriolis coupling with the C $^1\Pi_u$ state, which also dissociates to H(2p), is very inefficient [72]). The calculated branching ratio obtained through this mechanism is (35a):(35b) = 0.27. With off-resonance excitation an oscillating branching ratio, as a function of wavelength, is predicted [72].

Photodissociation of CH_3I is another well-studied problem, since it yields preferentially I atoms in the electronically excited fine structure level:

$$CH_3I + h\nu \rightarrow I^* \ (^2P_{1/2}) + CH_3 \tag{36a}$$

$$\rightarrow I \ (^2P_{3/2}) + CH_3. \tag{36b}$$

The branching ratio is of great importance for the iodine laser operation on the $I^* \rightarrow$ I transition (7603 cm^{-1}). A measurement by molecular beam photofragment spectroscopy [73] gave, from two well-resolved peaks in the I atom TOF distribution, a ratio (36a):(36b) = 3.5:1, at 2662 Å. This process has been treated quantum-mechanically [74], representing the molecule by a pseudo-triatomic, linear (H_3)-C-I model system. The absorption occurs initially to the $^3Q_0$ state which correlates with I^* + CH_3. The branch (30b) is explained by coupling with another surface, $^1Q_1$, which dissociates to ground state products. A branching ratio (36a):(36b) = 2.4:1 is calculated.

Very recently, the isotope effect in the photodissociation of HOD has been calculated [75]:

$$HOD + h\nu \rightarrow H + OD \tag{37a}$$

$$\rightarrow D + OH. \tag{37b}$$

Quantum as well as classical calculations, with quite similar results, were made from threshold, $h\nu$ = 2.7 eV, to $h\nu$ = 3.5 eV. Large effects are predicted: The ratio (37a):(37b) is ~ 6 at threshold, then decreases to 2-3. This large isotope effect is essentially due to the faster acceleration of an H fragment, compared to D. Unfortunately no experimental data are so far available.

7.2 Large Molecules

Bond-selective fission of large molecules is a much sought-after goal of photochemistry. An especially spectacular success in this area has been the recently reported UV induced rupture of the comparatively strong (67.7 kcal) C-Br bond in the CH_2BrI molecule in preference over the chemically similar, but weaker (55.1 kcal) C-I bond [76], processes (38a) and (38b):

$$CH_2BrI + h\nu \rightarrow CH_2I + Br \tag{38a}$$

$$\rightarrow CH_2Br + I \tag{38b}$$

$$\rightarrow CH_2 + Br + I \tag{38c}$$

$$\rightarrow CH_2 + IBr. \tag{38d}$$

Raman-shifted excimer laser light pulses at 210 nm produced fragments which were detected using TOF mass spectrometry. CH_2I^+, but no CH_2Br^+ at all was observed, demonstrating complete "branching" in favour of (38a) compared to (38b). The Br atoms from (38a) were also detected at the proper arrival time corresponding to that of CH_2I^+, while no I^+ from (38b) was observed (only I^+ from a secondary process, as well as Br^+ and I^+ from (38c)). The interpretation is that 210 nm absorption excites an $n(Br) \rightarrow \sigma^*(C-Br)$ transition to a repulsive state. The corresponding $n(I) \rightarrow \sigma^*(C-I)$ transition would occur around 270 nm. However, excitation in the "correct" absorption band will not necessarily guarantee the desired fragmentation pathway, as similar, but unsuccessful experiments with C_2F_4BrI showed. It is thus the electronic nature of the primary excitation, not just the photon energy, which determines the outcome. Possibly it is the geometry of CH_2BrI which makes it such a good candidate for selective dissociation: The C-Br and C-I bonds are roughly perpendicular to each other, so that intramolecular dipole-dipole energy transfer from the excited C-Br bond to the competing C-I bond is inhibited.

8. CONCLUSION

In this review a great variety of, but by no means all, branching processes have been covered. The emphasis has been on the broadest features of simple chemical processes, addressing questions such as: Which products are predominantly formed from given reactants? Are they in their electronic ground states? What is the fraction of charged products? In each case only one or two typical systems have been included, although it is hoped that the selection made does justice to most of the important developments. Also the questions of rotational/vibrational product excitation, fine structure level, Λ doublet populations and other finer details, which strictly also belong under the concept of "branching", had to be generally left out of consideration. From the immense number of very important studies of such internal state distributions only a few have been included which shed light on different possible reaction paths to the same product. The restriction to "chemical" processes is also clearly arbitrary. It means, for example, that electron impact on molecules, with the attendant branching between many excitations, fragmentation and ionization channels, is here considered as "not chemical". Finally "larger" systems have not been included, although they provide many beautiful examples of chemically important branching. A case in point is the quasi-equilibrium theory of mass spectra, which predicts the branching between the various competing fragmentation channels of a parent ion.

With all these limitations a survey has been attempted of the many different interpretations which have in the past been given of the striking phenomenon of branching. Not surprisingly, no unifying concept has emerged which explains "everything". It will be seen, however, that advances have been made on a broad front, from the original simplistic models (which still have their merits!) towards a faithful rendering by quasi-classical or even quantal scattering methods, or, in a well circumscribed sense, by statistical treatments. Still, many problems remain unsolved, especially where non-adiabatic processes are involved.

ACKNOWLEDGEMENTS

I am indebted to many colleagues and friends with whom I have enjoyed discussions on their work: N. Aristov, M. Baer, W. Hack, Z. Herman, A. Kowalski, Th. Krümpelmann, R. Martin, H. Morgner, B. Müller, R. Schinke, Ch. Schlier and J. Wolfrum. In some cases they have made unpublished results available to me. Work from our laboratory was supported by the Deutsche Forschungsgemeinschaft.

REFERENCES

[1] R. Grice, Adv. Chem. Phys., 30 (1975) 247.

[2] J.C. Whitehead, D.R. Hardin and R. Grice, Mol. Phys., 25 (1973) 515.

[3] S.M. Lin, J.C. Whitehead and R. Grice, Mol. Phys., 27 (1974) 741.

452

[4] W. Hack, H. Gg. Wagner, K. Hoyermann, Ber. Bunsenges., 82 (1978) 713.

[5] T.H. Dunning, Report 1984/85, Theor. Chem. Group, Argonne Natl. Lab., Argonne, Ill.

[6] H. Schacke, K.J. Schmatjko and J. Wolfrum, Ber. Bunsenges., 77 (1973) 248.

[7] K.J. Schmatjko and J. Wolfrum, 16th Int. Symp. on Combustion, The Combustion Institute (1977), p. 819.

[8] K.J. Schmatjko and J. Wolfrum, Ber. Bunsenges., 82 (1978) 419.

[9] K.J. Schmatjko and J. Wolfrum, Ber. Bunsenges., 79 (1975) 696.

[10] D. Brandt and J.C. Polanyi, Chem. Phys., 35 (1978) 23.

[11] J.C. Polanyi and W.J. Skrlac, Chem. Phys., 23 (1977) 167.

[12] J.C. Polanyi and J.L. Schreiber and W.J. Skrlac, Farad. Disc. Chem. Soc., 67 (1979) 66.

[13] W.H. Breckenridge and N. Umemoto, J. Chem. Phys., 80 (1984) 4168.

[14] D.A. Dixon, D.L. King and D.R. Herschbach, Farad. Disc. Chem. Soc., 55 (1973) 375.

[15] D.A. Dixon and D.R. Herschbach, J. Am. Chem. Soc., 97 (1975) 6269.

[16] D.L. King, D.A. Dixon and D.R. Herschbach, J. Am. Chem. Soc., 96 (1974) 3328.

[17] A. Persky, J. Chem. Phys., 59 (1973) 5576.

[18] M.J. Berry, J. Chem. Phys., 59 (1973) 6229.

[19] D.S. Perry and J.C. Polanyi, Chem. Phys., 12 (1976) 37.

[20] K. Tsukiyama, B. Katz and R. Bersohn, J. Chem. Phys., 83 (1985) 2889.

[21] M.S. Fitzcharles and G.C. Schatz, J. Phys. Chem., 90 (1986) 3634.

[22] F.S. Klein and L. Friedman, J. Chem. Phys., 41 (1964) 1789.

[23] M.A. Berta, B.Y. Ellis and W.S. Koski, J. Chem. Phys., 44 (1966) 4612.

[24] S. Chivalak and P.M. Hierl, J. Chem. Phys., **67** (1977) 4654.

[25] Z. Herman and R. Wolfgang, in Ion-Molecule Reactions, J.L. Franklin, Ed., Butterworth, London (1972), p. 553.

[26] P.M. Hierl, J. Chem. Phys., **67** (1977) 4665.

[27] J.L. Elkind and P.B. Armentrout, J. Phys. Chem., **89** (1985) 5626.

[28] P.B. Armentrout, in Structure, Reactivity and Thermochemistry of Ions, P. Ausloos and S.G. Lias, Eds., Reidel Publ., Dordrecht 1986.

[29] B. Müller and Ch. Ottinger, to be published.

[30] G. Ochs and E. Teloy, J. Chem. Phys., **61** (1974) 4930.

[31] R.K. Preston and J.C. Tully, J. Chem. Phys., **54** (1971) 4297.

[32] D. Gerlich, Dissertation, Freiburg 1977.

[33] Ch. Schlier and U. Vix, Chem. Phys., **95** (1985) 401.

[34] Ch. Schlier, priv. comm.

[35] D. Gerlich, U. Nowotny, Ch. Schlier and E. Teloy, Chem. Phys., **47** (1980) 245.

[36] J.R. Krenos, R.K. Preston, R. Wolfgang and J.C. Tully, J. Chem. Phys., **60** (1979) 1634.

[37] Ch. Schlier, U. Nowotny and E. Teloy, Chem. Phys., **111** (1987) 401.

[38] K. Tanaka, T. Kato and I. Koyano, J. Chem. Phys., **75** (1981) 4941.

[39] F.A. Houle, S.L. Anderson, D. Gerlich, T. Turner and Y.T. Lee, J. Chem. Phys., **77** (1982) 748.

[40] S. Chapman, J. Chem. Phys., **82** (1985) 4033.

[41] M. Baer and H. Nakamura, J. Chem. Phys., **87** (1987) 4651.

[42] W. Frobin, Ch. Schlier, K. Strein and E. Teloy, J. Chem. Phys., **67** (1977) 5505.

[43] Ch. Ottinger and S. Zimmermann, Chem. Phys., **53** (1980) 293.

[44] Ch. Ottinger, J. Reichmuth and S. Zimmermann, Int. J. Mass Spec. and Ion Physics, **38** (1981) 379.

454

[45] A.A. Wu and Ch. Schlier, Chem. Phys., **28** (1978) 73.

[46] Z. Herman and I. Koyano, Farad. Trans. Chem. Soc. 2, **83** (1987) 127.

[47] Z. Herman, K. Tanaka, T. Kato and I. Koyano, J. Chem. Phys., **85** (1986) 5705.

[48] N.G. Adams and D. Smith, in Reactions of Small Transient Species, A. Fontijn and M.A.A. Clyne, Eds., Acad. Press, (1983) p. 311.

[49] A. Kowalski and M. Menzinger, 1987, to be published.

[50] M. Menzinger, in Gas Phase Chemiluminescence and Chemiionization, A. Fontijn, Ed., Elsevier, (1985) p. 25.

[51] C.T. Rettner and R.N. Zare, J. Chem. Phys., **77** (1982) 2416.

[52] Ch. Ottinger, in Gas Phase Chemiluminescence and Chemiionization, A. Fontijn, Ed., Elsevier, (1985) p. 117.

[53] D.M. Manos and J.M. Parson, J. Chem. Phys., **63** (1975) 3575 and **69** (1978) 231.

[54] P. Dagdigian, Chem. Phys. Lett., **55** (1977) 239.

[55] N.A. Sondergaard, I. Sauers, A.C. Jones, J.J. Kaufman and W.S. Koski, J. Chem. Phys., **71** (1979) 2229.

[56] Ch. Ottinger and J. Reichmuth, J. Chem. Phys., **74** (1981) 928.

[57] J. Reichmuth, Diplomarbeit, Max-Planck-Institut für Strömungsforschung, Göttingen 1981.

[58] Ch. Ottinger, in Gas Phase Ion Chemistry, Vol. 3, M. Bowers, Ed., Acad. Press (1984) p. 249.

[59] B. Friedrich and Z. Herman, Chem. Phys., **69** (1982) 433.

[60] P. Rosmus and J. Murrell, priv. comm.

[61] A. Yencha, Electron Spectroscopy Vol. V, Acad. Press, (1984) p. 197

[62] A. Benz and H. Morgner, Mol. Phys., **58** (1986) 223.

[63] O. Leisin, H. Morgner and H. Seiberle, Mol. Phys., **56** (1985) 349.

[64] H. Morgner, priv. comm.

[65]   A.D. Isaacson, A.P. Hickman and W.H. Miller, J. Chem. Phys., 67 (1977) 370.

[66]   K. Johnson, R. Pease, J.P. Simons, P.A. Smith and A. Kvaran, Farad. Trans. Chem. Soc. 2, 82 (1986) 1281.

[67]   P.J. Kuntz, M.H. Mok and J. Polanyi, J. Chem. Phys., 50 (1969) 4623.

[68]   M.S. de Vries, G.W. Tyndall and R.M. Martin, J. Chem. Phys., 81 (1984) 2352.

[69]   T. Krümpelmann and Ch. Ottinger, Chem. Phys. Lett., 140 (1987) 142.

[70]   T. Krümpelmann and Ch. Ottinger, to be published.

[71]   J.E. Mentall and P.M. Guyon, J. Chem. Phys., 67 (1977) 3845.

[72]   J.A. Beswick and M. Glass-Maujean, Phys. Rev. A, 35 (1987) 3339.

[73]   S.R. Riley and K.R. Wilson, Farad. Disc. Chem. Soc., 53 (1972) 132.

[74]   M. Shapiro, J. Phys. Chem., 90 (1986) 3644.

[75]   V. Engel and R. Schinke, submitted to J. Chem. Phys., (1987).

[76]   L.J. Butler, E.J. Hintsa and Y.T. Lee, J. Chem. Phys., 84 (1986) 4104; L.J. Butler, E.J. Hintsa, S.F. Shane and Y.T. Lee, J. Chem. Phys., 86 (1987) 2051.

THE M + X₂ REACTIONS: PARADIGMS OF SELECTIVITY AND SPECIFICITY IN ELECTRONIC MULTI-CHANNEL REACTIONS

THE $M + X_2$ REACTIONS: PARADIGMS OF SELECTIVITY AND SPECIFICITY IN ELECTRONIC MULTI-CHANNEL REACTIONS

Michael Menzinger
Department of Chemistry
University of Toronto
Toronto
Ontario M5S 1A1

ABSTRACT. The dynamically rich alkaline earth + halogen reactions constitute paradigms for the dynamics of many covalent/ionic multi-channel reactions. This paper first reviews relevant experimental results on the $M + X_2$ dynamics. An attempt is subsequently made to relate these results to the network of relevant PES's making use of configuration and state correlation diagrams, locating the place and the geometry of the region where diabatic processes are likely, and predicting steric and alignment effects. Some specificity and selectivity effects observed experimentally are naturally explained in this way. In particular, the low chemiluminescence yield observed generally is shown to be due to the dynamics in the neighbourhood of a conical intersection at the outer covalent/ionic crossing.

1. INTRODUCTION

Reactions that occur adiabatically on a single potential energy surface (PES) are generally better understood than their diabatic, multi-PES, multi-channel counterparts whose dynamics are generally richer and the relevant PES's and their mutual interactions are harder to come by than the ground state PES alone. Yet this type of multi-channel and multi-PES reactions include many natural and industrial processes of great importance, e.g.: bioluminescence, photosynthesis and photochemical syntheses, to name only a few. Although the mechanisms of some of these complex processes may be known in rough outline, a detailed understanding in terms of consumption and disposal of energy and angular momentum and of alignment of reactant and products can be expected only for much simpler atom-diatom systems [1].

It is with this goal in mind that the MX_2 systems whose many product and reactant channels are schematically shown in Fig. 1, are proposed [1] as the paradigms for covalent/ionic multi-PES reactions. Their advantages lie on the one hand in the convenience with which experiments can be done and on the other in the simple structure of the fragment atoms and molecules and in the conceptual clarity of the PES's arising from them. The purpose of this paper is: (A) to document experimental findings, as far as they are relevant to the models presented, and build an inventory of dynamical information on all reaction channels (see Fig. 1), and (B) to develop a qualitative global model [2] of the multi-PES reactions, based on the Coulombic nature of the ionic electron configurations, on certain principles of electronic structure and on group theory.

457

J. C. Whitehead (ed.), Selectivity in Chemical Reactions, 457–479.
© 1988 *by Kluwer Academic Publishers.*

458

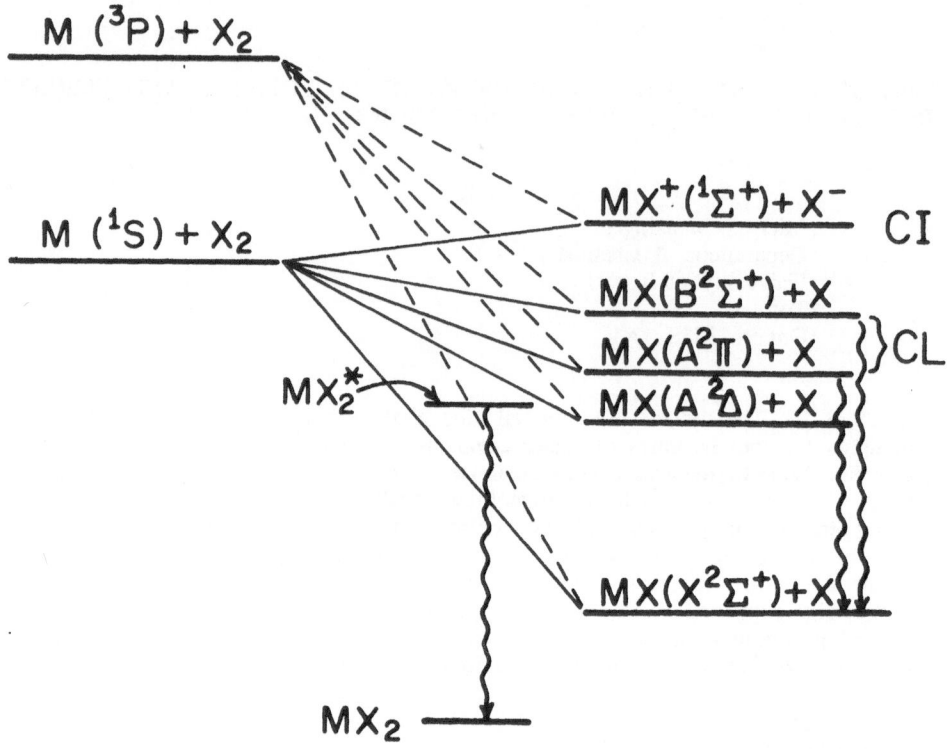

FIGURE 1. This schematic indicates the multitude of product channels (ground state $X^2\Sigma^+$, metastable $A'^2\Delta$, chemiluminescent $A^2\Pi$, $B^2\Sigma^+$, [the higher states $C^2\Pi$, $D^2\Sigma^+$,.. have been omitted for simplicity], and chemi-ions) that can arise from ground state ^1S and metastable ^3P reactant atoms. Emission from the transition state MX_2^* is indicated. The full and broken lines symbolize reaction channels, not PES's.

An extended correlation diagram for electronic configurations (Section 3.1) allows one to identify the major changes in electronic structure along the reaction coordinate and to identify the dominant attractive and repulsive forces and their domain of action, or briefly "to tell the story of the individual reaction channels". A detailed group theoretical analysis [2] of the harpooning regions in the adiabatic representation provides valuable information about the connectivity and the topology of the adiabatic PES's, and it yields insights into the alignment preferences and steric restrictions of the reaction channels [5]. This global model is described in Sections 3.1,2,3 where it is also shown to be in essential accord with many of the experimental findings. So far this model is qualitative, and work is in progress to quantify the ideas and to empirically construct full 4D PES's.

The MX_2 reactions have much in common with the closely related alkali-halogen reactions [3,4,5], from which they differ merely by one valence electron which makes the alkaline earth ions M^+ isoelectronic with the alkali atoms A, and endows them and their monohalides M^+X^- with low lying "chemiluminescent" electronic states and with a sufficiently low ionization potential for chemi-ions $MX^+ + X^-$ to be an open reaction channel [2]. Metastable $M(^3P,^1D)$ beams are generated with ease [6] and studies of discharge-excited $M(^3P)$ [6,7] and of laser excited and aligned $M(^1P)$ have been performed [8]. There is an ample literature [1,9] on the dynamics of X_2 reactions with metal atoms other than group IIA (e.g. Cu, Al, Ga, In, Ge, Si, Sn), and there is good reason to believe that a network of PES's similar to the one described in Section 3.1 also governs the dynamics of these reactions. The details of the adiabatic PES's, however, depend crucially on the symmetries of the atomic and ionic species which will generally not be the same as these for Ca/Ca$^+$. In a broader sense even, many organic and metallo-organic reactions, initiated by photochemical excitation, are believed to be initiated by a similar covalent-to-ionic transition and the present analysis may be relevant to them.

The interest in chemiluminescent (CL) processes in the early 1970's was fuelled by the need for efficient chemical light sources and above all for reactions capable of pumping an electronic transition laser. This programme did not meet with success since the CL yields of many reactions studied [1,10] were disappointingly low, and electronic inversions similar to those found in the decomposition of 1,2 dioxetanes [11] have not been found. It is timely therefore to inquire after the dynamical reasons for the monotonically decreasing, quasi-statistical electronic state distributions that have taken so many researchers by surprise. The answer is given in Sections 3.2 and 3.3 in terms of symmetry arguments.

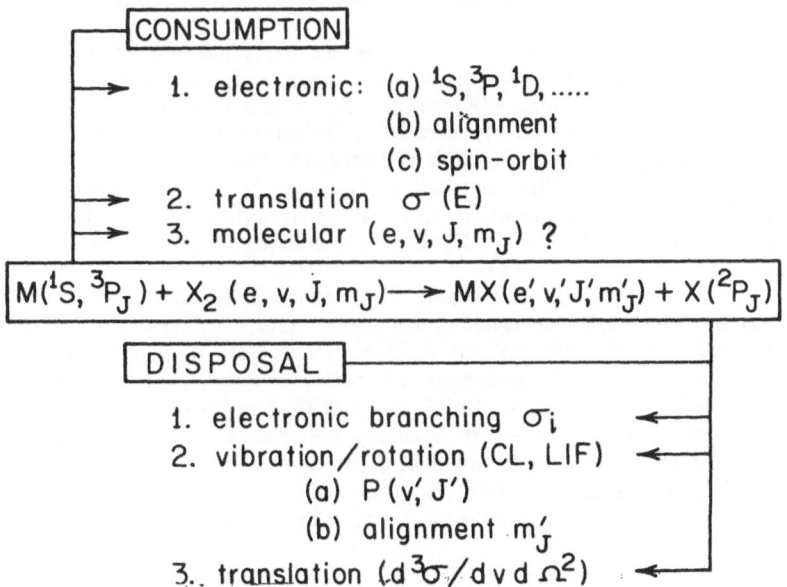

FIGURE 2. Experimental possibilities for studying the dynamics of MX_2 reactions.

2. EXPERIMENTAL RESULTS

These are discussed in the order of increasingly detailed information content. Many of the questions outlined in Fig. 2 have been addressed, *viz.* translational and electronic energy consumption, the reactivity of the $M(^3P_J)$ spin-orbit states and the effect of aligning $Ca(^1P)$, and studies of electronic, vibrational and rotational energy disposal, recoil velocity vector distributions of $MX(X^1\Sigma^+)$ and MX^+, and measurements of the alignment of MX products through CL polarization measurements [1,2,5]. Not yet addressed have been the questions of the effects of reactant molecular excitation, in particular the role of X_2 vibration, rotation and alignment, and of disposal into halogen atom J, m_J fine structure states.

2.1 Excitation Functions $\sigma_i(E)$

Excitation functions for chemiluminescence and chemi-ion production were measured by crossing a supersonic seeded halogen beam with an effusive metal beam and collecting the photon current using interference filters and the total ion current [12]. The resulting functions are presented in Fig. 3 such as to compare the CL and CI channels within a given system.[1] All four curves have qualitatively the same shape: a rapid drop at low energy to a constant value at higher energy.[2] Secondly, one finds that the chemiluminescence functions drop off more gradually than do the chemi-ionisation functions.

This shape provides evidence for harpooning as Fig. 4 illustrates. For a "primitive harpooning" potential characterized by an abrupt transition from covalent to ionic curve one would expect to find a constant reaction cross section πR_c^2 where R_c is the (absorbing sphere) crossing radius. Covalent-ionic configuration interaction however rounds off the avoided crossing. The extra attraction outside R_c causes some trajectories with impact parameters $b > R_c$ to be captured by the Coulomb potential, which is therefore responsible for the rapid rise of the cross section at low energy. The dashed curve of Fig. 4 represents the result of a model calculation [13] where the potential V(R) for $R > R_c$ is approximated by $-C_6/R^6$. The total reaction cross section clearly contains information about the potential at the avoided covalent/ionic intersection, and it demonstrates that the electron transfer controls the total reaction rate. The quantitative difference between CI and CL cross sections noted above will be shown in Section 3.2 to reflect the electron transfer to the ground state and first excited state configurations of the $M^+ + X_2^-$ ion pair (i.e. outer vs. inner harpooning).

It is of interest to compare these dynamically controlled exoergic channel cross sections (Figs. 3,4; exoergicities are $\Delta E = -1.47, -1.32$ eV for CaF(B) and CaF$^+$, and $-0.71, -0.46, -0.68$ eV for BaCl(A,B) and BaCl$^+$, respectively) with the excitation functions for endoergic ion pair production [14] presented in Fig. 5 ($\Delta E = +0.31, 0.78$ eV for SrCl$^+$ and CaCl$^+$, respectively). The experimental points are fitted to a high degree of accuracy by product phase space theory, as given by the solid lines. The dramatic shift from dynamical, entrance channel control to statistical, exit channel control in going from exoergic to endoergic channels reflects nicely the dominance of dynamical and statistical factors in terms of which the cross

---

[1] It should be noted that all four channels (CaF(CL, CI) and BaCl(CL and CI)) are exoergic.
[2] The functions are mutually normalized at high energy.

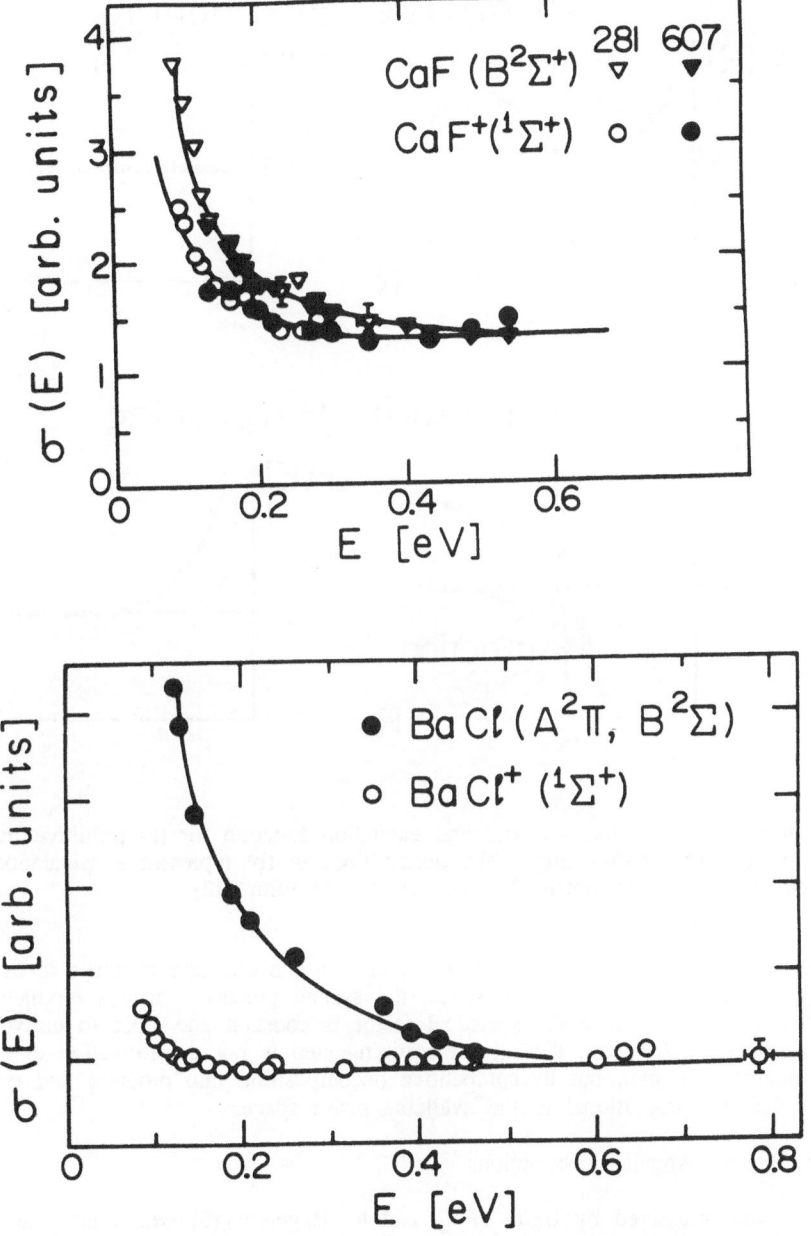

FIGURE 3. Chemiluminescence and chemi-ionization excitation functions [12] for the Ca + F$_2$ → CaF(B$^2\Sigma^+$) and CaF$^+$ and Ba + Cl$_2$ → BaCl(A$^2\Pi$, B$^2\Sigma^+$) and BaCl$^+$ systems. The CL and CI curves are normalised at high energy.

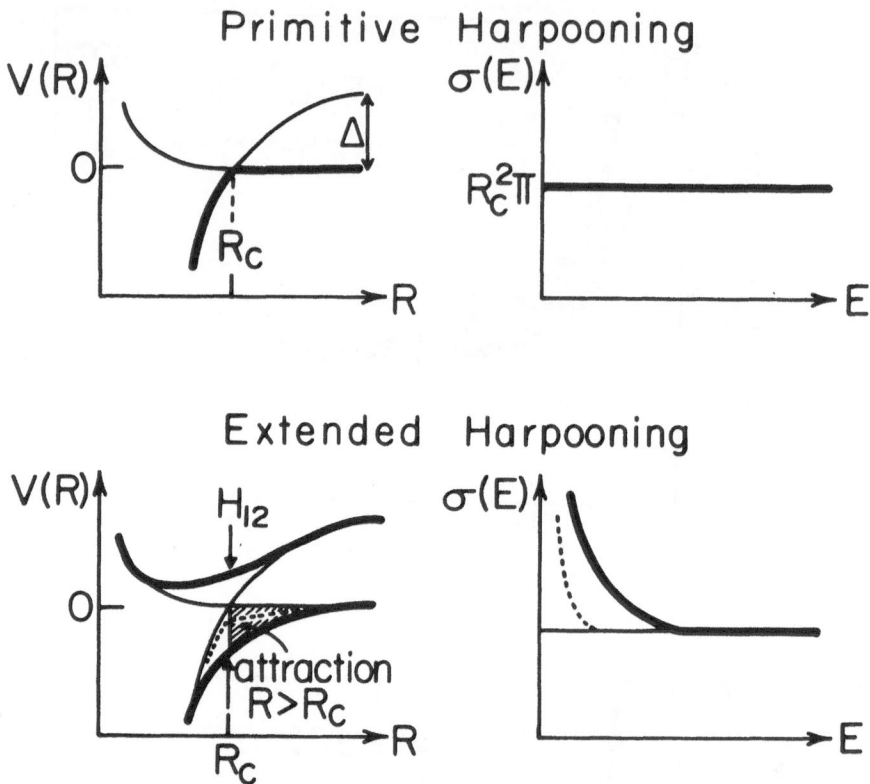

FIGURE 4. Interaction potential and excitation function for (a) primitive harpooning and (b) extended harpooning. The dotted lines in (b) represent a calculation in which the long range potential is of the $-C_6/R^6$ form [13].

section can be expressed: the only bottleneck in the exoergic channels is the initial electron transfer. After its occurrence, the system proceeds towards products with a high probability. Hence the statistical factor is constant and close to unity. In the endoergic case, however, the rate determining step is not the formation of the intermediate ion pair, but its competitive decomposition into products and reactants at rates that are proportional to the available phase space.

2.2 Product Angular Distributions

These were measured by Herm *et al.* and by Engelke [15] who found the angular distributions for the scattered $MX(X^2\Sigma^+)$ products to be anisotropic and peaked forward (w.r.t. metal beam). A small fraction of backward scattered products are thought to derive from low impact parameter head-on collisions. As expected, the ground state channel too appears to be initiated by harpooning, as are the CL and CI channels.

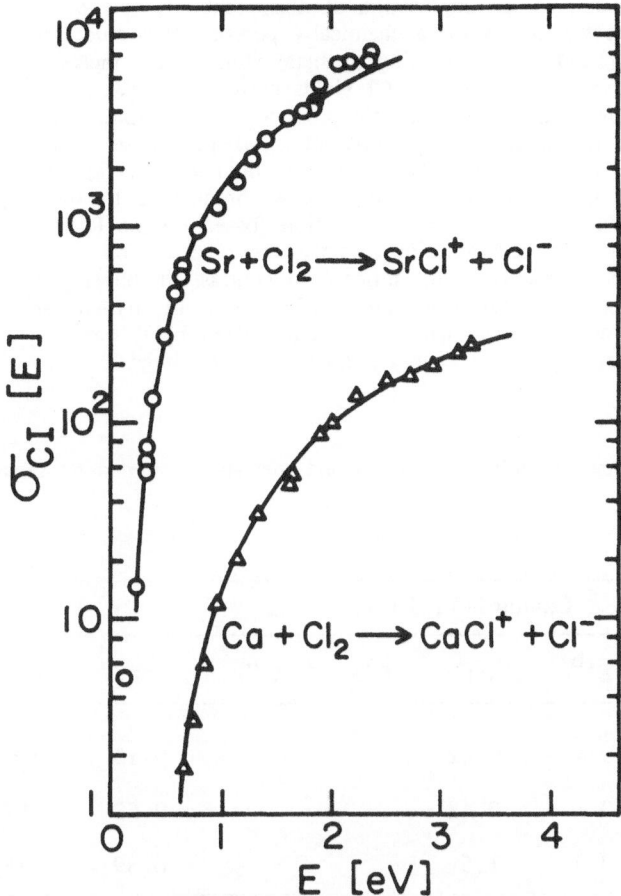

FIGURE 5. Chemi-ionization excitation functions for endoergic channels. The curves represent data fits by phase space theory [14].

2.3 Electronic Branching

Since the discovery of the Ba + Cl_2 chemiluminescence [16] it was found consistently in many reactions of metal atoms with halogen and oxygen containing molecules [1] that electronic product state distributions decrease monotonically with available energy in a manner suggesting qualitatively statistical behaviour and that the electronic inversions necessary for laser action [10] are elusive. CL yields for the reactions of ground state $M(^1S)$ are maximally a few percent (for Ca + F_2) and more typically a fraction of a percent [1,17]. This is also typical for the reactions of metal atoms with the oxygen-donors O_3, NO_2 and N_2O [1] and it is appropriate at least and at last to search for an explanation of disappointingly low quantum yields that have frustrated the search for a chemically pumped electronic transition laser since more than a decade ago.   Exciting themetal atom to the metastable ^{3}P and ^{1}D states usually enhances the CL and CI yields considerably [7], as Table 1 shows, without however removing the monotonic product state distribution and introducing an electronic inversion.   The question arises whether these product state distributions, statistical as they may appear qualitatively, do contain some dynamical bias.   An information theoretical analysis of the data is summarized by the surprisal plots, Figs. 6,7.   The deviations from prior expectation (based on the RRHO approximation) are considerable, and most importantly, the plots exhibit striking nonlinearities in 5 out of 7 systems.   This reflects, in contrast to the surprisal plots for vibrational and rotational disposal which are usually linear, the dynamical individuality of the electronic channels which are characterized by different exoergicities and by different potential energy surfaces and couplings.

TABLE 1. Electronic channel yields Y_i (in %) and metastable attenuation cross sections σ_T (in Å^2)

	Channel Yield Y_i[a]					
	$X^2\Sigma^+$[b]	$A^2\Pi$	$B^2\Sigma$	$C^2\Pi$	ions[d]	σ_T
Mg(^{3}P) + F$_2$	99.4	0.42			0.15	103
+ Cl$_2$	99	0.48			0.52	102
+ Br$_2$	98.1	1.5			0.39	118
Ca(^{1}S) + F$_2$	97.4	2.0	0.44		0.14	60
Ca(^{3}P) + F$_2$	94	5.3	0.44	0.07	0.21	109
+ Cl$_2$	65	31.0	1.0	1.6[c]	1.3	128
+ Br$_2$	74	18.0	6.5	0.54	0.67	143

[a] $Y_i = 100\ (\sigma_i/\sigma_T)$
[b] dark channels. For Ca reactions this includes the unknown $A'^2\Delta$ state.
[c] formed predominantly from Ca(^{1}D).
[d] calculated, assuming formation from M(^{3}P).

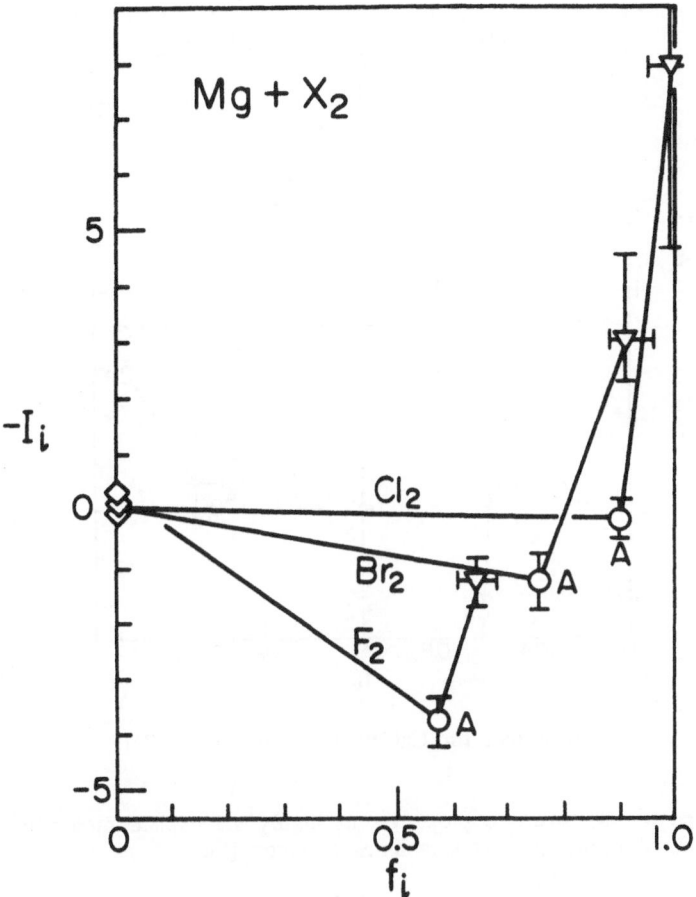

FIGURE 6. Surprisal plot for $Mg(^3P) + X_2$ reactions [7b].

In Section 3.2 the explanation will be given of why the MX(X) channel always dominates over the chemiluminescence channels, and an attempt will be made in 3.3 to rationalize, from a dynamical point of view, the monotonically decreasing branching among the CL channels.

2.4 Vib-Rotational Consumption and Disposal

CL and LIF spectra are the ideal sources of information on vibrational, and to a lesser extent of rotational, energy disposal. Extracting reliable population analyses from the generally congested spectra is however more demanding on one's

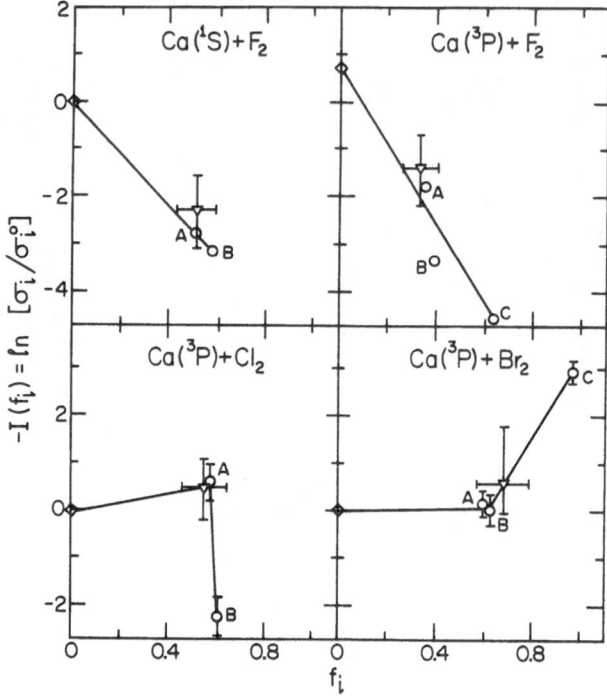

FIGURE 7.   Surprisal plot for $Ca(^3P) + X_2$ reactions [7c].

knowledge of spectroscopic properties of highly v,J excited molecules and on computational effort, than is sometimes assumed [5].

2.4.1   <u>Comparison: Same Reactant - Different Product States</u>.   Results from inversions [18] of the $CaF(B^2\Sigma^+ - X^2\Sigma^+)$ band in the CL and LIF spectra produced by the reactions

$$Ca(^1S) + F_2(X^1\Sigma^+_g) \longrightarrow CaF(B^2\Sigma^+) + F(^2P)$$

$$\longrightarrow CaF(X^2\Sigma^+) + F(^2P)$$

are shown in Fig. 8.   One notes a unimodal, monotonically falling v′ distribution in the $B^2\Sigma^+$ state.   The ground state $X^2\Sigma^+$ however possesses a bimodal distribution (truncated at $f_v = E'_v/E_{tot} = 0.7$ because of spectral overlap) with a strongly inverted broad peak centred beyond v′=45 and a monotonically falling component at low v′.   The latter has been interpreted [18] as stemming from microscopic branching due to the availability of competing pathways involving different PES's, as predicted by the model (Fig. 9) presented [2,5] in Section 3.1.   Even if the low f_v tail of the bimodal distribution were due to vibrational relaxation, one should notice the qualitative difference - monotonically falling vs. inverted - of the principal vibrational features of the two channels compared here.

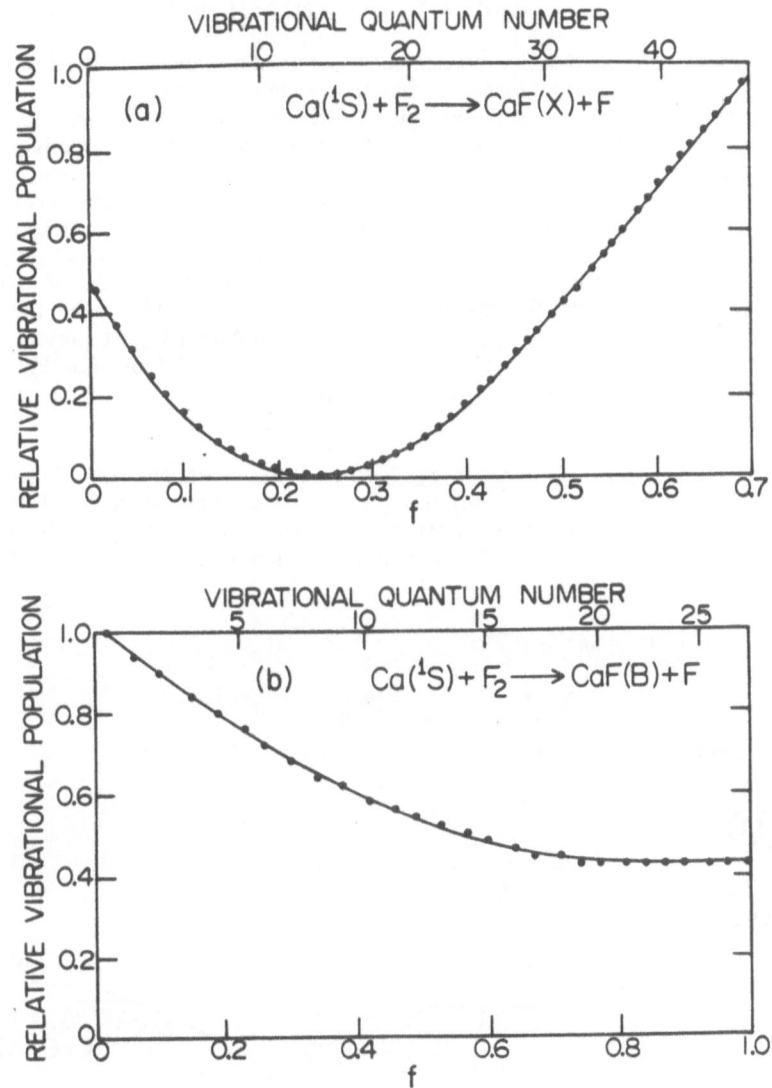

FIGURE 8. Vibrational distribution for the ground state $X^2\Sigma^+$ and chemiluminescence channel $B^2\Sigma^+$, determined by inversion of LIF and CL spectra [18].

2.4.2 <u>Comparison: Different Reactant - Same Product States</u>. CL spectra for the ground state and metastable reactions

$$Ca(^1S) + F_2 \longrightarrow CaF(A^2\Pi, \ B^2\Sigma^+) \text{ and}$$

$$Ca(^3P) + F_2 \longrightarrow CaF(A^2\Pi, B^2\Sigma^+)$$

have shown [19] a peculiar phenomenon: while the $CaF(A^2\Pi)$ band from the $Ca(^3P)$ reaction is dramatically broadened due to population of very high v-states as compared to that of the less exoergic $Ca(^1S)$ reaction ($-\Delta E$ values for the former and for the latter are 3.36 eV and 1.46 eV, respectively) as one might expect on energetic grounds alone, the spectral envelopes of the $CaF(B^2\Sigma^+)$ band emitted from the ^1S and the ^3P reactions are very similar to each other. Obviously, the vibrational state distributions (Fig. 8b) in $CaF(B^2\Sigma^+)$ for the two reactions resemble each other closely despite the 1.89 eV greater exoergicity of the $Ca(^3P)$ reaction. The fraction of the energy released into product vibration is $<f_v> = 0.62$ for the ^1S reaction [18]. For the ^3P reaction, one obtains $<f_v> = 0.27$ under the assumption of exactly equal emission spectra (vibrational distributions). At present it is unclear which feature of the PES's accounts for this reluctance of the $Ca(^3P)$ reaction to dispose of its extra energy by internal excitation.

2.4.3 <u>To be done: Dependence on $X_2(v,J)$ States</u>. An analysis of the PES for the harpooning process $Li + F_2$ [4] has predicted a dramatic increase of the crossing distance R_c depending on X_2 internuclear distance and hence on vibrational state of the halogen molecule. The CL and total reaction rates of the $Ba + (Cl_2, Br_2)$

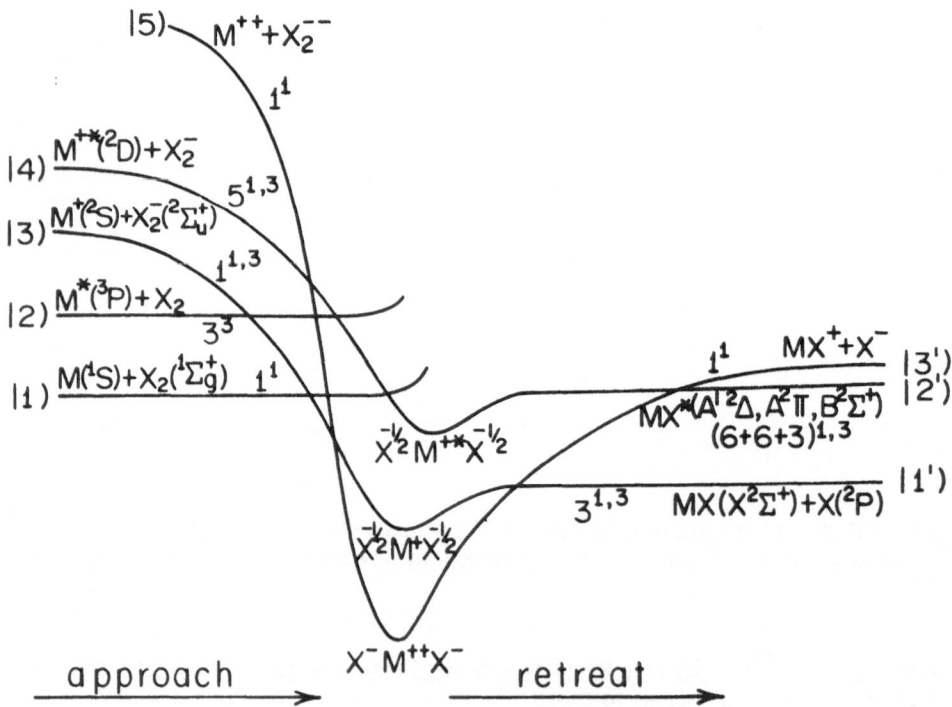

FIGURE 9. Correlation diagram for electronic configurations for M=Ca,Sr.

reactions were measured [12] in a beam-gas experiment as a function of scattering gas temperature but no enhancement of rate was discernible in this work in which only the lowest v-states are populated. Preparation of high (v,J) states, e.g. by stimulated emission pumping, appears feasible for studying the vibrational and rotational [20] state selectivity.

2.5 Alignment and Steric Effects

The dependence of $CL(A^2\Pi, B^2\Sigma^+)$ and CI yields on the alignment of $Ca(^1P)$ was studied [8], using polarized laser excitation, for the reactions

$$Ca(^1P) + Cl_2 \longrightarrow CaCl(A^2\Pi, B^2\Sigma^+) + Cl$$

$$\longrightarrow CaCl^+ + Cl^-.$$

The findings that the $A^2\Pi$ and to a lesser extent the $B^2\Sigma^+$ chemiluminescence are enhanced by perpendicular alignment of the p-orbital with respect to the collision axis in the LAB frame, while CI is diminished show the complementary nature of the CL and CI channels, which will be discussed in detail in Section 3.2.
   The polarization of the dispersed $CaF(B^2\Sigma^+ - X^2\Sigma^+)$ emission from the $Ca(^1S)$ + F_2 reaction has been analysed [23] and yielded information about the product alignment. A calculation based on a modified version of the DIPR DIP model is in general qualitative agreement with experiment.
   In Section 3.2 it will be shown that the low CL yield of the $M + X_2$ reactions is due to the pronounced steric requirements.

2.6 Reactivity of Spin-Orbit States

Dagdigian summarizes the measurements of his co-workers [21] of the relative rates of reaction of the $Ca(^3P_{0,1,2})$ fine structure states in this volume. For the present context it suffices to note another example of the complementarity of the $X^2\Sigma^+$ ground state and the CL channels: The CL cross sections are ordered J=2 > 1 > 0, whilst the ground state cross sections are ordered just in reverse, J=0 > 1 > 2. This will be seen in Section 3.1 to be in accord with inner and outer harpooning, respectively.

3. A GLOBAL MODEL FOR THE MX_2 REACTIONS [2,5]

3.1 Correlation of Electron Configurations

The facts that the MX(X,A,B...) products are ionically bound and that chemi-ions MX^+ presumably derive from the doubly ionic configuration $M^{++}X^-$ allows one to construct the configuration-correlation diagram shown in Fig. 9. Along each of the unbroken curves to which are affixed reactant ׀i) and product labels ׀f'), the electronic configuration remains invariant and only the nuclear arrangement changes. More detailed information on the relevant PES's and their steric features is obtained in Section 3.2 by examining group theoretically the states to which these configurations give rise [2,5].
   It states that $MX(X^2\Sigma^+)$ derives from the unexcited ion pair $M^+(^2S) + X_2^-$

($X^2\Sigma_u^+$) and that the first excited state[3] of the ion $M^+(^2D)$ gives rise, by splitting of the fivefold degeneracy of the ^2D state in the Coulomb field of the X^- ion, to the two lowest CL states $A^2\Pi$ and $B^2\Sigma^+$, as well as to a metastable $A'^2\Delta$. Although the latter has not yet been observed as a reaction product, presumably due to its long radiative lifetime (Einstein A factors for electric quadrupole transitions are *ca.* 10^6 times larger than for electric dipole transitions), its existence in BaCl has been established spectroscopically [22]. The chemi-ions $MX^+ = M^{++}X^-$ derive from the doubly ionic configuration ⌐5).

The ionic curves ⌐3), ⌐4), ⌐3'), ⌐5) are meant to represent Coulomb or Rittner potentials at long range, and they are drawn in the interaction region simply to interpolate between products and reactants, taking into account the bond strength of MX_2 and its excitation energy. The covalent curves ⌐1), ⌐2) are thought to be exponentially repulsive.

Figure 9 allows one to trace the changes of electron configuration along the reaction coordinate and to draw qualitative conclusions about the force field in which the nuclei move. The ground state channel ⌐1)-⌐1') is initiated by an electron transfer at the crossing ⌐1)*⌐3) which we call outer harpooning, at which point the system transits to the ionic ⌐3)-⌐1') configuration which leads directly to products. While the doubly ionic configuration ⌐5)-⌐3') is high in energy asymptotically, the Coulomb attraction causes it to drop below all the other configurations and form the ground state of MX_2 at close range. The crossings ⌐3)*⌐5) and ⌐1')*⌐3') at which this occurs afford a second reaction path in the ground state channel: transition from singly to doubly ionic curves, sampling of the deep Coulomb well of the intermediate, and return to the ⌐1') configuration as the products retreat. The low v'-portion of the experimental vibrational distribution, Fig. 8, has been thought to arise from this alternative dynamical pathway [18].

The chemi-ion channel ⌐1)-⌐3') is initiated similarly by outer harpooning with the possibilities of passing through the strong interaction region either on the doubly ionic or on the singly ionic configuration. The chemi-ions separate, climbing slowly the long-range Coulomb potential, whenever a sufficient amount of energy flows into the MX^+ - X^- coordinate. Otherwise the complex oscillates slowly and may decay statistically. The exoergic and endoergic excitation functions, Figs. 3 and 5, discussed in Section 2.1 are subject to these dynamics. We begin to describe the chemiluminescence channels starting on the product side. Although the crossing ⌐2')*⌐3') is formally avoided in the adiabatic representation, it lies at very large distances (*ca.* 30 Å) due to the small asymptotic separation of ionic and chemiluminescent products and their interaction vanishes. This means that the CL flux must come along the excited ion configuration ⌐4)*⌐2') which must be accessed at the inner harpooning crossing ⌐1)*⌐4). Thus, CL reaction requires a diabatic passage through the outer harpooning crossing followed by inner harpooning electron transfer, involving the simultaneous excitation of the remaining valence electron on Ca^+ from 4s to 3d orbitals. This process will be re-examined in the following Section.

The different reaction channels of metastable $M(^3P)$ and of ground state $M(^1S)$ have in common that they are initiated by covalent/ionic transitions. Only the $M(^3P)$ to chemi-ion channel stands out since it involves a change of spin multiplicity. The consistently low chemi-ion yields (Table 1) are probably due to the low probability of the spin flip.

---

[3] Ca is assumed to be the representative metal M in Fig. 11. In the cases of Mg and Ba, the first excited ion states are $Mg^+(^2P)$ and $Ba^+(^2D)$.

3.2   Local State Correlations in Harpooning Regions: Steric Factors

A more detailed analysis [2,5] is summarized in Fig. 10, the local state correlation diagram constructed for the quadrangle of harpooning intersections $|1)*|3)$, $|1)*|4)$, $|2)*|3)$, $|2)*|4)$.   It illustrates the connectivity of the relevant PES's[4] as a function of collision geometry ($C_{\infty v}$, C_s, C_{2v}, arranged horizontally) and as a function of the alignment possibilities of the p-orbital belonging to $M(^3P)$ relative to the axes (x,y,z) of MOL frame (p_x,p_y,p_z, arranged in columns).   This analysis will provide the general interpretation of the experimentally observed alignment dependences.

Conical intersections at the outer harpooning regions manifest themselves through identical symmetry species (avoided crossing) in one or two point groups and different species (symmetry enforced crossing) in another point group.   They are important primarily for the CL channels since they facilitate access to the inner harpooning region.   This is particularly striking for the ground state reaction $M(^1S)$ + X_2, where an important conical intersection occurs in C_{2v} geometry.   Here the ion pair state belongs to the B_2 representation and the covalent state to A_1.   The outer harpooning region $|1)*|3)$ lies close enough to make the intersection fairly strongly avoided in C_s causing harpooning to occur with a high probability at low collision energies and random geometry (C_s, $C_{\infty v}$).   The reactive flux is thus siphoned off towards the ground state and chemi-ion channels, except in perpendicular C_{2v} collisions where the conical intersection allows the system to pass through this protective shell without an electron transfer and to penetrate towards inner harpooning and chemiluminescence.   Figure 11 illustrates this protective outer harpooning shell and the narrow stereospecific slit into which only trajectories within a narrow cone surrounding C_{2v} can penetrate.   It is the low statistical weight of this symmetric geometry that is responsible for the low CL yields.

Figure 13b illustrates schematically that the attractive contribution to the harpooning potential arising from covalent-ionic configuration interaction is much more pronounced near the inner than at the outer harpooning region, due to exponential orbital overlap.   Since this extra attraction at $R > R_c$ is responsible for the low energy drop of the harpooning cross sections (Fig. 4) the different strength of configuration interaction at the two harpooning regions accounts for the more gradual falloff of the chemiluminescence cross sections relative to those for chemi-ion production (Fig. 3).

Clearly there exists a complementarity between those channels initiated by outer harpooning (ground state and chemi-ions) and by   inner harpooning (chemiluminescence): high cross sections for one type of process implies low cross sections for the other and *vice versa*.   This has been confirmed in experiments with spin-orbit selected $Ca(^3P_J)$ [21] and with aligned $Ca(^1P)$ [8] beams.

The experimentally observed alignment dependence of the CL and CI yields in the $Ca(^1P)$ + Cl_2 system [8] discussed in Section 1.5   can be discussed in the context of Fig. 10 since the relevant intersections for $Ca(^1P)$ are topologically equivalent to those of $Ca(^3P)$.   It was found [8] that perpendicular alignment of the p-orbital relative to the LAB-fixed collision axis enhances, as illustrated in Fig. 12, formation of $CaCl(A^2\Pi)$ and to a lesser extent of $CaCl(B^2\Sigma^+)$.   The correlation

---

4 Only the lowest PES's relevant in low energy collisions have been considered. The full multiplicity of the configurations is affixed to the reactant and product configurations in Fig. 9. Eg.: $3^{1,3}$ indicates 3 PES's each of singlet and of triplet multiplicity for $|1')$.

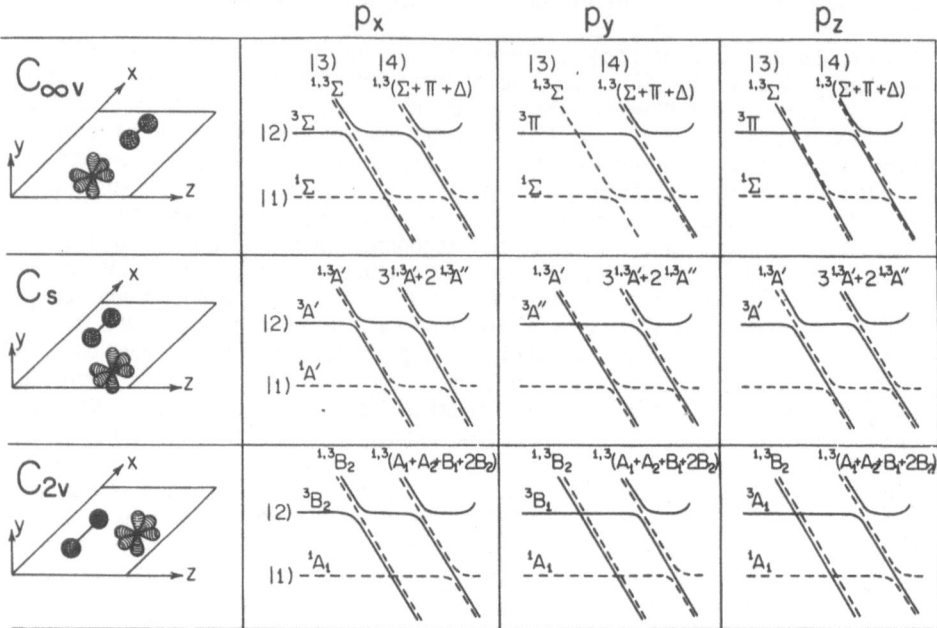

FIGURE 10. Local state correlation diagram in the quadrangle of harpooning intersections indicated in Fig. 9. Three collision geometries are arranged horizontally, while the three orthogonal alignments of the p-orbital belonging to $M(^3P)$ are arranged in columns.

diagram indicates that passage through the outer crossing and access to the CL channels is guaranteed for perpendicular p_y alignment in all collision geometries as well as for p_z alignment in $C_{\infty v}$ and C_{2v}. As Fig. 12b indicates this implies that collisions with low impact parameter are favoured by the CL channels, since for head-on collisions perpendicular alignment in the LAB frame transforms [24] into perpendicular alignment in the MOL frame, as required.

3.3 Branching among Chemiluminescence Channels

The preceding discussion has touched upon the mechanism by which chemiluminescence is initiated, but says nothing about the branching among CL states. It was noted above that the distributions over product states are invariably monotonically falling, suggesting interpretations based on statistical thinking. The very fervour with which electronic inversions have been searched for in many different CL reactions [1,10] since more than a decade ago highlights only the great disappointment and surprise to find results that were ultimately not surprising in the information theoretical sense of the word. It is therefore timely to enquire after the dynamical reasons for this quasi-statistical electronic energy disposal.

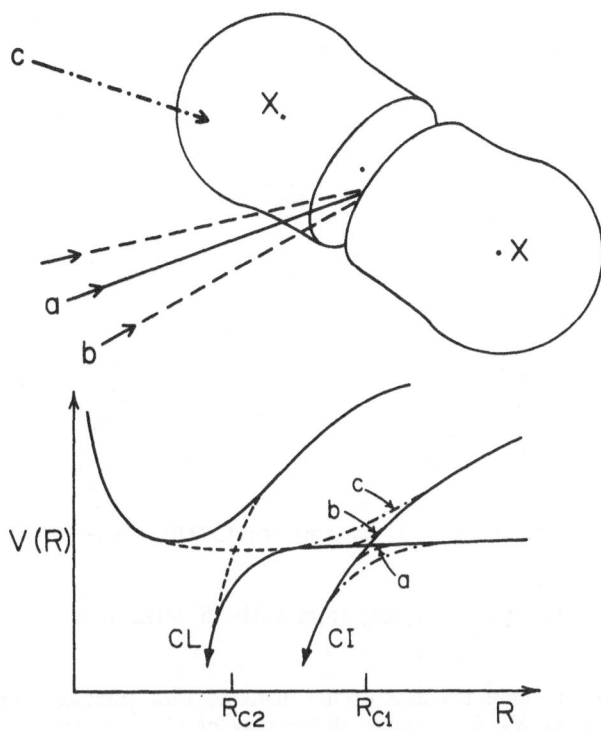

FIGURE 11. The conical intersection of the $^1A_1$ and $^1B_2$ surfaces at outer harpooning in C_{2v} geometry provides adiabatic access to inner harpooning. For C_s collisions the avoided outer harpooning crossing acts as a protective shell against inner harpooning. The low CL yield stems from the low statistical weight of C_{2v} collisions.

The problem is complex theoretically since the coupling of various angular momenta - nuclear (orbital and rotational), electronic (orbital and spin) angular momenta - comes into play. To make some progress we follow the widely used decoupling scheme of Born and Oppenheimer and assume nuclear and electronic motion to be independent, focussing attention primarily on locally adiabatic processes. Separating the electronic orbital angular momentum from the rest, it is still not possible to construct a unique model due to the vector coupling (Clebsch Gordan series) of angular momenta of the atomic $X(^2P)$ and molecular $MX(A'^2\Delta, A^2\Pi, B^2\Sigma^+)$ products.

We present here a limiting model[5] which assumes simply that in the course of the CL rearrangement, the m_L projection of the electronic angular momentum of $Ca^+(^2D)$ onto the $Ca...X_2$ axis is conserved and becomes the projection 1 of the

[5] The ideas for this model arose from conversations with Prof. R.N. Dixon.

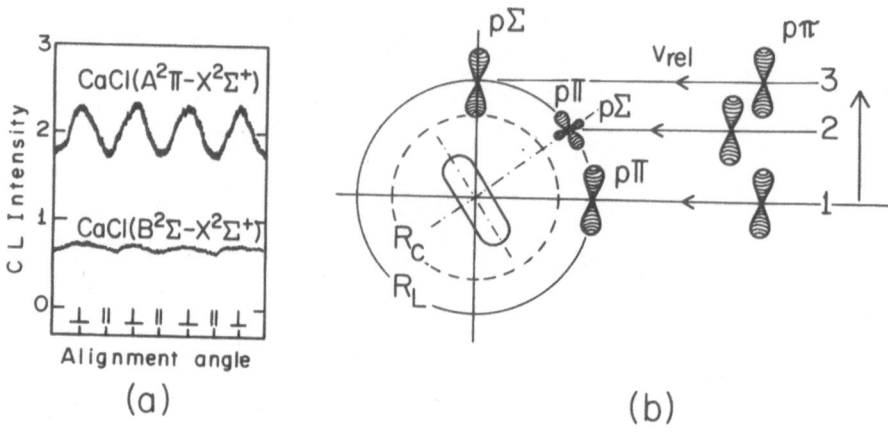

FIGURE 12. Dependence of CL intensity for $Ca(^1P) + Cl_2$ on the laser-induced alignment of $Ca(^1P)$.
(a) experimental results [8].
(b) schematic illustrating recoupling from LAB to MOL frames at R_L [24].

angular momentum along the axis of the diatomic CaX product. In the electric field of the distant anion X⁻, the fivefold degeneracy of $Ca^+(^2D)$ splits into a singly degenerate $^2\Sigma^+$ and two doubly degenerate $^2\Pi$ and $^2\Delta$ states. Accordingly, the atomic states of the excited $Ca^+(^2D)$ ion correlate with the product molecular states $B^2\Sigma^+$, $A^2\Pi$, $A'^2\Delta$ respectively. In the limit of low overlap, valid at the $|2)^*|4)$ intersection but less so at $|1)^*|4)$, crystal field theory predicts the relative ordering of the three states to be $^2\Delta < ^2\Pi < ^2\Sigma$. The modified correlation diagrams for the harpooning regions, for the ground state $M(^1S) + X_2$ and metastable $M(^3P) + X_2$ reactions, shown in Figs. 13 and 14 respectively, were constructed on the basis of this assumption.

To begin with the reaction of $M(^1S)$, Fig. 13 shows the previously mentioned conical intersection in C_{2v} geometry through which the reactive flux penetrates to the inner harpooning zone. There, the $A^2\Pi$ state is accessed adiabatically and with ease. The $B^2\Sigma^+$ state, on the other hand, cannot be accessed in C_{2v} by electronic coupling alone because even in the case of a diabatic transition between the two $^1A_1$ states, the $^1B_2$ state would interact with the $^1A_1$ reactant state only by vibronic coupling. A relatively facile avenue towards $B^2\Sigma^+$ involves the sterically unlikely· $C_{\infty v}$ collision and a diabatic transition at outer harpooning. In sterically unrestricted C_s, two diabatic transitions at outer and at inner harpooning make the occurrence of this event highly unlikely. This illustrates the severe dynamical restriction against the formation of $B^2\Sigma^+$, as compared to the relative ease of forming $A^2\Pi$ via the conical intersection. A typical experimental result is available for the $Ca(^1S) + F_2$ system [7] where the CL yields are Y(A) = 2% and Y(B) =

0.44% (see Table 1). The metastable $A'^2\Delta$ state cannot be accessed at all from ground state reagents.

For metastable $M(^3P)$ reactions, however, the branching model is given in Fig. 14. There are more pathways for each product state than in the case described by Fig. 13. The discussion begins with C_s, the statistically most favoured collision

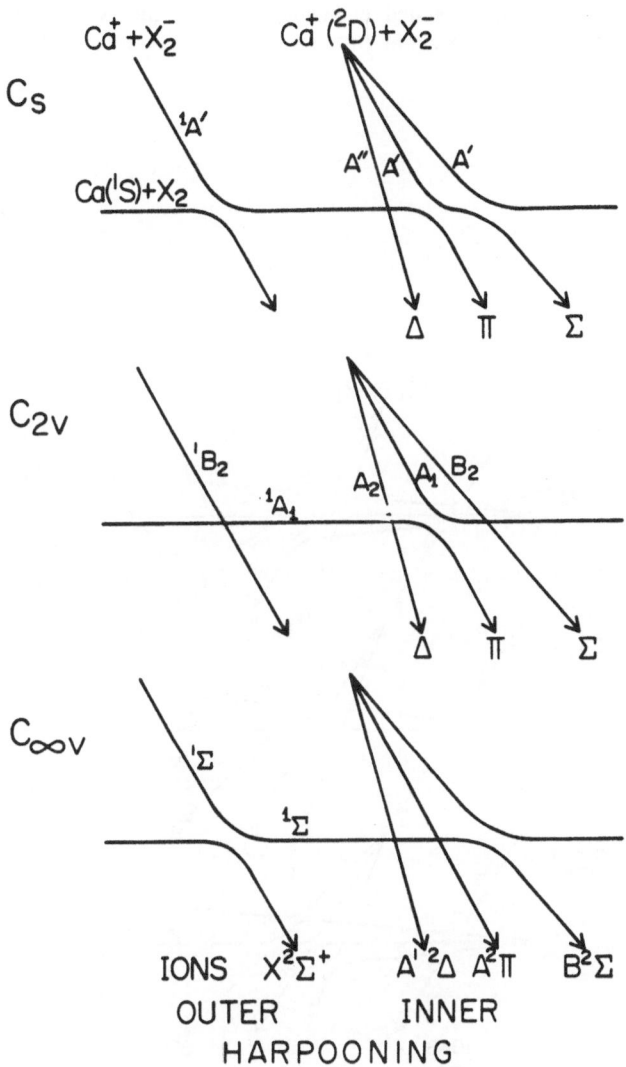

FIGURE 13. Chemiluminescence branching: local state correlation diagram for the harpooning regions for reaction of ground state $M(^1S)$, based on a model for angular momentum partitioning described in the text.

geometry. One notes, for A" symmetry, corresponding to alignment of the p-orbital perpendicular to the collision plane, an adiabatic avenue towards the elusive metastable $A'^2\Delta$ product. This avenue may be easier to travel than that yielding $A^2\Pi$, since in the latter diabatic transitions may deviate reactive flux towards $B^2\Sigma^+$ at the avoided outer harpooning intersection, although another diabatic transition at the outer harpooning from the PES leading to $X^2\Sigma^+$ may compensate for this loss of $A^2\Pi$ flux. All that can be concluded from this simple analysis is that $A'^2\Delta$ and $A^2\Pi$ constitute the principal excited channels, and that the $A'^2\Delta$ state which is hitherto unknown [22] as a reaction product, represents a major channel in the

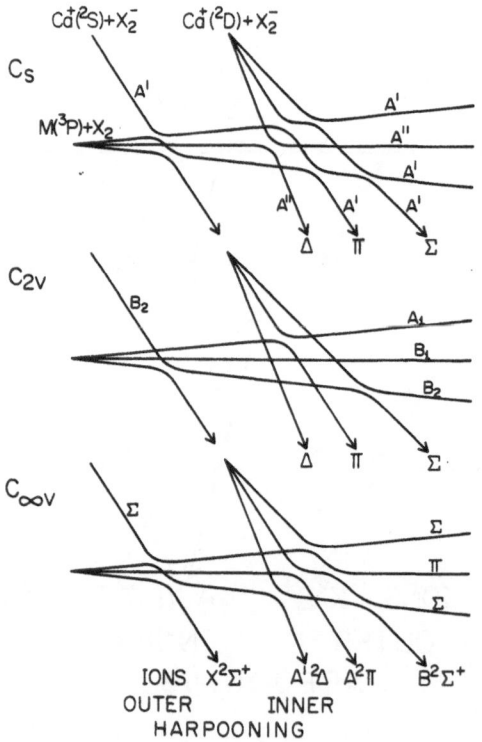

FIGURE 14. Chemiluminescence branching for $M(^3P) + X_2$. See Fig. 13.

reaction of metastables. In contrast to these facile adiabatic routes towards $A'^2\Delta$ and $A^2\Pi$, formation of $MX(B^2\Sigma^+)$ may occur in C_s and in C_{2v} geometry only *via* diabatic transitions at the inner and outer harpooning regions, respectively, and adiabatically in collinear $C_{\infty v}$ collisions of low statistical weight. Again this analysis confirms the experimental finding that the chemiluminescence rate decreases rapidly with increasing energy $A^2\Pi > B^2\Sigma^+$. The possibility that the $A'^2\Delta$ is slightly inverted relative to the $X^2\Sigma^+$ state makes the search for this missing link all the more interesting. Also, improved measurements of the kinetic energy dependences [12] for the individual CL channels might provide some evidence for the mechanisms described here.

The foregoing coupling scheme and CL model is admittedly oversimplified. Yet it provides what appears to be the first successful attempt at explaining on purely dynamical grounds the monotonically decreasing population of the lowest CL states.

4. CONCLUSION

We have shown that it is possible to obtain from qualitative considerations alone a global view of the relevant potential energy surfaces of the MX_2 reactions and of their role in the dynamics of individual electronic channels. We owe this to the neat covalent/ionic nature of the relevant configurations, which makes it easy to predict the locations of the all-important covalent/ionic surface crossings and to use group theory to obtain the topology of the network of PE surfaces and on their basis to make predictions about steric and alignment effects. The present analysis is of considerable generality since a similar version is appropriate for many other covalent/ionic systems whose fragments have low lying electronic states. This analysis would however not be possible for covalent/covalent systems where the intermolecular forces are governed by electronic overlap to a much greater extent than in MX_2, and the location or even the occurrence of (avoided) surface crossings can generally not be predicted without electronic structure calculations.

The present model shows some success with respect to the taxonomy of the reaction channels, the interpretation of the excitation functions in terms of outer and inner harpooning, a qualitative idea of the shapes of the relevant PES's, and last, but not least, with respect to the natural way by which it allows one to deal with steric and alignment problems. An advantage of the model is that it can be quantified and extended to the full dimension of the problem - a program which is currently under development at the University of Toronto.

ACKNOWLEDGMENT

This work was done under the aegis of the NSERC of Canada.

REFERENCES

[1] M. Menzinger, Adv. Chem. Phys., **42** (1980) 1.

478

[2] M. Menzinger, in : Gas-phase Chemiluminescence and
     Chemiionization, Ed.: A. Fontijn; Elsevier, Amsterdam, 1985.

[3] J.L. Magee, J. Chem. Phys., 8 (1940) 687.

[4] G.G. Balint Kurti, Mol. Phys., 25 (1973) 393.

[5] M. Menzinger, Polon. Phys. Acta, in press.

[6] P.J. Dagdigian, Chem. Phys. Lett., 55 (1977) 239.

[7] A. Kowalski and M. Menzinger, Chem. Phys. Lett., 78 (1981) 461.
     A. Kowalski and M. Menzinger, J. Chem. Phys., 78 (1983) 5612.
     A. Kowalski and M. Menzinger, J. Phys. Chem., submitted.

[8] C. Rettner and R.N. Zare, J. Chem. Phys., 77 (1982) 2416.

[9] D.M. Manos and J.M. Parson, J. Chem. Phys., 63 (1975) 3575.
     T. Ishikawa and J.M. Parson, J. Chem. Phys., 79 (1982) 4261.
     W.J. Rosano and J.M. Parson, J. Chem. Phys., 85 (1986) 2644.

[10] J.I. Steinfeld, Ed., Electronic Transition Lasers, MIT Press,
     Cambridge, MA., 1975.

[11] T. Wilson, in: MTP Int. Rev. of Sci., Phys. Chem., Series 2,
     vol. 2. Ed.: A.D. Buckingham and D.R. Herschbach; Butterworths,
     London; 1976.

[12] D.J. Wren, Ph.D. Thesis, Univ. of Toronto, 1978.

[13] M. Menzinger and D.J. Wren, Chem. Phys. Lett., 81 (1981) 599.

[14] H.J. Meyer, Th. Schultze and U. Ross, Chem. Phys., 90 (1984) 185.

[15] S.M. Lin, C.A. Mims and R.R. Herm, J. Chem. Phys., 58 (1973) 327.
     F. Engelke, Dissertation, Universität Freiburg, 1974.

[16] C.D. Jonah and R.N. Zare, Chem. Phys. Lett., 9 (1971) 65.

[17] D.J. Eckstrom, S.A. Edelstein and S.W. Benson, J. Chem. Phys.,
     60 (1973) 2930; G.A. Capelle, C.R. Jones, J. Zorskie and
     H.P. Broida J. Chem. Phys., 61 (1974) 4777; R.S. Bradford,
     C.R. Jones, L.A. Southall and H.P. Broida, J. Chem. Phys., 62
     (1975) 2060.

[18] M. Corbett, M.G. Prisant and M. Menzinger, Chem. Phys. Lett.,
     122 (1985) 365.

[19] M. Menzinger and M. Corbett, unpublished data.

[20] H.J. Loesch, Chem. Phys., **104** (1986) 213; *ibid.* **112** (1987) 85;
W. Grote, M. Hoffmeister, R. Schleysing, H. Zerhau-Dreihöfer
and H.J. Loesch, paper in this volume.

[21] P.J. Dagdigian and M. Campbell, Chem. Rev., **87** (1987) 1.
P.J. Dagdigian, paper in this volume.

[22] H. Martin and P. Royen, Chem. Phys. Lett., **97** (1983) 127.

[23] M.G. Prisant, C. Rettner and R.N. Zare, J. Chem. Phys.,
81 (1984) 2699; and Chem. Phys. Lett., **88** (1982) 271.

[24] I.V. Hertel, Adv. Chem. Phys., **45** (1981) 341.

THE ROLE OF SELECTIVE CHEMICAL REACTIONS IN LASER TECHNOLOGY

J.J. Ewing
Spectra Technology Inc.
Bellevue
Washington
USA

ABSTRACT. Selective chemical reactivity has played a long and vital role in the development of laser technology. Some recent examples of selective reactivity as applied to the understanding of rare gas halide excimer lasers and the search for novel chemically driven lasers are discussed. The future of selective chemistry as a component of laser development is projected.

1. INTRODUCTION AND HISTORICAL PERSPECTIVE

The discovery and understanding of gaseous molecular lasers has both utilized and spurred fundamental research in selective chemistry. The results of these efforts are incorporated into a broader field of laser device technology. Laser technology focuses on development of laser hardware using comprehensive understanding of all of the various phenomena that effect the design of devices. This paper will examine the role of selective chemistry in laser technology, both from a historic and future perspective.

Chemically excited infrared lasers were the first major "application" that derived from the early days of molecular dynamics research [1,2]. The significant research investment in the area of infrared chemiluminescence paralleled a substantially larger investment in coupling this chemical physics understanding to fluid mechanics and optical physics, as well as device technology. The results have been both large scale laser devices and a better fundamental understanding of the role of vibrational excitation as a product of reaction.

In parallel to these efforts on purely chemical systems, electrically excited lasers were being developed. In this area of laser technology, charged particle processes play an equally important role in energy transfer and selective reaction chemistry. The discovery of efficient and scalable ultraviolet lasers required the significant breakthrough of developing an understanding of a whole new class of molecular species, the rare gas halide excimers [3]. Spectroscopic and kinetic characterization of these systems was the tip of the iceberg in developing these lasers. The science content of this novel set of electrically driven systems has been quite large since subtle and complex mechanisms involving excited and ionic species have had to be addressed. Developing these lasers into useful practical devices has involved a number of technological and engineering challenges. The bulk of

J. C. Whitehead (ed.), Selectivity in Chemical Reactions, 481–495.
© *1988 by Kluwer Academic Publishers.*

development recently has focused on the coupling of plasma kinetic processes to circuit design, and with other aspects of high voltage, high speed electrical engineering. By improving overall efficiency, truly reliable ultraviolet sources for industrial photochemical applications, such as photolithography, should eventually be realized. Despite the engineering flavour of the laser technology discipline, there are still major research unknowns in chemical kinetics, some of which are discussed here.

The future prospects for selective chemistry research as an aspect of laser development are not as bright as several years ago. With the exception of the quest for a visible chemical laser, most new laser research is oriented towards systems that are patently nonmolecular. The leading candidate for very high power applications is the free electron laser (FEL). In this device, the interaction of a relativistic electron beam with a periodic magnetic structure produces coherent radiation. Such devices on paper can have substantially higher power and efficiency than electric molecular lasers. For moderate power applications, advances in solid state lasers, nonlinear optical conversion processes, and tunable solid state media offer the prospect of broadly tunable compact sources. At low powers, diode lasers and diode laser arrays are gaining increasing application and hold out the promise, when used with solid state media, of versatile tunable sources.

2. RARE GAS HALIDE LASERS

2.1. Previous Development

The development of rare gas halide lasers is continuing even though the first operation of these devices as lasers occurred over 12 years ago. The unravelling of the chemical kinetics and spectroscopic details of this rich class of molecules has proceeded in parallel with the technological development of these devices. This is not surprising since the first demonstration of lasing action and the observation of the emission spectra were almost simultaneous events. In fact, the deep UV (193 nm) ArF laser was discovered and producing over 100 J of laser output before its spectrum was recorded and understood [4].

The drive for developing such lasers was sparked by a very simple observation. If an efficient and scalable visible-ultraviolet laser could be built it would, in principle, be able to produce high intensity radiation in smaller spot sizes at longer distances compared to the then available high power infrared lasers: HF, DF, and CO_2. However, conceptualizing and developing such new short wavelength lasers required a fairly broad based combination of kinetic, spectroscopic, and laser technological skills. Elucidating the optical engineering and high voltage technology requirements, developing basic spectroscopic understanding, and developing a set of gas kinetic rate coefficients for potential candidates were each important in their own right. However, to achieve this breakthrough in laser technology, all of the preceding concerns had to be taken together and worked into a unified conceptual entity. Several parallel independent efforts also enhanced intellectual competition and cross fertilization.

The rare gas halide species are the brightest ultraviolet light sources available. They all share common molecular heritage, which is that their excited states are ionic in nature and are short-lived excited state analogs to the ground state alkali halides [5] (see Fig. 1). For example, the XeCl laser at 308 nm, which tends to be the most widely used of these devices, has an upper laser level that is a set of excited states corresponding to bound Xe^+Cl^- ion pairs. Because of quantum considerations arising from the two Xe$^+$ free ion states, three ion-pair-like species are generated. In fact, the collisional intercoupling rates of the various excited states [B(1/2), C(3/2), and D(1/2)] are still the subject of considerable speculation and very few clean experimental measurements are available. The binding energy relative to the free ion pair quantum states is roughly equal to that of the ground alkali halide states of the corresponding alkali halide, viz. XeCl* ≈ RbCl. Also, the most allowed emissions of the rare gas halide molecules, the so-called B → X or parallel transitions, have oscillator strengths of about 0.1 as one would expect for a charge transfer transition. That is, a charge hops from the halide negative ion to the positive rare gas ion to produce a rare gas/halogen atom pair along with ultraviolet light. For the case of XeCl and XeF lasers, this emission actually results in a weakly bound rare gas halide species which is dissociated in subsequent collisions. Despite its importance to the 351, 353 nm laser performance, the "debottlenecking" rate coefficients for specific vibrational and rotational states of XeF are not adequately understood.

The alkali halide nature of these species provides a ready first order interpretation of their kinetics of formation. The early observation of rare gas halide emission as products of metastable rare gas reactions [6,7], e.g.,

$$Xe* + Cl_2 \rightarrow XeCl* + Cl$$

led to a fairly complete characterization of a number of these reactions. Large rate coefficients and, for simple cases, unit branching fractions into the excited states, were all consistent with an alkali-like reaction mechanism involving a long range charge transfer or harpooning.

The dominant chemistry for an electrically excited laser mixture is ion chemistry rather than neutral chemistry. Both relativistic

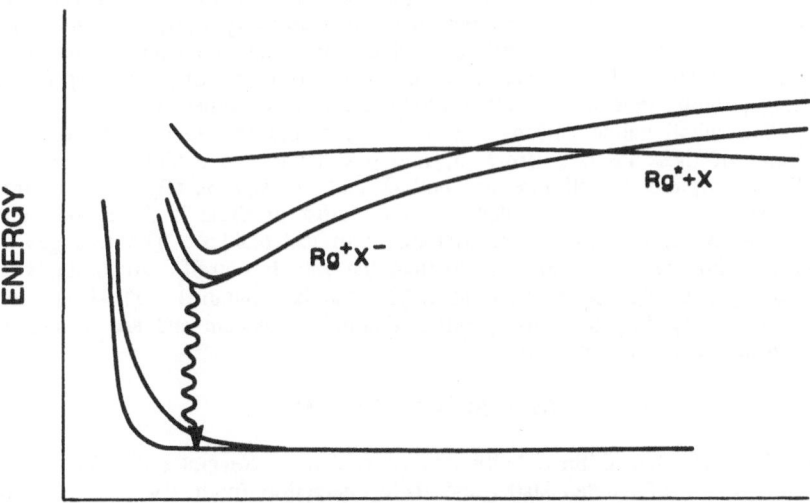

FIGURE 1. Schematic of typical rare gas halide energy diagram.

e-beams and the common self-sustained or avalanche discharges produce the excited states by first producing ions which then very rapidly combine to make the ion pair excited state:

$$e' + Xe \rightarrow Xe^+ + e_s + e'$$

$$e_s + Cl_2 \rightarrow Cl^- + Cl$$

$$Xe^+ + Cl^- + M \rightarrow XeCl^* + M$$

where e' is a relativistic or fast electron and e_s is a secondary or slow electron.

It is curious that the most used Cl source for an XeCl laser (either e-beam or discharge excited) is HCl and to first order neither of the above reaction schemes is relevant. The reaction $e + HCl \rightarrow H + Cl^-$ is 100 times too slow to allow for sufficient laser gain to be generated. Moreover, the harpoon mechanism from xenon metastables does not produce XeCl* emission in high yield. The above considerations are not true for a Cl_2 based XeCl laser, but materials handling and background absorption make this a less desirable source of Cl.

The HCl based XeCl laser is thought to work because HCl attaches to electrons as it is vibrationally excited by the electron bath. Significant enhancements in the attachment cross section for $v=1$ and $v=2$ imply that HCl must be first vibrationally excited before formation of the lasing species can proceed.

The ion chemistry nature of these species is also the cause of one of their principal drawbacks: non-saturable absorption which limits excited state extraction efficiency. Several sources of loss are known. The reagents used in the lasers, such as F_2 in a KrF laser; the halide ionic intermediates, (F^-, Cl^-, etc.), diatomic rare gas ions (Ar_2^+, Kr_2^+, etc.) and excited diatomic rare gas dimers (Ar_2^*, Kr_2^*, etc.) all absorb in the ultraviolet. A significant fraction of the early laser modelling of these systems focused on the understanding and control of the formation of these intermediates. The total background loss corresponds typically to 5 to 10% of the gain. By choosing kinetic conditions which maximize the gain to loss ratio, significant improvements in extraction efficiency (number of excited species extracted as laser photons ratioed to number produced) can be achieved.

The alkali halide analogy was a valuable tool to begin to unravel the mysteries of these lasers. However, rare gas halides do exhibit some processes which are common to all rare gas halides and are significantly different from other laser species or the alkali halides. The foremost of these is three body quenching and the subsequent formation of triatomic rare gas halides. The rare gas dimer ions were well known to be more stable than the free ion itself. Accordingly an ion pair rare gas halide species such as $Kr_2^+F^-$ can be imagined. These species are lower in energy than the corresponding diatomic molecule and are formed by termolecular reactions such as

$$KrF^* + Kr + M \rightarrow Kr_2F^* + M$$

which have substantial three body rate coefficients. Several other features, such as the multiplicity of excited states and their interaction (both VV and VE processes), lower level dynamics, and details of the role of power flow from electrons to electronic excited states differentiate these laser molecules from their alkali halide cousins.

As the media data base for these lasers has developed, so have our modelling tools. Nowadays multidimensional kinetic and electrodynamic codes are coupled together to do sophisticated laser designs. Here the effort is to turn all of our knowledge of chemistry, plasma processes, and pulsed power high voltage design into integrated engineering design tools. We are well along on that course despite an incomplete knowledge of some of the more fascinating reactive details [8].

However, laser technology runs in cycles in which a large kinetic or spectroscopic flavour is superseded by one with a large dose of some other relevant engineering technology. Following pioneering vibrational luminescence and excited state kinetics work, chemical laser technology focused on fluid mechanics and optical resonator design. In CO_2 laser development, plasma physics and pulsed power scaling became more important once the state specific kinetics were understood to first order. In excimer laser development, fast, reliable, high voltage electric circuitry and development of optics for the hard UV became dominant themes paralleling kinetic development. In the case of all of these lasers, some new application, desired mode of operation, or funding agency realization, may lead to a desire for more basic data and a refocusing on measurements of basic kinetic properties. Such a resurgence has recently happened for excimer lasers and is illustrated by recent work performed at Spectra Technology [9,11].

2.2 New Kinetic Measurements in Rare Gas Halides

Because of the fact that many kinetic formation and quenching processes are understood and the sources of background loss are reasonably well known, the prediction of laser output energy performance and efficiency is understood within certain ranges. However, the mechanisms in these laser systems are not without uncertainties. These uncertainties derive from some basic puzzles related to laser scaling, for example, the lower energy density for XeCl lasers relative to KrF or XeF. Recently we began measurements of some fundamental media properties to provide further anchoring points for laser models. The time dependence of the electron density in e-beam excited XeCl, KrF, and XeF lasers has been measured [9]. Curiously, even though electrons dominate the laser media chemistry, the density of this key reagent had never been quantified, but only predicted by laser models. In bulk terms, the net process in one of these lasers is the dissociation of HCl or other halogen source by electron and neutral collisional processes with the intermediate step of producing UV emitting laser species. Thus, fuel burnup clearly should be predicted by models as this is the simplest of chemical kinetic features of these lasers. The rate of removal of HCl has now been measured [10]. The Xe excited states have always been presumed to be formed, but their transient density has never been quantified. Recently measurements using hook interferometry have been performed to add these species to other code anchor points [11].

The electron density in a typical rare gas halide laser is calculated to be on the order of 10^{14} to 10^{15} electrons per cm^3 and is dependent on the pump power density, the type of plasma (i.e., e-beam excited or discharge), and laser mixture. This estimate can be developed from a steady state argument noting that the rate of ionization must balance the attachment rate after an initial transient excitation, viz.,

$$\frac{dn_e}{dt} = R_i - n_e k_a [halogen] = 0 \text{ at steady state}$$

The ionization rate is proportional to the energy deposition rate, usually a directly measured or inferred quantity.

The electron density experiments of Kimura *et al.* [9], used 10 micron interferometry to measure electron density. A CO_2 laser illuminates a quadrature interferometer in which one arm contains two passes through the 35 cm long laser chamber. One fringe shift at 10 microns corresponds to an electron density of 3 x 10^{14} per cm^3 for the conditions of the experiment. The use of a quadrature scheme provides for the elimination of ambiguities in counting the number of fringe shifts when an extremum is passed through as the plasma induced refractive index changes. The experimental layout used is shown in Fig. 2. Electron densities for XeCl plasmas (Ne/Xe/HCl mixtures) and XeF plasmas (Ne/Xe/F_2) are shown in Figs. 3 and 4. For each case the initial halogen density is varied to generate families of electron density curves ranging from nominal laser gas mixtures (0.1-0.15% halogen) to very dilute or rich halogen mixtures.

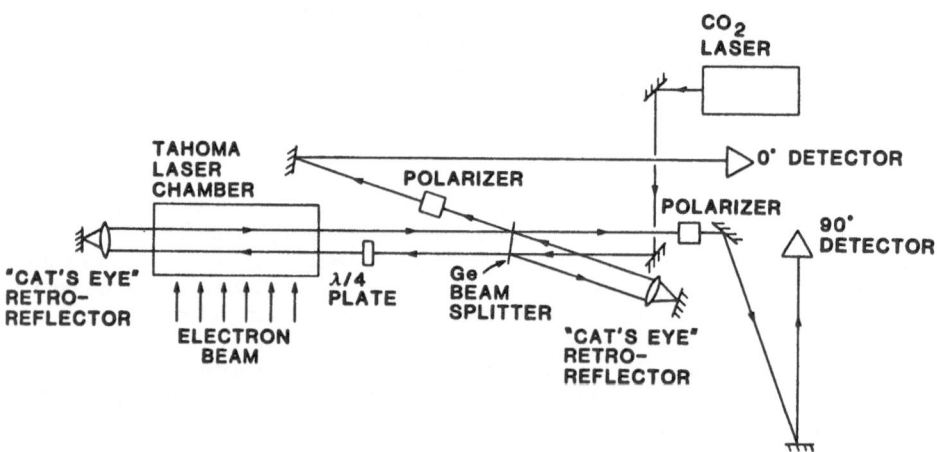

FIGURE 2. Schematic of quadrature interferometer system used to measure the time-dependent electron density.

One very striking result of the HCl based laser mixtures is the initial overshoot in electron density followed by a relaxation down to densities roughly predicted by steady state model predictions where attachment and electron production are in local equilibrium. The initial overshoot is due to the fact that ground vibrational state HCl does not attach rapidly. A degree of vibrational excitation is needed to turn on the attachment process. In contrast, the F_2 based lasers do not show this initial transient. F_2 effectively attaches when it is in v=0.

We see for both the fluorine and chlorine based laser mixtures that densities of order 2-3 x 10^{14} electrons per cm^3 are observed in steady state. This was at some variance with the commonly accepted computer models at the time. In the course of readjusting electron rate constants it was found that laser energy output was not terribly sensitive to (or a very good measure of) the electron density. Towards the end of the pulse (or sooner for dilute halogen mixtures) we see that the electron density increases rapidly. This is due to

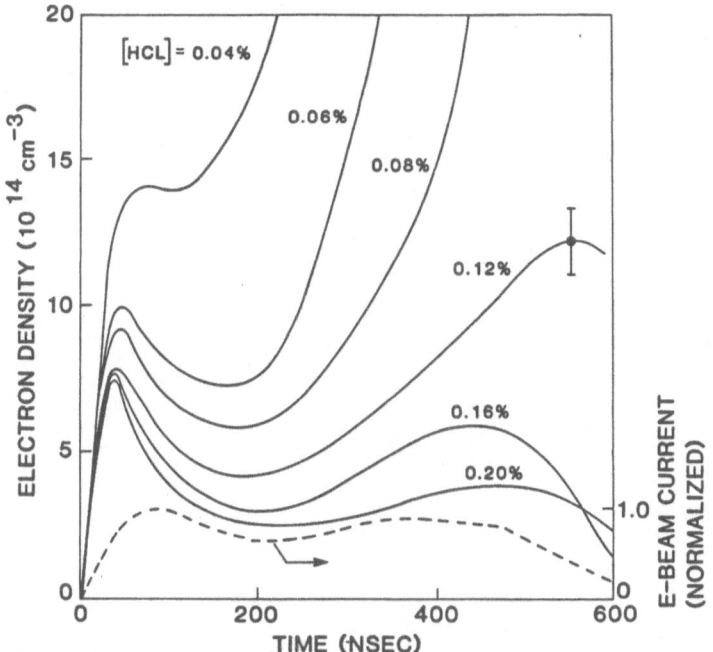

FIGURE 3. Typical electron densities of XeCl mixtures (1.5% Xe, balance Ne to 3000 torr) for various initial HCl concentrations. The dashed curve is a normalized trace of a typical electron beam current waveform. Data points are not shown; however, a representative error bar is indicated.

the burnup of the attaching halogen. As the halogen disappears the mechanism, which controls electron density, changes to one of recombination rather than attachment. The disappearance of attacher and the increase in electron density appears with a contemporaneous drop in XeCl* emission. This is due to both decreased production efficiency of XeCl* and increased quenching of this excited state, believed to be very fast with electrons.

Our models of these processes still need improvement however. The rate of growth of electron density and the actual electron density predicted towards the end of the pulse in the burnout phase is still off by approximately a factor of two. Understanding this phenomenum is a key to increasing the energy density of XeCl lasers.

The measurement of electron density suggests HCl is depleted such that roughly 2.4 torr of HCl is burned up in 400 nsec for 180 kW/cm^3 pump power. Or equivalently 1 torr is burned up per 30 J per litre deposited energy. To check this scaling and provide another model comparison point, the Spectra Technology, Inc. (STI) researchers measured the HCl removal dynamics. A Raman shifted Nd:YAG pumped dye laser was tuned to the HCl overtone band at 1.8 μm. Because of intense background absorption, intrapulse measurements could not be made. Rather, the remaining HCl was measured at 2 μs after termination of the

488

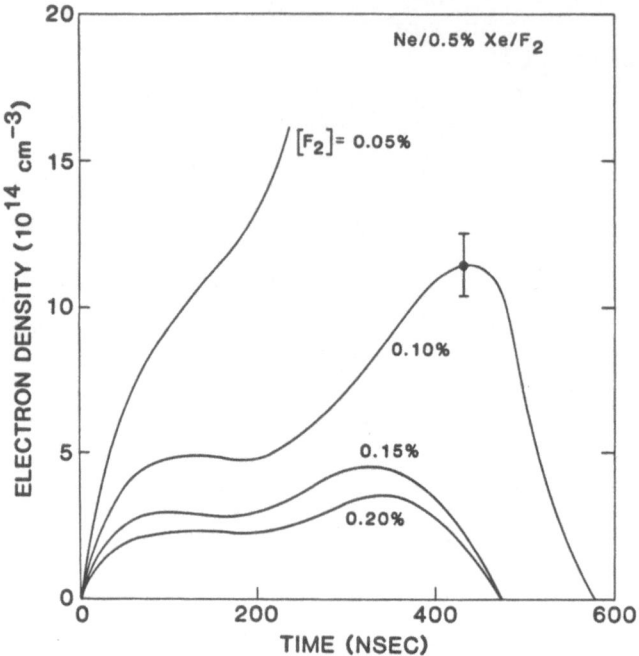

FIGURE 4. Typical electron densities of XeF mixtures (0.5% Xe, balance Ne to 2888 torr) for various initial F_2 concentrations. Data points are not shown; however, a representative error bar is indicated.

e-beam pulse. The HCl removal data is consistent with a burnup of all the HCl for a 2.4 torr mixture pumped for 400 nsec at 180 kW/cm^3. This is consistent with the electron density runaway shown in Fig. 3. For mixtures where less than complete burnup occurs, the models underpredict the rate of HCl disappearance. Processes such as:

$$e + HCl \rightarrow H + Cl + e$$

are very important in discharge lasers where the electron density and the electron temperature (10 eV) are higher. Their importance is traditionally thought to be low in e-beam pumped excimers where secondary electron temperatures are thought to be of the order of 1 eV. The implication of this data is that there are basic electron impact phenomena in HCl that are not known and/or that the electron temperature or distribution predicted by the models is not entirely correct.

A measurement of the electron temperature or distribution function is a worthwhile experiment we hope to undertake sometime in the future. In the interim, we have measured another media property that relates to electron density and temperature, and is a principle component of the rare gas excitation scheme. Time dependent density measurements of the lowest xenon excited states in XeCl lasers

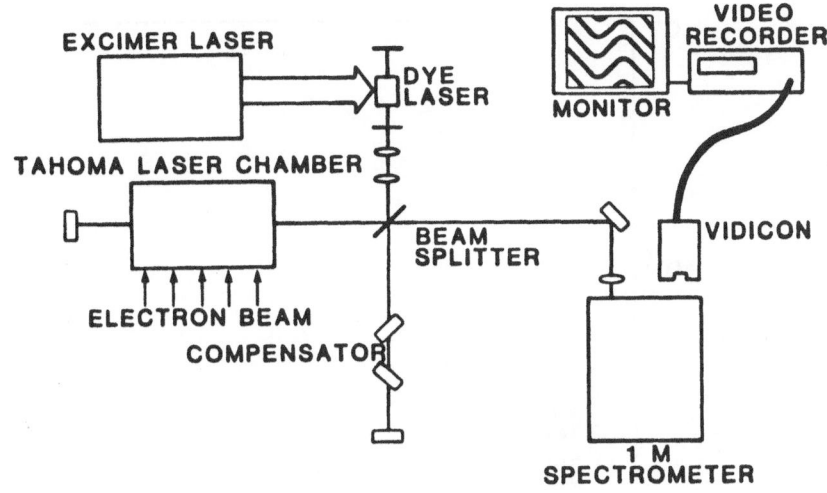

FIGURE 5. Schematic of hook interferometry system used to measure the xenon metastable density.

have been recently completed. This measurement uses the pulsed laser hook interferometry technique. Fig. 5 shows how a simple excimer pumped broadband dye laser is used to illuminate an interferometer. A spectrally resolved interferogram is recorded with a video camera and the well-known "hook" pattern is observed for the spectral region near an atomic transition. This method is very well suited to media where atomic lines are optically thick. From a knowledge of the oscillator strength and the media length, a population difference is measured. By changing the pulse delay of the probe dye laser with respect to the e-beam pulse, the atomic excited state density differences can be resolved in time.

By measuring a large number of density differences actual densities can be inferred. This has not been done yet for these measurements, but estimates of the upper Xe level densities suggest that the values of density difference are probably within roughly 30% of the actual lower level density. Densities of the two lowest xenon excited states are of order 3×10^{14} and 1.7×10^{14} per cm^3 for canonical 0.16% HCl laser mixtures. For mixtures lean in HCl, the xenon excited state density "runs away" as the HCl burns out (see Fig. 6). Preliminary modelling efforts show that the steady state densities are about what is expected, but the magnitude and time dependence of the Xe* run away are not as expected. The preceding are examples of some of the rich chemistry and kinetics that are still unknown about the rare gas halide lasers. Many other examples abound, but they are either not yet the subject of current research or there is still a significant amount of additional research needed. These include the kinetics of specific (v, J) states in XeF (a bound-bound laser transition), any quantitative understanding of ArF (one of the more potentially useful lasers), measurements of halogen burnup phenomena in KrF lasers, electron temperatures or distributions, the energy to produce an ion pair in actual laser mixtures, and a host of other challenging puzzles. On top of these "chemical physics" challenges there still remain numerous

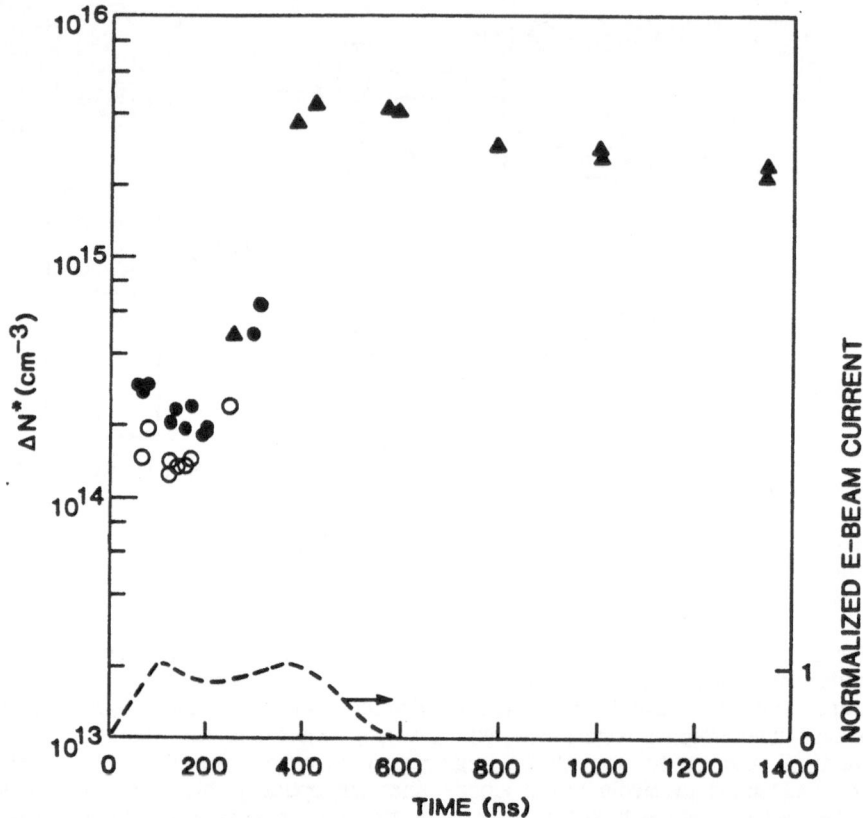

FIGURE 6. Time-dependent population density differences of a 3000 torr mixture of 98.46% Ne/1.5% Xe/0.04% HCl at 294 K. Closed, open circles, and closed triangles correspond to transitions at 823.2 nm ($6s[3/2]_2^o$ - $6p[3/2]_2$), 828.0 nm ($6s[3/2]_1^o$ - $6p[1/2]_0$), and 840.9 nm ($6s[3/2]_2^o$ - $6p[3/2]_1$), respectively. The dashed curve is a normalized trace of the e-beam current waveform.

device technology issues dealing with the coupling of electric power to the gas, optical coherence, short pulse generation, scaling, parasitic behaviour, and so forth. We have come a long way, but there are still interesting questions going unanswered.

2.3 A Chemically Pumped Rare Gas Halide Laser

The ultimate molecular laser is, of course, one with no recourse to electrons, ions, or discharges but is purely chemically pumped. These chemical schemes have worked well for infrared lasers and for the $O_2(^1\Delta)$ - iodine system. However, the search for visible or UV lasers pumped chemically still entices researchers. At STI, we have been working on one scheme which is attractive in its conceptual simplicity, but not as simple as it at first seemed. We are attempting to understand the generation of XeF radiation by azide chemistry.

The proposed scheme is based on the concept of using a known laser medium pumped by highly energetic reagents that are premixed. It turns out that all of the known visible or ultraviolet lasers that do not bottleneck are ones that require fairly robust pumping conditions to achieve lasing action. Our scheme would produce excited N_2 and then a lasing upper level by energy transfer to species such as vapour phase dyes, HgBr, or excimers. The attractiveness of such a scheme would be enhanced by a chain reaction mechanism. One possible mechanism is an F atom initiated $N_3/N_2(A)$ sequence listed below:

$$hv + F_2 \rightarrow 2F$$

$$XeF_2 \rightarrow 2F + Xe$$

$$F + HN_3 \rightarrow HF + N_3$$

$$N_3 + N_3 \rightarrow N_2(A,B) + 2N_2$$

$$N_2(A,B) + XeF_2 \rightarrow XeF^* + F + N_2$$

$$XeF^* + hv \rightarrow Xe + F + 2hv.$$

Luminescence in XeF_2/HN_3 mixtures is now under investigation [12]. We have, in fact, observed XeF emissions from premixed XeF_2/HN_3 mixtures. The observed chemiluminescence is clearly from XeF (see Fig. 7a), although various other emissions are observed as well.

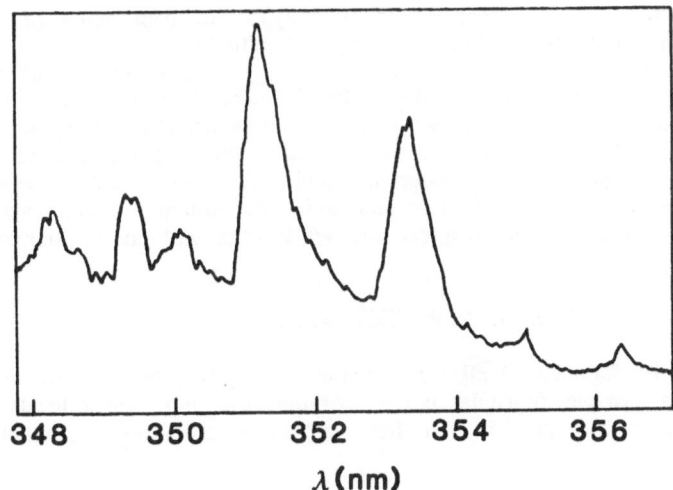

FIGURE 7A. XeF* chemiluminescence spectrum from premixed XeF_2/HN_3 mixtures.

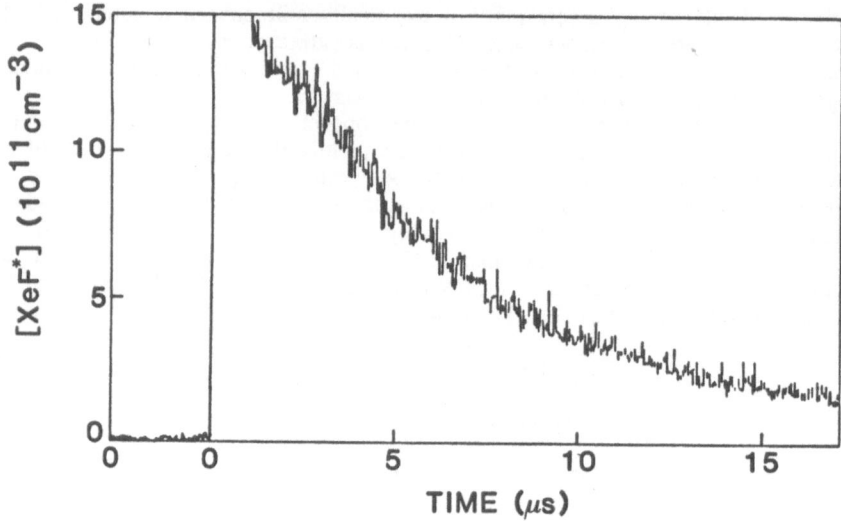

FIGURE 7B. XeF* chemiluminescence decay from premixed XeF_2/HN_3 mixtures. The photolysis source was a KrF laser, whose 5 eV photon energy is insufficient to produce XeF* from XeF_2 by a one photon process. The time dependence of the XeF emission shows a many microsecond fluorescence pulse that clearly is due to azide/nitrogen chemistry (see Fig. 7b).

At this stage we know more questions than answers. For example, separate measurements of $N_2(A)$ + XeF_2 reaction kinetics in low pressure flowing afterglows show that $N_2(A)$ does not produce XeF* [13,14]. Our measurements thus suggest that XeF* must be made from higher lying N_2 excited states, perhaps $N_2(B)$ or possibly others. There is uncertainty in the rates, products, and yields of the N_3 + N_3 reaction. It is not clear that the photolysis flux dependence of XeF* chemiluminescence is as simple as proposed initially.

The proposed scheme is really a prototype. Our experiments are focusing on improved spectroscopic/kinetic diagnostics. We need to define the yield of electronically excited N_2. If we can make large densities of $N_2(A)$ or $N_2(B)$, we may be able to find other laser species that could be pumped more favourably than XeF, for example, HgBr(B) comes to mind. We need to find acceptors for that energy which could possibly lase and preferably sustain a chain reaction. The search for a high power chemical visible/UV laser will not be simple.

3. THE FUTURE OF LASER TECHNOLOGY

Although development of all lasers continues to grow, there is an unmistakable trend away from gaseous molecular lasers. At the ultra high power levels, the focus is shifting away from gas lasers to free electron lasers. This is paralleled by a trend

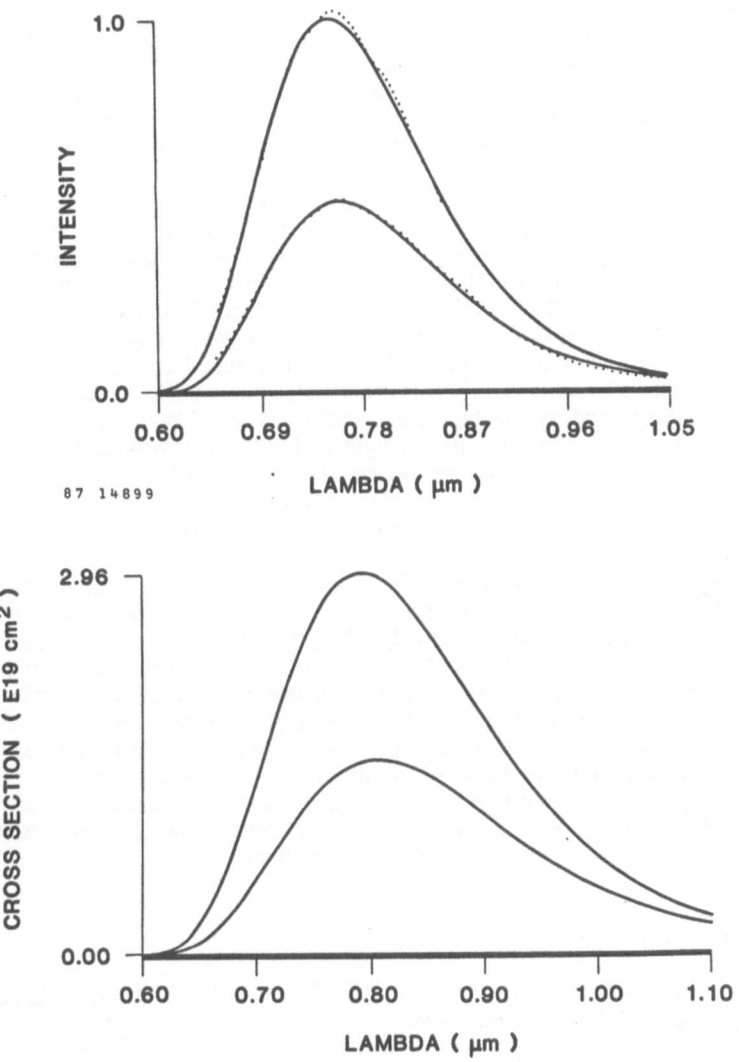

87 14899

FIGURE 8. Experimental and theoretical fluorescence and cross-section curves for Ti:Sapphire. Top: Experimentally measured fluorescence data (dots) and theoretical curve (solid line) for the π and σ emission spectra of Ti:Sapphire. Bottom: Theoretical cross-section curve for Ti:Sapphire using scaling based on a radiative lifetime of 3.9 μs.

toward solid state systems at low and moderate powers. Such trends have quantifiable merit. For example, the tunable solid state medium titanium doped sapphire is broadly tunable and with sufficient development of frequency doubling and nonlinear mixing schemes could be used to cover all of the wavelengths in the visible, ultraviolet, and near infrared. Ultimate powers would be limited, but the thought of dispensing with all dye lasers some day is intriguing to laser vendors as well as laser users.

We have been developing Ti:Sapphire lasers over the last several years. This very simple d^1 configuration ion can be optically pumped with green radiation, typically from a doubled Nd:YAG laser. Its emission cross-section shown in Fig. 8 can support lasing from 750 nm to beyond 1050 nm. The potential uses include remote sensing of atmospheric constituents such as H_2O vapour using differential absorption lidar (DIAL) techniques. The challenge of this laser medium has been the growth of high quality crystalline materials that have low background loss near the 800 nm emission peak. This "materials problem" has been resolved over the last few years and reasonable size crystals can be grown to produce laser rods or slabs capable of producing 10 J outputs. Now the performance limiting parameter is the green pump source, and the size and efficiency of such a pump laser.

The development of such solid state converters is spurred on by the parallel development of GaAlAs laser diodes and arrays of these diode lasers. GaAlAs laser diodes can pump the 1.06 μm Nd lasers (YAG or glass) as well as more esoteric materials such as Tm:Ho:YAG which lases at 2.1 μm. While laser diodes are becoming more common, the Holy Grail of high power solid state development is inexpensive diode arrays. Currently, the cost of a diode array to pulse excite a 1 J Nd:YAG laser is roughly 100 times the cost of a simple flashlamp pump arrangement. However, just as transistors and then micro chips have replaced vacuum tubes, the solid state laser champions claim that the time will come when flashlamps are passé, Nd lasers are 10% wall plug efficient, and tunable solid state lasers will do most of the work of dyes, excimers, and so forth.

As this day approaches, the decreased size and increased versatility of laser technology will provide a wide range of scientific tools that the chemical physics community can use for a variety of applications.

REFERENCES

[1]   K.G. Aulant, D.H. Maylotte, P.D. Pacey and J.C. Polanyi, Phys. Letters A, 24 (1967) 208.

[2]   J.V. Kasper and G.C. Pimentel, Phys. Rev. Letters, 14 (1965) 352.

[3]   For reviews of this field and the chemical kinetics involved see, J.J. Ewing, 'Excimer Lasers' in Laser Handbook, M.L. Stitch, editor (North Holland Amsterdam, 1979).

[4]   J.M. Hoffman, A.K. Hays and G.C. Tisone, Appl. Phys. Letters, 28 (1976) 538.

[5]   'Alkali Halide Analogy', J.J. Ewing and C.A. Braun, Phys. Rev. A, 12 (1975) 129.

[6] M.F. Golde and B.A. Thrush, Chem. Phys. Letters, **29** (1974) 486.

[7] J.E. Velazco, J.H. Kolts and D.W. Setser, J. Chem. Phys., **65** (1976) 3468.

[8] C.H. Fisher, M.J. Kushner, T.E. DeHart, J.P. McDaniel, R.A. Petr and J.J. Ewing, Appl. Phys. Letters, **48** (1986) 23, see also M.J. Kushner, A.L. Pindroh, C.H. Fisher, T.A. Znotins and J.J. Ewing, J. Appl. Phys., **57** (1985) 2406.

[9] W.D. Kimura, D.R. Guyer, S.E. Moody, J.F. Seamans and D.H. Ford, Appl. Phys. Letters, **49** (1986) 1569 and **50** (1987) 60.

[10] D.R. Guyer, W.D. Kimura and S.E. Moody, 'HCl Burnup and Electron Densities in E-Beam Pumped XeCl', in Proceedings of the International Conference on Lasers. '86, Orlando, Florida, 1986, edited by R.W. McMillan and W.B. LaCina (STS Press, McLean, in press).

[11] F. Kannari, W.D. Kimura, J.F. Seamans and D.R. Guyer, 'Xenon Metastable Density Measurements in Electron Beam Pumped XeCl Laser Mixtures', to be published.

[12] R.D. Mead, C.H. Fisher, M.A. Kushner, W.H. Pence and J.J. Ewing, 'A Novel Short Wavelength Chemical Laser Scheme', to be published.

[13] R.A. Young, J. Blauer and R. Bower, J. Chem. Phys., **87** (1987) 3708.

[14] D.W. Setser, private communication.

STATE SELECTIVE CHEMILUMINESCENT REACTIONS FOR CHEMICAL LASER APPLICATIONS

Stephen J. Davis and Anne M. Woodward
Physical Sciences Inc.
Dascomb Research Park
Andover
MA 01810
USA

1. INTRODUCTION

An important, but as yet unrealized, application of selective chemiluminescent reactions is the development of chemically driven short wavelength (visible) laser sources. Although this problem has been investigated since the early 1970's, no laser demonstrations have been reported. There are several unique features that such a device could offer. These include efficient, high power output for a variety of research applications. In addition, since the laser medium would likely be a diatomic molecule, such a device would be quasi-tunable. Consequently these devices might offer alternatives to dye lasers or could be used as pump sources for dye lasers. They could therefore find widespread use as kinetic and spectroscopic tools. Short wavelength chemical lasers also might offer unique advantages for medical applications.

In this paper we present a brief review of the field and attempt to give an historical perspective of prior attempts to develop a visible chemical laser. Finally we discuss a specific example from recent results which demonstrate several of the critical parameters that will likely be common to any such laser device.

The concept of using a chemical reaction to produce population inversions between two energy levels is conceptually simple and was beautifully demonstrated with the advent of infrared chemical lasers (e.g., HF) that operate on vibrational transitions in the ground electronic state. For example, the reaction of F with H_2 produces population inversions between several vibrational levels and lasing is obtained from $\Delta v = -1$ transitions:

$$\text{(Excitation)} \qquad H_2 + F \rightarrow HF(v) + H \qquad\qquad (1)$$

$$\text{(Lasing)} \qquad HF(v) + h\nu \rightarrow HF(v-1) + 2h\nu. \qquad\qquad (2)$$

Unfortunately, the HF laser, which is the shortest wavelength vibrational laser possible, operates at about 2.7 microns. Extension to visible wavelengths will require transitions between electronic energy levels. The relative simplicity and notable success of vibrational lasers led to the pursuit of visible wavelength lasers based upon simple chemical displacement processes similar to reaction (1), and early investigations used this approach. Before we discuss these early studies, it is

J. C. Whitehead (ed.), Selectivity in Chemical Reactions, 497–513.
© *1988 by Kluwer Academic Publishers.*

instructive to consider some parameters that will be relevant to any short wavelength chemical laser. We begin by examining a key component of optical gain, the stimulated emission cross section, σ_{SE}. In general, the stimulated emission cross-section for emission from an upper level u to a lower level 1 is given by Eq. 3

$$\sigma_{SE}(\nu) = \frac{\lambda^2 A_{u1}}{8\pi} g(\nu) \tag{3}$$

where

λ = wavelength of the transition

A_{u1} = Einstein A coefficient

and $g(\nu)$ = line shape function.

At visible wavelengths and under conditions applicable to chemical excitation $g(\nu)$ is usually Doppler (Gaussian). Thus we can write σ_{SE} at line centre as:

$$\sigma_{SE} = \left[\frac{4 \ln 2}{\pi}\right]^{1/2} \frac{\lambda^2 A_{u1}}{8\pi \, \Delta\nu_D} \tag{4}$$

But $\Delta\nu_D = 7.16 \times 10^{-7} \; \nu_0 \left[\frac{T(K)}{M(AMU)}\right]^{1/2}$

Thus $\sigma_{SE} = 1.74 \times 10^{-6} \; \lambda^3 \; A_{u1} \left[\frac{M}{T}\right]^{1/2} \tag{5}$

From Eq. 5 we recognize an important feature that makes visible lasers inherently more difficult to operate than infrared lasers. For equal A coefficients the stimulated emission cross section scales as λ^3. For example, comparable A coefficients for transitions at 500 nm and 3000 nm would lead to a stimulated emission cross section ratio $\sigma(500 \text{ nm})/\sigma(3000 \text{ nm}) \sim 0.005$ (assuming all other conditions were constant).

A second commonly used parameter, the small signal gain coefficient α, is given by the product of the population inversion density and σ_{SE}

$$\alpha = \left[N_u - \frac{g_u}{g_1}N_1\right]\sigma_{SE} \tag{6}$$

where g_u/g_1 is the degeneracy ratio for the states u and 1.

Since most chemiluminescent reactions produce excited diatomic products rather than excited atoms, molecular species are prime visible chemical laser candidates. However, each molecular electronic state contains a complete set of vibrational and rotational levels. Consequently, the population density of an excited electronic level is diluted over many sublevels. This can be illustrated by the expression for σ_{SE}

that applies to a single v', J' → v",J" emission line in a diatomic molecule.

$$\sigma_{SE} = \frac{8\pi^3}{3hc} \left[\frac{4 \ln 2}{\pi}\right]^{1/2} \left\{\frac{\nu_0}{\Delta\nu_D}\right\} |Re|^2 \frac{q_{v',v''} \; S_J \; F_v \; F_J}{2J+1} \tag{7}$$

where

ν_0 = frequency at line centre

$|Re|^2$ = transition moment

$q_{v',v''}$ = Franck-Condon factor for the v' → v" band

S_J = rotational line strength

F_v = fraction of population in upper vibrational level v'

F_J = fraction of population in upper rotational level J'.

We assume the optimistic case that the vibrational and rotational manifolds of the upper electronic state can be collisionally relaxed on a time scale short with respect to the upper electronic state lifetime. In that case the v',J' manifold is thermalized. In this limit, typical values are: $F_v \sim 0.8$, $F_J \sim 0.02$, $q_{v',v''} \sim 0.2$, and $S_J/2J+1 \sim 0.5$; this implies a dilution factor of nearly three orders of magnitude. In other words, for comparable radiative lifetimes and emission wavelengths, the single line stimulated emission cross section for a diatomic molecule will be ~1000 times smaller than for an atomic emitter.

The arguments presented above illustrate some of the difficulties to be encountered in any visible-chemical laser demonstration. Next we discuss some of the early attempts at developing these novel laser systems. Because HF vibrational lasers had been successfully demonstrated they provided a logical model for a visible chemical laser and, as we discuss below, considerable effort was expended in this direction.

Perhaps the most celebrated example of these early attempts to produce a visible laser that somewhat modelled HF devices was the BaO molecule. In the reaction

$$Ba + N_2O \rightarrow BaO^* + N_2 \tag{8}$$

intense emission on the (A,A' → X) bands near 500 nm is observed. The original observation of the Ba + N_2O chemiluminescence was by Ottinger and Zare in a beam gas experiment [1]. Subsequent flow tube experiments by Jones and Broida [2] demonstrated remarkably high photon yields (20 to 30 percent) when the reaction was run in the presence of a few Torr of Ar. Later work by Zare and co-workers [3] showed that only 2.5 percent of the total reactive cross section of Ba + N_2O produced chemiluminescence. It was becoming clear that the initial reaction between Ba and N_2O produced a dark, non-radiating state that was subsequently collisionally mixed with the radiating states. Hsu and Pruett [4] provided convincing evidence that the reaction of Ba + N_2O forms predominantly highly vibrationally excited ground state BaO. They found that over 50 percent of the BaO ground state formed had energy in excess of the lowest electronically excited states, and they concluded that intramolecular energy transfer from these high v" levels was the source for the high photon yields. Other reaction systems of this type (metal atom plus oxidizer) are thought to behave in a similar fashion. Thus, efficient, direct partitioning of energy into states that could be used as upper laser levels had not been demonstrated. An excellent collection of papers of the research during this period (mid to late seventies) can be found in Refs. 5 and 6.

The high photon yields observed for BaO demonstrate the inadequacy of photon yields as a quantitative figure of merit for screening potential laser systems. Indeed, high yields were not indicative of an efficient primary pumping process. The most common definition of photon yield in these early metal oxide studies was

$$\text{Photon yield} \equiv \frac{\text{number of photons emitted per second}}{\text{number of metal atoms injected per second}}.$$

Complete consumption of the metal atoms was assumed. The principal difficulty of the photon yield as a quantitative measurement is that it is very system-dependent, i.e., quenching, wall losses, and other processes influence its observed value. Photon yields shed little insight into the inherent specificity of a reaction. There has also been some confusion between photon yields and the reaction branching ratio. As we show below, the branching ratio is a more fundamental quantity. Consider the exchange reaction

$$A + BC \begin{cases} \xrightarrow{\ k^* \ } AC^* + B \\ \xrightarrow{\ k \ } AC + B \end{cases} \qquad (9)$$

We assume that the reaction is sufficiently exothermic that electronically excited AC^* can be produced. The respective rate coefficients for production of excited and ground state products are k^* and k. The branching ratio is defined as:

$$\text{Branching Ratio} \equiv \frac{k^*}{k + k^*}.$$

A branching ratio of unity tells us that the reaction produces only AC^* while the opposite is true for a branching ratio of zero. The branching ratio in effect describes the inherent efficiency of excited state production by a particular reaction. While quenching might limit the usable number density of excited states, one should strive to find reactions that have efficient branching to excited states.

In addition, a laser system must use "natural" reactants, i.e., the elegant and sophisticated reactant state selection preparation techniques discussed elsewhere in this volume do not offer practical solutions to product state selection. For a practical laser device, one must identify and utilize reaction schemes that yield electronically excited products. One is faced with the seemingly impossible task of testing an astronomically large number of possible reaction schemes. A useful starting point is to invoke spin and angular momentum selection rules that are constrained by the symmetry properties of the reactants and products.

Much of the chemiluminescence research has been guided by these correlation rule arguments [7]. Since many reacting systems are of the type A + BC → AB + C, the spin and orbital angular momentum of the reactants and products can be used to predict possible reaction pathways. Although violations of these correlations have been observed, in general, spin correlation is more rigorous. In brief, if the total spin of a potential product state is equal to an allowed value of the total spin of the reactants then that state is an allowed reaction product. As an example consider the reaction H + NF$_2$

$$NF_2(^2B) + H(^2S) \rightarrow HF(^1\Sigma) + NF(^1\Delta). \qquad (10)$$

Energetically both ground state NF($^3\Sigma$) and excited state NF($^1\Delta$) are allowed, but by spin correlation only singlet products can be produced. Experimental observations have shown that this reaction produces >95 percent NF($^1\Delta$) [8-10]. The only other electronically excited molecular species that has been efficiently produced via chemical reaction is the analogous ($^1\Delta$) state in O$_2$. This is accomplished by reacting Cl$_2$ in a slightly basic solution of H$_2$O$_2$. Interestingly, both NF(a$^1\Delta$) and O$_2$(a$^1\Delta$) lie fairly low in energy, T$_e$[NF(a)] = 12004 cm^{-1} and Te[O$_2$(a)] = 7918 cm^{-1}. Both also have extremely long radiative lifetimes: τ[O$_2$(a)] = 3600 s and τ[NF(a)] = 5.6 s. As exemplified by O$_2$(a) and NF(a) it appears that the most efficient direct chemical production of excited state occurs for those excited states that are metastable (by spin constraints) with respect to spontaneous radiation to the ground state. If the energetically accessible product excited and ground states have the same spin the selectivity is apparently lost.

An extremely metastable excited state is not itself a suitable laser candidate because the optical gain is directly proportional to the radiative rate. One is faced with a dilemma: If the only states that can be efficiently produced are extremely long lived, how can one hope to build a laser? A possible solution to this problem is found in the only electronic transition chemical laser yet demonstrated, the chemical oxygen iodine laser (COIL). In the COIL chemically produced, highly metastable O$_2$($^1\Delta$) resonantly transfers energy to atomic iodine, and an inversion is produced between the $^2P_{1/2}$ and $^2P_{3/2}$ iodine levels and atomic lasing occurs at 1.315 μm. This laser was predicted by Derwent and Thrush [11] in 1972 and was demonstrated by McDermott et al. in 1977 [12]. This device has been described in numerous papers [12-16], and we do not discuss it further. It clearly demonstrates the concept of a transfer-laser and may serve as a model for future visible lasers using this two step approach.

2. EXPERIMENTS ON A CANDIDATE SYSTEM

As an example of a potential visible chemical laser based upon energy transfer, we consider the interhalogen molecule IF. In this regard the (B → X) system of IF has received considerable attention in recent years. Indeed several papers contained in this volume describe details of chemical excitation processes in IF.

The radiative and collisional properties of IF(B) are nearly perfect for chemical laser development. Clyne and McDermid [17] found that the radiative lifetime (τ) is ~7 μs for 0 < v' < 9. This translates into large stimulated emission cross sections (σ_{SE}) for many v' → v" transitions. For example, σ_{SE} = 7.2 x 10^{-17} cm^2 for the 0 → 5 transition. This leads to a small signal gain (γ) of γ = 0.23 percent/cm on the R(20) line for only 1 mTorr of IF(B) at 300 K. In addition to radiative lifetimes that offer large optical gains, the equilibrium internuclear separation for the B state is 10 percent greater than that for the X state. Consequently, the largest Franck-Condon factors $(q_{v',v''})$ for IF(B → X) terminate on high v" facilitating the production of population inversions between v' = 0 and high v" (v"=4,5,6). Interestingly, even though there is a relatively large potential shift in IF, the emission spectrum for the entire IF(B → X) system is relatively narrow (440 to 750 nm), and, as a result, the B → X emission oscillator strength is not so severely diluted as it is in other molecules possessing shifted potentials.

Davis and co-workers clearly demonstrated that the IF(B → X) system will lase if a suitable chemical pump source can be found [18,19]. In a pulsed optically pumped laser containing 10 Torr of He they found that IF lased on the (v',J') → (v",J") = (0,22) → (4,21) band independent of which (v',J') level was pumped over the range v'= 0 → 6. Thermalization in IF(B) by He dominated all kinetic processes. Since most conceivable chemical excitation mechanisms will populate at least several v',J' levels, the observation of lasing from a thermalized distribution is evidence that relatively non-selective chemical excitation can be channelled into a single (v" = 0) level enhancing both the probability and the efficiency of lasing. Energy transfer experiments [20-22] have shown that thermalization can be rapidly attained, e.g. V-T collisions in IF(B) occur in single, $|\Delta v| = 1$, steps with rates that exceed electronic quenching by at least two orders of magnitude for all of the rare gases and N_2. Rotational relaxation occurs in large ($\Delta J \sim 10$) steps at nearly gas kinetic rates [22].

While the pulsed IF studies [18] answered critical questions concerning the kinetics of the upper laser level, the recent CW optically pumped demonstration by Davis and co-workers [19] has provided the definitive proof that relaxation within the v,J manifold in IF(X) can keep the terminal laser level essentially empty.

Although the above results clearly demonstrate the potential of IF, lasing by chemical excitation remains an unproven concept. Several schemes have been postulated. For example, energy transfer has been observed from $O_2(^1\Delta)$ [23,24], $NF(^1\Sigma)$ [25], $N_2(A^3\Sigma)$ [26] and "active" nitrogen [27]. In the remainder of this article we present recent results on $O_2(^1\Delta)$ transfer obtained in our laboratories. Singlet O_2 is of interest because efficient chemical production schemes are known as a result of the recent COIL investigations.

There have been only two reported investigations in which excited oxygen was used to excite IF(B). The first was by Clyne et al. [23] in 1972 when they reacted I + F + $O_2(^1\Delta)$. Although they proposed no definitive mechanism, they favoured the process

$$O_2(^1\Delta) + I(^2P_{3/2}) \rightarrow I(^2P_{1/2}) + O_2(^3\Sigma) \qquad (11)$$

$$I(^2P_{1/2}) + F + M \rightarrow IF(B) + M \qquad (12)$$

where M was a rare gas atom. $I(^2P_{1/2})$ and F might be expected to have an efficient three-body formation rate to IF(B) because they are the separated atom correlations of IF(B).

In more recent experiments Whitefield, Shea, and Davis [24] examined chemiluminescence from three reactions

$$I_2 + F_2 \qquad (13)$$

$$I_2 + F_2 + O_2(^1\Delta) \qquad\qquad (14)$$

$$I_2 + F + O_2(^1\Delta). \qquad\qquad (15)$$

Energy level diagrams of IF and O_2 are shown in Fig. 1. It is clear that neither $O_2(^1\Delta)$ nor $O_2(^1\Sigma)$ has sufficient energy to pump IF(X,v" = 0) directly to IF(B). Sequential collisions of $O_2(^1\Sigma)$ and $O_2(^1\Delta)$ with IF(X,v" = 0) could energetically transfer $\Delta E \sim 21,100$ cm^{-1}, which would produce IF(B,v'=5). Such a model involving sequential O_2* collisions requires an intermediate energy reservoir in the IF manifold. There are two likely candidates: IF(X,v" >> 0) and IF(A'$^3\Pi_2$). Since E-V energy transfer from $O_2(^1\Sigma)$, promoting IF(X,v" = 0) to IF(X,v" >> 0) would be expected to be improbable due to Franck-Condon considerations, it is an unlikely candidate.

In order that $O_2(^1\Delta)$ pump IF(X) to IF(A'), IF(X) must be vibrationally excited. Since IF(A') has $T_e \sim 13,250$ cm^{-1} (Ref. 28), only IF(X,v" ⩾ 9) can be

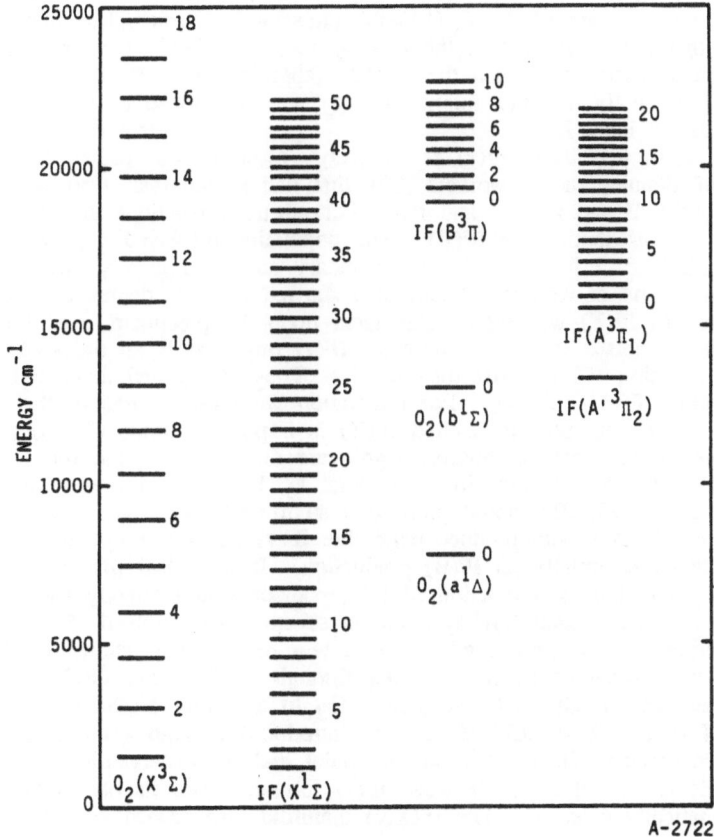

A-2722

FIGURE 1.   Energy level diagram of IF and O_2.

pumped to IF(A') by $O_2(^1\Delta)$ collisional transfer. Obviously, if T_e is higher then even more IF vibrational excitation is needed. It seems reasonable to expect the IF(A') can be pumped to IF(B) by an energy transfer collision from $O_2(^1\Delta)$. The efficiency of IF(B) production could critically depend upon the formation rate of IF(A'). Indeed, the rates for IF(A') formation could be considerably different depending upon whether $O_2(^1\Delta)$ or $O_2(^1\Sigma)$ is involved.

From the above considerations one can postulate two plausible models for IF(B) excitation by $O_2(^1\Sigma,^1\Delta)$. The first involves $O_2(^1\Sigma)$ pumping of IF(X,v" = 0) to IF(A') followed by promotion to IF(B) via an $O_2(^1\Delta)$ energy transfer collision. The second possible mechanism would use two $O_2(^1\Delta)$ molecules. As described in subsequent sections it is well known that the reactions F + IX → IF + X (X = I,Cl,Br), all partition energy into vibration of the IF(X) product [29]. It is energetically possible for $O_2(^1\Delta)$ to promote high lying IF(X,v) to the A' state. A secondary $O_2(^1\Delta)$ collision could produce IF(B).

Although no experimental evidence existed to support the $O_2(^1\Sigma)$ mechanism, the results of Whitefield et al. [24] were consistent with the IF(X,v) mechanism. They found that the IF(B → X) emission was most intense when the three reactants (I$_2$ + F + O$_2$*) were mixed at one spatial position in the flow tube. When the I$_2$ and F were mixed upstream of the $O_2(^1\Delta)$ the emission was much less intense. This suggested that the IF formed in the F + I$_2$ reaction initially contained internal energy that relaxed before it reached the $O_2(^1\Delta)$ (approximately 10 cm downstream). It is likely that this internal energy was partitioned as vibrationally excited IF which we label IF(v).

The catalyst to investigate IF(v) as an energy reservoir was provided by the detailed work of Wanner and co-workers [29], that had shown that IF(v) is produced in reactions between F and I$_2$, ICl and IBr. For all three reactions they observed bimodal vibrational distributions in IF(X). The population displayed a peak at v" = 0 then rapidly diminished for v' ⟨ 5. A secondary peak at higher v" was also observed. Wanner and co-workers [29,30] also showed that the degree of internal vibrational energy in IF(X) was highly dependent upon the precursors used for IF(X) production. They reported that the fraction of IF(v) produced is an order of magnitude greater using ICl and IBr than for I$_2$. They also found that CF$_3$I + F produces very little IF(v) (Ref. 30). For illustration we present some of their results in Figs. 2 and 3. If vibrationally excited IF(X) is important in the $O_2(^1\Delta)$ pumping process then one might expect to observe significant differences in the IF(B) production efficiencies for the three iodine donors: I$_2$, IBr, ICl. For example, based on the results of Ref. 29, IBr should produce a significant enhancement over I$_2$. CF$_3$I on the other hand should produce negligible IF(v) and our proposed IF(v) model would predict essentially no IF(B) production. It was clear that a systematic study of IF(B) production as a function of IX precursor would directly address several aspects of the proposed $O_2(^1\Delta)$ pumping scheme that involved IF(v).

The experiments were performed in the flow tube apparatus shown in Fig. 4. Fluorine atoms were provided by passing CF$_4$ through a microwave discharge, and $O_2(^1\Delta)$ molecules were produced by subjecting O$_2$ to a similar discharge. The iodine donor IX (where X = I,Cl,Br,CF$_3$) was introduced through a movable injector, thus the distance between the IX + F mixing point and the observation port was variable. In this manner the time between IF(v) formation and its observation point could be systematically changed. The IF(X,v) manifold was probed with a Molectron N$_2$ laser pumped dye laser using

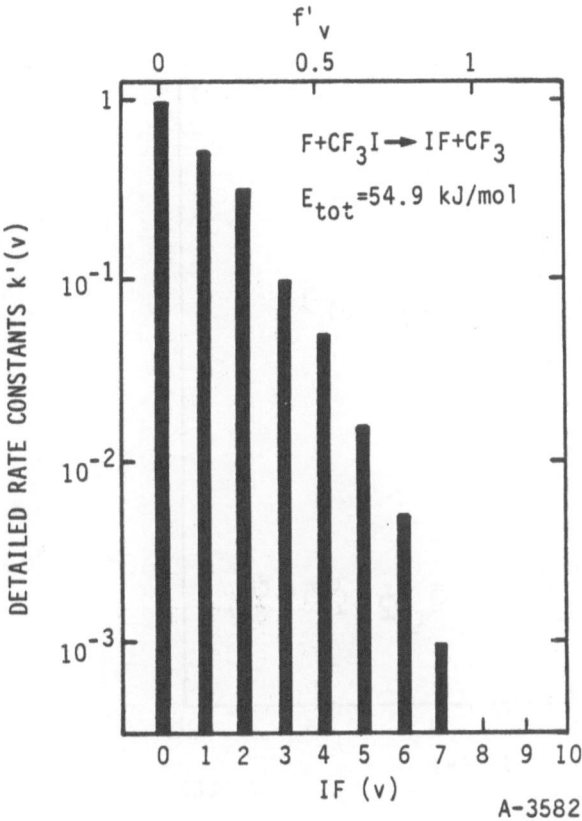

FIGURE 2. Plot of detailed rate constants, k_v, for partitioning of vibrational energy in IF(X) formed from F + CF$_3$I. Data taken from Ref. 30.

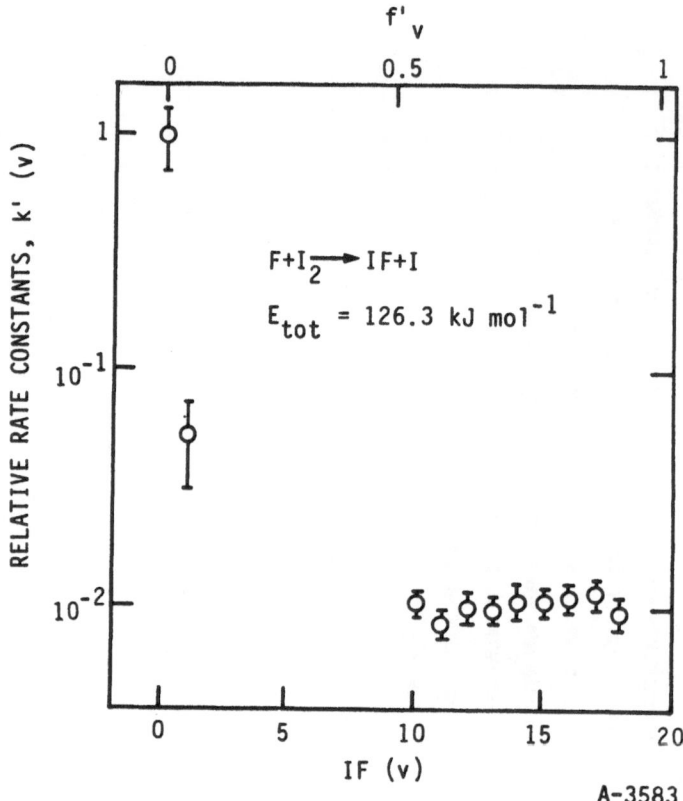

FIGURE 3. Plot of detailed rate constants, k_v, for partitioning of vibrational energy in IF(X) formed from $F + I_2$ reaction. Data taken from Ref. 29.

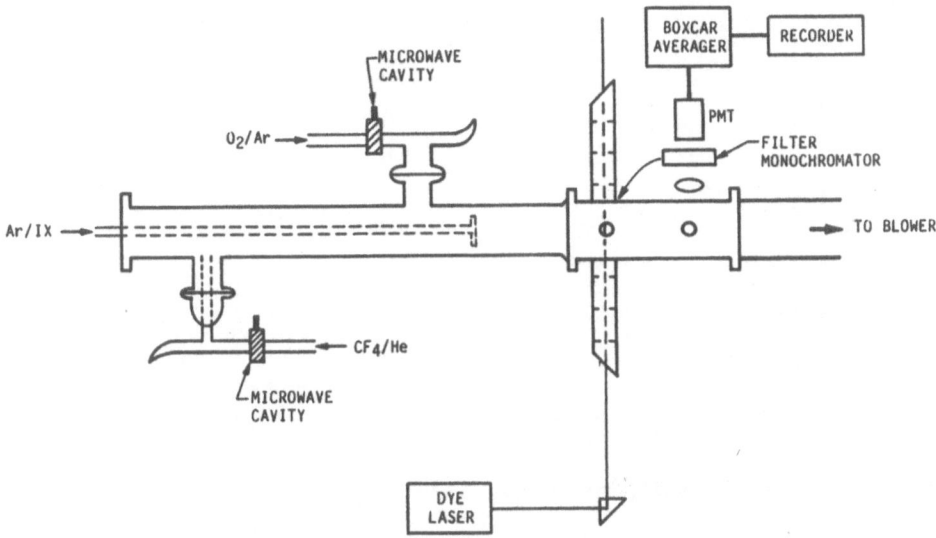

FIGURE 4. Block diagram of apparatus used for LIF studies. Flow tube reactor showing axial cross injector.

standard LIF techniques. Not shown in Fig. 4 is the detection system for monitoring IF(B → X) chemiluminescence which consisted of a 0.3 m monochromator and a photon counting system.

The model involving IF(v) is described by Eqs. 16 through 18:

$$IX + F \rightarrow IF(v) + X \tag{16}$$

$$IF(v) + O_2(^1\Delta) \rightarrow IF(A') + O_2 \tag{17}$$

$$IF(A') + O_2(^1\Delta) \rightarrow IF(B) + O_2. \tag{18}$$

Chemiluminescence techniques and laser induced fluorescence (LIF) were used to test this mechanism. The formation of IF(v) from reaction (16) was confirmed using LIF. Population distributions in IF(X,v) were mapped out, and $O_2(^1\Delta)$ was added to the flow. The subsequent production of IF(B) was monitored using chemiluminescence as a monitor. In this manner the causal relationship between IF(v) and IF(B) production was demonstrated. Some studies also probed the dependence of IF(B) production upon $[O_2(^1\Delta)]$.

In Fig. 5 we show representative LIF results using CF_3I and I_2 as iodine donors. Under flow tube conditions ($P_T \sim 1.0$ Torr) and a reaction time of 1 ms, we see that I_2 produced easily detectable IF(v) (v" = 8,9,10,11 are shown), while CF_3I produced none of these v" levels. This is consistent with Wanner's molecular beam results. We completed a series of these runs using ICl, IBr, I_2, and CF_3I

508

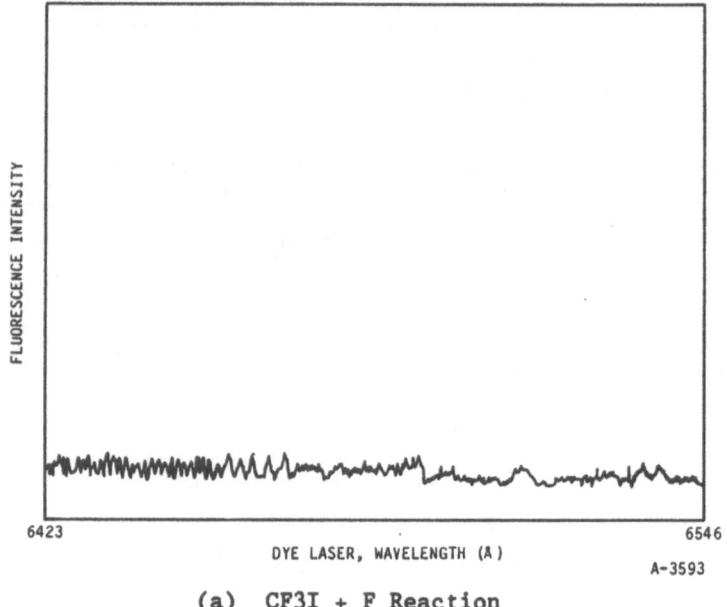

DYE LASER, WAVELENGTH (A)

A-3593

(a) CF3I + F Reaction

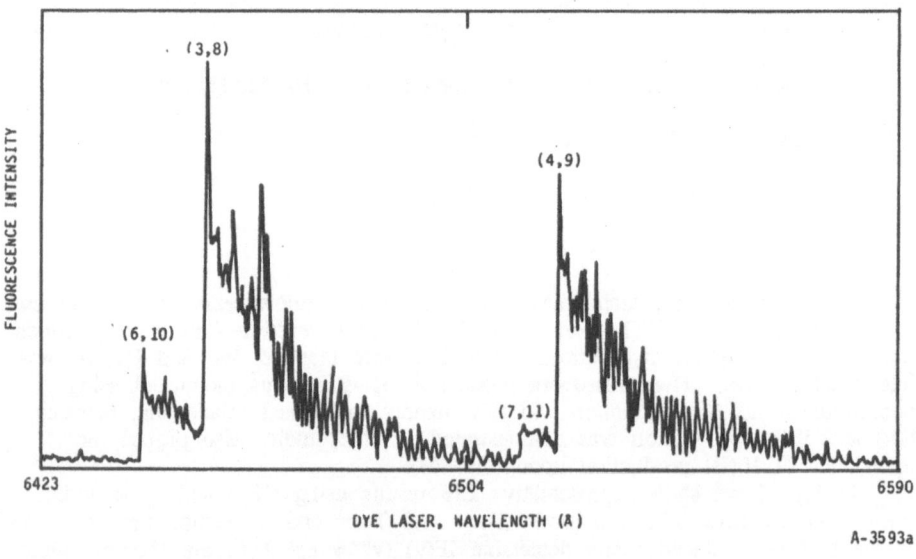

DYE LASER, WAVELENGTH (A)

A-3593a

(b) I_2 + F Reaction

FIGURE 5. Comparison of LIF excitation spectra using CF_3I + F (upper panel) and I_2 + F (lower panel) reactions when dye laser was probing hot vibrational levels in IF(X).

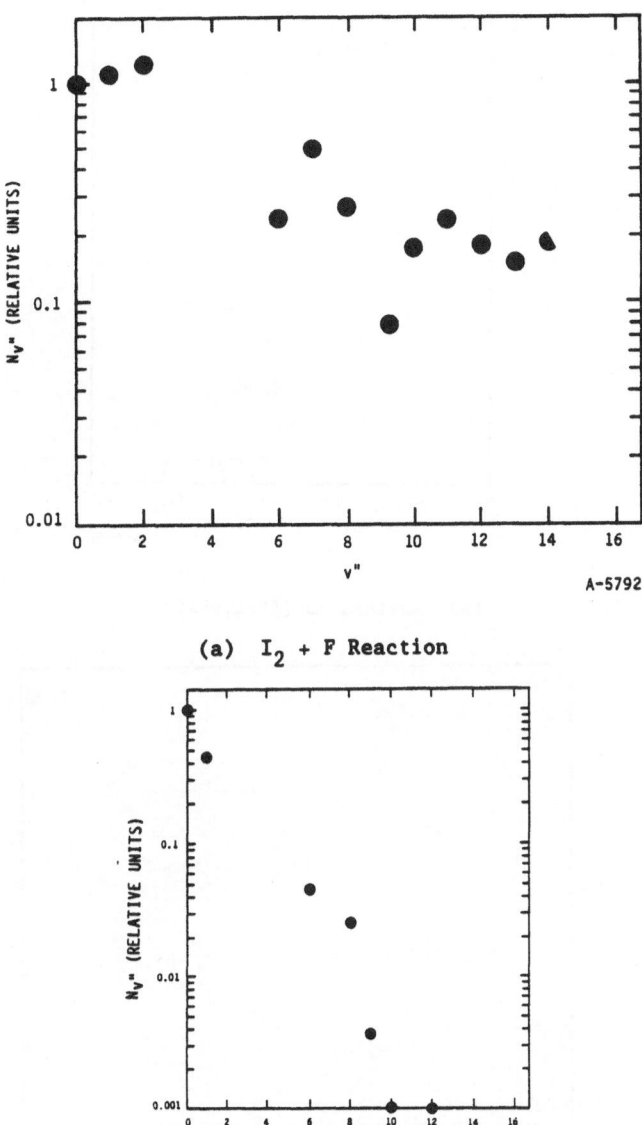

(a) I_2 + F Reaction

(b) ICl + F Reaction

FIGURE 6. Vibrational distribution of IF(X) produced from IX + F reaction. Flow tube pressure was 1 Torr.
Upper panel: I_2 + F reaction. Lower panel: ICl + F reaction

510

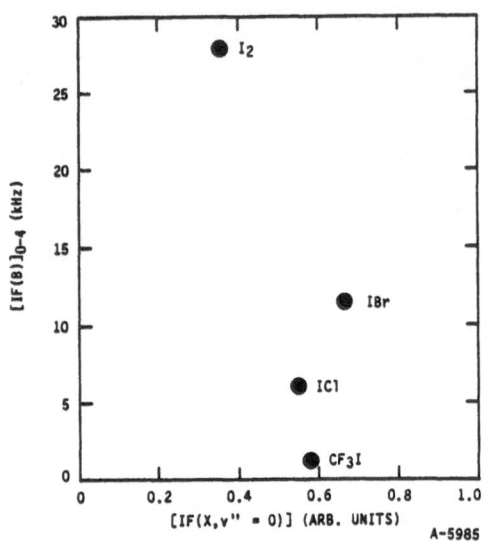

(a) Absissa is [IF(X,v"=0)

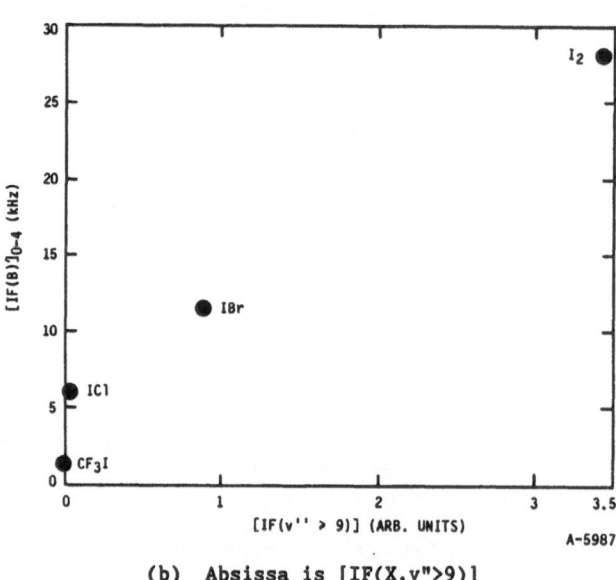

(b) Absissa is [IF(X,v"$\geq$9)]

FIGURE 7. [IF(B)] produced by O_2* pumping as a function of IF(v) at constant $O_2(^1\Delta)$. Labels indicate IX species used.
Upper panel: Abscissa is [IF(X,v"=0)]
Lower panel: Abscissa is [IF(X,v"$\geq$9)]

and extracted vibrational distributions in IF(X) produced by each IX donor. Consequently we were able to produce well characterized flows of IF(v) to which $O_2(^1\Delta)$ could be added. Distributions produced using I_2 and ICl are shown in Fig. 6. When equal flows of $O_2(^1\Delta)$ were added to each of the IF(v) flows, IF(B → X) chemiluminescence was produced. By measuring the chemiluminescence intensity as a function of the IF(v) distribution we were able to correlate the IF(B) production with IF(v).

From the IF energy levels shown in Fig. 1 we predict that only v" ≳ 9 should be excitable to IF(A') via $O_2(^1\Delta)$ E-E transfer. Our LIF results showed that the fraction of [IF(v ≳ 9)]/[IF(v=0)] was dependent upon the iodine donor in the order I_2 > IBr > ICl > CF_3I. In Fig. 7 we see that there is no apparent correlation between [IF(X,v"=0)] and IF(B) production, while a strong correlation exists when [IF(B)] is plotted versus [IF(X,v" ≳ 9)]. The non-zero intercept for CF_3I is attributed to a small contribution from $O_2(^1\Sigma)$ excitation of IF(X,v" = 0) to IF(A'). (While $O_2(^1\Sigma)$ does participate in IF(B) excitation its role is apparently minor and it is not discussed further here. These experiments will be published in detail elsewhere). These results provide strong evidence for the participation of IF(v) as an energy reservoir in the excitation of IF(B) by $O_2(^1\Delta)$. Although the IF(A') intermediate state could not be observed in the present work the proposed model is consistent with our results. It is clear that IF(v) opens reaction channels that would otherwise be inaccessible.

3. CONCLUDING REMARKS

The results described above illustrate some important aspects of selective chemical excitation of electronic states. First, the production of IF(v) from IX + F reactions represents the fairly general phenomenon of ground state vibrational excitation in simple A + BC exchange reactions and is analogous to HF laser chemistry. Efficient excitation of excited electronic states is only achieved by energy transfer from a metastable species, i.e. $O_2(^1\Delta)$. The O_2-IF system provides an interesting example in which these two processes are both utilized. The vibrational excitation in IF(v) opens pathways for $O_2(^1\Delta)$ transfer that would otherwise be energetically inaccessible. Whether this system can eventually be developed into an actual laser system remains to be seen. It does however provide interesting insights into selective chemical excitation of electronic states. In this brief survey we have attempted to discuss important issues relevant to chemical excitation of electronic states. The dominant obstacle that has delayed the development of short wavelength chemical lasers is the paucity of reactions that efficiently produce excited electronic states, and this remains the area needing the most attention.

ACKNOWLEDGEMENTS

A portion of this work was supported by the Air Force Weapons Laboratory, Kirtland AFB, NM under contract number 86-C-0017.

REFERENCES

[1] Ch. Ottinger and R.N. Zare, Chem. Phys. Lett., **5** (1970) 243.

[2] C.R. Jones and H.P. Broida, J. Chem. Phys., **60** (1974) 4369.

[3] C.R. Dickson, S.M. George and R.N. Zare, J. Chem. Phys., **67** (1977) 1024.

[4] Y.C. Hsu and J.G. Pruett, J. Chem. Phys., **76** (1982) 5849.

[5] L.E. Wilson, S.N. Suchard and J.I. Steinfeld, Electronic Transition Lasers, MIT Press, Cambridge, MA (1976).

[6] L.E. Wilson, S.N. Suchard and J.I. Steinfeld, Electronic Transition Lasers II, MIT Press, Cambridge, MA (1977).

[7] K.E. Shuler, J. Chem. Phys., **21** (1953) 624.

[8] J.M. Herbelin and N. Cohen, Chem. Phys. Lett., **20** (1973) 605.

[9] C.T. Chea and M.A.A. Clyne, J. Photochem., **15** (1981) 21.

[10] R.J. Malins and D.W. Setser, J. Phys. Chem., **85** (1981) 1342.

[11] R.G. Derwent and B.A. Thrush, J. Chem. Soc. Faraday Trans. II, **68** (1972) 720.

[12] W.E. McDermott, N.R. Pchelkin, D.J. Benard and R.R. Bousek, Appl. Phys. Lett., **32** (1978) 469.

[13] D.J. Benard, W.E. McDermott, N.R. Pchelkin and R.R. Bousek, Appl. Phys. Lett., **34** (1979) 40.

[14] R.F. Heidner, D.E. Gardner, G.I. Segal and T.M. El-Sayed, J. Phys. Chem., **87** (1983) 2348.

[15] R. Bacis and S. Churassy, Gas Flow and Chemical Lasers, Springer Verlag, New York, (1987).

[16] J. Bachar and S. Rosenwaks, Appl. Phys. Lett., **41** (1982) 16.

[17] M.A.A. Clyne and I.S. McDermid, J. Chem. Soc. Faraday Trans. II, **74** (1978) 1644.

[18] S.J. Davis, L. Hanko and R.F. Shea, J. Chem. Phys., **78** (1983) 172.

[19] S.J. Davis, L. Hanko and P.J. Wolf, J. Chem. Phys., **82** (1985) 4831.

[20]   P.J. Wolf, J.H. Glover, L. Hanko, R.F. Shea and S.J. Davis, J. Chem. Phys., **82** (1985) 2321.

[21]   P.J. Wolf and S.J. Davis, J. Chem. Phys., **83** (1985) 91.

[22]   P.J. Wolf and S.J. Davis, J. Chem. Phys., **87** (1987) 3492.

[23]   M.A.A. Clyne, J.A. Coxon, and L.W. Townsend, J. Chem. Soc. Faraday Trans. II, **68** (1972) 2134.

[24]   P.D. Whitefield, R.F. Shea, and S.J. Davis, J. Chem. Phys., **78** (1983) 6793.

[25]   A.T. Pritt, D. Patel and D.J. Benard, Chem. Phys. Lett., **97** (1983) 471.

[26]   L.G. Piper, W.J. Marinelli, W.T. Rawlins and B.D. Green, J. Chem. Phys., **83** (1985) 5602.

[27]   L.G. Piper, W.J. Marinelli, B.D. Green, W.T. Rawlins, H.C. Murphy, M.E. Donahue and P.F. Lewis, PSI Report TR-460 (Air Force Contract No. F29601-83-C-0051) (February 1985).

[28]   J.P. Nicolai and M.C. Heaven, J. Chem. Phys., **87** (1987) 3304.

[29]   T. Trickl and J. Wanner, J. Chem. Phys., **78** (1983) 6091.

[30]   L. Stein and J. Wanner, J. Chem. Phys., **72** (1980) 1128.

THE CHARACTERISATION OF THE MECHANISM OF IF(B) PRODUCTION IN FLUORINE/ IODIDE SYSTEMS

D. Raybone, T.M. Watkinson and J.C. Whitehead
Department of Chemistry
Manchester University
Manchester M13 9PL
U.K.

ABSTRACT. It is shown that it is possible to characterise the mechanism of IF(B) formation in low pressure (< 2 mbar) gaseous fluorine/iodide mixtures by inspecting the form of the IF(B) vibrational state distribution. A wide range of systems have been studied, but these can be divided into just three classes each with a distinct mechanism of IF(B) formation.

1.   INTRODUCTION

The search for a visible chemical laser system has recently been concentrated on the interhalogen molecules and in particular IF. Clyne and McDermid [1,2] suggested in 1977 that an electronic transition laser could be made operating on the $B(^3\Pi O^+) \rightarrow X(^1\Sigma^+)$ system of IF and Davis et al. [3,4] have demonstrated pulsed and CW laser operation on the B → X system using optical pumping of ground-state IF produced by the gas-phase reaction of F_2 with I_2 at ~ 1-27 mbar. Because of the relative shift in the B and X state potential curves in IF, it should be possible to develop a true chemical laser by direct production of excited IF(B) provided that a sufficiently high concentration of IF(B) can be achieved (> 1 x 10^{-3} mbar) [4]. Various chemiluminescent reactions directly producing IF(B) have been investigated [5-13] and it has also been shown that high yields of IF(B) can be produced in various fluorine/iodide systems on adding $O_2(^1\Delta_g)$ [14,15], NF($^1\Sigma$) [16,17], $N_2(^3\Sigma_u^+)$ [18] and vibrationally-excited HF($^1\Sigma^+$) [19]. In this paper, we show that by inspecting the forms of the IF(B) vibrational state distribution produced in a wide range of systems it is possible to categorise the distributions into a small number of groups each of which has a common mechanism for IF(B) formation.

2.   RESULTS

The IF(B) vibrational state distributions resulting from a variety of experiments performed by various groups are shown in Fig. 1 and the experimental conditions are given in the figure caption. All these experiments were conducted at very similar pressures (0.5-2.3 mbar) where the effect of collisional relaxation on the IF(B) vibrational distribution during the radiative lifetime (~ 7 μs) would be expected

J. C. Whitehead (ed.), Selectivity in Chemical Reactions, 515–523.
© 1988 by Kluwer Academic Publishers.

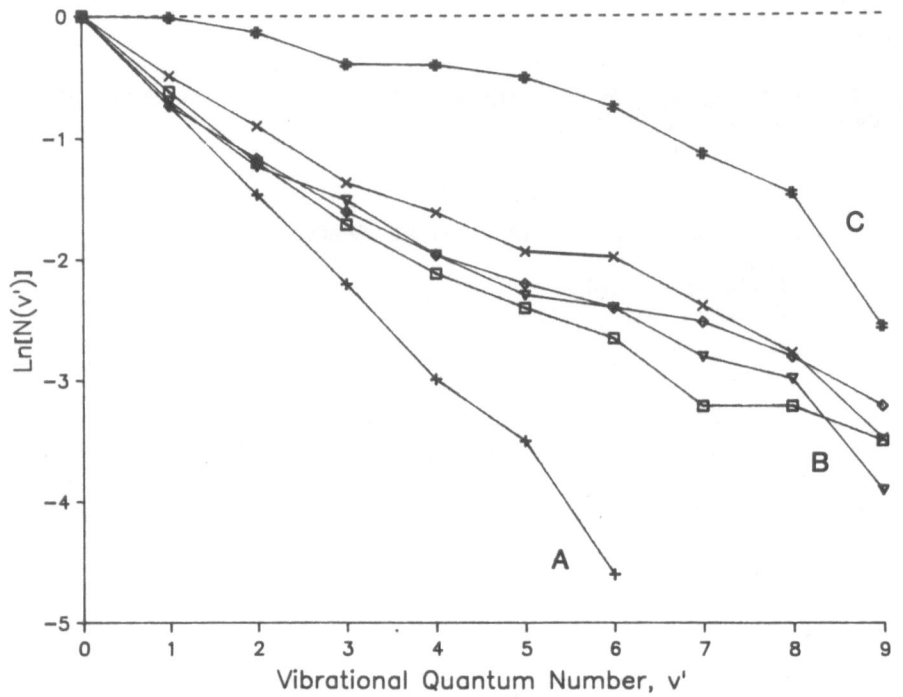

FIGURE 1. Logarithmic plots of the relative IF(B) vibrational populations resulting from a range of experiments

Symbol	System	Pressure/mbar	Ref.
+	$F_2 + I_2$	0.6	this work
∇	$F + I_2$	0.6	this work
X	$F + I_2 + O_2(^1\Delta)$	2.3	[14]
□	$F + Cl_4$	0.7	[11]
◊	$F + BiI_3$	0.6	[13]
#	$IF(X) + N_2(A)$	0.5	[18]

to be very similar in all cases. Studies on the quenching and relaxation of IF(B) by various gases [20-22] have shown that vibrational relaxation of IF(B) is typically 100 times more efficient than electronic quenching. Vibrational relaxation proceeds predominantly *via* a $\Delta v = -1$ cascade process with a cross section that is $\leqslant 1\%$ of gas kinetic. At the pressures studied here ($\leqslant 2$ mbar) some vibrational relaxation will take place but not to the extent that would cause thermalisation. Thus, these distributions are likely to be nascent or near-nascent [14]. We can classify the distributions into three distinct groupings. In the first (A) the distribution is characterised by a Boltzmann temperature of ~ 800K. The second group (B) has a vibrational distribution which is approximately Boltzmann (T_{vib} ~ 1200K) for $v' \leqslant 4$ but has an excess population for $v' > 4$. These distributions also show a pronounced "kink" at $v' = 6$ or 7. The final class (C) involves a very hot non-Boltzmann distribution. In both classes B and C the IF(B) vibrational distributions extend out to v' = 9. Levels above v' = 9 are absent due to predissociation of the B state of IF [2].

3. DISCUSSION

3.1 Case A

This 800K Boltzmann distribution is found in the low pressure reactions of F_2 with I_2 [7,14] where at even lower pressures (~ 0.01 Torr) emission is also observed from the long-lived A $(^3\prod(1))$ state of IF. It was suggested by Valentini *et al.* [23] that the stable trihalogen molecule I_2F which they observe in their crossed molecular beam studies of the endoergic reaction $F_2 + I_2$ might provide the route for IF^* (A or B) production in the F_2/I_2 flow system by the following steps

$$F_2 + I_2 \rightarrow I_2F + IF \qquad (1)$$

$$F + I_2F \rightarrow IF^* + IF. \qquad (2)$$

The presence of the I_2F intermediate in flowing systems has recently been confirmed [24].

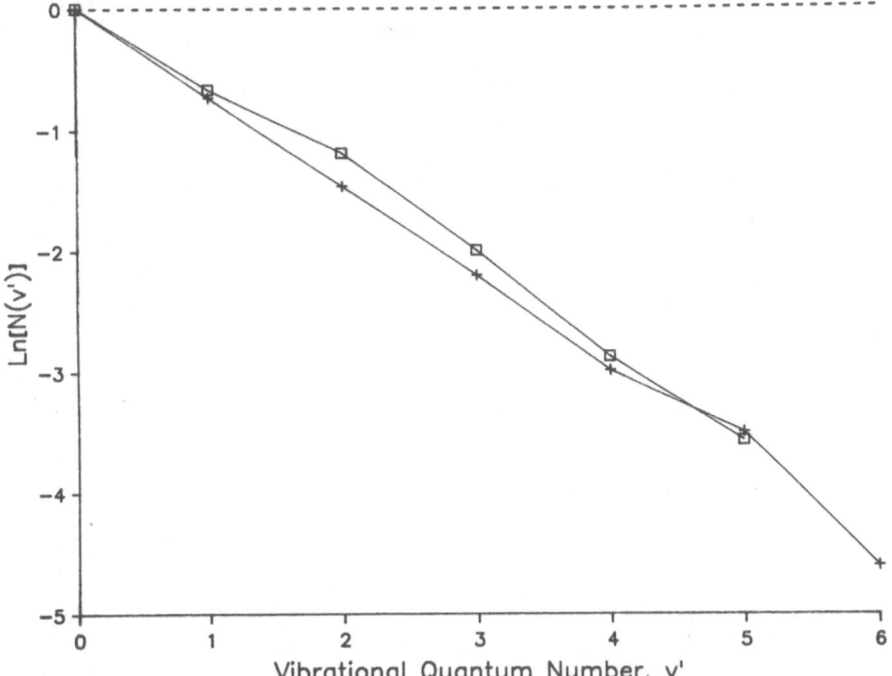

FIGURE 2. A comparison of the IF(B) vibrational state distributions resulting from the low pressure (0.6 mbar) flow cell reaction of $F_2 + I_2$ (+) and from the crossed molecular beam reaction $F + I_2$ (□).

IF* (A and B) has also been observed in the crossed molecular beam reaction F + I$_2$ [6]. This IF(B) vibrational state distribution is essentially identical to those measured in the low pressure F$_2$/I$_2$ flow systems (Fig. 2) and can be taken to indicate that in the beam reaction IF* is also produced by the consecutive reactions (1) and (2). This point is discussed in the following paper by Trautmann *et al.* [25] and it is thought that undissociated F$_2$ in the F atom beam source is the precursor of I$_2$F.

3.2 Case B

This very characteristic vibrational distribution was first observed in a F + I$_2$ flame to which metastable O$_2$($^1\Delta$) had been added [14,15]. It was suggested that IF(B) was produced by sequential collisional excitation ("energy laddering") of vibrationally-excited ground-state IF(X) from the reaction

$$F + I_2 \rightarrow IF(X) + I \tag{3}$$

in collisions with O$_2$($^1\Delta$)

$$IF(X) + O_2(^1\Delta) \rightarrow IF^{**} + O_2(^3\Sigma) \tag{4}$$

$$IF^{**} + O_2(^1\Delta) \rightarrow IF(B) + O_2(^3\Sigma), \tag{5}$$

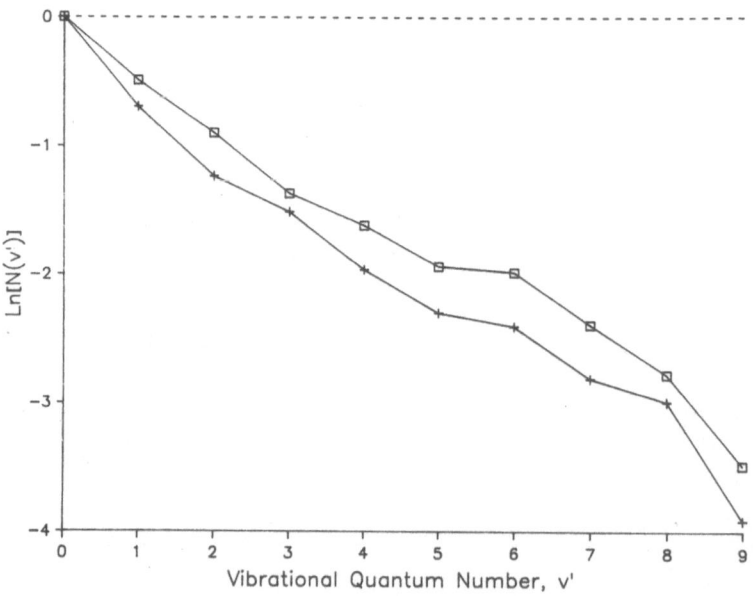

FIGURE 3. A comparison of the IF(B) vibrational state distributions resulting from the low pressure flow cell reactions of F + I$_2$ with (□) and without (+) the addition of O$_2$($^1\Delta$).

where IF** might be some electronic state of IF intermediate in energy between X and B, probably the A' state. However, we have shown [11-13] that identical IF(B) vibrational distributions can be produced from a wide range of F atom reactions with various iodides (I$_2$, organic and inorganic) <u>without</u> the addition of any O$_2$($^1\Delta$). This is clearly illustrated in Fig. 3 in which the IF(B) vibrational distributions from the reaction F + I$_2$ with and without O$_2$($^1\Delta$) are compared. The total IF(B) yield is greater when O$_2$($^1\Delta$) is added, but the form of the IF(B) vibrational distributions (both having a kink at v' = 6) is identical suggesting the same mechanism of IF(B) formation in both cases. A similar comparison has been made for IF(B) produced in the reaction of F + BiI$_3$ with and without the addition of O$_2$($^1\Delta$) [13].

Clearly, these distributions are characteristic of systems where there are fluorine atoms present, but there has to be an alternative explanation to that suggested by Davis et al. [14,15] {Reactions (3-5), where IF(B) is produced by collisional excitation of IF(X) by O$_2$($^1\Delta$)} in order to rationalise our observation that O$_2$($^1\Delta$) is not necessary to produce such distributions. We have suggested [26] that in addition to F atoms, the other key species responsible for IF(B) production is electronically-excited iodine atoms I*(^{2}P$_{1/2}$). These can be produced in such systems by a reaction of the form

$$F + RI \rightarrow RF + I^* \qquad (6)$$

which has an effective rate constant of 3.4 x 10^{-13} cm^3 molec^{-1} s^{-1} for R = CH$_3$ [27,28] and of 1.3 x 10^{-11} cm^3 molec^{-1} s^{-1} for R = I [29,30]. The increased yield of IF(B) upon the addition of O$_2$($^1\Delta$) to a fluorine/iodine system comes from the efficient pumping [31] of ground state iodine atoms, I(^{2}P$_{3/2}$), which are known [26] to be present in these systems, to give an increase in the I*(^{2}P$_{1/2}$) concentration

$$I(^2P_{3/2}) + O_2(^1\Delta) \rightarrow I^*(^2P_{1/2}) + O_2(^3\Sigma) \qquad (7)$$

The mechanism by which IF(B) is produced from I* and F atoms is not yet established. We have suggested [26] that IF(B) might be formed by a two or three body recombination process

$$I^* + F \rightarrow IF(B). \qquad (8)$$

Such a process would favour the formation of high vibrational levels of IF(B) which when partially relaxed, might give a vibrational distribution of the form B. In general, the rates of two-body or three-body recombination reactions would not be sufficient to account for the observed yields of IF(B). One possibility is that an exciplex of the excited iodine atom with the iodide, (I*...RI), might be formed [32] and the subsequent reaction of this exciplex with a fluorine atom would enhance the overall rate for IF*(B)

$$I^* + RI \rightarrow (I^*...RI) \qquad (9a)$$

$$F + (I^*...RI) \rightarrow IF^* + RI. \qquad (9b)$$

The concerted mechanism of reactions (9a) and (9b) could be regarded as an extreme of three-body recombination with an enhanced cross section due to the formation of an intermediate complex.

Another mechanism might be that IF(X) produced in the reaction

$$F + RI \quad \rightarrow \quad IF(X) + R \qquad\qquad (10)$$

is pumped to IF(B) in a sequence similar to reactions (4) and (5) but by $I^*(^2P_{1/2})$ rather than $O_2(^1\Delta)$

$$IF(X) + I^* \quad \rightarrow \quad IF(A') + I \qquad\qquad (11)$$

and

$$IF(A') + I^* \quad \rightarrow \quad IF(B) + I. \qquad\qquad (12)$$

The intermediate IF(A') state has recently been identified [33] in matrix studies as the metastable $^3\Pi(2)$ state lying at an energy of 13250 cm^{-1}. The excitation energy of an iodine atom (7603 cm^{-1}) requires that either the IF(X) in reaction (10) is already vibrationally-excited (v'' > 9) or that an additional pumping step is needed to produce vibrationally-excited IF(X). Reaction (12) has energetics such that near-resonant energy transfer will excite IF(A', v' = 0) to IF(B, v' = 5) [33] which might account for the excess populations in levels v' $\geqslant$ 4 compared with the Boltzmann populations of the lower v' levels and also for the "kink" in v' $\sim$ 6.

Clearly, there is a need to directly determine the IF(B) vibrational population distributions that result from the recombination process (8) (there is evidence [34] that levels up to v' = 9 are produced) and from the collisional pumping of IF(X) by I^* to definitively establish which mechanism for IF(B) production is correct. However, it is clear that IF(B) is not produced in these systems directly via collisional excitation of IF(X) by $O_2(^1\Delta)$. As yet, there have not been any IF(B) photon yield measurements for the F atom/iodide systems other than F + I_2 + $O_2(^1\Delta)$ [14] where a value of 0.3% was measured indicating an IF(B) concentration of $\sim$ 7 x 10^{-5} mbar, significantly less than that required for laser action [4].

3.3 Case C

In this category, a very hot non-Boltzmann IF(B) vibrational distribution is obtained. Examples of such distributions are found in systems where excited species such as metastable electronically-excited $N_2(A^3\Sigma_u^+)$ [18] and $NF(b^1\Sigma)$ [16] and vibrationally-excited $HF(X^1\Sigma^+)$ [19] collide with IF(X). In all cases, the excitation is thought to come from a single step E → E or V → E energy transfer process

$$IF(X) + M^* \text{ or } \neq \quad \rightarrow \quad IF^*(B) + M. \qquad\qquad (13)$$

The efficiency of this process decreases in the sequence N_2^*, NF^*, $HF^{\neq}$ where the rate constants for collisional excitation of IF(B) are 8.0 x 10^{-11}, 1.2 x 10^{-12} and 3.3 x 10^{-14} cm^3 $molec^{-1}$ s^{-1}, respectively [18,35,19]. This shows some correlation with the amount of excess energy of the collision partner compared with the IF (B ← X) excitation energy (3.5 eV for N_2^*, $\sim$ 0 for NF^* and $\sim$ 0 for $HF^{\neq}$, where the effective vibrational levels are not known and the excitation may even be multi-step).

For the case of excitation of IF(X) by $NF(b^1\Sigma)$ Herbelin [36] has developed a model to explain the IF(B) vibrational distribution. The model assumes that the dominant forces in the energy transfer come from long-range interactions in which a resonant energy transfer between NF^* and IF is followed by a charge transfer in which the resulting ion pair (NF^- - IF^+) then separates to give NF(X) and IF(B). The final IF(B) vibrational distribution depends on the IF(B-X) Franck-Condon factors and the initial IF(X) vibrational distribution. Because the IF(B-X) Franck-Condon factors favour transitions from low vibrational levels of the X state into high vibrational levels of the B state, it can be shown that NF(b) preferentially pumps IF(X) into high vibrational levels of IF(B) as is observed experimentally.

It would appear, however, that near resonant transfer of the energy stored in metastable species (which can be chemically prepared) such as $NF(b^1\Sigma)$ is not a particularly effective way of producing IF(B) [35]. This suggests that this mechanism does not appear to be a promising route for channelling chemical energy into visible laser output.

4. CONCLUSIONS

We have shown that for low pressure gas systems (< 2 mbar), it is possible to categorise the mechanism of IF(B) formation by inspecting the form of the IF(B) vibrational state distribution. At these pressures, the effects of vibrational relaxation are minimised and the measured distributions are near to their nascent forms and hence characteristic of their mode of formation. We find that all the systems studied so far can be divided into three broad categories. These are bimolecular chemiluminescent reactions of F_2, creation of IF(B) by fluorine atoms and spin-orbit excited iodine atoms and direct collisional excitation of IF(X) to IF(B) by an energy rich electronically or vibrationally excited species. Such a categorisation should assist the development and identification of promising schemes for a visible chemical IF (B-X) laser.

This work was supported by USAFOSR (grant no. AFOSR-85-0039).

REFERENCES

[1] M.A.A. Clyne and I.S. McDermid, J. Chem. Soc. Faraday Trans. 2, **73** (1977) 1094.

[2] M.A.A. Clyne and I.S. McDermid, J. Chem. Soc. Faraday Trans. 2, **74** (1978) 1644.

[3] S.J. Davis, L. Hanko and R.F. Shea, J. Chem. Phys., **78** (1983) 172.

[4] S.J. Davis, L. Hanko and P.J. Wolf, J. Chem. Phys., **82** (1985) 4831.

[5] G. Schatz and M. Kaufmann, J. Phys. Chem., **76** (1972) 3586.

[6] T. Trickl and J. Wanner, J. Chem. Phys., **74** (1981) 6508.

[7] J.W. Birks, S.D. Gabelnick and H.S. Johnston, J. Mol. Spec., **57** (1975) 23.

[8] C.C. Kahler and Y.T. Lee, J. Chem. Phys., **73** (1980) 5122.

[9] R.C. Estler, D. Lubman and R.N. Zare, Faraday Discuss. Chem. Soc., **62** (1977) 317.

522

[10] R.D. Coombe and R.K. Horne, J. Phys. Chem., 83 (1979) 2435.

[11] H.S. Braynis, D. Raybone and J.C. Whitehead, J. Chem. Soc. Faraday Trans. 2, 83 (1987) 627.

[12] H.S. Braynis, D. Raybone and J.C. Whitehead, J. Chem. Soc. Faraday Trans. 2, 83 (1987) 639.

[13] D. Raybone, T.M. Watkinson and J.C. Whitehead, J. Chem. Soc. Faraday Trans. 2, 83 (1987) 767.

[14] P.D. Whitefield, R.F. Shea and S.J. Davis, J. Chem. Phys., 78 (1983) 6793.

[15] P.D. Whitefield, J. Photochem., 25 (1984) 465.

[16] A.T. Pritt, D. Patel and D.J. Benard, Chem. Phys. Lett., 97 (1983) 471.

[17] H. Cha and D.W. Setser, J. Phys. Chem., 91 (1987) 3758.

[18] L.G. Piper, W.J. Marinelli, W.T. Rawlins and B.D. Green, J. Chem. Phys., 83 (1985) 5602.

[19] G.W. Tregay, J.W. Raymonda, H.M. Thompson and T.E. Furner, Chem. Phys. Lett., 123 (1986) 458.

[20] P.J. Wolf, J.H. Glover, L. Hanko, R.F. Shea and S.J. Davis, J. Chem. Phys., 82 (1985) 2321.

[21] P.J. Wolf and S.J. Davis, J. Chem. Phys., 83 (1985) 91.

[22] P.J. Wolf and S.J. Davis, J. Chem. Phys., 87 (1987) 3492.

[23] J.J. Valentini, M.J. Coggiola and Y.T. Lee, Faraday Discuss. Chem. Soc., 62 (1977) 232.

[24] H.V. Lillenfeld and G.R. Bradburn, in Gas Flow and Chemical Lasers (Springer-Verlag, 1987).

[25] M. Trautmann, T. Trickl and J. Wanner, 'IF (A-X, B-X) Chemiluminescence of Fluorine-Iodide Systems in a Crossed Molecular Beam Experiment', the following paper in this volume.

[26] D. Raybone, T.M. Watkinson and J.C. Whitehead, Chem. Phys. Lett., 135 (1987) 170.

[27] R.S. Iyer and F.S. Rowland, J. Phys. Chem., 85 (1981) 2488.

[28] T.V. Venkitachalam, P. Das and R. Bersohn, J. Am. Chem. Soc., 105

(1983) 7452.

[29] E.H. Appelman and M.A.A. Clyne, J. Chem. Soc. Faraday Trans. 1,
      71 (1975) 2072.
[30] H. Brunet, Ph. Chauvet, M. Mabru and L. Torchin, Chem. Phys.
      Lett., **117** (1985) 371.
[31] R.G. Derwent and B.A. Thrush, Faraday Discuss. Chem. Soc., **53**
      (1972) 162.

[32] E. Gerck, Opt. Comm., **41** (1982) 102.

[33] J.P. Nicolai and M.C. Heaven, J. Chem. Phys., **87** (1987) 3304.

[34] M.A.A. Clyne, J.A. Coxon and L.W. Townsend, J. Chem. Soc. Faraday
      Trans. 2, **68** (1972) 2134.

[35] R.A. Young, J. Blauer, R. Bower and C.L. Lin, J. Chem. Phys., **87**
      (1987) 4634.

[36] J. Herbelin, Chem. Phys. Lett., **133** (1987) 331.

IF(A-X, B-X) CHEMILUMINESCENCE OF FLUORINE-IODIDE SYSTEMS IN A CROSSED MOLECULAR BEAM EXPERIMENT

M. Trautmann[+], T. Trickl[++], J. Wanner
Max-Planck-Institut für Quantenoptik
8046 Garching
Federal Republic of Germany

Very weak visible IF(A-X) and (B-X) chemiluminescence was observed in connection with the F + I_2, ICl, IBr reaction systems as studied in a single-stage molecular beam apparatus with two crossed effusive nozzles. In an earlier communication [1] we reported a first attempt at spectral resolution of the light emission associated with the F + I_2 system. There we attributed the spectrum to the reaction of F atoms with the trihalogen radical I_2F. The existence of this light-emitting reaction step had been verified in a molecular beam experiment by Kahler and Lee, yet without recording a spectrum of the emitter. In their experiment the trihalogen radicals, e.g. XIF (X=Cl,I), were formed by reactions of supersonically seeded F_2 above distinct translational threshold energies according to: F_2 + XI → XIF + F. It could be shown that this step including the subsequent reaction of F + XIF → XF + IF obeys bimolecular dynamics and hence proceeds as an elementary reaction [2]. We were able to extend our earlier work with a more sensitive detection system [3]. Hence at a collision energy of ~3.2 kJ mol^{-1} it was possible to obtain a vibrational product state analysis of the IF(B) state. The chemiluminescence spectrum for the reaction F + I_2F is depicted in Fig. 1a.

Fig. 1b shows the corresponding IF(B) vibrational product state distribution. The population limit closely corresponds to the enthalpy according to ΔH_0^0 = -248.5 kJ mol^{-1}. The distribution is characterized by a "vibrational temperature" T_v ~800 K. The latter observation is essentially identical with results of Whitehead *et al.* for a low-pressure I_2/F_2 flame [4], where the reaction mechanism of Lee *et al.* is also anticipated. In our experiment the above reaction sequence is most likely induced by the approximately 5% undissociated F_2 internally excited in the microwave discharge.

[+] Wilhelmsgymnasium, München, [++] Department of Chemistry, University of California, Berkeley, CA 94720, USA.

J. C. Whitehead (ed.), Selectivity in Chemical Reactions, 525–529.
© *1988 by Kluwer Academic Publishers.*

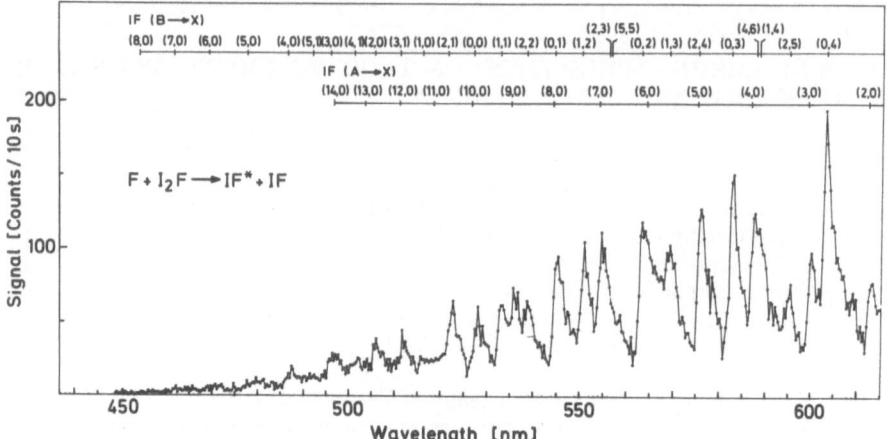

FIGURE 1a.   Chemiluminescence spectrum of the reaction $F + I_2F$ as observed from a crossed molecular beam experiment of $F(F_2) + I_2$.

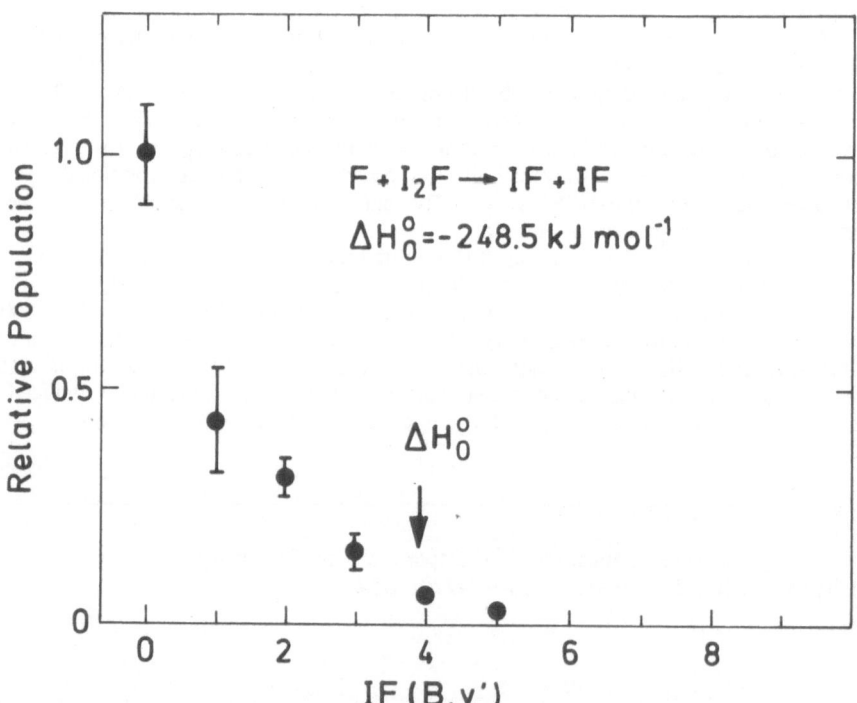

FIGURE 1b.   The IF(B) vibrational product state distribution in agreement with results of Whitehead *et al.* for a low-pressure I_2/F_2 flame confirms the work of Kahler and Lee [2].

Much weaker chemiluminescence spectra were recorded from the systems F + ICl and IBr only at iodide flows approximately one order of magnitude higher (F_2 = 18 SCCM in all experiments; I_2 = 1.4; ICl = 17; IBr ~10 SCCM). Under these conditions a further increase in iodide flow or addition of Ar resulted in a strong nonlinear increase in chemiluminescence intensity without changing the spectral distribution, whereas at iodide flows of ~2 SCCM the overall chemiluminescence signal showed a linear flow dependence. The spectra for the F/ICl and F/IBr systems are shown in Fig. 2a.

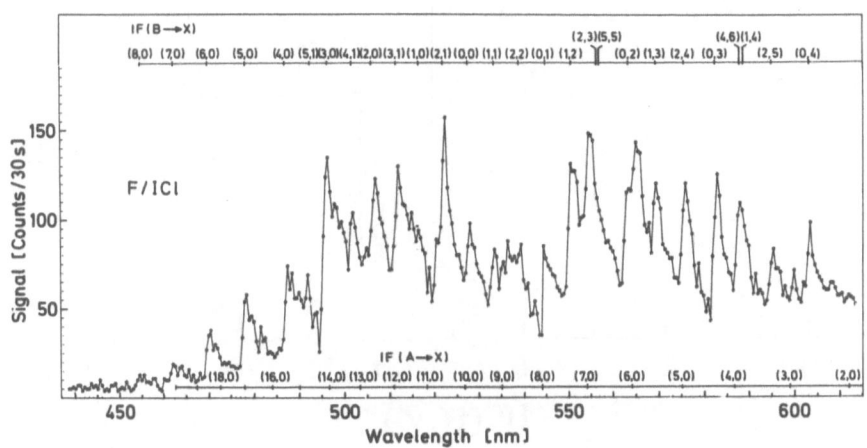

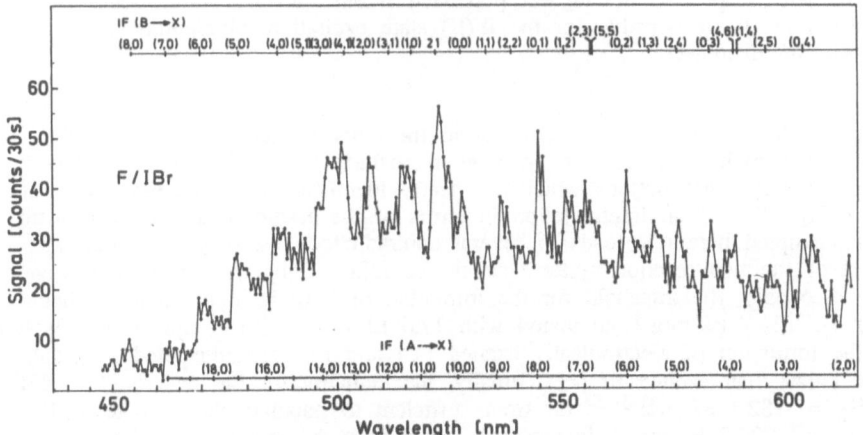

FIGURE 2a. Chemiluminescence spectra for the F/ICl and F/IBr systems under enhanced flow conditions.

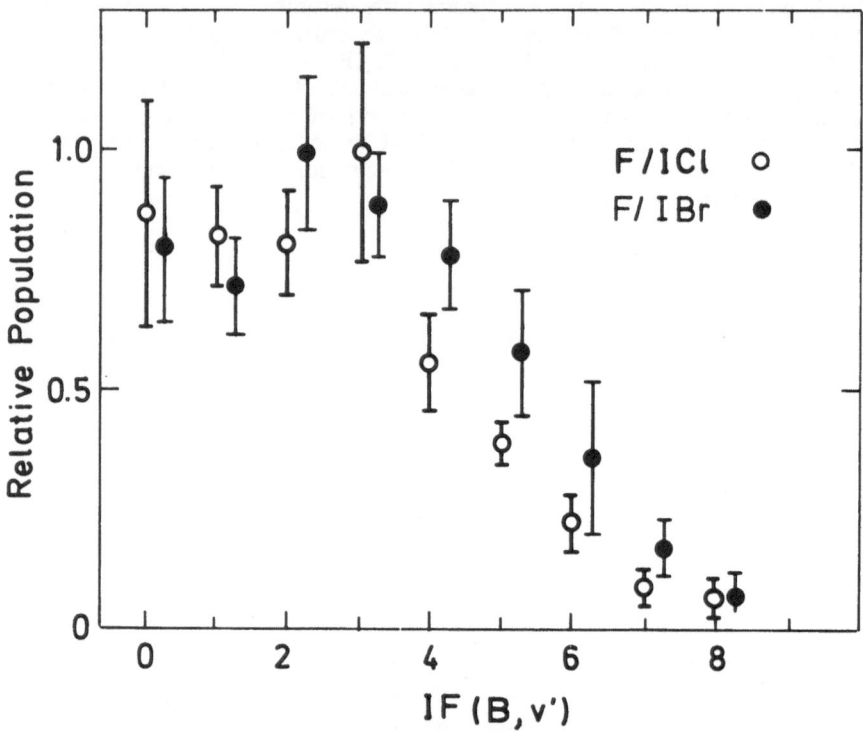

FIGURE 2b. The non-statistical IF(B) vibrational product state distributions are normalized with respect to the highest populated B state level. A different mechanism must be responsible for the IF(B) state excitation which may not depend on the XIF intermediate.

They are similar and significantly differ from the emission due to the F + I_2F reaction as shown in Fig. 1a. The population analysis for both systems resulted in more non-statistical distributions which are nearly identical within the error bars, as shown in Fig. 2b. If an interpretation in terms of the Kahler and Lee mechanism is again attempted here this leads to several contradictions. Firstly the different energetics of the two reaction systems should be reflected by the chemiluminescence spectra. Secondly, the threshold for the formation of ClIF is significantly higher than for I_2F (24.7 kJ mol^{-1} compared with 17.6 kJ mol^{-1}) [2]. Thirdly, one would expect the formation of electronically excited FCl and FBr as anticipated by Kahler and Lee rather than excited IF [2]. Finally, the enthalpy for the reaction F + ClIF with ΔH_0^o = -182.1 kJ mol^{-1} is far from sufficient to populate the B state of IF where at least 226.7 kJ mol^{-1} is required. (Although the stability of BrIF is not exactly known, the argument also holds for the F + BrIF reaction).

Hence for the F/ICl and F/IBr systems under given conditions there must be a different mechanism responsible for the IF(B) state excitation which may not depend on the XIF intermediate.

REFERENCES

[1] T. Trickl, J. Wanner, J. Chem. Phys., **74** (1981) 6509.

[2] C.C. Kahler, Y.T. Lee, J. Chem. Phys., **73** (1980) 5122.

[3] The apparatus used is identical to the one described by
     M. Trautmann, J. Wanner, S.K. Zhou and C.R. Vidal, J. Chem.
     Phys., **82** (1985) 693.

[4] D. Raybone, T.M. Watkinson, and J.C. Whitehead, 'The characterisation
     of the mechanism of IF(B) production in fluorine/iodide systems',
     preceding paper in this volume.

UNEXPECTED PROPERTIES OF ATMOSPHERIC REACTIONS

B.A. Thrush
University of Cambridge
Department of Physical Chemistry
Lensfield Road
Cambridge CB2 1EP
U.K.

ABSTRACT. Numerical modelling of the chemistry of stratospheric ozone requires precise laboratory data on many free radical reactions under atmospheric conditions. Although these processes occur under conditions where Maxwell-Boltzmann energy distributions obtain, the wide range of pressures and temperatures (down to 200 K or below) covered has yielded many unexpected features. The rates of many reactions differ widely from those which might be predicted on the basis of transition state theory. In particular a number of atom transfer reactions involving HO and HO_2 exhibit parallel second and third order rate expressions with fairly similar small or negative temperature coefficients. These can be understood in terms of hydrogen-bonded or perhaps covalently bonded intermediates, but the quantitative interpretation of such pressure dependences is beyond the scope of theories which assume the randomisation of internal energy.

1. INTRODUCTION

The realisation in the 1970's that the chlorofluorocarbons used as aerosol propellants, refrigerants and foam blowing agents might eventually deplete stratospheric ozone [1] came soon after a similar concern about the nitrogen oxides which are released directly into the stratosphere from high flying aircraft [2]. These effects were predicted on the basis of modelling calculations which included the rates of vertical transport of species through the stratosphere and laboratory data on the rate coefficients of the reactions which were considered to be important in the stratosphere. Although several workers independently calculated that continuous release of chlorofluoromethanes at 1973 levels would lead to an eventual ozone depletion of about 10% with a time scale close to 100 years, it was quickly realised that this agreement arose from similar assumptions as to the rates of the chemical and transport processes and that many of the crucial reaction rates were highly uncertain [3]. At that time the models included the rate coefficients of about 50 chemical reactions, few of which were known to ± 20% (and then at room rather than stratospheric temperatures) and many of which were guessed. Although the sensitivity of the predicted ozone depletions to the values adopted for the individual rate coefficient varied greatly, their combined effect made the predicted ozone depletion uncertain by more than a factor of three [4]. The long time-scale

J. C. Whitehead (ed.), Selectivity in Chemical Reactions, 531–542.
© 1988 by Kluwer Academic Publishers.

predicted for the response of stratospheric ozone to such perturbations meant that any legislation limiting the use and release of halocarbons would have to be based on model calculations rather than on observation of ozone depletion. This predicated three main approaches to the problem;

(i)   precise laboratory measurements of individual rate coefficients under the appropriate conditions of temperature and pressure, and the determination of reaction products,

(ii)   the identification of important chemical processes or species not included in the model,

(iii)   measurement of important trace species in the stratosphere as a test of the models.

The present paper concentrates on the first of these topics, because the need to measure the rate coefficients of many elementary reactions to an accuracy of 10% has established some unpredicted features such as the pressure dependence of the second-order rate coefficients for a number of bimolecular reactions involving hydrogen atom transfer. However, we will consider first the second topic which is important to the background and development of the main theme. It also emphasizes our current inability to predict the rates of elementary bimolecular transfer reactions even to an order of magnitude.

2. CHEMISTRY OF THE STRATOSPHERE

Any discussion of this topic inevitably begins with the familiar Chapman mechanism [5]

$$O_2 + h\nu \ (\lambda < 242nm) \rightarrow O + O \qquad \text{Slow} \qquad (1)$$

$$O + O_2 + M \rightarrow O_3 + M \qquad \text{Fast} \qquad (2)$$

$$O_3 + h\nu \rightarrow O_2 + O \qquad \text{Fast} \qquad (3)$$

$$O + O_3 \rightarrow O_2 + O_2 \qquad \text{Slow} \qquad (4)$$

Here the processes (2) and (3) rapidly interconvert the odd-oxygen species O and O_3 which are formed by process (1) and destroyed by reaction (4). This latter process only removes about one-quarter of the stratospheric ozone (odd oxygen) and the remainder is largely accounted for by catalytic cycles of the type

$$X + O_3 \rightarrow XO + O_2$$

$$XO + O \rightarrow X + O_2$$

$$\overline{O + O_3 \rightarrow O_2 + O_2}$$

which are equivalent to reaction (4). Here the principal species X are NO, Cl, H and HO. The dominant processes involved in the first two cases are illustrated in

Figs. 1 and 2.   In parallel with the ozone destruction cycle involving NO_x

$$NO + O_3 \rightarrow NO_2 + O_2 \tag{5}$$

$$O + NO_2 \rightarrow NO + O_2 \tag{6}$$

$$\overline{}$$

$$O + O_3 \rightarrow O_2 + O_2$$

the rapid photolysis of NO_2 in daylight generates a null cycle in which ozone is not destroyed

$$NO_2 + h\nu \rightarrow NO + O \tag{7}$$

$$O + O_2 + M \rightarrow O_3 + M \tag{2}$$

$$NO + O_3 \rightarrow NO_2 + O_2 \tag{5}$$

The importance of NO_2 photolysis is further enhanced by the rapid reactions

$$HO_2 + NO \rightarrow HO + NO_2 \tag{8}$$

$$ClO + NO \rightarrow Cl + NO_2 \tag{9}$$

which couple together the ozone distribution cycles with X = NO, HO and Cl and by which the XO species in the latter cycles can enter a null cycle based on NO_2 photolysis.

The comparable importances of the transfer reactions given above may appear somewhat surprising because the relative abundances of the trace species in the stratosphere are $O_3 \gg O \gg NO \gg HO_2$, HO, ClO, Cl. Furthermore, it is generally assumed that atom transfer reactions show a clear correlation between activation energy and exothermicity. Activation energy is the dominant factor in determining the rate coefficients of these reactions at the stratospheric temperatures of 200-250 K.

Table 1 shows that there is no clear correlation between the Arrhenius parameters and exothermicities of the oxygen atom transfer reactions which are important in the stratosphere, this divergence is made more remarkable by the recent work of Barnett, Marston and Wayne [7] who obtained a rate coefficient of 1×10^{-16} cm^3 molecule^{-1} s^{-1} for the highly exothermic reaction

$$N(^4S) + O_3(^1A_1) \rightarrow NO(^2\Pi) + O_2(^3\Sigma_g^-) + 526 \text{ kJ mol}^{-1}. \tag{13}$$

The explanation for this lack of correlation appears to lie in the factor governing the energies of the transition states. Reactions (4) and (13) are overall spin and symmetry allowed and their low rates must be attributed to the higher energies of their respective triplet and quartet transition states as compared with the doublet surfaces over which the other reactions can proceed. In the absence of Arrhenius parameters it cannot be excluded that reaction (10) proceeds non-adiabatically via a doublet surface. Reactions (8) and (9) must pass through intermediate states corresponding to pernitrous acid and chlorine nitrite both of which have been prepared by Knauth at Kiel; they therefore go via attractive potential surfaces.

534

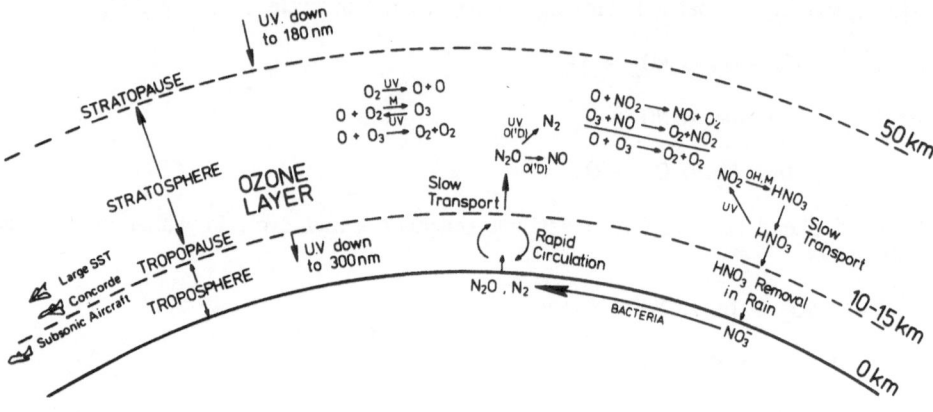

FIGURE 1. Ozone formation and removal in the unperturbed stratosphere. (Taken from Ref. 3).

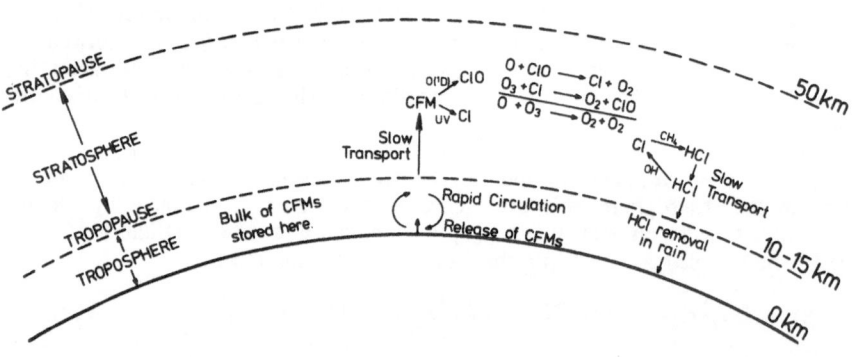

FIGURE 2. Simplified mechanism for effect of chlorofluoromethanes on stratospheric ozone. (Taken from Ref. 3).

TABLE 1. Arrhenius Parameters of some Oxygen Atom Transfer Reactions (Taken from Ref. 6).

Reaction Number	Reaction	$-\Delta H$ /kJ mol^{-1}	E_a /kJ mol^{-1}	log A/s
4	$O + O_3 \rightarrow O_2 + O_2$	393	17.1	-11.1
10	$H + O_3 \rightarrow HO + O_2$	313	3.9	-10.5
5	$NO + O_3 \rightarrow NO_2 + O_2$	200	11.4	-11.7
11	$Cl + O_3 \rightarrow ClO + O_2$	163	2.1	-10.6
12	$HO + O_3 \rightarrow HO_2 + O_2$	162	7.8	-11.8

$$HO_2 + O_3 \rightarrow HO + O_2 + O_2 \qquad (14)$$

Reaction (14) (which is important for ozone destruction at night) is unusual in that it is a bimolecular process generating three species; i.e. it involves a large increase in entropy. It might therefore be expected to have a large pre-exponential factor as do the second order dissociations of small molecules. The very recent measurements by Howard's group [8] give a curved Arrhenius plot represented by $k = 3.2 \times 10^{-13} \exp. (-1730/T) + 1.2 \times 10^{-15}$ cm^3 s^{-1} corresponding to a very small A factor. The temperature independent term may well be associated with tunnelling because experiments using $H^{18}O_2$ showed that $(75 \pm 10)\%$ of HO is formed by a hydrogen atom transfer mechanism.

The dominant reactions forming and removing nitrogen oxides and chlorine radicals in the stratosphere were given in Figs. 1 and 2. There it can be seen that the number of hydroxyl radicals present controls the balance between ozone destruction by nitrogen oxides and by chlorine radicals because the reaction

$$HO + NO_2 + M \rightarrow HNO_3 + M \qquad (15)$$

removes nitrogen oxides but the reaction

$$HO + HCl \rightarrow H_2O + Cl \qquad (16)$$

regenerates chlorine radicals from inactive HCl. For this reason, the reactions which control the concentration of HO radicals in the stratosphere are of particular importance. These are summarised in Fig. 3. Some of these processes show the unexpected pressure dependences which are discussed in more detail below.

In addition to the better measurement of individual reaction rates, much effort was also put into 'unknown' chemistry:- the identification of species and reactions which might be important in the stratosphere but which had not been included in the model calculations of the mid- to late 1970's. Here there was a clear precedent in

$$O(^1D) + H_2O,\ CH_4,\ H_2 \longrightarrow HO,\ H$$

$$H_2O + h\nu \longrightarrow HO,\ H$$

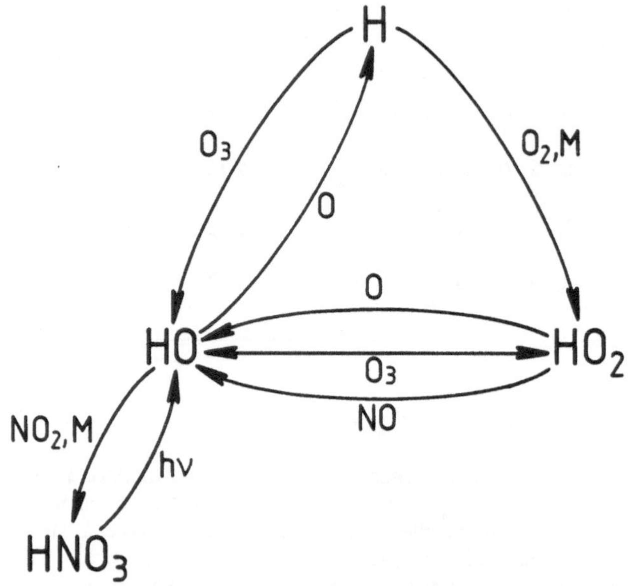

$$HO + HO_2 = H_2O + O_2$$

$$HO_2 + HO_2 = H_2O_2 + O_2$$

$$H + HO_2 = H_2 + O_2$$

FIGURE 3. Reactions controlling H, HO and HO_2 in the stratosphere.

peroxyacetylnitrate (PAN) which was identified as an active irritant in Los Angeles smog before it was synthesised in the laboratory [9]. Significantly, the additional species identified as having potential importance in stratospheric chemistry resemble PAN in being highly oxidised species which have limited thermal stability and which absorb light only weakly above 280 nm. These properties reflect the highly oxidising nature of the atmosphere, and the relative stability of such molecules in the lower stratosphere which is cold (200-230 K) and protected by ozone and oxygen from light below 280 nm. Each of the three species concerned:- pernitric acid, chlorine nitrate and hypochlorous acid is formed by the cross-combination of the less reactive radicals (XO species) in main ozone destruction cycles.

$$HO_2 + NO_2 + M \rightarrow HO_2NO_2 + M$$

$$ClO + NO_2 + M \rightarrow ClONO_2 + M$$

$$HO_2 + ClO \rightarrow HOCl + O_2$$

The stability of these species and their relatively long photochemical lifetimes by day (4 hours for HO_2NO_2 and $ClONO_2$) means that their inclusion in atmospheric models reduces the predicted rates of ozone destruction. As Fig. 4 shows the inclusion of such species in atmospheric models has led to considerable changes over the last few years in the predictions for future ozone depletion by halocarbons and by nitrogen oxides from stratospheric flight.

The first numerical models of atmospheric chemistry were one-dimensional in that they considered only vertical transport and neglected latitudinal and longitudinal variations. Initially they assumed a uniform twelve hour day because ozone formation and destruction essentially cease at night when the concentration of atomic oxygen rapidly drops to zero. The inclusion of the semi-stable molecules mentioned above made it necessary to use diurnal models which include the variation of the concentration of species over a period of 24 hours. In recent years much effort has gone into the development of two-dimensional models which consider the latitudinal variation of species and the corresponding circulations. The eventual achievement of a three-dimensional model of the earth's atmosphere which can also include enough chemistry to be realistic is a goal for the next decade.

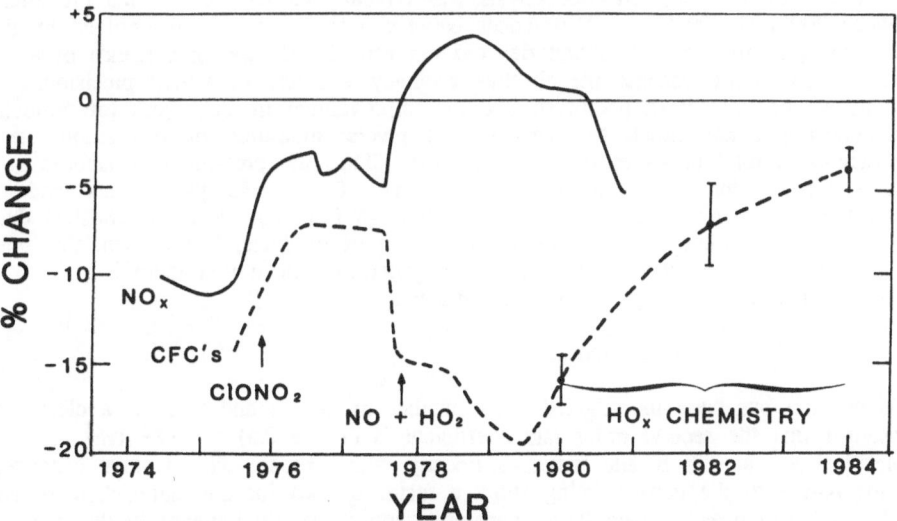

FIGURE 4. Predictions of ozone reduction due to continued release of chlorofluorocarbons or to a large fleet of supersonic aircraft.

538

3. LABORATORY STUDIES

3.1 The HO + CO Reaction

The first simple atom transfer reaction for which a pressure dependence was reported was that between hydroxyl and carbon monoxide.

$$HO + CO \rightarrow CO_2 + H \qquad (17)$$

However, early results gave somewhat conflicting results and for a time it was believed that the reaction was accelerated only by molecular oxygen. However, it is now clear that the rate coefficient of this reaction is increased to different extents by the addition of chemically inert gases [10] as are the reactions considered below. For [M] = air, $k_{17} = 1.5 \times 10^{-13} + 3.7 \times 10^{-33}$ [M] cm^3 s^{-1} at 298 K [6], the latter term being within the range expected for a third order combination reaction, albeit one which yields a product which decomposes (HOCO). This reaction is of atmospheric importance because it governs the lifetime of ^{14}CO generated by cosmic rays and can therefore be used to estimate [HO].

When this reaction was first studied directly in the laboratory, an accuracy of ± 30% was considered good for the determination of the rate coefficient of a rapid radical reaction, and a pressure dependence of the rate coefficient which was smaller than this could easily be overlooked unless the reproducibility of the experiments was high. Furthermore, the two main techniques used have very different pressure ranges which do not overlap. Discharge-flow systems are most satisfactory at total pressures below 5 Torr, a range over which pressure dependences are not detectable. Conventional flash-photolysis was typically carried out at total pressures between 200 and 800 Torr. Differences between rate coefficients determined by the two techniques in different laboratories did not provide convincing evidence of a pressure dependence because the absolute accuracy was less than their precision. The advent of laser flash photolysis and the improvement in techniques for retrieving and averaging small signals has permitted the precise measurement of reaction rate coefficients at total pressures down to 20 Torr. Thus the pressure dependences discussed below have a sound experimental basis. These techniques do not lend themselves to precise product analysis, but although it is not clearly established that the reactions considered below only yield one set of products, it is known that one channel predominates in each case and it is generally agreed that there is no evidence of a change of products with pressure.

3.2 The HO_2 + HO_2 reaction

This reaction has been investigated by a number of groups and there is a clear consensus that the second order rate coefficient is $(1.7 \pm 0.5) \times 10^{-12}$ cm^3 molecule^{-1} s^{-1} at 298 K and increases linearly with pressure with a slope dependent on the nature of the carrier, being about a factor of two for one atmosphere of air [11]. H_2O produces a much larger apparently more specific increase in the rate coefficient; this is probably due to the formation of a hydrogen bonded complex $HO_2 \cdot H_2O$ which is more reactive than HO_2, but the kinetics have not been investigated in sufficient detail [12]. Although there is some evidence that ca. 1% of the reaction yields H_2, the dominant path at all pressures is

$$HO_2 + HO_2 \rightarrow H_2O_2 + O_2. \qquad (18)$$

The reaction exhibits a negative temperature coefficient both at low and high pressures, and for M = air the rate coefficient can be expressed as

$$k_{18} = 2.3 \times 10^{-13} \exp(590/T) + 1.7 \times 10^{-33} \exp(1000/T) \, [M] \, cm^3 \, molecule^{-1} \, s^{-1}.$$

The second term in this expression is reasonable for a three body combination reaction between molecules of the complexity of HO_2. However, there is no chemical or spectroscopic evidence for the formation of a stabilised H_2O_4 species and estimates of the stability of the H_2O_4 molecule are based on an O-O chain structure which would decompose to $H_2 + 2O_2$ rather than $H_2O_2 + O_2$. The negative temperature coefficient at low pressures again suggests a reaction proceeding over an attractive potential surface, albeit with a low collisional efficiency.

The most obvious explanation of these parallel processes involves the formation of an H_2O_4 species which after stabilisation has an increased probability of decomposing to $H_2O_2 + O_2$ as against redissociating to $HO_2 + HO_2$.

In terms of a simple Hinshelwood - Lindemann type mechanism this would be

$$HO_2 + HO_2 \underset{k_b}{\overset{k_a}{\rightleftharpoons}} H_2O_4^* \overset{k_c}{\longrightarrow} H_2O_2 + O_2$$

$$\downarrow k_d \, [M]$$

$$H_2O_4 \longrightarrow H_2O_2 + O_2.$$

Because stabilised or partly stabilised H_2O_4 species must have a higher probability of decomposition than redissociation the barrier for the former (c) must be lower than for the latter (b). In terms of conventional transition state theory k_c should therefore be much larger than k_b and the overall rate coefficient

$$k = \frac{k_a(k_c + k_d[M])}{k_b + k_c + k_d[M]}$$

would not exhibit the observed linear pressure dependence which requires that $k_b \gg k_c$. It is therefore hardly surprising that attempts to explain the $HO_2 + HO_2$ reaction in terms of conventional transition state theory have been unsuccessful. Because all the reactions which exhibit this behaviour involve the species HO_2 and/or HO which are capable of forming strong hydrogen bonds, the explanation almost certainly lies in the formation of a hydrogen bonded complex $HO_2...HO_2$ in which the rate of decomposition (k_c) is determined by the rate of flow of energy from the H-O chemical bond into the O...H hydrogen bond. The observed pressure dependence gives $k_c \sim 5 \times 10^9 \, s^{-1}$ and $k_b \sim 5 \times 10^{11} \, s^{-1}$ which are wholly plausible.

3.3. The HO + HO_2 Reaction

The process

$$HO + HO_2 \rightarrow H_2O + O_2 \qquad (19)$$

is two orders of magnitude faster than reactions (17) and (18) at low pressures and a significant acceleration at atmospheric pressure would not be expected. Although DeMore [13] has reported that k_{19} is pressure dependent, the main evidence for this dependence has been the fact that the rate coefficients determined at low pressures by different groups are consistently lower than those found by other groups around atmospheric pressure [6]. This is not a convincing argument and recent unpublished measurements at low pressures by C.J. Howard give higher rate coefficients,

$$k_{19} = 4.6 \times 10^{-11} \exp (230/T) \, cm^3 \, s^{-1}$$

which suggests that the rate coefficient may not increase by more than 20% at one atmosphere pressure. More work is clearly needed on this difficult system.

3.4   The HO + HNO$_3$ Reaction

Because of its importance in atmospheric chemistry, the reaction has been studied repeatedly by several groups of workers mainly using flash photolysis. A major difficulty has been the determination of the concentrations of HNO_3 present, particularly because it has a low vapour pressure at temperatures below 250 K which correspond to stratospheric conditions. Experiments under different conditions provide no evidence for any reaction path other than

$$HO + HNO_3 \rightarrow H_2O + NO_3. \qquad (20)$$

The most recent assessment of the experimental rate coefficients for this reaction gives a rate expression of the form

$$k_{20} = k_0 + \frac{k_f k_g [M]}{k_f + k_g [M]}$$

where $k_0 = 7.2 \times 10^{-15} \exp (785/T)$, $k_f = 4.1 \times 10^{-16} \exp (1440/T)$ and $k_g = 1.9 \times 10^{-33} \exp (725/T) \, cm^3 \, s^{-1}$ for [M] = air [6]. This implies a simple bimolecular reaction plus a three body process (represented by a Hinshelwood - Lindemann mechanism) occurring in parallel and yielding similar products. However the algebraic form of the pressure dependence given above is formally the same as that derived for the $HO_2 + HO_2$ reaction and it is more logical to regard the HO + $HNO_3 \rightarrow H_2O + NO_3$ reaction as proceeding via a common energised species the stabilisation of which causes the forward rather than the reverse path to predominate. Again the energised species is almost certainly hydrogen bonded.

4.   CONCLUSIONS

The unexpected pressure dependences discovered in atmospheric atom transfer reactions involving HO and HO_2 probably have their origin in the slow exchange of vibrational energy between hydrogen bonds and the much higher frequency H-O chemical bonds. This phenomenon has much in common with many other current studies involving limited rates of energy flow:- intramolecular energy transfer.

photo-dissociation of van der Waals molecules, desorption from surfaces, etc., although it is less susceptible to detailed study.

This phenomenon will clearly be more important at low temperatures, a regime of great current interest in atmospheric chemistry in connection with the Antarctic "ozone hole" where temperatures as low as 160 K are encountered in the lower stratosphere in winter. It is exceptionally difficult to study many important reactions in the laboratory at such temperatures because of the low vapour pressure of the species involved. Indeed, an important factor in the chemistry of the Antarctic lower stratosphere may well be the condensation of nitric acid on ice crystals at temperatures below 180 K which could result in heterogeneous reactions which would again be very hard to study in the laboratory. Furthermore, our understanding of the temperature dependence of the rates of simple bimolecular reactions with small activation energies is not such that we could confidently extrapolate measurements to temperatures down to, say, 250 K as far as 160 K. Past experience of shorter extrapolations to 220 K have exposed the risks involved.

This review has concentrated on the information which has been obtained from very careful studies of the kinetics of stratospheric reactions. These studies have almost always involved spectroscopic measurements of the rates of disappearance of an atom or free radical under pseudo-first-order conditions, i.e. in the presence of excess of the other reagent. Product analyses have rarely been carried out because of the small amounts of product formed from the low concentrations of transient species employed. An important but rather neglected area of selectivity in atmospheric reactions is the possible occurrence of parallel reaction paths leading to different products. In some cases the alternative reaction path could have a significant effect on predicted ozone depletions. For instance, it is generally assumed that the reaction $ClO + HO_2$ yields exclusively $HOCl + O_2$ rather than the energetically feasible $HCl + O_3$. Similarly the photolysis of HOCl is believed to yield exclusively $HO + Cl$ rather than $HCl + O$. If HCl were formed in either of these processes its relatively high stability in the stratosphere would reduce the proportion of chlorine present as Cl and ClO and hence decrease the predicted ozone depletions.

REFERENCES

[1]   M.J. Molina and F.S. Rowland, Nature, **249** (1974) 810;
      F.S. Rowland and M.J. Molina, Rev. Geophys. Space Phys.,
      13 (1975) 1.

[2]   H.S. Johnston, Science, **173** (1971) 517.

[3]   Halocarbons: Effects on Stratospheric Ozone, Nat. Acad. of
      Science, Washington, D.C. (1976).

[4]   Chlorofluoromethanes and the Stratosphere, Ed. R.D. Hudson,
      NASA Ref. Publ. **1010** (1977).

[5]   S. Chapman, Phil. Mag., **10** (1930) 369.

[6]   Chemical Kinetics and Photochemical Data for Use in
      Stratospheric Modelling. Evaluation Number 7, JPL Publ. **85-37**,
      NASA (1985).

542

[7]   A.J. Barnett, G. Marston and R.P. Wayne, J. Chem. Soc., Faraday
      Trans. 2, 83 (1987) 1453.

[8]   A. Sinha, E.R. Lovejoy and C.J. Howard, J. Chem. Phys., 87
      (1987) 2122.

[9]   Photochemistry of Air Pollution, P.A. Leighton, Academic
      Press, New York (1961).

[10] G. Paraskevopoulos and R.S. Irwin, J. Chem. Phys., 80 (1984)
      259.

[11] S.P. Sander, M. Peterson, R.T. Watson and R. Patrick, J. Phys.
      Chem., 86 (1982) 1236.

[12] R.-R. Li, M.C. Sauer, Jr. and S. Gordon, J. Phys. Chem., 85
      (1981) 2833.

[13] W.B. DeMore, J. Phys. Chem., 86 (1982) 121.

STATE SELECTIVITY IN LIGHT EMISSION FROM FLAMES

David R. Crosley, Karen J. Rensberger and Richard A. Copeland
Molecular Physics Department
SRI International
Menlo Park
California 94025
USA

ABSTRACT. Chemiluminescence in flames is the result of energetic and highly specific reactions involving free radicals. If understood and linked to the major combustion chemical mechanism, it can be of use for monitoring the progress of that chemistry. Although there is firm evidence for certain reactions to form electronically excited OH and CH in hydrocarbon flames, the generally accepted chemiluminescence mechanisms cannot explain all of the observations, indicating the importance of other pathways. Some of the evidence for other pathways can be found in the highly nonequilibrium internal level distributions seen in flame chemiluminescence. An interpretation of these observed distributions in terms of the nascent results from the elementary reactions involves consideration of state-specific electronic quenching and energy transfer. We present some recent observations on chemiluminescent emission from CH and OH in hydrocarbon flames on a low-pressure flat flame burner, and summarize pertinent experiments in flames and other systems on excited state quenching and energy transfer in these two radicals.

1. INTRODUCTION

The light emitted as a result of chemiluminescent reactions of free radicals forms what we familiarly think of as a flame, even though those reactions constitute but a negligible pathway for the overall conversion of fuel and oxidant to combustion products. There have been numerous early studies of flame luminescence, as exemplified by Gaydon's book [1] on the topic, but in recent years the concentration on laser-based optical methods has generally superseded such studies. Nevertheless, an understanding of the process of chemiluminescence from flames can have practical benefits, for its use as a monitor of the progress of the flame chemistry under conditions where probing by lasers cannot be carried out. We can cite several such recent examples: the measurement of CH, C_2 and OH chemiluminescence to deduce spatial and temporal regimes of heat release in pulse combustors [2]; the use of CH emission to conditionally sample flame fronts in gas turbine engines [3]; and the observations of emission from C_2 and from both the $B^2\Sigma^-$ and $A^2\Delta$ states of CH as markers of shock fronts in detonation waves [4]. It can be noted, interestingly, that in this last study the radiation from these two states did not coincide temporally in the shock. Additionally, the chemiluminescence can be of direct significance such as

J. C. Whitehead (ed.), Selectivity in Chemical Reactions, 543–554.
© *1988 by Kluwer Academic Publishers.*

its use in detecting rocket plumes through observations in visible and ultraviolet spectral regions [5].

Despite the many observations of these phenomena, there remain many unanswered mechanistic questions. Nearly all of the identifiable light emission from flames is from small free radicals, and the formation of the radicals in electronically excited states demands energetic precursors. Thus the reactants are also free radicals; many of the likely reactions form the highly stable CO molecule as one of the products. Only fairly recently [6] were any mechanistic pathways established for the production of excited OH and CH in one low-pressure flame, using absorption measurements of ground state radical concentrations; but, for at least CH, even that mechanism has been questioned in a different series of experiments [7].

In these reactions, there is often considerable energy deposited into internal degrees of freedom of the emitting, electronically excited radical. An illustrative example is shown in Fig. 1, which exhibits a Boltzmann plot of the rotational populations in the $v'=0$ level of the $A^2\Sigma^+$ state of the OH radical in the primary reaction zone of an atmospheric pressure methane/oxygen flame. The distribution appears bimodal, described by two "temperatures". At lowest N', the (poorly determined) value of 1700 K is not far from the rotational temperature of ground state OH, determined by absorption measurements. The highly nonequilibrium distribution at higher N', however, is far hotter than any possible flame temperature. It is the result of the nascent distribution from the chemiluminescent reaction forming [8] OH* (possibly CH + O_2), together with rotational energy

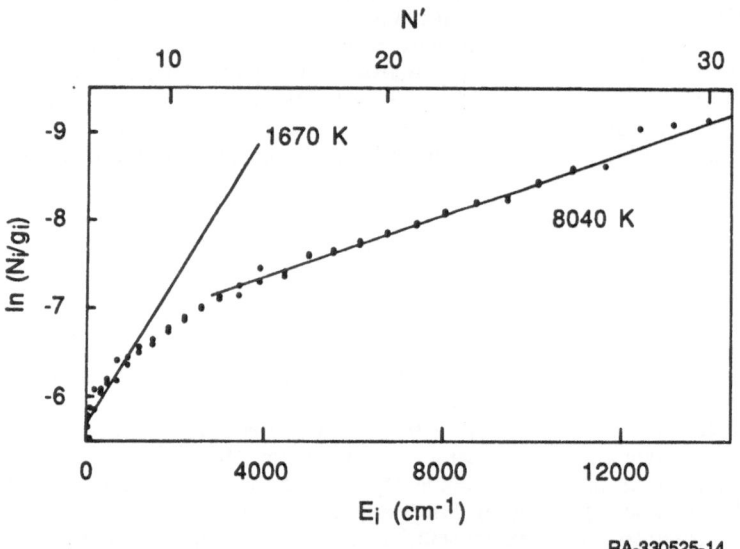

RA-330525-14

FIGURE 1. Rotational level Boltzmann plot: the logarithm of the state population divided by its degeneracy vs. its energy. Note that negative values increase to the top. This is for emission from the $v'=0$ level of the $A^2\Sigma^+$ state of the OH molecule, in the reaction zone of a stoichiometric, atmospheric pressure CH_4/O_2 flame. The distribution is bimodal, with the populations at low-N' close to that of the ground state OH rotational temperature, plus a nonequilibrium high-N' component reflecting the chemiluminescent reaction forming OH in the flame.

transfer, and rotationally state-specific collisional quenching which in this flame is the dominant removal mechanism from the excited state [9]. Similar distributions have been observed in other hydrocarbon flames [1,10].

It would be useful to relate the formation of the chemiluminescent, emitting radicals to the progress of the primary combustion reactions for a variety of flame conditions. In doing so, relationships among excited and ground state radical concentrations form important clues. So also do the particular state distributions (electronic, rotational and vibrational) populated by these reactions in the flames. Deducing those nascent distributions from the measured ones demands knowledge of collisional quenching and internal energy transfer in the flame environment. For several years we have been studying these collisional processes for a variety of small free radicals important in combustion, and have recently begun observations of the chemiluminescence itself in a low-pressure burner system. In this paper, we will concentrate on two species: OH and CH. We will describe some of the chemiluminescence results, which raise new questions (but do not yet resolve them!) concerning the mechanistic pathways, and shall summarize current knowledge of quenching and energy transfer for these radicals in flame environments.

2. CHEMILUMINESCENCE OF CH AND OH

It was long ago suggested [1] but only more recently established [11,12] that in hydrocarbon/oxygen flames the major, perhaps sole, means of formation of OH* is the reaction between CH and O_2. The evidence supporting this conclusion [11] comes from the constancy over various operating conditions, in a low-pressure C_2H_2/O_2 flame, of the relationship $[OH^*]nT^{1/2}/[CH][O_2]$, where n is the total density; this is as expected from a simple steady state balance between this formation reaction and collisional removal of the excited state. Further support is found in the results from a room temperature discharge flow system [12] of O + C_2H_2, which are in agreement with a complex computer model of the reaction network. This reaction yields rotationally hot distributions, as seen in Fig. 1, contrasting for example with the nonequilibrium vibrational distributions of $A^2\Sigma^+$ OH in a H_2/O_2 flame [1] (formed in that case via inverse predissociation).

Several spatial profiles of the intensities of chemiluminescence of CH* and OH* and laser-induced fluorescence (LIF) of CH and OH from the low-pressure burner experiments [13] are shown in Figs. 2 and 3. A sintered, porous disc flat flame burner of 5 cm diameter is housed in an evacuable chamber, permitting flames to be stabilized down to about 3 Torr. Measurements of excited state chemiluminescence are made with a small monochromator, here operated with a typical 10 nm bandpass and a wavelength fixed to the center of the emission band being monitored. The spatial resolution is governed by the slit width and is about 0.4 mm. Ground state measurements are made using LIF with an excimer-pumped dye laser, frequency doubled to detect OH or used directly in the ultraviolet or blue for exciting the CH B and A states respectively. Here, the pulsed laser produces pulsed LIF, which is time-resolved to discriminate against the flame emission.

Fig. 2 shows the spatial profiles for a stoichiometric propane/oxygen flame at 6.5 Torr. There are several salient features. First, the OH* emission profile peaks higher off the burner (i.e., later in the flow) than the ground state CH profile. The O_2 concentration, although not measured here, must continually decrease as a function of distance from the burner. Over this region, the temperature (measured

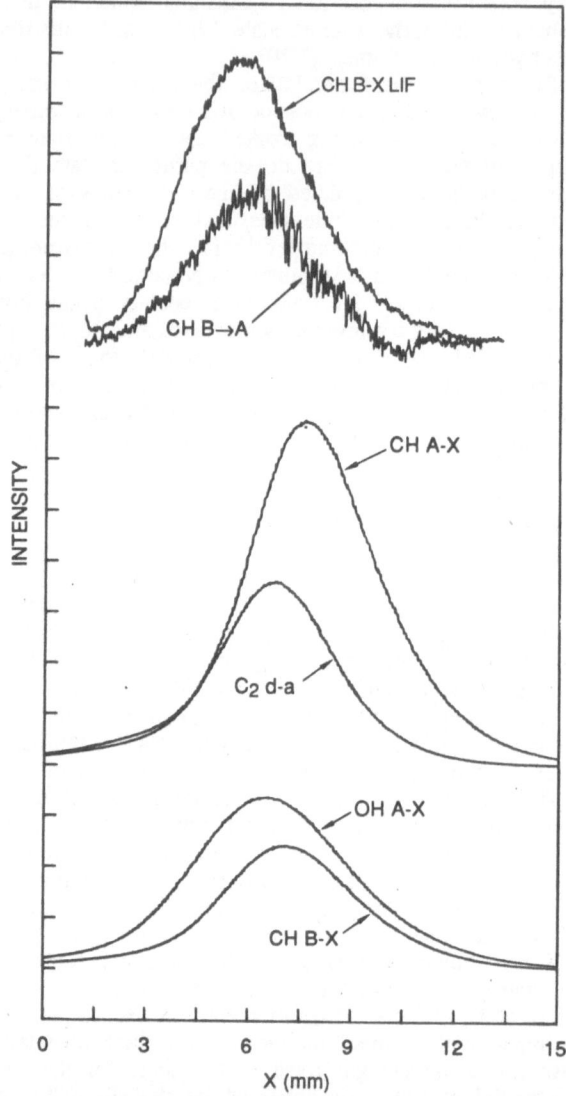

FIGURE 2. Spatial profiles of LIF and chemiluminescence for several flame radicals, taken in the low-pressure burner. The flame is C_3H_8/O_2 at a pressure of 6.5 Torr. Note that intensities, not radical concentrations, are plotted; the relationship between the two involves a quantum yield which is however nearly constant throughout this flame.

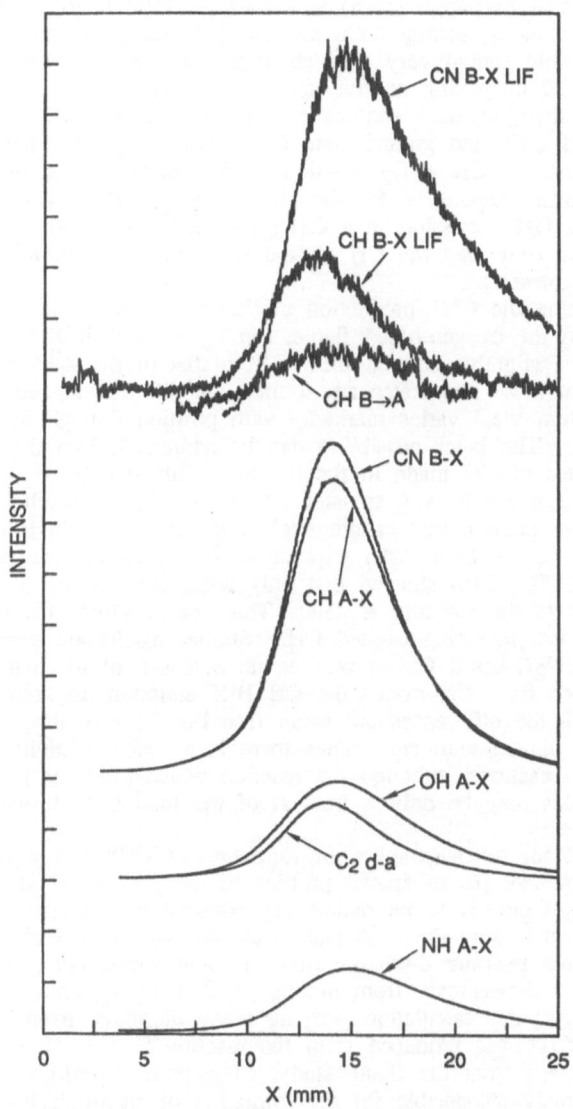

FIGURE 3.   As Fig. 2 for a C_3H_8/N_2O flame.

separately in CH rotational excitation scans) varies from ~1500 K to 2000 K. Using temperature-dependent quenching cross sections [9], described below, we expect that the OH quantum yield should vary only about 5% through the flame; direct measurements [14] of CH quenching on the same burner show that its quantum yield too is constant. Thus these intensity profiles can be taken as equivalent to concentration profiles of OH* and ground state CH. Taken together, they do not match that expected from the CH + O_2 reaction. Consequently, there must here be some additional mechanism responsible for the production of OH*. This is also evident from the strong OH* emission in a C_3H_8/N_2O flame (Fig. 3). Again, the hydroxyl emission peaks later than the CH ground state, even though here there is little molecular oxygen present.

Evidence concerning the CH* production mechanism comes from the state selectivity observed. In the oxygen-based flame, Fig. 2, the CH $B^2\Sigma^-$ chemiluminescence has a slightly different profile from that of the $A^2\Delta$ emission. This suggests either that these two states are formed via different reactions, or else that the emission quantum yield varies markedly with position through the flame, differing for each state. The latter possibility can be addressed from the quenching study [14] on both states of CH made in the burner. Although the B state quenches more rapidly and the B → A transfer rate varies slightly (perhaps 10%) with burner position (see below), that cannot explain the difference in profiles. This can be concluded from an energy transfer experiment in an atmospheric pressure methane/oxygen flame [15], which showed that only about 20% of the B-state molecules are quenched to the emitting A-state. Thus we conclude that there are (at least) two mechanisms for producing excited CH. Similar results are seen for methane flames. (The N_2O-based flames provide no evidence of this nature because strong chemiluminescence from CN masks the CH B-X emission, as seen in Fig. 3). On the other hand, the differences are small (see Fig. 2) and disappeared when the propane flame was diluted with N_2. Thus there is a state selectivity indicating at least two production reactions, although the reaction which produces the difference in A and B state profiles may be only a fraction of the total CH* formation mechanism.

Previous evidence for the mechanism of formation of CH* comes from two experiments. A measurement [6] of spatial profiles in low-pressure C_2H_2/O_2 flames showed the ratio $[CH^*]/[C_2][OH]$ to be remarkably constant with variation in many different flame parameters. Only the $A^2\Delta$ state was observed, however. In a more recent study [7] in a low pressure discharge flow at room temperature, emission spatial profiles, measured downstream from mixing of O + C_2H_2, were compared with the results of a computer calculation, varying many discharge parameters. Here it was concluded that CH* was produced from the reaction O + C_2H, not the reaction C_2 + OH deduced from the flame study. The present profiles suggest that neither mechanism is solely responsible for the formation of electronically excited CH.

Another pertinent observation is that of a third excited state of CH, $C^2\Sigma^+$, which lies some 6300 cm^{-1} or 9000 K higher than $B^2\Sigma^-$. It is not seen in low pressure discharges [7,16] of acetylene and oxygen but is observed in atmospheric pressure flames of those two reactants [16]. This is significant in that the energetics of the two proposed CH* formation reactions are considerably different; O + C_2H has just enough energy at threshold to populate the v'=0 level of $B^2\Sigma^-$ whereas OH + C_2 is energetic enough to directly produce $C^2\Sigma^+$. We find the argument in Ref. 7, that the C-state is formed in flames from vibrationally excited C_2H, to be questionable, in view of the very large amounts of excess energy required above threshold.

3. COLLISIONAL QUENCHING OF CH AND OH IN FLAMES

It can be seen from the above that it is necessary to account for collisional effects on the emission quantum yields when interpreting chemiluminescence profiles to deduce mechanistic information. In this section, we summarize the results of a series of experiments on quenching and energy transfer in electronically excited OH and CH, which are pertinent to such flame studies.

We have made OH measurements in discharge flow cells and a laser pyrolysis/laser fluorescence (LP/LF) apparatus [16] for T = 230 to 1400 K, and on CH in the burner, at T ~ 1700 K. The measurements of quenching were made from the pressure dependence of the direct time decay of LIF signals; an example of such a decay trace from the CH experiments is shown in the upper panel of Fig. 4. Energy transfer has been studied using fluorescence scans following laser excitation of an individual upper state level.

The OH room temperature results showed a key finding [17,18]: that the quenching cross section σ_Q varied markedly with rotational level N', for some 20 collision partners studied. This state-specific effect can be seen because individual rotational levels are excited by the laser, and they do not rotationally thermalize before undergoing a significant amount of quenching. In some cases, σ_Q drops as much as twofold as N' increases from 0 to 5. This has important implications for interpreting distributions such as in Fig. 1, because the OH in the flame also does not become rotationally thermalized before quenching [19]. On the other hand, neither the rotational level dependence at high N', nor the temperature dependence of the N'-dependence is known, so at present the effects of this variation of σ_Q can only be estimated. Recent experiments in the burner [14] have shown the quenching of both the B- and A-states of CH vary with N', although the variation in the flames is less marked, 20% at most. The degree varies with flame; it is flat for CH_4/O_2 and decreases similarly for C_3H_8/O_2 and C_2H_2/O_2. This can be ascribed to the different collisional environments in these flames. We can make analogy with both the OH results [18] as well as with recent room temperature studies [20] of σ_Q for $A^3\Pi_i$ NH, which show a degree of rotational level dependence similar to that for OH. We then expect a large N' variation for CO_2 and little for H_2O, although this does not offer a full explanation of the variation with flame.

Room temperature σ_Q values for OH are large, ranging from 10 to 100 Å^2, indicating [18] that attractive forces are involved in the collision. (The rotational level dependence is then ascribed [18] to anisotropies in this attractive surface). A governing attractive interaction is in accord with studies [21] of the temperature dependence of OH quenching, measured in both the flow and LP/LF systems. In all cases, σ_Q decreases as the temperature increases. There is a further decrease in the cross section averaged over a thermal rotational distribution, because of the shift with increasing temperature to higher N', having lower state-specific σ_Q. This must also be taken into account in interpreting chemiluminescence data, which exist over a range of temperature near the flame front.

The average cross section for a rotationally thermal distribution in the v'=0 level of the $A^2\Delta$ state of CH has also been measured at elevated temperature in the LP/LF apparatus [22]. Here, a comparison could be made with cross sections determined from a photodissociation method at room temperature [23]. For the four colliders common to both experiments, σ_Q increased from 25% to tenfold, in going from 300 K to 1300 K. This behaviour is markedly different from that of OH. This indicates a quenching mechanism for CH involving some kind of barrier or repulsive surface.

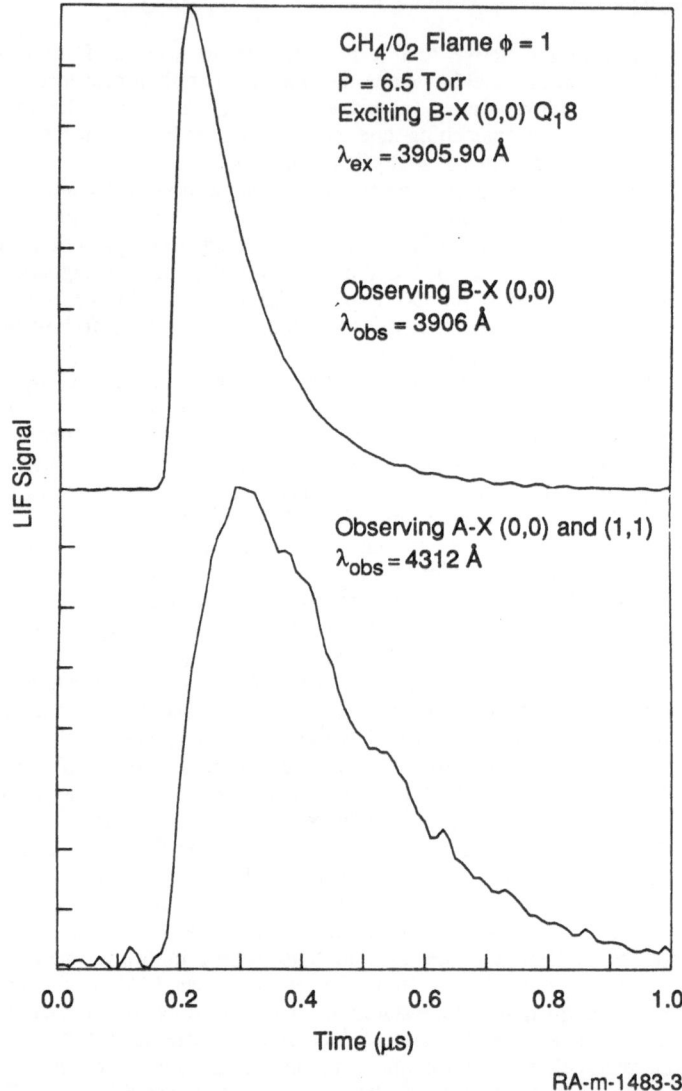

CH$_4$/O$_2$ Flame ϕ = 1
P = 6.5 Torr
Exciting B-X (0,0) Q$_1$8
λ_{ex} = 3905.90 Å

Observing B-X (0,0)
λ_{obs} = 3906 Å

Observing A-X (0,0) and (1,1)
λ_{obs} = 4312 Å

LIF Signal

Time (µs)

RA-m-1483-3

FIGURE 4. Time dependence of fluorescence from the CH molecule in a CH$_4$/O$_2$ flame at 6.5 Torr. The laser, with a pulse length of 10 ns, excites the B$^2\Sigma^-$ state. The upper panel shows the fluorescence from this state; the decay time is governed by both quenching and radiation. The bottom panel shows the fluorescence from A$^2\Delta$, populated by collisional energy transfer from B. Its rise time is the same as the B decay time, and this trace decays due to quenching and radiation from A.

Quenching of both the $B^2\Sigma^-$ and $A^2\Delta$ states of CH has also been studied in the low-pressure burner [14]. Unlike the flow or LP/LF systems, flames do not offer an environment consisting of only one collision partner. Nonetheless, some collider-specific cross sections can be obtained for important flame gases. Operating in an H_2/O_2 flame at about 1500 K (with trace amounts of CH_4 added to produce CH), σ_Q for H_2O could be measured. Results for N_2 and CO_2 were obtained by diluting flames with these chemically inert gases, and extrapolating the quenching rate to 100% added diluent. An example of this method is shown in Fig. 5. For these two colliders the quenching cross sections are larger than those from the LP/LF experiments. The LP/LF experiments were performed at 1300 K while these flame measurements are at 1700 K. This thus shows an increase in cross section with increasing temperature for these two colliders, for which no room temperature measurements exist. For H_2O, this represents the first quenching determination for this important collision partner. The cross sections are collected in Table 1.

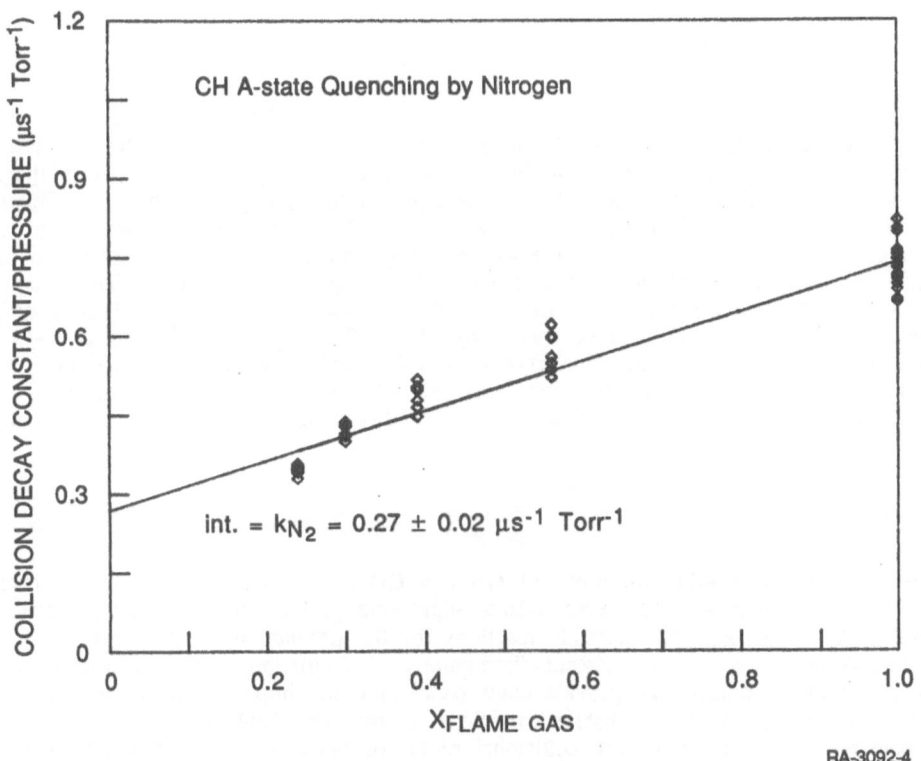

FIGURE 5. Plot of the collisional rate constant in CH_4/O_2 flames diluted with N_2, as a function of mole fraction of flame gas. X=0.0 corresponds to 100% N_2, so that the intercept is the rate constant due to nitrogen quenching.

TABLE 1.

CH $A^2\Delta$ and $B^2\Sigma^-$ collisional cross sections/Å^2.

Collider Flame	N_2	CO_2	H_2O
CH B	3.2	3.6	9.0
CH A	2.3	3.6	6.8
LP/LF (Ref. 22)			
CH A	1.4	2.1	-

The decay trace shown in the upper part of Fig. 4 is obtained following laser excitation of a specific N' level in the v'=0 level of the $B^2\Sigma^-$ state of CH in the burner. Here, the spectrometer which viewed the fluorescence was tuned to 390 nm, the location of the (0,0) band of the B-X system. If the spectrometer wavelength is changed to 413 nm, the trace in the lower panel results. This is fluorescence in the (0,0) and (1,1) bands of the A-X system, produced by CH molecules which have been collisionally transferred from the B to the A state. Note that the risetime of the A-X emission is the same as the decay of the B-X trace, whereas the A state dies out more slowly, having smaller σ_Q for all flame conditions and the three individual collision partners for which measurements were made. Spatial profiles showing the fluorescence from A caused by collisional transfer from B are given in Figs. 2 and 3.

4. SUMMARY

Observations on chemiluminescence of OH and CH in hydrocarbon flames have been summarized, and new experimental results suggesting possible multireaction pathways have been described. The possible reactions for the production of CH* have very different energetics, and the nascent distributions of populations within internal levels of the emitting radical may provide clues concerning the important pathway. A proper interpretation of the observed rotational distributions must include considerations of the competing collisional routes of quenching and rotational energy transfer. Experiments performed over a large temperature range have shown that for these two radicals the quenching cross section depends markedly on both rotational quantum state and temperature.

ACKNOWLEDGMENTS

M.A. DeWilde of the Ballistic Research Laboratory took the OH emission scan resulting in Fig. 1; M.J. Dyer, N.L. Garland, J.B. Jeffries and G.P. Smith took part in the quenching and energy transfer studies on OH and CH described in the paper. We thank all of these researchers for their help and numerous discussions. The assembly of this summary was funded by the Gas Research Institute; the research described has received support from GRI, the Air Force Aeropropulsion Laboratory, NASA and the Department of Energy.

REFERENCES

[1] A.G. Gaydon, The Spectroscopy of Flames, Second Edition, Chapman and Hall, London, 1974.

[2] J.M. Corliss, D.D. Paul, R.H. Barnes and W.A. Ivanic, Proceedings of the International Gas Research Conference, in press, 1987; D. Reuter, B.R. Daniel, J. Jagoda and B.T. Zinn, Comb. Flame, 65 (1986) 281.

[3] L.P. Goss, B.G. MacDonald, D.D. Trump and G.L. Switzer, AIAA Paper 831480, Eighteenth Thermophysics Conference, Montreal, Canada, June 1983.

[4] K. Saito and I. Murakami, Comb. Flame, 34 (1979) 331.

[5] W.L. Shackleford, JANAF Exhaust Plume Technology Meeting, San Antonio, Texas, May 1985; R.B. Lyons, J.B. Wormhoudt and C.E. Kolb, Prog. Astron. Aero., 83 (1982) 128.

[6] E.M. Bulewicz, P.J. Padley and R.E. Smith, Proc. Roy. Soc., A315 (1970) 129.

[7] J. Grebe and K.H. Homann, Ber. Bun. Phys. Chem., 86 (1982) 587.

[8] An asterisk is used to denote an electronically excited molecule.

[9] N.L. Garland and D.R. Crosley, Twenty-First Symposium (International) on Combustion, The Combustion Institute, Pittsburgh, 1987, in press.

[10] M.A. DeWilde and D.R. Crosley, NBS Special Publication 561, (1979) 1171.

[11] R.P. Porter, A.H. Clark, W.E. Kaskan and W.E. Browne, Eleventh Symposium (International) on Combustion, The Combustion Institute, Pittsburgh, 1967, p. 907.

[12] J. Grebe and K.H. Homann, Ber. Bun. Phys. Chem., 86 (1982) 581.

[13] K.J. Rensberger and R.A. Copeland, unpublished results.

[14] K.J. Rensberger, M.J. Dyer and R.A. Copeland, Appl. Opt., to be published.

[15] N.L. Garland and D.R. Crosley, Appl. Opt., **24** (1985) 4229.

[16] G.P. Smith, P.W. Fairchild, J.B. Jeffries and D.R. Crosley, J. Phys. Chem., **89** (1985) 1269.

[17] I.S. McDermid and J.B. Laudenslager, J. Chem. Phys., **76** (1982) 1824.

[18] R.A. Copeland, M.J. Dyer and D.R. Crosley, J. Chem. Phys., **82** (1985) 4022.

[19] G.P. Smith and D.R. Crosley, Eighteenth Symposium (International) on Combustion, The Combustion Institute, Pittsburgh, 1981, p. 1511.

[20] N.L. Garland and D.R. Crosley, J. Chem. Phys., to be published.

[21] R.A. Copeland and D.R. Crosley, J. Chem. Phys., **84** (1986) 3099; J.B. Jeffries, R.A. Copeland and D.R. Crosley, J. Chem. Phys., **85** (1986) 1898; G.P. Smith and D.R. Crosley, J. Chem. Phys., **85** (1986) 3896.

[22] N.L. Garland and D.R. Crosley, Chem. Phys. Lett., **134** (1987) 189.

[23] C.J. Nokes and R.J. Donovan, Chem. Phys., **90** (1984) 167.

CONCLUDING REMARKS

R.J. Donovan
Department of Chemistry
University of Edinburgh
West Mains Road
Edinburgh EH9 3JJ

1. INTRODUCTION

Our task at this workshop has been to "review, critically discuss and identify new directions in the area of reactivity and production of excited species in gas phase neutral and ionic collisions", in the context of chemical selectivity. It is now my task to try to identify the key areas in which we have made progress and to summarise our findings. Inevitably as we have covered so much material during the week, I shall not be able to comment on all of the contributions. My remarks will be gathered under five main headings which formed recurrent themes in our discussions.

2. REAGENT STATE SELECTION, ORIENTATION AND ALIGNMENT

Enormous advances have been made over the past decade in the experimental techniques at our disposal for state selection and orientation of reactants, and we have seen some of the most important and impressive applications of these techniques at this workshop. The exploration of selectivity in chemical processes requires that at least in some of our experiments we can carefully state select and orient the reagents, and similarly identify the states and orientation of the products. Fully detailed studies are still rare but this workshop gives abundant evidence of the progress that has been made. Professor Bernstein has long been a practitioner in this area and it was fitting that he should start the workshop with a review of state selection and orientation: his message was clear, the techniques are now available and well characterised and we were given many elegant examples to prove the point.

The alignment of atomic orbitals and the effect that this has on the rate of electronic curve crossing processes was discussed by Professor Leone. This work is still at an early stage but it will allow future study of alignment dependent effects on competing reactive and energy transfer pathways. A simple model, in which the initially aligned atomic orbital locks into one configuration at a specific distance from the collision partner, provided a useful vehicle for our discussions in this area. More sophisticated quantum models will no doubt be required for a full description, however as noted by Professor Alexander during our discussions, quantum mechanics gives us the truth but little physical insight!

J. C. Whitehead (ed.), Selectivity in Chemical Reactions, 555–560.
© 1988 by Kluwer Academic Publishers.

The most penetrating insight into the dynamics of elementary chemical reactions comes from the measurement of vector correlations as we can then see the stereochemistry of the dynamics. Three vector (j,k,j') correlation measurements for the Ba + N$_2$O reaction were presented by Professor Stolte. In this reaction the orientation of the N$_2$O was specified while chemiluminescence from the BaO* product gave information on product alignment: the results suggest that concerted motions couple the entrance and exit channels. Similar detailed studies of reactions involving rotor accelerated and electronically excited rare gas atoms and photofragment vector correlations, were presented by Professor Simons. Further riches were provided by Professor Wolfrum in studies of selective vibrational and translational excitation of oriented reactants: multiple pathways were also explored and we sought ways of distinguishing the main routes from the side routes and of diverting the reactive flux into the desired route. This theme was further amplified in the carefully chosen examples presented by Professor Ottinger.

With so much detailed information on the dynamics of chemical reactions becoming available we seriously have to face the question of how to optimise our approach and concentrate on the key points required to advance our understanding on a broad front. In this context I regard the development of a new "imaging" photodissociation spectrometer (Parker) as very important. This impressive instrument will greatly accelerate the rate at which photofragmentation data can be collected and will facilitate rapid survey work to identify key systems for detailed study.

3. VAN DER WAALS COMPLEXES AND CLUSTERS

In this section we moved towards the condensed phase and saw how van der Waals clusters are formed and the novel reactions that can occur in this new physical state that lies between the gas and liquid phases. Starting with the simplest type of van der Waals complex we considered reactions involving species of the type BrH...OCO, where photolysis causes the hydrogen to recoil from the halogen atom towards the CO$_2$ molecule. In this type of experiment the kinetic energy of the hydrogen atom, the internal energy of CO$_2$, the impact parameter and the collision geometry are all well defined and thus the product scattering can be studied in detail. The initial hope, that the hydrogen atom would not be further involved in the scattering process, has turned out to be false but this "failure" may yet prove to be of value. Secondary scattering from the halogen may hinder our approach to understanding the reactive scattering of H atoms with CO$_2$ but it provides an interesting chemical environment, one that starts to resemble the conditions met in the liquid phase where multiple encounters are normal.

Our view was further broadened by Dr. Jouvet's discussion of the reaction that occurs when complexes such as Xe...Cl$_2$ are excited. From gas phase studies we know that reaction can occur if either the noble gas atom or the halogen molecule is excited. In both cases the reaction involves harpooning with the formation of Xe$^+$ and X$_2^-$, followed by direct stripping to yield XeX*(B,C). Recent work by Apkarian and co-workers [1] has shown that this type of reaction can also occur in the liquid phase (ie. in liquid Xe) and that the initial excitation process is probably best thought of as a simple charge transfer transition (ie. Xe...Cl$_2$ + hν → Xe$^+$...Cl$_2^-$). This description is probably also valid for the simpler van der Waals complexes and we saw evidence for this in the excitation spectra presented by Dr. Jouvet. This system provides an ideal opportunity to study a reaction in detail in both the gas and liquid phases, and <u>all</u> "phases" in between, by observing reaction with clusters of the type $Xe_n...Cl_2$ (n=1 to ∞).

The reactions of large "solvated" ions, with up to 10^2 Ar atoms per cluster were then explored by Dr. Stace. It was clear from this and from the brief presentation by Professor Bernstein that the reactions of ions in clusters differ very significantly from those in the gas phase and that alternative chemical pathways become available with partially solvated species.

This is an exciting area where we can expect to see rapid developments. Cluster reactions of both neutral and ionic species will open up new and selective pathways that differ from those in the gas or liquid phases.

4. POTENTIAL ENERGY SURFACES

Potential surfaces pervaded the whole of our discussions, however it was useful to have a special section in which to concentrate on the accuracy and availability of potential surfaces. The $H + CO_2$ and $OH + CO$ surface again featured strongly: our understanding of this potential surface is reaching a useful stage of maturity but the apparent conflict between theory (Schatz) and experiment (Bernstein, Zewail, *et al.*) over the lifetime of the $HOCO^\dagger$ complex shows that further work is needed in at least one of these areas.

The diatomics-in-molecules (DIM) formalisms were reviewed in detail by Professor Kuntz who also brought the encouraging news that a comprehensive suite of programmes for DIM calculations will soon be available. The application of DIM techniques was well illustrated by Professor Grice who pointed to the involvement of charge transfer (ionic) surfaces in a number of reactions involving halogen species, even where these surfaces lay well above the main surface on which reaction occurs. Configuration interaction is clearly important in these systems and my own feeling is that we will find charge transfer surfaces playing a wider and increasing role in our discussions in the future.

Difficulties arise in systems with more than four atoms, but as pointed out by Professor Murrell, we rarely need to know all of the details of the 3N-6 dimensional surface and we can settle for sections of the surface of lower dimensionality. Unfortunately, quite subtle differences between surfaces can produce profound dynamical effects: furthermore several of the surfaces in use at present are "flawed" (Schatz) and care is needed in drawing conclusions on the dynamics of reaction when using such surfaces.

5. ELECTRONICALLY EXCITED STATES

Despite a large and growing volume of work in this area, the chemistry of electronically excited states remains essentially in its infancy. As every atom, molecule and ion has an infinite number of electronic states it is clear that we have barely scratched the surface of this area as yet and we can confidently predict that many surprises lie ahead. One only has to think of the way in which the discovery of the reactions between excited noble gas atoms and halogen molecules appeared so unexpectedly, despite a long gestation period (during which the evidence from oscillatory continuum emission stared us in the face), to realise that we are probably stumbling over similar novel chemistry without realising it at this meeting! As with

558

most things in life serendipity* will play an important role. We have seen that even simple reactions such as that between $O(^1D_2)$ + H_2 still hold surprises and Professor Sloan has presented evidence for the involvement of an excited $^1A_2(^1A")$ hypersurface which provides a path for direct stripping in this reaction. Thus, while the main flux for reaction occurs over the ground state surface, with insertion of $O(^1D)$ into the H-H bond (ie. complex formation), direct reaction can also occur *via* the $^1A_2$ state. This view is further substantiated by theoretical work (Schatz).

Reactions leading to electronically excited products have also featured prominently in our discussions and Dr. Whitehead's interesting work on the formation of IF*(B) in a wide range of chemical reactions and its relative resilience to quenching, further substantiates the importance of this species for energy storage and as a potential intermediate for the operation of future chemical laser systems. The involvement of other "dark states" is also intriguing in this context although it is clearly more difficult to identify and monitor such species.

Processes involving spin-orbit excited states were thoroughly reviewed by Professor Dagdigian. We saw that states with large spin-orbit coupling behave adiabatically, as expected, however even for weakly spin-orbit coupled states a large degree of adiabatic behaviour is observed. It had generally been assumed that rapid quenching would scramble states with small spin-orbit coupling energies and that any selectivity would be lost. Clearly selectivity remains in some cases and it will be interesting to explore this area in further detail.

6. APPLICATIONS OF CHEMICAL SELECTIVITY

Dr. Ewing gave us an interesting view of the interplay between pure and applied science. His discussion of *total system engineering* amply illustrated the interdependence of the two and the integrated approach that is needed to produce major new laser systems. The need for the discovery of "new chemistry" with which to develop new lasers was emphasised and our views on the prospects for this varied widely. As you will have gathered, I personally remain convinced that we have barely scratched the surface of chemistry and that excited state species, though difficult to handle efficiently, will provide major new avenues in the future.

Our knowledge of atmospheric chemistry has advanced dramatically over the past decade or so, as both existing knowledge and new discoveries have been built into the models which summarise our understanding of this complex "system". Professor Thrush gave us a thorough review of our present standing in this area and of the extent to which selectivity plays a role in channelling atmospheric processes. Unexpected features were highlighted and it became clear that much fundamental work still remains to be done, particularly on pressure dependent rate coefficients and reactions involving hydrogen-bonded species. It is interesting to note in this context that hydrogen-bonding is responsible for much of the *selectivity* in biological processes.

Many aspects of atmospheric oxidation are echoed in flame chemistry, albeit at much higher temperatures, and we were given an interesting view of this area by Dr. Crosley. Our discussions centred mainly on processes involving OH, CH, NH and their excited states, the latter being frequently observed in emission from

---

* serendipity can be defined as looking for a needle in a haystack and finding the farmer's daughter.

flames. Chemiluminescence from flames has long been a source of fascination but we have a long way to go before we can fully understand even the main combustion processes that are involved. Nevertheless, considerable progress has been made in understanding fundamental energy transfer processes for excited species such as OH(A$^2\Sigma^+$) and CH(A$^2\Delta$) over a wide temperature range. It is worth recalling that the spectroscopic description of hydrides has been greatly facilitated by the united atom model and recent work [2] on collisional processes involving CH(A$^2\Delta$) suggests that the same model may be useful in this context also. Figure 1 illustrates this point by comparing rate data for collisional removal of CH(A$^2\Delta$) and CH(X$^2\Pi$) with that for the united atom N(^{2}D). It is clear that there are strong parallels in behaviour between these three species. The most surprising feature is that the ground state of CH is the most reactive, while N(^{2}D) is consistently least reactive: this suggests that there is a small entry barrier on the surfaces involving CH(A$^2\Delta$) with an even higher barrier for the N(^{2}D) surfaces. Some care is needed in using such analogies as diatomic species have internal rovibrational states that are absent for atoms. Thus some enhancement in the rate of removal of hydrides, due to chance resonances (energy transfer channels), can be expected but the broad picture should still hold.

    Ab initio studies (Farantos) of the collisional deactivation of OH(A$^2\Sigma$) by CO indicate that complex formation is important and that for quenching by He there is a surprising propensity for rotational changes with $\Delta J = \pm 2$.

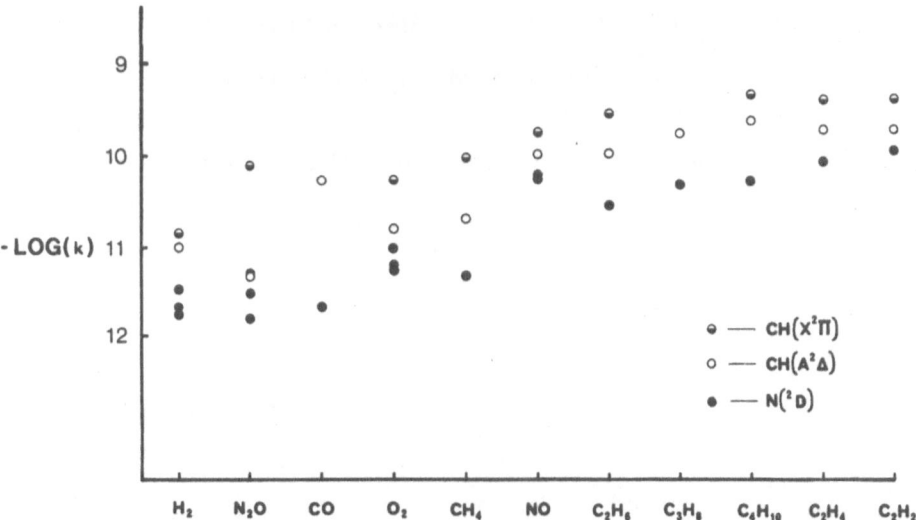

FIGURE 1. Correlation of rate data for CH(A$^2\Delta$), CH(X$^2\Pi$) and N(^{2}D) with the collision partners indicated along the horizontal axis. Rate constants (cm^3 molecule^{-1} s^{-1}) are presented in logarithmic form (base 10): in some cases rate constants for N(^{2}D) have been determined by more than one group and all data have been included (further details are given in Ref. 2). A broad correlation is seen, with the data for N(^{2}D) lying below that for CH(A$^2\Delta$), which in turn lies below that for CH(X$^2\Pi$). The datum for removal of CH(X$^2\Pi$) by CO has not been included as this process is third order.

7. EPILOGUE

The material presented during this workshop illustrates very well the dramatic advances that have been made over the past few years in both experimental techniques and theoretical methods. These advances are likely to continue as the polarisation and coherence properties of lasers, the applications of jet-cooling techniques and cluster formation are exploited in experimental studies, and the development of parallel processing and other computer hardware facilitates theoretical work. The main challenge for the future will be to choose chemical systems which probe the widest fundamental principles. It is all too easy to become enthralled by the apparent beauty of a system that has only local significance. Like the orienteer we need to get to the mountain-tops to see our paths clearly and thereby reduce the chance of wasting our energy by following blind valleys. This workshop has provided us with a panoramic view of chemical selectivity and through its publication, this view will hopefully reach a wider audience.

Finally, on behalf of all of us here, I would like to thank Dr. Whitehead for his very considerable efforts in organising this workshop and for providing such a stimulating and convivial environment. Both he and the organising committee are to be congratulated on the excellent blend of science that they have achieved.

REFERENCES

[1]   L. Wiedeman, M.E. Fajardo and V.A. Apkarian, Chem. Phys. Lett., **134** (1987) 55.
        M.E. Fajardo, V.A. Apkarian, A. Moustakas, H. Krueger and E. Weitz, to be published.

[2]   C.J. Nokes and R.J. Donovan, Chem. Phys., **90** (1984) 167.

ORGANISING COMMITTEE

J.N.L. Connor (Manchester, U.K.)
P.J. Dagdigian (Johns Hopkins, U.S.A.)
S. Stolte (Katholieke Universiteit, Nijmegan, Neth.)
J.C. Whitehead, Director (Manchester, U.K.)

LIST OF PARTICIPANTS

M.H. Alexander (Maryland, U.S.A.)
V. Aquilanti (Perugia, Italy)
R.B. Bernstein (UCLA, U.S.A.)
P.R. Brooks (Rice, U.S.A.)
M. Costes (Bordeaux, France)
J.N.L. Connor (Manchester, U.K.)
D.R. Crosley (S.R.I. International, U.S.A.)
P.J. Dagdigian (Johns Hopkins, U.S.A.)
S.J. Davis (Physical Sciences Inc., U.S.A.)
R.J. Donovan (Edinburgh, U.K.)
F. Engelke (Bielefeld, W. Germany)
J.J. Ewing (Spectra Technology, U.S.A.)
S.C. Farantos (Crete, Greece)
A. Gonzalez Ureña (Madrid, Spain)
R. Grice (Manchester, U.K.)
D.M. Hirst (Warwick, U.K.)
Ch. Jouvet (Orsay, France)
P.J. Kuntz (Hahn-Meitner-Institut, Berlin, W. Germany)
S.R. Leone (JILA and Colarado, U.S.A.)
H. Loesch (Bielefeld, W. Germany)
M. Menzinger (Toronto, Canada)
J.N. Murrell (Sussex, U.K.)
Ch. Ottinger (Göttingen, W. Germany)
D.H. Parker (Santa Cruz, U.S.A.)
N. Sadeghi (Grenoble, France)
G.C. Schatz (Northwestern, U.S.A.)
J.P. Simons (Nottingham, U.K.)
J.J. Sloan (Waterloo, Canada)
A.J. Stace (Sussex, U.K.)
S. Stolte (Nijmegan, Neth.)
B.A. Thrush (Cambridge, U.K.)
J. Wanner (Garching, W. Germany)
J.C. Whitehead (Manchester, U.K.)
J. Wolfrum (Heidelberg, W. Germany)

INDEX

565